Electric Drives

Concepts and Applications

Second Edition

ABOUT THE AUTHOR

Vedam Subrahmanyam received his MSc (Engineering) degree in Electrical Machine Design from Madras University in 1963 and his PhD from IIT Madras in 1973. He joined IIT Madras in 1966 and served as a Professor in the department of Electrical Engineering. Prior to this, he had taught for two years at Visweswaraya Regional College of Engineering, Nagpur, and Regional Engineering College, Warangal, as a faculty member. A popular academician, he has guided several research students at the MTech and PhD levels.

Prof. Subrahmanyam has published around 60 articles in national and international journals and has also presented papers at international conferences. He has attended programmes in Germany and worked as a faculty member at the University of Dar-es-Salaam for two years. He is the author of well-acclaimed books like *Power Electronics: Devices, Converters, Applications* and *Thyristor Control of Electric Drives*, the latter published by McGraw-Hill.

Electric Drives

Concepts and Applications

Second Edition

Vedam Subrahmanyam

Former Professor
Department of Electrical Engineering
Indian Institute of Technology
Chennai

McGraw Hill Education (India) Private Limited

Published by McGraw Hill Education (India) Private Limited,
Registered Office: Anjana Complex No: 5/90 A, Butt Road, St. Thomas Mount, Chennai – 600016, Tamil Nadu, India.

Electric Drives, 2e

Seventeenth reprint 2023

ISBN-13: 978-0-07-070199-1
ISBN-10: 0-07-070199-7

18 19 20 21 22 NIC 27 26 25 24 23

Managing Director: *Lalit Singh*
Head—Production: *Pankaj Manohar*

Typeset at ACE PRO India Pvt. Ltd., Ghaziabad 201014, and printed and bound in India at Nice Printing Press, 3, Rashid Market Extn., Delhi 110051.

Cover Printer: SDR Printers.

Visit us at: www.mheducation.co.in
Toll free in India: 1800 103 5875
Write to us at: info.india@mheducation.com
CIN: U80302TN2010PTC111532

Dedicated to my beloved mother late Smt Seshamma

Contents

Preface

Even in the 21st century electric drives continue to be widely used in the industry. During the last one and a half decades, since the publication of the first edition, research in electric drives has been very active. Significant improvements in computing have given rise to technologies that make design and control of electric drives efficient. Electric drives have a number of advantages and are widely used in the industry. Among these advantages are reliability and versatility of speed control, especially with a dc drive. Speed control of electric motors has become possible with the advent of the thyristor. Optimisation and automation have, however, improved the static and dynamic behaviour of these drives. Further developments in the area of control of static apparatus such as converters and inverters using thyristors have made these drives more reliable and accurate in operation which in turn, has led to a wider application of these drives.

ABOUT THE BOOK

This book is written to cover the subject at a basic level in order to familiarize the reader with the problems of the drives and possible solutions for them. Before going into the details of closed-loop control and associated design problems connected with dynamics of the drives, knowledge of conventional drives for variable speed operation is necessary. The dynamics of these conventional systems are discussed to give an idea as to how these can be improved with closed-loop systems. The availability of thyristor power converters has made an impact on the area of drives with sophistications such as fast dynamic response, accurate speed control, etc. A detailed discussion of these converters and their features has also been provided. The features of ac and dc drives are discussed in detail. The control aspects of the drives have been elaborated upon so that improvements in both the static and dynamic behaviour of the drives can be achieved with proper design of the controllers.

The text covers cutting-edge research and development in effective self-control of electric motors. Sensorless control using advanced control technologies like Artificial Intelligence including Fuzzy Logic and Neural Networks are covered. It discusses topics like state estimation using sophisticated filtering techniques in significant detail. Concepts and applications of cycloconverter fed synchronous motors are also explained.

The book provides an exhaustive and comparative study of all drives, both conventional and those fed from static converters. It also discusses the utility of static drives for these applications.

TARGET AUDIENCE

The book is intended to serve as a textbook for basic courses in drives covering fundamentals of the drives, as well as advanced courses dealing with the design of closed-loop speed control and design of controllers. The focus on applications makes it useful to practicing engineers.

NEW TO THIS EDITION

While maintaining core chapters from the previous edition, the second edition brings the readers abreast to state-of-the-art design of electric drives and their latest applications. This edition introduces Permanent Magnet Synchronous Motors, their mathematical modeling and applications. The other new topics introduced are Brushless Motors, Three Phase Induction Motor Drives, Current Source Inverter Control (CSI Control), Voltage Source Inverter Control (VSI Control) and Application of Electric Devices in IGBT, MOSFET and BJTs. The pedagogy is refreshed with inclusion of new problems in the form of Solved examples, Review Questions and Objective Type Questions.

SALIENT FEATURES

- In-depth coverage of thyristor power converters and drives employing these converters
- Highlights closed-loop control and dynamics of electric drives
- Detailed discussion on industrial applications, utility and technical problems of electric drives
- Comparative analysis of various drives along with the relevant discussion
- Application of Microprocessor in electric drives presented in separate chapter
- Excellent pedagogy including
    - 65 solved problems
    - 190 review questions
    - 115 objective-type questions

CHAPTER ORGANIZATION

Chapter 1 deals with an introduction to the drives, their classification, and their dynamics. A brief description of methods of speed control and braking of the motors using conventional methods is given in **Chapter 2**. **Chapter 3** gives a review of all the power converters used for drives. The methods of improving the performance are also discussed. **Chapter 4** deals with ac and dc drives. The former uses induction and synchronous motors. All the technical problems of the drives are discussed where they are operated on static power converters. The problems of motor heating on different load cycles are discussed in **Chapter 5**. **Chapter 6**

discusses the control aspects of electrical drives, the design of controllers to improve performance. **Chapter 7** discusses typical applications of electrical drives in the industry. **Chapter 8** presents Microprocessors and Control of Electric Drives in detail.

ACKNOWLEDGEMENTS

First of all, I would like to acknowledge the support and patience of my beloved wife, Leelavathi, during the process of writing and revising this book. Also, the reviewers of this edition deserve a special mention for their efforts in scrutinizing the manuscript and providing helpful inputs. Their names are given below.

G K Singh	*Indian Institute of Technology(IIT)* *Roorkee, Uttarakhand*
Vivek Deep Verma	*BBD Institute of Technology* *Lucknow, Uttar Pradesh*
Ranjan Behera	*Indian Institute of Technology(IIT)* *Patna, Bihar*
S B L Seksena	*National Institute of Technology(NIT)* *Jamshedpur, Jharkhand*
K N Pawar	*BSD College of Engineering,* *Dhule, Maharashtra*
N Kumaresan	*National Institute of Technology(NIT)* *Tiruchirapalli, Tamil Nadu*
B Vinod	*PSG College of Technology* *Coimbatore, Tamil Nadu*
B K Murthy	*National Institute of Technology(NIT)* *Warangal, Andhra Pradesh*
Subbarayudu Duggisetty	*GPR Engineering College* *Kurnool, Andhra Pradesh*
Gita Ramadas	*RMK Engineering College* *Kavaraipettai, Tamil Nadu*

Finally, my wholehearted appreciation and gratitude is due to the editorial and production staff of Tata McGraw-Hill, especially Shalini Jha, Suman Sen, Manish Choudhary, Sohini Mukherjee and P L Pandita for their constant support and encoragement during various stages of revision of the text.

VEDAM SUBRAHMANYAM

FEEDBACK

Constructive criticism and suggestions are always welcome for the enhancement of the book, all of which can be sent to the publisher's email at tmh.elefeedback@gmail.com, mentioning the title and author's name in the subject line. Feel free to report any piracy spotted by you as well.

Introduction

Industrial loads require operation at any one of a wide range of speeds. Such loads are generally termed as variable speed drives. These drives demand precise adjustment of speed in a stepless continuous manner over the complete speed range required. The loads may be constant torque loads, requiring constant torque at all speeds. The torque may also be a function of speed, e.g. in a fan type or pump type load the torque is proportional to the square of the speed. These loads are driven by hydraulic, pneumatic or electric motors. An industrial drive has some special features when driven by electric motors. The speed–torque characteristic of the motor can be very easily modified to suit the load characteristic. It has a sufficient overload capacity and can be overloaded for short periods without affecting the life of the motor. The motors can be brought to operation without any warming up period.

One of the important features of an industrial drive is its four quadrant operation in the speed–torque plane. An electric motor can operate in all the four quadrants of the *V–I* plane, corresponding to the mechanical quantities, speed and torque. The normal motoring operation is in the first quadrant. Sometimes the drive motor is required to maintain a constant speed, especially when there is a tendency for the load to get accelerated beyond the safe speed. Electric motors have a natural tendency to become generators automatically if the speed of the motor goes beyond the no-load value. The energy responsible for acceleration is successfully returned to the mains, thus maintaining constant speed. This operation corresponds to the second quadrant. An electric motor can operate in the reverse direction of rotation with suitable changes in its supply connections. This operation corresponds to the third quadrant. Operation is also possible in the fourth quadrant, where the motor acts as a brake. This can be done by suitably modifying the characteristic, e.g. by inserting a resistance in the rotor circuit of an induction motor. A suitable torque developed by the motor drives the load at the desired speed, providing the braking action. An electric motor thus adapts itself for four quadrant operation. Rapid acceleration of the drive and frequent reversing, if required, can be easily accomplished when an electric motor is employed.

Another feature of drives employing electric motors is smooth speed control over a wide range, e.g. the speed control of a dc motor in Ward Leonard control when a variable voltage is applied to the armature. This had not been possible with ac motors till the advent of thyristor. The development of compact thyristor power converters has made possible the smooth speed control of both dc and ac motors. The efficiency of the drive in the complete speed range also improves. Electric motors have a good starting torque and can be started on load.

Industrial drives demand very precise variations in speed. This can be easily accomplished by means of an electric motor, up to an accuracy of 1%. An electric drive is easy to maintain and can even be operated in a contaminated atmosphere.

Because of these features industrial drives employ electric motors ac or dc.

REQUIREMENTS OF AN ADJUSTABLE SPEED DRIVE

An electric drive should satisfy the following requirements:

i. *Stable operation* The speed–torque characteristic of the motor should be such that stable operation of the drive is assured in all the four quadrants over a wide range of speeds. The drive must be controllable with regenerative braking to maintain constant speed if the load overhauls the motor. The motor must have stable operation in the fourth quadrant.

ii. The drive motor should also have a good transient response. The drive should return to its original operating condition very quickly in case of disturbance. If there is a step change in torque or speed the drive must attain its new operating point quickly without any large overshoots. It must operate with stability, not have a sluggish transient response, not be very oscillatory, and have a suitably chosen damping.

The frequency and transient responses may be corrected with a proper design of the controllers in the closed loop controls. Stability considerations of electric motors and the optimum designs for improving the transient performance are discussed in subsequent chapters.

FORMS OF DRIVE MOTORS

The possible forms of drive motors are

i. dc motors fed from dc supply
ii. dc motors fed from ac supply
iii. ac motors fed from ac supply

The speed control of a motor can be accomplished either in the conventional manner or by using thyristor power converters. Solid state drives using thyristor power converters are gaining popularity over drives employing conventional methods due to their reliability, compactness and capability for controlling speed. Thyristorised drives employ both dc and ac motors. The dc motors can be driven from an ac supply with a thyristor power converter interposed. The capacity of this converter to operate both as a rectifier and an inverter makes regeneration possible. A converter fed drive therefore has a natural tendency to operate in the first and second quadrants. With suitable changes in the power circuit of the converter operation can be extended to the third and fourth quadrants also. If operation is needed only in the first quadrant a half controlled converter with better characteristics may be used. Power converters, such as choppers, may be used to feed (operate) a dc motor from a dc supply.

Till the advent of thyristor power converters, variable frequency to an ac induction, synchronous motor had been provided by rotating the motor–generator set for the purpose of speed control. The development of inverters providing variable voltage, variable frequency sources has widened their applicability in controlling the speed of ac motors over a wide range. Both dynamic and regenerative braking are possible. Conventional armature voltage control, slip energy recovery schemes and rotor resistance control are now being replaced by static converters. These drives and discussed in the following chapters.

Thyristor power converters have made possible the speed control of synchronous motors, which had otherwise been of the constant speed type. Self controlled synchronous motors with constant margin angle control have a steadystate and dynamic behaviour comparable to that of a dc motor.

AC DRIVES AND DC DRIVES—A COMPARISON

DC motors are very versatile for purposes of speed control. Smooth speed control is possible by varying the armature voltage and/or field current, but they suffer from disadvantages imposed by the commutator. Sparking at the brushes limits both the highest speed of operation and the design capacity of the motor. Also. the ripple content of the motor current due to processes of rectification impairs the commutation capability of the motor. Induction motors, which had been constant speed type till the advent of thyristor power converters, posed problems in the area of speed control. When controlled using conventional methods, only stepped speeds are possible with a limited range of speed control. The power factor and efficiency are poor. With thyristor power converters providing a variable voltage, variable frequency supply the speed control of induction motors has become straightforward, making them viable competitors to dc motors. Current research is oriented towards the sophisticated control of induction motors such that the characteristics of dc motors can be realised. Further, the static and dynamic characteristics of synchronous motors can be improved by self control making them similar to those of a dc motor. A comparison of dc and ac drives is given in Table 1.

From the table it is clear that developments in the area of solid state device technology are providing the thyristor at increasingly more economical prices, thus making ac drives more popular than their dc counterparts.

TRENDS IN DRIVE TECHNOLOGY

With thyristor power converters achieving popularity in the area of adjustable speed drives, the operational problems of motors on these converters have started posing problem. The ac motor has a non-sinusoidal voltage or current while operating on inverters while a dc motor has a ripple superimposed on its current. These lead to additional losses and torque pulsations which are objectionable at low speeds. The trend is now towards designing the motor as well as the inverter such as to minimise these problems. The design of a dc motor is being improved to provide a sufficiently good commutating capability, while that of the inverters is being improved to provide voltage with the least harmonic content. PWM techniques have been developed, and the thyristors are being replaced by transistors.

Table 1 *Comparison of dc and ac drives*

DC drives	*AC drives*
The commutator makes the motor bulky, costly and heavy. Sparking at the brushes makes it environmentally unsuitable in certain locations. The highest speed and design rating are limited due to commutation. The commutator requires frequent maintenance.	The problems are not there. Motors are inexpensive, particularly the squire1 cage motor. Speed and design rating have no upper limits. Motor is reliabie, requires little maintenance and can be used in all locations. Economical in some applications.
The converter technology is well established. The power converter is simple and inexpensive.	The inverter technology is still king developed. The power circuit of the converter and its control are complex.
Line commutation of the converter.	Forced commutation is usedwith induction motors. Sometimes machine commutation may be used with synchronous motors.
The line conditions are very poor, i.e. poor power factor, harmonic distortion of the current.	For regenerative drives the line power factor is poor. For non-regenerative drives the power factor is better.
Fast response and wide speed range smooth control.	Response depends upon the type of control. With solid state converters the speed range is wide. With conventional methods it is stepped and limited.
Small power/weight ratio.	Large power/weight ratio.
Cost does not depend on the solid state converter.	Solid state converter employed also decides the cost.

Vector controlled induction motors, and margin angle controlled synchronous motors are employed in high performance drives having precise speed and torque control. These also have a very good static and dynamic response.

Solid state drives employ closed loop control both for speed and torque control. Modern control methods use state space techniques. Methods of stabilising the drives and improving their transient performance are being developed.

Digital control using microprocessors is slowly replacing the analog control of the drives, which had limitations due to component drift and temperature. Microprocessor control has several features and advantages. Modern control techniques to the implementation of sophisticated drives (which are very difficult to realise with hard wired systems) are possible, as are diagnosis, monitoring, warning etc.

1 Characteristics of Electric Motors

1.1 INTRODUCTION

Variable speed drives in the industry employ electric motors as their drive motors mainly because they enjoy several specific advantages, such as overload capacity, smooth speed control over a wide range, capability of operating in all the four quadrants of the speed-torque plane, etc.

Till the advent of thyristors and thyristor power converters the dc motor had been very popular in the area of adjustable speed drives, even though it suffered from the disadvantages imposed by the presence of a mechanical commutator. Thyristor power converters capable of providing variable voltage, variable frequency supplies have now made ac motors increasingly popular.

Industrial loads have different types of speed-torque characteristics. Once the speed-torque of the load to be driven is determined a proper motor has to be selected for driving the load. Some of the factors that can influence the choice of a motor to drive the load are:

i. the available ratings, capital and running costs involved
ii. the limits of speed range, hardness of speed control and speed regulation
iii. the efficiency during variable speed operation
iv. controllability
v. braking requirements
vi. reliability of operation
vii. starting requirements
viii. power/weight ratio
ix. power factor
x. capability of operating on a load factor or duty cycle
xi. availability of supply
xii. effects of supply variations
xiii. loading of the supply, and
xiv. environmental effects.

Therefore, a knowledge of the behaviour of electric motors and with regard to the factors listed above is required. It may also be possible to modify the system to improve its performance and make it more economical and efficient.

A short survey of the characteristics of electric motors, the conventional methods of speed control, braking and starting are discussed in this chapter.

1.2 CHARACTERISTICS OF DC MOTORS

DC motors are of the rotating armature type. The armature winding is a closed winding through the commutator. The armature is supplied through the brushes which are placed along the neutral axis on the commutator. The field system is stationary. The mmf produced by the field is along the magnetic axis while the current flowing through the armature produces an mmf directed along the brush axis. The two mmfs are in space quadrature and occur simultaneously in the motor. They react with each other and develop a torque under whose action the armature rotates. A voltage is induced in the armature called the back emf. The direction of rotation can be determined by the left hand rule. The mmfs and the direction of rotation of the armature are illustrated in Fig. 1.1.

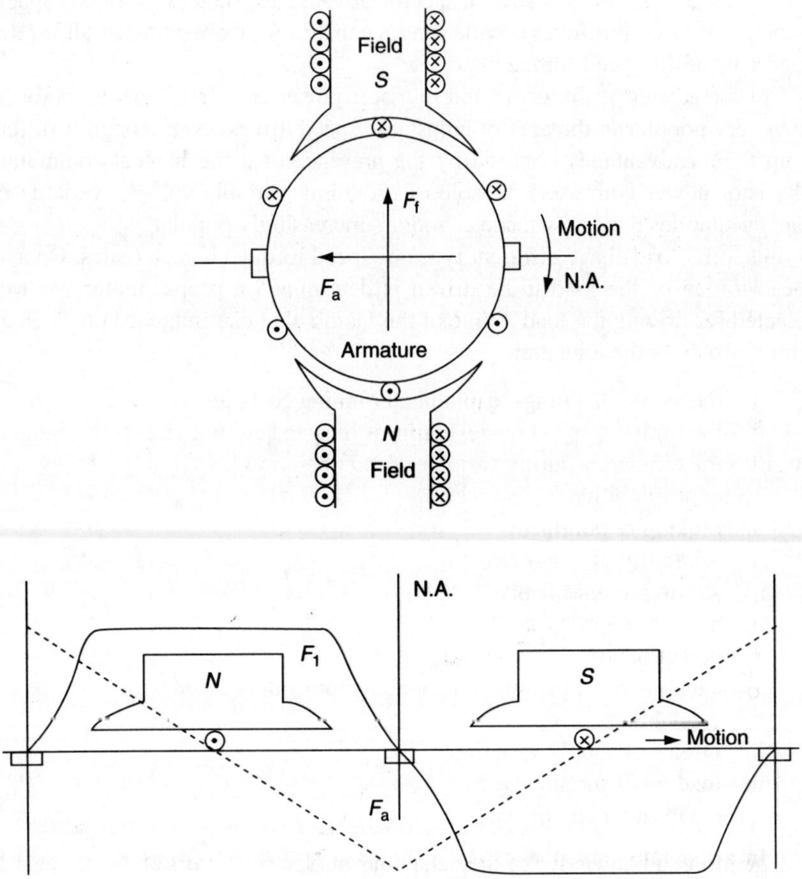

Fig. 1.1 *The magnetomotive forces of a dc machine*

A special feature of dc motors is that it is possible to connect the field and armature windings in several ways so as to achieve a variety of speed-torque characteristics. The motors are classified depending upon the type of connection between the armature and the field.

In a separately excited dc motor the armature and field are excited by independent voltage sources. In a dc shunt motor the field winding is connected in parallel with the armature, both of which are supplied from the same source. In series motors the armature and field are connected in series and are obviously supplied from the same source. In a compound motor both series and shunt fields are present. The performance of the motor is determined by the relative strengths of the series, shunt or separately excited field windings and by the orientation of the fluxes produced by them. Different types of dc motor connections are shown in Fig. 1.2.

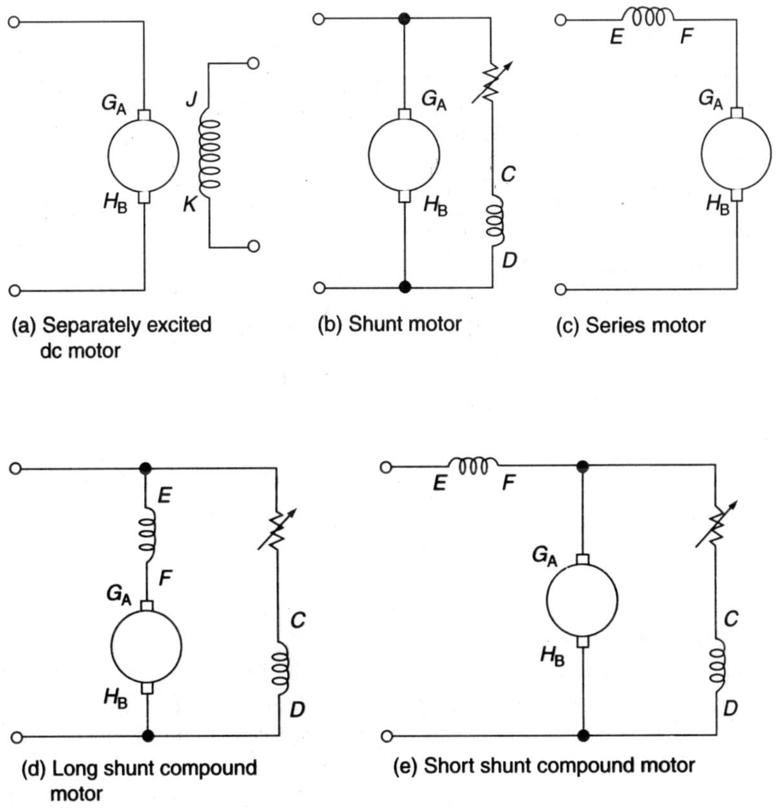

Fig. 1.2 *Types of dc motors*

In a separately excited motor, it is possible to control both armature voltage and field current, so as to control the speed over a wide range in a smooth manner. Speeds ranging from zero to base speed may be obtained at constant torque by armature voltage control. Speeds above base speed are possible at constant power output by a weakening of the flux.

Shunt motors operate at almost constant speed from no load to full load when operated from constant voltage mains. With a series connection of the field winding with armature an inverse relation between the speed and torque may be achieved. By properly adjusting the relative field strengths of the series and shunt windings of a compound motor a speed-torque curve with desired speed regulation may be obtained.

The speed-torque characteristic of an electric motor is very significant since it decides the application of the motor.

1.2.1 Speed-Torque Characteristic of a Separately Excited dc Motor

The circuit equation of a dc motor whose armature, having a total of Z conductors, is wound for $2P$ poles (the brushes divide the winding into $2a$ parallel paths), is

$$V_a = E + I_a r_a \tag{1.1}$$

Where E is the back emf of the armature given by

$$E = \frac{\phi Z N}{60} \frac{2P}{2a} = K_e \phi N = K_t \phi \omega \tag{1.2}$$

where ω is the angular velocity given by $\omega = (2\pi N/60)$, K_e and K_t can be easily identified.

From Eqs (1.1) and (1.2) we get

$$N = \frac{V_a - I_a r_a}{K_e \phi} \tag{1.3}$$

The torque developed by the motor is given by

$$T_d = \frac{1}{2\pi} 2P\phi \frac{I_a}{2a} Z = K_t \phi I_a \tag{1.4}$$

Substituting for I_a in Eq. (1.3) from Eq. (1.4) we have

$$N = \frac{V_a}{K_e \phi} - \frac{T_d}{K_t K_e \phi^2} r_a \tag{1.5}$$

When $T_d = 0$ the coriesponding speed $N_0 = V_a/(K_e\phi)$ is the no-load speed. The motor speed decreases as the torque developed increases, resulting in a drooping characteristic. The speed-torque curves are shown in Fig. 1.3. The figure clearly shows a speed drop of 2 to 3% as the torque varies from no-load to full load.

In dc machines the armature mmf reacts with the field mmf this reaction is known as the armature reaction. When the effects of the armature reaction are neglected, the flux per pole of the motor is constant and is independent of load. In normal construction the brushes are placed in the neutral zone. The armature reaction, though cross magnetising, is followed by demagnetisation due to saturation. The effect of demagnetisation on the field flux due to armature flux is clearly

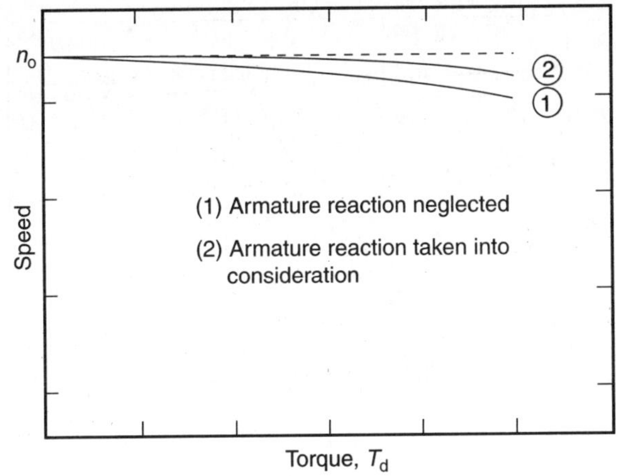

Fig. 1.3 *Speed versus torque of a separately excited dc motor*

shown in Fig. 1.3. The speed drop from no-load to full load decreases, improving speed regulation.

The effect of additional resistance in the armature circuit is depicted in Fig. 1.4. Speeds in the range of zero to base speed may be obtained. With a suitable value of r_a very slow speeds are possible, at the cost of efficiency.

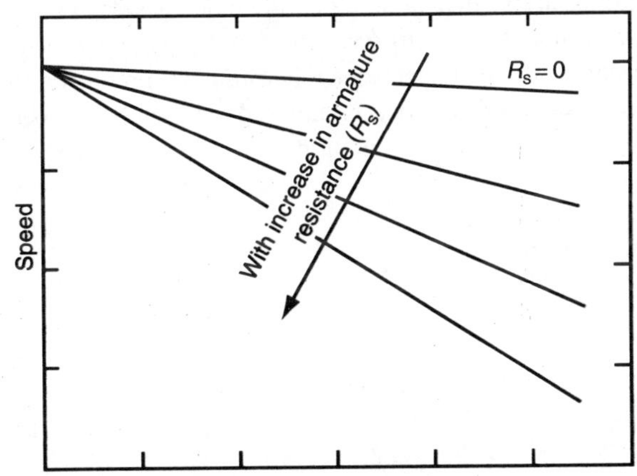

Fig. 1.4 *Torque developed, T_d*

The speed-torque curves for a smooth variation of armature voltage are shown in Fig. 1.5. They move along the *Y*-axis (speed axis) following changes in the armature voltage. The field winding of the motor is supplied from a separate source. The smooth variation of armature voltage brings about speed control in the zero to base speed range very efficiently. The motor operates in a constant torque mode. This method of controlling the speed of a dc motor using variable voltage to the armature is employed in Ward Leonard control.

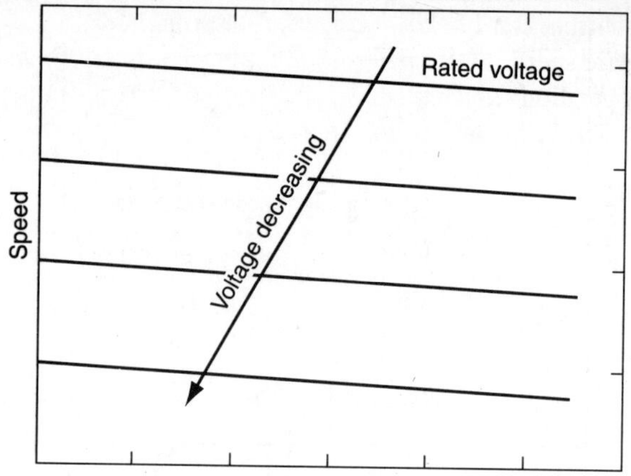

Fig. 1.5 *Torque developed, T_d*

The effect of the variation of field current on the speed-torque characteristic is shown in Fig. 1.6. The field current is decreased to achieve speeds above base speed when the armature voltage reaches its rated value. The flux weakening mode is best suited for constant power applications, since the armature current may be maintained at its rated value. The torque decreases. In the flux weakening mode the motor cannot be used to drive constant torque loads as the motor draws increased currents as the speed increases. This mode is employed to obtain speeds in the range of base speed to twice base speed. The highest speed attainable by flux weakening is limited by commutation.

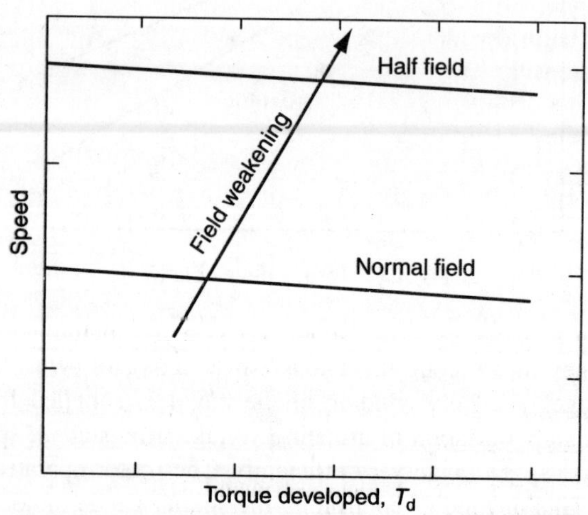

Fig. 1.6 *Effect of field weakening on speed-torque characteristic*

The armature current can remain constant in the complete range of speeds from zero to twice base speed. Constant torque and constant power modes are shown in Fig. 1.7. The operation depicted in Fig. 1.7 is possible with shunt motors also by means of a variable resistance in the field circuit.

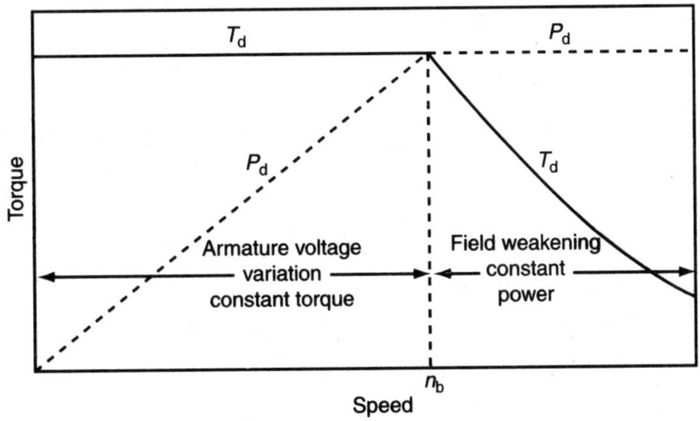

Fig. 1.7 *Constant torque and constant power modes of a separately excited motor*

1.2.2 Modifications to the Speed-Torque Characteristic of a dc Shunt Motor

The preceding discussion shows that armature voltage variation gives creeping speeds. The simple rheostatic method provides a characteristic with little hardness and little stability. Ward Leonard control (smooth variation of voltage), on the other hand, produces a flat characteristic with reasonable hardness and stability, but high initial cost. A simple method with low initial cost, to obtain crawling speeds with sufficient hardness, is depicted in Fig. 1.8. The conventional rheostatic control with a resistance in series with the armature is modified by shunting the armature with a low resistance. By varying the values of series and shunt resistances the speed-torque characteristics can be made to have any desired shape.

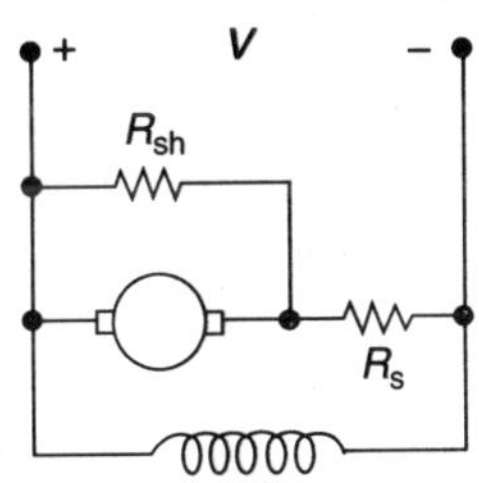

Fig. 1.8 *Modification to speed-torque characteristic by shunting armature*

In the simple rheostatic control using only a series resistance, the voltage across the armature at no-load is V. The no-load speed is decided by V, whatever be the value of R_s. If the armature is shunted by R_{sh} the voltage across the armature becomes less than V even at no load. The no-load speed decreases to the desired value with proper values of R_s and R_{sb}. The smaller the value of R_{sh}, the lesser is the voltage across the armature at no-load. Eventually, the no-load speed decreases. The value of R_{sh} is also effective in making the characteristic flat.

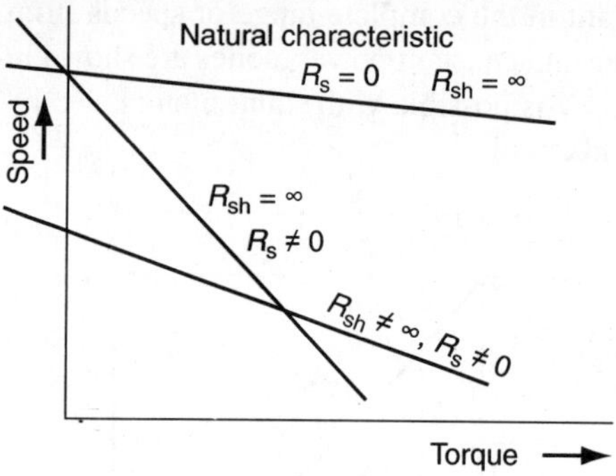

Fig. 1.9 *Typical speed-torque curve*

Typical speed-torque characteristics are shown in Fig. 1.9 in which the natural characteristic of the shunt motor and the characteristic with simple rheostatic control are shown. This modification may be used if stable low speed operation is required. It can be employed for accurate stopping of the drive. By changing the value of R_{sh} the speed can be reduced to a very low value and then suitable mechanical braking may be applied to have accurate stopping.

Referring to Fig. 1.8, we have

$$V_a = E + I_a r_a + I_s R_s \tag{1.6}$$

$$V_a = I_{sh} R_{sh} + I_s R_s \tag{1.7}$$

$$I_s = I_a + I_{sh} \tag{1.8}$$

Using these equations we have

$$V_a = I_s(R_s + R_{sh}) - I_a R_{sh} \tag{1.9}$$

from which

$$I_s = \frac{V_a + I_a R_{sh}}{(R_s + R_{sh})} \tag{1.10}$$

Also from Eqs (1.8) and (1.7)

$$I_s = I_a + \frac{V - R_s I_s}{R_{sh}} \tag{1.11}$$

Using these relations in Eq. (1.6) we have

$$N = \frac{V}{K_e \phi} - \frac{R_{sh}}{(R_{sh} + R_s)} - \frac{I_a}{K_e \phi}\left(r_a + \frac{R_{sh} R_s}{R_{sh} + R_s}\right) \tag{1.12}$$

Substituting for I_a in terms of T_d we have

$$N = \frac{V}{K_e \phi}\left(\frac{R_{sh}}{R_{sh} + R_s}\right) - \frac{T_d}{K_t K_e \phi^2}\left(r_a + \frac{R_{sh} R_s}{R_{sh} + R_s}\right) \tag{1.13}$$

The speed-torque characteristic is shown in Fig. 1.9. The following points are clear from the figure:

i. The no load speed ($T_d = 0$) decreases to $\frac{V}{K_e \phi}(\frac{R_{sh}}{R_{sh} + R_s})$ as the value of $\frac{R_{sh}}{R_{sh} + R_s} < 1$. Smaller the value of R_{sh}, smaller is this value. The slope also decreases if R_{sh} is small. The hardness is thus improved and stable operation is assured when compared to simple rheostatic control.

ii. Smoothness of speed control depends on how R_{sh} and R_s are varied. The speed control is stepped, as the resistances can be varied in a stepped manner.

iii. Speeds below base speed are possible. The no-load speed itself changes following variations in R_{sh}. A sharp drop in the no-load speed may be observed when R_{sh} is decreased. Speed control is achieved by varying the value of R_s. The method is equivalent to making the field stronger and gives results similar to those obtained by increasing the field current at a given armature current.

iv. The method is suitable for constant torque loads, so that the armature current is at its rated value.

v. The method is suitable if accurate stopping is required.

vi. It is not economical for continuous operation. The losses in R_{sh} and R_s make the system inefficient. The method can be employed if stable creeping speeds are required for short periods.

1.2.3 Speed-Torque Characteristics of Series Motor

The field winding is connected in series with the armature (Fig. 1.2(b)). The armature current and field current are the same. The operating speed-torque characteristic may be deduced for a series motor in the same way as described before for shunt motor/separately excited motor using the basic equations (Eqs 1.3 and 1.4). The speed and armature current are related by the equation

$$N = \frac{V}{K_e\phi} - \frac{I_a r_a}{K_e\phi} \tag{1.14}$$

whereas T_d and I_a are related by

$$T_d = K_t \phi I_a \tag{1.15}$$

In a series motor $I_f = I_a$ and hence the field flux

$$\phi = f(I_a) \tag{1.16}$$

For low values of armature current causing no saturation we have

$$\phi = K_f I_a \tag{1.16a}$$

using which we have

$$T_d = K_t K_t I_a^2 \tag{1.17}$$

substituting for I_a in Eq. 1.14, we have

$$N = \frac{V}{C_1\sqrt{T_d}} - \frac{r_a}{K_f K_e} \tag{1.18}$$

where $C_1 = K_e/\sqrt{K_f K_t}$

The speed-torque characteristic of an unsaturated series motor is represented by AB in Fig. 1.10. As the torque increases, the speed drops rapidly to B.

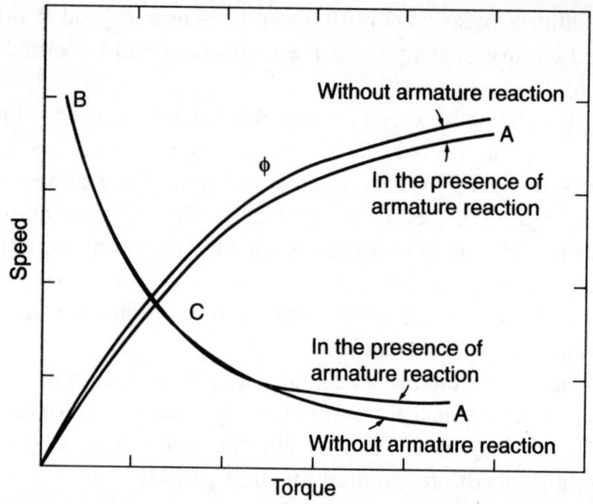

Fig. 1.10 *Speed-torque characteristic of a dc series motor*

When the armature current is greater than the value corresponding to point *B*, saturation sets in and the actual value of flux must be substituted in the equation to get the characteristic. However, to understand the trend of the curve it can be assumed that ϕ is constant. Under this assumption the speed-torque characteristic of saturated series motor is

$$N = \frac{V}{K_e\phi} - \frac{T_d r_a}{K_e K_t \phi^2} \tag{1.19}$$

and *BC* in Fig. 1.10 represents the characteristic. *AC* therefore, is a typical speed-torque curve of a dc series motor. Also, the effect of armature reaction is predominant only when the armature current is large. The speed drop is restricted due to a decrease in the field flux. The effect of armature reaction is also depicted in Fig. 1.10.

Effect of Variation of Armature Voltage The armature voltage variation to a series motor can be achieved by means of an additional resistance in the armature circuit or by using a thyristor power converter. On doing this the torque-speed curve moves towards the right when the voltage is increased. For a given torque developed the motor speed increases with an increase in voltage. The speed-torque curves for different armature voltages (for different armature resistances) are shown in Fig. 1.11. Of the two methods employed for the variation of the voltage the use of additional resistance has greater losses, leading to poor efficiency.

Field Weakening in Series Motor Speeds above base characteristic are normally obtained by means of field weakening. In series motors this is achieved by a diverter coil connected across the field or by a series-parallel combination of field

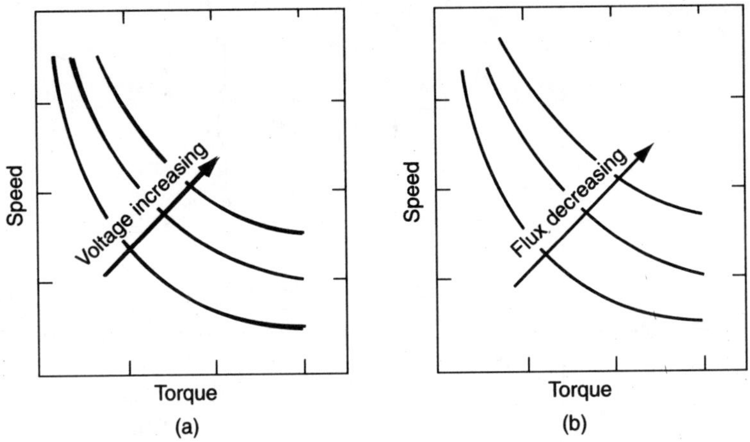

Fig. 1.11 *Torque-speed curves of a series motor (a) at different voltages (b) at different field currents*

coils. Following field weakening the armature current increases to develop a constant torque or a given armature current develops a reduced torque. The armature reaction further demagnetises the field flux, thus enhancing the effect. Due to field weakening the characteristic moves towards the right.

The speed-torque characteristic of the motor shows a steep fall in the speed as the load increases. This mode of operation is suitable for driving the loads of constant power.

1.2.4 Desirable Modifications of the Speed-Torque Characteristic of Series Motors

The light load (no-load) operation of series motors is not possible due to very high speeds. Connecting an additional resistance in series with the armature does not help in reducing no-load speeds. Speeds are reduced only under loaded conditions. However, the load requirement may be such that even at light loads the motor may be required to run at low or medium speeds. The normal methods of controlling the speeds of dc series motors do not give satisfactory solutions. Some special types of connection may be required to operate the lightly loaded series motors at low and medium speeds.

Several such connections are shown in Fig. 1.12.

A resistance connected across the armature and field provides a speed control downward from its base speed (Fig. 1.12(a)). The voltage drop across the series resistance causes a speed drop and the torque developed at this speed is reduced by diverting the current through R_{sh}. In this connection the field and armature currents are the same. Hence it fails to provide a light load operation at low or medium speeds. However, the shift of the speed-torque characteristic towards the left is more than that achieved with simple conventional rheostatic control.

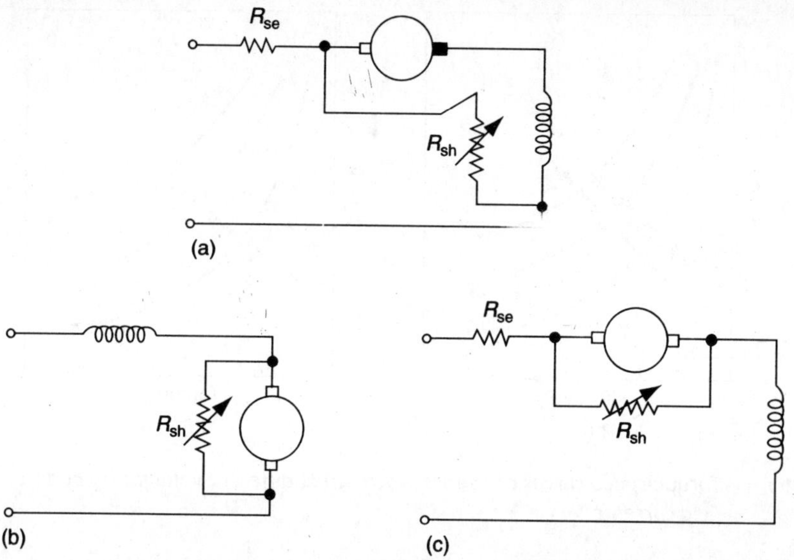

Fig. 1.12 *Some possible methods of shunting armature and field of a dc series motor to modify its speed-torque characteristic*

A better connection is shown in Fig. 1.12(b) where a resistance is shunted across the armature. Thus, only a part of the line current flows in the armature and the field current is maintained at the line value. By properly selecting the shunting resistance, a desirable speed-torque characteristic may be obtained. The higher field current stabilises the speed of lower values and the smaller armature current develops a lower torque. Very low speed operation is made possible by further connecting a resistance in series with the field. The shift of the characteristic towards the left is more due to R_s. Typical speed-torque curves are shown in Fig. 1.13. The characteristics show a greater hardness, being flat over a range of speeds. The method, however, is not economical due to losses in the resistances and is suitable for driving the load at low speeds for short intervals of time. A wide range of speeds below the base speed is not attainable with this method.

When medium speeds are required, the shunting resistance may be disconnected. The motor then operates with series resistance R_s. When R_s is short circuited the motor runs on its natural characteristic. It is possible to operate the motor from no-load to full load at very low, medium and high speeds.

While lowering the load using a hoist, sometimes, the empty cage may have to be driven at low, medium or high speeds. High speed operation may be required for rapid lowering. A suitable connection for controlling the speed in such cases is shown in Fig. 1.14. The scheme is called the potentiometer lowering circuit. In this connection, the field of the motor is connected across the armature with a limiting resistance (R_{sh}) in series. R_a is the resistance in series with the armature. A variable resistance R_{se} is connected in series with the motor. The armature of the

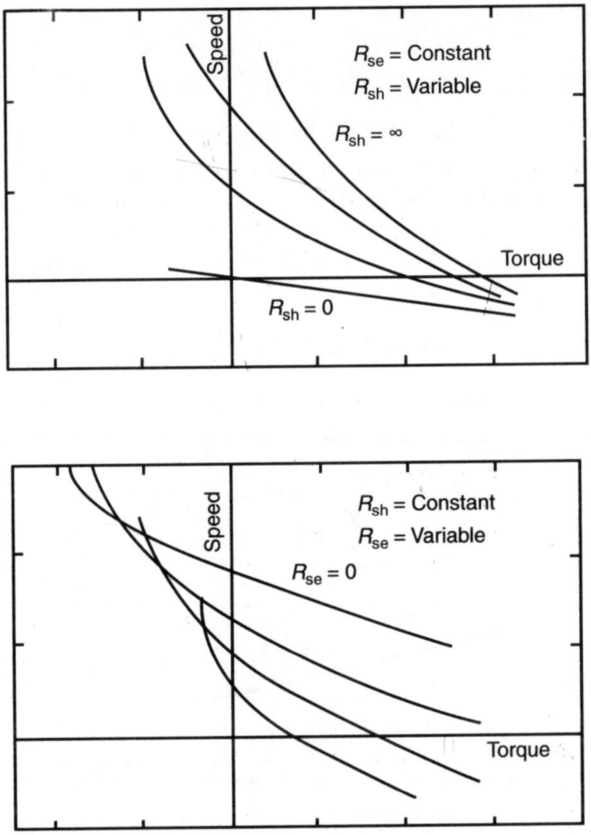

Fig. 1.13 *Typical speed-torque curves of connection (c) of Fig. 1.12*

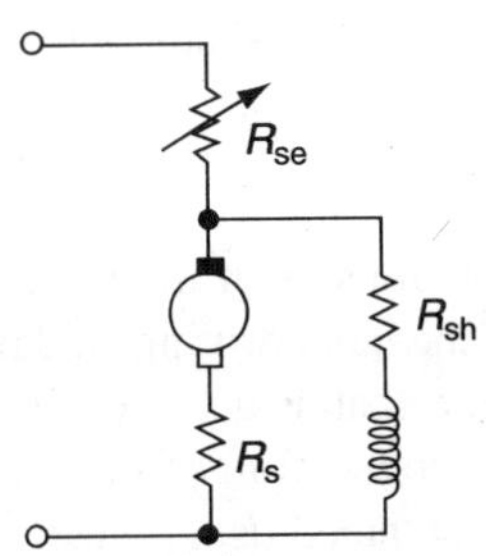

Fig. 1.14 *Connections of a dc series motor as a shunt motor*

dc motor gets its supply from the potential divider formed by R_{se} and the combined resistance of the field and R_{sh}. The motor speed-torque characteristic can be derived from the following equations.

The armature voltage

$$V_a = V_s - IR_{se} \qquad (1.19a)$$

The field current

$$I_f = \frac{V_a}{(R_{sh} + R_f)} \qquad (1.19b)$$

The armature current

$$I_a = I - I_f \text{ for normal connection}$$

The back emf

$$E_b = V - I_a R_a \qquad (1.19c)$$

The torque developed

$$T_d = \frac{E_b I_a}{\omega_r} \tag{1.19d}$$

The motor has a shunt characteristic. The derived speed-torque curves may be obtained by varying the resistance in series with the field. Reverse rotation is achieved by reversing the supply to the armature. A set of speed torque obtained by this connection is shown in Fig. 1.15, which indicates the possibility of operation in the fourth quadrant. The series field in this connection carries rated current

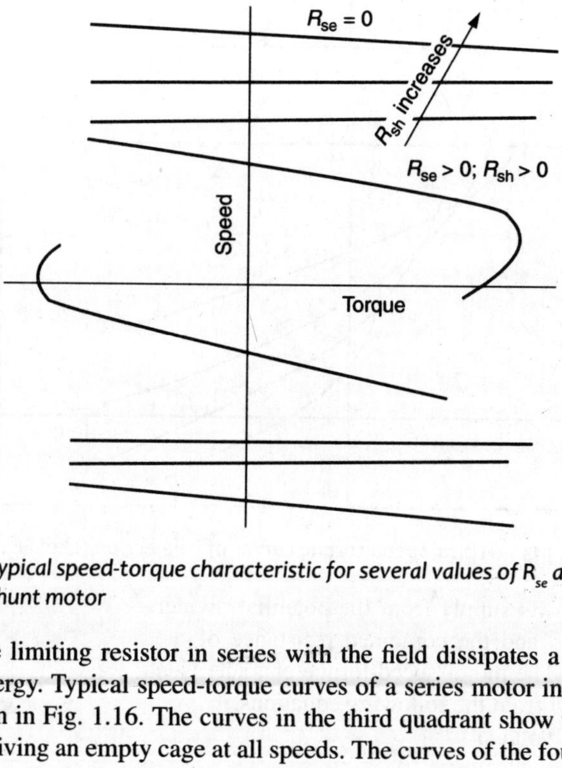

Fig. 1.15 *Typical speed-torque characteristic for several values of R_{se} and R_{sh} in a shunt motor*

and hence the limiting resistor in series with the field dissipates a considerable amount of energy. Typical speed-torque curves of a series motor in this application are shown in Fig. 1.16. The curves in the third quadrant show the operation of the hoist driving an empty cage at all speeds. The curves of the fourth quadrant show that heavy loads may be lowered at these speeds. These may be obtained by the connection of Fig. 1.17. The speed-torque curves in the fourth quadrant may be obtained using the connection of Fig. 1.17(a).

1.2.5 Speed-Torque Characteristics of Compound Motors

A schematic of a compound motor is shown in Fig. 1.18. There are both series and shunt fields (separately excited fields). The orientation of the series flux may be such as to aid or oppose the shunt flux in the magnetic circuit. In the former case, with the additive nature of the component fluxes, the motor is said to be a cumulatively compound motor. In the latter case, it is called a differential compound motor. Again, depending upon the position of the series field the motor may be long shunt

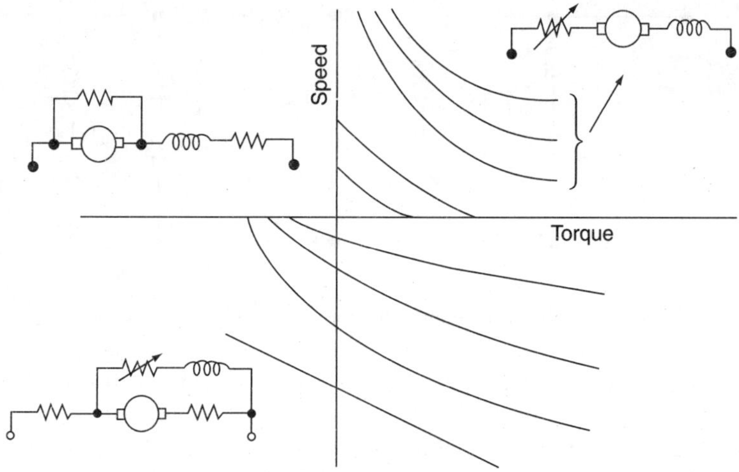

Fig. 1.16 *Modifications of series motor connections for hoist applications*

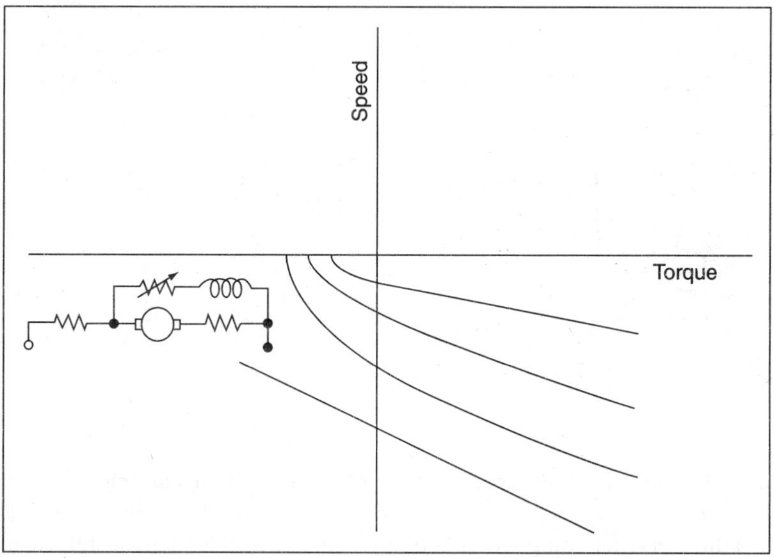

Fig. 1.17 *Series motor connections for speed-torque curve in fourth quadrant*

or short shunt. In the former, the series field is in series with the armature, whereas in the latter it is in series with the line. These are shown in Fig. 1.18.

The torque-speed characteristic of a compound motor may be obtained with reference to the following equation

$$N = \frac{V}{K_e \phi} - \frac{T_d (r_a + r_s)}{K_e K_t \phi^2} \tag{1.19e}$$

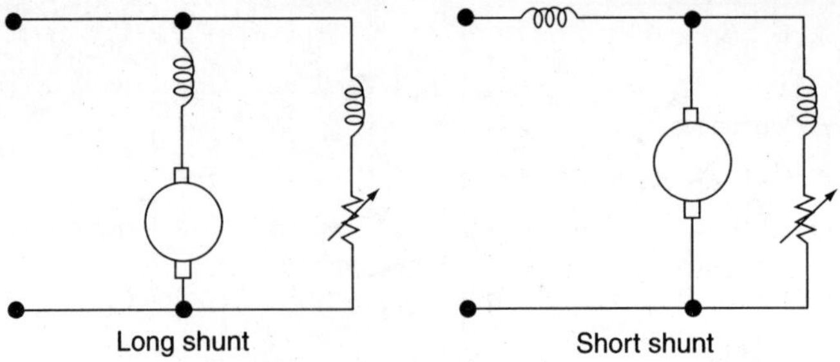

Fig. 1.18 *Compound motor connections*

In a cumulatively compound motor the flux increases with load. Therefore, the speed falls as the load increases. The curve has a drooping nature and shows less hardness than that of a shunt motor. However, the speed drop is limited by saturation and armature reaction. A typical characteristic is shown in Fig. 1.19.

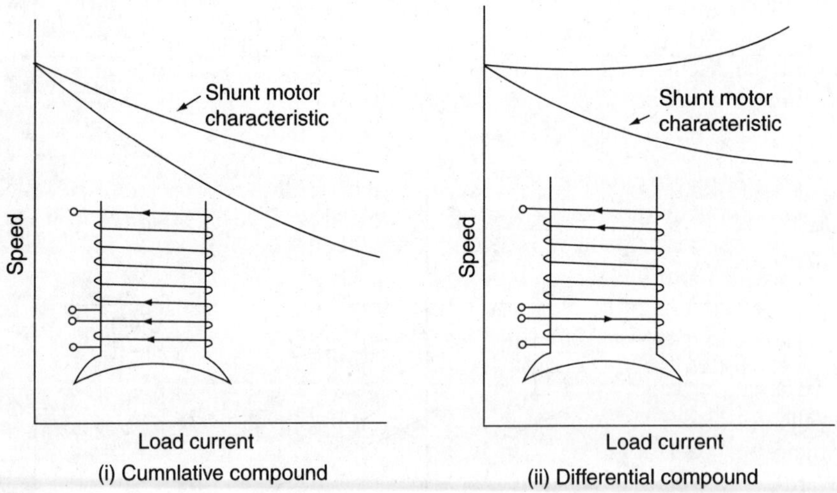

Fig. 1.19(a) *Typical load current versus speed of compound motors*

In a differentially compound motor the air gap flux decreases with load, causing an increase in the speed as load increases. There may be no saturation effects. The armature reaction demagnetises the air gap flux causing a further increase in speed. It is less stable than a cumulatively compound motor. Typical speed-torque curves are shown in Fig. 1.19.

1.2.6 Speed Control of dc Motors

In the foregoing sections we discussed the torque-speed characteristics of dc motors, bringing out the effects of armature voltage variation and field current on

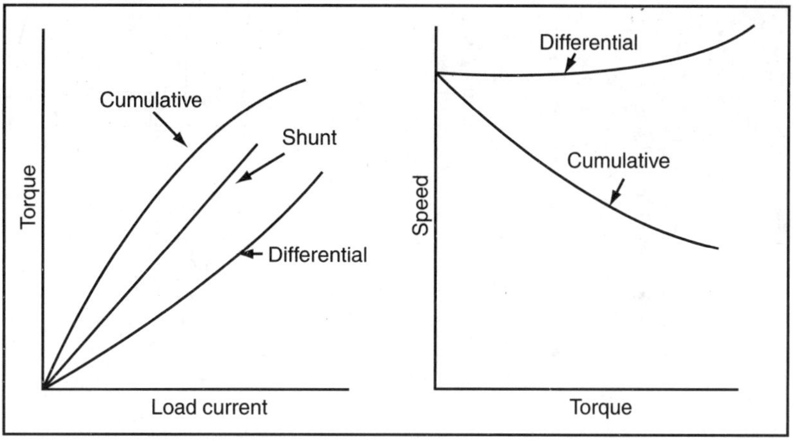

Fig. 1.19(b) *Torque versus armature current and Torque versus speed curves of a compound motor*

them. The discussion also applies to the study of the methods of speed control of dc motors, which may be summarised as follows:

i. Variation of armature voltage by inserting an additional resistance in the armature circuit.
ii. Smooth variation of applied voltage using conventional Ward Leonard control or static Ward Leonard control employed for separately excited dc motors.
iii. Variation of field current either by inserting a resistance in the field circuit or by varying the field voltage using thyristor power converters.

The conventional Ward Leonard method of speed control, where the voltage to the drive motor is supplied by a variable voltage generator, suffers from low efficiency, need for periodic maintenance, bulk and size restrictions, etc. Static Ward Leonard control using either phase controlled rectifiers or choppers has now become very popular.

A comparison of these methods bringing out their salient features may prove worthwhile. The comparison given in Tables 1.1 and 1.2 is based on the following aspects:

i. Limit or range of speed control (Hardness of speed control)
ii. Smoothness of speed control
iii. Economics of speed control
iv. Stability of operation
v. Direction of speed control
vi. Permissible load.

Table 1.1 *Comparison of the speed control methods of shunt/separately excited dc motor*

Basis of comparison	*Significance*	*Armature resistance control*	*Field control*	*Ward Leonard control*
1. Limit, range and hardness of speed control	Signifies the ratio of $\frac{N_{max}}{N_{min}}$ constancy of speed with load	The speeds in the range of zero or almost crawling to base speeds are possible. The slope of the characteristic increases and the curve becomes less hard with increase in the resistance. The range of speed control depends on the load.	The speed control is obtained in the range base to twice base speed. Speeds below base speed require field currents above the rated value. Hence, in view of the field rating it is not employed for speeds below base speed. The upper speed limit by field weakening is decided by the armature reaction and commutation. The hardness is maintained and is almost the same as that at rated field current and rated voltage if the motor is compensated otherwise the armature reaction affects the hardness.	An elegant method for smooth variation of speed over a wide range. Speed reversal is possible by reversing the generator field. The speed control in both the directions of rotation up to base speed. The field weakening may be employed to increase the speed above base value. Hardness of the speed-torque curve does not materially change. The char characteristics obtained with variation of voltage arc all parallel to the normal characteristic. The characteristic is sufficiently hard. The hardness is affected by armature reaction, armature resistance. In static Ward Leonard control the converter voltage regulation affects the hardness of speed torque curve.

2. Smoothness of speed control	Signifies the number of steady speeds in the operating range. Larger the number of steps, smoother the speed control.	In large power drives speed control is obtained by varying the armature resistance using contactors. The control is normally stepped and is smoother with larger number of steps. If the resistance is varied smoothly by means of a motor the speed control would be smooth.	The field circuit is a low power one. The resistance in the field circuit can be smoothly varied. This method offers a very smooth speed control of speeds above the base speed.	Speed control is achieved by controlling the generator field. The power required for control is low. A smooth variation of voltage makes a smooth speed control possible. The smoothness with which the generator field can be varied decides the smoothness of speed control. In static Ward Leonard control the smoothness of speed depends upon the fine control of voltage using a rectifier or chopper.
3. Economics of speed control	Decided by the efficiency of the motor, maintenance and overall cost	The power loss in the armature resistance reduces the efficiency. The efficiency is very poor, particularly at low speeds. The power input is constant while driving constant torque loads. As the speed is lowered a larger part of the input is wasted in the resistance. The efficiency is poor at lowspeeds, it is approximately equal to operating speed/base speed. Longer operating periods at low speeds are not economical. Frequent replacement of resistance may be required.	The method is economical. Power to be handled is only 5 per cent. The efficiency is not very much affected, because the field losses decrease and rotational losses increases. The method is well suited for constant power loads.	High initial cost both for conventional and static Ward Leonard controls. Efficiency is better than resistance control. Owing to several rotating machines, overall efficiency may decrease. The bulk of the system and need for considerable space are the disadvantages. Static systems have better efficiency and are compact.

(*Continued*)

Basis of comparison	*Significance*	*Armature resistance control*	*Field control*	*Ward Leonard control*
4. Stability of operation	The capacity of the drive to return to its original state or a state of new operating conditions following a disturbance in speed or torque	The motor operation is stable due to the drooping nature of the speed-torque curve. The operating conditions affect stability. The hardness of speed control affects the stability. Harder the characteristic more stable is the operation. The stability is decided by the torque driven at given speed.	The weak flux conditions in the machine are prone to increased degree of armature reaction. At high speeds the cumulative effects of weak flux and armature reaction may cause stability problems. The stability considerations also place a limit on the upper speed limit.	Stability is very good at speeds below base speed. At higher speeds obtained by field weakening the stability is a matter of concern.
5. Direction of speed control.	Signifies the posibility of speed control below and above base speeds.	Only speeds below base speed up to zero are possible, because of the motor's inherent resistance and additional resistance. Speed control is possible only in the downward direction from base speed.	Only speeds above base speed are possible due to field rating. The upper limit is set to twice the base speed in view of commutation, stability and centrifugal forces. Speed control is possible only in the upward direction from the base speed.	Speed variation below base speed by armature voltage control and above base speed by field control.

6. Permissible load	Signifies whether a machine is overloaded or underloaded in its speed range and thus utilisation of the machine.	Suitable for constant torque loads. The drive has low efficiency at low speeds. As the motor draws heavy current this is not suitable for constant power at low speeds. The underloading of the motor does not occur as the armature current is maintained at its rated value.	Suitable for constant power loads. Torque is a parabolic function of speed. The current can be maintained at its rated value. The method is not suitable for constant torque loads as the armature current drawn is excessive. If the armature current is limited it leads to underutilisation of the motor.	Economical and suitable for constant torque loads when armature voltage is varied and constant power loads when field current is controlled. With pulsating loads a flywheel may be used to equalise the load.
7. Others		Special cooling may be required at very low speeds for the motor as well as the resistance. The self ventilation of the motor becomes poor.	At high speeds it presents poor commutation due to increased reactance voltage. Stability problems due to armature reaction. Mechanical design of the armature for running at high speeds may be required.	Requires no starting equipment. For speeds above base speeds due consideration must be given to commutation and stability.

Table 1.2 *Speed control of dc series motors*

Basis of comparison	*Armature resistance control*	*Excitation control*	*Armature voltage control*	*Shunting armature and field or armature alone*	*Shunting the field to the armature*
1	2	3	4	5	6
1. Limit of speed control	Speed control in the range 2:1 or 3:1. Depends on load, the hardness of the speed–torque characteristic decreases with increase in armature resistance.	Control is achieved by shunting the field by a diverter. Effectively, the method is field weakening. Hardness of the speed–torque curve also decreases. Speed control range is 2:1.	accomplished by series parallel control of motors. The range of speed control depends upon the number of motors. This range can be widened by/means of additional resistances.	When both armature and field are shunted, low speeds at light loads are not possible. To get low speeds at light loads the armature alone is shunted. The series motor may be made to operate at no-load with definite low speed. Speed ranges of 3:1 to 5:1 may be obtained. The hardness of the characteristic is improved at light loads.	The speed torque can be modified and desired characteristic may be obtained. The series motor is made quite useful in this connection. Very good speed range is possible by properly varying resistances.
2. Smoothness of speed control	Speed control is stepped. The variation of resistance is accomplished by means of contractors.	The field current handled is of the same order as the armature current. Speed control is stepped here also.	Stepped speed control, Even with additional resitances. However the step can be decreased in size.	Smoothness depends upon the shunting resistor and its variation. Normally contactors are used and therefore control is stepped.	The resistance variations are stepped, as is the speed control.

3. Economics of speed control	Heavy power loss in the resistance. Resistance needs replacement. However it is popularly employed in cranes and traction as the method is simple. Capital cost may be less.	Power loss takes place in the shunting resistor. This loss is of the order of the field copper loss and the method is efficient, The shunting value of the resistance is of the order of the field value.	No power is wasted if no resistances are used. Very good η. Use of several motors has advantages: 1. Time of starting and braking decreases due to decrease in intertia. 2. Reliability of operation. At least one motor is operating. Continuity of operation is assured. 3. Due to space limitation several motors may be preferred to a single one. If resistances are used, η is poor.	Power loss in shunting resistors make the method limited to short time duty. The method is not economical.	The resistances carry nearly full load value. Poor efficiency due to power loss.
4. Stability of operation	The characteristic becomes less hard as the resistance increases. The stability of operation also becomes somewhat poor.	At very weak fields the armature reaction may result in unstable operation.	Stable operation	Operation is stable at both light loads and light speed.	Operation may not be stable at every point in the high speed operation obtained by field weakening. At low speeds the hardness improves, thereby improving stability.

(*Continued*)

Basis of comparison	*Armature resistance control*	*Excitation control*	*Armature voltage control*	*Shunting armature and field or armature alone*	*Shunting the field to the armature*
1	2	3	4	5	6
5. Direction of speed control	Speeds below the base speed are possible.	The speeds can be varied in the upward direction from the base speed due to flux weakening.	Speeds below base speed.	Very low speeds with light loads. Speeds well below base speed are possible. At these speeds the characteristic is sufficiently hard.	Speed variation in both directions, below and above base speed.
6. Permissible load	To drive constant torque loads so that the current drawn is within the rated value. Due to heavy losses and low efficiency the method is advisable only for intermittent duty.	To drive constant power loads if a wide range of speeds is required. With constant torque the current drawn increases and range of speed control decreases. Economical and advantageous with constant output; torque decreases at high speeds.	Speed control at constant torque so that no motor is overloaded and every motor is fully utilised.	When used for constant torque loads speed ranges of 3:1 to 5:1 are obtained. To limit the series field to its rated value the motor must have lower torques at lower speeds.	Constant torque operation when the speed is below base rated speed. Constant power operation at high speed. Suitable for crane drives requiring rapid lowering of empty cage.

1.3 CHARACTERISTICS OF A THREE-PHASE INDUCTION MOTOR

The performance characteristics of a three-phase induction motor can be derived using the approximate equivalent circuit shown in Fig. 1.20(a). In the circuit

V_1 is the applied voltage
r_1, x_1 are the stator resistance and leakage reactance per phase respectively
r'_2, x'_2 are the rotor resistance and leakage reactance referred to the stator
x_m is the mutual reactance between stator and rotor
I_1, is the stator current
I'_2 is the rotor current referred to the stator, and
s is the slip of the motor.

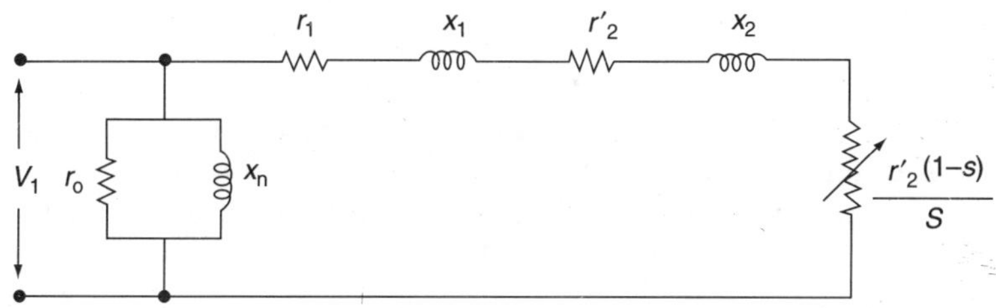

Fig. 1.20(a) *Approximate equivalent circuit of a three phase induction motor*

The phasor diagram of the motor is shown in Fig. 1.20(b).

In an induction motor the power transferred to the rotor (P_d), rotor copper loss (P_{cu2}) and mechanical power developed (P_m) are in the ratio of $1 : s : (1 - s)$. The torque developed by the motor

$$T_d = m_1 \frac{P_m}{2\pi n_r} Nm \qquad (1.20)$$

But using the relations $P_m = P_d(l - s)$ and $n_r = n_s(l - s)$ we have

$$T_d = m_1 \frac{P_d}{2\pi n_s} Nm \qquad (1.21)$$

Further using $P_d = \frac{P_{cu2}}{s}$

$$T_d = \frac{m_1}{2\pi n_s} \; \frac{\text{rotor copper loss}}{s} \qquad (1.22)$$

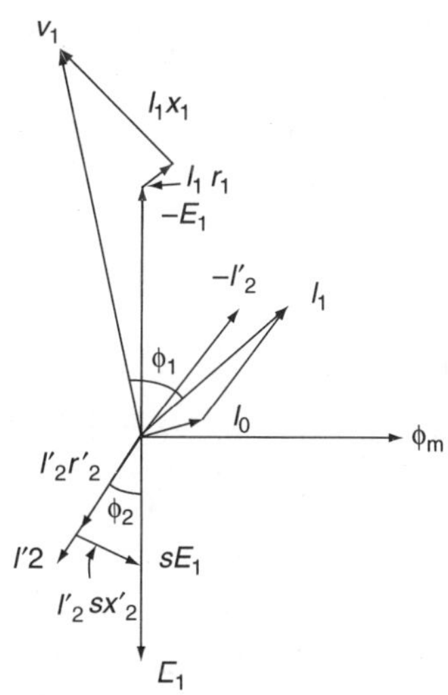

Fig. 1.20(b) *Phasor diagram of a three phase induction motor*

From the equivalent circuit

$$I_2' = \frac{V_1}{\sqrt{\left(\left(r_1 + \frac{r_2'}{s}\right)^2 + (x_1 + x_2')^2\right)}} \tag{1.23}$$

The torque developed is therefore

$$T_d = \frac{m_1}{2\pi n_s} \frac{V_1^2}{(r_1 + r_2'/s)^2 + (x_1 + x_2')^2} \frac{r_2'}{s} \tag{1.24}$$

At very small slips (operating region of the motor) $\left(r_1 + \frac{r_2'}{s}\right) \gg (x_1 + x_2')$ and $r_2'/s \gg r_1$ leading to

$$T_d = \frac{m_1}{2\pi n_s} \frac{V_1^2}{r_2'} (s) \tag{1.25}$$

This shows that the torque developed is directly proportional to the slip. In a similar way, at large slips the torque varies in inverse proportion to the slip. At intermediate slips the torque developed needs to be calculated. The speed (slip) torque characteristic is shown in Fig. 1.21.

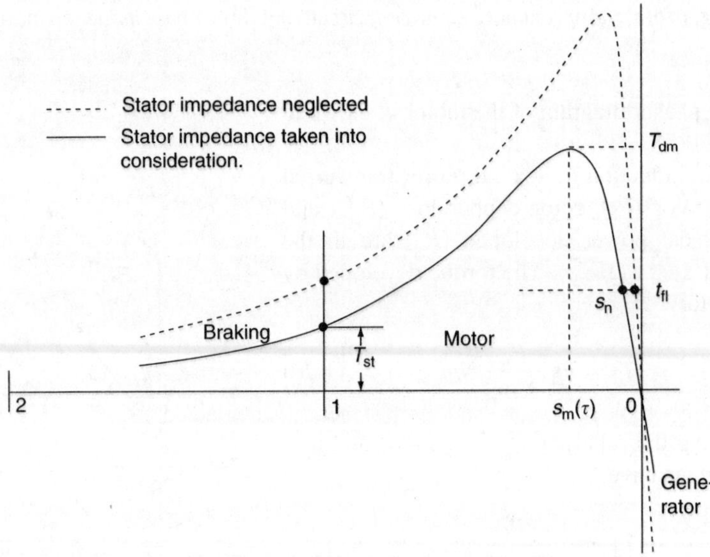

Fig. 1.21 *Typical speed torque curve of a three phase induction motor*

From the trend of the characteristic it can be seen that there is a maximum value for the torque. The slip at which this maximum torque occurs is given by

$$s_{m(T)} = \pm \frac{r_2'}{\sqrt{r_1^2 + (x_1 + x_2')^2}} \tag{1.26}$$

The maximum torque is

$$T_{dm} = \frac{m_1}{2\pi n_s} \frac{V_1^2}{2\left[r_1 \pm \sqrt{r_1^2 + (x_1 + x_2')^2}\right]} \tag{1.27}$$

Sometimes the stator impedance is neglected or the stator impedance drop is compensated to operate the motor at constant flux at all slips. The applied voltage is increased such that the induced voltage. E_1/f is constant (equal to the value at rated voltage). The applied voltage varies as a function of frequency such that E_1/f remains constant. The stator impedance can be assumed to be zero since it has no effect. In such a case the relations are

$$T_d = \frac{m_1}{2\pi n_s} \frac{E_1^2}{(r_2'/s)^2 + (x_2')^2} \frac{r_2'}{s} \tag{1.28}$$

$$s_{m(T)} = \pm \frac{r_2'}{x_2'} \tag{1.29}$$

$$T_{dm} = \frac{m_1}{2\pi n_s} \frac{E_1^2}{2x_2'} \tag{1.30}$$

The speed-torque curve for this case is also shown in Fig. 1.21.

Note the following features of a typical speed-torque curve (Fig. 1.21):

i. At exactly synchronous speed $s = 0$ the torque developed is zero ($T_d = 0$). This can be expected because there are no induced currents due to zero relative speed.
ii. Full load torque (T_{fl}) corresponds to the rated slip (s_{fl}).
iii. T_{dm} is the maximum torque at the slip $s_{m(T)}$
iv. T_{st} is the starting torque at $s = 1$.

The torque developed at any slip s expressed as a fraction of maximum torque is given by

$$\frac{T_d}{T_{dm}} = \frac{2(1 + as_m)}{s/s_m + s_m/s + as_m} \tag{1.31}$$

where $a = r_1/r_2'$. When the stator resistance r_1 is neglected

$$\frac{T_d}{T_{dm}} = \frac{2}{s/s_m + s_m/s} \tag{1.32}$$

The operation of the motor in the range of slips $0 - s_m$ is stable. When the motor is operating in this range any disturbance in the operating point by change of either speed or torque is damped out and the motor returns to its original operating point or attains a new one. For stable operation the torque developed must increase when the speed falls, i.e.,

$$\frac{dT_d}{dn_s} \text{ should be negative}$$

The operation of the motor in the range s_m to l is unstable. In this range the curve has a positive dT_d/dn_s, i.e. torque decreases when speed falls.

The characteristic is almost linear at very small slips (in the stable operating region). This linearity continues till the break down torque point for the case of operation with constant flux. The characteristics shown in Fig. 1.21 are redrawn in the same figure.

For slips greater than unity, the operation is in the fourth quadrant. The rotation of the rotor and the rotating magnetic field are in opposite direction. The torque developed is a braking torque, tending to stop the motor. This can occur in two ways:

i. The phase sequence of the supply to the motor is reversed while it is running.
ii. A negative torque is applied to the shaft.

The motor is operated as a brake in the range of slips ($s > 1$) to make it drive the load at constant speed while lowering the load. The torque is positive, whereas the direction of rotation is reversed. By a suitable resistance of the rotor the point of operation is shifted to the quadrant of operation so that the load is lowered at constant speed.

The torque-speed curve extends to the second quadrant, representing a negative torque in the forward direction of rotation. This occurs if the speed of the rotor is greater than the synchronous speed. Any tendency of the rotor to accelerate beyond synchronous speed is arrested by a generating torque. In this mode of operation all the kinetic energy connected with increase in speed is returned to the mains. The maximum (break down) torque depends on the following:

It varies as the square of the applied voltage.
It decreases with stator impedance.
Its value is independent of the rotor resistance.
Its value decreases with an increase in the rotor leakage reactance.

1.3.1 Desirable Modifications to the Speed-Torque Characteristic of an Induction Motor

Additional Rotor Resistance From Eq. 1.27 it can be observed that the maximum torque is independent of rotor resistance. However, the slip at which the maximum torque occurs changes with rotor resistance. When the rotor resistance is increased, so is the slip for maximum torque, and the stable operating slip range of the motor increases. Typical characteristics of an induction motor for different values of rotor resistance are shown in Fig. 1.22. From the figure it is seen that the starling torque can be increased by increasing the rotor resistance. The maximum torque occurs at starting if the rotor resistance is increased to a value.

$$r_2' = \sqrt{r_1^2 + (x_1 + x_2')^2} \tag{1.33}$$

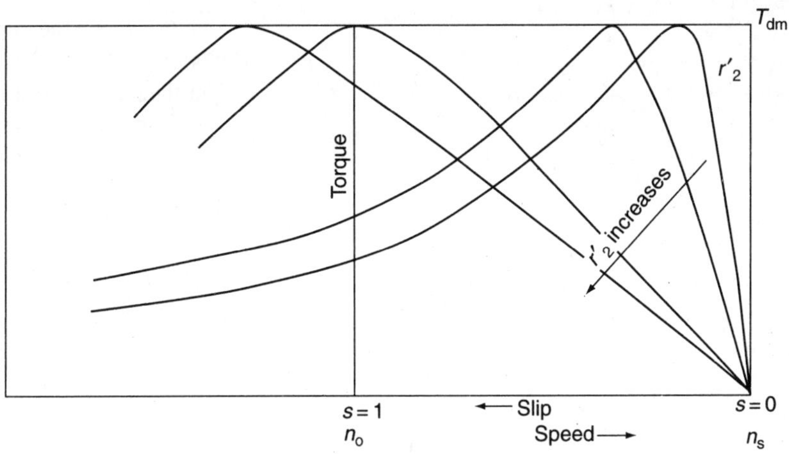

Fig. 1.22 *Effect of variation of rotor resistance on the speed-torque curve*

If the stator impedance is neglected the rotor resistance needs to be increased to a value equal to the rotor leakage reactance. If the rotor resistance is increased beyond this value the starting torque decreases. The breakdown torque occurs at slips greater than one (in the braking region). The starting current decreases and the starting power factor is better at increased values of rotor resistances. The full load slip changes, facilitating speed control in a limited range when the rotor resistance is varied. However, efficiency is impaired at high rotor resistances due to increased losses. Rotor heating is present in an inherently high-resistance rotor.

In short, the starting performance of the motor is improved with large rotor resistances while the running performance is impaired. To get the advantages of a high rotor resistance at starting, an additional resistance is connected in the rotor circuit of the wound rotor induction motor and slowly cut off as the rotor accelerates. At rated speed the motor operates on its natural characteristic. The connections are shown in Fig. 1.23(b).

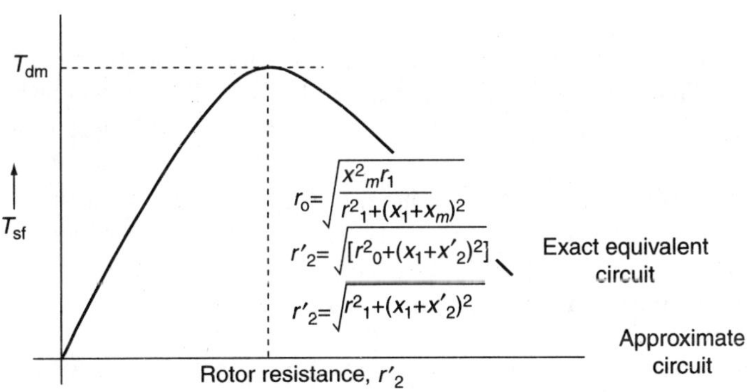

Fig. 1.23(a) *Starting torque as a function of rotor resistance*

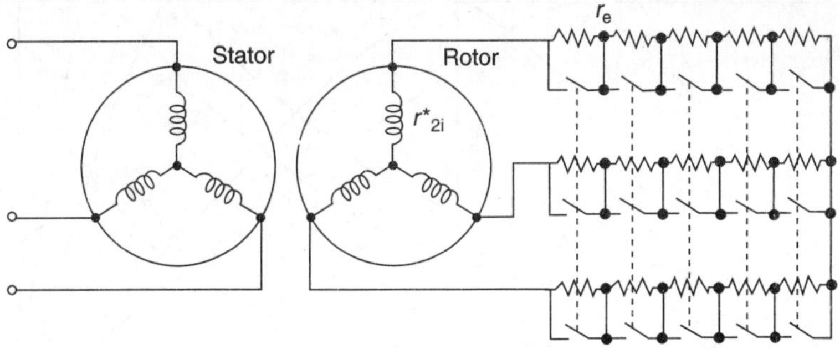

Fig. 1.23(b) *Slip ring rotor with additional rotor resistance. Rotor resistance* $r_2 = r_{2i} + r_e$

However, connecting an additional resistance is not possible in squirrel cage motors. Special rotor constructions, such as double cage and deep bar rotors are employed. At staring, due to high rotor frequency the current distributes in the outer cage of a double cage rotor or in the top portion of the bar in the case of a deep bar rotor. The effect of high resistance is thus achieved. As the motor speeds up, the rotor frequency decreases and the current distributes in both the cages of the double cage rotor or in the complete bar in the deep bar rotor. The effective resistance is small and running performance is improved. Typical torque-speed curves are shown in Fig. 1.23.

Variation of Applied Voltage The speed-torque characteristic of an induction motor can be modified by varying the applied voltage, Typical speed-torque characteristics of the motor when supplied from variable voltage at rated frequency are given in Fig. 1.24(a). They are based on the fact that the induction motor torque (at a given slip) varies as the square of the voltage. The slip for maximum torque is independent of voltage. The full load torque occurs at different slips when the voltage is varied. This renders the speed control of induction motors feasible over a limited range by supply voltage variation However, the torque capability of the motor decreases at low voltages, because of reduction in the air gap flux. The power factor decreases. The motor draws heavy currents to develop a given torque at low voltages. The current drawn at different voltages is shown in Fig. 1.24(a), along with the torque developed at rated current at different voltages.

Figure 1.24(b) shows the advantages of high resistance in the rotor when the applied voltage is varied to modify the speed-torque characteristic. Besides increasing the range of speed control, the current drawn by the motor at low voltages can be limited by a proper choice of rotor resistance.

Pole Change Motors The speed-torque curve of an induction motor can be modified by an armature winding reconnected to give different sets of poles. When

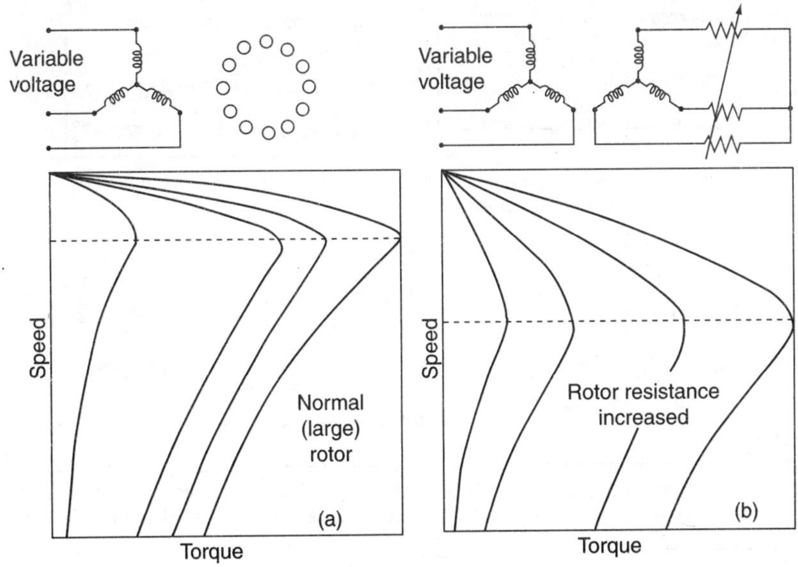

Fig. 1.24 *Effect of voltage variation on speed-torque curve*

the number of poles changes, so does the speed. The type of connection decides the permissible loading at constant torque or constant power. This method is suitable for squirrel cage motors as their rotors can adopt to any number of poles. No reconnection of the rotor winding is required. If, on the other hand, a slip ring rotor is used, it must be reconnected to different sets of poles. The consequent pole winding is employed for reconnection. The coil pitch effectively changes at different speeds.

Each phase has a winding split into halves. These are connected either in series or in parallel, to effectively change the number of poles. The possible combinations are shown in Fig. 1.25. Constant power operation is provided by the series-delta connection for high speeds and parallel-star for low speeds. At high speeds, a low torque is developed so that the power is constant. Voltage per half is $V/2$ in the high speed connection and $V/\sqrt{3}$ at low speeds.

Constant torque operation is possible at both speeds if parallel-star is used for high speed operation and series-delta for low speed. In this case, the voltage per half is $V/\sqrt{3}$ at high speeds and $V/2$ at low speeds. Series-parallel and series-star connections for high and low speeds respectively make variable torque loading possible. The half winding voltages are $V/2\sqrt{3}$ and $V/\sqrt{3}$ respectively.

Different pole changing connections and typical speed-torque curves are depicted in Fig. 1.25(a), (b). Several possible connections for high and low speeds are given in Fig. 1.25(c) and compared in Table 1.3.

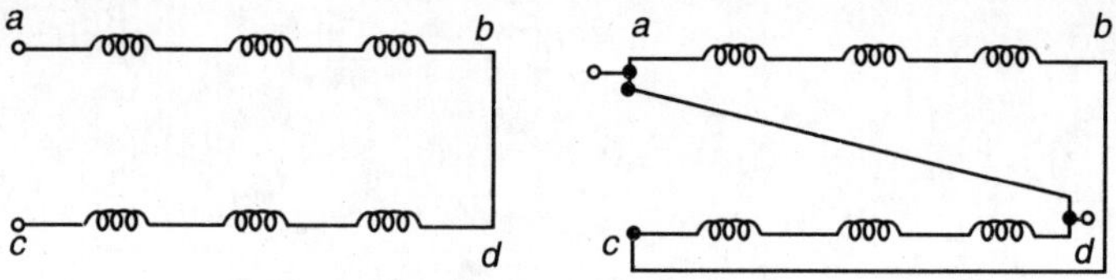

Series parallel connections of winding to produce a field of required number of poles (6-pole field).

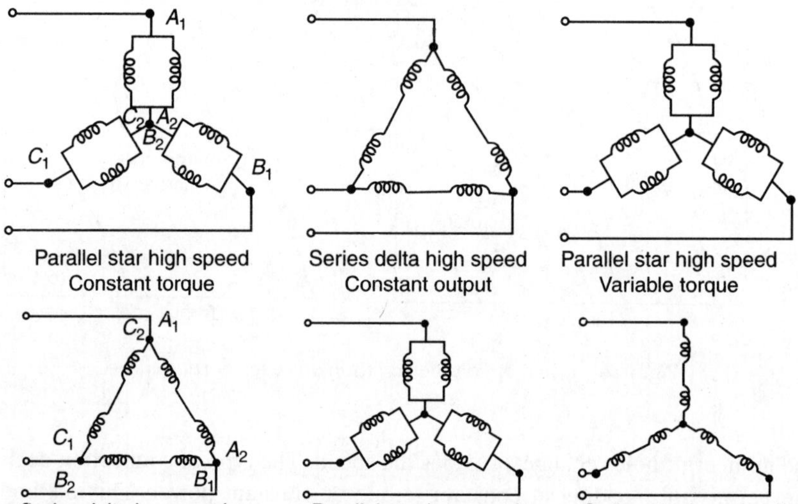

Fig. 1.25(a) *Pole change connections*

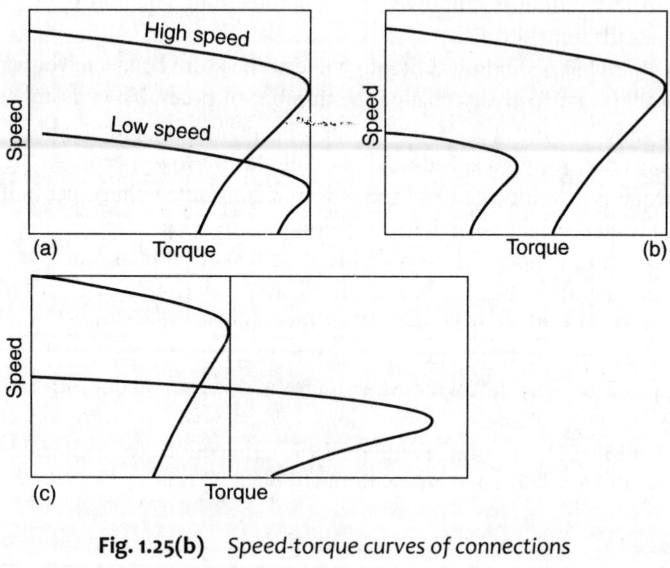

Fig. 1.25(b) *Speed-torque curves of connections*

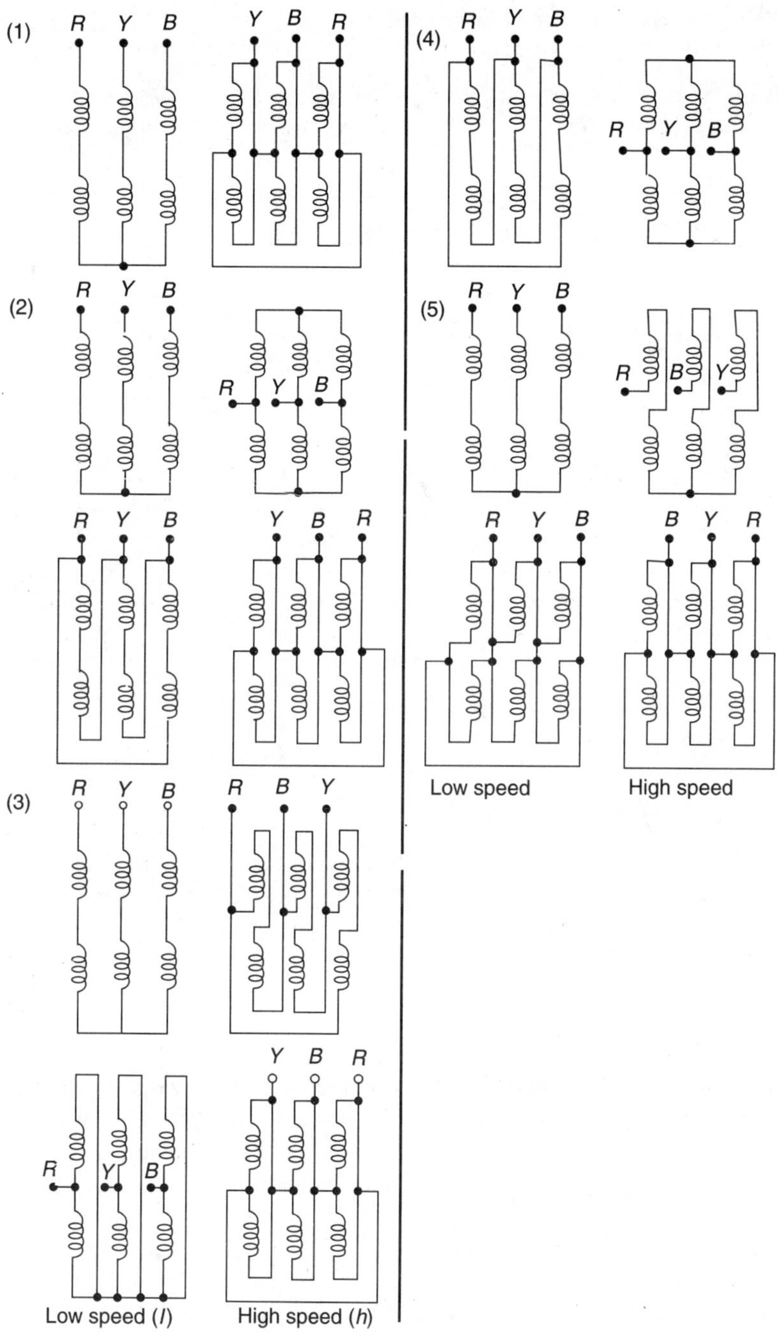

Fig. 1.25(c) *Pole change connections*

Table 1.3 *Comparison of pole change windings*

Connection Fig.	*Low speed subscript 'l'*		*High speed subscript 'h'*		*Flux density B_l/B_h*	*Type of operation*
	Half winding connection	Phase connection	Half winding connection	Phase connection		
1.	Series	Star	Parallel	Delta	0.58	Variable torque
2.	Series	Star or delta	Parallel	Star-star or delta	1.00	Constant torque
3.	Series or parallel	Star	As with low speed operation	Delta	1.16	Constant torque
4.	Series	Delta	Parallel	Star-star	1.73	Constant horse power
5.	Series or parallel	Star or delta	As with low speed operation	Star or delta	2.00	Constant horse power

No. of poles on high speed operation $2P$

No. of poles on low speed operation $2 \times 2P$

$$\frac{T_l}{T_h} = \frac{B_l}{B_h} \quad \frac{P_l}{P_h} = \frac{T_l}{T_h} \cdot \frac{n_l}{n_h} = \frac{n_l}{n_h}\frac{B_l}{B_h}$$

Slip Power Recovery Schemes The modification of the speed-torque characteristic using a variable rotor resistance has the major disadvantage of poor efficiency, thus making it uneconomical. Continuous low speed operation is not possible due to overheating of the rotor. These low speeds can be very effectively achieved with reasonable efficiency using slip energy recovery schemes. The slip power which is wasted in the external resistance in the rotor circuit is returned to the mains in these schemes.

The conventional methods of slip power recovery employ rotating machines, such as rotary converters, alternators, dc machines, etc. in the rotor circuit to convert the power at slip frequency to power at line frequency. Some typical conventional schemes, known as Scherbius and Kramer controls, are shown in Fig. 1.26.

When these methods are employed, the motor may be operated to drive both constant torque and constant power loads. These are illustrated in Fig. 1.27(a) and (b), in principle. In Fig. 1.27(a) the rotor power at slip frequency is converted to line frequency by means of a slip converter. If the slip power converter allows power flow in both directions, the motor may be operated both at sub and super synchronous speeds. This scheme is used to drive constant torque loads. In sub

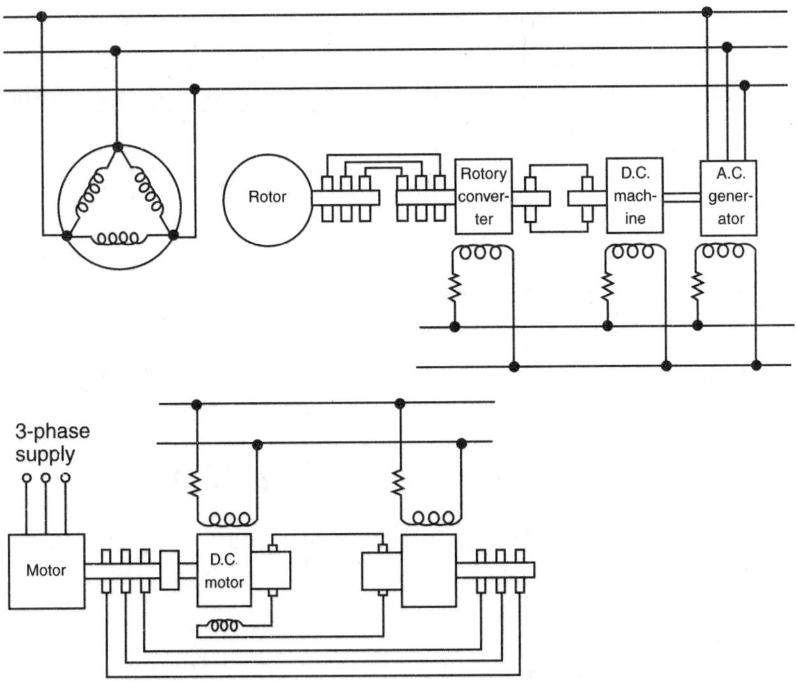

Fig. 1.26 *A typical conventional slip energy recovery schemes*

synchronous operation the slip power is converted to line frequency and fed to the mains. In supersynchronous operation the power at line frequency is converted to slip power and fed to the motor. One significant feature of this modification is that the developed torque is proportional to rotor current under the assumption of constant flux in the motor. The speed torque curves for this scheme are depicted in Fig. 1.28(a). The desired modification of the torque speed curves shown in Fig. 1.28(a) is achieved by controlling the slip power converter to match the motor voltage at a given slip. The control of the converter is represented by the parameter a. Increase in a increases the voltage on the rotor side of the slip power converter causing a speed drop, a may be fixed for no-load conditions. It may be varied in a closed loop control to maintain constant speed. When it is fixed at no-load value the motor has a drooping speed-torque characteristic. For example a can be the firing angle of the line-side converter in the case of static slip power schemes.

The scheme shown in Fig. 1.27(b) uses the slip power to drive an auxiliary machine. In this case the slip power converter is coupled to the rotor of the induction motor and gets power from it. Here too both sub and super synchronous speeds are possible. In subsynchronous operation the auxiliary machine converts slip power to mechanical power. In supersynchronous operation the additional power is fed to the rotor windings through the slip converter from the auxiliary machine. The connection maintains constant power. The speed-torque characteristics

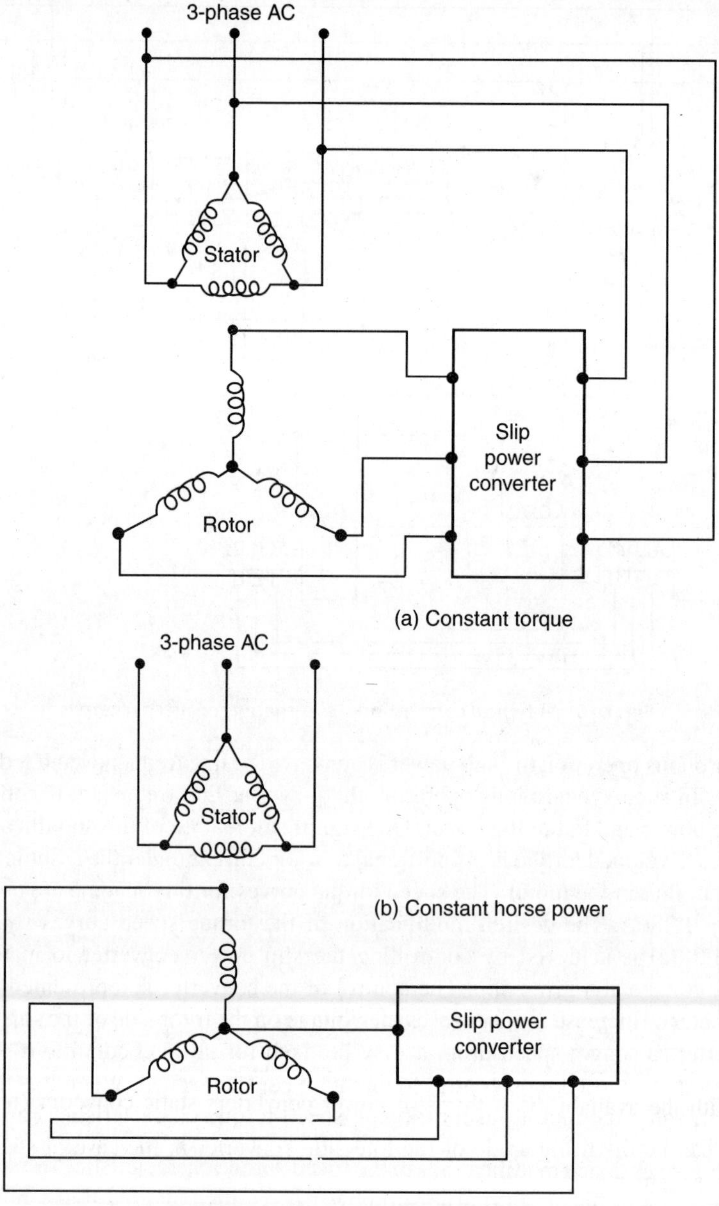

Fig. 1.27 *Slip energy recovery schemes*

of the motor are shown in Fig. 1.28(b). Here also a is a parameter of the slip converter chosen in a manner as to cause the speed control as described above, when varied in a given manner. The slip power is handled by the main motor shaft. The torque decreases with an increase in speed.

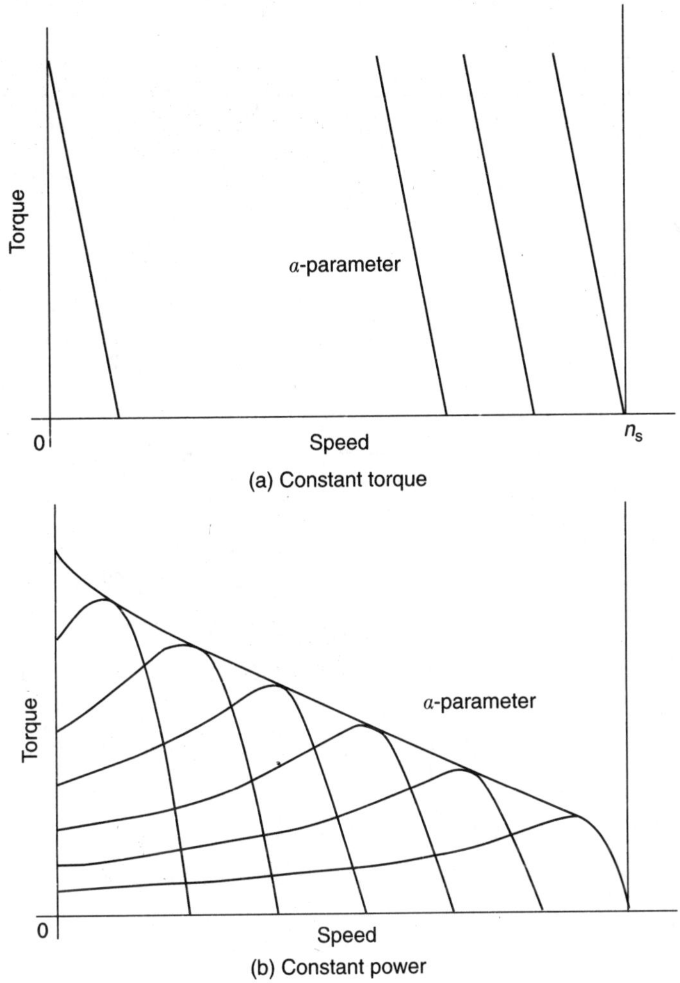

Fig. 1.28 *Speed-torque for slip energy recovery schemes*

With the availability of thyristor power converters static converter (rectifier-inverter) cascades are being used in the rotor circuits of induction motors to get the abovementioned modifications to the speed-torque characteristic. The schemes are depicted in Fig. 1.29(a). The slip power is rectified and fed to the line commutated converter which feeds the power to the mains. The speed-torque curves obtained by the variation of firing angle of the inverter, are shown in Fig. 1.29(b). A cycloconverter may also be used in the rotor circuit.

Injection of Voltage Into the Rotor Circuit The torque-speed characteristic of an induction motor can be modified by injecting a voltage into the rotor circuit (wound rotor) of an induction motor. The voltage injected must be at slip frequency.

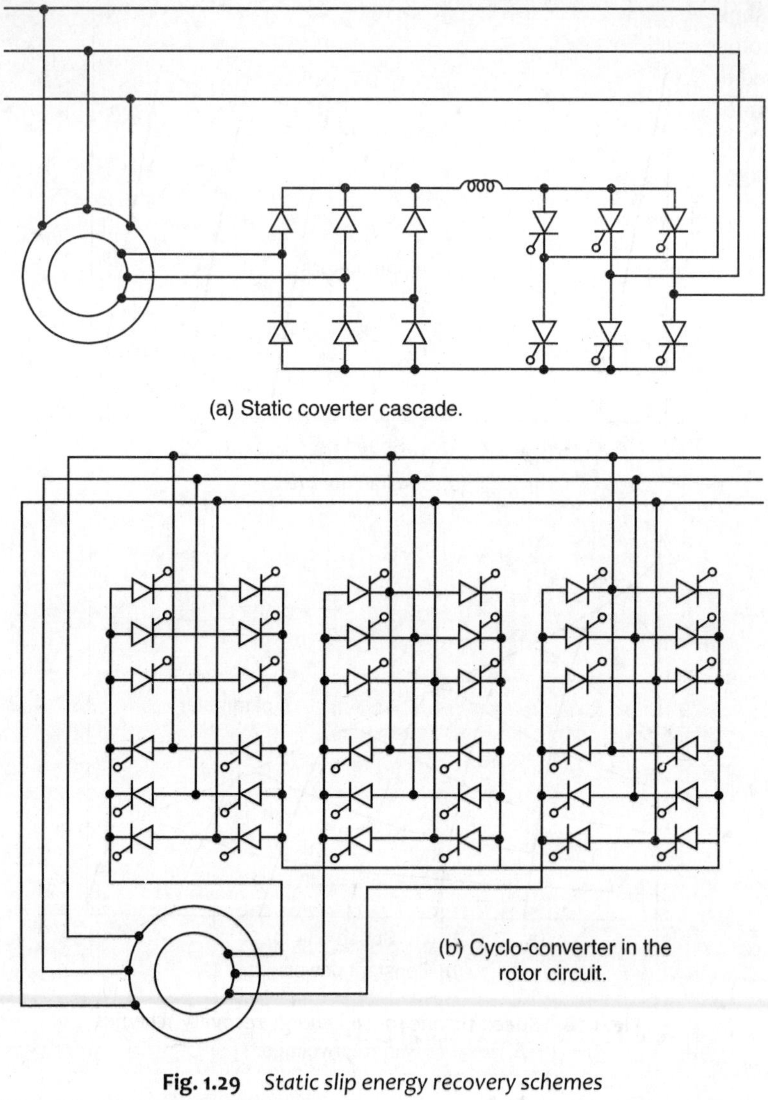

(a) Static coverter cascade.

(b) Cyclo-converter in the rotor circuit.

Fig. 1.29 *Static slip energy recovery schemes*

If the voltage injected opposes the rotor voltage, the effective rotor current decreases, which instantly affects the torque. The reduced torque cannot drive the load. The rotor speed decreases to a value which ensures sufficient induced rotor voltage and hence rotor current to drive the load. If, on the other hand, the injected voltage aids the rotor voltage, it results in an increased rotor current. The increased torque developed accelerates the rotor to a speed at which sufficient rotor current flows to drive the load. The speed torque curves for the two cases are shown in Fig. 1.30. For comparison, the torque-speed curve of a short circuited rotor with zero injected voltage is also shown. From the figures it can be inferred that it is possible to change the torque capability of the motor by changing the injected voltage. When the injected voltage opposes the rotor current torque capability decreases, whereas it increases when the injected voltage aids the rotor voltage.

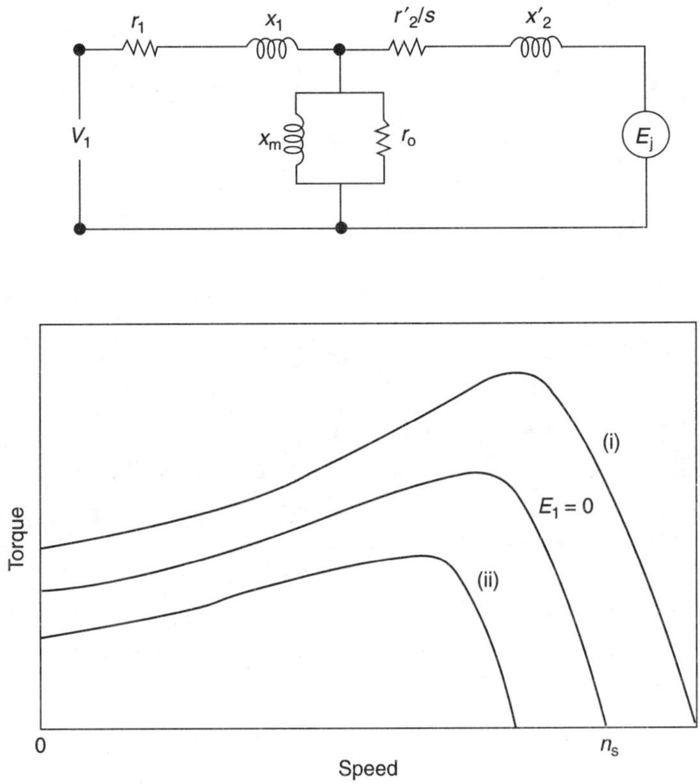

(i) E_j is phase with rotor voltage.

(ii) E_j is out of phase by 180° with rotor voltage.

Fig. 1.30 *Speed control by injection of rotor voltage*

Variation of Supply Frequency The speed of a synchronously rotating magnetic field is a function of supply frequency. Therefore, by varying the supply frequency the synchronous speed and hence the rotor speed can be varied. To avoid saturation due to an increase in the flux at low frequencies, the applied voltage to the motor is also varied so that the flux remains constant at its rated value at all frequencies. To achieve this a simple method is to vary both voltage and frequency so that *V*/*f* is constant. The torque-speed curves with constant *V*/*f* are depicted in Fig. 1.31. There is a depletion of torque at low frequencies. The motor has reduced torque capability and overload capacity. This is because of the dominant effect of stator resistance at low frequencies. The resistance drop become appreciable as compared to the applied voltage. This causes a depletion of flux, whose constancy cannot be maintained at low frequencies. The torque developed with *V*/*f* constant is

$$T_d = \frac{pm_1}{2\pi}\left(\frac{v_1}{f_1}\right)^2 \frac{f_2 x_m^2 / r_2'}{\left[r_1 + \frac{f_2}{r_2' f_1}(x_m^2 - x_{11}x_{22})\right]^2 + \left[x_{11} - \frac{f_2 r_1}{f_1 r_2'} x_{22}\right]^2} \tag{1.34}$$

where $x_{11} = x_m + xx;\ x_{22} = x_2' + x_m$

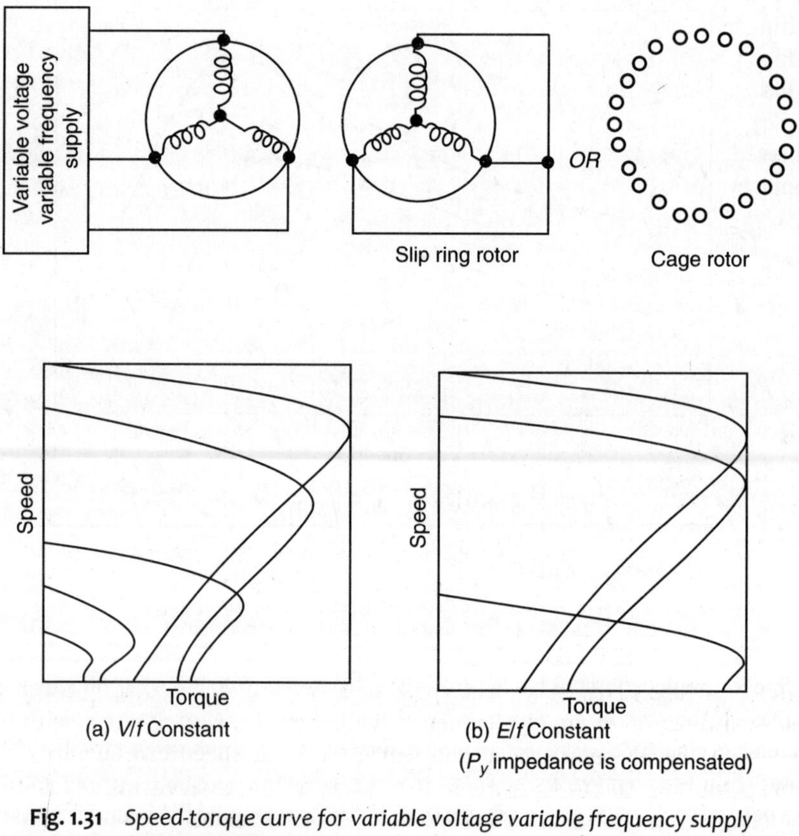

Fig. 1.31 *Speed-torque curve for variable voltage variable frequency supply (V/f constant)*

To have the same torque and overload capacity at all frequencies it is necessary to compensate for the stator (resistance) drop in order to keep *E*/*f* constant. *V*/*f* is no longer constant since it increases as the frequency decreases. The torque developed in this case is given by

$$T_d = \frac{pm_1}{2\pi}\left(\frac{E_1}{f_1}\right)^2 \frac{f_2 r_2'}{r_2'^2 + (2\pi f_2 L_{2\sigma}')^2} \tag{1.35}$$

where $L_{2\sigma}'$ is rotor leakage inductance.

The torque-speed curves for constant *E*/*f* are shown in Fig. 1.31(b).

With *V*/*f* (constant) control the starting torque increases with a decrease in the frequency, up to a certain value. Below this value of frequency the starting torque decreases. This effect is considered similar to that achieved by changing rotor leakage reactance. As the frequency decreases, the rotor leakage reactance. As the frequency decreases, the rotor leakage reactance decreases. Effectively, an increase in the rotor, resistance relative to leakage reactance takes place. Therefore, the starting torque increases up to a certain frequency, where the rotor leakage reactance equals the rotor resistance. If the frequency is decreased further the starting torque decreases. The variation of starting torque with frequency is shown in Fig. 1.31(c). However, with constant *E*/*f* control the starting torque increases as the frequency decreases up to a value decided by the parameters. If the frequency is further decreased the starting torque decreases. The acceleration may be achieved at constant torque and armature current by varying the stator frequency from a low value keeping *E*/*f* constant.

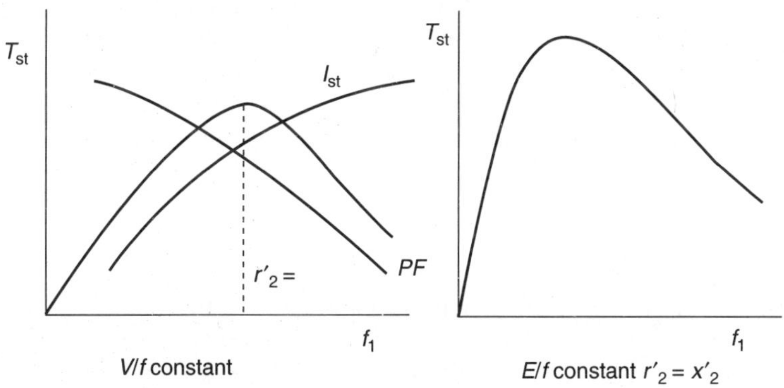

Fig. 1.31 *(c) Starting torque as a function of stator frequency*

Speed-torque characteristics above the base characteristic are obtained by increasing the supply frequency beyond the rated value. The flux in the motor decreases, since the voltage cannot be increased beyond the rated value. The motor is said to be operating in the flux weakening mode.

The torque-speed curves run parallel to each other at all frequencies. They extend to the second quadrant, showing that regeneration is possible.

The starting of the motor can be easily accomplished using a variable voltage, variable frequency supply. This decreases the starting current, giving a reasonably good accelerating torque at a good power factor even with low resistance cage motors.

1.3.2 Speed Control of Induction Motors

A three-phase induction motor is essentially a constant speed motor. It is not possible to achieve smooth speed control of the motor over a wide range, when supplied from a conventional three-phase constant voltage, constant frequency supply. Thyristor power converters have made variable frequency and variable voltage supplies possible. These are employed to get a smooth speed control of induction motors over a wide range.

The methods of modification of speed-torque characteristics discussed are more or less the methods of speed control also. Thyristor power converters are being employed widely in adopting the methods of speed control, e.g. a chopper to control the rotor resistance, ac voltage controller to vary supply voltage, static converter cascades for slip energy recovery, etc. A comparison of these methods is given in Table 1.4.

1.4 CHARACTERISTICS OF SYNCHRONOUS MOTORS

Synchronous motors are constant speed motors. The speed of the motor is decided by the number of poles and frequency. Compared to an induction motor, it is very sensitive to sudden changes of load. This causes a hunting of the rotor and finally leads to stability problems. It has no starting torque and requires starting equipment to bring it to its rated speed. When it is running at its rated speed the field is excited. The damper windings on the field poles help in damping the hunting and providing the starting torque. The motor can be operated at different power factors by changing the excitation. Overexcited synchronous motors operate at leading power factors whereas underexcited ones operate at lagging power factors. They are reasonably efficient. Their efficiency and ability to correct the power factor by varying the excitation make synchronous motors attractive in large power applications. They are preferred as constant speed drives in the industry.

The phasor diagram of a synchronous motor is shown in Fig. 1.32. The theory of these motors has been developed on the basis of synchronous reactance, which takes care of leakage reactance and armature reaction. A salient pole machine, which has a non-uniform air gap, is described by direct and quadrature axis reactances. Variation of the armature current of the motor when its excitation is varied is described by V-curves when the motor develops a given power. The variation of excitation brings about the following:

i. change in armature current
ii. change in line power factor
iii. slight change in the load angle.

Table 1.4 *Speed control of three-phase induction motors*

Basis of comparison	*Signifi-cance*	*Variable voltage at constant frequency*	*Rotor resistance control*	*Variable voltage variable frequency*	*Slip power recovery*	*Pole change motors*	*Injection of voltage into rotor*
1. Range of speed control		With normal squirrel cage induction motor the speed control range is limited. To increase the range a high resistance rotor is required. A wound rotor induction motor adopts itself very well to this, as an external resistance can be included in the rotor circuit. Inferior speed regulation. At low slips speed control is not effective for normal motor.	Variable resistance in the rotor circuit provides speed control over a limited range. Mainly suitable for slip ring motors. Using a suitable value of rotor resistance the operating point may be made to occur in the braking region so that the acceleration of the motor by the load may be avoided. Range of speed control depends upon rotor resistance and load speed control range can be increased by having voltage control on stator side. Poor speed regulation. Hardness decreases with increase in resistance.	Speed control can be obtained from 1.2 times the base speed to almost zero speed. A speed range of 20:1 is possible can also be used to start the motor with good starting performance. Low starting current at good starting torque and power factor. The characteristic has sufficient hardness.	Speed range depends on the system. Larger speed control range is possible. Speeds below base speed are possible. With cycloconverter speeds above base speed are also possible. Speed ranges 1:3 to 1:4 may be possible.	Speed range of 1:2	A wide range of speeds is possible by injecting voltage into the secondary at slip frequency.

(*Continued*)

Basis of comparison	*Significance*	*Variable voltage at constant frequency*	*Rotor resistance control*	*Variable voltage variable frequency*	*Slip power recovery*	*Pole change motors*	*Injection of voltage into rotor*
2. Smoothness of speed control		Depends on how the stator voltage is varied. Speed control may be smooth in the specified range.	As the variation of rotor resistance is accomplished in steps, only stepped speed control is possible. A smooth variation may be possible with a chopper controlled resistance in the rotor.	A smooth speed control is possible over the complete speed range.	Speed control is smooth in the given range.	Stepped speed control.	Smooth speed control.
3. Economics of speed control		Motor operates at very low values of flux. Rotor currents are high to drive the torque. Efficiency is poor. Economical and simple for low power drive where efficiency is not a criterion. With ac voltage controllers the method is simple. At large slips, which are possible with large resistance rotors, the operation is inefficient.	When employed with wound rotor the external resistance may be connected in the rotor circuits power is dissipated in the external resistance. Rotor heating is not present. With squirrel cage motors a high resistance rotor is required. Besides poor η rotor heating also there. Not suitable or economical for continuous operation.	Operates with reasonably good efficiency at all speeds. The capital cost is more. Rotating machines were used. But at present power electronic equipment are used for providing the required supply. This equipment is costly.	The initial cost of the equipment is decided by the range of speed control. Efficiency is high. Running cost is low. Rotating machines like commutator machines, dc machines, synchronous machines and static converters are used in the rotor circuit.	No initial cost. The armature may have to be reconnected for different sets of poles. Regeneration may be possible when speed changes from high to low values.	Special commutator motor is required. Thus initial cost is more.

4. Stability		Depends on the load speed torque curve. Stability problem may occur at very low voltages.	Stable operation. Region of stable operation increases.	Operation may show low frequency instability when speed control is accomplished with static converters.	Stable	Stable	Stable
5. Direction of speed control		Speeds below base speed. Loss of torque at low speeds.	Speeds below base speed. Low speed operation is inefficient.	Speeds below and above base speed are possible. Flux weakening occurs at speeds above syn. speed.	Speeds below base speed if dc link conversion is employed. With a cycloconverter speeds below and above base speeds are possible.	Two sets of speed are possible.	Speeds below and above synchronous speeds are possible.
6. Permissible load.		Constant torque loads cannot be driven at low speeds due to heavy currents. Suitable for fan type loads where $T \alpha \omega^2$.	Constant torque loads can be driven as the rotor current remains constant at every r_2' to fully utilise the motor.	Constant torque loads up to base speed and constant power loads beyond base speed.	Constant torque and constant power drives are available.	The connections may be made to give constant torque or constant power at all speeds. Variable torque operation is also possible.	Shunt motor characteristic for constant torque loads below synchronous speed and constant power loads above syn. speed.

(*Continued*)

Basis of compariso.	*Signifi-cance*	*Variable voltage at constant frequency*	*Rotor resistance control*	*Variable voltage variable frequency*	*Slip power recovery*	*Pole change motors*	*Injection of voltage into rotor*
7. General		Suitable for low power drives. Power factor is poor.	Used for fan type loads and also for loads where the power reduces with decrease in speed.	Very popular drive.	The power factor is poor. Special methods are employed to improve the p.f. Transformer is connected between the line and converter for this purpose. Efficiency is good.	P.f. is different at different speeds.	Power factor control may also be employed.

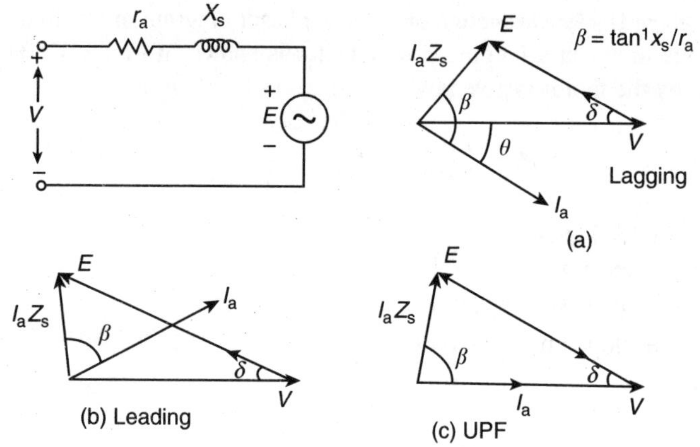

Fig. 1.32 *Phasor diagram based on synchronous impedance*

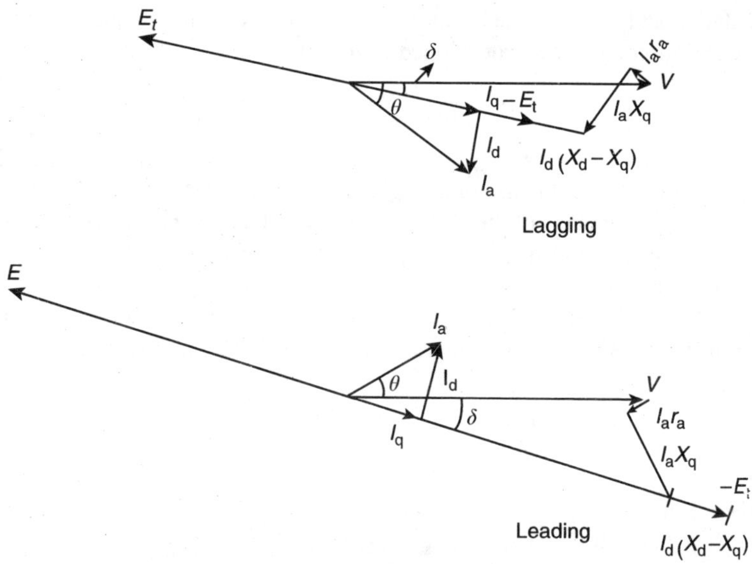

Fig. 1.32(a) *Phasor diagram of salient pole synchronous motor*

However, there are minimum and maximum excitations for a given power developed.

An increase in the mechanical load at constant excitation would tend to retard the rotor. The angle by which the rotor tends to fall behind the no-load position is called the load angle. In the process of attaining a final position the rotor undergoes oscillations which are damped by damper windings.

Power Developed by Synchronous Motor The phasor diagram of a cylindrical rotor synchronous motor at a lagging power factor is shown in Fig. 1.32. The power developed by the motor is given by

$$P_d = \frac{E_t V}{Z_s} \cos(\theta - \delta) - \frac{E_t^2}{Z_s} \cos \theta \tag{1.36}$$

where δ is the load angle

V is the terminal voltage

E_t is the induced voltage

The torque developed by the motor is

$$T_d = \frac{P_d}{2\pi n_s} \tag{1.37}$$

The power developed depends on the excitation. An increase in the excitation results in an increase of P_d. Consequently, the load angle decreases for a given power developed. The overload capacity of the motor increases with an increase in excitation and the machine becomes more stable. If the resistance of the armature is negligible, the power developed is given by

$$P_d = \frac{E_t V \sin \delta}{X_s} \tag{1.38}$$

For a salient pole rotor the use of a single synchronous reactance gives unreliable results. The performance of the motor is determined by the use of two reactances, namely direct axis and quadrature axis reactances (X_d, X_q), The former being greater than the latter.

The phasor diagram of an overexcited salient pole synchronous motor is shown in Fig. 1.32(b). Neglecting armature resistance the power developed is given by

$$P_d = \frac{E_t V}{X_d} \sin \delta + \frac{V^2}{2X_d X_q} (X_d - X_q) \sin 2\delta \tag{1.39}$$

and the torque developed

$$T_d = \frac{P_d}{2\pi n_s}$$

When compared with round rotor machines, the following differences are clear:

i. The power (torque) developed by a salient pole rotor has an additional component due to saliency, which depends upon the difference of the two axes reactances. This is called reluctance power (torque). A salient pole rotor develops more power for a given load angle. This means the power per degree of load angle is greater in a salient pole rotor than that in a round rotor when excitation is the same in both the cases (the two motors are otherwise identical).

ii. The maximum torque in a salient pole rotor occurs at a torque angle which is less than the corresponding one of a round rotor motor.
iii. Torque is available in a salient pole rotor even at zero excitation.

The power angle characteristics of both types of rotor are shown in Fig. 1.33, for different excitations. The torque developed in a synchronous motor is directly proportional to the applied voltage unlike in an induction motors where it is proportional to the square of voltage. Hence it is less sensitive to voltage variations.

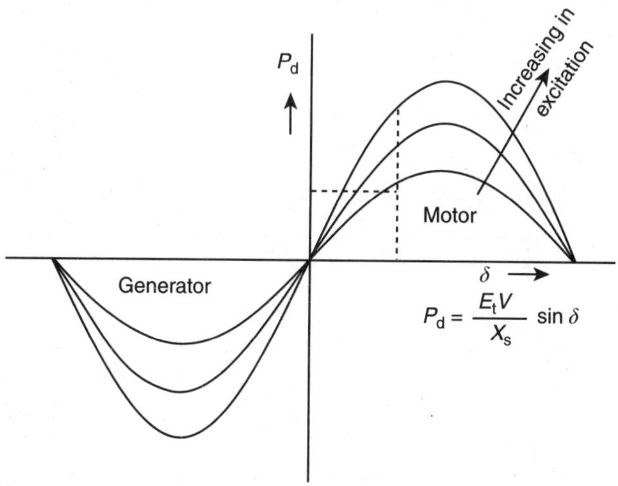

Fig. 1.33(a) *Power-angle diagrams of a synchronous motor at different excitations*

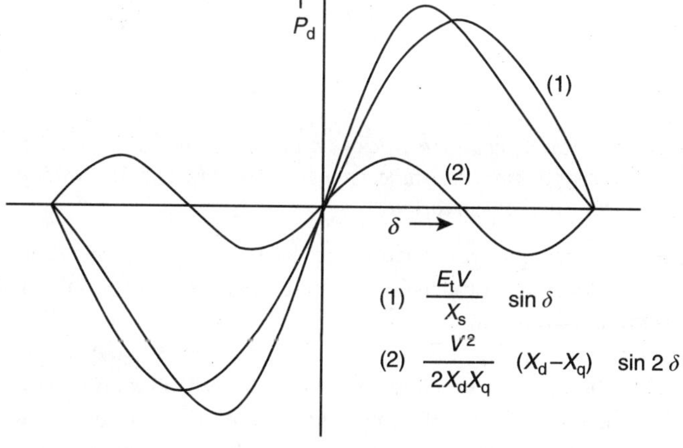

Fig. 1.33(b) *Power-angle characteristic of salient pole motor*

The torque-speed characteristic of a synchronous motor is shown in Fig. 1.34. The characteristic is parallel to the torque axis since the motor is of constant speed type.

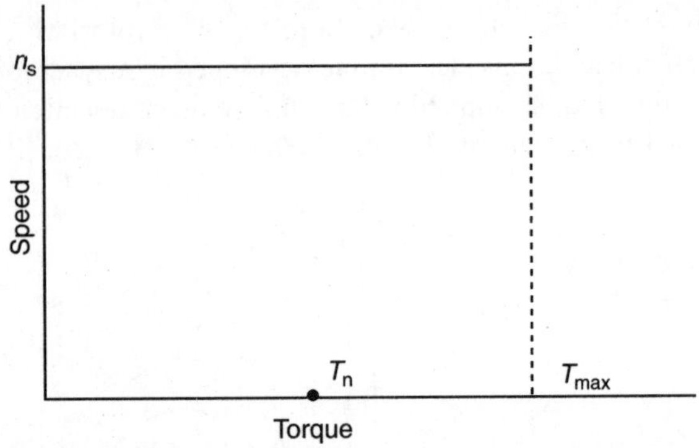

Fig. 1.34(a) *Speed torque characteristic of a synchronous motor*

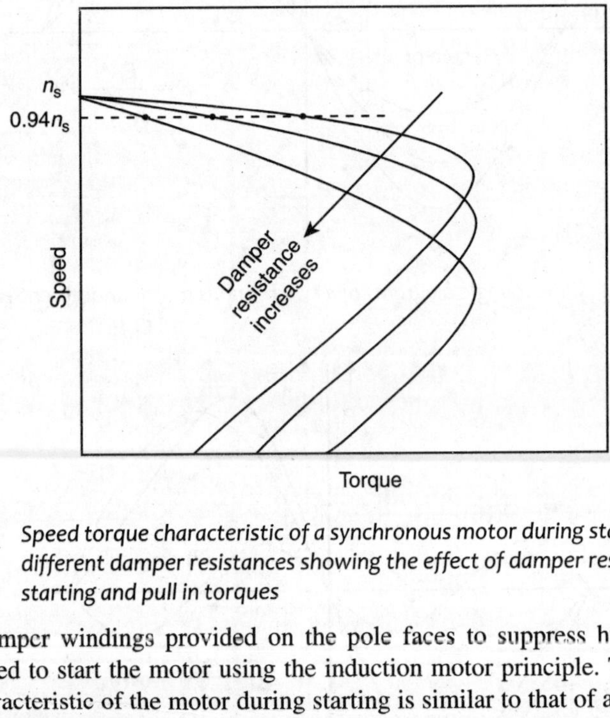

Fig. 1.34(b) *Speed torque characteristic of a synchronous motor during starting for different damper resistances showing the effect of damper resistance on starting and pull in torques*

The damper windings provided on the pole faces to suppress hunting may also be used to start the motor using the induction motor principle. The torque-speed characteristic of the motor during starting is similar to that of an induction motor, and is depicted in Fig. 1.34 for different damper resistances. To get a better starting torque the damper winding must have a high resistance. However, this inhibits their primary function of damping the oscillations, since a low resistance damper is more effective for this task. A judicious choice of resistance is required, depending upon the application of the motor.

1.4.1 Desirable Modifications to the Speed-Torque Curve of the Motor

The speed of a synchronous motor can be changed only by varying the frequency, i.e. the speed-torque characteristic can be modified only by a variation of the supply frequency. To avoid possible saturation, the voltage is also varied as a function of the frequency. Variable voltage, variable frequency supplies are used to achieve modifications in the speed-torque curve.

Thyristor power converters provide this supply to the motor. Three modes of operation may be identified:

i. V/f and E/f constant
ii. V and E/f constant
iii. V/f and E constant

In the first mode of operation the motor operates at constant flux, disregarding the effects of resistance. The current drawn by the motor is independent of stator frequency. Thus speed-torque curves similar to those of a shunt motor are possible. The performance characteristics are shown in Fig. 1.35. In the second mode of operation the motor possesses a series motor characteristic, but the effects of saturation are present. The current drawn at low frequencies is high. The performance characteristics are shown in Fig. 1.36. The third mode of operation gives a mixed performance, as depicted in Fig. 1.37.

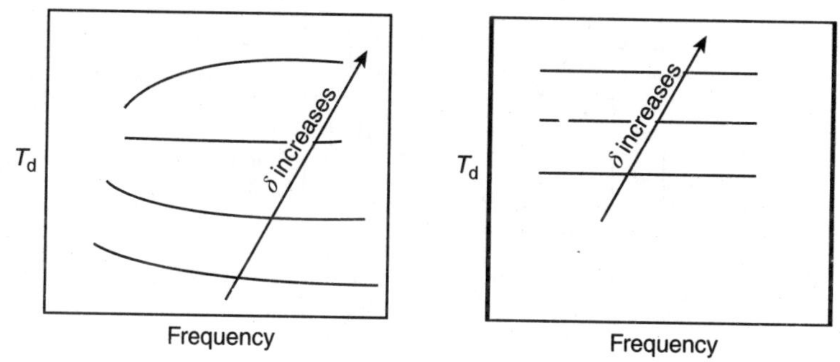

Fig. 1.35 *(a) V/f and E/f constant (b) current-feeding*

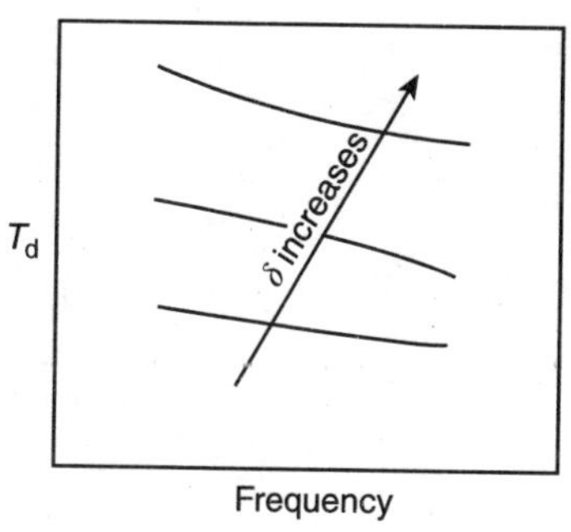

Fig. 1.36 *V/f and E constant*

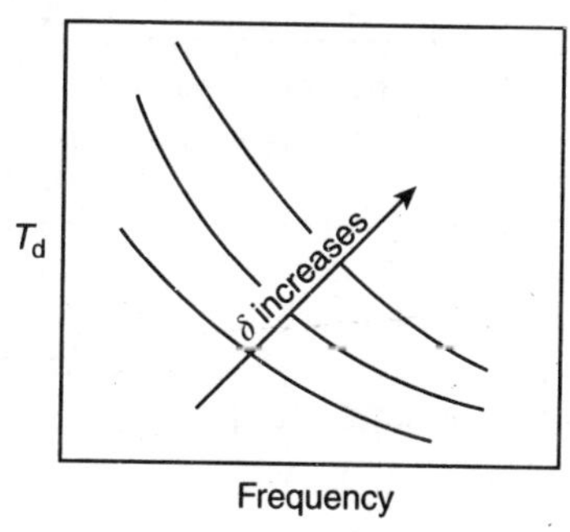

Fig. 1.37 *V and E/f constant*

1.4.2 Speed Control of Synchronous Motors

With the availability of thyristor power converters, synchronous motors are becoming increasingly popular as variable speed drives. Two types of control are possible for synchronous motors, when fed from thyristor power converters.

i. Separate control
ii. Self control

The type of control employed affects both the dynamic and the steady-state performance of the motor. In the former (Fig. 1.38), the motor is fed from a variable frequency supply, the frequency being controlled externally from a crystal oscillator. The motor has the normal synchronous motor operation with all its stability and hunting problems. In self control (Fig. 1.39), the frequency of the input is decided by the rotor speed or stator voltages. A rotor position sensor is used to control the inverter firing pulses. By the time the rotor moves by two pole pitches, all the thyristors in the inverter are fired once in the sequence. Thus the input frequency and rotor speed are related. The firing pluses may be derived by sensing the stator voltages also. With self control the motor has good

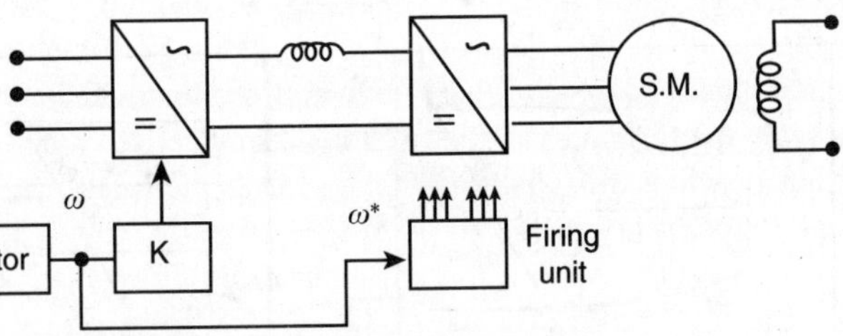

Fig. 1.38 *Separate control using crystal oscillator (converter can be either cycloconverter or dc link inverter)*

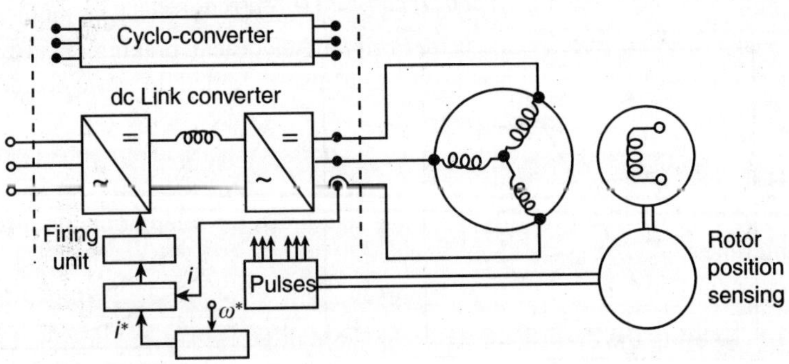

Fig. 1.39 *Self control of synchronous motor (using either cycloconverter or dc link converter)*

stability as well as good dynamic performance. The motor acquires dc motor characteristics. The inverter with rotor position sensing or induced voltage sensing is equivalent to a six-segment commutator. A self controlled synchronous motor can replace a dc motor, which has limitations due to its mechanical commutator. In this mode of operation it is also called a commutator-less motor (CLM). The motor may be fed from a VSI or CSI or cycloconverter. When fed from a CSI the motor may be overexcited to make use of the machine voltages for commutation.

The input voltage waveform and current to the synchronous motor are non-sinusoidal. The time harmonics of the waveform result in torque pulsations and armature heating. These effects are minimal with a cycloconverter In the case of CSI feeding, voltage spikes are present. These affect the motor insulation and voltage rating of the inverter.

1.5 BRAKING OF ELECTRIC MOTORS

While operating electrical drives it is often necessary to stop the motor quickly and also reverse it. In applications like cranes or hoists the torque of the drive motor may have to be controlled so that the load does not have any undesirable acceleration, e.g. in the case of lowering of loads under the influence of gravity. The speed and accuracy of stopping or reversing operations improve the productivity of the system and quality of the product. For both the applications stated above, a braking torque is required, which may be supplied either mechanically or electrically. In the former case, the frictional force between the rotating parts and brake drums provides the required braking. Mechanical equipment, such as brake linings and brake drums are required. On the other hand, in electrical braking a braking torque which opposes the motion of the rotating member is developed during the braking operation. This is achieved by suitably changing the electrical connections of the motor. The motor operates on a speed-torque characteristic depending upon the method of braking employed. Whether mechanical or electrical, the braking of the drive should be such as to stop the motor at the specified point of time and location, for reasons of safety.

A comparison of electrical and mechanical braking is given in Table 1.5, to bring out the effectiveness and superiority of electrical braking. However, in hoists a stand by mechanical brake system is also provided, to avoid accidents in case of power failure.

From the preceding discussion it is clear that electrical braking is preferable However, in view of the severe operating cycle of the motor, it should be employed only when it is highly desirable to control the retardation and limit the braking time.

Electric braking is of two kinds, depending upon the application of the drive:

i. To bring the driving motor completely to rest in a given amount of time and at exactly specified points. While doing so, the K.E. is fed back to the mains.

Table 1.5 *Comparison of electrical and mechanical braking*

	Mechanical	*Electrical*
Maintenance	Mechanical brakes require frequent maintenance, like adjustment of brakes replacement of brake linings. They are prone to wear and tear.	Very little maintenance dust free operation due to absence of mechanical equipment.
Utilisation of energy of rotating parts	The energy of the rotating parts is wasted as heat in friction. Heat generation during braking.	The energy of the rotating parts can be converted to electrical energy which can be utilised or returned to the mains. This happens during braking or while preventing the drive from attaining undue acceleration.
	Depending upon the conditions the braking may not be smooth.	Braking is smooth, without snatching.
	Brake shoes, brake linings, brake drum are required.	Equipment of higher rating than the motor rating may be required in certain types of braking.
	This braking can be applied to hold the system at any position.	Cannot produce holding torque. Requires electrical energy for operation.

ii. To restrict the speed to safe values. This arises normally while lowering loads using a hoist or crane. While maintaining constant speed the excess of K.E. and P.E. of the load tending to accelerate the motor is fed back to the mains.

1.5.1 Methods of Braking of Electric Motors

Electric braking may be accomplished in any of the following three ways:

i. Regenerative braking
ii. Dynamic or rheostatic braking
iii. Counter current braking.

We have seen that electrical braking makes it possible to convert the K.E. of the rotating parts to electrical energy. This energy can be returned to the mains or dissipated in an external resistance. The braking is called regenerative when the energy is returned to the mains. It is called dynamic or rheostatic braking when the energy is dissipated in a resistance. In either case the machine operates as a generator. Electric machines are capable of smooth transition from motor to generator action.

The load forces the motor in some applications to accelerate beyond no-load speed. The induced voltage becomes greater than the supply voltage to which the motor is connected and the current flows from the machine to the supply. It develops a torque to oppose the motion and hence to control the acceleration. The K.E. and P.E., minus the losses of the motor, are returned to the mains and the motor runs at constant speed. Thus, regenerative braking eliminates any tendency of the load to accelerate the motor.

Regenerative braking is also possible if the terminal voltage can be instantly decreased. The machine can be braked to zero speed. Reconnection of the motor is not necessary for regeneration. On the other hand, in dynamic braking the motor must be switched to the load resistance keeping the field constant.

Counter current braking is accomplished by reconnecting the supply to the armature of the motor so that the motor draws a current to develop a torque to oppose its already existing rotation. The motor acts as a brake and comes to rest very fast but has a tendency to accelerate in the reverse direction. If reversal is not required the supply to the motor must be cut off at zero speed. In the case of dc motors this is achieved by reversing the polarity of the supply voltage to the armature, while for ac motors the phase sequence is altered. The method is inefficient because of power loss in the resistors used for limiting the current, due to interconnection. The mechanical energy is converted to heat; there is additional power input from the supply. This method of braking is also employed to maintain a constant speed when the load tries to accelerate the rotor to high speeds, e.g. in the case of induction motors a suitable resistance of the rotor shifts the point of operation to the fourth quadrant.

1.5.2 Characteristics of DC Motors During Braking

Regenerative Braking This type of braking is possible in the case of drive motors where the speed can go beyond no-load speed or the terminal voltage can be momentarily decreased, e.g. in Ward Leonard control by decreasing the excitation of the generator. The armature current reverses and braking takes place.

Shunt Motors The speed and armature current are related by the equation

$$N = \frac{V}{K_e\phi} - \frac{I_a r_a}{K_e\phi} \tag{1.40}$$

where

$$I_a = \frac{V - E}{r_a} \tag{1.41}$$

During braking I_a is negative because $E > V$. The torque developed $T_b = K_t\phi\, I_a$ is negative, and hence is a braking torque. During braking

$$N = \frac{V}{K_e\phi} + \frac{T_b r_a}{K_e K_t \phi^2} \tag{1.42}$$

The speed-torque curve of the motor extends into the second quadrant. It passes through the no-load speed point with the same slope as in the first quadrant (motoring). The characteristics with different armature resistances are shown in Fig. 1.40.

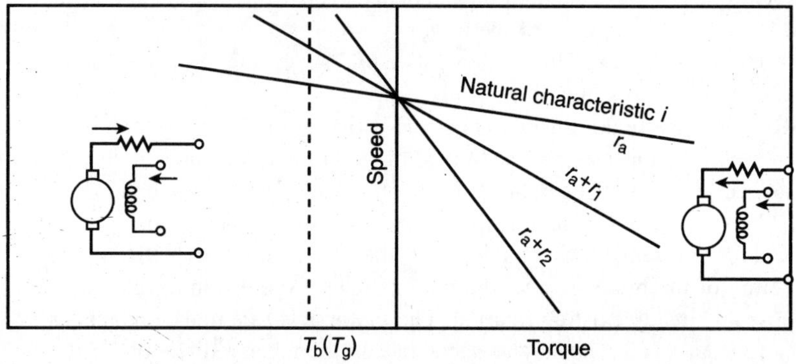

Fig. 1.40 *Regenerative braking: Torque-speed curve of a separately excited (shunt) motor*

Any tendency of the motor to accelerate to speeds beyond no-load speed is offset by means of the torque T_b. This occurs, for instance, in a hoist lowering the load which overhauls the motor and tries to accelerate it. The transition takes place to the second quadrant and the torque developed maintains constant speed. The slope of the characteristic can be varied by an additional resistance in the armature. The speed-torque curves in the second quadrant, as affected by a smooth variation of applied voltage, are shown in Fig. 1.41. The braking can be accomplished here at constant torque T_b, as shown in the figure. The armature current is also constant.

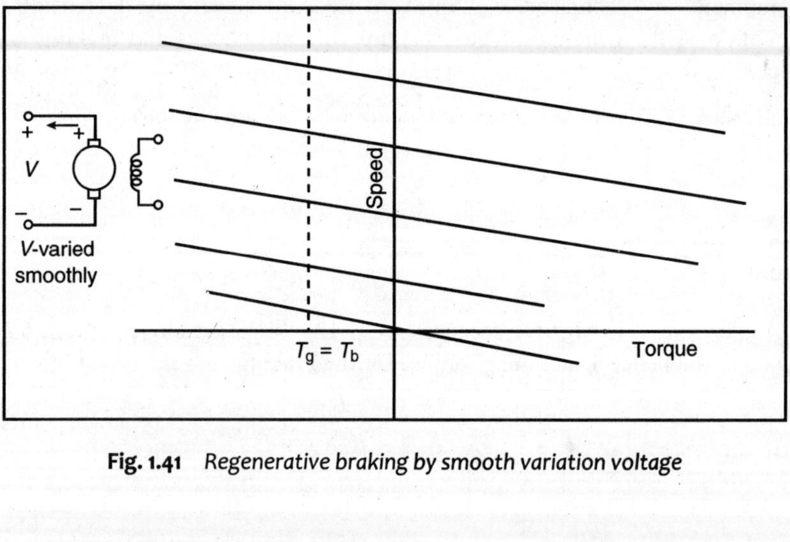

Fig. 1.41 *Regenerative braking by smooth variation voltage*

This kind of situation also occurs when the hoist is raising an empty cage. The counterweight moving under the influence of gravity tries to accelerate the motor and a transition to the second quadrant takes place. The torque developed is T_b, which maintains the speed constant. The K.E. trying to accelerate the rotor is regenerated to the mains.

Series Motors The nature of the speed-torque curve of a dc series motor is such that it does not extend to the second quadrant. As no-load is approached, the speed increases asymptotically to the speed axis. This implies that an increase in speed of a dc series motor is followed by a decrease in the armature current and field flux. The induced emf cannot be greater than the terminal voltage. Regeneration is not possible in a plain dc series motor since the field current cannot be made greater than the armature current.

However, in applications such as traction and hoists, where series motors are used extensively, regeneration may be required. For example, in locomotives moving down a gradient a constant speed descent may be necessary and in hoist drives the speed needs to be limited whenever it becomes dangerously high. The regeneration in such cases is achieved by separately exciting the field, as shown in Fig. 1.42. The motor has characteristics similar to those of a separately excited motor.

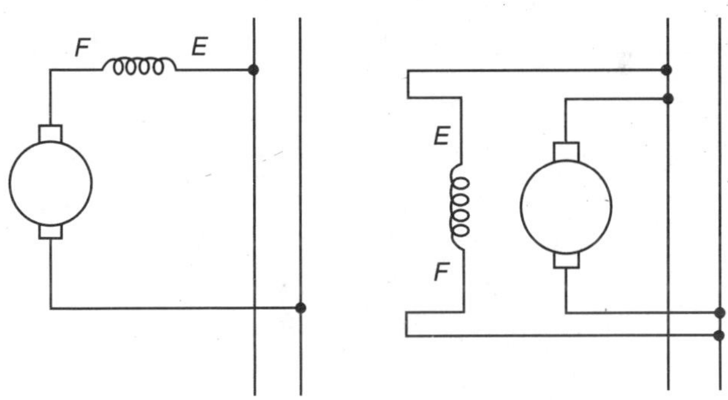

Fig. 1.42 *Reconnection of a series motor as separately excited one for regeneration*

Dynamic Braking Regenerative braking is not possible if it is impossible for the motor speed to be greater than the no-load speed or if the armature voltage is constant and cannot be varied smoothly. Dynamic braking is employed in such cases where the K.E. of the rotating parts is dissipated in an external resistance.

Shunt Motors The connections for dynamic braking are shown in Fig. 1.43. The motor operates at its rated voltage. When braking is required, the armature is switched on to an external resistance R_e. The field remains connected to the supply with full excitation, and the induced voltage in the armature has the same polarity. The armature current reverses and flows in a direction opposite

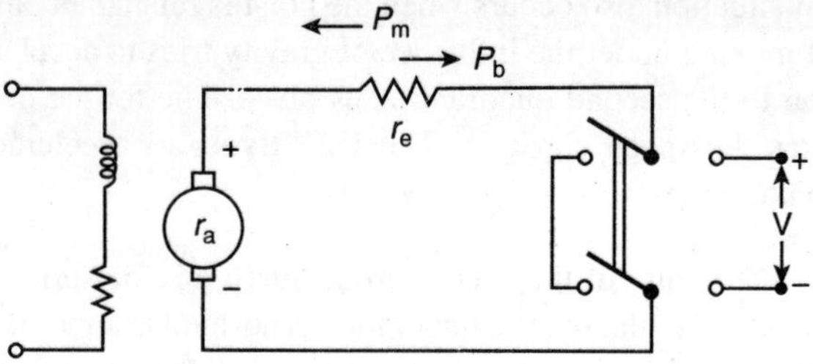

Fig. 1.43(a) *Connections for dynamic braking. P_m—Power flow during motoring P_b—Power flow during braking*

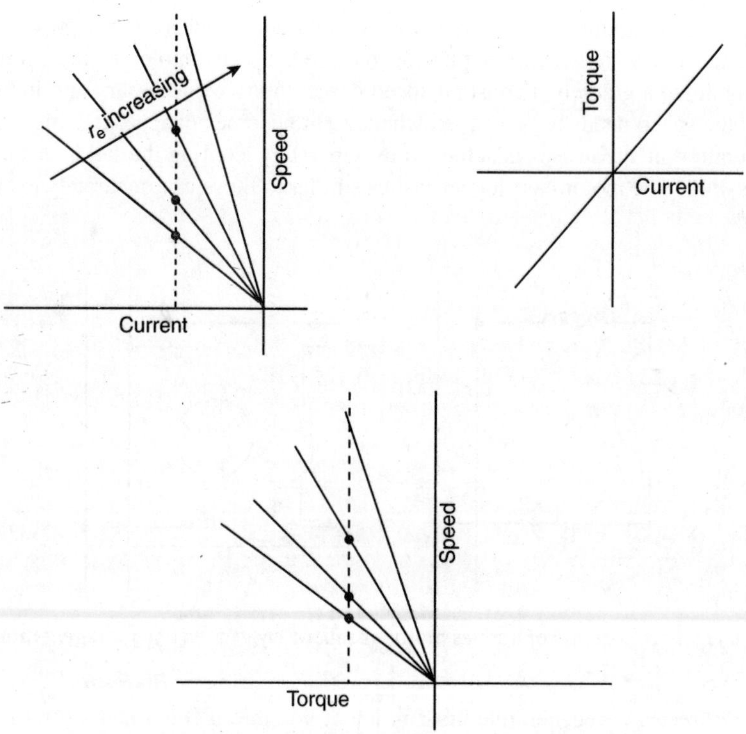

Fig. 1.43(b) *Braking characteristics*

to the current during motoring (Fig. 1.43), developing a braking torque. Even though the motor is braked by generator action the method is not similar to regenerative braking. The braking is effective and the motor stops very fast if the field is available at its full value, for which reason it is separately excited. If the field is shunt excited, the field current falls with speed leading to very poor braking below critical speed.

If E is the induced voltage, the armature current during braking is

$$I_a = \frac{-E}{(r_a + R_e)} = \frac{-K_e \phi N}{(r_a + R_e)} \tag{1.43}$$

The torque developed

$$T_d = K_t \phi I_a = T_b = \frac{-K_t}{(r_a + R_e)} \tag{1.44}$$

If separate excitation is employed the speed-torque characteristic is a straight line, as shown in Fig. 1.43. The slope of the line $-K_tK_e\phi^2N/(r_a+R_c)$ decreases with an increase in R_e. This shows that the braking torque decreases with an increase in the armature resistance, which increases the time of braking. A suitable value of R_e can be chosen such as to obtain stopping in the required time. The method is adopted for non-reversing drives where regeneration is not possible.

Series Motors When dynamic braking is employed the armature current would reverse. Obviously the field mmf also reverses, causing demagnetisation. To avoid this, the field connections are reverse connected before the series combination of armature and field is switched on to the braking resistance. The machine is then able to self excite in this case. The connections of a series motor during braking are shown in Fig. 1.44(a). The speed-torque curves during braking are shown in Fig. 1.44(b), in the second quadrant. The torque during braking is

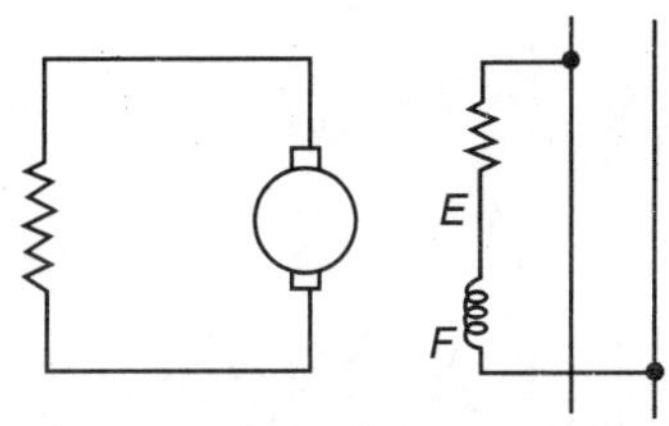

Fig. 1.44(a) *Methods of connecting a series motor for dynamic braking*

$$T_d = T_b = -K_t K_e \frac{\phi^2 N}{(r_a + R_e)} \tag{1.45}$$

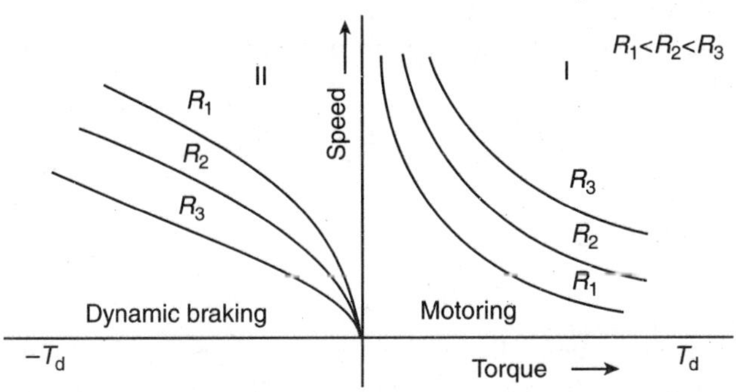

Fig. 1.44(b) *Torque-speed curves of a series motor with dynamic braking*

At the instant of initiating the braking, the current is more and hence the flux builds up. The torque developed is approximately proportional to the square of the armature current. At this instant the braking effect is more and there may be a jump in the torque developed, causing an objectionable shock to the load.

In case this torque is objectionable, dynamic braking is employed by separately exciting the field (Fig. 1.44(c)). This braking is similar to that in shunt motors and has already been discussed.

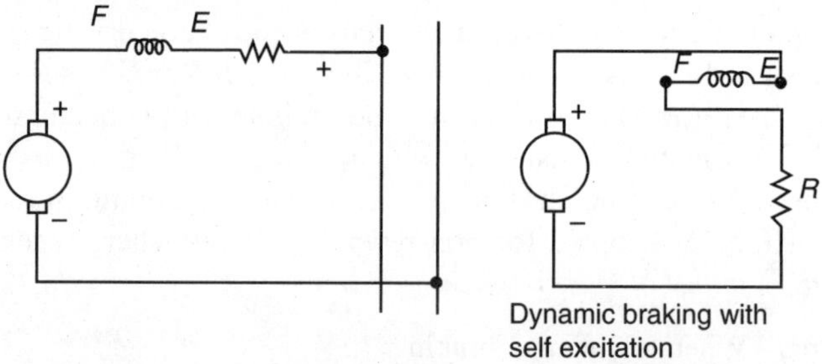

Fig. 1.44(c) *Dynamic braking with separate excitation*

Counter Current Braking This braking is employed in the following cases:

i. For quick stopping of the motor.
ii. For reversing drives requiring a short time for reversal.
iii. In cranes and hoists the motor is switched on to raise the load at the instant when the associated gearing operates in the direction to lower the load.

Shunt Motor The counter current braking in a shunt motor is accomplished by reversing the supply to the armature, while the connection to the field winding remains unchanged. When the polarity of the armature terminals of the running motor is reversed, the applied voltage and back emf reinforce each other in the circuit. Consequently, a current flows in a direction opposite to the one existing during normal motoring. The braking current may be limited by a resistance in the armature circuit. As this current flows only for a short duration of braking time, the starter resistances may be used. The torque-speed characteristic is given by

$$N = \frac{V}{K_e\phi}\,\frac{1}{(r_a + R_e)} + \frac{T_b(r_a + R_e)}{K_t K_e \phi^2} \tag{1.46}$$

The torque developed is

$$T_d = T_b = +K_t\phi I_a \tag{1.47}$$

If there is a load torque T_L, the total braking torque is $T_b + T_L$. In Fig. 1.45(b), the curve (1) indicates the speed-torque curve of the motor and curve (2) indicates the speed-torque curve during braking. In both cases of braking and acceleration

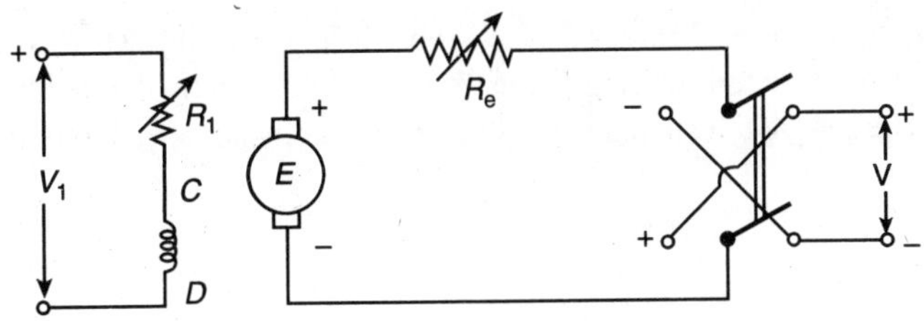

Fig. 1.45(a) *Connections for reverse current braking plugging*

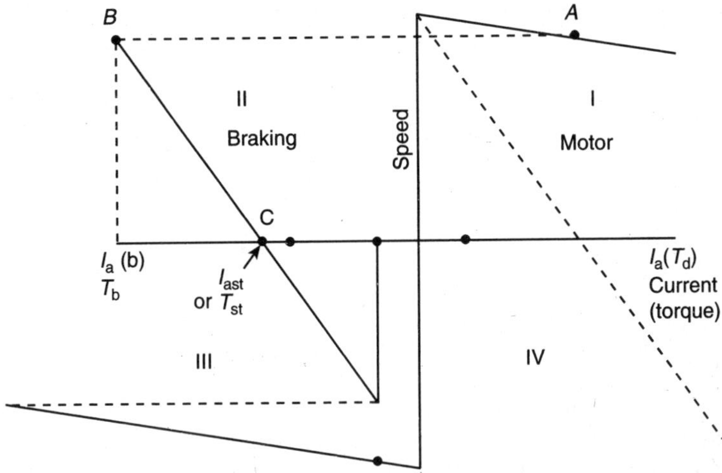

Fig. 1.45(b) *Speed versus armature current (torque) during plugging*

the same resistance R_e is connected. Let the motor be operating at Point *A* before plugging takes place. The operation is in the first quadrant. When the armature is reversed for braking, it may be assumed that the speed remains the same due to inertia (the limiting resistor is in the circuit). The point shifts to *B* in the second quadrant. The developed torque added to the load torque retards the rotor and the speed decreases along *BC*. If speed reversal is not required the motor must be switched off when the speed reaches zero at *C*.

If the motor is not switched off at zero speed, the developed torque accelerates the rotor in the reverse direction. Now the polarity of the induced voltage changes. The applied voltage and induced voltage oppose each other. The motor operates in the third quadrant. The series resistance can be short circuited, so that the motor operates on its natural characteristic.

Plugging or reverse current braking may be employed to lower the load at constant speed. If the friction torque is greater than the load torque the motor is stationary. In the opposite situation the motor accelerates in the direction of the load. At this moment the armature supply is reversed, with a suitable resistance to limit the current. The torque developed retards the load being lowered and runs it at constant speed. The torque-speed curves are as shown in Fig. 1.45(c). The point B on the curve corresponds to the stalling point. The current drawn by the motor is $(V + E)/(r_a + R_e)$. The torque developed lowers the load at constant speed to point C.

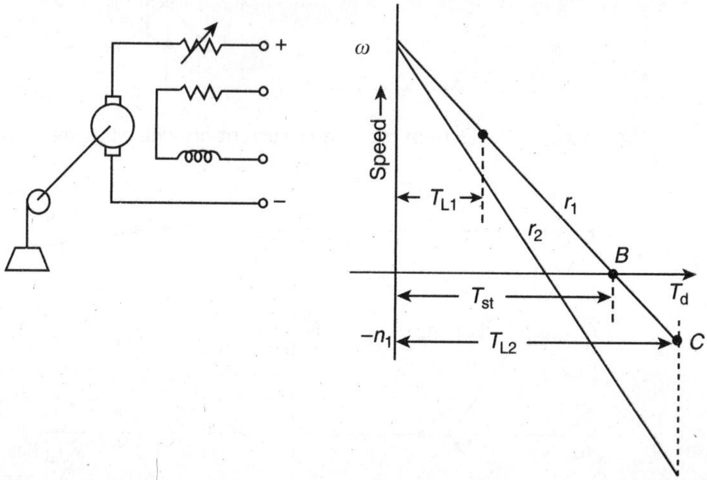

Fig. 1.45(c) *Speed-torque curve of a dc separately excited motor with counter current braking while lowering a load. Controlled lowering of load by counter torque*

Series Motors When the supply to the armature of a dc series motor is reversed, care must be taken to see that the field current retains its direction, for effective braking. The circuit conditions of a series motor are as shown in Fig. 1.46. When

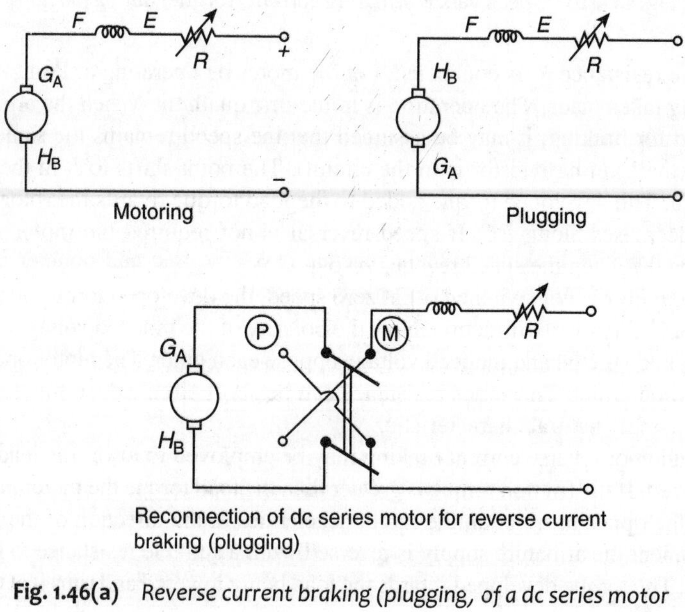

Fig. 1.46(a) *Reverse current braking (plugging, of a dc series motor*

the supply terminals are reversed, the polarity of the field must also be reversed so that there is no demagnetisation. To limit the current during braking, an external resistance may be required in the armature circuit. Compared to dc shunt motors the braking torque of dc series motor falls very rapidly with speed and below a certain speed braking may not be effective, because the field flux also decreases with speed. In the case of shunt motors the flux is constant in the complete braking region. A typical speed-torque curve of a series motor during braking with external resistance is shown in Fig. 1.46(b) The curve of the motor with zero resistance is also shown. The braking torque depends on the load the motor is drivint at the instant of braking. The braking of the motor by this method is effective only if the load torque is less than the short circuit developed torque. Alternatively, the field may be separately excited during braking. The braking characteristics in this case are obviously similar to those of a separately excited motor.

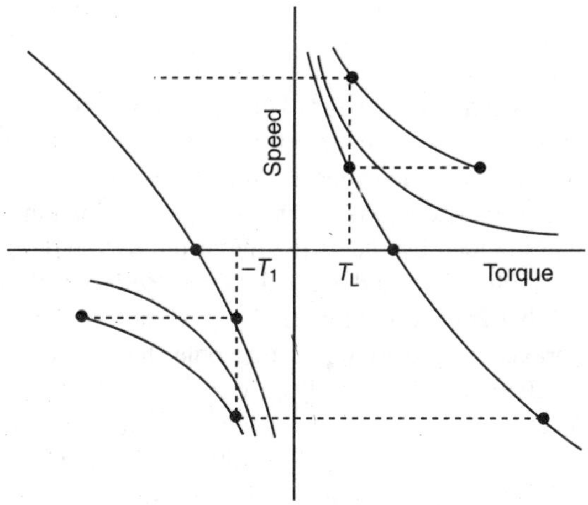

Fig. 1.46(b) *Torque-speed curves with counter current braking*

1.5.3 Braking of Induction Motors

The three types of braking, namely regenerative, dynamic and counter current braking can be accomplished with induction motors also.

Regenerative Braking When the rotor of an induction motor runs faster than the stator field the slip becomes negative and the machine generates power. Therefore, whenever the motor has a tendency to run faster than the rotating field, regenerative braking occurs and the K.E. of the rotating parts is returned to the mains. The speed-torque curve extends to the second quadrant (Fig. 1.47). The speed of the motor decreases. The braking torque makes the motor run at constant speed, arresting its tendency to rotate faster. Due to the effects of stator resistance, the maximum torque developed during regeneration is greater than the maximum torque during motoring. In hoists and cranes the drive motor has a tendency to

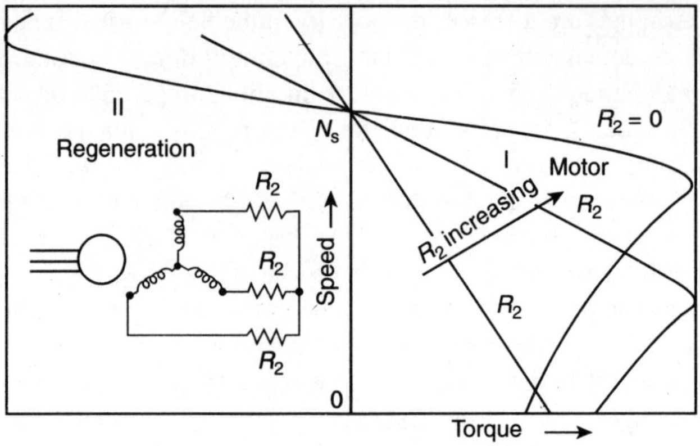

Fig. 1.47 *Speed-torque curves of a three phase induction motor during regeneration R_2 additional rotor resistance*

run faster than the synchronous speed. This situation occurs when the hoist is raising an empty cage. Due to the counterweight, the case may acquire dangerous speeds. The transition takes place almost automatically and a torque is developed to arrest the acceleration and regeneration takes place. This kind of operation is also possible when the load overhauls the motor during the descent of the load. Automatic regeneration arrests undue acceleration. Rotor resistance control may be employed to get better braking torque.

Regenerative braking is also possible with a pole change motor when the speed is changed from high to low. It can easily be accomplished in a variable frequency drive also. By decreasing the frequency of the motor momentarily, the synchronous speed decreases and conditions favourable to regeneration take place. As the motor speed decreases, the frequency is continuously reduced so that braking takes place at constant torque and stator current, till the motor comes to zero speed.

During regenerative braking there is a possibility of dangerous speeds if the operating point during braking falls in the unstable portion of the characteristic. This happens if the load torque is greater than the breakdown torque of the motor. The torque developed cannot brake the motor and undue acceleiation takes place. This possibility can be eliminated by means of a high resistance in the rotor.

Dynamic Braking Dynamic braking is employed to brake a non-reversing drive. The stator is transferred from ac mains to dc mains (Fig. 1.48(a)). The dc flowing through the stator sets up a stationary field. This induces rotor currents which produce a torque to bring the rotor to rest quickly. The torque developed and the retardation during braking may be controlled by the amount of dc power. Additional resistances r_1 and r_{2e} in the stator and rotor circuits respectively control the dc excitation and braking torques.

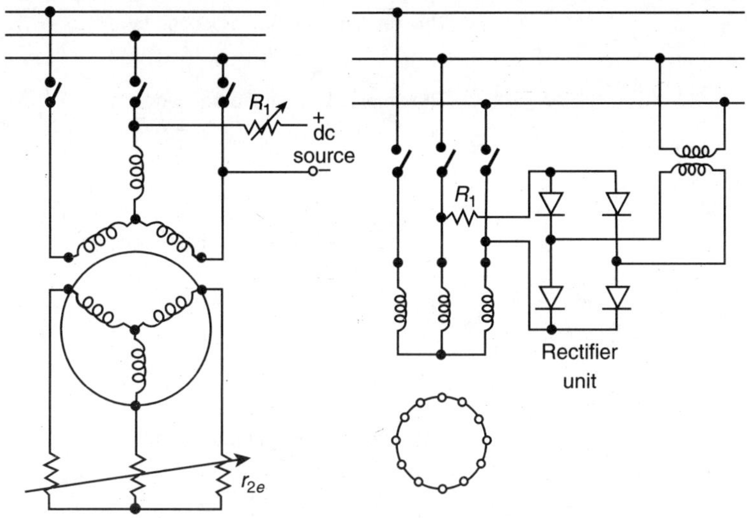

Fig. 1.48(a) *Induction motor connections for dynamic braking*

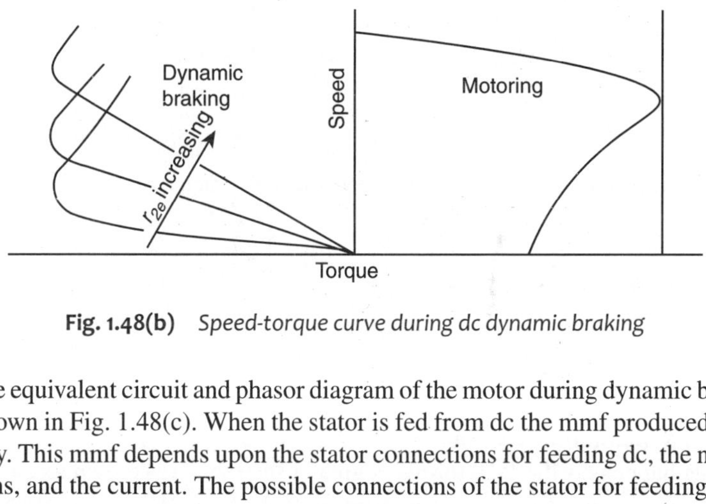

Fig. 1.48(b) *Speed-torque curve during dc dynamic braking*

The equivalent circuit and phasor diagram of the motor during dynamic braking are shown in Fig. 1.48(c). When the stator is fed from dc the mmf produced is stationary. This mmf depends upon the stator connections for feeding dc, the number of turns, and the current. The possible connections of the stator for feeding dc are

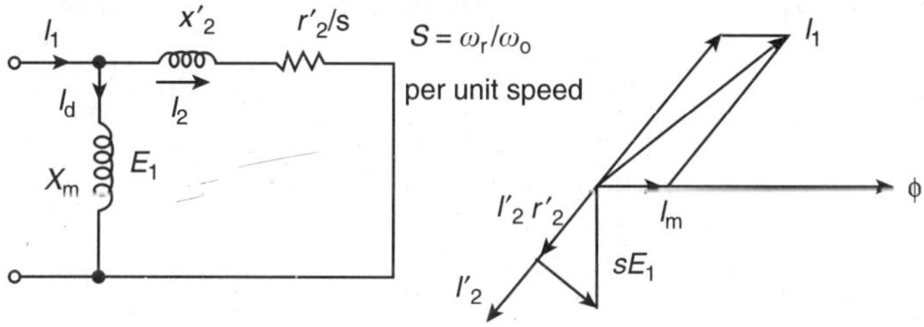

Fig. 1.48(c) *Equivalent circuit and phasor diagram during dynamic braking*

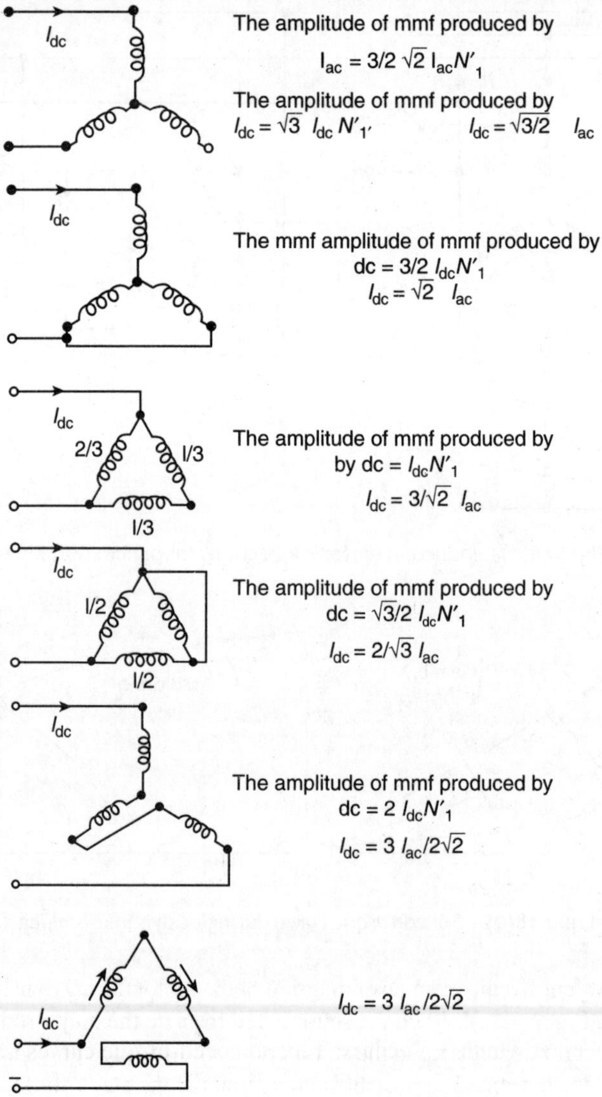

Fig. 1.49 *Possible connections of stator for dc dynamic braking*

shown in Fig. 1.49. The dc equivalent current can be determined by equating the mmfs produced by this equivalent current and the ac current.

The equivalent primary current is responsible for the magnetisation and the secondary current for the torque. The torque is given by

$$T = \frac{3}{2\pi n_s} I_2'^2 \frac{r_2'}{s} \tag{1.48}$$

where s is p.u. slip.

From the equivalent circuit we have

$$I_2' = I_{\text{eq}} \frac{x_{\text{m}}}{\sqrt{(r_2'/s)^2 + (x_2' + x_{\text{m}})^2}} \tag{1.49}$$

Also, from the phasor diagram we have

$$I_{\text{eq}}^2 = I_{\text{m}}^2 + I_2'^2 + 2I_{\text{m}}I_2' \cos \phi_2 \tag{1.50}$$

from which I_{m} can be determined. The value of x_{m} is given by

$$x_{\text{m}} = E_{\text{m}}/I_{\text{m}} \tag{1.51}$$

The torque is

$$T = \frac{3}{2\pi n_s} I_{\text{eq}}^2 \frac{x_m^2 (r_2'/s)}{(r_2'/s)^2 + (x_{\text{m}} + x_2')^2} \tag{1.52}$$

The torque slip curve can be determined using this expression. The maximum torque occurs at a slip

$$s_{\text{mb}} = \frac{r_2'}{\left(x_{\text{m}} + x_2'\right)} \tag{1.53}$$

and the maximum braking torque is

$$T_{\text{mb}} = \frac{3}{2\pi n_s} I_1^2 \frac{x_{\text{m}}^2}{2\left(x_{\text{m}} + x_2'\right)} \tag{1.54}$$

The characteristic can also be drawn using the relation

$$\frac{T}{T_{\text{mb}}} = \frac{2}{s_{\text{mb}}/s + s_{\text{mb}}/s} \tag{1.55}$$

The braking torque is proportional to I_1^2. However saturation plays its own role and there is a depletion of the torque. The effect of r_2' is similar to the effects during motoring, i.e. it does not change the value of T_{mb}, but changes the value of the speed at which T_{mb} occurs. Typical speed-torque curves are shown in Fig. 1.48.

The values of I_1 and r_2' are controlled to provide the desired braking. The former is limited by R_1. This method is applied commonly to brake the motors driving active loads. Dynamic braking is employed in conjunction with automatic control. The induction motor is more popular in hoists than the dc motor, due to this feature.

The methods of feeding a dc supply to the stator are depicted in Fig. 1.50. It may be provided by a dc supply using a limiting resistor R_1 in the circuit for controlling dc excitation. Torque control is achieved by rotor resistance variation. Alternatively, an ac supply may be rectified by means of a diode rectifier and the resulting dc may be fed to the motor.

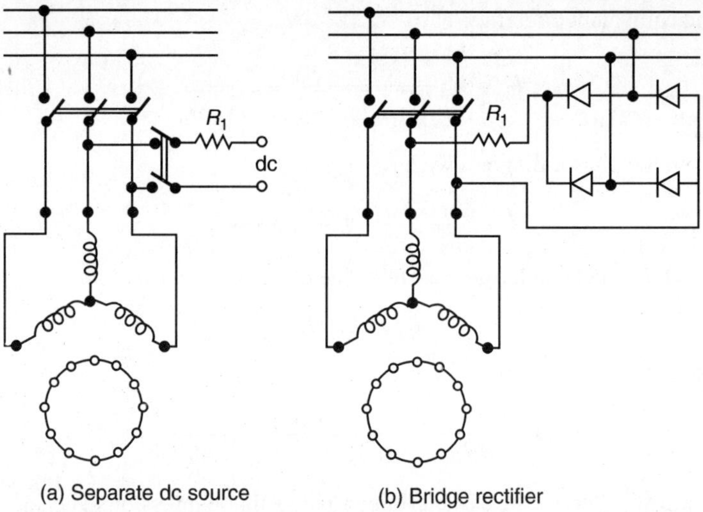

Fig. 1.50 *Methods of feeding dc to the stator for dynamic braking*

In ac dynamic braking the stator is switched to a capacitance bank. The machine runs as a self excited induction generator. All the mechanical energy is dissipated as electrical energy in the rotor resistance. This method is uneconomical, due to the cost of the capacitors.

Counter Current Braking By changing the phase sequence of the input to an induction motor, the direction of the stator field can be reversed. In practice this is done by interchanging the supply to any two terminals of the motor (Fig. 1.51(a)). A braking torque is developed and the motor comes to rest very fast. The motor must be switched off from the mains when zero speed is approached. Else the torque developed accelerates the motor in the reverse direction. This method is also called plugging.

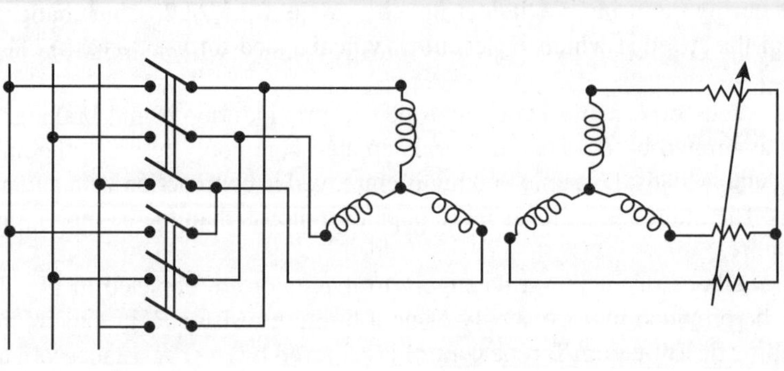

Fig. 1.51(a) *Connections for plugging of a three phase induction motor with rotor resistance for current limiting and braking torque*

When the motor is plugged, the induced voltage E in the armature and the applied voltage V aid each other and the current during braking is caused by $E + V$. This may result in very high currents, which are limited by the high rotor resistance, and also effectively increase the braking torque.

The speed-torque curve of an induction motor can be modified by varying the rotor resistance. The maximum torque point can be made to occur in the range of slips 1–2, where the torque developed tends to brake the rotor. This torque can also be used to arrest the tendency of the rotor to accelerate due to one reason or the other (e.g. load overhauling the motor or a hoist raising the empty cage). A high resistance is introduced in the rotor, so that the operating point shifts to the fourth quadrant. The braking torque developed prevents any acceleration of the rotor, and the rotor works at uniform speed (Fig. 1.51(b)).

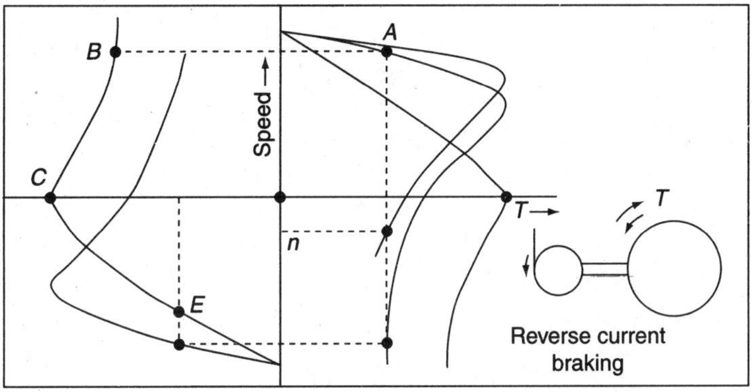

Fig. 1.51(b) *Speed-torque curves during plugging*

If the motor is operating at a slip s at the instant of plugging, the total braking torque is the sum of the plugging torque at $(2 - s)$ and the load torque

$$T_P = T_{\text{p}} + T_{\text{L}} \tag{1.56}$$

T_{p} may be controlled by a variable rotor resistance which limits the braking current.

1.5.4 Braking of Synchronous Motors

The methods that are employed for braking synchronous motors are:

i. Regenerative braking while operating on a variable frequency supply
ii. Rheostatic braking
iii. Plugging.

Regenerative Braking When the motor operates as a variable speed drive motor utilising a variable frequency supply, it can be regeneratively braked and all the K.E. returned to the mains. As in an induction motor, regeneration is possible if the synchronous speed is less than the rotor speed. The input frequency is gradually decreased to achieve this at every instant. The K.E. of the rotating parts is returned to the mains. The braking takes place at constant torque. With a CSI and cycloconverter, regeneration is simple and straightforward. With VSI an additional converter is required on the line side.

Rheostatic or Dynamic Braking A synchronous motor is switched on to a three-phase balanced resistive load after disconnecting it from the mains, keeping the excitation constant. To achieve greater braking torque for effective braking, the excitation may be increased. The terminal voltage and current (change) decrease as the speed decreases. At very low speeds the resistance effect becomes considerable. The value of resistance affects the speed at which the maximum torque occurs. It can ideally be made to occur just before the stopping of the motor.

The braking current at any instant is given by

$$I_{br} = \frac{E}{\sqrt{r_1^2 + (\omega L_s)^2}} \tag{1.57}$$

where $E = \omega L_{af} I_f / \sqrt{2}$ is the induced voltage.

In the above equations r_1 = stator resistance per phase

L_s = synchronous inductance per phase

L_{af} = mutual inductance between armature and field

I_f = field current

$$\text{The braking torque} = \frac{3I_{br}^2 r_1}{\omega} = K\frac{\omega}{r_1^2 + (\omega L_s)^2} \tag{1.58}$$

The speed at which the T_{br} is maximum can be obtained as $\omega_m = r_1/L_s$. By proper choice of r_1, the maximum braking torque can be made to occur just before stopping.

Plugging The braking of a synchronous motor by plugging has serious disadvantages. Very heavy braking current flows causing line disturbances. The torque is also not effective. However, if the motor is synchronous induction type it ran be braked effectively by plugging only if the machine is working as an induction motor.

1.5.5 Energy Relations During Braking of Electric Motors

In conventional methods of braking, such as rheostatic braking and plugging, it is necessary to know the energy wasted, so as to satisfactorily carry out the design of the braking equipment.

i. When a shunt motor is braked dynamically using a resistance the energy dissipated in the resistance is equal to the K.E. of the motor:

$$W_{br} = \frac{1}{2}J\omega_1^2 \tag{1.59}$$

Where J is the moment of inertia and ω_1 is the speed at which the braking is initiated.

ii. During counter current braking (plugging) the energy dissipated is $\frac{3}{2}J\omega_1^2$. If speed reversal is required, it is $2J\omega_1^2$. The extra energy drawn from the mains is due to the application of the voltage in the reverse direction.

iii. For a three-phase induction motor which is braked by dc in the stator

$$W_{br} = \frac{1}{2} I_a^2 r_1 t_{br} + \frac{1}{2} J\omega_s^2 \tag{1.60}$$

where t_{br} is the braking time, which can be estimated from the dynamics of the motor during braking as

$$t_{br} = \frac{J\omega_s}{T_{dm}} \frac{1 - s_1^2}{2s_m} + s_m \text{In}\left(1/s_1\right) \tag{1.61}$$

where s_1 is the slip at the instant of braking.

iv. When an induction motor is plugged, the energy wasted during braking is $\frac{3}{2}J\omega_s^2$. If it is allowed to reverse, the energy is $2J\omega_s^2$. If the stator losses are also added, these are $\frac{3}{2}J\omega_s^2\left(1 + r_1/r_2\right)$ and $2J\omega_s^2\left(1 + r_1/r_2'\right)$ respectively.

v. For a synchronous motor the energy dissipated during rheostatic braking is

$$W_{br} = 3I_{br}^2 r_1 t_{br} + \frac{1}{2} J\omega_s^2 \tag{1.62}$$

Knowing the various torques occurring in the motor during dynamic braking the dynamic behaviour of the motor can be established.

1.6 STARTING OF ELECTRIC MOTORS

Transient processes involved with the starting of the (drive) motor in a variable speed drive require a detailed study. The electric motor and connected load accelerates to the rated speed from rest under the influence of the starting torque. The transient operation during starting is satisfactory if a sufficiently good starting torque is developed with a reduced value of starting current, to accelerate the rotor in the desired amount of time.

The need to limit the starting current arises due to heavy dips of the voltage of the motor following starting peaks of current. The starting equipment used should be capable of minimising these dips in the voltage to a tolerable value, so that the other equipment on the network is not affected.

The starting current affects the motor also. High starting currents heat up the rotor. If starting is frequent, the heating needs to be reduced or limited. In dc machines high starting currents produce sparking at the brushes. For good commutation the starting currents must be limited. In converter operation additional harmonics of current affect the commutation.

These problems of commutation in dc motors due to ripple may be solved by (properly) increasing the pulse number of the power converter, and modifying the design of the motor itself, e.g. laminating the interpoles.

The starting torque must produce uniform acceleration. Acceleration time must be reduced with a view to improving the productivity and to reduce the energy lost during starting.

1.6.1 Methods of Starting Electric Motors

The purpose of starting equipment in an electric motor is to limit the starting current and to provide a reasonably good starting torque, if possible, so that the motor accelerates in the desired period to the rated speed. For dc motors the starting current is limited by using an additional resistance in series with the armature. The motor is switched on with full field. This is effectively reduced voltage starting. Thyristor power converters used for speed control may also be used for the purpose of starting, since the voltage is smoothly variable and starting losses are absent.

Induction motors are started by any of the following methods:

i. Direct on line starting
ii. Low voltage starting
iii. Rotor resistance starting
iv. Low frequency starting
v. Special rotor construction.

A comparison of these methods is given in Table 1.6.

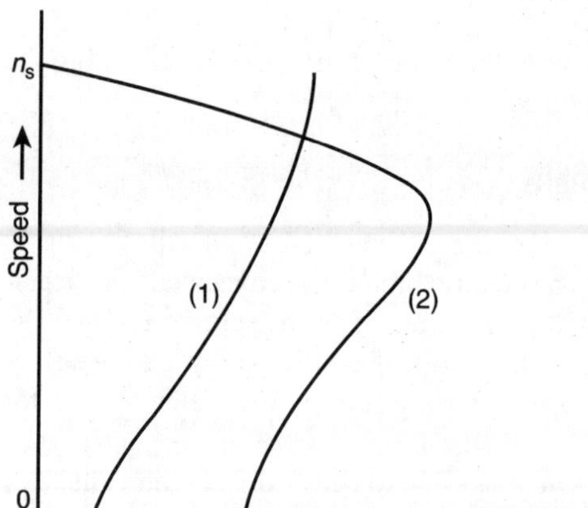

Fig. 1.52 *Comparison of starting a synchronous motor on mains and variable frequency supply. (1) Mains starting (2) Low frequency starting*

Synchronous motors are not self starting, but are started by an auxiliary motor or using the induction motor principle. The damper windings are used as starting windings. To make a synchronous motor self starting, a wound rotor is employed. It is short circuited with an additional resistance during starting and fed from a dc when accelerated to synchronous speed. A synchronous motor can also be started using a variable frequency inverter system. The application of this kind of starting is discussed in Chapter 7. A comparison of variable frequency starting with mains starting is given in Fig. 1.52.

Table 1.6 *Comparison of starting methods of induction motors*

Basis of comparison	*Direct on starting*	*Low voltage starting*	*Rotor rheostat*	*Low frequency starting*	*Special rotor*
General	Small squirrel cage motors up to 2 kW are started using full rated voltage. This is not suitable for motors of higher rating. The starting current depends upon blocked rotor impedance. The saturation effects on leakage reactance will increase the starting current. The high starting current causes dips in line voltage as well as heating of the motor if the acceleration time is large.	Squirrel cage motors are started normally with reduced voltage to limit the starting current. The associated motor heating and voltage dips are controlled. After the motor attains 80% of rated speed the full voltage is applied. Reduced voltage starting may be accomplished using an autotransformer, a star/delta switch or a series reactor.	Employed for slip ring induction motors. The resistance which provides the desired starting performance is inserted in the rotor circuit. A suitable resistance may also be chosen to provide fast acceleration. The resistance is slowly cut off as the motor accelerates so that the acceleration takes place at constant torque. The resistance is fully cut off when the motor reaches rated speed.	This can be employed if the motor is fed from a frequency converter. A variable frequency converter may be developed only for starting. The motor frequency is slowly varied such that the slip frequency is constant. This maintains the input current at a constant value and acceleration takes place at constant torque. Both VSI and CSI may be used for starting the motor. The starting dynamics are improved. The current and acceleration during starting may be maintained at the desired value by controlling slip frequency. If used only for starting the system is costly.	Squirrel cage motors have double cage or deep bar to improve starting conditions. The starting current decreases whereas torque increases. Direct on line starting may be used. The running performance of the motor is reasonably good. However, depending upon the cage parameters the full load efficiency and power factor may be poorer than plain cage motors. The desired performance at starting may be achieved by designing the motor with proper choice of cage parameters.

(*Continued*)

Basis of comparison	*Direct on starting*	*Low voltage starting*			*Rotor rheostat*	*Low frequency starting*	*Special rotor*
		Auto transformer	*Star/ delta*	*Series reactor*			
Starting current	I_{st}, = (6 to 8)I_{fl}	$x = \frac{V_x}{V_r}$			Rotor resistance may be designed to obtain the desired values of starting torque and current	Characteristics are similar to those of rotor rheostat starting.	300 to 400% of full load current
Starting torque	$\left(\frac{I_{st}}{I_{fl}}\right)^2$ full load slip (1.5 to 2 times full load torque)	$I_{st} = x^2$ times that of DOL starting x^2 times that of DOL starting	1/3 times DOL starting 1/3 times that of DOL starting	x times that of DOL starting x^2 times that of DOL starting		V/f is kept constant and the acceleration is achieved at constant slip frequency.	300% full load torque
Starting power factor	Poor				better	better	reasonably good power factor

1.6.2 Energy Relations During Starting

In order to select a proper starting equipment, e.g. the starting rheostat in a shunt motor, it becomes necessary to determine the energy loss during starting.

i. The energy loss during starting is the K.E. of the rotating parts at final speed. The same energy is drawn from the supply. Therefore the total electrical energy drawn from the supply during starting is

$$W_e = J\omega_s^2$$

where ω_s is the final speed at no-load.
When started against a load of torque T_L the total loss in the armature circuit is

$$W_e = J\left(\omega_s\omega_L - \frac{\omega_s^2}{2}\right) + T_L F \tag{1.63}$$

where ω_L is the speed with load T_L and $T_L F$ is the energy loss due to the load torque and is given by

$$F = \omega_0 t_{st} - \int_0^{t_{st}} \omega dt$$

In the case of dc series motors the energy loss depends on R, and is given by

$$W_e = \frac{JR}{K_t}\omega_0 + \frac{T_L R}{K} t_{st} \tag{1.64}$$

ii. In the case of an induction motor, an additional rotor resistance is employed. The energy loss in the rotor resistance is the K.E. of the rotating parts. Part of the energy is dissipated in the stator resistance also. The total energy lost during starting is

$$W_e = \frac{J\omega_s^2}{2}\left(1 + \frac{r_1}{r_2'}\right) \tag{1.65}$$

An increase in rotor resistance decreases the energy loss in the stator resistance, whereas it does not affect the loss in the rotor resistance itself. This decreases the time of acceleration.

When the motor is started on load the loss dissipated is greater than at no-load, and is given by

$$\Delta W_{st} = J\omega_s\left(1 + \frac{r_1}{r_2'}\right)\frac{T_d}{T_d - T_L} s\, ds \tag{1.66}$$

where T_d and T_L are functions of slip. The integration of ΔW_{SL} from $0 - \omega_s$ gives the total energy loss. When squirrel cage motors are started directly from the line, there is minimum energy dissipation because the torque developed is large with full voltage. With low voltage starting, T_d decreases and the energy wasted increases, even though starting current decreases. This may be attributed to the increase of acceleration time at low voltage starting.

In squirrel cage motors the entire energy is lost in the machine itself, whereas in slip ring rotors an external resistance may be used for the dissipation of energy, thus minimising the motor heat but increasing the starting time.

1.6.3 Starting Dynamics of Electric Motors

The starting, of electric motors is normally done with graded resistances which are cut off slowly as the motor accelerates. The grading is based on two limits between which starting current is allowed to vary. The dynamics during starting may be necessary to actually find out the values of these resistances.

DC Shunt Motor When a shunt motor is accelerated under load, the equations are

$$V = K_t\omega + I_a r_a \tag{1.67}$$

$$T_d = K_t I_a = J\frac{d\omega}{dt} + T_L \tag{1.68}$$

Using these relations we have

$$\frac{d\omega}{dt} + \frac{\omega}{\tau_m} = \frac{\omega_0 - \omega_L}{\tau_m} \tag{1.69}$$

where ω_0 is the no-load speed

ω_L is the speed drop under load

τ_m is the mechanical time constant.

If the motor is accelerated from ω_1, to ω_2 the solution of this equation may be obtained as

$$\omega = \omega_2 + (\omega_1 - \omega_2)e^{-t/\tau_m} \tag{1.70}$$

If the acceleration is from zero to ω_0 we have

$$\omega = \omega_0\left(1 - e^{-t/\tau_m}\right) \tag{1.71}$$

During starting, the current drops as the motor speeds up, due to the building up of back emf. The armature current drops exponentially as

$$i = I_L + (I_1 - I_L)e^{-t/\tau_m} \tag{1.72}$$

where I_1 is the starting current and I_L the final value. If the acceleration is at no-load

$$i = I_1 e^{-t/\tau_m} \tag{1.73}$$

The time of acceleration can be determined using these relations. If a multi-stepped starter is used, the time taken for the current to drop from I_1 to I_2 is

$$t_x = \tau_m \text{In}\left(\frac{I_1 - I_L}{I_2 - I_L}\right) \tag{1.74}$$

The mechanical time constant is different at different steps, and is given by

$$\tau_{mx} = \frac{JR_x}{K_t^2} \tag{1.75}$$

$$R_x = \frac{V - E_x}{I_1} \tag{1.76}$$

The value of E_x is obtained by using the value of speed, which increases exponentially. Finally, the total accelerating time

$$t_x = \sum_{r=1}^{n} t_{2r} \tag{1.77}$$

Induction Motor We can analyse the transient conditions in a three-phase induction motor with a view to determining the time of acceleration. It is well known that by properly adjusting the rotor resistance, maximum torque may be made to occur at starting. But analysis shows that this does not give the minimum time of acceleration, which is obtained if the rotor resistance is selected such that the maximum torque occurs at a slip of 0.407.

The torque developed at any slip is given by

$$T = \frac{2T_{dm}}{s/s_m + s_m/s} \tag{1.78}$$

Assuming the acceleration takes place at no-load, the torque developed accelerates the rotor

$$T_d = J\frac{d\omega}{dt} \tag{1.79}$$

Also, we have

$$\omega = \omega_s(1 - s) \tag{1.80}$$

assuming the no-load speed is synchronous speed. Using the above equations we have

$$\frac{2T_{dm}}{s_m/s + s/s_m} = -J\omega_s\frac{ds}{dt} \tag{1.81}$$

form which

$$dt = \frac{-\tau_m}{2}\left[\frac{s}{s_m} + \frac{s_m}{s}\right]ds \tag{1.82}$$

where $\tau_m = \dfrac{J\omega_s}{T_{dm}}$ is the mechanical time constant of the motor. The motor starts from rest and runs up to no-load speed. The slip varies from 1 to s. Integrating Eq. 1.82 between these limits we have

$$t_{st} = \frac{\tau_m}{s^2}\frac{1 - s^2}{2s_m} + s_m \text{In}\left(\frac{1}{s}\right) \tag{1.83}$$

If the final slip is assumed to be $s = 0.03$, the starting time t_{st} is

$$t_{st} = \tau_m \left[\frac{1}{4s_m} + 1.5s_m \right] \tag{1.84}$$

The minimum value of t_{st} is obtained by

$$\frac{dt_{st}}{ds_m} = 0 \tag{1.85}$$

which gives $s_m = 0.407$. The ratio t_{st}/τ_m as a function of s_m is shown in Fig. 1.53. To accelerate the rotor in the time t_{st} the effective torque is

$$T_{\text{eff st}} = \frac{J\omega_s}{t_{st}} = \frac{T_{dm}s_m}{0.25 + 1.5s_m^2} \tag{1.86}$$

This shows again that the starting time is a minimum if the starting torque is 0.81 T_{dm}. This happens again if T_{dm} occurs at $s_m = 0.407$. An induction motor can be made to accelerate in minimum time if the starting torque is 0.81 times the maximum torque and the maximum torque occurs at $s_m = 0.407$.

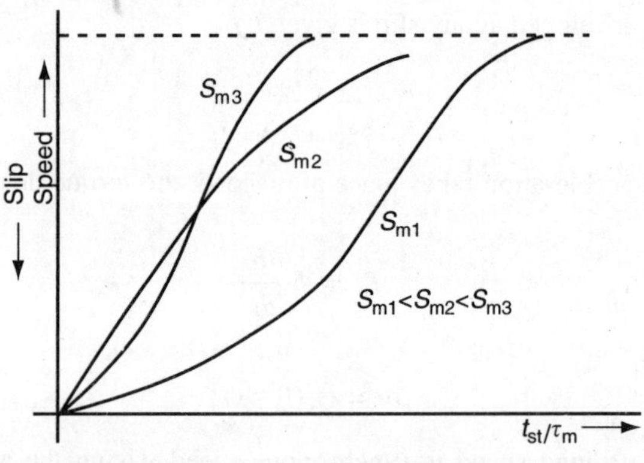

Fig. 1.53(a) *The starting time(t_{st}/τ_m) as a function of slip, the slip for maximum torque being a parameter*

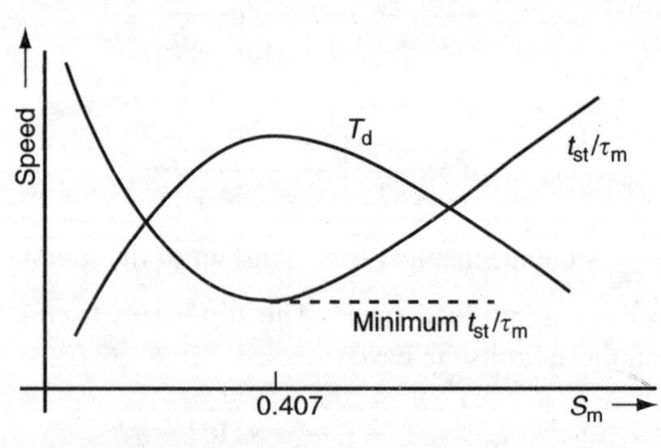

Fig. 1.53(b) *Motor torque and starting time as functions of slip for maximum torque*

Worked Examples

1.1 *A 500 V de shunt motor with constant field drives a load whose torque is proportional to the square of the speed. When running at 900 rpm it takes an armature current of 45 A. Find the speed at which the motor runs if a resistance of 8Ω is connected in series with the armature. The armature resistance may be taken as 1Ω.*

Solution In a shunt motor the torque developed

$$T_d = KI_a$$

As the load torque is proportional to the square of the speed

$$T_L = K_L N^2$$

$$\text{As} \quad T_d = T_L$$

$$\left(\frac{N_1}{N_2}\right)^2 = \frac{I_{a1}}{I_{a2}}$$

The back emf $E_b \propto N_1$

Therefore $\left(\frac{E_{b1}}{E_{b2}}\right)^2 = \frac{I_{a1}}{I_{a2}}$

$$E_{b1} = V - I_{a1}r_a = 500 - 45 = 455 \ V$$

$$E_{b2} = 500 - 9I_{a2}$$

Substituting

$$\frac{455^2}{(500 - 9I_{a2})^2} = \frac{45}{I_{a2}}$$

Cross multiplying and simplifying,

$$I_{a2}^2 - 167.9I_{a2} + 3086.4 = 0$$

Solving, we get $I_{a2} = 21$ A.

$$E_{b2} = 500 - 9 \times 21 = 500 - 189 = 311 \text{ V}$$

$$\text{The new speed} = \frac{311}{455} \times 900 = 615.2 \text{ rpm}$$

1.2 *A 400 V, 750 rpm, 70 A dc shunt motor has an armature resistance of 0.3Ω. When running under rated conditions, the motor is to be braked by plugging with armature current limited to 90 A. What external resistance should be connected in series with the armature? Calculate the initial braking torque and its value when the speed has fallen to 300 rpm. Neglect saturation.*

Solution The speed of the motor = 750 rpm = 78.5 rad/s

The back emf = 400 – 70 × 0.3 = 379 V

$$\text{The torque developed} = \frac{379 \times 70}{78.5} = 338 \text{ Nm}$$

When plugged,

the voltage causing the current $= 400 + 379 = 779$ V

$$\text{current} = 90 \text{ A}$$

$$\text{resistance} = \frac{779}{90} - 0.3 = 8.36\ \Omega$$

The braking torque at the instant of braking

$$= \frac{379 \times 90}{78.5} = 434.5 \text{ Nm}$$

$$\text{At 300 rpm the back emf} = 379 \times \frac{300}{750}$$

$$= 15.1.6 \text{ V}$$

$$\text{Current} = \frac{400 + 151.6}{8.356} = 63.725 \text{ A}$$

$$\text{Torque} = \frac{151.6 + 63.725}{2\pi \times 5} = 307.66 \text{ Nm}$$

1.3 *A 250 V dc series motor drives a fan, the load torque being proportional to the 1.5th power of the speed. At a certain speed the motor takes 40 A The machine resistance is 0.6Ω. Find the extra resistance needed to reduce the speed to one half of the original speed. Saturation may be ignored.*

Solution The torque developed in a dc series motor $= T_d \propto \phi I_a \propto I_a^2$ as there is no saturation

$$\frac{T_1}{T_2} = \frac{I_{a1}^2}{I_{a2}^2} = \frac{N_1^{1.5}}{N_2^{1.5}}$$

$$\text{But } \frac{N_1}{N_2} = 2$$

$$\frac{I_{a1}^2}{I_{a2}^2} = 2\sqrt{2} = 2.828$$

$$I_{a2} = \frac{40}{\sqrt{2.828}} = 23.79 \text{ A}$$

$$\frac{E_{b1}}{E_{b2}} = \frac{I_{a1}N_1}{I_{a2}N_2} = \frac{2 \times 40}{23.79} = 3.363$$

$$E_{b2} = \frac{E_{b1}}{3.363} = \frac{250 - 40 \times 0.6}{3.363} = 67.2 \text{ V}$$

$$250 - 23.79\,R = 67.2$$

$$R = 7.684\Omega$$

Extra resistance required $= 7.084\Omega$

1.4(a) *A 440 V, 80 amp, 1200 rpm dc shunt motor has armature and field resistances of 0.55Ω and 110Ω respectively. The shunted armature connection is used to obtain a no-load speed of 600 rpm and full load speed of 300 rpm (full load torque) Determine the necessary value of series and shunting resistances. Calculate also the line current and overall efficiency of the system.*

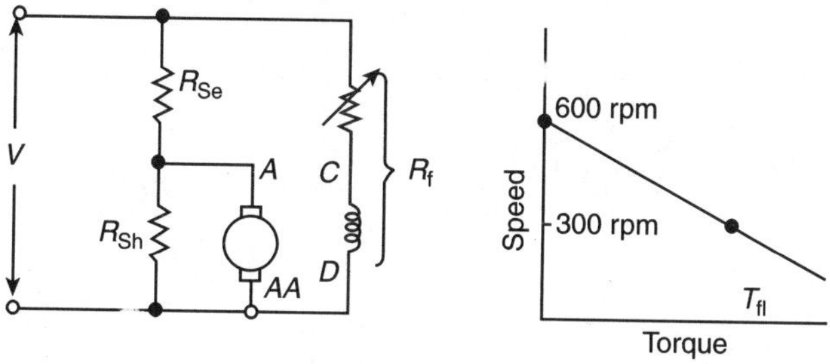

Fig. P1.4 Connections and torque speed characteristics of example 1.4

Solution With the arrangement shown, the no-load speed = 600 rpm

$$\omega_r = 62.8 \text{ rad/s}$$

Induced voltage

$$= K\omega_r$$

The value of K can be obtained from the ratings as

$$E = 440 - 80 \times 0.55 = 396 \text{ V}$$

$$\text{Rated speed} = 1200 = 125.6 \text{ rad/s}$$

$$K = \frac{396}{125.6} = 3.153$$

The induced voltage with R_{sc} and R_{sh} is

$$E_1 = 3.153 \times 62.8 = 198 \text{ V}$$

$$\text{The armature voltage} = V\left(\frac{R_{sh}}{R_{sh} + R_{sc}}\right) = \frac{V}{(1+a)}$$

Since the current through the armature is zero

$$\frac{V}{1+a} = 198 \, V$$

$$1 + a = \frac{V}{198} = 2.22$$

$$\text{Full load torque} = KI_a = 3.153 \times 80 = KI_a = 252.24 \text{ Nm}$$

At 300 rpm the motor develops full load torque. As the torque is constant armature current at this speed $I_{n2} = 80$ A

The back emf is

$$E_2 = 198 \times \frac{300}{600} = 99 \, V$$

$$99 = V_2 - 80 \times 0.55$$

$$V_2 = 99 + 44 = 143 \, V$$

The supply current

$$I_s = \frac{143}{R_{sh}} + 80$$

Also

$$440 = I_s R_s + 143$$

$$= R_s\left(\frac{143}{R_{sh}} + 80\right) + 143$$

Solving $R_s = 1.532\Omega$
$R_{sh} = 1.256\Omega$

$$\text{The line current} = 193.854\text{A} + \frac{440}{110} = 197.854$$

$$\text{Input power} = 85.3 + 1.76 = 87.06 \text{ kW}$$

$$\text{Output} = 252.24 \times 2\pi n_s = 7920 = 7.920 \text{ kW}$$

$$\text{Efficiency} = 9.1\%$$

1.4(b) *If in the previous problem* $R_s = 2\Omega$ *and* $R_{sh} = 1.5\Omega$, *determine the no-load and full load speeds.*

Solution

$$a = \frac{R_s}{R_{sh}} = \frac{2}{1.5} = 1.333$$

$$No - load\ voltage = \frac{440}{1.333} = 330\ V$$

$$\text{No-load speed} = \frac{330}{3.153} = 104.6622 \text{ rad/s}$$

$$= 1000 \text{ rpm}$$

Full load speed $I_a = 80\text{A}$

$$\text{The supply current} = \frac{V_2}{R_{sh}} + 80$$

$$\frac{V - V_2}{R_s} = \frac{V_2}{R_{sh}} + 80$$

$$V_2\left[\frac{1}{R_s} + \frac{1}{R_{sh}}\right] = -80 + \frac{V}{R_s} = -80 + 220 = 140$$

$$V_2[0.5 + 0.6667] = 140$$

$$V_2 = 120V$$

$$E_2 = 120 - 80 \times 0.55 = 76\ V$$

$$N_2 = 1000 \times \frac{76}{330} = 230.30 \text{ rpm}$$

1.5(a) *A 250 V dc series motor has an armature resistance of 0.4Ω. From its magnetisation curve at 480 rpm, the armature voltage is 125 V at 40 A of field current when running as a motor. Determine the speed at which it will run while drawing a current of 40 A.*

Solution Induced voltage of the motor $= 250 - 0.4 \times 40$

$$= 234\ V$$

$$\text{The speed of the motor} = 480 \times \frac{234}{125} = 898.56 \text{ rpm}$$

$$\text{The torque developed} = \frac{234 \times 40}{2\pi n_s} = \frac{234 \times 40}{94.05}$$

$$= 99.52 \text{ NWm}$$

1.5(b) *In part (a) determine the value of resistance to be included in series with the motor so that the motor runs at 480 rpm. Determine the torque. The motor current is 40 A.*

Solution The back emf = 125 V

$$250 - 40(0.4 + R_e) = 125$$

$$R_e = 2.725\Omega$$

$$\text{The torque developed} = \frac{125 \times 40}{2\pi \times 480/60} = 265.4 \text{ Nm}$$

1.6 *The series motor in the previous problem has a resistance of 2.725Ω in series with the field and 3.5Ω in parallel with the armature. Determine the speed of the motor when the supply current is 40 A. Assume the field has a resistance of 0,15Ω.*

Assume a linear magnetisation characteristic.

Solution Voltage across the motor $= 250 - 40(2.725 + 0.15)$

$$= 135 \text{ V}$$

The current through shunting resistor = 38.57 A

Armature current = 1.43 A

$$\text{Back emf} = 135 - 1.43 \times 0.25 = 134.6425 \text{ V}$$

$$\text{Speed of the motor} = \frac{125}{134.64} \times 480 = 446 \text{ rpm}$$

$$\text{Torque} = \frac{134.64 \times 1.43}{2\pi \times 446/60} = 4.128 \text{ Nm}$$

1.7 *The series motor of Example 1.6 is shunted with a resistance of 3.5Ω, while a resistance of 2.725Ω is connected in series with the line. Determine the speed and torque of the motor. Assume the motor current to have a value = 40 A*

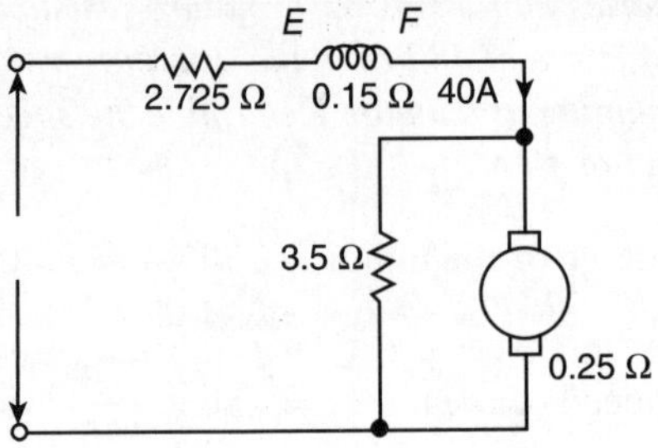

Fig. P1.7 *Pertaining to example 1.7*

Solution Back emf at 40 A = 125 V and speed = 480 rpm
Let the voltage across the motor be V_m

$$I_L = \frac{250 - V_m}{2.725} \quad I_r = \frac{V_m}{3.5} \quad I_a = I_L - I_r$$

$$40 = \frac{250 - V_m}{2.725} - \frac{V_m}{3.5}$$

$$= \frac{250}{2.725} - V_m\left(\frac{1}{2.725} + \frac{1}{3.5}\right)$$

$$= \frac{250}{2.725} - V_m 0.653$$

$$= 91.743 - 0.653V_m$$

$$V_m = 79.24\ V$$

$$E_m = 79.24 - 40 \times 0.4 = 63.24\ V$$

$$\text{Speed} = \frac{63.24}{125} \times 480 = 242.8 \text{ rpm}$$

$$\text{The supply current} = I_L = \frac{250 - 79.24}{2.725} = 62.66\ A$$

$$\text{Input} = 15.67\ kW$$

$$\text{Output} = 63.24 \times 40 = 2.53$$

$$\text{Efficiency} = 16.14\%$$

$$\text{Torque} = \frac{2.53 \times 10^3}{2\pi n_s}$$

$$= \frac{2.53 \times 10^3}{25.41} = 99.56 \text{ Nm}$$

1.8 *The series motor connections are as shown in Fig. P1.8 to lower the empty container of a hoist. If the current through the field circuit is 40 A determine the speed and torque. The motor details may be taken from the previous example. The supply to the armature is reversed during lowering.*

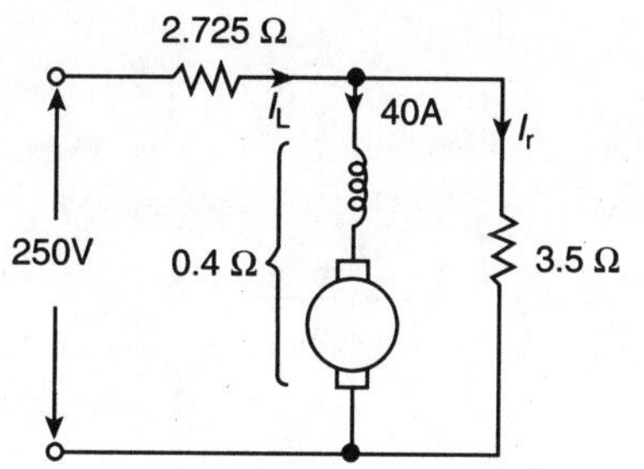

Fig. P1.8 *Pertaining to example 1.8*

Solution Motor voltage

$$V_m = 0.95 I_f$$
$$= 40 \times 0.95 = 38 \text{ V}$$

Armature current $= -(I - I_f)$

$$I_a = I_f - \frac{V - V_m}{2} = 40 - \frac{250 - 38}{2}$$
$$= 40 - 106 = -66\ A$$
$$E_m = -38 + 66 \times 0.55 = -22\ V$$
$$\text{Speed} = -\frac{22}{125} \times 480 = -84.48$$
$$\text{Torque} = \frac{-22 \times -66}{2\pi \times (-84.48/60)} = -164.2 \text{ Nm}$$

1.9 *The maximum torque of a 3-phase squirrel cage induction motor is 2.5 times the full load torque, and the starting torque is 1.2 times the full load torque. For negligible stator resistance, compute the*

i. *slip at maximum torque*
ii. *full load slip*
iii. *starting current in terms of full load value.*

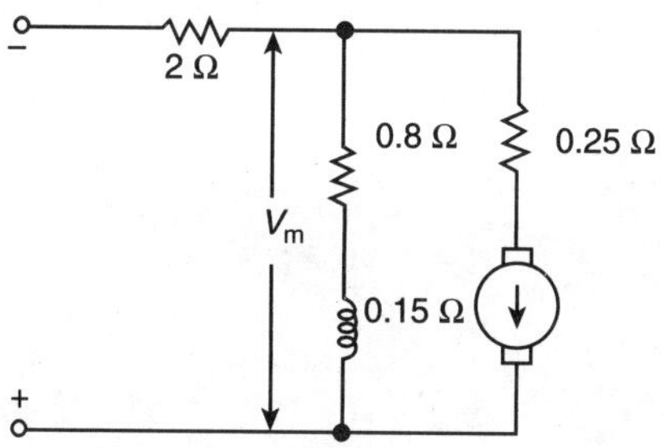

Fig. P1.9 *Pertaining to example 1.9*

Solution The ratio of full load torque to maximum torque

$$\frac{T_d}{T_{dm}} = \frac{2s\, s_m}{s^2 + s_m^2}$$

also

$$\frac{T_{st}}{T_{dm}} = \frac{2s_m}{1+s_m^2}$$

$$\frac{T_{st}}{T_d} = \frac{2s_m}{1+s_m^2}\,\frac{s^2+s_m^2}{2ss_m}$$

$$\frac{1}{2.5} = \frac{2ss_m}{s^2+s_m^2}$$

$$\text{i.e. } s^2 + s_m^2 = 5ss_m$$

$$1.2 = \frac{2s_m}{1+s_m^2}\,\frac{5ss_m}{ss_m} = \frac{10s_m}{1+s_m^2}$$

$$1 + s_m^2 = 8.33s_m s^2 - 0.6006s + 0.0144 = 0$$

$$s_m^2 - 8.33s_m + 1 = 0 \qquad s = 0.0.25$$

$$s_m = 0.120$$

1.10 *A 3-phase squirrel cage induction motor has a starting current eight times the full load value. The full load slip is 4%. Compute the starting torque as a percentage of full load torque if the motor is started*

i. *direct on line*
ii. *by a star/delta starter*
iii. *using an autotransformer to limit the starting current to three times the full load value. What is the line current as a percentage of full load value?*

Solution Direct on line starting torque $= \left(\frac{I_{st}}{I_{fl}}\right)^2 \times s_{fl}$

$$= 64 \times 0.04 = 2.56$$

$$\text{Star/}\Delta\text{ starter} = \frac{2.56}{3} = 0.8533$$

Autotransformer

$$T_{st} = 9 \times 0.04 = 0.36$$

$$I_{1L} = \frac{9}{8} I_{fl}$$

$$I_L = \frac{3}{8}\frac{I_{st}}{8} = 9/64 I_{st}$$

$$x = \frac{3}{8} \times 3I_{f1}\,\frac{9}{8}$$

1.11 *If in the previous example, the line current is to be limited to three times the full load value, determine the phase current and torque.*

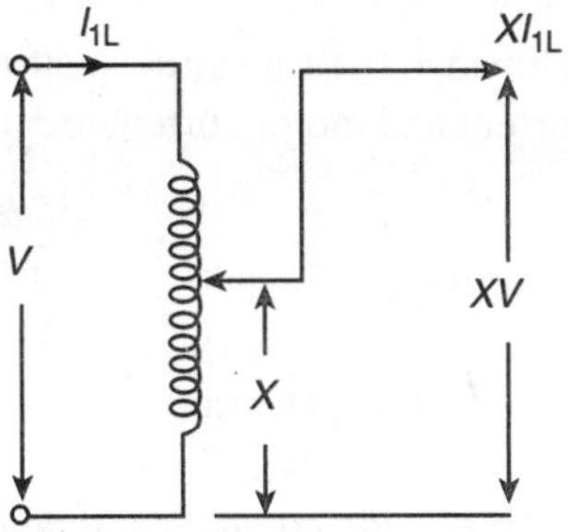

Fig. Pl.11 *Pertaining to example 1.11*

Solution

$$I_{1L} = \frac{x^2}{8} I_{f1}$$

$$3I_{f1} = \frac{x^2}{8} I_{f1}$$

$$x^2 = 24 \quad x = 4.9$$

$$T_{st} = (4.9)^2 \times 0.04 = 0.96 \quad \text{tap} = 61.25\%$$

$$\left(\frac{I_{st}}{I_{f1}}\right)^2 \times 0.025 = 1.2$$

The starting current = 6.93 times full load value.

1.12 *A three-phase, cage induction motor takes at normal voltage a starting current of 5 times the full load value and its full load slip is 4%. What ratio of autotransformer would enable the motor to be started with not more than twice full load current drawn from the supply? What would be the starting current under this condition and how would it compare with that obtained using a stator resistance starter under the same limitations?*

Solution Starting torque with direct on line

$= 5^2 \times 0.04 = 1.00$ times full load torque

If an autotransformer is used for starting with $x\%$ tapping, the motor current is limited to $\frac{5x}{100} \times I_{f1}$

The corresponding primary current would be such that

$$V_p I_p = \left(\frac{x}{100}\right) V_p \frac{5x}{100} \times I_{f1}$$

But $I_p = 2I_{f1}$

Therefore

$$2I_{f1} = \frac{5x^2}{(100)(100)} I_{f1}$$

$$x = \sqrt{\left(\frac{2}{5}\right)} \times 100 = 63.3\%$$

The starting torque developed = 0.4 times full load torque. If a stator resistance starter is used, the line current and motor current are the same. Therefore the starting torque

$$= \left(\frac{2}{5}\right)^2 (5)^2 \times 0.04$$

$$= 0.16 \text{ times the full load torque.}$$

1.13 *A 500 V, 3-phase, 50 Hz star-connected induction motor has the following equivalent circuit parameters:*

$$r_{1'} = 0.13\Omega,\ r_2 = 0.32\Omega,\ x_1 = 0.6\Omega$$
$$x_2 = 1.48\Omega,\ r_m = 250\Omega,\ x_m = 20\Omega$$

all referred to the stator side. The full load slip is 5%. The machine is to be braked from full load speed by plugging, after inserting a resistance of l.5Ω per phase referred to the stator. Determine the initial braking torque. Neglect mechanical losses and use the approximate equivalent circuit.

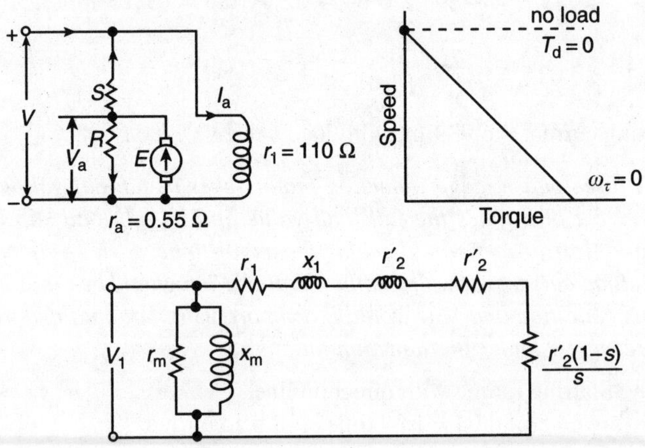

Fig. P1.13 *Equivalent circuit of induction motor*

Solution The equivalent circuit is shown in Fig. PI. 13.
The impedance Z_2 of the motor

$$Z_2 = 0.13 + j0.6 + \frac{0.32}{0.05} + j1.48$$

$$= 0.13 + j0.6 + 6.4 + j1.48 = 6.53 + j2.08 = 6.853\angle 17.67°$$

$$I_2 = \frac{500}{\sqrt{3} \times 6.853} = 42.13 \text{ A}$$

$$\text{Torque} = \frac{3}{2\pi n_s}\frac{I_2'^2 r_2}{s} = \frac{3}{2\pi n_s} \times (42.13)^2 \times 6.4 = \frac{1}{2\pi n_s} 34.07 \text{ Nm}$$

$$= 34.07 \text{ syn. kW}$$

At the instant of plugging

$$s_b = 2 - s = 1.95$$

$$Z_b = 0.13 + j\,0.6 + \frac{1.82}{1.95} + j\,1.48 = 1.063 + j\,2.08 = 2.336\Omega$$

$$\text{Current} = \frac{500}{\sqrt{3} \times 2.336} = 123.6\ A$$

$$\text{Torque} = \frac{3}{2\pi n_s}(1236)^2 \times 0.93 = 42.662 \text{ syn kW}$$

The braking torque = 77.293 kW

Assuming four poles, this is 492.3 Nm

Problems

1.1 A 50 hp, 220 V, 1200 rpm dc shunt motor has an efficiency of 85% at full load. The shunt field resista: ce is 110Ω and the armature circuit resistance is 0.06Ω. This motor is required to drive a load requiring 120% of the full load motor torque at a speed of 550 rpm. Determine the additional armature resistance required to operate the motor.

1.2 A 50 hp, 220 V, 1200 rpm shunt motor has an armature re istance of 0.06Ω. At rated speed and output the motor takes a current of 180 A. The field current is 2 A The field flux is reduced to 70% of its original value by using a field rheostat while the torque remains the same. Find the new speed and efficiency, assuming that the rotational losses vary as the square of speed.

1.3 The armature of a dc shunt motor has a resistance of 0.08Ω. The armature current drawn by the motor at a speed of 1400 rpm is 190 A The motor drives a load having a torque speed characteristic given by

$$T_1 = 0.65\omega \text{ Nm}$$

where ω is rad/s. If the rated voltage of the motor is 230 V determine the torque and speed of the motor.

1.4 (a) Determine the additional armature resistance required to reduce the speed to 600 rpm in Problem 3.

(b) Determine also the voltage to be applied to the motor for the conditions of 4(a). Assume the field is separately excited from a 230 V supply.

1.5 A 600 V, 50 hp, 600 rpm dc series motor has an armature and series field resistance of 0.2Ω. The full load current of the motor is 215 A. The armature voltage is varied to control the speed of the motor. Determine the voltage required to reduce the speed to 450 rpm if the torque driven by the motor is given by the relation

$$T_L = K\omega^2 \text{ Nm}$$

where ω is the speed in rad/s.

1.6 A series motor having a combined series and field resistance of 2Ω draws a current of 8 A from a 240 V supply and runs at a speed of 400 rpm.

(a) Determine the torque developed, assuming a linear magnetisation curve.
(b) Determine the value of resistance to be connected in series with the armature circuit if the speed has to be reduced to 200 rpm keeping the torque at its value in (a).
(c) What would be the speed of the motor to develop (3/4)th of the torque in (a) if the resistance connected is double the value in (b)?

1.7 The magnetisation curve of a dc series motor, obtained by separately exciting the field is given in the following table, at 800 rpm. The combined armature and field resistance is 0.2Ω when connected as a series motor Rheostatic braking is employed to limit the speed of the motor to 400 rpm against a load torque of 15 Nm. Determine the current drawn by the motor. The moment of inertia of the motor is 100 kgm². Determine also the braking time.

Table P1.7

Current I_f *(A)*	10	20	30	40	50	60	70	80
EMF (V)	115	228	343	434	514	560	602	629

1.8 A 220 V dc shunt motor with constant field drives a load whose torque is proportional to square of the speed. When running at 800 rpm, it takes an armature current of 40 A. Find the speed at which the motor runs if a resistance of 12Ω is connected in series with the armature. The armature resistance may be taken as in.

1.9 A 200 V dc shunt motor has a no-load speed of 700 rpm with rated field current. The resistance of the armature is 0.9Ω. The rated current of the motor is 20 A. A resistance of 6Ω in series and 3Ω in parallel, is connected with it. Determine the current drawn and speed at full load torque.

1.10 A dc shunt motor takes 70 A from a 220 V supply and runs at 1000 rpm, when delivering an output of 11 kW. The armature circuit resistance is 0.15Ω. Determine the additional resistance required in the armature to reduce the speed to 800 rpm, if
(a) the load torque is proportional to the square of the speed
(b) the load torque is constant
(c) Determine the loss in the resistance in both the cases.

1.11 The armature resistance of a dc shunt motor is 0.6Ω. The motor draws an armature current of 180 A at a speed of 1250 rpm and operates on a supply system of 220 V It drives a load torque having a characteristic.

$$T_1 = 0.7\omega \text{ Nm}$$

where ω is in rad/s. Determine the torque and speed of the motor. Determine also the torque and speed when the load has a characteristic.

$$T_1 = 78.69\sqrt{\omega} \text{ Nm}$$

1.12 A220Vdcshuntmotorrunsat700rpm with full field. The armature resistance is 0.5Ω. The armature current required to give rated torque is 20 A. The speed of the motor is controlled by resistances connected in series and parallel with the armature, equal to 6Ω and 5Ω respectively. Determine the speed-torque characteristic. What current would the motor draw to develop rated torque? What is the efficiency?

1.13 A 20 hp, 1200 rpm, 220 V shunt motor has an armature resistance of 0.2Ω and field resistance of 124Ω. The series and shunt resistances connected are 0.05 and 2Ω respectively. Find the speed of the motor required to develop the rated torque. Assume its efficiency as 85%.

1.14 Consider a 20 hp, 1200 rpm, 220 V shunt motor having an armature resistance of 0.2Ω and field resistance of 124Ω. An extra resistance of 10Ω is connected in the field circuit. Find the speed at which the machine runs while developing the rated torque. Assume the efficiency of the motor without extra field resistance as 85%.

1.15 A series motor having a combined series field and armature resistance of 1.5Ω drives a fan load having a torque speed characteristic given by

$$T_1 = KN^2 \text{ Nm}$$

where N is in rpm. At 220 V the motor runs at 350 rpm talking a current of 8 A. The speed of the fan has to be increased to 500 rpm by increasing the armature voltage. Determine the voltage and current at the new speed. Neglect saturation.

1.16 A series motor working on a 500 V system runs at a rated speed of 1000 rpm and develops rated torque with a current of 95 A. The armature resistance is 0.15Ω and series field resistance is 0.05 ohm.

Assuming that the magnetic circuit is unsaturated, determine:

(a) the speed of the armature when the supply current is 40 A and the armature voltage is 450 V.

(b) the speed when the motor develops 50% of the full load torque with a diverter resistance of 0.05Ω connected in parallel with the field. The supply voltage is 500 V.

(c) The machine is supplied with rated voltage and has a resistance of 0.15Ω in series with the armature. Determine the speed when the motor develops 75% of rated torque.

1.17 A 250 V shunt motor with a constant field current of 3.2 A runs at a no-load speed of 600 rpm. The rotational inductance is 0.8 H, armature resistance is 0.5Ω and the system inertia is 5 kgm^2. Determine the braking times and currents if the motor is to be retarded to (a) 300 rpm (b) standstill, when braked with an external resistance of 4.5Ω (i) dynamically (ii) by reverse-current.

1.18 The 250 V shunt motor problem 1 runs at no-load at a speed of 600 rpm. Find the braking time for the motor to be retarded to 300 rpm regeneratively (a) by halving the armature voltage, (b) by reducing the field circuit resistance to half its value at 600 rpm. The field inductance is 4.5 H. Neglect armature inductance.

1.19 A 240 V dc shunt motor has an armature resistance of 0.06Ω and an emf constant of 2.2 V rad/s. The motor runs at a speed of 1000 rpm. It is overhauling a load with a torque of 200 Nm

(a) Calculate the resistance to be inserted to lower the load at 1000 rpm.

(b) If regenerative braking is employed, determine the minimum speed at which the motor can hold the load.

1.20 A 200 V, 25 kW dc shunt motor runs at its rated speed of 1500 rpm. Reverse current braking is employed. The armature resistance is 0.15Ω and the efficiency is 87%. Determine

(a) the resistance required in series with the armature to limit the initial braking current to 2.5 times the rated value.

(b) the initial braking torque.

(c) the armature current and torque when the speed has fallen to 500 rpm.

1.21 A 240 V dc shunt motor with field and armature resistances of 110Ω and 0.05Ω has the following magnetisation curve at 1000 rpm

I_f (A)	0.25	0.5	0.75	1.0	1.5	2.0	2.5
E (V)	40	80	117	142	168	184	215

Determine the speed torque curve of the motor when it is braked dynamically by a resistance of 2.50.

1.22 A series motor having a combined field and armature resistance of 1.5Ω between the terminals drives a load for which the torque is proportional to square of the speed. At 250 V the motor draws a current of 15 A and runs at 400 rpm. If the speed is to be decreased to 200 rpm, determine the value of resistance required in the armature circuit for the limiting cases of (a) linear variation of flux with field current, (b) constant flux due to saturation.

1.23 A series motor has an unsaturated field having armature and field resistances of 0.5 and 0.7Ω respectively. It drives a load for which the torque is proportional to the cube of the speed. The motor, while operating on a 250 V supply, takes 45 A and runs at 300 rpm. Determine the voltage required to raise the speed to 450 rpm.

1.24 A series motor having a combined series and field resistance of 1.5Ω drives a load whose speed-torque characteristic is given by

$$T_1 = KN^2 \text{ Nm}$$

where N is in rpm. At 200 V it runs at a speed of 300 rpm taking a current of 10 A. The speed of the load has to be increased to 500 rpm; determine the voltage and current at the new speed by (a) neglecting saturation (b) assuming the flux is almost constant.

1.25 A 400 V, 4 pole, 50 Hz motor delivers its rated load of 10 kW at 1450 rpm. It has a rotor resistance per phase of 0.3Ω. If an extra resistance of 2Ω is connected in every phase, determine the speed of the motor. Assume constant torque.

1.26 A 3-phase induction motor has a full load slip erf4%. It has resistance and standstill reactances per phase of 0.02 and 0.1Ω respectively. The motor drives a constant torque load. The speed has to be reduced to 50% of rated speed.

(a) Determine the percentage reduction in stator voltage.

(b) Repeat the problem if $T_d \propto w^2$

1.27 The speed of a slip ring induction motor is varied by means of a resistance in the rotor circuit. The motor drives a load whose torque is proportional to square of the rotor speed. The details of the motor are: 3-phase, 440 V, 50 Hz, 12 pole, 492 rpm, stator and rotor connected in star. The motor delivers a power of 200 k\V at rated speed. What resistance must be added to the rotor circuit to reduce the speed to 200 rpm? Assume the rotor resistance is 0.02Ω/phase. The stator to rotor turns ratio is 1.2. Neglect stator impedance and rotational losses of the motor.

1.28 The rotor resistance and standstill reactance of a wound rotor induction motor are 0.15Ω and 1.0Ω per phase respectively. The motor has a turns ratio of 1.2 The applied voltage per phase of the 8 pole stator is 220 V at 50 Hz. Determine additional rotor resistance to produce maximum torque at (a) starting, (b) a speed of 350 rpm.

1.29 (a) Show that in a 3-phase induction motor

$$\frac{T_{dm}}{T_d} = \left[\frac{s/s_m + s_m/s}{2}\right]$$

where T_{dm} is breakdown torque
s_m is the slip for T_{dm}
T_d is the torque at slip s

(b) A 3-phase, 50 Hz induction motor has a starting torque of 1,8 times full load torque and a breakdown torque of 2.5 times full load torque, while operating on its rated voltage rated frequency mains. Determine the speeds for maximum as well as rated torque.

1.30 A 3-phase, 4 pole, 50 Hz induction motor has a full load slip of 5% while operating at rated voltage and rated frequency. The rotor resistance and standstill reactance are r_2 = 0,02Ω and x_2 = 0.15Ω respectively. Find the speed of the motor when the voltage is reduced by 30% and the torque delivered by the motor is the rated full load torque.

1.31 A 3-phase 220 V, 50 Hz, six pole induction motor has the following parameters:

$r_1 = 0.25\Omega \quad x_1 = 0.4\Omega$

$r_2' = 0.15\Omega \quad x_2' = 0.16\Omega$

$X_m = 20\Omega$

For a slip of 3%, determine the stator current, output power, torque and efficiency of the motor. Assume constant losses = 350 W.

1.32 A 3-phase, 8 pole, 50 Hz induction motor has rotor parameters of r_2 = 0.15Ω and $x_{1:}$ = 0.6Ω.
Neglecting stator parameters and rotational losses, determine the speed of the motor when a rated voltage of 234 V is applied. The load torque is proportional to square of the rotor angular velocity. The constant of proportionality is 0.0136. If the voltage applied is reduced by 25%, determine the speed.

1.33 A 3-phase, 4 pole, 440 V, 50 Hz slip ring induction motor has a full load slip of 3%. The stator to rotor turns ratio = 1.8. The rotor resistance and standstill reactance are 0.2Ω and 0.8Ω respectively. Determine

(a) the maximum torque
(b) the slip and speed at which maximum torque occurs
(c) the resistance to be connected in the rotor circuit to give a starting torque of 0.8 times maximum torque.

1.34 An 8 pole, 500 V, star-connected, 3-phase induction motor has a stator impedance of (0.0615 + j0.21)Ω and an equivalent rotor standstill impedance of (0.065 + j 0.214)Ω. The motor has a breakdown torque equal to 2.5 times the full load torque. The motor is running at its rated speed, delivering rated torque. The motor is braked with an extra resistance of 1.75Ω/phase Determine the initial braking current and torque for (a) dc dynamic braking and (b) plugging. If the moment of inertia of the motor is 10 kg m^2, determine the time taken for the motor to come to standstill from rated speed for both of the above cases.

1.35 A 220 V shunt motor runs at a speed of 1000 rpm with a constant field resistance of 110Ω. The armature resistance and rotational inductances are 0.8 and 1.0 H respectively. The motor is dynamically braked with an external resistance of 5Ω. Determine the time taken for the motor to attain speeds of 500 rpm and 200 rpm. Determine also the braking currents. The inertia of the rotating parts is 6 kg m^2.

1.36 The shunt motor of Example 35 runs at a no-load speed of 1000 rpm. The motor is braked by means of plugging. Determine the braking time and current to reduce the speed to (a) 500 rpm and (b) standstill.

1.37 A 240 V, 20 hp, 1150 rpm dc shunt motor has an armature resistance of 0.3Ω. The armature takes a current of 36.7 A for a given load. The rotational inductance is 1.0 H and the inertia 5 kg m². Neglecting armature inductance, find the braking time for speed reduction to 500 rpm regeneratively (a) by halving the armature voltage, and (b) by doubling the field current.

1.38 A 3-phase, 415 V, 6 pole, 50 Hz induction motor has the following parameters: r_1 = 1Ω/phase, r_2' = 1.5Ω/phase, x_1 = 2.2Ω/phase, x_2' = 2.2Ω/phase, all values referred to stator. The full load speed of the motor is 960 rpm. The motor is braked with an external resistance of 2Ω/phase (referred to stator) inserted in the rotor circuit. Determine the initial braking torque for (a) dc rheostatic braking, and (b) plugging.

1.39 If the moment of inertia of the motor in Problem 35 is 4 kg m² determine the time taken by it to come to rest when plugged at a speed of 960 rpm.

1.40 The following data refer to a 12 pole, 420 V, 50 Hz, 3-phase induction motor: r_1 = 2.95Ω, r_2' = 2.08Ω, x_1 = 6.82Ω, x_2' = 4.11Ω, per phase. The full load slip of the motor is 3%. The stator is star-connected. The motor is dynamically braked. Determine the value of additional rotor resistance needed to limit the current to three times the full load value at the instant of braking? What is the braking torque? With this value of resistance in the rotor circuit the motor is braked by plugging. Determine the current and torque at the instant of plugging.

Multiple-Choice Questions

1.1 An unsaturated dc shunt motor runs at its rated speed when rated voltage is applied to it. If the supply voltage to the motor is reduced by 25% the speed of the motor
(a) increases by 25%
(b) remains the same
(c) decreases by 25%
(d) increases only slightly by an amount less than 25%

1.2 Speed control of dc shunt motors by means of field weakening is suitable for
(a) constant power operation
(b) constant torque operation
(c) constant torque and constant power operation
(d) variable torque and variable power operation

1.3 The polarity of the supply voltage to a dc motor running at rated speed is reversed. Then
(a) the motor reverses its direction of rotation
(b) the motor comes to a dead stop
(c) the motor speed increases slightly
(d) the motor continues to run in the same direction

1.4 The series field of a cumulatively compounded motor is short circuited when it runs at rated speed. The speed of the motor
(a) remains the same
(b) increases
(c) decreases
(d) becomes zero

1.5 In an unsaturated dc series motor
(a) torque is proportional to armature current
(b) torque is proportional to field current
(c) torque is proportional to the square root of the armature current
(d) torque is proportional to the square of the armature current.

1.6 The starting torque of a three-phase induction motor
(a) increases with rotor resistance
(b) decreases with rotor resistance
(c) increases with rotor resistance up to a certain value and decreases as the rotor resistance increases further
(d) does not depend on rotor resistance

1.7 In a three-phase induction motor the maximum torque is
(a) directly proportional to the voltage
(b) inversely proportional to the standstill rotor reactance
(c) directly proportional to the standstill rotor reactance
(d) inversely proportional to the rotor resistance

1.8 A three-phase induction motor is started by means of a star/delta switch. The starting current is
(a) three times the current with DOL
(b) 1/3 times the current with DOL
(c) $\sqrt{3}$ times with current with DOL
(d) $1/\sqrt{3}$ times the current with DOL

1.9 A three-phase induction motor has negligible stator impedance. When the motor has constant torque load, its speed is controlled by rotor resistance The slip of the motor
(a) varies directly with rotor resistance
(b) varies inversely with rotor resistance
(c) does not depend on rotor resistance
(d) varies directly with rotor standstill reactance.

1.10 A variable speed three-phase induction motor drive obtained by varying the applied voltage is suitable for
(a) driving constant torque loads
(b) driving fan type loads where $T \propto N^2$
(c) driving loads where $T \propto 1/N^2$
(d) driving all types of loads

1.11 An ideal synchronous motor has no starting torque because
(a) the rotor is made up of salient poles
(b) the relative velocity between the stator and rotor mmfs is zero
(c) the relative velocity between the stator and rotor mmfs is not zero
(d) The rotor contains dc winding

1.12 The locus of constant power factor points on a set of V-curves is called
(a) power circle
(b) excitation circle
(c) compounding curves
(d) open circuit characteristic

References

1.1 Clayton, A.E., *The performance and design of direct current machines.*
1.2 Chillikin, M., *Electric drive*, Mir Publishers, Moscow.
1.3 Say, M.G., *The performance and design of alternating current machines'*, Sir Isaac Pitman Ltd., London, 1958.
1.4 Pillai, S.K., *A first course on electrical drives*, Wiley Eastern Ltd., 1982.
1.5 Fitzgerald, A.E. and C. Kingsley Jr., *Electric machinery*, McGraw-Hill Book Company Ltd., London, Tokyo, 1961.
1.6 Say, M.G. and E.O Taylor, *Direct current machines*, Pitman Publishing Company Ltd., London, 1980.

2 Dynamics of Electric Drives

2.1 INTRODUCTION

An industrial drive system basically consists of a mechanical working equipment or load, which has to be kept in motion to turn out mechanical work, equipment to do this job, called the prime mover, and a transmission to transfer energy from the prime mover to the mechanical load. Transmission equipment such as a gearing or belt may be used to match the speeds of the prime mover and the load. The transmission may also be required sometimes to convert rotatory to linear motion and vice versa. Thus, a combination of a prime mover, transmission equipment and mechanical working load is called a drive.

I.C. engines, steam engine, turbines or electric motors may be used as prime movers. However, in industrial drives electric motors are predominantly employed for this purpose due to their inherent advantages, such as overload capacity, efficiency, better dynamic and transient behaviour, availability in various sizes and designs compatible to load requirements, etc. An electric drive can be defined as a drive, using an electric motor as a prime mover, and ultimately converting electrical energy to mechanical energy. The electric motors used may require some types of control equipment to achieve speed control and/or torque control. These controls make the motor work on a specific speed-torque curve, and may be operated using open loop or closed loop control.

The advantages and drawbacks of electric drives can be summarised as follows:

i. Availability of electric drives over a wide range of power from a few watts to mega watts.
ii. Ability to provide a wide range of torques over a wide range of speeds. DC motors are very versatile in this aspect. However, with the advent of thyristors and thyristor power converters, ac motors are also now capable of giving a smooth speed control.
iii. Adaptability to almost any type of environmental or operating conditions, such as natural, forced ventilation, totally enclosed, submerged in liquids, exposed to explosive, or radioactive environment, etc.
iv. No hazardous fuel is required. No exhaust gases are emitted to pollute the environment. The noise level is also low.

v. Electric drives have an overload capacity which can be made use of in selecting a smaller motor for short time duties. The efficiency is very good.

vi. The speed control of these motors is straightforward. Using a proper control they can be made to operate on a desired speed-torque curve to suit the mechanical load. Gearless coupling to the mechanical load, especially for low-speed rolling mills is possible. A smooth transition from one set of operating conditions to another is possible, with a high quality of dynamic performance.

vii. An electric drive is capable of operating in all four quadrants of the speed-torque plane, i.e. motoring and braking in either direction of rotation. Regenerative braking, in which the kinetic energy of the rotating parts is advantageously returned to the mains is possible only with electric drives.

viii. The motor can develop a steady torque because of symmetry, on a balanced sinusoidal supply. The operation is quiet. However, non-sinusoidal supplies to the motors when fed from converters may cause some torque pulsation due to the time harmonics of the supply voltage/current which may become objectionable particularly at low speed.

ix. Electric motors are available in a variety of design ratings to make them compatible to any type of load.

x. The drive can be started and accelerated to the design speed at very short notice. Full load may be applied almost immediately. There is no need for refuelling or warming up of the motor, and it requires little servicing. However, electric drives do have certain drawbacks:

They require a continuous power supply, particularly in vehicle propulsion if there is no power rail available. The power supply equipment needs to be carried on board, requires a lot of space, and is bulky.

Problems of saturation of iron and cooling make the electric motors have a lower power/weight ratio.

2.2 CLASSIFICATION OF ELECTRIC DRIVES

Electric drives are normally classified into three groups, based on their development, namely group, individual and multimotor electric drives.

If several groups of mechanisms or machines are organised on one shaft and driven or actuated by one motor, the system is called a *group drive* or *shaft drive*. The various mechanisms connected may have different speeds. Hence the shaft is equipped with multistepped pulleys and belts for connection to individual loads. In this type of drive a single machine whose rating is smaller than the sum total of all connected loads may be used, because all the loads may not appear at the same time. This makes the drive economical, even though the cost of the shaft with stepped pulleys may seem to be high.

This method is rarely used in modern drive systems and has become of historical interest, because of the following disadvantages:

i. The efficiency of the drive is low, because of the losses occurring in several transmitting mechanisms.
ii. The complete drive system requires shutdown if the motor requires servicing or repair.
iii. The location of the mechanical equipment being driven depends on the shaft and there is little flexibility in its arrangement.
iv. The system is not very safe to operate.
v. The noise level at the work spot is high.

If a single motor is used to drive or actuate a given mechanism and it does all the jobs connected with this load, the drive is called an *individual drive*. For example, all the operations connected with operating a lathe may be performed by a single machine. If these operations have to be performed at different speeds, transmission devices may be required. The efficiency may become poor over several operations, due to power loss. In some cases it is possible to have the drive motor and driven load in one unit.

In a *multimotor drive* each operation of the mechanism is taken care of by a separate drive motor. The system contains several individual drives, each of which is used to operate its own mechanism. This type of drive finds application in complicated machine tools, travelling cranes, rolling mills, etc. Automatic control methods can be employed and each operation can be executed under optimum conditions.

2.3 BASIC ELEMENTS OF AN ELECTRIC DRIVE

The above discussion makes clear that an electric drive system basically consists of a mechanical load to which the required mechanical motion is imparted through a transmission drive usually equipped with gears or pulleys. Gearless transmission is possible sometimes, in which case there exists a direct coupling between the motor and load. The system also comprises certain controls for the motor for precise adjustment of the speed-torque curve as demanded by the mechanical load. The elements of a typical variable speed electric drive are depicted in Fig. 2.1.

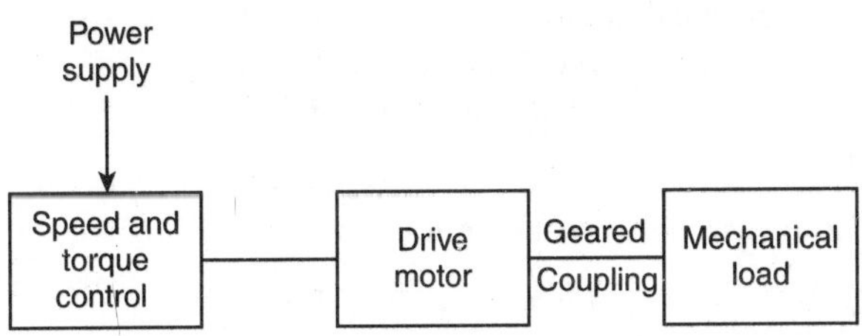

Fig. 2.1 *Elements of an electric drive*

The function of the control equipment is to set the desired speed or torque precisely. Until the advent of thyristors and associated power converters the speed control of motors had been achieved by means of contactors and relays which include or cut off the resistors. The development of power converters has made the control of motors quite straightforward. With this help the speed of ac motors is also smoothly variable. The elements of a drive system when converters are employed are shown in Fig. 2.2.

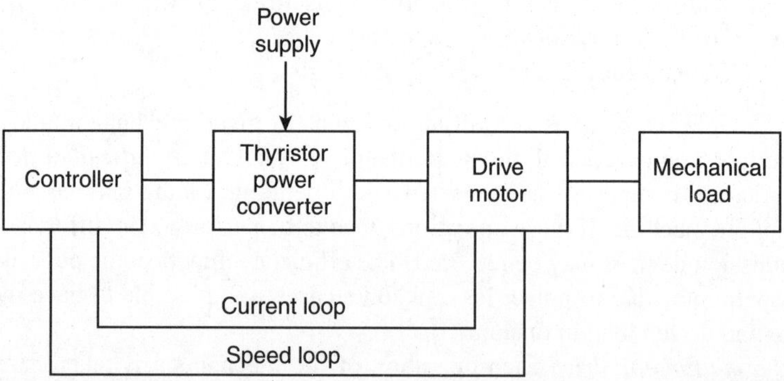

Fig. 2.2 *Elements of an electric drive using a static thyristor power converter*

Among these elements, the mechanical load and its characteristics are normally specified by means of its load diagram and torque-speed curve. A motor and its controls have to be selected to suit the given power supply and drive the load. We now discuss various types of loads and their characteristics in detail.

Mechanical System The mechanical system is coupled to the motor by means of a transmitting device. The motor has to develop a torque as required by the mechanical work to be carried out to drive the load, and the mechanical losses occurring in the system. A mechanical system is specified by a speed-torque curve. The motor while driving this mechanical load must provide enough torque to drive the load against losses like friction and to accelerate the load toque to the desired speed. Hence the load torque required by the load at the shaft has the following components.

i. Torque component to overcome friction and windage which accompanies mechanical motion.
ii. Torque required to accelerate the load to the desired speed.
iii. Torque required to do the prescribed mechanical work, i.e. to run to the load at the desired speed.

The load torque seen by the motor at the shaft

$$T_L = T_{fw} + T_a + T_w \tag{2.1}$$

Mechanical motion is accompanied by frictional forces between the surfaces undergoing relative motion. Frictional forces of both the load and the transmission equipment have to be considered. Also, the electromagnetic torque developed by the motor has to provide for its own friction and windage before appearing at the shaft to drive the load. The torque to overcome the friction should be delivered by the motor to keep the mechanism in motion when the system is on no load.

The friction existing in a mechanical system may be classified as follows:

i. *Viscous friction:* In this type of friction the torque required is directly proportional to the speed of rotation

$$T_B = B\omega = B\frac{d\theta}{dt} \tag{2.2}$$

where B is a constant of proportionality. This is depicted in Fig. 2.3. It occurs in well lubricated bearings and the laminar flow of lubricating liquids.

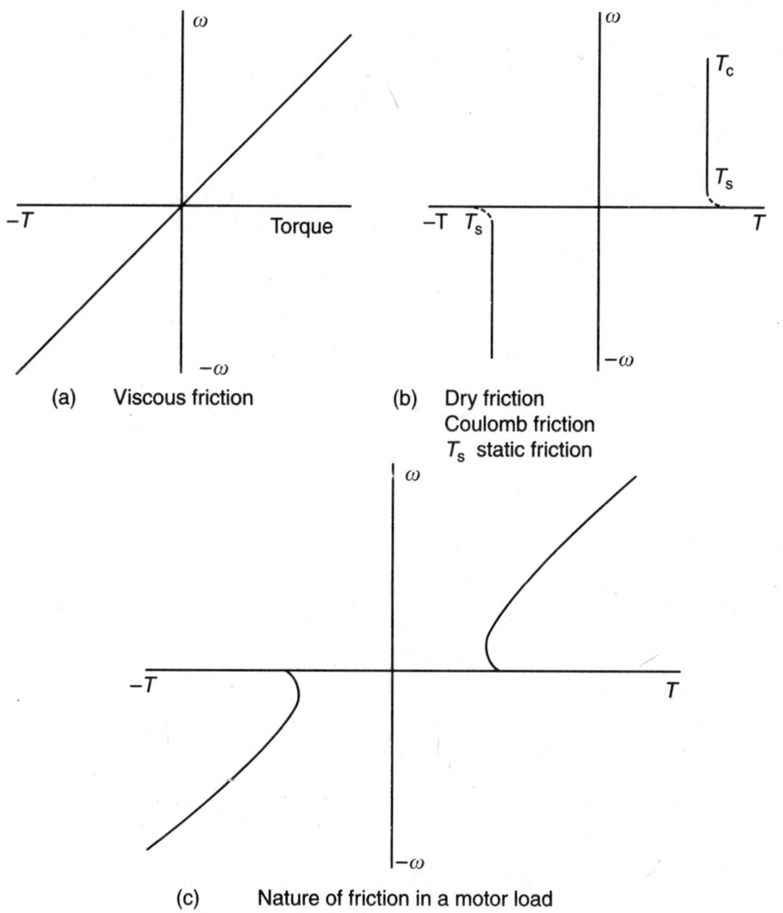

Fig. 2.3 *Types of friction in drive systems*

ii. Coulomb friction: The torque required is independent of speed in this type of friction. It acts as a load torque in either direction and is also called dry friction. Viscous friction changes to coulomb friction at very low speeds.
iii. Static friction or stiction: Occurs due to the sticking nature of the surfaces. This is generally very small and can be neglected.

Windage Torque The torque required by the load when the air surrounding the rotating parts moves. It is normally proportional to the square of the speed. However, at normal speeds of operation it may be considered equivalent to viscous friction and the value of B may be taken to contain both friction and windage.

The working mechanism must be accelerated and brought to the desired speed. The motor should provide a torque at the shaft capable of accelerating the rotating parts against their inertia. Inertia, as seen by the motor shaft, includes the inertia of the mechanism as well as of transmission. The inertia of the rotating parts of the motor is taken care of by the internal torque developed by the motor before it appears as a shaft torque. Taking J as the inertia of the mechanical system,

$$T_a = J\frac{d\omega}{dt} = J\frac{d^2\theta}{dt^2}\ \text{Nm} \tag{2.3}$$

A torque component may exist sometimes due to the torsional elasticity of the shaft. This has to be considered under transient conditions and is given by

$$T_e = K\theta_e \tag{2.4}$$

i.e. it is proportional to the torsion angle of coupling. The constant of proportionality K is called the stiffness of coupling. Normally the shafts are perfectly stiff and this component does not exist. However, if it does, it represents a store of potential energy where T_a represents kinetic energy. Under ideal conditions without friction, there exist oscillations leading to the shaft breaking.

The useful working torque required to do the mechanical job is a function of speed

$$T_w = T(\omega) \tag{2.5}$$

Therefore the total torque required at the shaft

$$T_L = J\frac{d\omega}{dt} + B\omega + T(\omega) \tag{2.6}$$

Considering the acceleration of the motor against its inertia and friction, the electromagnetic torque developed is

$$T_d = T_L + J_m\frac{d\omega}{dt} + B_m\omega \tag{2.7}$$

Let us now study the various types of load torques that occur in industrial practice.

The load torque may be constant at all speeds, as shown in Fig. 2.4 by curve (a).

This type of torque is represented by a compressor load and the spee1-torque curve is as shown by the curve (b). This drive is unidirectional.

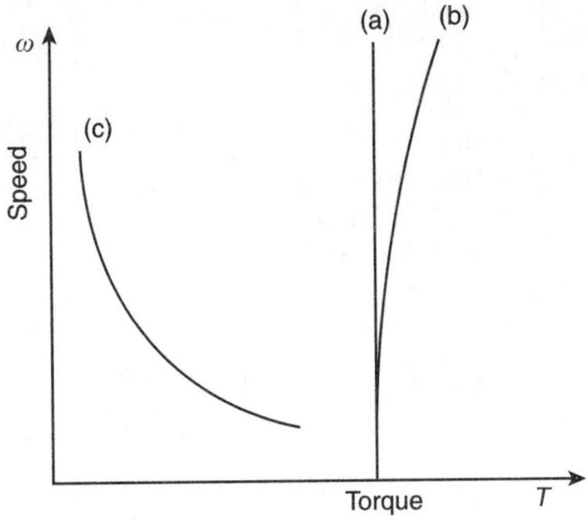

Fig. 2.4 *(a) Constant torque load*
(b) Compressor load characteristic
(c) Constant power curve (load)

Another type of load has its torque proportional to the square of the speed, as shown in Fig. 2.5.
This speed-torque curve is found with pump or fan type loads.

$$T_w = K\omega^2 \quad (2.8)$$

The power developed at the motor shaft

$$P = K\omega^3 \quad (2.9)$$

This load is also unidirectional. It is required to run at constant speeds for a longer period of time or at several speed settings or over a range of speeds.

Another type of mechanical load requires constant power at all speeds

$$P_w = T_w\omega \quad (2.10)$$

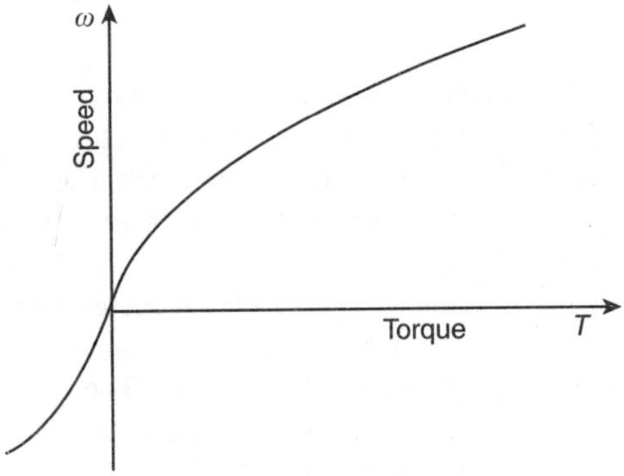

Fig. 2.5 *Fan type load characteristic* $T_L \propto \omega^2$

P constant at all ω. The torque-speed characteristic of such a load is a rectangular hyperbola. This load is found with steel rolling mills, papers mills, etc. It occurs in transportation also. The torque-speed curve is shown in Fig. 2.4 (c). With certain types of loads, like winch drives, a constant torque is required when the mechanism is under standstill conditions. The direction of rotation may need to be reversed. These occur in ships when it is required to hold the ship in a particular location or to warp it through a lock gate.

From the above types of loads one can easily see that some operate only in one direction with no reversal of speed. In the speed-torque plane these are represented in the first quadrant. The forward driving of the mechanism corresponds to an operation in the first quadrant where motor draws electrical power to drive the mechanical load coupled to it.

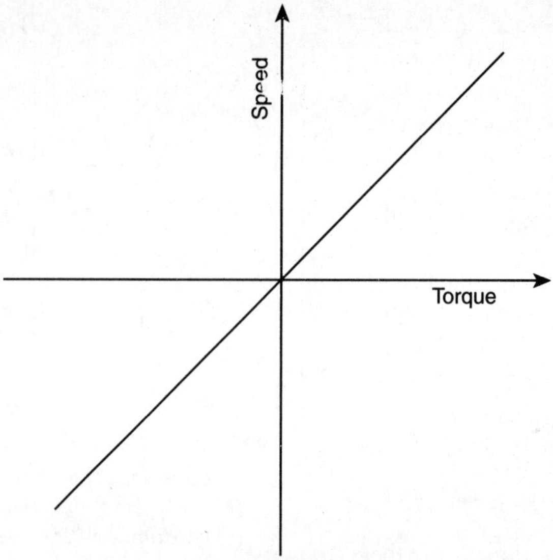

Fig. 2.6 *Load torque proportional to speed*

The mechanical load sometimes requires braking to bring it to rest quickly. During braking, which torque is achieved either by electrical or mechanical method is applied in the negative direction. Electrical methods are dynamic braking, eddy current braking, regenerative braking and plugging. In dynamic braking, the machine driving the load acts as a generator and the kinetic energy of the rotating parts is converted to electrical energy and dissipated in the resistances. In eddy current braking this energy is dissipated in eddy current losses or resistance losses in a specially constructed machine. The energy of the rotating masses, instead of being dissipated, is returned to the mains in regenerative braking. Another kind of electrical braking is called plugging or reverse current braking. A motor acts as an electric brake. The motor develops a negative torque whose nature is to produce a torque which would oppose the direction of rotation already existing. In mechanical braking, the energy of the rotating parts of the system is dissipated as heat due to friction in the mechanical brake coupled to the system.

Operation in the second quadrant represents braking, because in this part of the torque-speed plane the direction of rotation is positive and the torque is negative. The machine operates as a generator developing a counter torque which opposes motion. The K.E. of the rotating parts is available as electrical energy which may be pumped to the mains or in dynamic braking, dissipated in some resistances.

In the third quadrant which corresponds to motor action in the reverse direction both speed and torque have negative values, while power is positive. Operation is similar to that in the first quadrant, with direction of rotation reversed.

In the fourth quadrant the torque and speed have opposite signs—positive and negative respectively. The operation can be examined in either of two ways.

The motion in this quadrant may be under the action of load itself. The motor tends to attain dangerously high speeds. The motor must develop a torque which opposes the acceleration due to load. The motor acts as a brake. This quadrant corresponds to braking in reverse motoring. The transition tc this may be also from the third quadrant if there is a tendency of the motor to accelerate in the reverse direction. The generator torque developed arrests the acceleration. This situation occurs when a hoist is lowering the load. These operations in all four quadrants are depicted in Fig. 2.7.

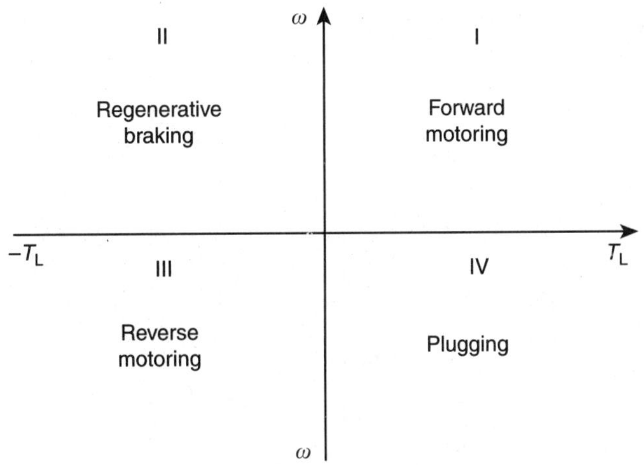

Fig. 2.7 *Four quadrant operation*

Compressor, pump and fan type loads require operation in the first quadrant only, since their operation is unidirectional. They are one quadrant drive systems.

Transportation drives require operation in both directions. The method of braking depends upon the conditions of availability of power supply. Braking may be electrical or mechanical. If regeneration is necessary, operation in all four quadrants may be required. If not, operation is restricted to quadrants 1 and 3. Dynamic braking or mechanical braking may be employed.

In hoist drives a four quadrant operation invariably occurs. The speed-torque curve is depicted in Fig. 2.8. The operations are shown in Fig. 2.9. The first quadrant represents the operation of the hoist while going up. The load in the cage has to be moved

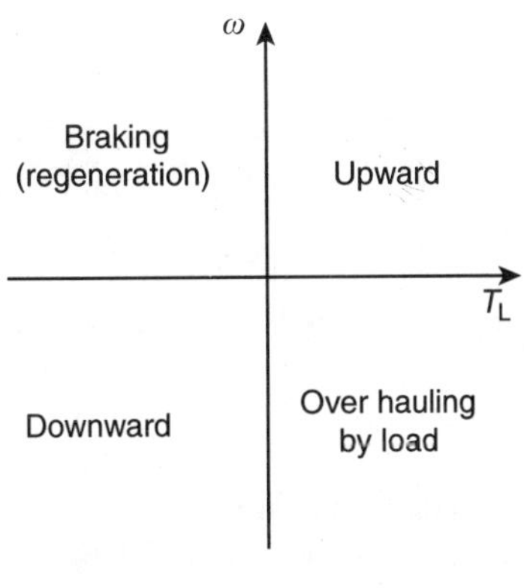

Fig. 2.8 *Speed torque diagram of hoist*

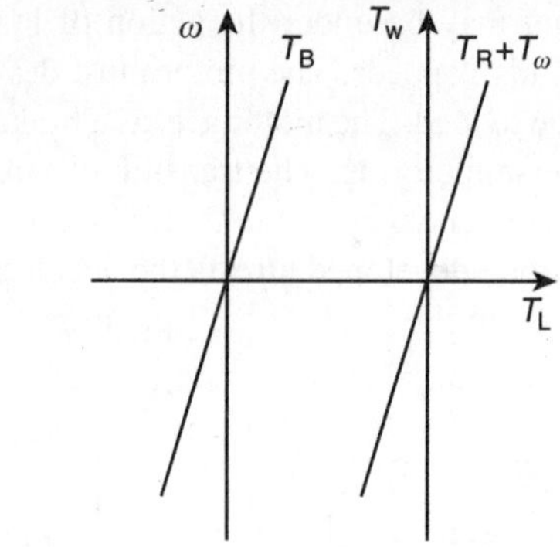

Fig. 2.9 (a) *Torque of hoist*

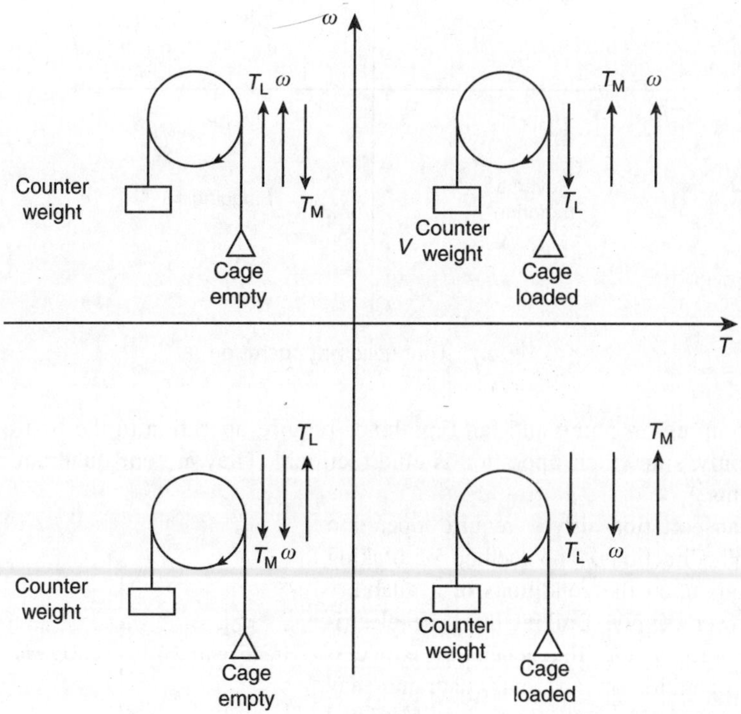

Fig. 2.9 (b) *Four quadrant operation of a motor driving hoist*

against gravity and a motoring torque occurs. In the second quadrant the cage movement remains upwards. The load torque is not there because of the absence of load. Under the action of the counter weight, very high speeds may result, causing regeneration. For operation in this quadrant, and as a safety arrangement, a mechanical brake is provided, which brakes the mechanism in case of power failure. It also helps to hold the suspended load stationary at any point and prevents excessive acceleration Dynamic braking may also be applied for quick stopping.

In the third quadrant the direction of rotating reverses; the load is lowered by the hoist. In case of acceleration of the load under gravity, there is a slight transition to the fourth quadrant, wherein regenerative braking or dynamic braking may occur, limiting the speed of motion.

The load moves downwards in the fourth quadrant. It overhauls the motor and may cause regeneration. Under gravity the acceleration might result in dangerous speeds. Therefore the motor is made to act as a brake and develop a torque to run the motor in the positive direction, opposite to the load.

Typical speed-torques of an induction motor and dc motor are shown in Fig. 2.10. Under steady conditions of the drive the operation is restricted to 1, 3, and 4 quadrants. Regenerative braking takes place in the second quadrant under transient conditions. Plugging may take place in the second quadrant for motoring operation in the third quadrant.

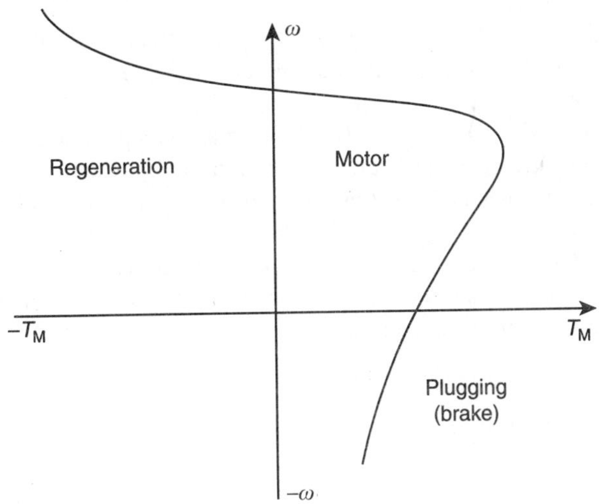

Fig. 2.10 *Typical speed-torque characteristic of an induction motor*

In some types of loads the load torque depends on the position of the load during motion. For example, a train moving up a gradient will have different torque compared to one moving down it, or on a level track or along a curve. Another example of such loads is the hoisting mechanisms where the weight of the rope affects the load torque, depending upon the position of the load. It may act as a load, requiring power from the drive motor, or as an effort moving the load in the required direction. This might cause a jerky operation of the hoist and needs to be compensated by tail ropes.

While selecting a motor for driving a load, it is necessary to consider the variation of load torque with both speed and time. The torque-speed curve normally decides the type of motor, whereas the variation of load with time decides the rating of the motor. Loads that occur in industrial practice can be classified depending upon their variation with time, which is specified by the load diagram. A load whose torque is independent of speed may be a continuous load or appearing intermittently in a periodic manner. A fan type load where $T \propto N^2$ may also be continuous or occurring intermittently. Hence loads may also be classified depending upon the duty they have to perform. The classification, according to variation with time, is as follows:

i. *Continuous constant loads:* These loads occur for a long time under the same conditions, e.g. fan type loads.
ii. *Continuous variable type loads:* The load is variable over a period of time, but occurs repetitively for a longer duration. It occurs in metal cutting lathes, conveyors, etc.
iii. *Pulsating loads:* Certain types of loads exhibit a torque behaviour which can be thought of as a constant torque superimposed by pulsations. They are present with reciprocating pumps and all loads having crank-shafts.
iv. *Impact loads:* Peak loads occur at regular intervals of time, e.g. in rolling mills, forging hammers, etc. Motors driving these loads are equipped with flywheels for load equalisation.
v. *Short time intermittent loads:* The load appears periodically in identical duty cycles, each consisting of a period of application of load and one of rest. Cranes and hoisting mechanisms are examples of this type of loading.
vi. *Short time loads:* A constant load appears on the drive for a short time and the system rests for the remainder. Battery charging and household equipment offer such loads.

These types of loads are depicted in Fig. 2.11.

Equivalent Systems The load may be coupled to the shaft of the driving motor either directly or through a transmission system comprising, gears, pulleys or belts. The latter is mainly to match the speeds of the motor and the load. If it is impossible to select a motor having the same speed as the load, the two nay be running at different speeds. In such cases they are coupled to each other by means of gears or belts. While analysing such systems it is necessary to refer all the torques and inertia to a common shaft, e.g. for convenience, the load torque and load inertia are referred to the motor shaft. This is equivalent to direct coupling between the motor and load. This is similar to referring the quantities of primary and secondary winding of a transformer to one side while analysing the same using the equivalent circuit.

The moments of inertia are referred to a common shaft on the assumption that the K.E. remains unchanged. Consider a motor coupled to a load of inertia J_2. The speeds of rotation of the machine and load are ω_1 and ω_2 respectively. The gear ratio is N_1/N_2, as shown in Fig. 2.12(a).

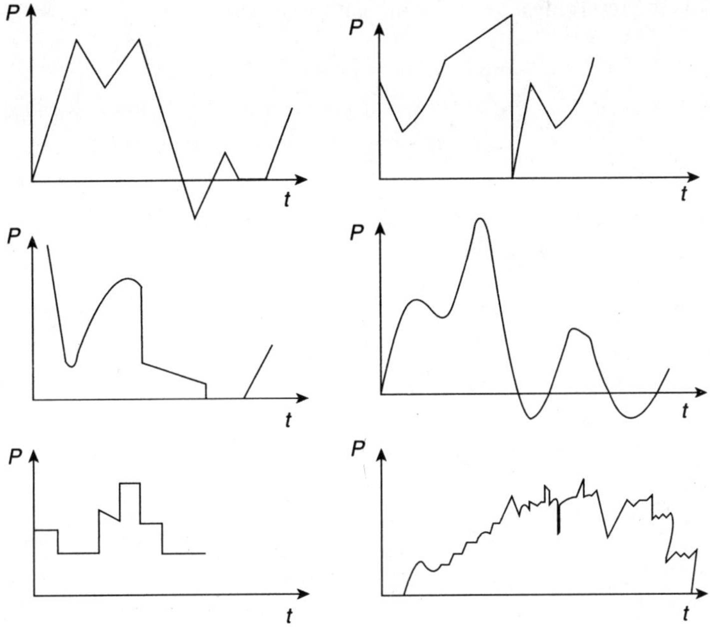

Fig. 2.11 *Some typical common time varying loads*

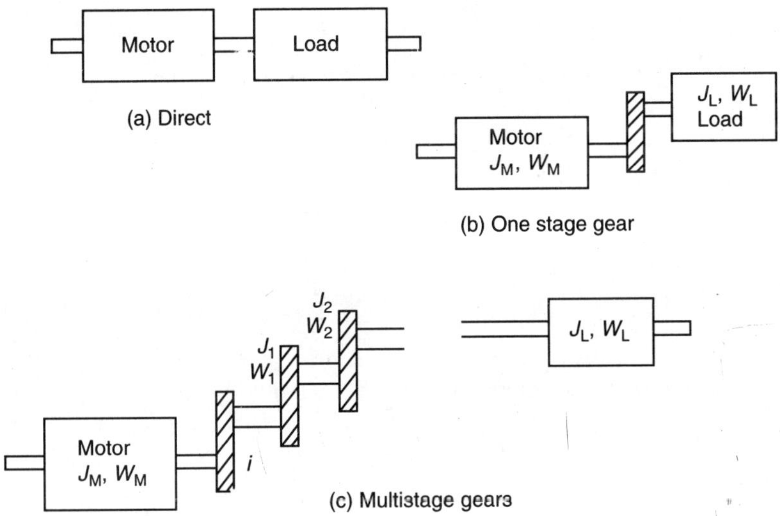

Fig. 2.12 *Transmission using gears*

If J_2' is the referred value of the moment of inertia

$$\frac{1}{2} J_2' \omega_1^2 = \frac{1}{2} J_2 \omega_2^2 \tag{2.11}$$

$$J_2' = J_2 \left(\frac{\omega_2}{\omega_1} \right)^2 \tag{2.12}$$

The torques are referred, based on the equivalence of power. Assuming an efficiency of transmission η, we have

$$T_2' \omega_1 = \frac{T_2 \omega_2}{\eta} \tag{2.13}$$

$$T_2' = T_2 \left(\frac{\omega_2}{\omega_1} \right) \frac{1}{\eta} \tag{2.14}$$

The speeds are referred using the relation

$$\omega_2' N_1 = \omega_2 N_2 \tag{2.15}$$

$$\omega_2' = \omega_1 = \omega_2 \left(\frac{N_2}{N_1} \right) \tag{2.16}$$

or

$$\frac{\omega_2}{\omega_1} = \frac{N_1}{N_2}$$

Similarly, friction B_2 can referred to the motor shaft as

$$B_2' = B_2 \left(\frac{\omega_2}{\omega_1} \right)^2 \tag{2.17}$$

Using the relation between the speeds and the gear ratio, we have

$$J_2' = J_2 \left(\frac{N_1}{N_2} \right)^2 \tag{2.18}$$

$$T_2' = T_2 \left(\frac{N_1}{N_2} \right) \frac{1}{\eta} \tag{2.19}$$

$$B_2' = B_2 \left(\frac{N_1}{N_2} \right)^2 \tag{2.20}$$

Therefore the load torque referred to the motor shaft is

$$T_2' = J_2 \left(\frac{\omega_2}{\omega_1} \right)^2 \frac{d\omega_1}{dt} + B_2 \left(\frac{\omega_2}{\omega_1} \right)^2 \omega_1 \tag{2.21}$$

$$T_2' = J_2' \frac{d\omega_1}{dt} + B_2' \omega_1 \tag{2.22}$$

If the system has several stages of transmission, as shown in Fig. 2.12(b), the referred values of torque, inertia and friction can be obtained as

$$T_L' = T_L \left(\frac{\omega_n}{\omega_2}\right)\frac{\omega_2}{\omega_1}\left(\frac{\omega_1}{\omega_m}\right)\frac{1}{\eta_1}\frac{1}{\eta_2}\frac{1}{\eta_3} \tag{2.23}$$

$$J' = J_n\left(\frac{\omega_n}{\omega_2}\right)^2\left(\frac{\omega_2}{\omega_1}\right)^2\left(\frac{\omega_1}{\omega_m}\right)^2 + J_2\left(\frac{\omega_2}{\omega_1}\right)^2\left(\frac{\omega_1}{\omega_m}\right)^2 + J_m \tag{2.24}$$

$$B' = B_L\left(\frac{\omega_n}{\omega_2}\right)^2\left(\frac{\omega_2}{\omega_1}\right)^2\left(\frac{\omega_1}{\omega_m}\right)^2 \tag{2.25}$$

If, in a mechanism, there is translational motion of a force driven by a rotating shaft, the quantities have to be referred to the rotating shaft and referred values of torque and inertia obtained. The torque as seen by the shaft when a force is moved upwards is determined on the basis of equal powers.

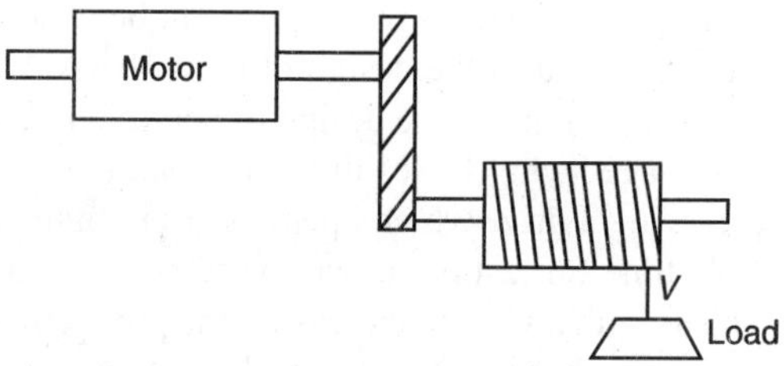

Fig. 2.13 *Rotary motion to translational motion*

Therefore, referring to Fig. 2.13,

$$T'\omega_1 = \frac{F_L \times V_L}{\eta_L} \tag{2.26}$$

where F_L is the force lifted, V_L is the velocity and η_L is the efficiency.

Therefore torque $$T' = \frac{F_L V_L}{\omega_1 \eta_L} \text{ Nm} \tag{2.27}$$

The inertia at the motor shaft is determined from the equivalence of the kinetic energies

$$\frac{1}{2} m V_L^2 = \frac{1}{2} J' \omega_1^2 \tag{2.28}$$

J' is the inertia referred to the shaft

$$= m\left(\frac{V_L}{\omega_1}\right)^2 = \frac{J'}{g}\left(\frac{V_L}{\omega_1}\right)^2 \tag{2.29}$$

It is therefore possible to reduce a complex drive system having several moments of inertia to a simpler system which is directly coupled to the motor. The dynamic performance of this system is a replica of the performance of the complex one.

2.4 DYNAMIC CONDITIONS OF A DRIVE SYSTEM

Dynamic or transient conditions occur in electric drive systems when the operating point changes from one steady state condition to another, following a change introduced in the system variables. These variables may be mechanical, such as speed, torque, etc. or electrical, such as voltage, current etc.

These conditions generally exist during starting, braking and speed reversal of the drive. The dynamic conditions arise in a variable speed drive when transition from one speed to another is required. The drive may also have transient behaviour if there are sudden changes of load, supply, voltage or frequency.

The energy storing elements in the drive system, such as mechanical inertia (J) and electrical inductance (L), are responsible for the delay in response following a disturbance and for variations in transient behaviour.

The investigations of the transient behaviour or dynamic conditions of a drive are significant in the design of controllers. A knowledge of the dynamic behaviour is essential for the design of control circuits, for the correct choice of motors and for reducing losses during the starting and stopping of drives. One aims at achieving a drive which operates at optimum speed and takes the minimum possible time for settling down to the new steady-state after the initial disturbance. This is with a view to increasing the productivity of the drive. Varying parameters such as voltage, frequency and machine constants also influences the transient behaviour and a knowledge of this variation is necessary for a suitable choice of these parameters.

The dynamic behaviour of a drive has a close relation to its stability. A drive is said to be stable if it can go from one state of equilibrium to another following a disturbance in one of the parameters of the system. Stability can be identified as either steady-state or transient. The conditions of stability depend on the operating point. Stability will be discussed in detail in later sections.

Transient conditions normally exist in a motor and drive system for a short interval of time. The change in thermal processes during this interval is not significant due to the large value of the thermal time constants compared to the electrical and mechanical time constants. Therefore, it can be assumed that changes in thermal conditions do not affect transient behaviour.

In drives which have infrequent starting and stopping as well as very simple control circuits, transient behaviour need not be considered.

The fundamental equation of motion giving a balance between the various torques in the drive may have to be considered while investigating the dynamic behaviour.

As has been stated earlier, any complex mechanical system driven by an electric motor can be simplified and represented by a single moment of inertia, friction and load, all referred to the motor shaft. A typical equivalent system is shown in Fig. 2.14. The dynamics of the drive can be investigated using the torque balance equation given by

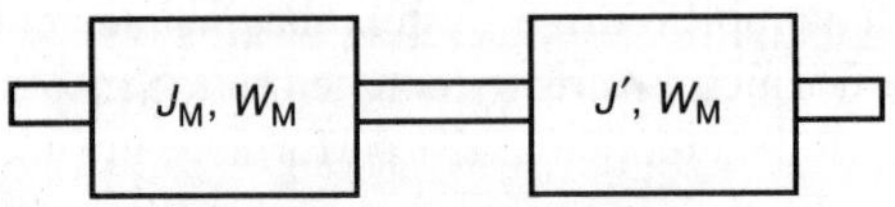

Fig. 2.14 *Equivalent systems*

$$T_d = J\frac{d\omega}{dt} + B\omega + T_L' \tag{2.30}$$

where $J = (J_M + J')$
$B = (B_M + B')$

The equation of motion considers the motor inertia J_M and the inertia of the load as seen by the shaft (J'). It also considers the friction of the motor B_m and friction of the load referred to the motor shaft (B'_1). T'_L is the load torque referred to the motor shaft. Considering $B\omega$ also as a part of the load torque, we can further simplify the equation

$$T_d = J\frac{d\omega}{dt} + T_L \tag{2.31}$$

or

$$J\frac{d\omega}{dt} = T_d - T_L$$

If the electromagnetic torque developed is greater than the load torque including the friction torque, the motor accelerates.

$$\text{i.e., if} \quad T_d > T_L; \quad \frac{d\omega}{dt} > 0, \quad \omega \text{ increases}$$

If the electromagnetic torque developed is less than the load torque including friction, the motor decelerates.

$$\text{i.e., if} \quad T_d < T_L; \quad \frac{d\omega}{dt} < 0, \quad \omega \text{ decreases}$$

This situation also occurs if T_d is negative, opposing the motion, as in the case of braking.

If both torque components are equal, the motor runs at constant speed.

$$\text{i.e., if} \quad T_d = T_L; \quad \frac{d\omega}{dt} = 0, \quad \omega = \text{constant}$$

As has been stated previously, T_L comprises load torque as well as friction. The load torques occurring in mechanical systems may be passive or active. If a torque always opposes the direction of motion of the drive motor it is called a passive torque. The nature of friction is to oppose the motion, whatever be its direction. This has to be properly taken care of in writing down the equation. If the sign of drive rotation changes, these torques also change. The torques occurring due to friction, metal cutting, etc. are passive torques.

Active torques, on the other hand, oppose the motion in one direction and aid it in the other. These are associated with the potential energy of the moving parts of the system. They may therefore be either positive or negative, depending upon whether they oppose or aid the motion. The motor has to provide sufficient torque when these torques oppose the motion. In the other direction they aid the motoring torque. As an example, the potential energy of the cage of a crane or hoist is associated with an active torque. While going up it resists the motion and while coming down it aids it. Active torques retain their sign, whatever the sign of rotation in the dynamic equation.

$J\frac{d\omega}{dt}$ is the inertial torque having a positive sign (opposes motion) during acceleration and negative sign (aids motion) during retardation. It exists only during a dynamic condition. However, the dynamic equation is determined by the signs of T_d and T_L.

The dynamic equation of the drive can be written based on the foregoing considerations, which may be summarised as:

T_d is positive during motoring

T_L is negative for passive torques during motoring

$$T_d - T_L = J\frac{d\omega}{dt}$$

The sign of $(T_d - T_L)$ determines the sign of $\frac{d\omega}{dt}$; If $(T_d - T_L)$ is positive acceleration takes place.

If the direction of rotation is reversed the equation of motion is

$$-T_d + T_L = -J\frac{d\omega}{dt}$$

or

$$(T_d - T_L) = J\frac{d\omega}{dt}$$

If T_L is an active torque

$(T_d - T_L) = J\frac{d\omega}{dt}$ when the potential energy due to T_L opposes T_d

$T_d + T_L = J\frac{d\omega}{dt}$ when it aids T_d

In general, the equation of motion can be written as

$$\pm T_d \mp T_L = J\frac{d\omega}{dt} \tag{2.32}$$

The sign of T_d depends upon whether it is motoring (+) or braking (–). T_L has a sign depending upon whether it is passive or active. Using this equation we can investigate the torque-speed of the system during dynamic conditions. The duration of dynamic behaviour can also be obtained.

2.5 STABILITY CONSIDERATIONS OF ELECTRICAL DRIVES

From the preceding discussion it is clear that a drive is in its state of equilibrium at constant speed if the developed motor torque is equal to the sum of load torque and friction. The basis of the investigation of the dynamics of a drive from one steady-state equilibrium condition to another has already been discussed in detail. Electrical drives have closed loop control systems. Investigations on the stability of drives must include all the controls available along with the motor and mechanical system. A system is in an equilibrium condition, if there is no disturbance. In the presence of a disturbance the equilibrium conditions are disturbed and the drive tries to take a new equilibrium position under the new input conditions. The system is said to be stable if sometime after the appearance of a disturbance it attains a new equilibrium condition. The system is said to be unstable if it comes to rest or has a continuous increase in speed following the disturbance. i.e. the system is unable to take up a new equilibrium position The disturbances may be external or internal to the drive.

The stability of a drive may be affected very much by the parameters of controllers, motors, etc. A system which is stable for one set of parameters may become unstable for another set. The stability investigations of a drive system having closed loop controls and represented by means of a block diagram will be discussed in detail in a later chapter dealing with the control techniques of drives.

The stability of a drive is influenced by the inertia of the rotating masses and inductances of the motor, in the same manner as the transient behaviour. Therefore in cases of sudden changes of parameter causing the drive to change its equilibrium state, the effects of these components cannot be ignored. The stability behaviour of the drive taking into account the effects of these parameters is called transient stability. On the other hand, if the changes from one state of equilibrium to another take place too slowly to have the effects of the above parameters, the stability conditions refer to steady-state stability.

Before discussing the stability characteristics of a drive in a closed loop or open loop operation it is first necessary to investigate the stability of the motor driving the load. The motors have both steady-state and transient stability conditions existing in them.

Stability Behaviour of Electrical Machines—Steady-state Stability The nature of response of an electrical machine between two states of equilibrium following a disturbance, which would also refer to the nature of its stability, can be obtained by solving the differential equations of the motor along with the dynamic equation of motion. The time response so obtained gives information regarding the transient response as well as stability. The motor is stable if the transient portion of the response dies down with time. As the equations are non-linear they are linearised using the method of small signals about an operating point. These equations give operating point stability. This tedious investigation is necessary to get an insight into the steady-state stability and to develop criteria for it in cases of disturbances. Stability studies can be easily done using the steady-state speed-torque curves assuming that in cases of disturbances the operating point moves along these curves.

The steady-state stability can be explained with reference to Fig. 2.15, in which some typical speed-torque curves of a motor and load are shown. Consider the situation at point A. A small decrease in speed is followed by an increase in the motor torque which accelerates the load and brings it back to the equilibrium

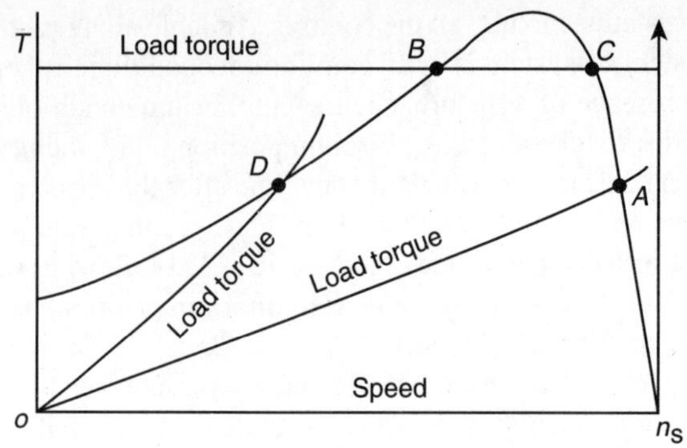

Fig. 2.15 *Steady-state stability of induction motor*

point. Note also that after the disturbance has occurred, the motor torque is greater than the load torque. In reaching the point of equilibrium, the drive may overshoot it, with further increase in speed. The motor torque decreases and becomes smaller than the load torque. Under these conditions no more acceleration is possible and the drive retards. It finally comes to a state of equilibrium after several oscillations. In the case of an induction motor the oscillations are effectively damped out by the induced currents in the squirrel cage. The motor successfully returns to its original point of equilibrium in case of a small increase in speed also. It is stable at point *A*. $T_L > T_d$ following an increase in speed which effectively brakes the motor.

Consider the operation at *B*. A small decrease in speed decreases the torque developed by the motor. Further, from the figure we see that the load torque is greater than the torque developed by the motor. This causes further retardation. The process being cumulative, the motor finally comes to rest. If there is a small increase in speed, the developed torque increases, which accelerates the load torque. This process is also cumulative and the speed goes on increasing—the motor does not return to its original operating point. Point *B* is unstable.

Following the reasoning given for point *A*, the operation at point *C* is stable. At point *D*, even though the torque developed decreases with decrease in speed it is able to accelerate the load, because $T_d > T_L$. The motor reaches its original point of operation. It is also a stable point.

Taking these discussions into consideration, it can be concluded that the motor is stable if the change in motor torque following a decrease in speed is such that it is greater than the load torque, or smaller than the load torque following an increase in speed.

In the foregoing discussions, the load torque is constant at the operating point. If in a motor-load combination a change in speed can bring about changes in the motor as well as load torques, stable operation can be envisaged only if the change in the load torque is greater (smaller) than the change in the motor torque in case of an increase (decrease) in speed. In the former case excess load torque has a tendency to brake the motor, whereas in the latter case excess motor torque has a tendency to accelerate the rotor so that the system is stable again.

These conclusions can also be drawn on the basis of the dynamic equation of motion applied to the operating point.

Let the motor be operating at a developed torque T_d and a load torque T_L. The speed of operation is ω. Let there be small variations of δT_L and δT_d following a speed variation $\delta\omega$. Before the disturbance occurs we have

$$J\frac{d\omega}{dt} = T_d - T_L \tag{2.33}$$

After the disturbance we have

$$J\left(\frac{d}{dt}(\omega + \delta\omega)\right) = (T_d + \delta T_d) - (T_L + \delta T_L) \tag{2.34}$$

From these equations we get

$$J\frac{d}{dt}(\delta\omega) = \delta T_d - \delta T_L \tag{2.35}$$

This can be written as

$$J\frac{d}{dt}(\delta\omega) = \left(\frac{dT_d}{d\omega}\right)\delta\omega - \left(\frac{dT_L}{d\omega}\right)\delta\omega \tag{2.36}$$

assuming that the increments are so small that they can be expressed as linear functions of speed.

Solving this equation we get

$$\delta\omega = (\delta\omega)_0 e^{-\frac{1}{J}\left[\frac{dT_L}{d\omega} - \frac{dT_m}{d\omega}\right]t} \tag{2.37}$$

where $(\delta\omega)$ is the initial disturbance in speed. The speed transient can be obtained from Eq. 2.37. The transient will die down to zero if

$$\left(\frac{dT_L}{dt} - \frac{dT_m}{d\omega}\right) > 0 \tag{2.38}$$

making the exponential term decrease with time. When this condition is satisfied, the motor returns to its original operating point and the machine is said to be stable. Else, the speed transient increases continuously and the machine is unstable.

From the condition given in Eq. 2.38 we can conclude that a machine is stable if its load speed-torque curves are such that for

i. a decrease in speed, the motor torque is greater than the load torque.
ii. an increase in speed, the load torque is greater than the motor torque.

From the condition it can also be seen that a decrease (increase) in speed must bring about a greater (smaller) change in the motor torque than in the load torque so that the machine can be accelerated (retarded) under the influence of the motor torque (load torque) to its original point of operation.

Transient Stability The steady-state stability criterion discussed above assumes that the deviations in speed and torque follow the steady-state speed-torque curves of the motor and load. Also, the changes are so slow that the energy storage elements, such as inertia and inductance, do not affect the variation while going from one state of equilibrium to the other. However, if the changes are very fast the effects of these parameters cannot be ignored. In that case the results of steady-state stability are optimistic. A machine which is stable in the steady-state may lose its stability under transient conditions when the effects of inertia are considerable. A motor can be loaded to its maximum capacity of torque or power when it is slowly or gradually loaded. For transient type of loads the motor cannot reach this limit because of kinetic energy of the rotating parts also has to be considered. Therefore the stability information during transient conditions is different from steady-state stability information. A machine which can be loaded to its maximum power under steady-state conditions may not be loaded to the same extent under transient loading conditions.

A knowledge of transient stability is essential for the proper design of flywheel and other associated controls.

In a synchronous motor drive, the steady-state speed and torque of the motor and load may be such that the steady-state stability criterion does not provide any reliable information regarding its stability. Therefore it is necessary to study its transient stability.

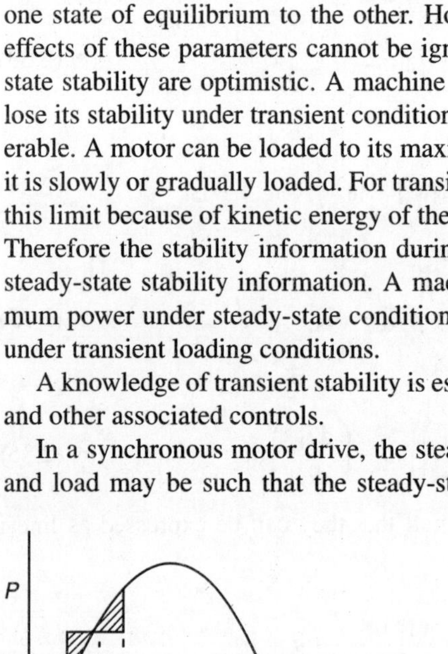

Fig. 2.16 *Power angle characteristics of a synchronous motor*

It is well known that a synchronous motor is very sensitive to sudden changes of load and prone to what is called hunting.

(The oscillatory behaviour of a rotor about a mean torque angle position is called hunting.) The nature of hunting can be discussed with reference to the power angle characteristic shown in Fig. 2.16.

The power developed by synchronous motor, neglecting armature resistance is

$$P_d = \frac{E_t V}{X_s} \sin \delta$$

assuming a cylindrical rotor. Under a steady-state this may reach a maximum value

$$P_{dm} = \frac{E_t V}{X_s} \quad \text{when} \quad \delta = \frac{\pi}{2}$$

which cannot, however, be reached under transient conditions. Let the machine be operating at a power angle equal to δ, giving a power of P. The torque balance equation may be replaced by the power balance equation due to the constancy of speed. The power developed in the motor is equal to the load power and damping power. The latter may be neglected while deriving the criteria for transient stability.

When a load is suddenly applied to the shaft of a synchronous motor there is a momentary slowing down of the motor. This causes the rotor to fall back from its old position in order to develop the required torque. Till the rotor attains its new position the rotor releases the kinetic energy to drive the increased load during retardation. The rotor angle increases to a value where sufficient torque is developed to drive the load. At this position rotor speed is less than synchronous speed. The rotor does not remain at this new equilibrium position. It has to be brought back to its synchronous speed by replenishing the kinetic energy released by the rotating mass. Therefore the torque angle increases further. The torque developed is more than the load torque. Excess torque accelerates the rotor and eventually reduces the torque angle. The rotor may overshoot the equilibrium position. The rotor thus oscillates like a torsional pendulum about the equilibrium position corresponding to new torque. The motor operation is stable when the torque angle changes within $\frac{\pi}{2}$. However, during oscillations or hunting of the motor there is a possibility that the motor has a torque angle greater than $\frac{\pi}{2}$. The motor may be stable or may lose its stability. In such cases the stability of the drive can be estimated using equal area criterion. Referring to Fig. 2.17, δ_o is the load angle before the load is applied. δ_m is the mean position of oscillation and δ_f is the final value of torque angle during oscillation. For the machine to retain stability and remain in synchronism the value of $A_2 \geq A_1$. If $A_2 < A_1$ the motor loses its synchronism.

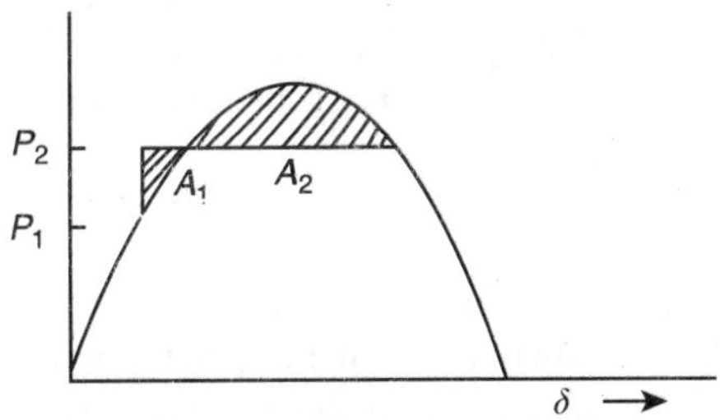

Fig. 2.17 *Equal area criterion*

Worked Examples

2.1 *Use the equal area criterion to determine the maximum load that can be suddenly thrown on to a synchronous motor without affecting stability assume that the motor is initially under no-load.*

Solution

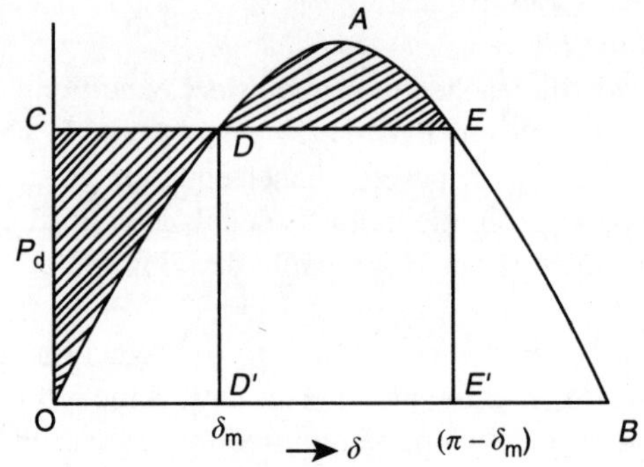

The power developed by the motor

$$P_d - P_{max} \sin \delta$$

where δ is the torque angle

As the motor is on no-load, $\delta = 0$. When the load is thrown on to the motor the corresponding value of δ_m should be such that

$$\text{area } OCD = \text{area } ADE$$

Area OCD = Area $OCDD'$ – Area ODD'. Let the load be P'_d.

$$P'_d \cdot \delta_m - \int_0^{\delta_m} P_{max} \sin \delta (d\delta) = \int_{\delta_m}^{\pi-\delta_m} P_{max} \sin \delta (d\delta) - (P'_d(\pi - 2\delta_e)).$$

After integrating and simplifying we get

$$P'_d = P_{max} \frac{1 + \cos \delta_m}{\pi - \delta_m}$$

But $P'_d = P_{max} \sin \delta_m$, substituting and simplifying we have

$$\tan \frac{\delta_m}{2} = \frac{1}{\pi - \delta_m}$$

Solution is by trial and error method to obtain δ_m. The value of $\delta_m = 46.2°$.

The value of load that can be thrown on the motor $P'_d = 0.722 P_{max}$.

This clearly shows that the equilibrium position of the rotor is well below 90° and transient overload capacity is 72.2% of the steady state overload capacity.

2.2 *A cylindrical rotor synchronous motor has 2p poles and a synchronous speed n_s rps. The moment of inertia is J kg.m². Show that the time period of natural undamped oscillation is given by*

$$t = 12.88 \left(\frac{J n_s}{2p \cdot E_s \cdot I_s} \right)^{1/2}$$

where E_s is the induced voltage and I_s is the short circuit current.

The period of oscillation of a 4,000 kVA, 6,600 V, 50 Hz, 4 pole, 3 phase synchronous machine having a synchronous reactance of 20% is found to be 1.25 seconds. The machine operates on infinite busbars. Determine the moment of inertia of the motor.

Solution Synchronising power $= \frac{3E_s^2}{Z_s} \sin\theta_s \cdot \sin\delta$; $\theta_s = 90°, \delta = \delta_0$ is very small at no-load. There $P_s = 3E_s \cdot I_s \cdot \delta = 3E_sI_s\left(\frac{2p}{2}\right)\delta_m$. Synchronising power per mechanical radian

$$K_s' = \frac{P_s}{\delta_m} = 3E_sI_s\left(\frac{2p}{2}\right)$$

Synchronising torque $K_s = \frac{3}{2}\,\frac{E_sI_s(2p)}{2\pi n_s}$

Undamped natural frequency of oscillation $= \frac{1}{2\pi}\sqrt{\frac{K_s}{J}}$ = period of oscillation

$$T = 12.88\sqrt{\frac{Jn_s}{(2p)E_sI_s}}.$$

2.3 *Show that the condition of stability for large angular deviations of a synchronous machine is given by*

$$\frac{d\delta}{dt} = 0$$

Using this derive the equal area criterion for transient stability of a synchronous motor.

Solution · The dynamic equation of motion of the motor is such that the mechanical power developed is used in accelerating the rotor and driving mechanical load. Therefore

$$P_m + P_{dyn} + P_L$$

The mechanical power developed comprises two parts: (a) damping power which varies linearly with change in load angle, (b) synchronising power which varies directly with load angle

$$P_m = P_d \frac{d\delta}{dt} + P_s(\delta)$$

The dynamic power $P_{dm} = -P_j; \frac{d^2\delta}{dt_2}$ and $P_j = J\frac{\omega}{P}$ where P is pairs of poles and $\omega_m = 2\pi\, n_s$. Therefore the equation of motion is

$$P_j \frac{d^2\delta}{dt^2} + P_d \frac{d\delta}{dt} + P_s(\delta) = P_L$$

Assuming that there is no damping

$$P_d = 0$$

We have

$$P_j \frac{d^2\delta}{dt^2} + P_s(\delta) = P_L$$

But $P_s(\delta) = P_m \sin \delta$. Using which we have

$$P_j \frac{d^2\delta}{dt^2} + P_m \sin \delta = P_L$$

from which

$$\frac{d^2\delta}{dt^2} = \frac{P_L - P_m \sin \delta}{P_j}$$

when the load is suddenly thrown on the motor the motor retards, releasing the kinetic energy of the rotating parts to meet the load requirement. When the rotor angle reaches the angle corresponding to the new load, the motor is at subsynchronous speed and an accelerating torque is developed to bring it to synchronous speed. Thus the rotor oscillates about a mean δ corresponding to the new load. The energy released from the rotating parts should be balanced by the accelerating torque. Otherwise the value of δ goes on changing, leading to instability. For stable conditions to be realised the load angle should stop changing at some instant of time during oscillation, i.e., $\frac{d\delta}{dt}$ must be zero. If in the variation of δ, finally $\frac{d\delta}{dt} = 0$, stability is assured.

$$\frac{d}{dt}\left(\frac{d\delta}{dt}\right)^2 = 2\frac{d^2\delta}{dt^2}\frac{d\delta}{dt} = 2\left(\frac{P_L - P_m \sin \delta}{P_j}\right)\frac{d\delta}{dt}$$

$$\frac{d\delta}{dt} = \sqrt{\int_{\delta_0}^{\delta} 2\,\frac{(P_L - P_m \sin \delta)}{P_j}\, d\delta}$$

But for stability $\frac{d\delta}{dt} = 0$. Therefore

$$\int_{\delta_0}^{\delta_f} \frac{2(P_L - P_m \sin \delta)}{P_j}\, d\delta = 0$$

concluding that the energy released from the rotating parts should be exactly balanced or replenished so that the motor attains synchronism at which $\frac{d\delta}{dt} = 0$. If the oscillation of the rotor is about a mean position δ_m we have

$$\int_{\delta_0}^{\delta_m} \frac{2(P_L - P_m \sin \delta)}{P_j}\, d\delta$$

This is the energy released by the motor given by area in Fig. 2.17

$$\int_{\delta_0}^{\delta_f} \frac{2(P_L - P_m \sin \delta)}{P_j}\, d\delta$$

is the energy replenished to bring the rotor to synchronous speed and it is the area A_2 Fig. 2.17. These two are equal for transient stability. Thus equal area criterion can be used to investigate transient stability. Also referring to figure if

$$\underbrace{\int_{\delta_m}^{\delta_f} \frac{2(P_L - P_m \sin \delta)}{P_j} d\delta}_{(A_2)} > \underbrace{\int_{\delta_0}^{\delta_m} \frac{2(P_L - P_m \sin \delta)}{P_j} d\delta}_{(A_1)}$$

the synchronous operation and hence stability is assured. If $A_2 = A_1$, the machine is just stable. If $A_2 < A_1$, the machine is unstable.

Problems

2.1 A gear train system used to drive a load is shown in Fig. P2. Assuming the ideal conditions of no back lash and no elastic deformation in the system determine the equivalent inertia and equivalent friction referred to
(a) motor shaft,
(b) load shaft.

2.2 Show that the torque to inertia ratios referred to the motor shaft and to the load shaft differ from each other by a factor of N, where N is the gear ratio. Show that the torque squared to inertia referred to the motor shaft or load shaft are the same.

2.3 Figure P2 shows a motor lifting a load by means of a winch. The weight lifted is 1500 kg at a velocity of 0.75 m/s. The motor runs at a speed of 1000 rpm. The inertia of the winch drum and motor are 1.8 kg.m^2 and 3.6 kg.m^2 respectively. Calculate the total load torque of the system referred to motor shaft.

2.4 Use the considerations of stabilit[illegible] show the equilibrium of operatio[illegible] the motor at points P_1, P_2, P_3 and [illegible] If there is a point of instability you suggest the modification to m[illegible] speed torque curve to make the op[illegible]tion stable? Based on the conclus[illegible] suggest a method to determine speed torque curve of the motor i[illegible] first, second and fourth quadrants

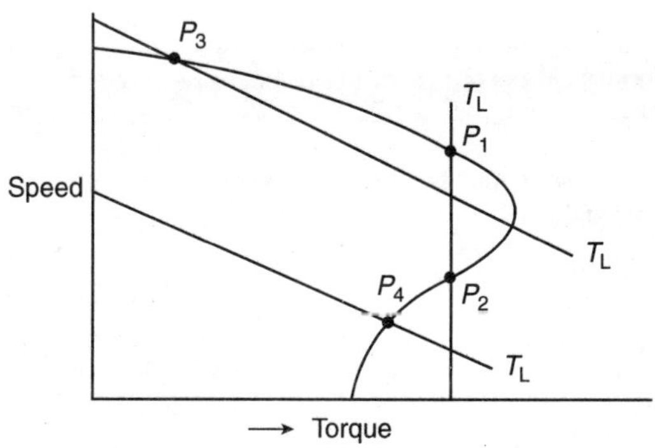

Fig. P.2

2.5 Use equal area of criterion to estimate the stability of a synchronous motor. Use the same to discuss the effects of damper windings on the oscillation of a synchronous motor.

2.6 Set up the dynamic equation of a synchronous motor during hunting and determine the expressions for natural frequency of oscillation and damped frequency of oscillation.
Determine the natural frequency of oscillation of a 50 Hz, 20 pole, 1000 kW, 6.6 kV synchronous motor having a total inertia of 2600 kg.m² and a torque angle of 30° when the power developed is 1000 kW. If the damping power coefficient is 2.5 kW/elec. degree, determine the damped frequency of oscillation.

2.7 A synchronous motor connected to an infinite bus drives a mechanical load. When the motor is loaded to its rated capacity the torque angle is 45°. The load is suddenly increased to a value which corresponds to a torque angle of 60°. Determine whether the motor is stable or not. Determine also the maximum additional load that can be suddenly thrown on to the motor without affecting stability.

2.8 A cylindrical rotor synchronous motor having an inertia J is operating on constant voltage constant frequency busbars of voltage V. Show that its natural period of oscillation

$$t_n = 9.10 n_s \sqrt{\frac{J}{f I_{sc} V}}$$

A 10 MVA, 11 kV, 3 phase, 6 pole, 50 Hz star connected synchronous motor operates on an infinite bus. Its short circuit current is 3 times the full load current. Its natural period of oscillation is 1.4 seconds. Determine the inertia of the rotating parts.

2.9 A 500 kW, 10 pole, 50 Hz synchronous motor has a torque angle of 35° on full load. Determine the natural frequency of oscillation if the moment of inertia is 1200 kg.m².

2.10 A 150 kW, 2300 V, 3 phase, 50 Hz, 28 pole synchronous motor is connected to constant voltage constant frequency busbars. The motor has the following data:
The moment of inertia = 460 kg.m²
Synchronising power = 12 kW/elec. degree.
Damping torque = 2500 Nm/mech/rad/sec. Set up the electrodynamic equation of motion. Determine the natural and damped frequencies of oscillations. If the rated load is switched on to the motor suddenly when it is on no-load determine the electromagnetic transients with and without damping torque.

Multiple-Choice Questions

2.1 A motor driving a passive load is said to be steady-state stable if

(a) $\frac{dT_l}{d\omega} - \frac{dT_M}{d\omega} = 0$

(b) $\frac{dT_l}{d\omega} - \frac{dT_M}{d\omega} < 0$

(c) $\frac{dT_l}{d\omega} - \frac{dT_M}{d\omega} > 0$

(d) All

2.2 Equal area criterion gives information regarding the

(a) transient stability of a synchronous motor

(b) steady-state stability of a synchronous motor

(c) steady-state stability of an induction motor

(d) transient stability of an induction motor

2.3 From the characteristics of load and motor torques given below, a typical example of unstable system is

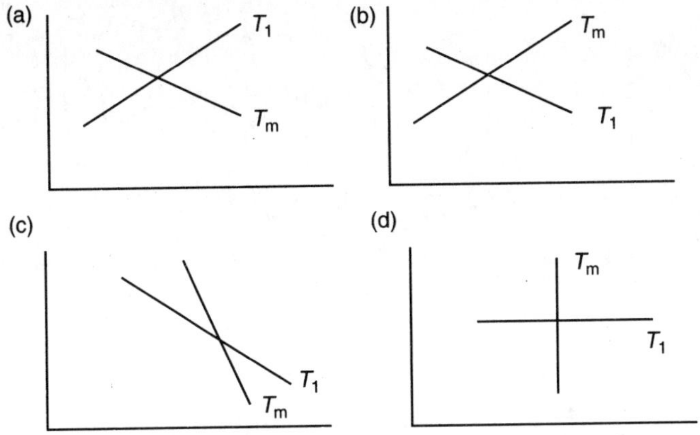

2.4 In a drive system all the mechanical quantities are referred to a single rotation shaft using the principle of
(a) Power invariance
(b) torque invariance
(c) conservation of momentum
(d) All (a), (b), (c)

2.5 Retardation test is employed
(a) to determine the losses of the motor
(b) to determine the moment of inertia of the rotating parts
(c) to determine the speed-torque curve of the motor
(d) to determine the overload capacity of the motor.

2.6 Active loads
(a) have the capacity to accelerate as well as decelerate the drive motor
(b) have the capacity only to oppose the motion trying to retard the motor
(c) have the capacity to provide accelerating torque only
(d) All the above

2.7 A typical active load is
(a) Hoist
(b) Blower
(c) Pump
(d) Lathe

References

2.1 Slemon, S.B. Dewan and A. Straughen, *Power Semiconductor Drives*.
2.2 Leonhard, W., *Control of Electrical Drives*, Springer Verlag, Berlin, 1985.
2.3 Chillikin, M., *Electric Drive*, Mir publishers, Moscow.
2.4 Kümmel, F., *Elektrische Antriebstechnik*, Springer Verlag, Berlin, 1971.
2.5 Deshpande, M. V., *Electric Motors: Applications and Control*, A.H. Wheeler, Madras, 1985.

3 Converters for Feeding Electric Motors

3.1 A GENERAL SURVEY OF CONVERTERS FOR FEEDING ELECTRIC MOTORS

The speed of a dc motor can be varied by varying the armature voltage or the field current, for which a variable dc supply is required. The speed of an ac motor, on the other hand, can be varied by varying its supply frequency. In order to achieve the rated torque capability of the motor it is necessary to operate it at rated flux (maintained constant), which can be obtained by varying the applied voltage also. Hence, a variable voltage, variable frequency supply is required to control the speed of an ac motor.

With rapid developments made in the area of thyristors, power converters using them have become very popular in the control of electric drives. The variable dc voltage needed to control the speed of dc motors can be obtained by means of a phase controlled line commutated converter. The basic theory and important fundamentals of these converters are well known from the days of mercury arc rectifiers. A variable frequency supply is possible by means of inverters employing forced commutation if the ac side of the converter is unable to provide the necessary reactive power for the converter. The basic principles of inversion have been known for long, but the devices were not popular because of the poor dynamic properties of mercury are rectifiers. With the advent of thyristors and developments in integrating circuits for the control of thyristors there has been a lot of development in the area of inverters making use of forced commutation. These find application in industry in the area of electric drives.

Converters are systems used to transform or control electrical energy using the phase control of thyristors. A thyristor in these circuits periodically provides conducting and non-conducting states alternately, which are used to control the output voltage or output frequency or both. Figure 3.1 summarises several possibilities of converting electrical energy using power converters. The arrows indicate the direction of power flow and the circles, the type of source (ac or dc).

Converters, when used to control the speed of a dc motor, are called upon to provide a variable voltage. This is possible by converting the existing ac to dc. The process is called rectification and the converter is a phase controlled line commutated rectifier. The phase control of thyristors using a control voltage makes it possible to have a variable voltage at the output terminals. Sometimes, the converter is required to allow power flow from a dc load to an ac supply, e.g. during the regenerative braking of a dc motor. The converter performs a function called synchronous inversion. A converter which can perform both rectification and inversion is called two quadrant converter. The speed of a dc motor can be varied very smoothly and continuously with it. The two quadrant representation is shown in Fig. 3.2.

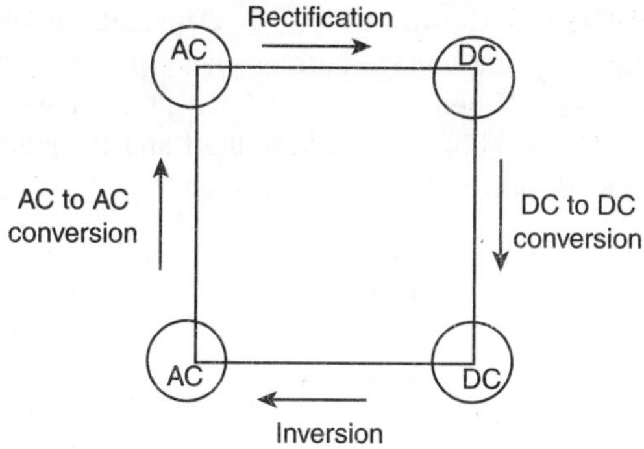

Fig. 3.1 *Possibilities of converting electrical energy using thyristor power converters*

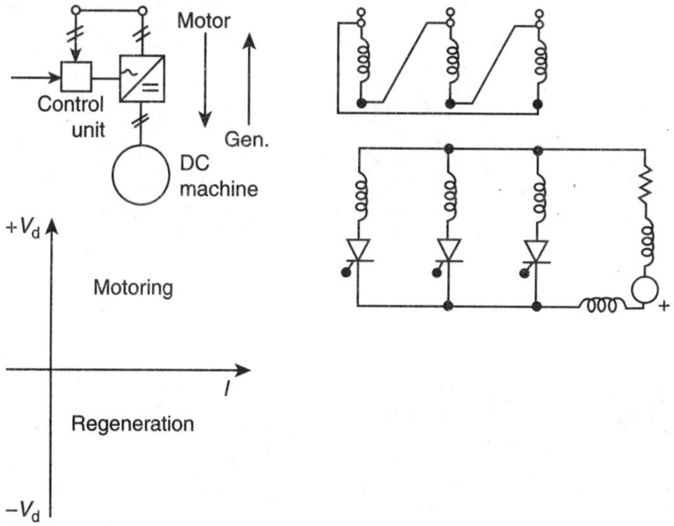

Fig. 3.2 *Two quadrant converter and its representation*

A dc motor is sometimes required to operate as a variable speed motor in both directions of rotation, with a possibility of regenerative power transfer to the ac supply system. The operation is called four-quadrant operation. Two two quadrant converters connected back to back provide this reversible drive. Four quadrant operation is shown in Fig. 3.3. Depending upon the polarity of voltage V_d and current I_d, the voltage-current plane of the dc system can be divided into four quadrants. When both V_d and I_d have the same sign, the power is delivered to the dc system. This is possible in the I and III quadrants. When they have opposite signs the power flows from dc to ac.

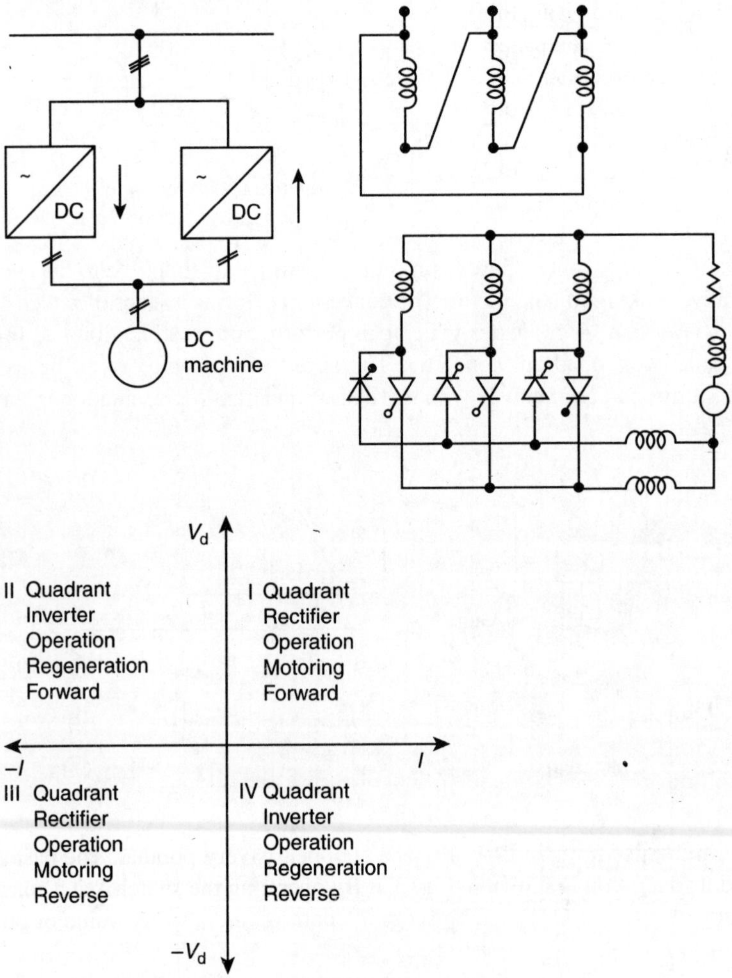

Fig. 3.3 *Four quadrant representation*

Certain applications of dc motor drives may require operation only in the first quadrant. In such a case the converter feeding the motor is called a one quadrant converter. As no regeneration is required, diodes can be used at some points as a substitutes for the thyristors. This, besides reducing the cost of the converter,

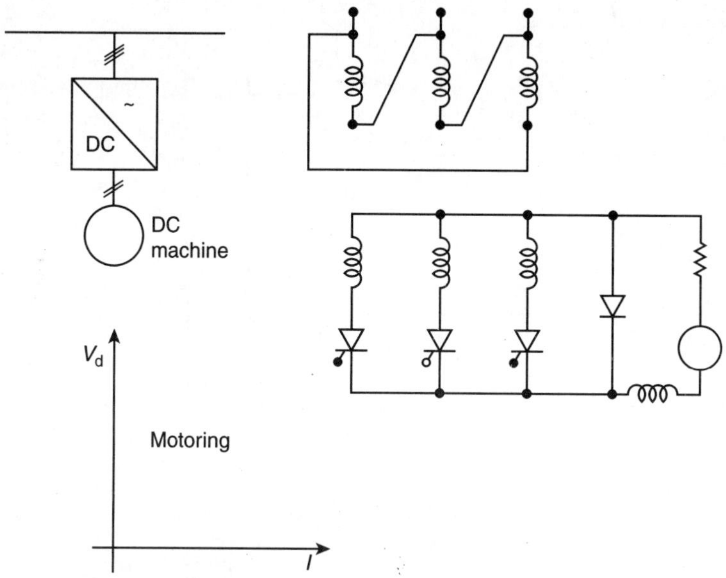

Fig. 3.4 *One quadrant converter*

improves its performance with respect to the power factor, and line harmonics. One quadrant operation of the converter is depicted in Fig. 3.4. A summary of the converters for controlling the speed of a dc motor from single phase and three-phase supplies is given in Fig. 3.5.

A variable dc voltage can be made available, to a dc motor from a constant dc voltage using a two stage conversion. A two stage dc to dc converter has an intermediate ac link. DC is first converted to ac by means of an inverter employing forced commutation. This ac is then rectified to variable dc. The schematic is shown in Fig. 3.6.

The conversion of a constant dc voltage to a variable dc voltage using a single stage conversion equipment, known as a chopper, is very popular. The basic circuit and its operation is shown in Fig. 3.7. By operating the switch *S* at a constant frequency of ON and OFF with appropriate control, the average value of output voltage can be varied. The switch S is called the thyristor chopper. The FWD allows the flow of load current when the switch S is open, especially when the load is inductive. When the FWD conducts, the energy stored in *L* is partly dissipated in *R*. By suitably selecting the chopper frequency and load inductance *L* the output current can be made to have a very small ripple content. A dc chopper is therefore suitable for controlling the speed of a dc motor efficiently. Four quadrant operation is possible. Figure 3.8 gives a summary of the operation of a dc motor on one-two- and four-quadrant choppers. The dc chopper circuit is very simple and straightforward.

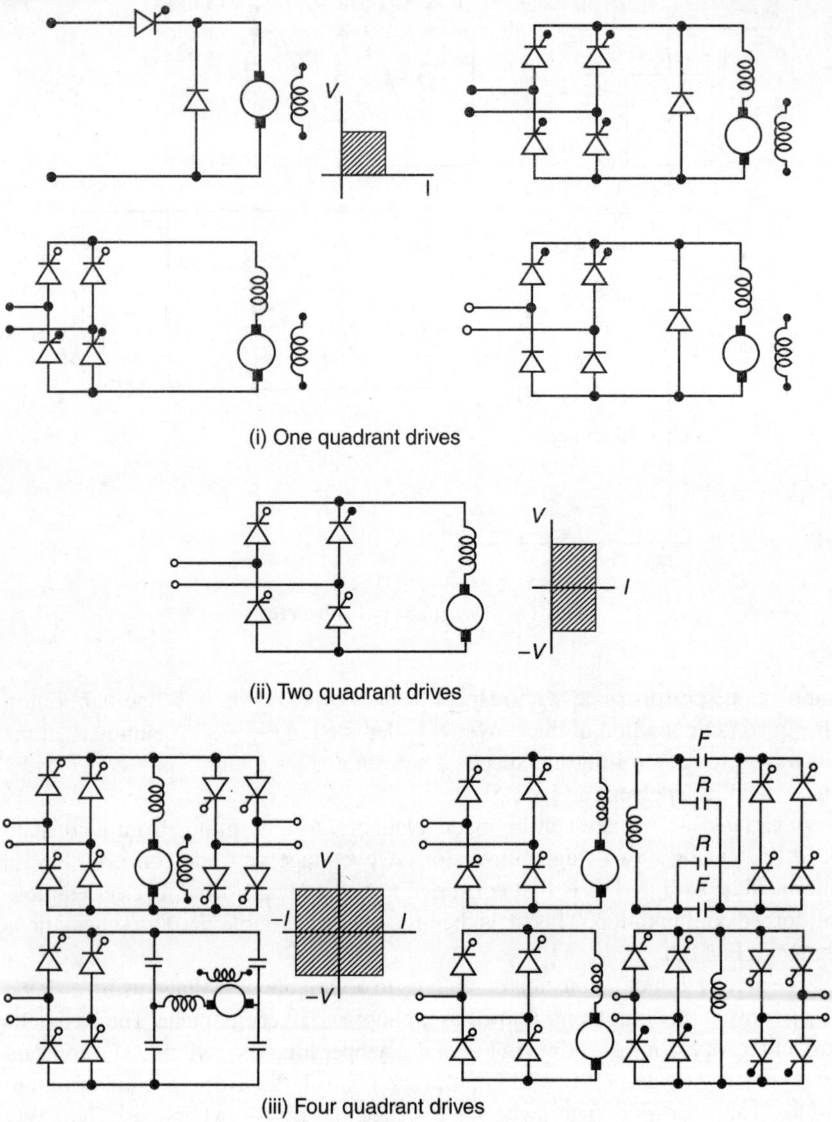

Fig. 3.5(a) *Summary of single phase converters for dc motor*

The block diagram for the control of an ac machine from a constant ac supply is shown in Fig. 3.9. The variation of stator voltage at constant frequency is obtained from a three phase voltage controller as shown in Fig. 3.9(a). In the other schemes shown in Fig. 3.9(b) and Fig. 3.9(c) both machine voltage and frequency are varied simultaneously so that air gap flux is constant. Both machine voltage and frequency can be independently varied using control voltage (V_{st1}) of line side

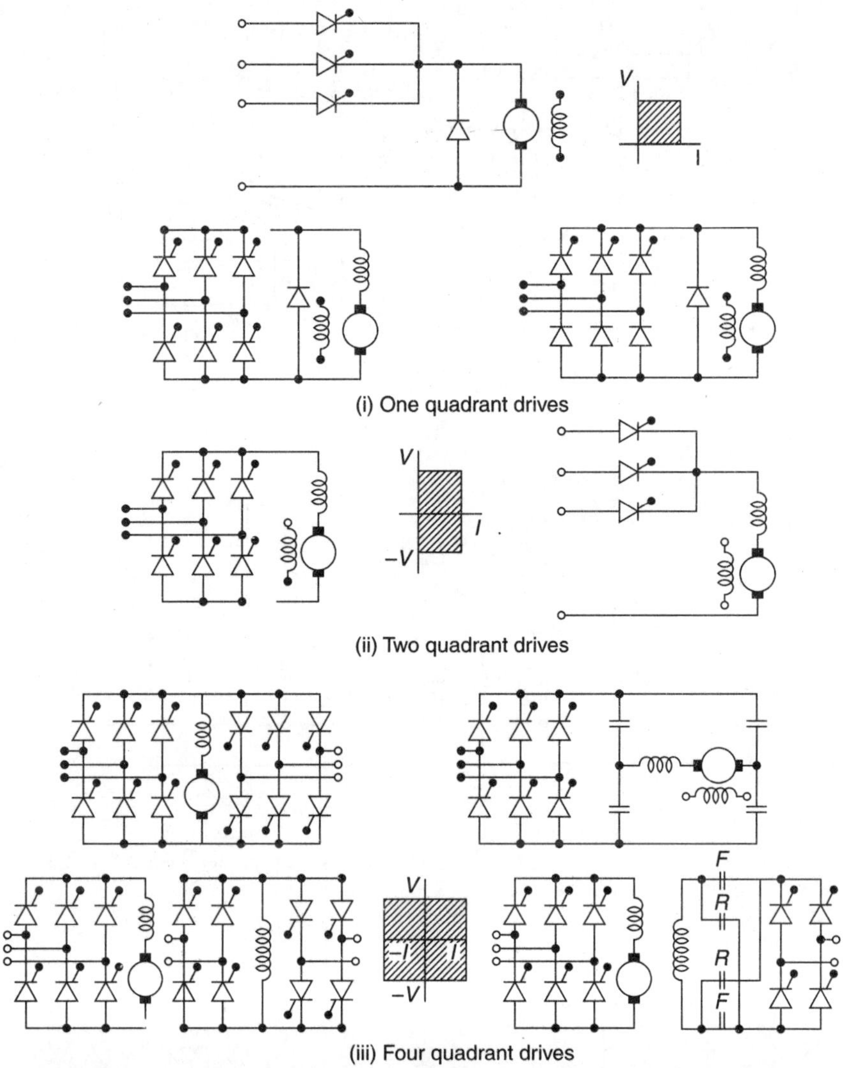

Fig. 3.5(b) *Summary of three-phase converters for speed control of dc motor*

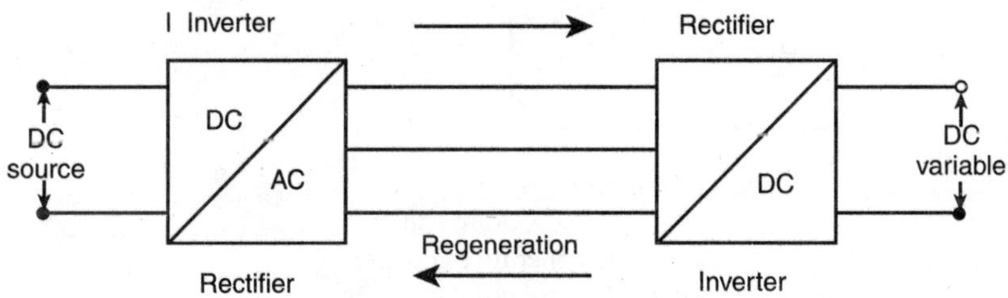

Fig. 3.6 *Two stage dc to dc conversion*

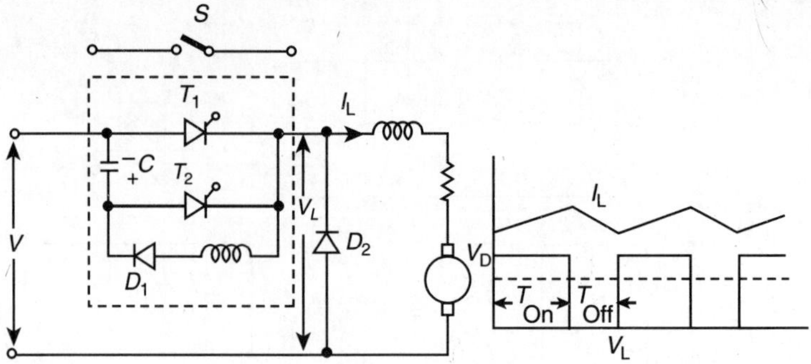

Fig. 3.7 *Basic circuit of a chopper*

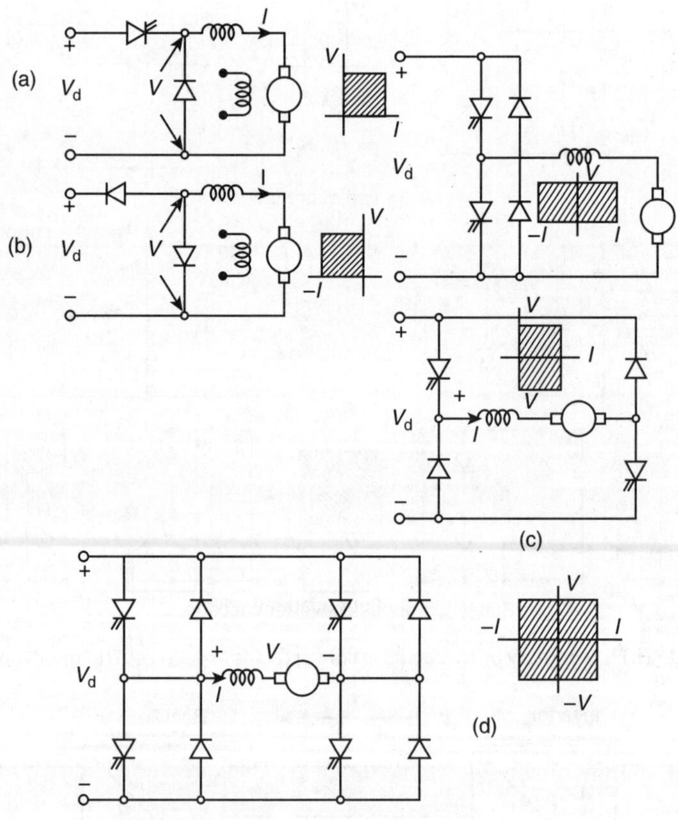

Fig. 3.8 *Summary of chopper-fed dc drives*
(a) One quadrant chopper
(b) Regenerative chopper
(c) Two quadrant chopper
(d) Four quadrant chopper

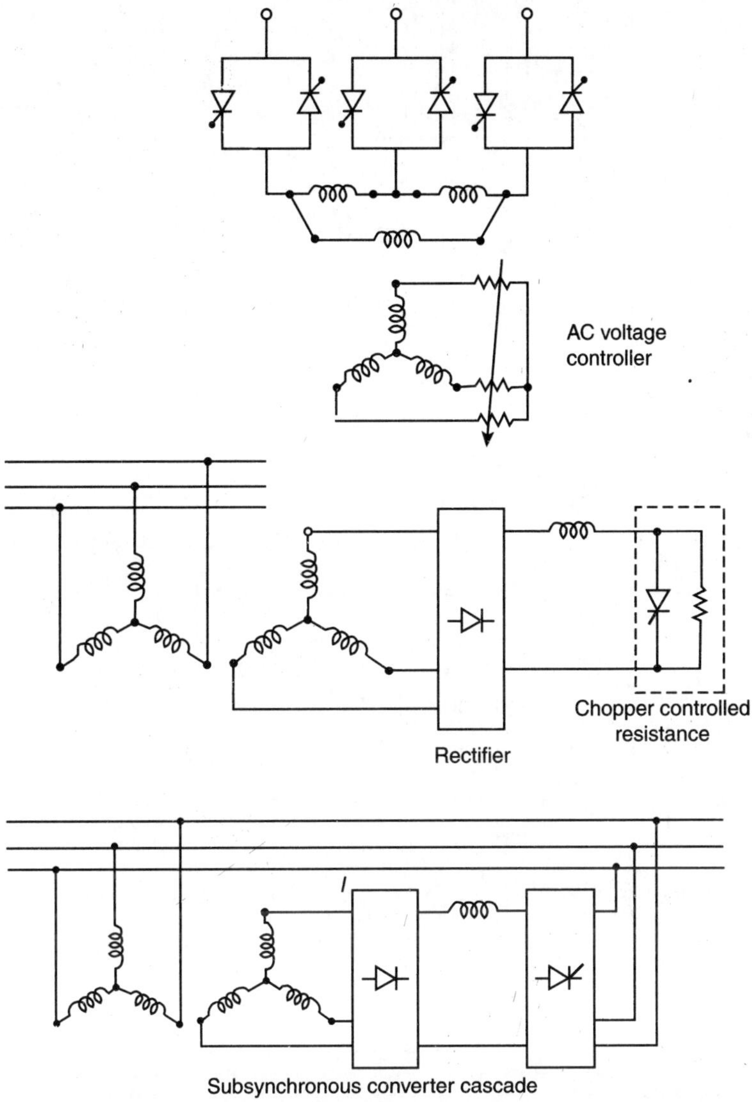

Fig. 3.9 (a) *Summary of slip ring induction motor speed control*

converter and the motor side converter (V_{st2}). They are varied so that optimum magnetic flux conditions exist in the motor corresponding to the rated torque. This provides an efficient operation of the motor, providing constant torque at every speed.

When speed control of the ac motor is required from a 3-phase system of constant frequency, the line commutated phase controlled converter and the force

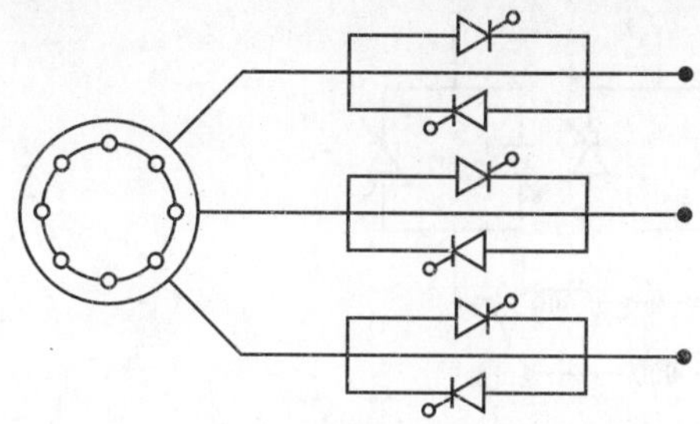

Fig. 3.9 (b) *Voltage control of sq cage motor*

commutated inverter are connected in series, as shown in Fig. 3.10. The interconnection of these converters is by means of energy storage elements which decouple the converters so that, the disturbances on the power system feeding the line side converter have no effect on machine behaviour. Such a converter is called a dc link converter. Both the converters are linked by means of a dc voltage or dc current. The former is called a voltage source inverter, with the motor impressed with voltages, and is shown in Fig. 3.11. The dc link voltage is applied to the motor phases alternately by controlling the inverter. The latter is called a current source inverter, and the motor is impressed with currents. The two differ in several aspects of performance also. In the motor operation, the line side

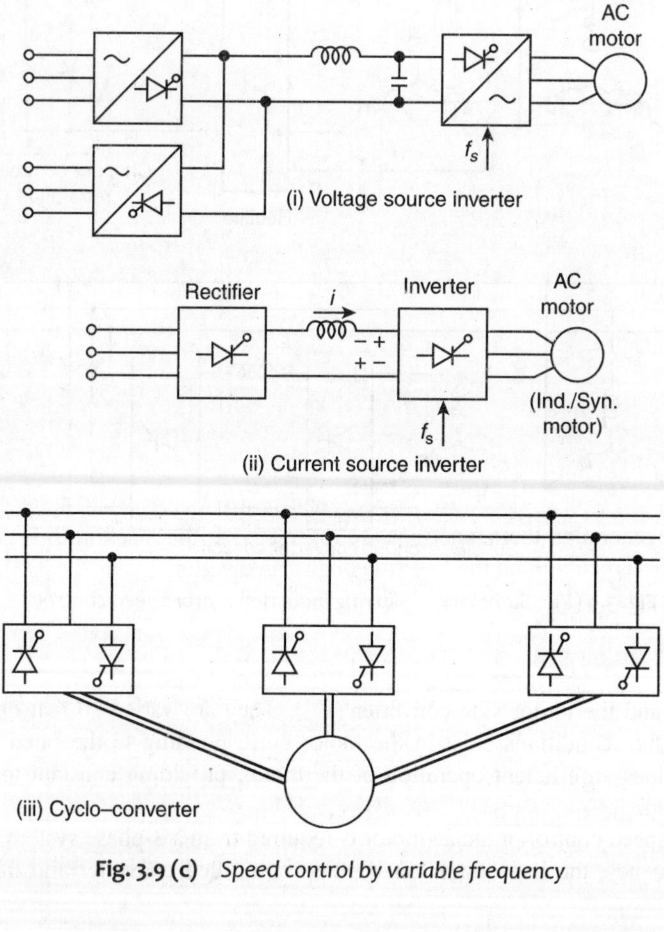

Fig. 3.9 (c) *Speed control by variable frequency*

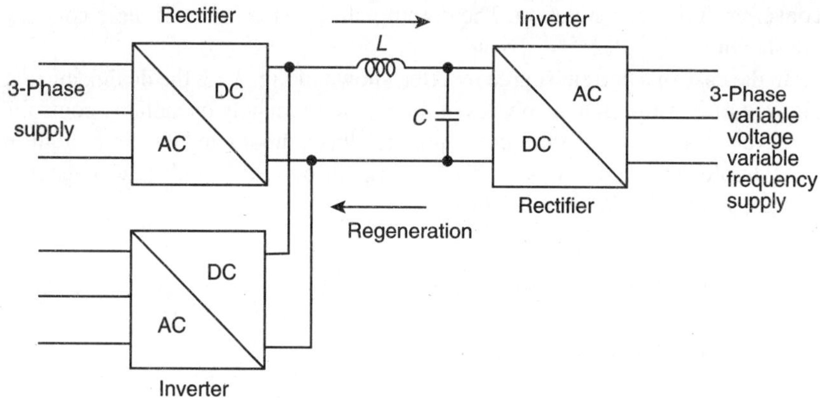

Fig. 3.10 *DC link converter*

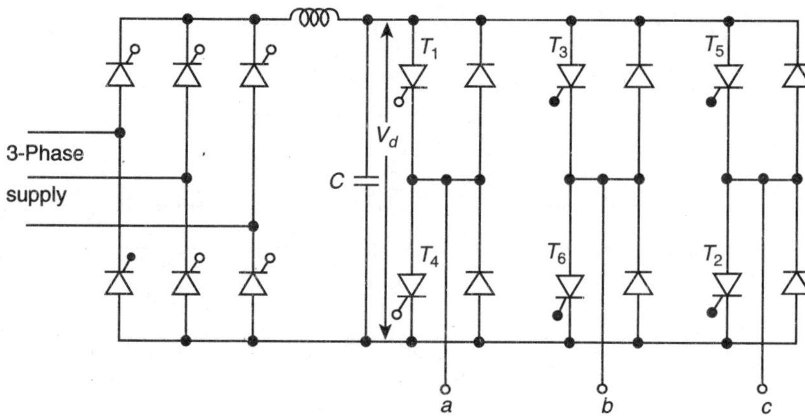

Fig. 3.11 *Voltage source inverter*

converter operates as a rectifier and the one on the machine side as an inverter. During regeneration their functions are reversed, i.e. the machine side converter operates as a rectifier and the line side one as an inverter.

An inverter requires voltage control so that the operating flux of the motor can ensure efficient operation. This can be achieved external to the inverter. The control angle of the line side converter is varied to provide the required voltage while the machine side converter is controlled to provide the required frequency. The inverter operates at variable dc voltage. The main drawback is the insufficient voltage available for commutation at low speeds, which consequently puts a limit on the lower speeds. Sometimes voltage control can be obtained in the inverter itself, using the principles of PWM or PSM. The inverter operates at constant dc link voltage, and is controlled to vary both the frequency and amplitude of the voltage. These are invariably voltage source inverters which require an additional converter during regeneration. The output voltage waveforms of these converters are shown in Fig. 3.12.

In the case of a current source inverter, shown in Fig. 3.13, the dc link current is allowed to flow through the phases of the motor alternately by suitably controlling the inverter. The inverter does not require feedback diodes and the configuration is very simple. Regeneration is simple and straightforward. The disadvantage is that it is not suitable for multimotor drives.

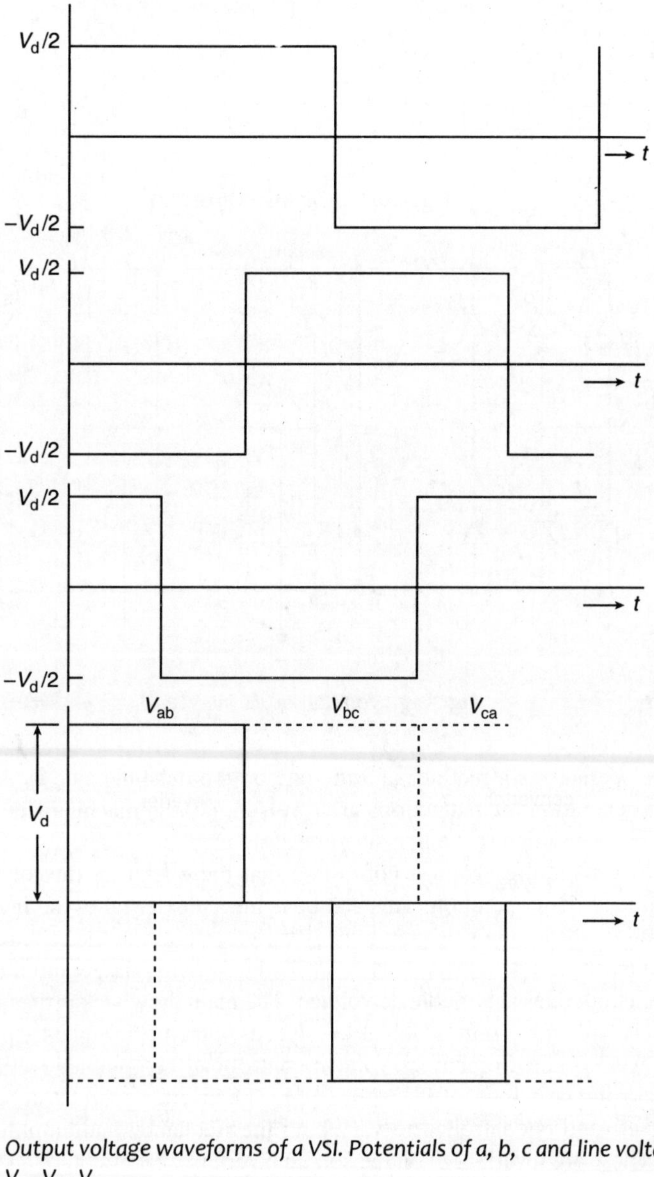

Fig. 3.12 *Output voltage waveforms of a VSI. Potentials of a, b, c and line voltages* V_{ab}, V_{bc}, V_{ca}

Variable frequency supply to an ac motor can also be provided by a cycloconverter which is a single stage (ac to ac) frequency converter having both voltage and frequency control. The output frequency of a cycloconverter can be at most 1/3 times the supply frequency. This gives a speed control in the range 0–33% of base speed. Line commutation can be made use of. The output voltage waveform has less ripple content. The schematic of a cycloconverter feeding a 3-phase ac motor is shown in Fig. 3.14.

Inverters can have natural commutation using load voltages if the load is capable of providing the necessary reactive power for the converter. Such a case occurs

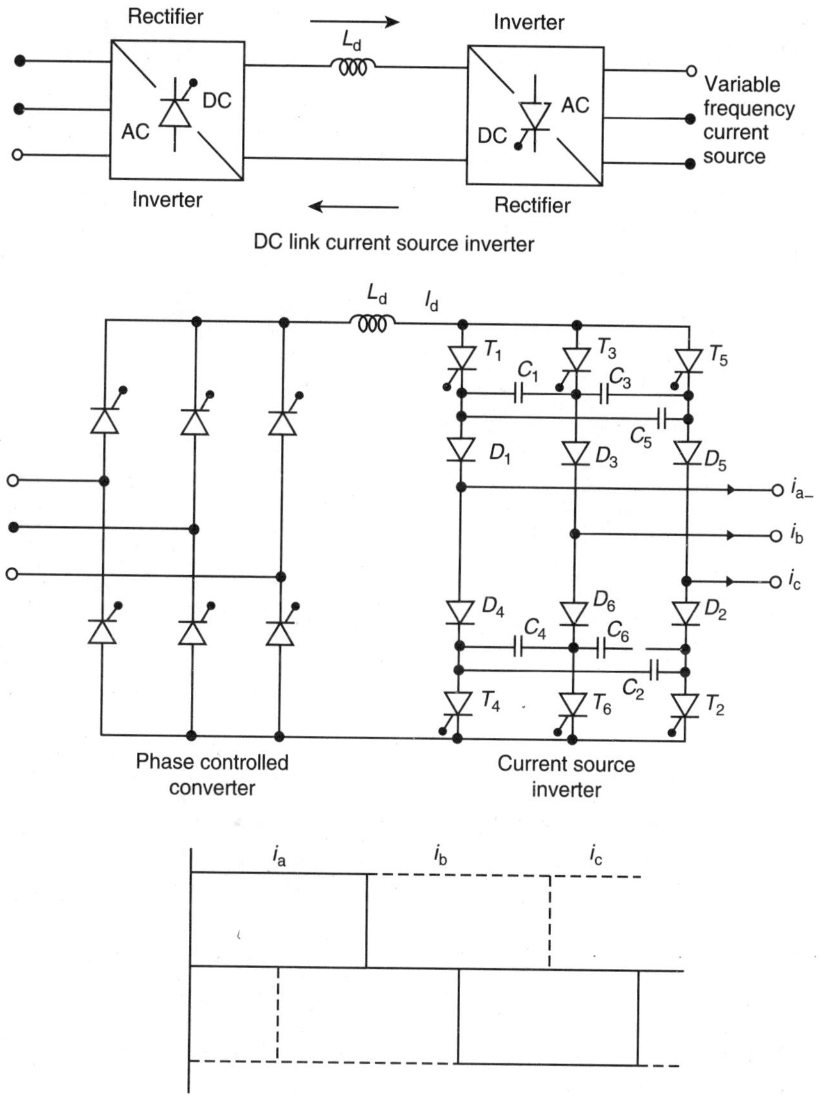

Fig. 3.13 *Current source inverter*

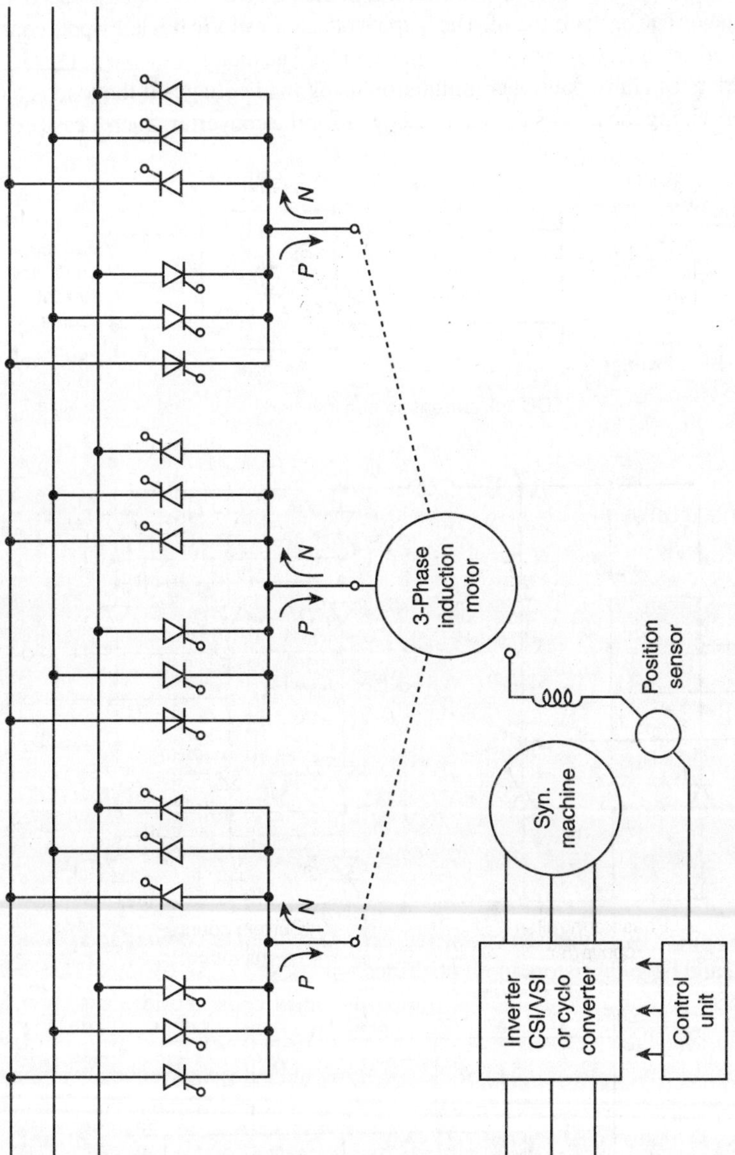

Fig. 3.14 *Cycloconverter feeding an ac motor*

when a variable frequency converter feeds a synchronous motor. An overexcited synchronous motor can provide the reactive power necessary for commutation and the machine voltages can be made use of to turn off the thyristors. A current source inverter or a cycloconverter feeding a synchronous motor employs machine commutation. The former does not require a commutation circuit and has a very simple configuration. The latter does not have frequency limitations and the speed range can go up to base speed. The firing signals to the thyristors of the converter are derived from the motor position. The motor in this case is called self controlled and has a behaviour similar to that of a dc motor.

3.2 PHASE CONTROLLED LINE COMMUTATED CONVERTERS

In these converters the commutation voltage, i.e. the voltage required to transfer current from one thyristor to the other, is provided by the supply lines to which the converter is connected. The classification of these converters is done in several ways. Depending on the direction of power flow, i.e. the type of energy conversion performed, they may be one quadrant or two quadrant converters. As has already been explained, a two quadrant converter allows power in both directions, and can perform both phase controlled rectification and inversion, i.e. ac to dc as well as dc to ac. The converter necessarily has thyristors in all positions and, is called a fully controlled converter. On the other hand, a one quadrant converter has a power flow from ac to dc and diodes can be used in a few positions of the converter. It can perform only phase controlled rectification, and is called a half controlled converter.

Converters are also classified according to the pulse number of the ac voltage superimposing the average dc voltage of the converter. Thus, we have

i. Two pulse converters
ii. Three pulse converters
iii. Six pulse converters
iv. Twelve pulse converters

Converters can be midpoint or bridge type converters depending upon their layout.

3.2.1 Two Quadrant Converters

Phase controlled converters which perform both phase controlled rectification and inversion are two quadrant converters. They are used to convert ac to dc and vice versa, and have thyristors in all positions.

These are further classified as midpoint converters and bridge type converters.

Two Pulse Mid Point Converter Two pulse converters are, in general, single phase converters. The pulse number of the converter indicates the frequency of the ripple voltage superimposing the average dc voltage at the terminals. The schematic of a two pulse mid point converter is shown in Fig. 3.15. In the figure v_{s1} and v_{s2} are midpoint to line voltages on the secondary side, which are out of phase by 180°.

$$v_{s1} = \sqrt{2}\, V_s \cos \omega t = -v_{s2} \tag{3.1}$$

The branch which has a higher voltage with respect to the midpoint can be made to conduct by giving a firing pulse to the thyristor in that branch. The midpoint

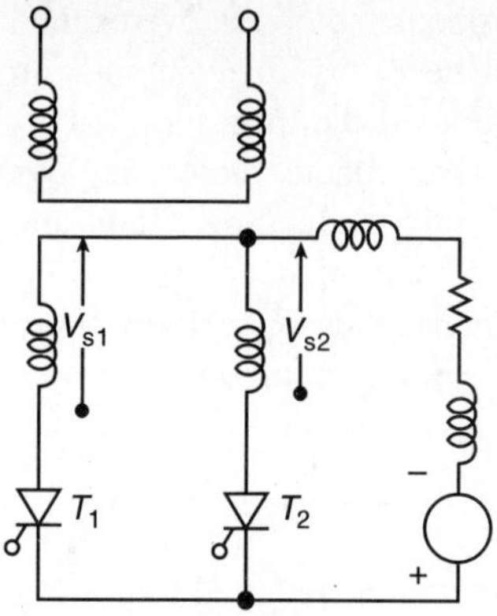

Fig. 3.15 2-Pulse mid point converter

of the secondary serves as the return for the current. The thyristor T_1 can be fired when υ_{s1} is positive and T_2 can be fired when υ_{s2} is positive with respect to the midpoint. The thyristor conducts the load current when its voltage is positive, once it is turned ON. The output voltage and current waveforms, assuming a highly inductive load are given in Fig. 3.16. If a thyristor, say T_1, is switched ON when the voltage υ_{s1} is positive the current through the load builds up. T_1 maintains conduction, depending upon the nature of the load, even in the period when υ_{s1} is negative. At this stage υ_{s2} becomes positive and thyristor T_2 takes over if a firing pulse is given. Assuming instantaneous commutation, performance, equations of the converter can be derived. Assuming a turns ratio of unity for the transformer and neglecting the commutation reactances, the average voltage at the dc terminals of a two pulse midpoint converter can be derived as

$$V_{dia} = 0.9V_s \cos a \tag{3.2}$$

where a is the firing angle of the converter. For firing delay angles in the range 0 to 90° the converter operates as a rectifier providing dc voltage at the load. Power

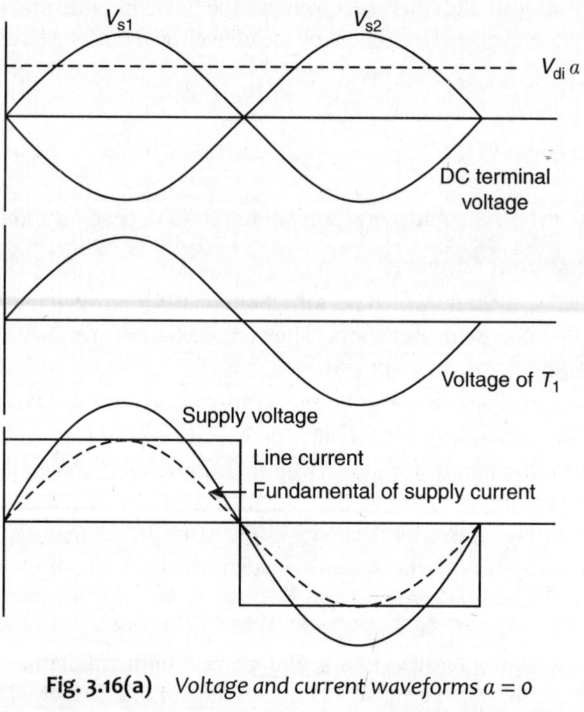

Fig. 3.16(a) Voltage and current waveforms $a = 0$

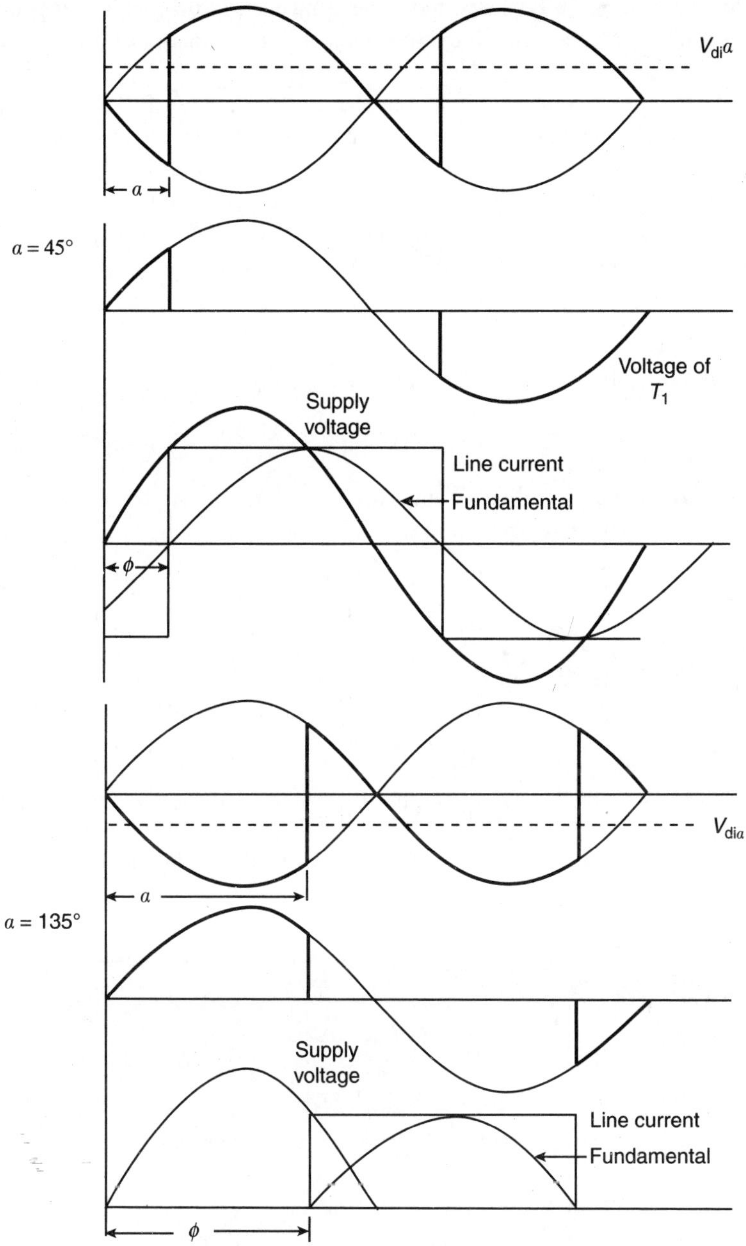

Fig. 3.16(b) *Voltage and current waveforms*

transfer takes place from ac to dc. For angles in the range 90° to 180° (theoretically) the converter operates in the inverting mode. If there is a dc source on the dc side, dc power can be converted to ac. However, in practice a firing angle of 180° cannot be reached due to overlap and the finite amount of time taken by the thyristor to go to a blocking state.

The average dc voltage is maximum at a firing angle of 0°, It decreases as the firing angle changes from 0 to 90°, and finally becomes zero when the firing angle is 90°. The voltage reverses its polarity for firing angles greater than 90° and increases with reversed polarity as a is increased beyond 90°. It reaches its maximum negative value at $a = 180°$ (Fig. 3.17).

The average value of the thyristor current

$$I_d = T_L = \frac{V_m}{\pi R} = 0.45\frac{V}{R} \tag{3.3}$$

The rms value of the thyristor current

$$I_{rms} = \frac{I_d}{\sqrt{2}} \tag{3.4}$$

Using these values of voltages and currents the ratings of the converter transformer can be obtained. The rating of the secondary winding is

$$P_s = 2V_s\frac{I_{dN}}{\sqrt{2}} = 1.57P_{di} \tag{3.5}$$

where $P_{di} = V_{dio}I_{dN}$

The primary rating is

$$V_L I_d = V_s I_d = 1.11P_{di} \tag{3.6}$$

The design rating of the transformer is

$$1/2\,(1.57 + 1.11)P_{di} \cong 1.35P_{di} \tag{3.7}$$

The increased rating of the transformers is due to the dc component of the current.

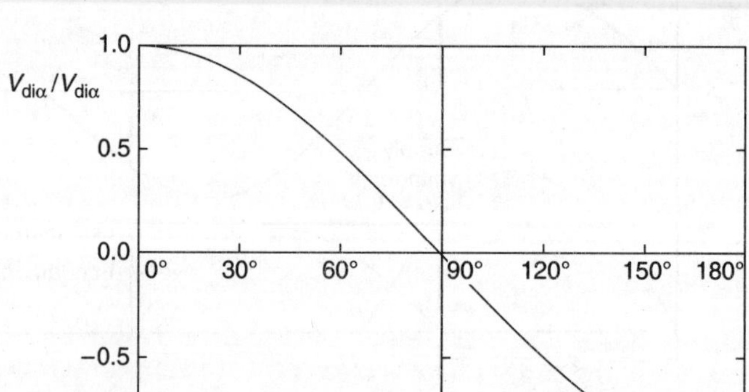

Fig. 3.17 *Control characteristic*

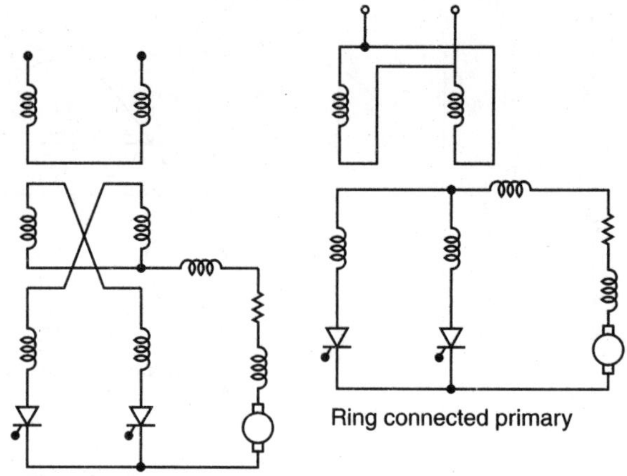

Fig. 3.18 *Arrangements of transformer for preventing dc magnetisation*

The peak forward or reverse voltages applied to the thyristors are

$$2V_m = 2\sqrt{2}\,V_s = 3.142V_{dio} \tag{3.8}$$

The premagnetisation of the transformer existing in the circuit of Fig. 3.15 can be avoided by the ring connection of the transformer shown in Fig. 3.18.

Overlap the current through the load builds up when a thyristor is turned ON and the voltage across it is positive. Say, for example, T_1 is fired when V_{s1} is positive. T_1 maintains its conduction till the next thyristor is fired at any instant during its positive voltage. V_{s2} becomes positive when V_{s1} becomes negative. Thyristor T_2 is fired. The current transfer takes place from T_1 to T_2. The preceding discussion has assumed instantaneous commutation. But in practice, due to the leakage reactance of the transformer, and the line inductances and additional inductances of the circuit (to protect the thyristors from d*i*/d*t*), the transfer of current is never instantaneous, but takes a definite amount of time. During commutation both the thyristors conduct. The current of the outgoing thyristor decreases and that of the incoming one increases. The process is complete when all the current has been transferred to the incoming thyristor. The angle of overlap is denoted by u. The voltage and current wave forms of the converter, taking overlap into consideration, are shown in Fig. 3.19.

The effect of overlap is to cause a kind of voltage drop at the output terminals. The average value of the dc voltage at the dc terminals of the converter is

$$V_{da} = 0.45V_s(\cos a + \cos(a + u)) \tag{3.9}$$

where $\cos(a + u) = \cos a - (I_d/\sqrt{2}I_k)$ with $\sqrt{2}I_k = \sqrt{2}\,V_s/X_k$

Therefore

$$V_{da} = V_{dia} - \left(\frac{V_{dio}}{2\sqrt{2}}\right)(I_d/I_k) \tag{3.10}$$

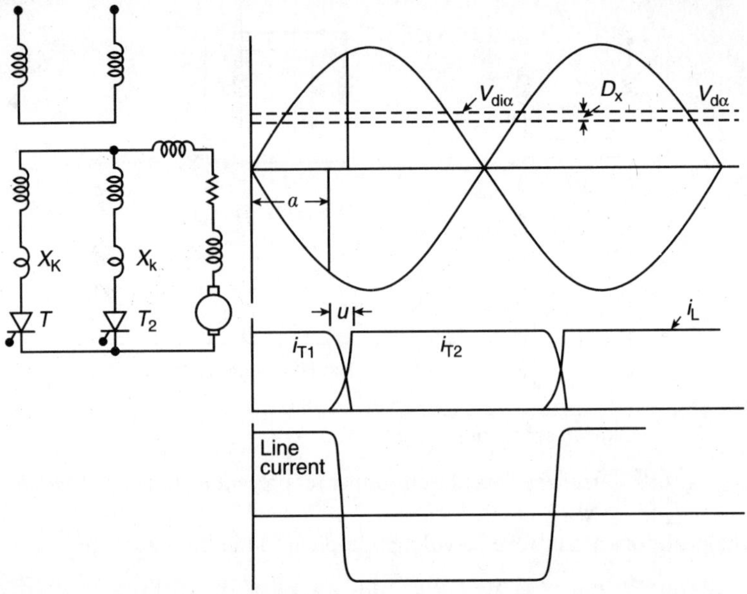

Fig. 3.19 *Current and voltage waveforms taking overlap into consideration*

Therefore, as the converter is loaded there is a reduction in the terminal voltage. This reduction is called voltage regulation. Besides overlap, the drops in the thyristors and circuit resistances contribute to voltage regulation. During overlap the rate of change of current causes a drop in the inductive reactances in series with the thyristors, which is the main cause of voltage regulation.

The mean dc voltage of a two pulse converter is superimposed by a ripple voltage of twice the supply frequency. The ripple content is minimum at $a = 0°$ and increases to a maximum at $a = 90°$. When a is increased further the ripple content decreases and falls to minimum when $a = 180°$.

When the load is purely resistive, the current in it becomes discontinuous. To explain this, note that the current is in phase with the voltage. When the load voltage falls to zero and the thyristor is reverse biased, conduction ceases. When the load has sufficient inductance one thyristor conducts for 180° and the other thyristor takes over before the load current falls to zero. Thus, conduction is made continuous. The load current becomes pure dc if the load has infinite inductance.

The mean value of the dc output voltage would be different for the cases of continuous and discontinuous conduction for the same firing angle being less for the former case. This is mainly because negative excursions of the voltage are possible across the load in the case of continuous conduction, as the load current is maintained even after the voltage has become negative. The back emf load on a converter is prone to discontinuous conduction. A resistive load inherently has discontinuous conduction. These cases are depicted in Fig. 3.20.

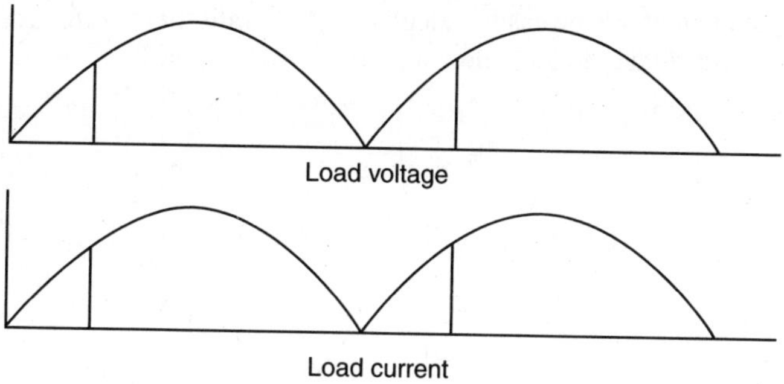

Fig. 3.20(a) *Discontinuous conduction resistance load*

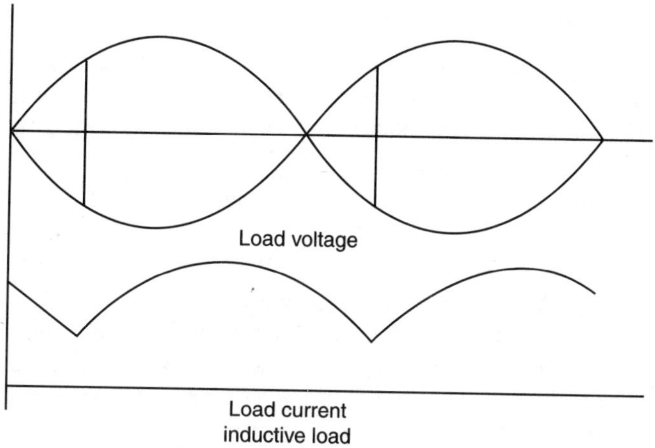

Fig. 3.20(b) *Effect of load inductance on continuous conduction*

The performance of a converter is characterised by the superimposed ripple content of ac voltage on the mean dc voltage. The effective value of the rth harmonic referred to V_{di} is

$$\frac{V_{rda}}{V_{di}} = \frac{\sqrt{2}}{r^2 - 1}(r^2 + (1 - r^2)\cos^2 a)^{1/2} \tag{3.11}$$

neglecting the overlap. When the effect of the overlap is taken into consideration, the effective value of the rth harmonic referred to V_{di} would be

$$\left.\frac{V_{mda}}{V_{di}}\right|_u = \frac{\sqrt{2}}{r^2 - 1}(f_1(a) + f_2(u) + f_3(a, u))^{1/2} \tag{3.12}$$

where $f_2(u) = \sin ru \sin u$

$$f_1(a) = 1 + \cos^2 a + r^2 \sin^2 a$$

$$f_3(a, u) = (r^2 - 1)\sin^2(a + u) + 2\cos ru(\cos a \cos a + u) + r^2 \sin a \sin(a + u)$$

The ripple content can be easily calculated as the ratio of the effective value of superimposed ac voltage to ideal dc voltage

$$W_u = \sqrt{\frac{\Sigma V_{rda}^2}{V_{di}}} \tag{3.13}$$

For a two pulse converter the ripple content is 48.2% for $a = 0°$ and 111.1% for $a = 90°$.

A smoothing inductance is necessary in the load circuit. This inductance serves two purposes:

i. to smooth the ripple content of output current
ii. to make conduction continuous in the load or to minimise the possibility of discontinuous conduction.

The value of L_d is normally determined such as to avoid discontinuous conduction rather than to smoothen out the ripple content. The layout of the smoothing inductance is rather large. The smoothing inductance required in the load circuit is

$$L_d = 3.18 \sin a \, \frac{V_{di}}{I_d} \text{ mH} \tag{3.14}$$

where V_{di} average value of dc voltage
a firing angle at which the smoothing is required
I_d is the dc current at which the conduction must be continuous

The performance of a converter is also characterised by harmonic currents on the ac side. The harmonic components present on the ac side do not contribute to any power transfer. On the other hand, they cause undesirable effects in converter operation and also reduce the power factor markedly. They may cause resonance effects due to the line inductance and capacitance. When overlap is neglected, the rms value of the harmonic current referred to the fundamental is

$$\left(\frac{I_{rL}}{I_{1L}}\right) = \frac{1}{r} \tag{3.15}$$

e.g. $r = 3$ 33.33%
$r = 5$ 20%
$r = 7$ 14.29% and so on.

For a single phase

$$I_{1L} = \frac{I_d V_{di}}{V_L} \tag{3.16}$$

The effective value of the line current expressed as the ratio of the fundamental

$$\frac{I_L}{I_{1L}} = \sqrt{1 + \Sigma \frac{1}{r^2}} \tag{3.17}$$

For a two pulse converter this ratio is 111.1%. This distortion of the input ac current is also seen by examining the fundamental content of the input current. The ratio of the fundamental component to the total rms current, $g = I_{1L}/I_L$. For a two pulse converter

$$g = 0.9 \tag{3.18}$$

When the overlap u is taken into consideration

$$\frac{I_{rL}}{I_{1L}} = \left(\frac{1}{r}\right)\frac{\sin(ru/2)}{(ru/2)} = \frac{I_{rL}}{I_{1L}} f(r,u) \tag{3.19}$$

The effect of overlap is to reduce the distortion on the ac side and decrease the rms value of a harmonic.

The reactive power required by a converter is also a significant factor in evaluating its performance. The fundamental displacement factor is the phase difference between the voltage and the fundamental of input current. From the waveforms of Fig. 3.16 the displacement factor is cos a. The total power factor on the input side is somewhat less than the displacement factor. It can be shown that the total power factor is given by

$$\text{Power factor} = g \cos a \tag{3.20}$$

For a two pulse converter, $g = 0.9$. The harmonics therefore effectively decrease the pf even though they do not contribute to power transfer.

The reactive power required by a converter is due to the phase control employed, as well as commutation. Unlike the active power which is decided by the fundamental only, the control reactive power is decided by the harmonics also. The fundamental displacement factor is the cosine of the control angle. When commutation is considered there is a certain overlap angle, because of which the current waveform shifts further to the right, increasing the angle of lag of the current. The overlap angle depends upon the firing angle. The reactive power required because of phase control is $V_{di} I_d \sin a$ and it increases as the firing angle increases, or the pf becomes poor. The reactive power due to commutation overlap is

$$Q_1 = V_{id} I_d \frac{2u_o - \sin 2u_o}{4(1 - \cos u_o)} \tag{3.21}$$

where u_o is the overlap angle at $a = 0°$. It can be shown that the reactive power required by the converter at a given firing angle a is

$$Q_1 = V_{id} I_d \frac{2u + \sin 2a - \sin 2(a+u)}{4(\cos a - \cos(a+u))} \tag{3.22}$$

The fundamental displacement factor, taking overlap into consideration is approximately $\cos(a + u/2)$ or $\cos(a + 2u/3)$ depending upon $60 \le a \le 90°$ or $0 \le a \le 30°$.

The characteristic curves of a midpoint converter are given in Fig. 3.21.

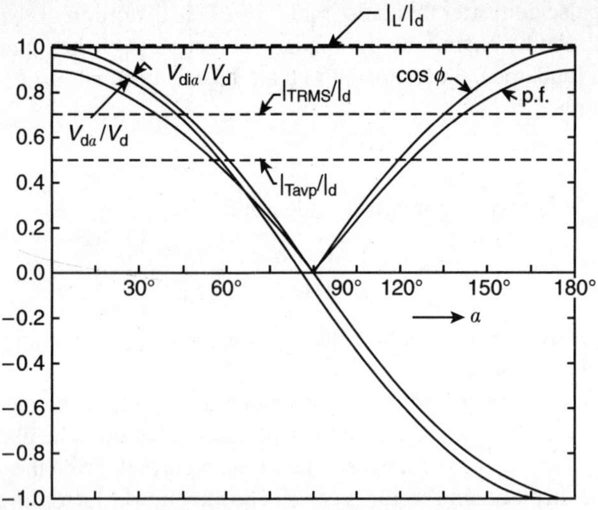

Fig. 3.21 *Characteristics of a 2-pulse mid point converter*

Two Pulse Bridge Converter A two pulse bridge converter is achieved as shown in Fig. 3.22 from two midpoint converters. They are connected in series on the dc side and in parallel on the ac side. It is also a single phase converter. At any given time, two diagonally opposite thyristors conduct, one thyristor acting as a return path for the current, e.g. T_1 conducts the current to load whereas T_4 returns the current to the supply. In the next half cycle T_2 and T_3 take over the jobs of conducting the current as described above. The common point of the transformer is not used as return any more and the input transformer can even be eliminated. The performance of a two pulse bridge converter is similar to that of a two pulse midpoint converter, but several differences between the two are worth noting. The voltage regulation at the dc terminals of the converter due to forward voltage drops of the thyristors is greater, as two thyristors conduct at any time. The control circuitry is slightly more complicated since two channels of gating pulses displaced by 180° are required. The channels must have outputs to provide the firing pulses to the conducting thyristors. The voltage and current waveforms of the converter are depicted in Fig. 3.23 under idealised conditions of instantaneous commutation, smooth dc current, etc.

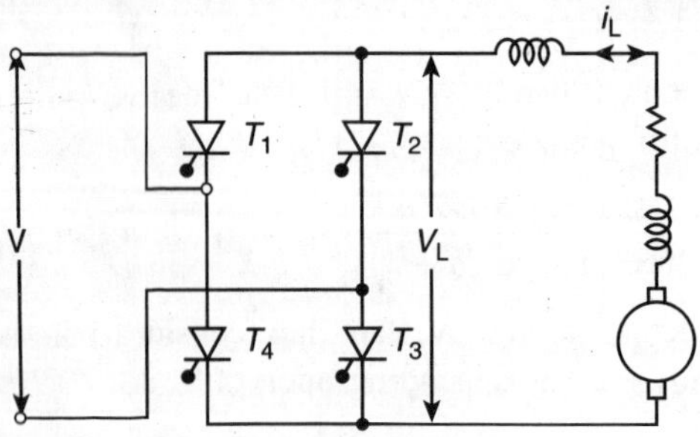

Fig. 3.22 *Two pulse bridge converter*

The average value of the dc voltage in the bridge converter is twice that in a midpoint converter, for the same inverse voltage of the thyristors. This means that the peak forward or reverse voltage of a thyristor in a bridge

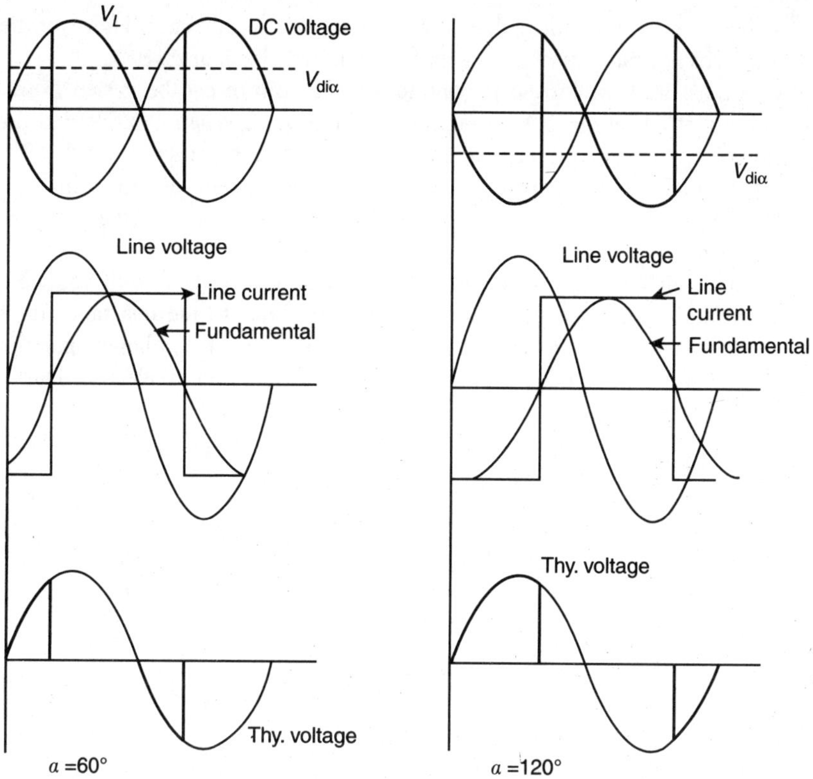

Fig. 3.23 *Voltage and current waveforms*

converter is just half its value in a midpoint converter, for the same dc voltage at the terminals.

The mean voltage at the dc terminals

$$V_{dia} = 0.9V \cos a \quad (3.23)$$

where V is the ac voltage. The peak forward or reverse voltage of the thyristors is equal to the peak value of the source voltage $= \sqrt{2}V$. In terms of V_{dio} it is equal to $\frac{\pi}{2}V_{dio}$. The average thyristor current

$$I_{AV} = 0.45\frac{V}{R} = 0.5I_d \quad (3.24a)$$

The rms value of the thyristor current

$$= I_d/\sqrt{2} \quad (3.24b)$$

If a converter transformer is used, its design rating is considerably smaller than that used with a midpoint converter, and is given by

$$P = 1.11P_{di} \quad (3.25)$$

This is because the secondary does not carry any dc in this case. However, the transformer is normally dispensed with as has already been discussed.

As long as the load current is continuous, the voltage can be varied from a maximum of 0.9 V at $a = 0°$ to a minimum of 0 at $\pi/2$. When a is retarded further the polarity of the voltage reverses. As the angle is increased towards 180°, the voltage increases in the reverse direction. It reaches a negative maximum at $a = 180°$. In practice, $a = 180°$ cannot be realized due to overlap and the finite turn off time of the thyristors.

The converter's voltage regulation, characterised by a reduction of voltage at the dc terminals, can also be attributed to (a) overlap, (b) the resistance drop, and (c) the device drop. The overlap (commutation) affects the voltage regulation in the same way as has been explained earlier for midpoint converters. During

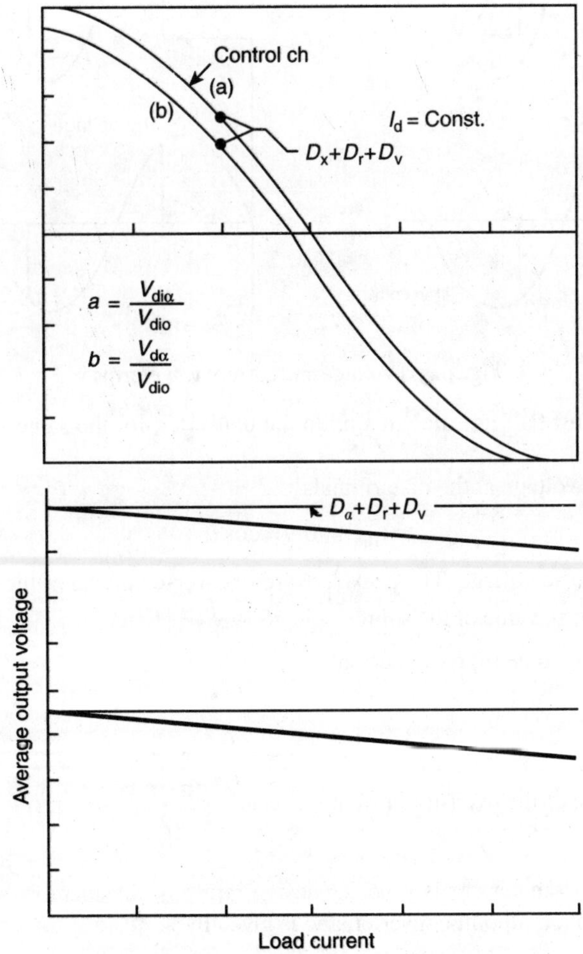

Fig. 3.24 *Voltage regulation of the converter*

commutation the rate of change of current causes a voltage drop across the reactance in series with the thyristors, e.g. transformer leakage reactance, line reactance and any inductive reactance in the circuit to protect from *di*/*dt* (Fig. 3.24)

$$V_{da} = 0.9V\cos a - 0.9V\frac{X_t}{\sqrt{2}} - I_d R - 2V_T \qquad (3.26)$$

where V ac supply voltage

- X_t per unit reactance in series with thyristor
- R circuit resistance
- I_d mean value of load current
- V_T forward drop of the thyristor at current I_d

The voltage current waveforms, taking overlap into consideration, are depicted in Fig. 3.25.

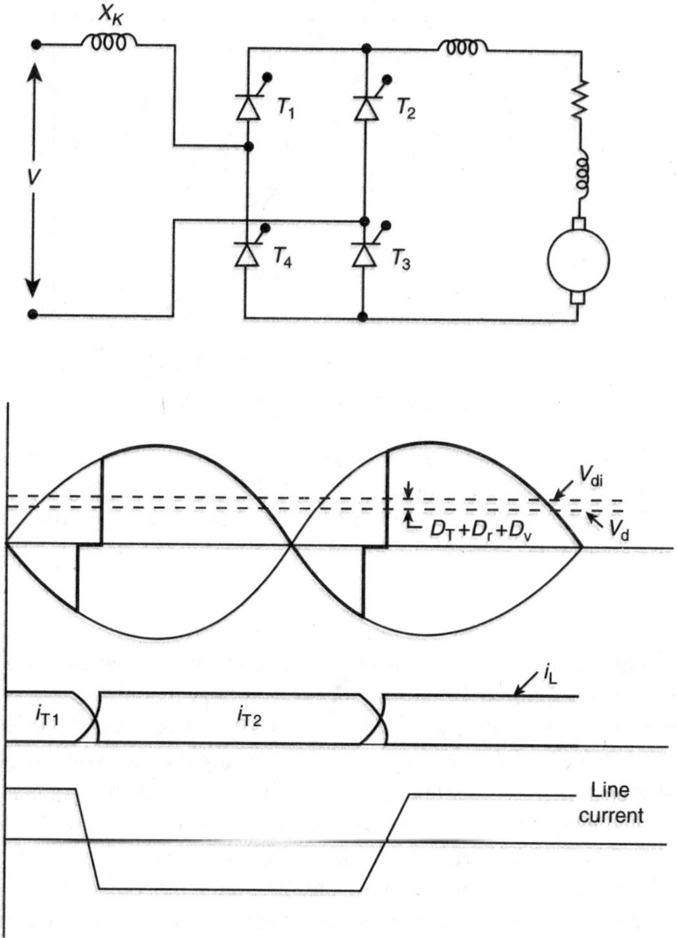

Fig. 3.25 *Voltage and current waveforms taking overlap into consideration* $a = 60°$

Equations 3.12–3.19 are also applicable to a bridge converter, taking the overlap into consideration. The performance of a bridge converter with respect to the ac ripple superimposing the dc voltage, harmonics in the input current, power factor, reactive power requirement, and discontinuous condition lay out of smoothing reactor, is the same as that of a midpoint converter.

Even though a large number of thyristors are required in the case of a bridge connection, the voltage rating is just half that of a midpoint converter. This may offset the cost of the converter. The characteristics of the converter are given in Fig. 3.26.

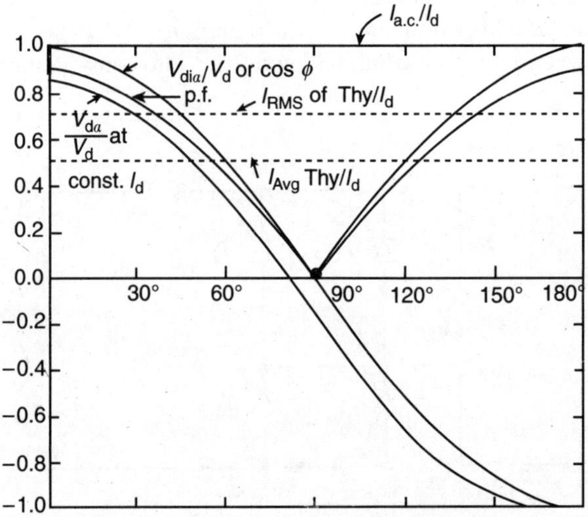

Fig. 3.26 *Characteristics of two pulse bridge rectifier*

Two pulse converters have a limited power capability, since they are basically single phase converters. The ripple content in the output voltage is large and the amount of inductance required to smooth it, as well as to avoid discontinuous conduction is rather large. Further, a two pulse midpoint converter requires a special type of converter transformer. They therefore find application only under special circumstances.

Three Pulse Midpoint Converter These are basically three-phase converters, and are very popular because a 3-phase supply is readily available. 3-phase converters have a greater power capability than single phase converters. With these converters the pulse number of the output voltage superimposing the mean dc voltage can be increased using suitable transformer connections. Increasing the number of pulses improves the converter performance with respect to the amplitude of the dc voltage as well as the magnitude of ripple content. The dc voltage of a 3-phase converter is more, with reduced ripple content. Also, the smoothing inductance becomes small.

A three pulse midpoint converter is the simplest form of a 3-phase converter. It is shown in Fig. 3.27. From the figure it is clear that the star point of secondary of the converter transformer is required to serve as a return path for the current. The

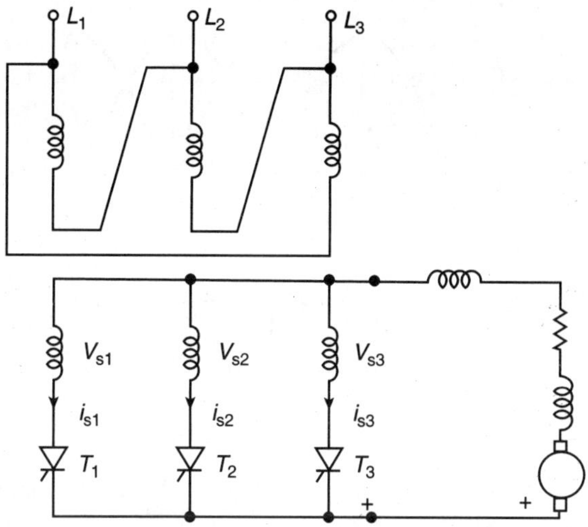

Fig. 3.27 *Basic circuit of 3-pulse mid point converter*

converter is also called a star point converter. The output voltage can be varied steplessly by varying the firing angle of the thyristor. A thyristor which has forward voltage across it can start conducting if it receives a firing pulse. The phase voltages V_{s1}, V_{s2} and V_{s3} are shown in Fig. 3.28. From the figure, the natural firing point of the thyristor can be identified. It is the instant at which the diodes would start conducting if the converter were uncontrolled, and is the point of intersection of the voltages. This instant, at which a thyristor is forward biased, occurs 30° after its voltage has crossed zero. A thyristor can always go into conduction if it receives a firing pulse when its phase voltage is greater than that of the outgoing one. The firing angle is reckoned from this point. The phase voltage having the largest instantaneous value can only appear across the load. If the firing turning-on takes place at the natural firing instant, the mean voltage at the output terminals is a maximum. The voltage and current waveforms for several firing angles are shown in Fig. 3.29. Each thyristor conducts for 120° and blocks for 240°.

The maximum reverse voltage across a thyristor is the line to line voltage, which is equal to $\sqrt{3}$ times the phase voltage.

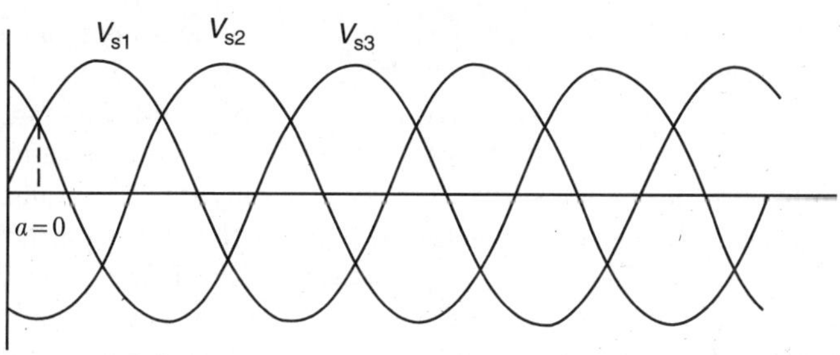

Fig. 3.28 *Phase voltage of secondary $\alpha = 0$ natural firing instant*

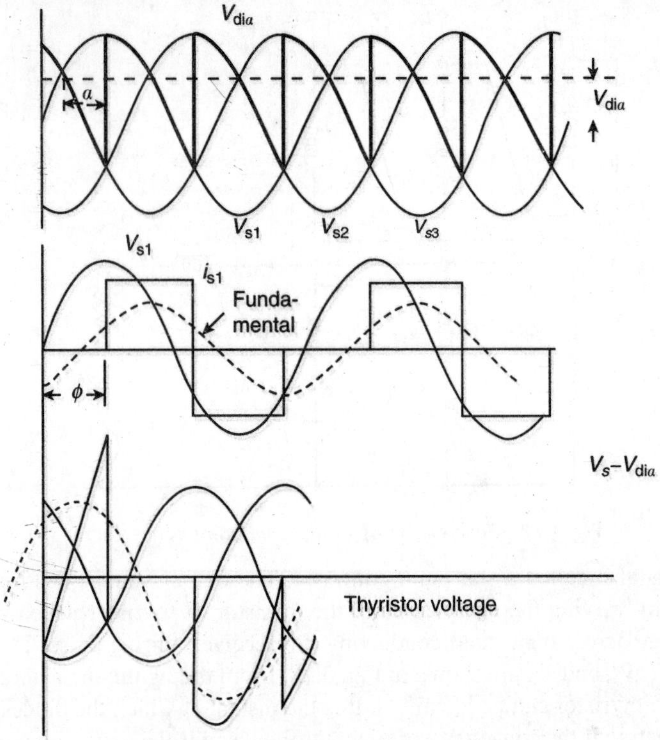

Fig. 3.29 *Voltage and current waveforms*

The average thyristor current is

$$I_d/3 = 0.39\frac{V}{R} \tag{3.27}$$

The rms value of the thyristor current $= 0.23\dfrac{V}{R}$

The mean output dc voltage at any firing angle is given by

$$V_{dia} = 1.17V\cos\alpha \tag{3.28}$$

assuming instantaneous commutation. This equation is valid for continuous conduction of load current, i.e. the current should not become zero when the voltage is zero, as happens in a resistive load or back emf load. With sufficient inductance in the load circuit, the load current flows even when the voltage is negative.

As the firing angle α increases from 0 to 90°, the output voltage falls from a maximum of 1.17V to zero very smoothly. The power flow takes place from ac to dc and the converter is in the rectifying mode. If the firing angle is increased further or retarded, the output voltage has a reversed polarity. Power flow can take place from dc to ac only if there is a source of dc voltage, e.g. a counter emf load, such as a dc motor load will be able to do this during regeneration. The converter is in the inverting mode. The output voltage increases progressively in the negative direction as a reaches 180°. Therefore for angles $0 < a < 90°$ the converter is in the rectifying mode and for $90° < a < 180°$ it is in the inverting mode. Normally the reverse voltage must exist across the thyristor for a time greater than its turn off time, so that it successfully goes into the forward blocking condition. The turn off time of a thyristor is of the order of 100 μs to 300 μs. When $a = 180°$ the conducting thyristor can never be blocked, as there is no time at all for it to regain its blocking state. The time for which the reverse voltage exists across a thyristor must be greater than ωt_q.

Also, the negative voltage occurring across the thyristor should continue even after the thyristors have under gone commutation for a definite amount of time t_q. For successful commutation there should be a marginal turn off angle, and $a = 180°$ cannot be realised. Figure 3.30 depicts the variation of output voltage as the firing angle is extended from 0–180°. (The control characteristic of the converter shown in Fig. 3.30 also shows the inverter limit.)

The finite angle of overlap, due to reactances on the line side of the converter, also affects the maximum firing angle for inverter operation.

Taking these factors into consideration, the maximum firing angle a_{max} is fixed at 150°.

Sometimes commutation failure occurs if the applied voltage is small. This increases the overlap which affects the inverter limit, making the lead angle of firing less than the sum of the overlap and turn off angle. Commutation troubles normally arise if circuit turn off is less than the turn off time of the thyristor. For successful commutation there must be a marginal quenching angle.

Under ideal smoothing conditions of dc load current the voltage V_{dia} is dropped across the resistance and the superimposed ripple content across the load inductance.

A close examination of the primary and secondary currents of transformers shows that there is no mmf balance because of the dc component of current in the secondary winding. The dc mmf premagnetises the core. Each leg of the transformer carries a dc flux, passing mainly through air. A heavy magnetising current is required by the transformer. Also, when the primary is star-connected there exists a third harmonic flux besides the dc flux. This produces additional losses and consequent heating of the transformer. It also induces additional voltages in the transformer windings. This harmonic flux does not exist in a Δ connection.

This dc and the third harmonic core flux can be completely eliminated by using a zig-zag connection, as shown in Fig. 3.31. In this connection the current of a thyristor is made to flow through the windings on different legs, so that the dc magnetisation and third harmonic flux get cancelled. However the rating of this transformer is 8% higher than one with a normal conncction.

Even though the fundamental component of current decides the power used up in the dc side, the rating of the transformer must be decided using the total or actual current flowing through the winding.

RMS value of the secondary current $\dfrac{I_d}{\sqrt{3}}$.

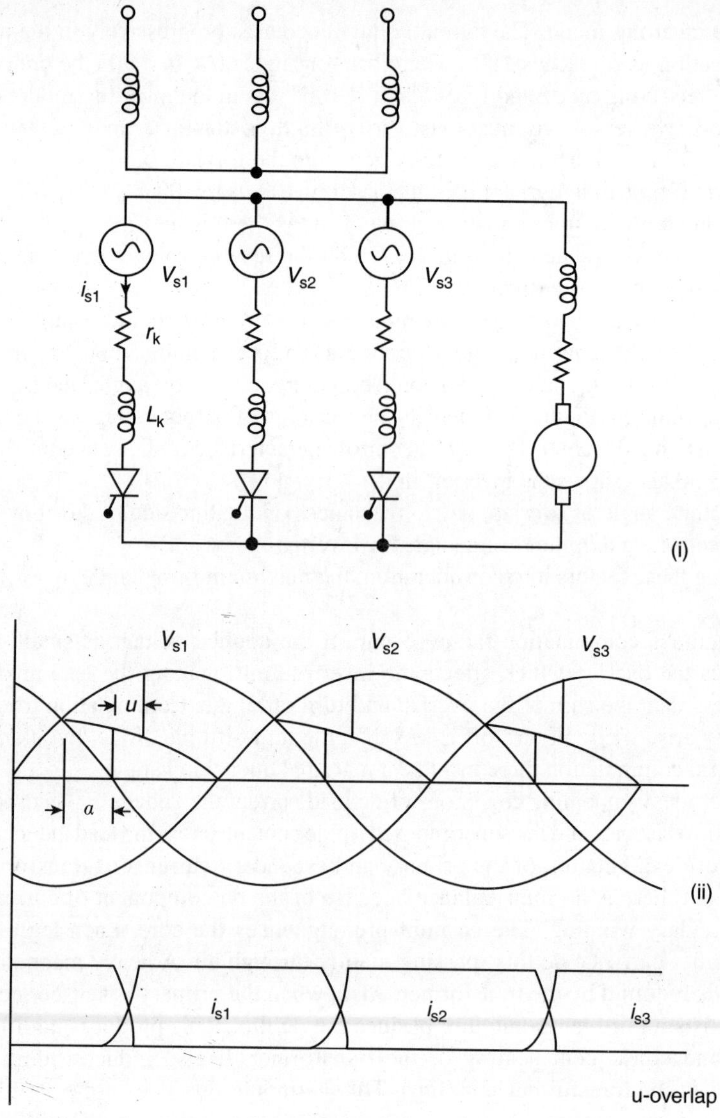

Fig. 3.30(a) *Effect of input reactance (i) Equivalent circuit of 3-pulse converter (ii) Voltage and current waveforms*

The rating of the secondary winding

$$P_s = 1.48P_{di} \tag{3.29}$$

where $P_{di} = V_{di} I_d$

The effective value of primary current $= \dfrac{\sqrt{2}I_d}{3}$

The rating of the primary winding = 1.21 P_{di}

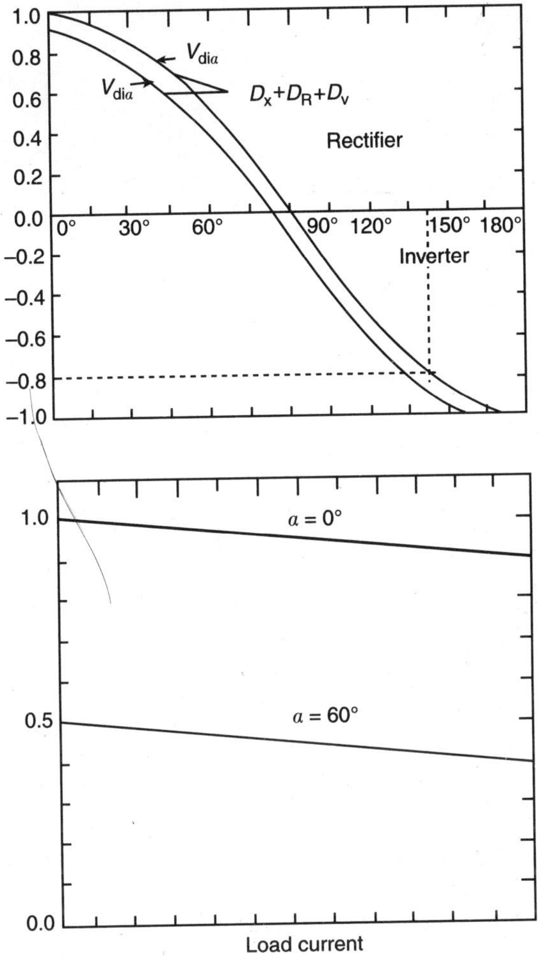

Fig. 3.30(b) *Control characteristics and voltage regulation of the converter*

$$\text{The design rating of the transformer} = \frac{1}{2}(P_s + P_p) = 1.35P_{di}$$

Overlap The commutation of the current from one thyristor to the other is never instantaneous. The incoming and outgoing thyristors conduct simultaneously during commutation. The period of simultaneous conduction is called overlap. The angle of overlap depends upon the transformer leakage reactance, line reactances and any other inductances in the circuit which limit the d*i*/d*t* of the thyristor. The overlap depends upon the load current and angle of firing. The dependence of overlap on angle of firing can be understood from the following equation

$$(1 - \cos u_o) = \cos a - \cos(a + u) \tag{3.30}$$

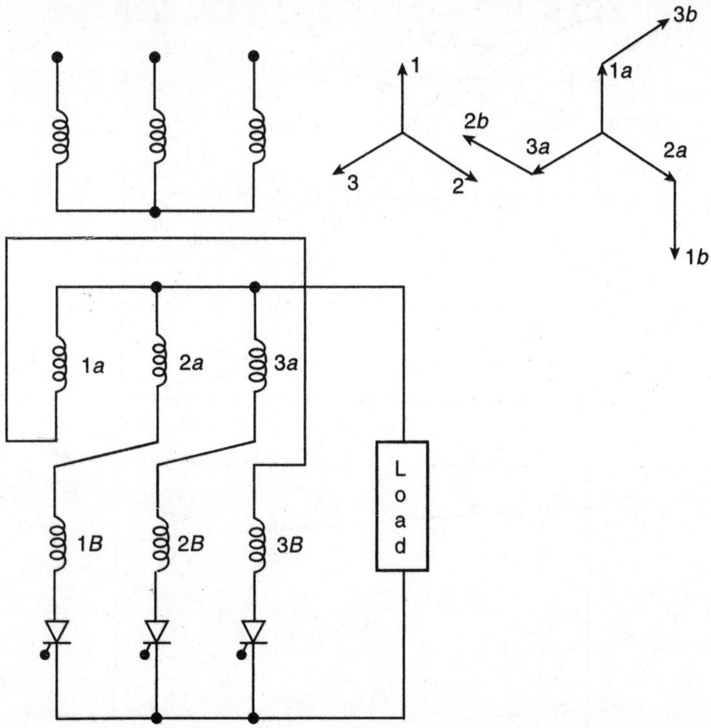

Fig. 3.31 *Zig-zag connection of secondary to eliminate premagnetisation*

where u_0 is the overlap angle at $a = 0$ and u is the overlap angle at a. The effects of total reactance (X_k) and total circuit resistance, R_k, which is negligibly small on the terminal voltage can be derived in general. Due to overlap there is a reduction in terminal voltage. This can be attributed to the voltage drops in the reactances when there is a change of current. The voltage regulation and associated equations can be determined using the equations given for single phase connections.

The voltage drop due to overlap

$$D_x = (V_{dio}/2)(\cos a - \cos(a + u)) \tag{3.31}$$

where $\cos(a + u) = \cos a - I_d/\sqrt{2I_k}$

with $\sqrt{2}I_k = \dfrac{\sqrt{2}V_s\sqrt{3}}{2X_k}$

$$D_x = \frac{I_d X_k}{2\pi/P}$$

The resistance drops and device drops in the forward direction also contribute to voltage regulation.

Therefore the terminal voltage

$$V_{da} = V_{di} \cos a - \frac{I_d X_k}{(2\pi/P)} - V_T \text{ — resistance drop} \quad (3.32)$$

or

$$V_{da} = \frac{V_{di}}{2}(\cos a + \cos(a + u)) - V_T \text{ — resistance drop}$$

The control characteristic, taking regulation into consideration, is shown in Fig. 3.30.

The mean dc voltage of the converter is superimposed by a non-sinusoidal ac voltage. It has a ripple of three times the supply frequency. The effective value of the rth harmonic referred to V_{di} is

$$\frac{V_{rda}}{V_{di}} = \frac{\sqrt{2}}{r^2 - 1}(r^2 + (1 - r^2)\cos^2 a)^{\frac{1}{2}} \quad (3.33)$$

$r = vP$ $v = 1, 2, 3$ and P is the pulse number. For $a = 0$

$$\frac{V_{rd}}{V_{di}} = \frac{\sqrt{2}}{r^2 - 1}$$

when the overlap is considered

$$\frac{V_{rda}}{V_{di}} = \frac{1}{2(r^2 - 1)}\sqrt{f_1(a) + f_2(u) + f_3(a, u)} \quad (3.34)$$

where $f_1(a) = 1 + \cos^2 a + r^2 \sin^2 a$

$f_2(u) = \sin ru \sin u$

$$f_3(a, u) = (r^2 - 1)\sin^2(a + u) + 2\cos ru(\cos a \cos(a + u)) + r^2 \sin a \sin(a + u))$$

Defining the ripple content as the ratio of the effective value of superimposed ac voltage to ideal dc voltage, (W_u) we have

$$W_u = \frac{\sqrt{V_{rda}^2}}{V_{di}} \quad (3.35)$$

For a three pulse converter, W_u is 18.3% for $a = 0$ and 65.5% for $a = 90°$.

As has already been discussed, the inductance L_d smooths out the ripple content in the load current and helps to make it continuous. Under ideal smoothing, the load voltage is dropped across the resistance and the ripple voltage across L_d. L_d is normally determined such as to avoid discontinuous conduction. For a three pulse converter

$$L_d = 1.26 \sin a \frac{V_d}{I_d} \text{ mH} \quad (3.36)$$

where a is the firing angle and I_d is the value at which the current becomes continuous.

The input current of a three pulse converter also has a harmonic content. The effective value of the rth harmonic is

$$I_{rL} = \frac{I_{1L}}{r} \tag{3.37}$$

which is 50%, 25% and 20% of the fundamental for $r = 2$, 4 and 5 respectively.

The effective value of the line current

$$I_L = I_{1L}\sqrt{1 + \Sigma \frac{1}{r^2}} = 1.21\, I_{1L} \tag{3.38}$$

The effect of overlap is to reduce the harmonic content

$$\frac{I_r}{I_{1L}} = \frac{1}{r} \cdot \frac{\sin ru/2}{ru/2} \tag{3.39}$$

The converter requires reactive power for phase control as well as for commutation. This reactive power is supplied from the lines and affects the power factor of the input current. The harmonics in the current waveform contribute to the reactive power. The fundamental displacement factor is the cosine of the angle between the fundamental of the input current and the voltage. This and the fundamental current contribute to the active power. The total power factor is affected by the harmonics. Therefore the fundamental displacement factor must be corrected properly to obtain this altered power factor. A close examination of the voltage and current waveforms reveals that the fundamental displacement factor is nothing but the cosine of the firing angle of the thyristors. For $a = 0$ the secondary phase voltage and the fundamental current are in phase and the displacement factor is unity. As a increases, this factor decreases. The effect of harmonics on the amount of reactive power can be considered by taking into consideration their effect on the power factors. The distortion factor g is defined as the effective value of the fundamental to the effective value of the total current (I_1/I). The total input power factor = $g \cos a$. For three pulse converters this is 0.827. As g is always less than one, the distortion decreases the total power factor.

The reactive power due to commutation can be determined using the following equation.

$$\frac{Q_1}{V_{di} I_d} = \frac{2u_o - \sin 2u_o}{4(1 - \cos u_o)} \tag{3.40}$$

The total reactive power can be determined as a ratio of the dc power at any control angle, using

$$\frac{Q_{1a}}{V_{di} I_d} = \frac{2u + \sin 2a - \sin 2(a + u)}{4(\cos a - \cos(a + u)} \tag{3.41}$$

The fundamental displacement factor taking overlap into consideration is approximately $\cos(a + u/2)$ or $\cos(a + 2u/3)$ depending upon $60° < a < 90°$ or $0 \le a \le 30°$.

Therefore, the total power factor $= g \cos(a + u/2)$ (3.42)

Transformer Utilisation in Controlled Rectifiers The expressions of device currents in midpoint converters discussed emphasize that the utilisation of devices as well as transformers decrease as the number of phases or pulses increases. The conduction angle decreases, adversely affecting the losses and current ratings of the devices. The ratio of the rms to average value of current increases, which increases the copper losses of the transformer. The utility of the transformer can be conveniently defined by the ratio of dc output power from the rectifier to the effective volt amperes of the transformer, or the design rating of the transformer, which is the average of the primary and secondary ratings of the windings. Under ideal conditions it may be unity, but is normally less than 1.

In the case of controlled rectifiers, the transformer utilisation decreases at reduced voltages in the same ratio as the power factor. This is evident from the voltage and current waveforms. We see that the firing angle variation introduces a phase shift in the current pulses, as compared to the uncontrolled case. This has no effect on the harmonics but changes the input displacement factor. The value of a also has no effect on the value of ac input current relative to dc output current. The reduction in power factor matches the reduction in output voltage and the necessary power balance is thus maintained. Therefore, at reduced voltage transformer utilisation reduces in the same ratio as the power factor.

Six Pulse Midpoint Converter The converter transformer converts the existing three phase system to a six phase one, since it has six secondary windings. The converter connections are shown in Fig. 3.32.

A thyristor having maximum voltage can go into conduction only if it has a firing pulse. The natural firing instant is the point of intersection of the two phase voltages in the sequence of phases. The firing angle of a thyristor is reckoned from this instant. The secondary voltages are shown in Fig. 3.32.

The current and voltage waveforms are shown in Fig. 3.32, and Fig. 3.33 assuming instantaneous commutation and ideal smoothing of dc. The average value of the dc voltage at the converter terminals is given by

$$V_{dia} = 1.35V \cos a \tag{3.43}$$

By varying the a, the output voltage can be varied steplessly. In the range of angles $0 < a < 90°$ the converter operates as a rectifier. In the range of angles $90° < a < 180°$ it operates as an inverter. There is a limit to the firing angle ($< 180°$) during inversion due to the reactances, as has already been discussed.

From Fig. 3.33, the effective value of the secondary current is given by

$$I_s = \left(\frac{1}{2\pi} I_d^2 \frac{\pi}{3}\right)^{1/2} = \frac{I_d}{\sqrt{6}} \tag{3.44}$$

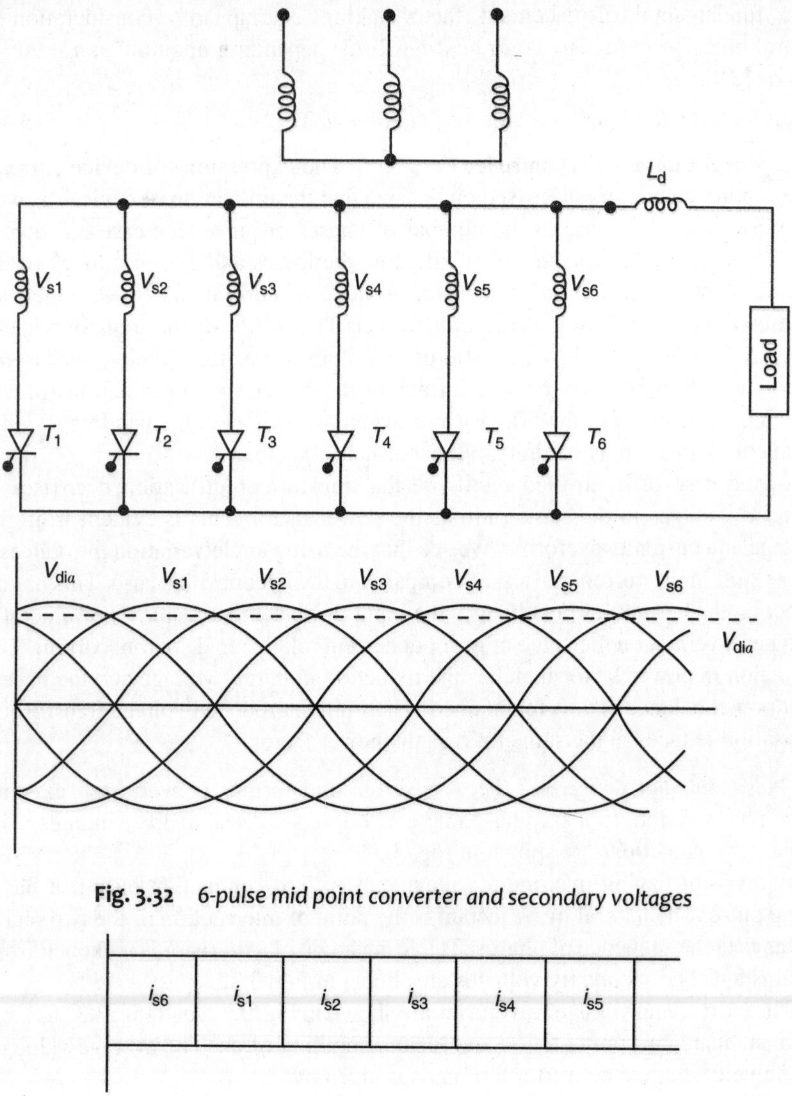

Fig. 3.32 *6-pulse mid point converter and secondary voltages*

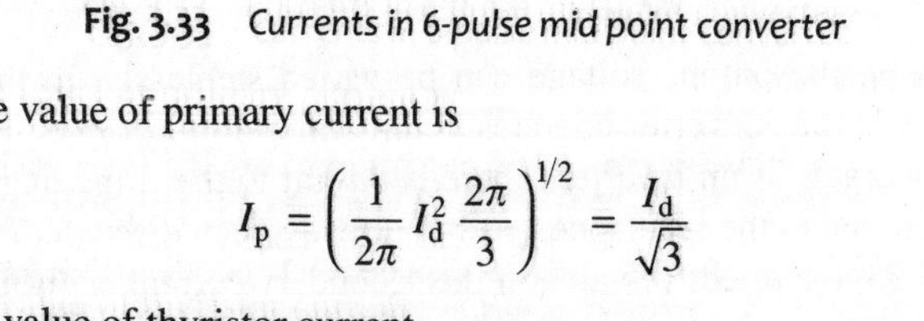

Fig. 3.33 *Currents in 6-pulse mid point converter*

The effective value of primary current is

$$I_p = \left(\frac{1}{2\pi} I_d^2 \frac{2\pi}{3} \right)^{1/2} = \frac{I_d}{\sqrt{3}} \tag{3.45}$$

The average value of thyristor current

$$= \left(\frac{1}{2\pi} \right) I_d \frac{\pi}{3} = \frac{I_d}{6} \tag{3.46}$$

The rating of the secondary

$$P_s = 6V_s \frac{I_{dN}}{\sqrt{6}} = \frac{6V_{dio}}{1.35} \frac{I_{dN}}{\sqrt{6}} = 1.81P_{di} \quad (3.47)$$

and the rating of the primary

$$P_p = 3V_s \frac{I_{dN}}{\sqrt{3}} = 1.28P_{di} \quad (3.48)$$

assuming a unity ratio transformer.

The design rating of the transformer

$$= 1/2(P_s + P_p) = 1.55P_{di} \quad (3.49)$$

The peak forward or reverse voltage of the thyristor $= \sqrt{2}V_s$.

The rms to average current ratio of the transformer is very large which is the main disadvantage of this connection. The thyristor conducts only for 60° in a period and blocks for the remaining 300°. Therefore the utilisation of the thyristor and secondary winding is somewhat poor.

The ripple content in the output voltage of the converter can be determined as a percentage of the maximum dc voltage using Eq. 3.13 It is 4.2% at maximum voltage ($a = 0°$, $a = 180°$) and 30.8% at zero voltage ($a = 90°$).

The load inductance, which serves the purpose of smoothing the output voltage and decreasing the possibility of discontinuous conduction is obtained by the equation

$$L_d = 0.296 \sin a \frac{V_{di}}{R} mH \quad (3.50)$$

Under ideal smoothing, the average output voltage is dropped across the resistance and the ripple content across the inductance.

The order of the harmonics present in the input current are 5, 7, 11, 13, etc. Their intensity is 20% for the 5th, 14.29%, for 7th harmonic etc. The total effective value of the rms current in the input as a percentage of the fundamental current is 104.72% and $g_1 = 0.955$.

The performance of the converter with respect to the power factor and reactive power is similar to that of a 3-pulse converter. The fundamental displacement factor is the cosine of the firing angle and the power factor is $g_i \cos a$. The 6-pulse converter gives a better performance than a 3-pulse converter.

The effects of overlap on the terminal voltage, harmonic content and inverter limit are the same as those in the case a three pulse converter.

The terminal voltage of the converter taking overlap into consideration is given by

$$V_{da} = V_{dio} \cos a - D_x \quad (3.51)$$

where

$$D_x = \frac{V_{dio}}{2}(\cos a - \cos(a + u))$$

with

$$\cos(a + u) = \cos a - \frac{I_d}{(\sqrt{2}I_k)}$$

Therefore

$$V_d = V_{dio}\cos a - \frac{V_{dio}}{2}\frac{I_d}{(\sqrt{2}I_k)} \tag{3.52}$$

substituting $I_k = \dfrac{V_k}{2X_k} = \dfrac{V_s}{2X_k}$

$$\frac{V_{dio}}{\sqrt{2}V_s} = \left(\frac{P}{\pi}\right)\sin\left(\frac{\pi}{P}\right)$$

We have

$$V_{da} = V_{dio}\cos a - \frac{I_d X_k}{(2\pi/P)}$$

Taking the resistance drop and device drop into consideration,

$$V_{da} = V_{di}\cos a - \frac{I_d X_k}{(2\pi/P)} - V_T - V_R$$

$$= \frac{V_{di}}{2}(\cos a + \cos(a + u)) - V_T - V_R \tag{3.53}$$

The harmonic content of the rth harmonic decreases in the ratio of $(\sin ru/2)/(ru/2)$ due to the overlap u. The dependence of the overlap angle on the firing angle is given by

$$1 - \cos u_o = \cos(a) - \cos(a + u) \tag{3.54}$$

The fundamental displacement factor is approximately $\cos\left(a + \dfrac{u}{2}\right)$.

The effect of u on the reactive power consumption can be determined using Eqs 3.40–3.41.

However, because of the disadvantages of poor utilisation of thyristor, transformer secondary winding (as effective thyristor conduction is only for 60°) and the high ratio of rms to average current of the transformer and thyristors, this convertor is rarely used.

Six Pulse Bridge Converter The conduction time of a thyristor in a six pulse connection can be increased when the converter is obtained by a series or parallel connection of two three-pulse midpoint converters. A six pulse bridge converter is obtained by connecting two three pulse converters in series on the dc side and in parallel on the ac side. The converter configuration is shown in Fig. 3.34. The three-pulse converter feeding current to the load is called the positive group and the other providing the return path for the current is called the negative group.

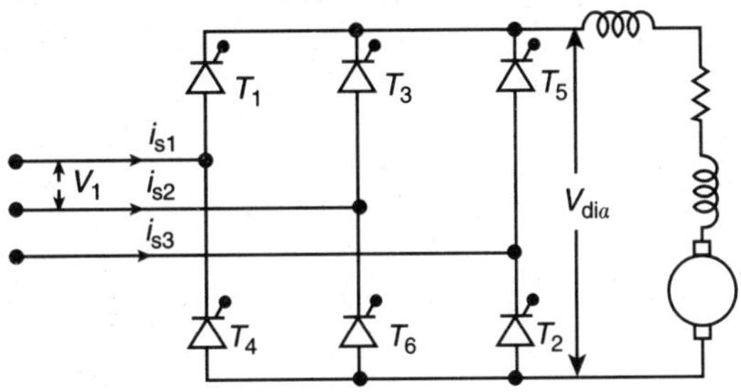

Fig. 3.34 *Six pulse bridge converter*

The star point is no longer necessary, and can be eliminated. There are some obvious advantages when compared to a three-pulse converter:

i. Because of series connection of the converters on the dc side, the mean output voltage is twice that of a 3 pulse converter for the same supply voltage. Consequently the power capability doubles.
ii. As no star point is required for the return path, a transformer can be avoided.
iii. The number of pulses is increased to six and the amplitude of the ac ripple is decreased.
iv. The dc component in the secondary of the transformer is completely eliminated. This aspect decreases the design rating of the transformer, if used.

Bridge connections are called two way circuits, since the transformer windings carry current in both directions. This is the reason for eliminating the dc magnetisation of core. On the other hand, midpoint connections have no such facility and thus are one way connections. The bridge connection has thyristors conducting for 120° which increases the utilisation of both the thyristors and the transformer, as compared to a six-pulse midpoint converter.

The current and voltage waveforms of three-pulse converters connected to form the bridge converter are shown in Fig. 3.35 for different values of the firing angle α. The output voltage of the bridge converter is the algebraic sum of the voltages of the component converters. The positive group has common cathode connection, and will have a thyristor with maximum positive anode voltage conducting. The negative group has a common anode connection and will have a thyristor with maximum negative cathode voltage, conducting. Thus the output voltage at the dc terminals has segments of three phase voltages. Due to the phase difference between the positive and negative group voltages, the output voltage

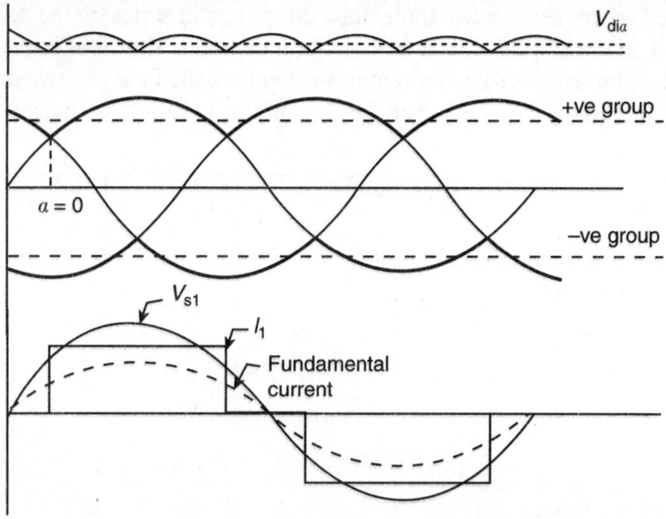

Fig. 3.35(a) *Voltage and current waveforms $a = 0$*

has a pulse frequency of 6f. The commutations occur alternately in the positive and negative groups. At all times, two thyristors, one in the positive group and the other in the negative group are in conduction. From Fig. 3.35 it is clear that the firing angle can be increased from 0 to 180°. Firing angle a = 0 is the natural firing instant, as has already been defined in the case of a three pulse converter. As the firing angle increases to 90°, the average dc voltage falls to zero. Power flow takes place from ac to dc and the converter is in the rectifying mode. When the angle increases further to 180°, the mean voltage increases from 0 to its maximum value with reversed polarity. If there is a dc source on the load side, power flow takes place from dc to ac and the converter is in the inversion mode. As has already been explained, there is an inverter limit, which is the maximum firing angle (< 180°), beyond which commutation failure occurs in the converter.

The mean voltage at the dc terminals can be shown to be

$$V_{dia} = 2.34V_s \cos a = 1.35V_L \cos a \tag{3.55}$$

where V_L is the line to line voltage. The control characteristic is shown in Fig. 3.36.

The peak forward or reverse voltage of a thyristor in a six-pulse bridge converter is the peak value of the line voltage, $\sqrt{2}V_L$. The average current carried by the thyristor is

$$\frac{I_d}{3} = 0.45\frac{V_L}{R}$$

The RMS current carried by the thyristor is

$$\frac{I_d}{\sqrt{3}} = 0.78\frac{V}{R} \tag{3.56}$$

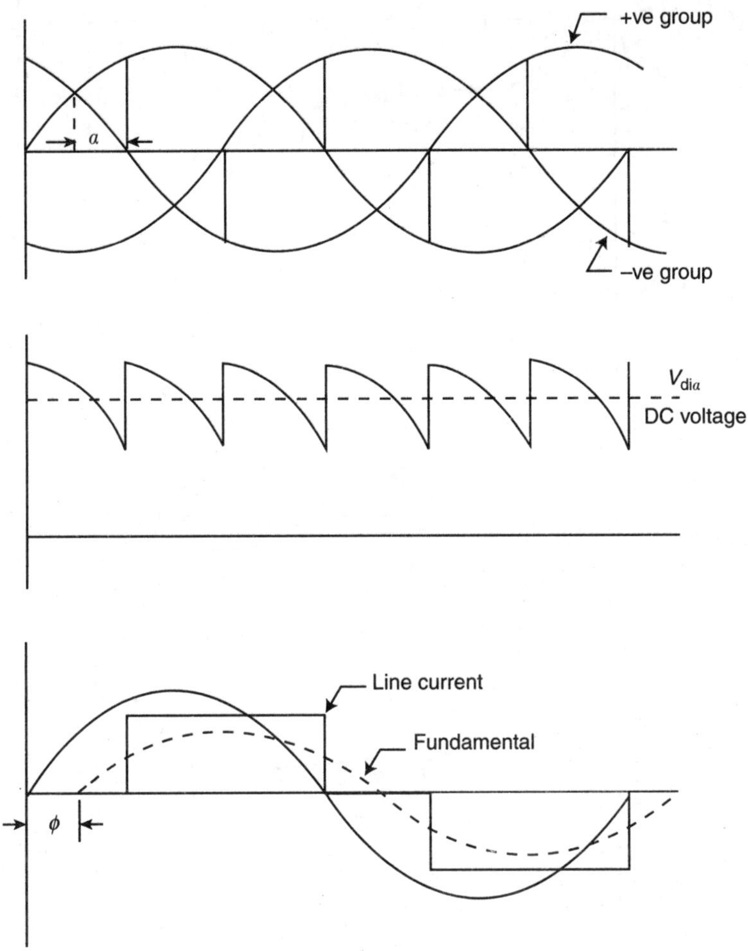

Fig. 3.35(b) *Voltage and current waveforms a = 30°*

The effective value of the secondary current is

$$I_s = \sqrt{\frac{1}{2\pi} I_d^2 \frac{4\pi}{3}} = I_d \sqrt{\frac{2}{3}}$$

$$P_s = P_p = 3V_s I_{sN} = 3\frac{V_{di}}{2.34} I_{dN}\sqrt{\frac{2}{3}} = 1.05 P_{di}$$

It is clear that a bridge circuit is superior compared to a three-pulse or six-pulse connection, in the design rating of the converter transformer.

Considering the effects of overlap, resistance and device drop, the terminal voltage of the converter can be determined in the same way as for a six-pulse converter. It is given by

$$V_{d\alpha} = V_{dio} \cos \alpha - D_x - 2V_T - V_R \tag{3.57}$$

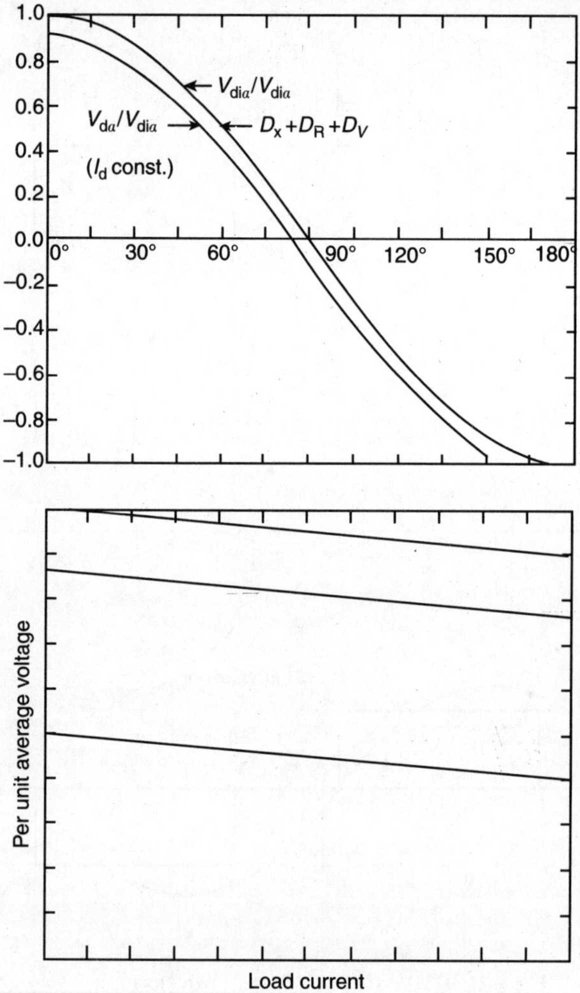

Fig. 3.36 *Control characteristics and voltage regulation*

where

$$D_x = \frac{1}{2} V_{dio}(\cos a - \cos(a + u))$$

with $\cos(a + u) = \cos a - \dfrac{I_d}{\sqrt{2} I_k}$

$$I_k = \frac{V_k}{(2X_k)} = \frac{\sqrt{2} V_L}{2X_k}$$

$$D_x = \frac{I_d X_k}{2\pi/3}$$

The ripple factor of the ac voltage, the smoothing inductance, harmonics in the line current, the power factor and the effects of overlap on these quantities can be determined in the same way as for a six pulse converter. The performance of the converter is shown graphically in Fig. 3.37.

Six channels of firing pulses are required, each separated by 60°. During continuous conduction a single pulse is sufficient, whereas in discontinuous conduction each thyristor must be gated for 60°, because an initial starting current is required.

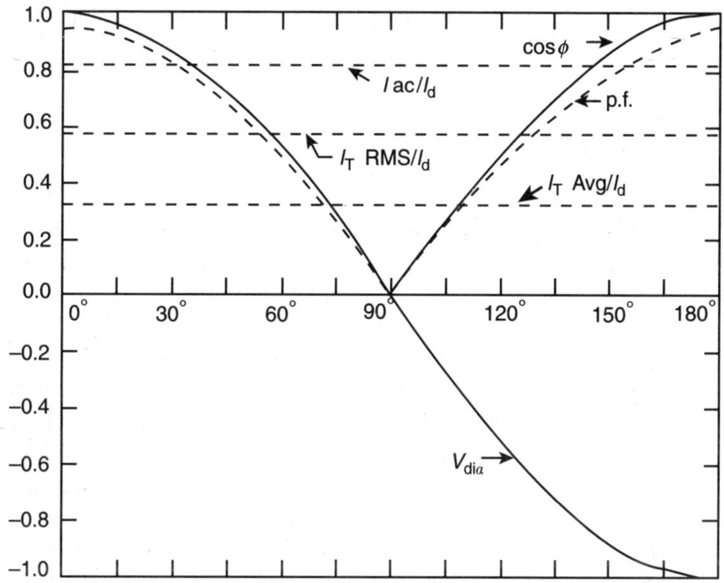

Fig. 3.37 *Performance characteristics of six pulse bridge converter*

Six Pulse Converter with Interphase Transformer The basic connection of a six pulse converter with an interphase transformer is shown in Fig. 3.38. This is obtained by connecting two three pulse converters in parallel on the dc side. The interphase transformer absorbs the difference of the instantaneous voltages of the two converters. The total load current is shared by the two converters and hence is the sum of the individual converter currents.

$$I_d = I_{dI} + I_{dII}$$

The output voltages of individual converters have a ripple content of frequency $3f$. Owing to the phase difference between the instantaneous values of output voltages, the net dc voltage has a reduced ripple content and a pulse frequency of $6f$. This means the output voltage is made up of six segments of dc

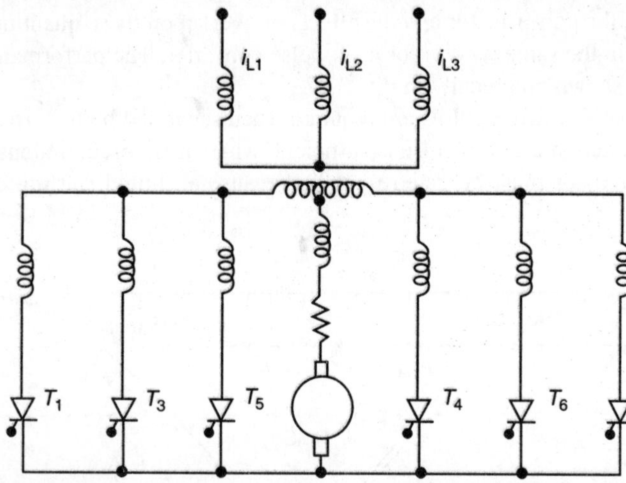

Fig. 3.38 *Six pulse converter with interphase transformer*

voltage. A thyristor conducts for 120° and there is, therefore, a good utilisation of the thyristor and transformers. The current and voltage waveforms for $a = 45°$ and $\alpha = 135°$ are depicted in Figs 3.39(a) and (b). From these figures it is clear that the average dc voltage of the converters is the same, but the instantaneous values are different. The latter differ because of the phase difference between the output voltages of the converters. The parallel connection of the two converters must be made through an interphase transformer, so that difference between the instantaneous voltages becomes a voltage drop across the reactor.

If V_{dI} and V_{dII} are the voltages of the individual converters and V_{tr} the voltage across the interphase transformer, we have (referring to Figs 3.38 and 3.39).

$$V_d = V_{dI} - \frac{V_{tr}}{2} \tag{3.58}$$

$$V_d = V_{dII} + \frac{V_{tr}}{2}$$

From these equations we have

$$V_d = \frac{V_{dI} + V_{dII}}{2} \tag{3.59}$$

This shows that the pulse number is six and the amplitude of the ripple is smaller than that of individual converters.

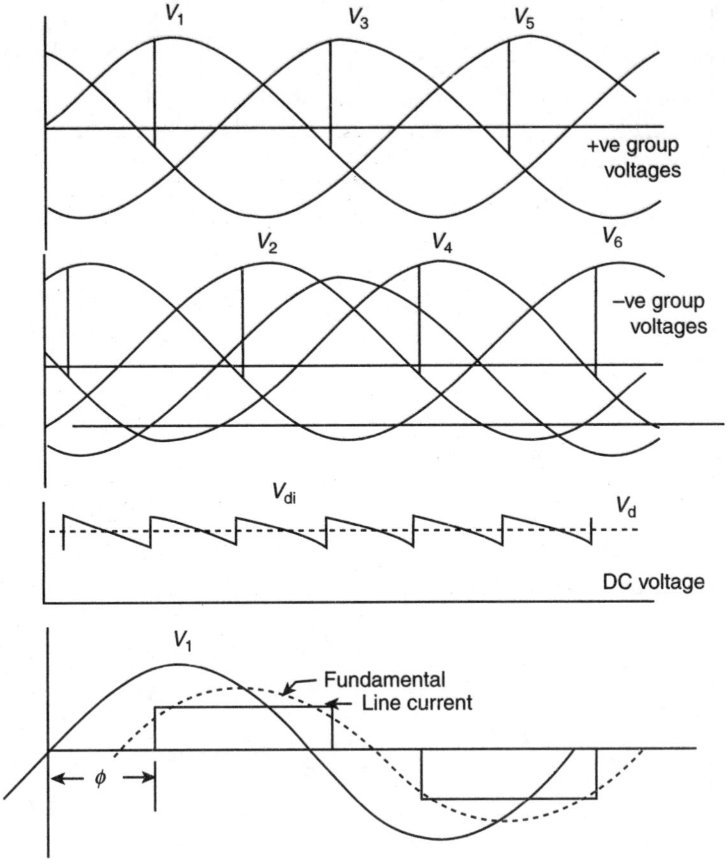

Fig. 3.39(a) *Voltage and current waveforms $a = 45°$*

From the equation we have,

$$\frac{V_{tr}}{2} = \frac{V_{dI} - V_{dII}}{2} \tag{3.60}$$

which is the voltage across one half of the interphase transformer. The voltage across the interphase transformer and the mean dc voltage are shown in Fig. 3.40 for $a = 90°$. The voltage across the interphase transformer is non sinusoidal. This results in a non-sinusoidal magnetising current of the transformer. It has a frequency of $3f$. The dc load current is superimposed by this magnetising current. On the other hand, for $\alpha = 0$ the magnetising current is sinusoidal. Because of the centre tap, the dc current flows simultaneously in each half of the transformer which avoids premagnetisation of the core. The peak value of the ac current must be very small compared to the load current, which should not fall below a critical value. This critical value is the peak value of the magnetising current. In case the load current is less than the critical value, the converter operates as a normal six

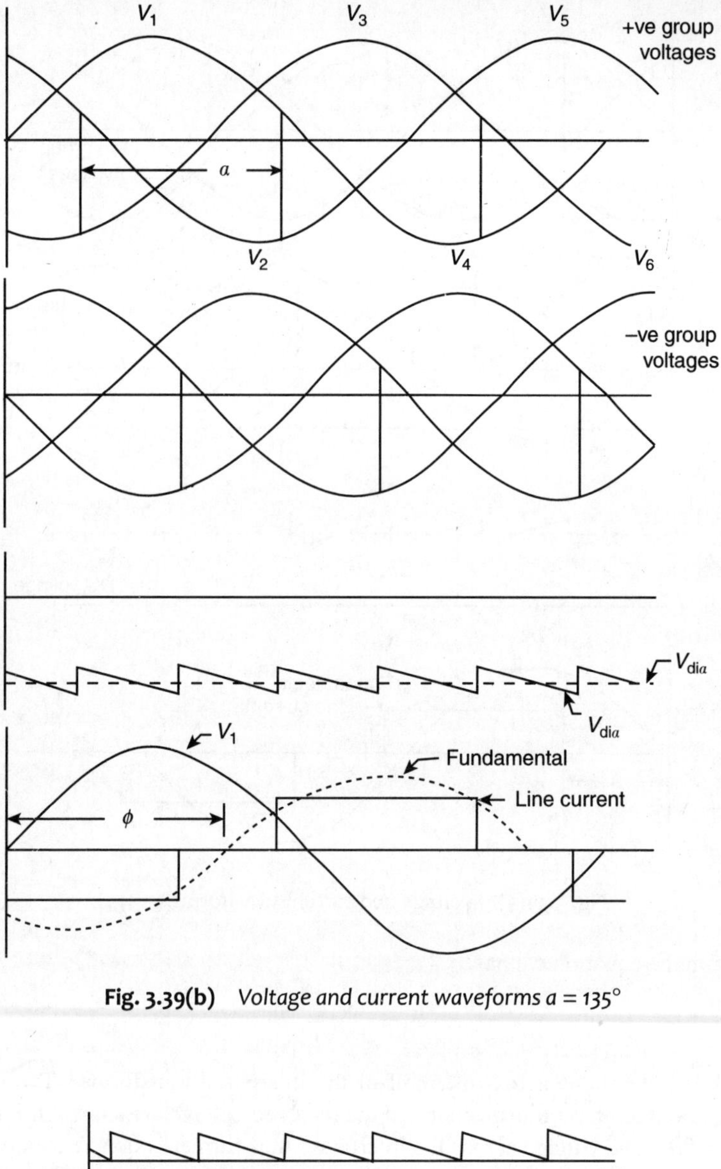

Fig. 3.39(b) *Voltage and current waveforms a = 135°*

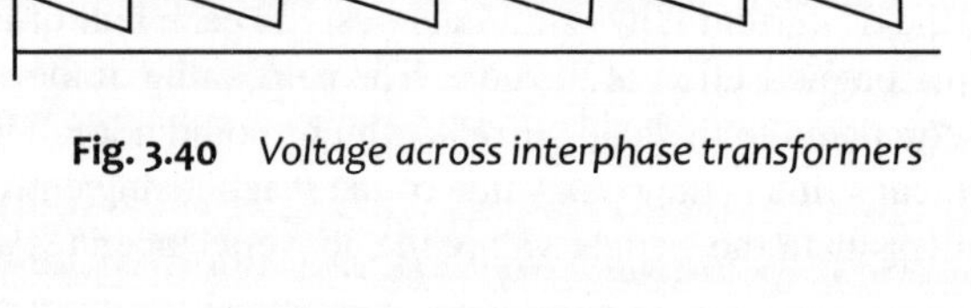

Fig. 3.40 *Voltage across interphase transformers*

pulse midpoint converter. Considering this fact, the mean value of the dc voltage of the converter is

$$V_{dia} = 1.35V_s \cos a \qquad I_d \le I_{cr} \tag{3.61}$$

$$V_{dia} = 1.17V_s \cos a \qquad I_{cr} \le I_d$$

The thyristors conduct for 60° in the former case and for 120° in the latter. Therefore, as the load current increases there is first a significant voltage drop of 15% due to transition, or a sudden rise if the load current is below I_{cr}.

The mean value of the dc voltage of the converter under ideal conditions with an interphase transformer is

$$V_{di} = 1.17V_L \cos a = V_{dio} \cos a \tag{3.62}$$

The converter operates as a rectifier in the range of firing angles $0 < a < 90°$ and as an inverter for $90° < a < 180°$.

Each converter group is an individual commutator group with three commutations independent of each other. There are six commutations, which are not simultaneous.

As has already been discussed, the terminal voltage falls when the thyristor conduction is increased to 120° using an interphase transformer. The terminal voltage of the converter is affected by the overlap in the same manner as in the other types of converters discussed previously. If the drop due to overlap be D_x we have

$$\cos u_o = 1 - 2D_x \tag{3.63}$$

and

$$2D_x = \frac{I_d X_k}{2\sqrt{2}V_s \sin(\pi/3)}$$

or

$$\cos(a + u) = \cos a - \frac{I_d}{\sqrt{2}I_k} \tag{3.64}$$

with

$$\sqrt{2}I_k = \frac{\sqrt{2}V_k}{2X_k} = \frac{\sqrt{2}V_L}{2X_k}$$

$$D_x = \frac{I_d X_k}{(2\pi/3)}$$

Other factors that influence the terminal voltage are the forward voltage drop of the device and the resistance drop. Therefore

$$V_d = V_{dio} \cos a - D_x - V_T - I_d R \tag{3.65}$$

The secondaries, carrying current in the same direction, are wound on the same core limb to avoid saturation.

$$\text{The peak value of the thyristor current} = \frac{I_d}{2} = 0.5I_d \tag{3.66}$$

Average value of the thyristor current = $0.167\ I_d$

RMS value of the thyristor current = $0.289\ I_d$

The peak forward or inverse voltage of the thyristor is $2.095\ V_{dio}$ or $\sqrt{2}V_L$. The reverse voltage occurs during inversion. The thyristors do not have a sudden rise of voltage in the forward direction.

Six channels of firing phases, separated by 60° each, with a single firing pulse in each channel are required.

The power factor, displacement factor, etc., describing the performance of the converter, are depicted in Fig. 3.41 in graphical form.

The secondaries share the load current. Two three pulse systems, each with half dc current have the secondary rating as a three-pulse connection. The rating of the secondary

$$P_s = 1.48P_{di} \tag{3.67}$$

The RMS value of the primary winding current is $I_d/\sqrt{6}$.

The primary rating is $P_P = 1.05P_{di}$.

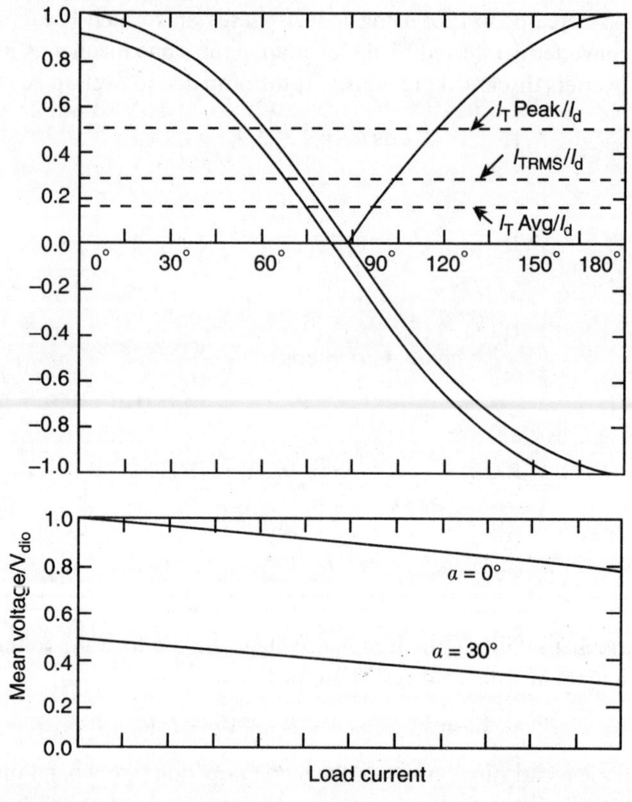

Fig. 3.41 *Performance of a 6-pulse converter with interphase transformer*

The design rating of the transformer is

$$\frac{1}{2}(1.48 + 1.05)P_{\text{di}} = 1.26P_{\text{di}} \tag{3.68}$$

The interphase transformer acts as a common smoothing inductance. The extra load inductance required may be small and there is little tendency to conduct discontinuously.

The interphase connection is used in applications where large dc currents are required at low voltage outputs.

Converters with Large Pulse Number It is clear from the preceding discussion that increasing the pulse number to six greatly improves the performance of the converter. The ac line current has only odd harmonics and the value of g increases to 0.96, indicating that the fundamental content is 96% of the total rms value of the current. A six-pulse converter is obtained by suitably connecting two three-pulse converters. The idea can be extended to increases the pulse number of the output voltage to 12 or 24 by suitable interconnections of six-pulse converters. It can be shown that connecting two six-pulse converters with 30° phase displacement results in a 12-pulse converter having an input current in which the lower fifth and seventh harmonics are absent. Also, increasing the pulse number decreases the output ripple which results in a reduced layout of smoothing inductance. The power factor improves, consequent to the improvement of g.

A 12 pulse converter, for higher voltage applications, is obtained by interconnecting two six pulse converters, as shown in Fig. 3.42. The input voltage to the

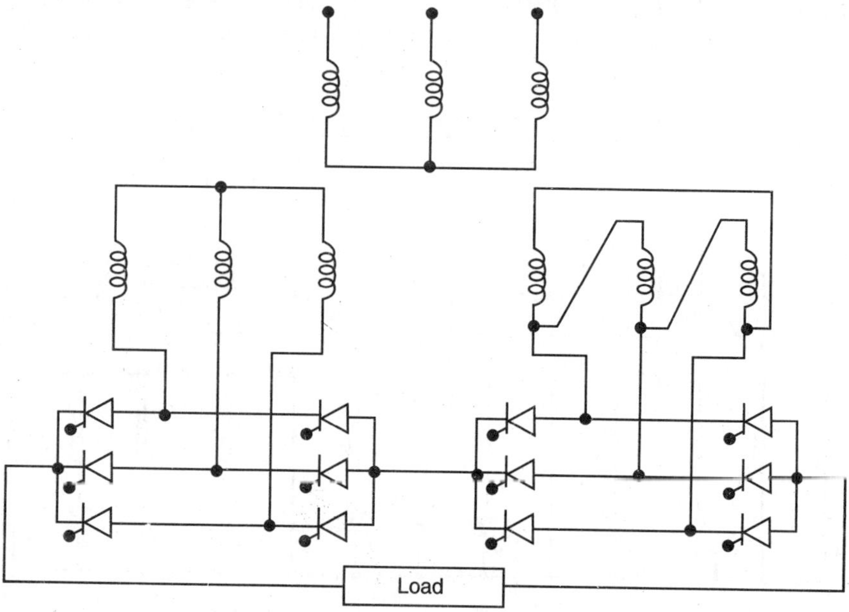

Fig. 3.42 *Twelve pulse converter*

converters should have a phase difference of 30°, which can be achieved in two ways.

i. The primary of one transformer is connected in star and that of the other in delta.

ii. The converter transformer has two secondaries, one of which is connected in star and the other in delta.

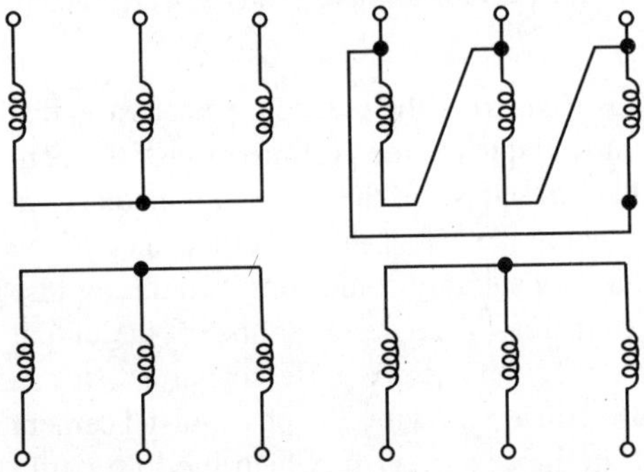

Fig. 3.43 *Alternative arrangement of transformer to achieve 30° phase difference between secondary voltages*

These connections are depicted in Fig. 3.43, 3.42.

Twelve pulse converters are also obtained by connecting two six pulse converters with an interphase transformer through another interphase transformer. 12 pulses are obtained due to the phase difference between the instantaneous values and output voltages of individual converters. The additional interphase transformer has almost the same design as other interphase transformer. The connections are shown in Fig. 3.44.

The advantages of 12 pulse converters are obvious. The ripple content of the ac voltage superimposing the mean value of the dc voltage is reduced greatly. The value of *g* increases (0.988) effectively improving the power factor. The line currents of 12 pulse converters are shown in Fig. 3.44(b). They are built up by the

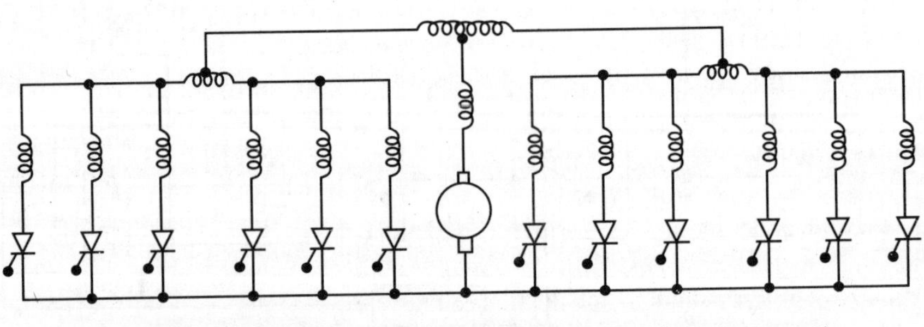

Fig. 3.44(a) *Twelve pulse converter in interphase transformer connection*

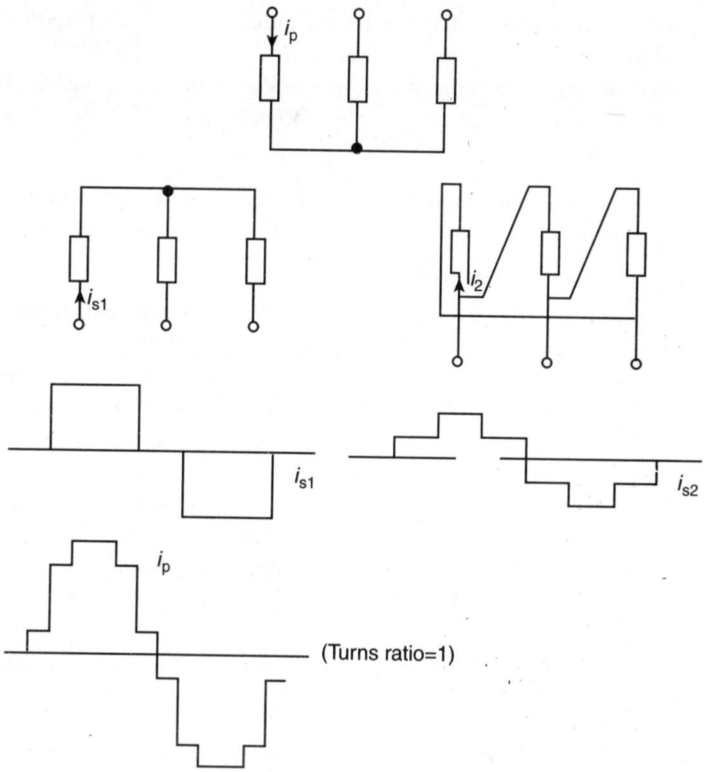

Fig. 3.44(b) *Line current of a twelve pulse converter*

individual primary currents shown in the figure. The harmonics present are $= 12K \pm 1$, $K = 1,2,3,\ldots$

Sometimes the advantages of increasing the pulse number are offset by the complexity of the transformer connection and by the difficulties of maintaining balance in the system. This happens when two 12 pulse converters are interconnected to form a 24 pulse one. Such difficulties do not encourage increasing the pulse number beyond 12 or at most 24.

General Comparison of Two Quadrant Converters The performance of various two quadrant converters can be compared. From the comparison, the following conclusions can be drawn:

i. The input power factor improves and the ripple content of the output voltage decreases as the number of pulses of the converter increases.
ii. The number of thyristors required in midpoint converters is just half the number required in a bridge connection. The current rating of the thyristors is half, whereas the voltage rating is twice that for a bridge converter.
iii. The reduction in the number of thyristors is normally offset by the cost of high voltage thyristors and the converter transformer. Therefore

midpoint converters may not be economical, when compared to bridge converters.

iv. The transformer is utilised well in a bridge converter. If the supply voltage, gives the output voltage appropriately, one can dispense with the transformer.

Midpoint converters are suitable for low voltage and high current use as the number of thyristors is a minimum. Bridge circuits are widely employed.

3.2.2 One Quadrant Converters

In certain drive applications, operation in the second quadrant, i.e. regeneration, is not necessary. The converter need not perform inversion. Such converters, operating only in one quadrant, are called one quadrant converters. They perform only rectification, i.e. convert ac to dc. When only one quadrant operation is required the converter can be made up of both diodes and thyristors. Normally half the number of devices are diodes and the other half are thyristors. They are therefore called half controlled or semi-controlled converters, and their overall cost is less.

One quadrant operation can also be obtained by having a free wheeling diode across the load, which can be used with both midpoint and bridge type converters. Semi controlled converters are possible only with bridge type converters. As we shall see later, the effectiveness of the free wheeling diode decreases as the pulse number increases. Compared to two quadrant converters, these have the following advantages:

i. The ripple content of the output voltage is decreased.
ii. The input power factor improves resulting in less reactive power consumption.

Two Pulse Half Controlled Bridge Converters A two pulse half controlled bridge converter is obtained by connecting a two pulse controlled midpoint converter in series with an uncontrolled one, as shown in Fig. 3.45. The controlled converter has thyristors T_1 and T_2 in a common cathode connection whereas the uncontrolled one has diodes D_1 and D_2 in a common anode connection. The former is a positive group and the latter a negative group. The bridge can also be obtained

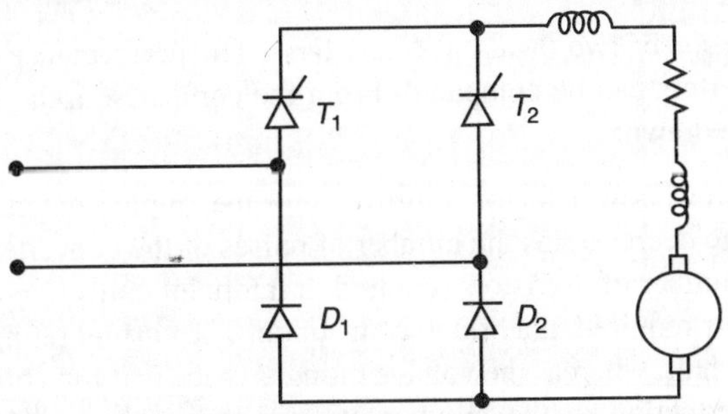

Fig. 3.45 *Two pulse half controlled bridge converter (symmetrical connection)*

with the uncontrolled converter in the positive group and the controlled one in the negative group.

The voltage of the controlled converter can be varied from positive maximum to negative maximum by changing the phase angle from 0 to 180°. The output voltage of the uncontrolled bridge is constant. The net voltage at the load terminals is the sum of the voltages of the controlled and uncontrolled groups. The average voltage varies from a maximum value to zero. The output voltage cannot reverse its polarity, due to the uncontrolled bridge. Only rectification is possible.

Taking the inverter limit of the controlled group into consideration, the phase angle can be varied from 0 to near about 180°. The average dc voltage varies from a maximum value to near about zero. Because of the inverter limit of the controlled converter, zero voltage is not possible.

The average value of the voltage at the dc terminals is given by

$$V_{dia} = \frac{V_{dio}}{2}(1 + \cos a) \tag{3.69}$$

where

$$V_{dio} = \frac{2\sqrt{2}V_L}{\pi}$$

The control characteristic is shown in Fig. 3.46, wherein the inverter limit is indicated.

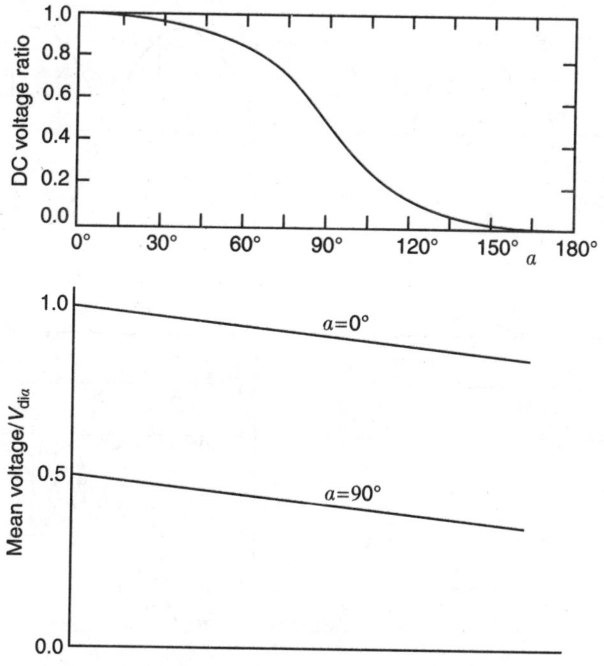

Fig. 3.46 *Control characteristic and regulation*

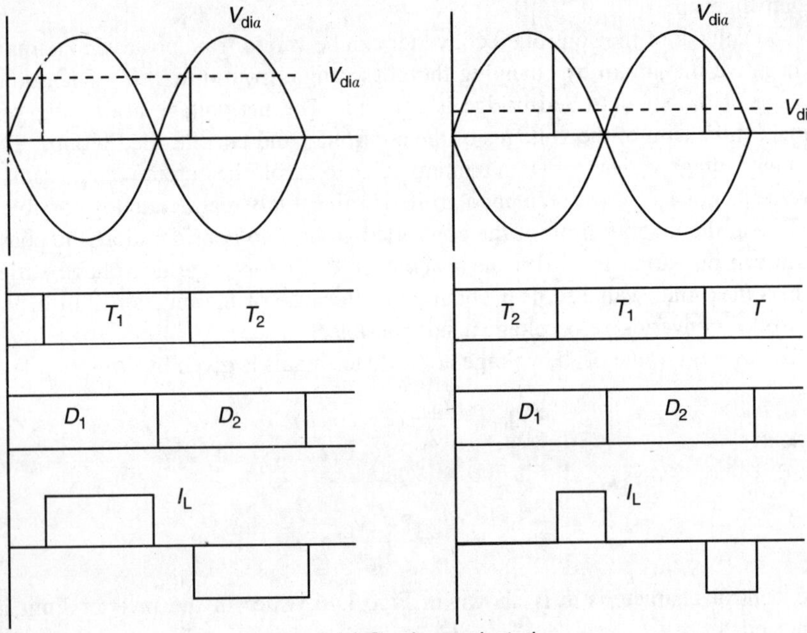

(a) Overlap neglected

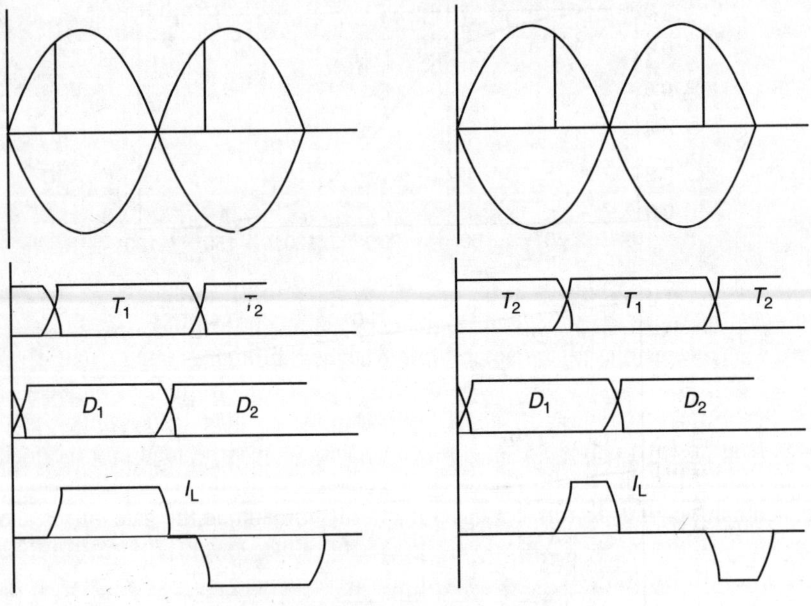

(b) Overlap considered

Fig. 3.47 *Voltage and current waveforms*

For inductive loads there is an inherent free wheeling action provided by the incoming diode and the reverse biased thyristor, until the incoming thyristor is turned on. Referring to Fig. 3.47, the thyristor T_1 and the diode D_2 conduct during the positive half cycle while T_2 and D_1 do so in the next half cycle.

There are also periods of time where T_1D_1 or T_2D_2 conduct simultaneously. The load current free wheels and mains are relieved from supplying the current when the devices placed diagonally opposite conduct, the supply voltage appears across the load. The current flows from the supply to the load. When the devices connected in series conduct, the load voltage is zero and the load current free wheels.

To start with let us assume that T_1 and D_2 are conducting. The portion of the positive half cycle appears across the load (a to π). At $\omega t = \pi$, D_1 gets forward biased and the current transfers from D_2 to D_1 D_2 to T_1 continues to conduct till T_2 is fired at appropriate firing angle $[a + \pi]$. Current transfers from T_1 to T_2. Now T_2 and D_1 conduct. The supply voltage appears across the load. Thus the thyristor is in conduction for 180°. There is an inherent free wheeling action because of the diodes, and this makes the voltage across the load zero. The negative excursions of the load voltage are prevented by the uncontrolled bridge.

The output voltage in the working range is affected by the device drops in the forward directions, as well as the resistance drops. The reactances in the line side of the converter are responsible for overlap. The current transfer takes place in a definite amount of time. The rate of change of current combined with the reactances causes a voltage drop. This can be determined in the same way as for a fully controlled converter. Therefore

$$V_d = V_{da} - D_x - I_d R - (V_T + V_D) \tag{3.70}$$

where

$$V_d = \frac{V_{dio}}{2}(1 + \cos a); \qquad D_x = \frac{V_{dio}}{2}(\cos a - \cos(a + u)) \tag{3.71}$$

For very small firing angles and highly inductive loads, it is not possible to turn off the converter in this connection by a sudden removal of firing pulses. The last thyristor fired remains in conduction indefinitely, making the input voltage appear across the dc terminals every other half cycle. For example, assume that at some instant of time T_1 and D_2 are conducting and instant the firing pulses are removed. When T_1 is reverse biased, the natural free wheeling through T_1 and D_1 maintains the current in the load and the load voltage is zero. If the current has not decayed to zero during this time, the thyristor T_1 gets a forward voltage and conducts as if $a = 0$. The positive half cycle of voltage appears across the load and the load current flows through T_1 and D_2 for the complete half cycle. This phenomenon is called half waving. To turn off the converter is to maintain the gate pulses and retard the firing angle such that the load voltage is reduced. Consequently the load current falls below the holding current of the thyristor.

The elimination of the negative excursions of dc voltage due to natural free wheeling is advantageous, in that the superimposed ac ripple on the average dc voltage decreases with reduced smoothing equipment.

It can also be observed from the current waveforms that the period of conduction of the input current pulses decreases as the firing angle is retarded, i.e. output voltage is reduced. Thus the reactive power requirements of a half controlled bridge are considerably less than those of the fully controlled one at reduced voltage. The rms value of the input current

$$I_s = I_d\sqrt{1-\frac{a}{\pi}} \tag{3.72}$$

The fundamental displacement factor is cos $a/2$. The saving in control reactive power is considerable, as the free wheeling action takes place even at $a = \pi$. However there is no saving in commutation reactive power.

The periods of current flow in the thyristor and diode are equal. The peak forward or reverse voltage of the devices is equal to

$$\sqrt{2}V_L \quad or \quad \frac{V_{dio}}{2\pi}$$

The harmonic components of the input current depend upon the control angle. They have different spectrum at different a. Therefore the value of g depends upon a. The total power factor is

$$\lambda = g\cos a/2 = \frac{2\sqrt{2}}{\pi}\frac{\cos^2 a/2}{\sqrt{(1-a/\pi)}} \tag{3.73}$$

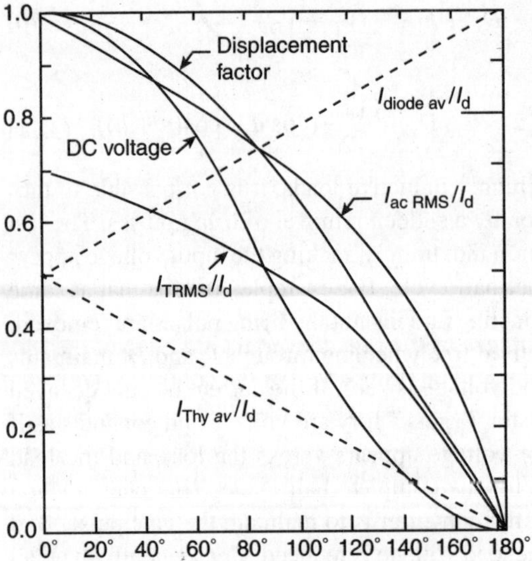

Fig. 3.48 *Performance characteristics*

The ripple content in the output voltage is reduced because of the inherent free wheeling action, which prevents the negative excursions of the output voltage. The amount of smoothing inductance in the load circuit is small it is only 57% of the value required with a fully controlled converter. The performance of the converter is summarised in Fig. 3.48.

The other possible half controlled two pulse bridge connection is as shown in Fig. 3.49. This is a symmetrical connection. The voltage and current waveforms of the converter are shown in Fig. 3.50.

The free wheeling is provided by the diodes D_1 D_2. The thyristor T_1 and diode D_2 conduct in the positive half cycle and T_2 and D_1 in the negative half cycle.

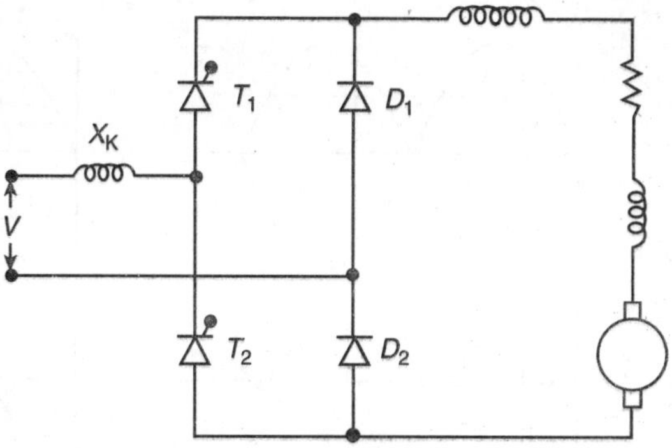

Fig. 3.49 *Two pulse half controlled asymmetrical bridge converter*

At the end of the positive half cycle, D_2 acquires a positive voltage and D_1D_2 free wheel the load current. This free wheeling prevents the negative portions from appearing in the output voltage. The operation and behaviour of the circuit is the same as the previous circuit as far as the output voltage, power factor, etc. are concerned. However there are two differences. First, the half waving of the circuit is not present. From the waveforms of the current, it can be seen that as a increases the period of conduction of the thyristor decreases and that of the diode increases. The devices can, therefore, be rated accordingly depending upon their maximum period of conduction. The period of the input current pulses decreases as a is delayed.

Half controlled 2-pulse converters of large power find application in tram cars in which single phase power is rectified to feed the traction motor. In these applications one tries to improve the reactive power saving without giving much importance to regenerative braking.

Three Phase Half Controlled Bridge Circuit This is obtained by a series connection of a 3 pulse controlled converter and a 3 pulse uncontrolled one. The three arms of the former consist of thyristors and the three arms of the latter comprise diodes. A typical circuit is shown in Fig. 3.51. Here the controlled converter is shown with a common cathode connection and forming the positive group. The uncontrolled one has a common anode connection and forms the negative group. It is also possible to have the converter with connections the other way around. The thyristors commutate at the phase angle at which they are fired. The diodes commutate at the natural firing instant $a = 0$. The thyristors conduct for 120° and are fired at intervals of 120°. The output voltages of the two converters are added to get the net output voltage at the dc terminals. The firing angle of the controlled converter ranges from 0 to 180° (under ideal conditions). The output voltage is varied from a positive maximum to a negative maximum. The average value of the voltage for the uncontrolled converter is fixed at the maximum value of the controlled one. The net voltage at the dc terminal varies from a positive maximum to zero. In practice, due to the inverter limit of the controlled converter, the voltage cannot go to zero.

A thyristor and a diode conduct at any given time the diode is forward biased at the natural firing instant. A thyristor conducts, even if it is reverse biased, until the next thyristor in the sequence is fired. Thus there is a natural free wheeling of the load current via the incoming diode and outgoing thyristor. The load voltage

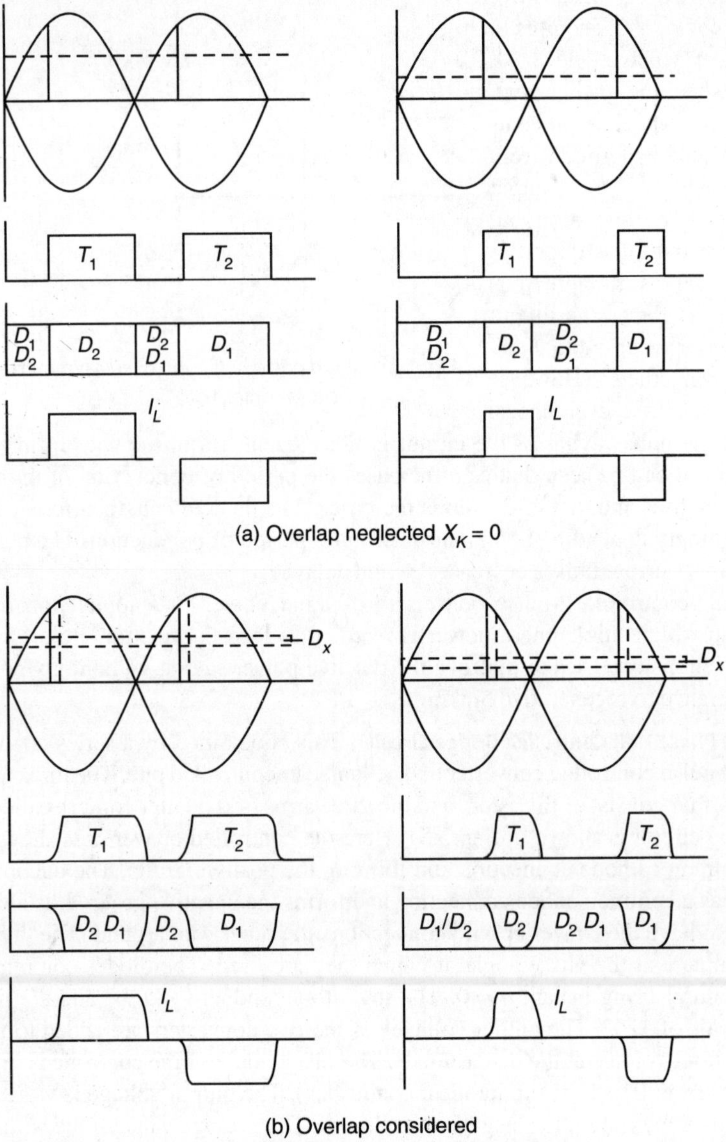

Fig. 3.50 *Voltage and current waveforms of a half controlled asymmetrical two pulse circuit*

is zero during the free wheeling period. Free wheeling due to conduction of diode does not allow negative excursions of load voltage. This reduces the ripple content in the output voltage. The ripple frequency of the output voltage at $a = 0$ is $6f$. For $a < 60°$, free wheeling does not take place since the voltage is always positive on the dc side. The negative instantaneous value does not occur. Free wheeling

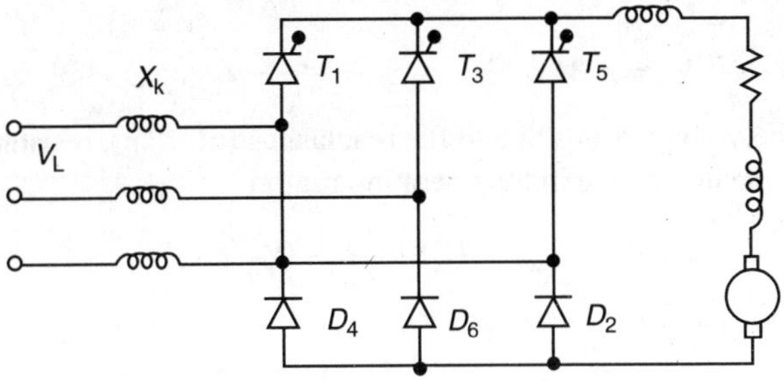

Fig. 3.51 *Three-phase half controlled bridge circuit*

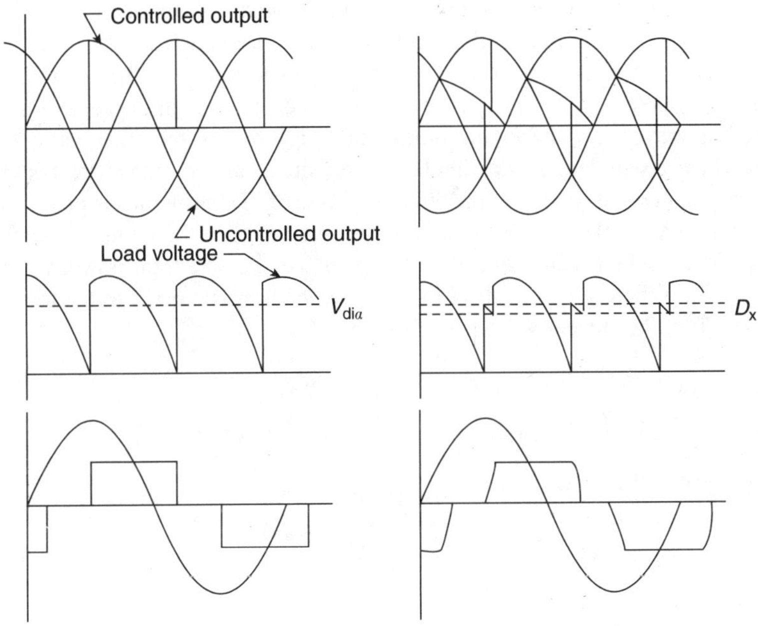

Fig. 3.52 *Voltage and current waveforms of half controlled 3-phase bridge converter $a = 60°$*

takes place only when $a \geq 60°$. The ripple frequency decreases to $3f$ at these firing angles ($a \geq 60°$). The ripple voltage is less at $a = 60°$ and it increases for $a > 60°$. Compared to a fully controlled converter, the smoothing inductance required is large at $a = 90°$, even though there is inherent free wheeling.

The average values of dc voltage can be obtained as (Fig. 3.52)

$$V_{dia} = \frac{V_{dio}}{2}(1 + \cos a) \text{ for } 0° \leq a \leq 180° \tag{3.74}$$

where $V_{dio} = \frac{3\sqrt{2}}{\pi} V_L = 1.35V_L$

considering the voltage drops due to the reactances (overlap), resistances and the device drops, the dc voltage in the operating region

$$V_a = V_{dia} - D_x - I_d R - (V_T + V_D) \tag{3.75}$$

where $D_x = V_{dio}/2\ (\cos a - \cos(a - u))$

$$\cos a - \cos(a + u) = \frac{I_d}{\sqrt{2}I_k} = \frac{2I_d X_k}{(2\pi/3)}$$

The voltage and current waveforms of the converter at different firing angles are shown in Fig. 3.52. The line current waveforms show that the period of the current pulse decreases in the line as the firing angle is increased. The effective values of the fundamental as well as the harmonics depend upon the firing angle. The value of g depends on the firing angle; it is not constant, as in the case of a fully controlled converter. All the harmonics can be referred to the mean value of dc current. Hence with larger delay angles, the effective value of line current is much smaller than at $a = 0$. However the effective value of the harmonics is much larger at greater delay angles, and becomes a large percentage of the input current.

The fundamental displacement factor is $\cos(a/2)$. The total power factor is $g\cos(a/2)$. Where g is the ratio of the fundamental rms current to the total rms current. From the current waveforms, we get

$$\lambda = (3/\pi)\cos(a/2) \quad 0 < a < 60° \tag{3.76}$$

$$\lambda = (\sqrt{6/\pi})\ (\cos^2 a/2)/\sqrt{1 - a/\pi} \quad 60° < a < 180°$$

The rms value of current

$$I_s = I_d\sqrt{2/3} \quad 0 \le a \le 60°$$

and

$$I_s = I_d\sqrt{1 - a/\pi}\ 60° < a \le 180° \tag{3.77}$$

There is a saving of control reactive power but not of commutation reactive power. The power factor as a function of V_{dia}/V_{dio} is depicted in Fig. 3.53. An improvement in the power factor can be seen in the range $0° < a < 180°$. The value of g is the same as for a fully controlled converter in the range $0 < a < 60°$.

The peak forward and reverse voltage of the thyristors and diodes is $\sqrt{2}V_L$ where V_L is the rms line to line voltage. The thyristor current is $0.45\ V_L/R$. The rms value of the thyristor and diode currents is

$$\frac{I_d}{\sqrt{3}} = 0.78\frac{V_L}{R} \tag{3.78}$$

For highly inductive loads and small firing angles the converter shows the phenomenon of half waving. This can be prevented by means of the FWD across the load.

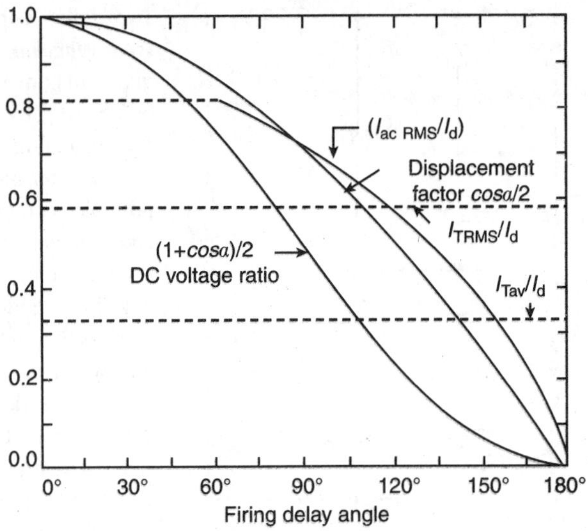

Fig. 3.53 *Performance of three phase half controlled bridge circuit*

General Features of Half Controlled Converters When only unidirectional applications are involved it is advantageous to use half controlled converters as they provide the following special features over two quadrant converters:

i. The converters are economical as half the positions are occupied by diodes.
ii. The firing circuit provides signals only to half the number of thyristors and therefore are simple and less costly.
iii. The performance of the converter on the line improves as the power factor improves. The control reactive power is less. This is because the period of conduction of the input current pulse decreases as a increases. The input current is zero when the voltage is zero. Thus the reactive power requirement becomes less. There is no saving in commutation reactive power.
iv. The voltage variation is between a maximum value and (near about) zero when the firing angle varies from 0 to 180°. However the inverter limit does not allow a to equal 180° and hence the voltage cannot go to zero. Zero output voltage can be obtained by supplying the component converter with different voltages. The converter transformer having two secondaries of appropriate turns feeds the converters. The voltage supplied to the controlled portion is 10% larger than that supplied to the uncontrolled one.
v. The amplitude of ripple decreases and hence the amount of smoothing inductance required is less. This is because natural free wheeling does not allow negative excursions of voltage.
vi. The ripple frequency is half that of a fully controlled converter (3-phase converters).

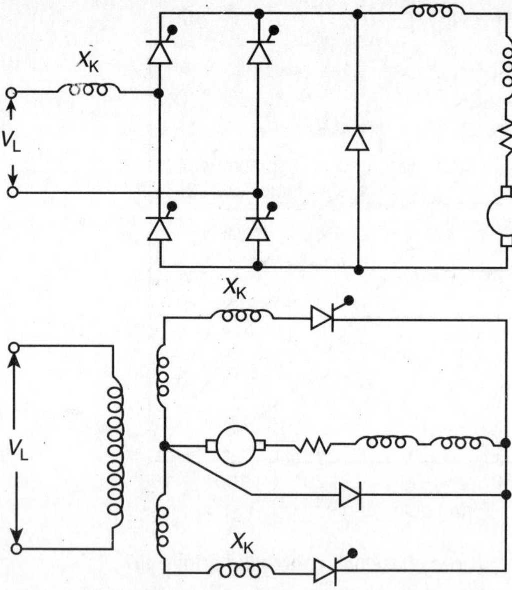

Fig. 3.54 *Two pulse converters with free wheeling diode*

Two Pulse Converters with Free Wheeling Diode One quadrant converters can also be obtained by connecting a free wheeling diode across the load terminals. This is possible with both midpoint and bridge converters, as shown in Fig. 3.54. The free wheeling diode provides a local path for the load current when the thyristors are reverse biased. The thyristor conducts only for (π – a), instead of π, similar to the case of an unsymmetrical half controlled bridge. The free wheeling of the load current does not allow negative swings of the load voltage, and the load voltage is zero. The dc output voltage is

$$V_{dia} = \frac{V_{dio}}{2}(1 + \cos a) \tag{3.79}$$

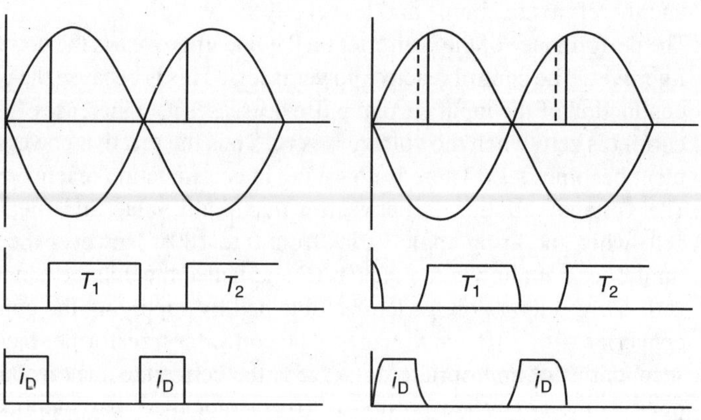

Fig. 3.55 *Voltage and current waveforms of 2-pulse converters with FWD*

The absence of negative swings in the output voltage improves the output voltage as far as its ripple is concerned. The associated smoothing inductance is also small. The reduction in the conduction angle of the input current ripple, as the firing angle increases, decreases the total reactive power demand. The commutation reactive power does not, however, change. The voltage and current waveforms are shown in Fig. 3.55.

The behaviour of these converters is same as that of symmetrical and unsymmetrical bridge connections. The free wheeling diode is effective in the complete range of firing angles, $a = 0$ to 180°.

Three Pulse Converter with Free Wheeling Diode A three pulse converter with a free wheeling diode across the load terminals is depicted in Fig. 3.56. The diode is meant to provide an alternative path for the load current, which would otherwise have flown through the converter. The diode becomes positively biased when the load voltage becomes negative, and starts conducting. At this time the thyristor of the converter is reverse biased and stops conduction. Thus, negative swings of the load voltage are avoided and it is zero during free wheeling.

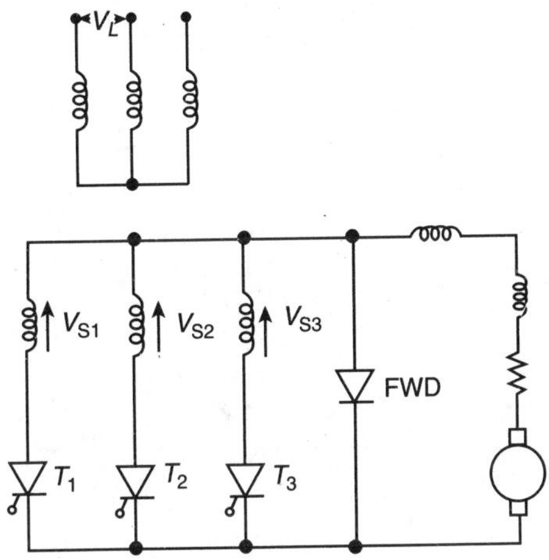

Fig. 3.56 *Three pulse converter with free wheeling diode*

Inversion is not possible. In the converter, the FWD becomes effective only for those firing angles for which the load voltage tends to become negative. For firing angles, up to $\pi/6$, the diode does not get a forward voltage as the load voltage has no tendency to become negative. For a 3 pulse converter the diode is effective for $a > \pi/6$.

Therefore $\quad 0 < a < \pi/6 \quad V_{dia} = V_{dio} \cos a \quad (3.80)$

$$\pi/6 < a < 5\pi/6 \quad V_{dia} = \frac{V_{dio}}{\sqrt{3}}(1 + \cos(a + \pi/6))$$

The effective variation of a is only from 0 to $5\pi/6$ The control characteristic is shown in Fig. 3.60. Until the free wheeling diode takes part in conduction, i.e. $0 < a < \dfrac{\pi}{6}$, the converter operates as a two quadrant one and there is thus no reactive power saving in this range of angles.

The terminal voltage is affected in the working range by the drops in the input reactances, resistance and device drops.

The free wheeling diode improves the performance in the same way as in the case of a half controlled rectifier. The input current pulses have a conduction angle which decreases with increasing *a*. During free wheeling, no current is drawn from the mains when the load voltage is zero. This helps in reducing the control reactive power requirement. The current in mid point converters flows from line to neutral. The input power factor and harmonics are the same as in the case of half controlled converters. The overlap due to commutation and the necessary reactive power for commutation is the same as in the fully controlled case. The advantages

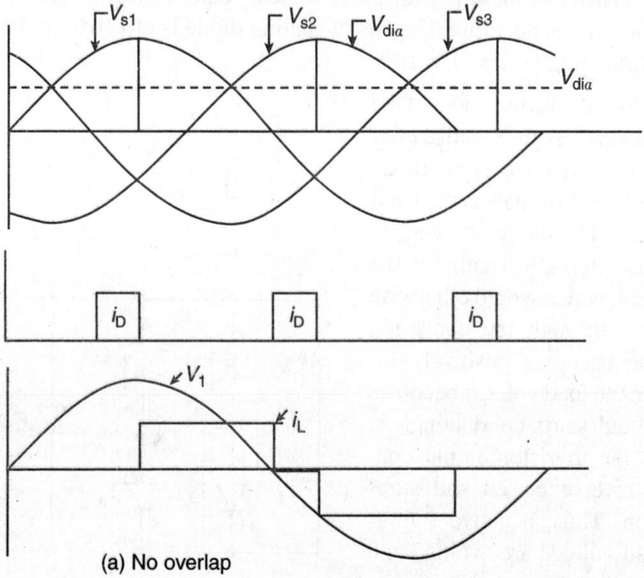

(a) No overlap

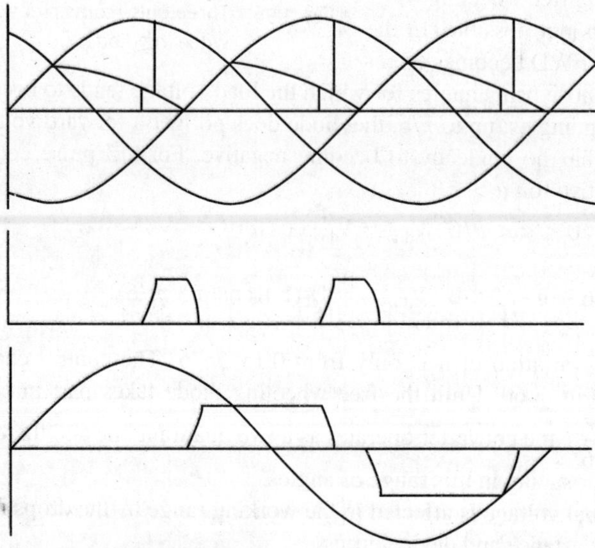
(b) With overlap

Fig. 3.57 *Voltage and current waveforms of a three-pulse converter with FWD* of the converter are available only when the diode participates in conduction, i.e. $a \geq 60°$. The voltage and current waveforms are shown in Fig. 3.57.

Bridge Rectifier with Free Wheeling Diode A 6 pulse bridge rectifier with a free wheeling diode across the load terminals is shown in Fig. 3.58. As has already been explained, the diode participates in conduction only when the instantaneous load voltage tends to become negative. This occurs for firing angles greater than 60°. For $a \leq 60°$ the bridge operates as a fully controlled one and the mean output voltage is

$$V_{dia} = \frac{3\sqrt{6}}{\pi} V_{ph} \cos a = \frac{3\sqrt{2}}{\pi} V_L \cos a = V_{dio} \cos a \tag{3.81}$$

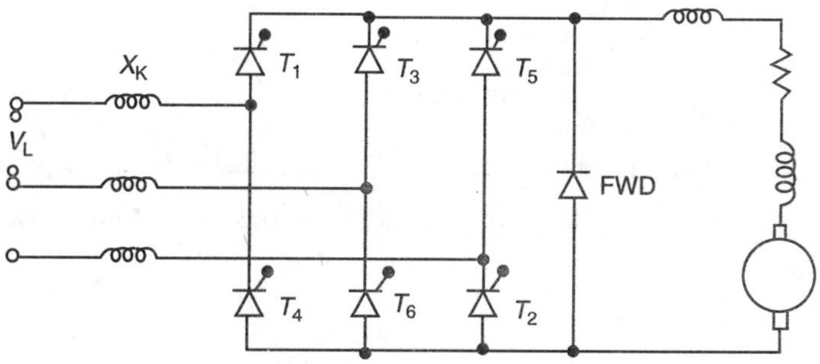

Fig. 3.58 *Three phase bridge rectifier with FWD*

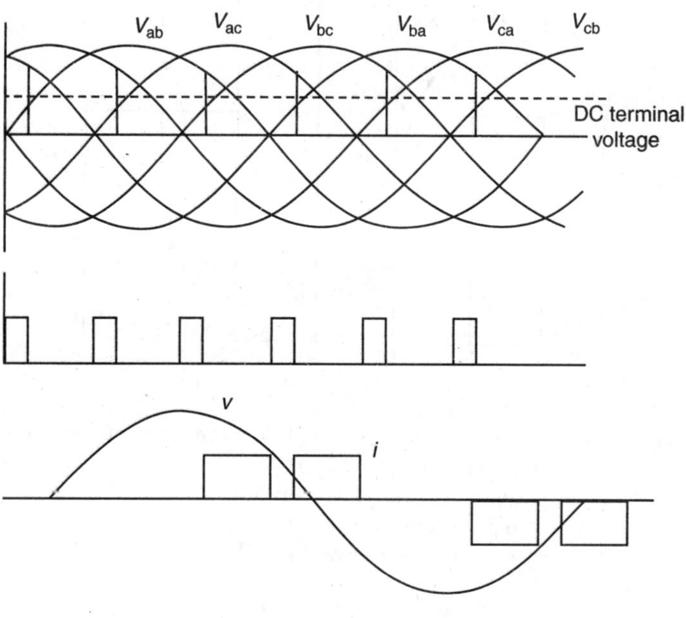

Fig. 3.59 *Voltage and current waveforms of three phase bridge rectifier with FWD* $\alpha = 75°$

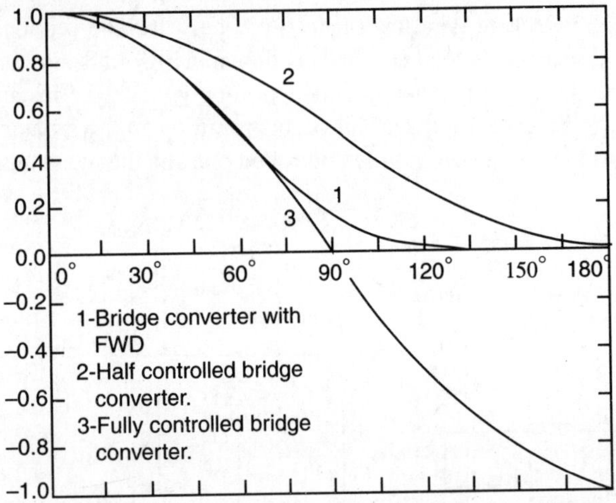

Fig. 3.60(a) *Control characteristics of bridge type converter with FWD.1) two-pulse*

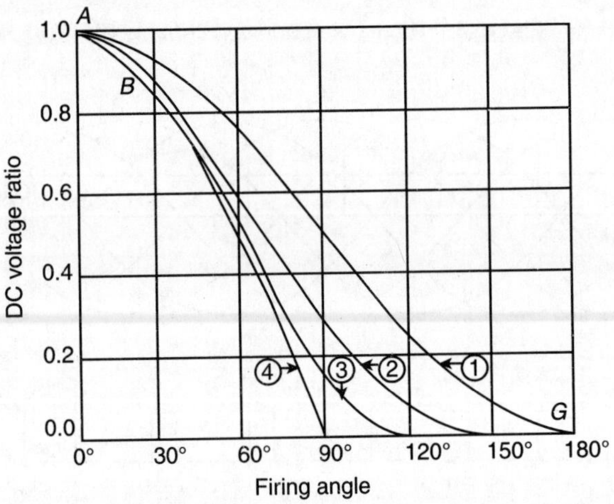

Fig. 3.60(b) *Control characteristics of bridge type converter with FWD*

1. 2-pulse converter with FWD
2. 3-pulse converter with FWD
3. 6-pulse converter with 2 FWD
4. 6-pulse converter with 1 FWD

For a > 60° and < 120° the free wheeling action of the diode does not permit negative excursions of the instantaneous voltage and hence the average value of voltage is

$$V_{dia} = \frac{3\sqrt{6}}{\pi} V_{ph}(1 + \cos(a + \pi/3)) \tag{3.82}$$

The control characteristic is shown in Fig. 3.60. There is a distinct difference between the output voltage of a half controlled converter and that of a fully controlled one with FWD. But for this difference, the improvement in the power factor, reactive power requirement, etc. during free wheeling are the same. These advantages are present only during the free wheeling action.

It is clear from the above discussion that as the number of pulses increases, the advantages of FWD get sharply reduced. For example, in a six pulse converter, with a view to saving reactive power, FWD is effective only after $a = 60°$. On the other hand, in two pulse converters it is effective in the complete range of firing angles, from 0 to 180°. The control reactive power is reduced but not the commutation reactive p‹ wer. The dc side voltage harmonics are smoothened, thereby reducing the amount of smoothing equipment. The ripple content of the unsmoothed dc voltage is reduced and a continuous dc is available with small smoothing inductances, thereby avoiding discontinuous conduction.

Series Connection of Converters—One in the Fully Controlled Mode and the other in the Uncontrolled Mode The converter connections are shown in Fig. 3.61. Operation is similar to that of a half controlled bridge circuit. Actually the principle of a half

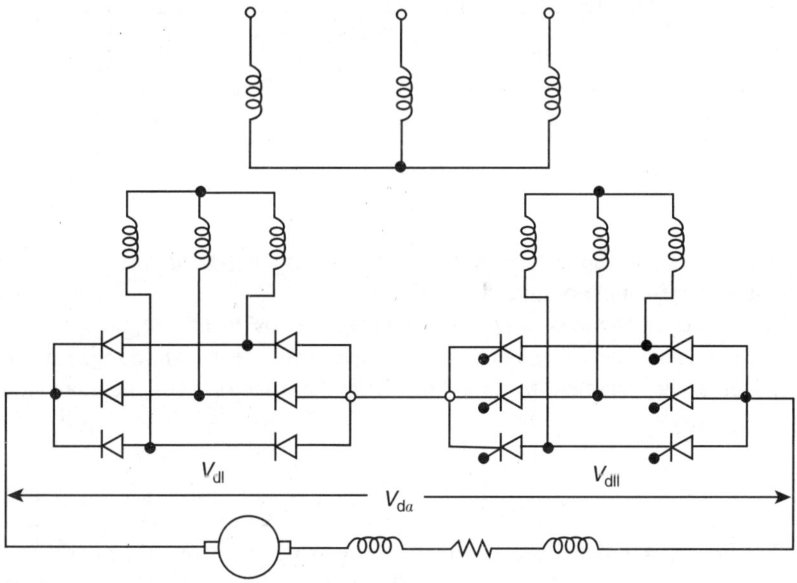

Fig. 3.61 *Series connection of fully controlled and uncontrolled converters*

controlled bridge circuit is extended to obtain the present circuit, by connecting two individual bridge circuits in series, one in the fully controlled mode and the other in the uncontrolled mode. This connection is preferred when high dc voltages are required. The output voltage

$$V_{dia} = \left(\frac{V_{dio}}{2}\right)(1 + \cos a) \tag{3.83}$$

$\frac{V_{dio}}{2}$ is the output voltage provided by the diode bridge and $\frac{V_{dio}}{2}\cos a$ is the voltage provided by the thyristor bridge. Also

$$\frac{V_{dio}}{2} = \frac{3}{2\pi}V_L = \frac{3\sqrt{3}}{2\pi}V_{ph} \tag{3.84}$$

The control characteristic is shown in Fig. 3.62. Voltage variation up to zero is not possible, due to the inverter limit of the controlled converter. Zero voltage can be obtained by supplying the two converters with different voltages, the supply

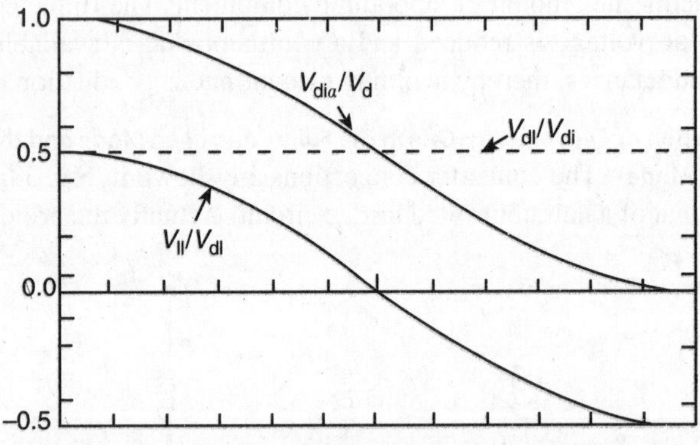

Fig. 3.62 *Control characteristic of converter in Fig. 3.61*

voltage to the controlled one being 10% greater. The inverter limit is fixed so that there is a margin angle of quenching of 10°.

The converter provides some saving in reactive power. The output voltage has a reduced ripple. The voltage and current waveforms are shown in Fig. 3.63. The fundamental displacement factor as seen from the waveforms on the input side is $\cos(a/2)$, i.e.

$$\cos\phi = \cos(a/2) = \sqrt{\frac{V_{dia}}{V_{dio}}} \tag{3.85}$$

The input harmonics depend upon the firing angle. Therefore the improvement in the total power factor is not quite the same as that in the displacement factor. The total power factor as a function of firing angle is shown in Fig. 3.64.

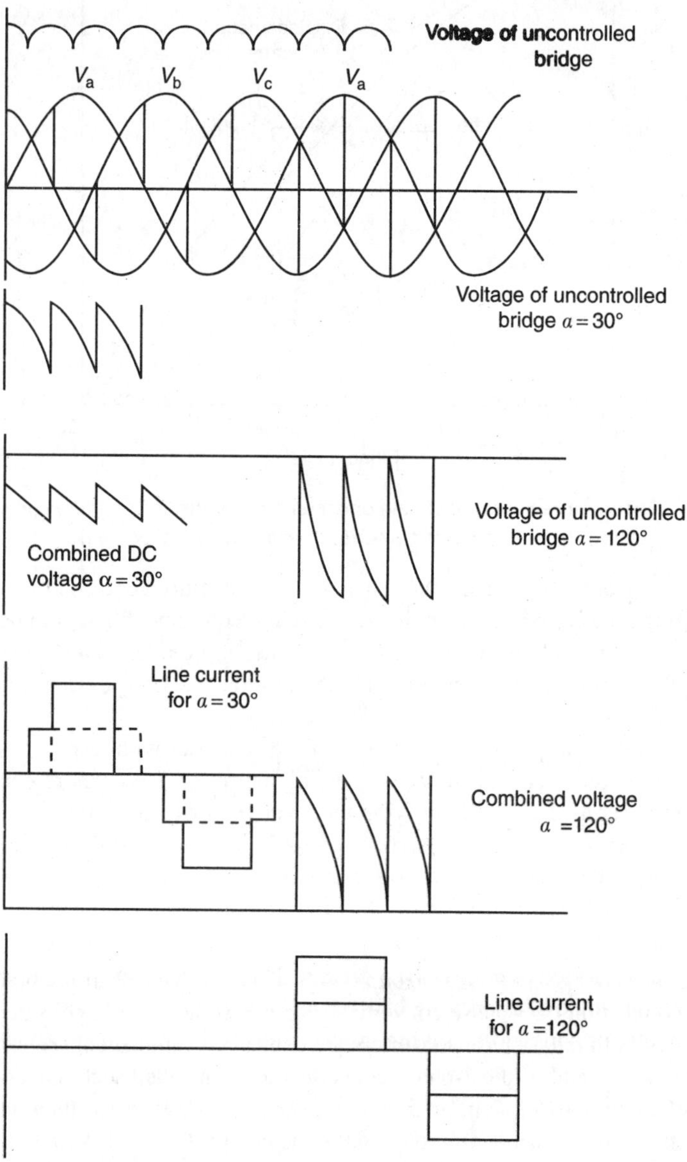

Fig. 3.63 *Voltage and current waveforms of series connected fully controlled and uncontrolled bridges*

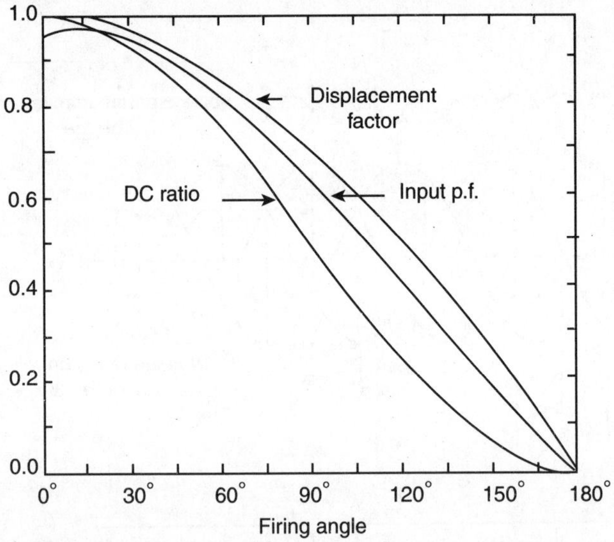

Fig. 3.64 *Performance curves of series connected half controlled six pulse converter (fully controlled converter in series with an uncontrolled one)*

Sequence Control of Converters We have shown that fully controlled converters take lagging reactive power from the mains. This requirement is less in half controlled converters and converters with a free wheeling diode, which however do not provide the process of inversion. In both cases the reactive power requirement increases as the angle of firing is retarded.

It is possible to improve the behaviour of the converter with respect to control reactive power requirement, by the sequential control of two or more converters connected in series. Besides improving the reactive power consumption converters with sequential control can also be used for high voltages at the output.

When several converters are connected in series and controlled one after another, it is called sequence control. The reactive power requirement of the converters in this method is considerably less than the sum of each individually.

1. Series Control of 3-phase Bridge Circuits A 3-phase bridge circuit is obtained by a series connection of 3-phase converters. For $a = 0$ the dc voltage is zero. The thyristors of both the 3-phase converters are controlled simultaneously.

On the other hand, if the 3-phase converters are controlled such that converter I operates as an inverter (Fig. 3.65) and converter II as a rectifier, then when the firing angle of the former is 180° and that of the latter is 0°, the dc voltage $V_{dia} = 0$. The individual converters require zero reactive power.

The converter operates as a rectifier if converter II is uncontrolled ($a = 0$) and converter I has a transition from an inverter to a rectifier. Only converter I requires reactive power for control. When inverter operation is required, converter I is always operated as an inverter and the firing angle of converter II is retarded from 0°.

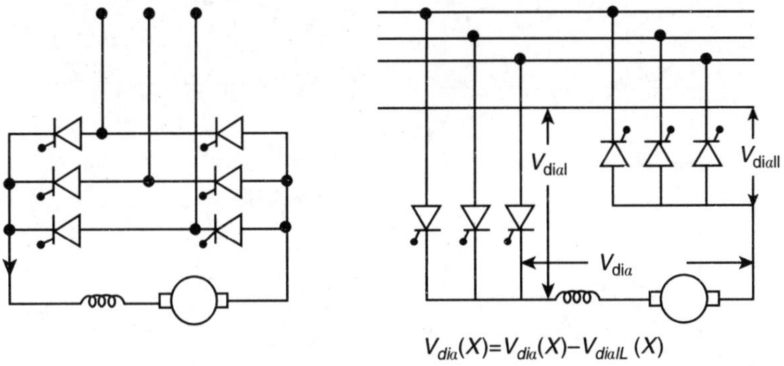

Fig. 3.65 *Three-phase bridge employing sequence control*

In sequence control, the reactive power never exceeds the value of the reactive power required when a single fully controlled converter is used. When two are used in series, the total reactive power is just half the reactive power required by one converter.

The operation, with respect to reactive power becomes more comfortable if the number of series connected converters increases. The ratio of the reactive power required with sequence control, to the reactive power required with a converter decreases. In view of the resulting complexity, however it is not advisable to use more than two converters in series for sequence control.

Saving is possible only for control reactive power and not for commutation reactive power. As the number of series connected converters increases the commutation reactive power increases, even though there is a saving in control reactive power.

The inverter limit should be given due consideration.

Two bridge type converters can also be used for sequence control. The individual bridges may have different initial overlap angles. The dc voltage at the output terminals of 3-phase bridge converters in sequence control is shown in Fig. 3.66a for rectifier operation and Fig. 3.66b for inverter operation.

2. *Series Connection of Two Bridge Converters with Sequence Control of Both* This arrangement has a behaviour similar to that of bridge circuit with sequence control of individual converters. It has an advantage of reduced ripple content because of 6-pulse operation.

3. *Series Connection of Two Bridge Converters with Sequence Control of Individual Converters* The arrangement is depicted in Fig. 3.67. The bridges are supplied from a transformer having two secondaries, each supplying one bridge. The reactive power requirement of this arrangement is less than that of a single converter having sequence control in its halves. The converter should be controlled with a definite margin angle of quenching, which decides its inverter limit. It is not worthwhile having more than two converters in series.

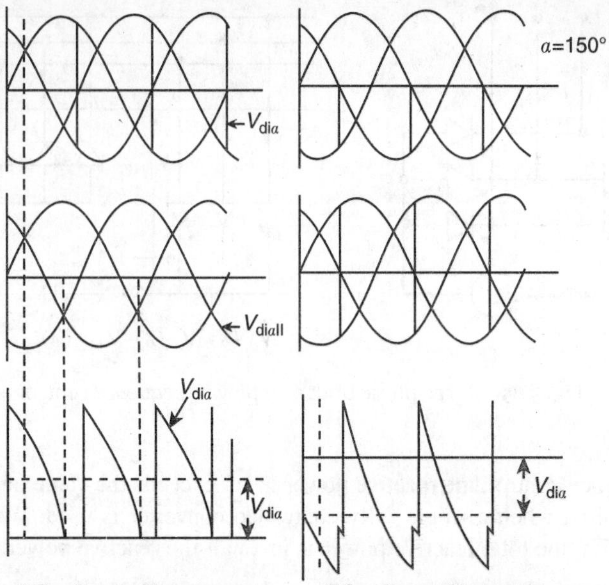

DC voltage with sequence control as a rectifier α
α_1=90°, α_{11}=0°, $V_{di\alpha}$=V_{di}/2

DC voltage with sequence control as an inverter
α_1=150°, α=97.7°, $V_{di\alpha}$=V_{di}/2

Fig. 3.66 *Voltage waveforms of dc voltage of a bridge with sequence control*

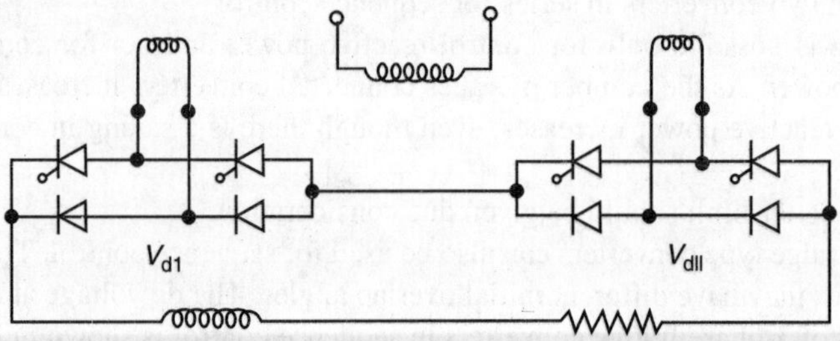

Fig. 3.67(a) *Sequence control of half controlled bridges*

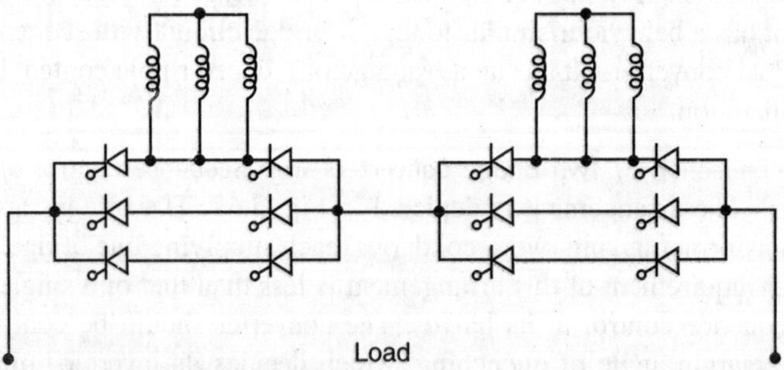

Fig. 3.67(b) *Sequence control of three phase controlled bridges*

Optional Free Wheeling Optional free wheeling in the bridges using thyristors is given in Fig 3.68.

A fully controlled bridge converter has a disadvantage of poor powerfactor at small control ratios. A half controlled bridge converter has a power-factor better than that of a fully controlled one at the same voltage ratio. However, half controlled converters do not possess the capacity of inversion. Thus half controlled converters have limited application as one quadrant ones. It is possible to use a fully controlled bridge circuit as half controlled one by controlling a set of thyristors a diodes by triggering them at their natural firing instant. When the inversion is required they are operated as thyristors and their firing angles are varied from 90° to 180°. Such a facility is called optional freewheeling. In Fig. 3.68, two quadrant converters are shown with encircled thyristors to provide optional freewheeling. These are controlled as diodes during rectification and as thyristors during inversion. Such a control is possible in converters with freewheeling diodes. Here the diode is replaced by a thyristor which is fired at its natural firing instant for freewheeling action. When

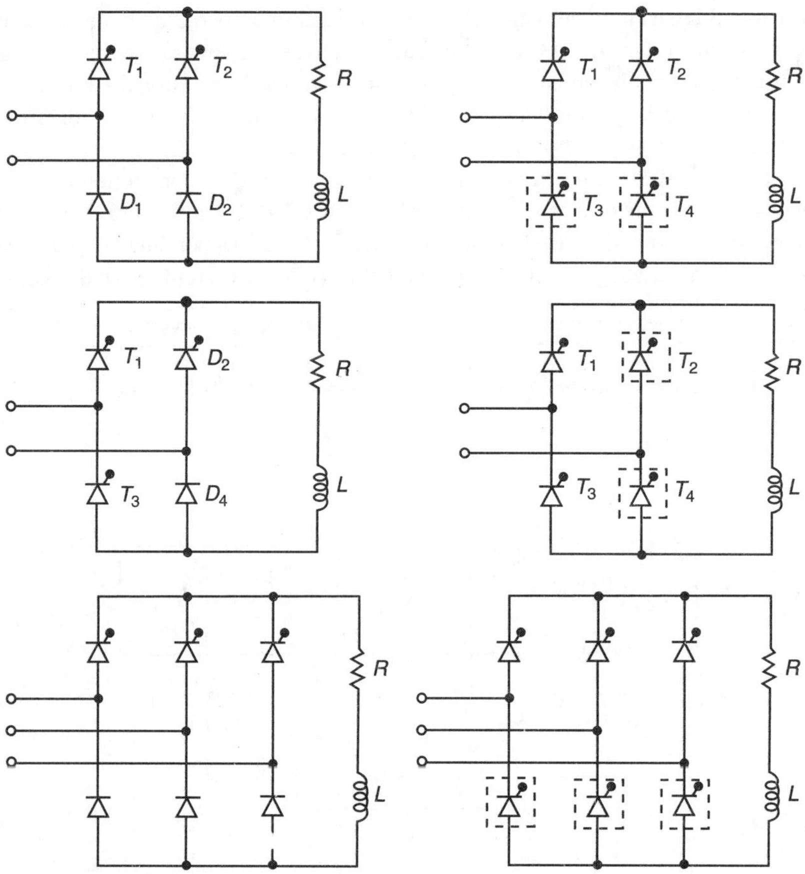

Fig. 3.68 *Optional free wheeling in the bridges using thyristors*

the inversion is required the control pulses to the thyristors are blocked and the converter operates as a two quadrant one. Optional freewheeling enables a converter operate in both rectification and inversion modes providing the advantages of free wheeling during rectification. Sequence control discussed in the foregoing is a kind of optional freewheeling providing the advantages during inversion also.

3.2.3 Four Quadrant Converters

The converters described in the previous sections are suitable either for one quadrant or two quadrant operation. In the former case stepless speed control is possible by changing the applied voltage. There is no regeneration. In the latter case both speed control and regeneration are possible. The converter operates in the inversion mode as well. The sign of the *dc* voltage changes and the current direction is specified by the thyristor. Operation is possible in the neighbouring quadrant in the V-I plane (speed–torque plane). The converter fed drive has regeneration capability.

Sometimes dual converter finds application where speed control and regenerative capability are available in both directions of rotation. It operates in the reverse direction. Such a converter can be realised by combining two 2 quadrant converters having different directions of current. A converter allowing operation in all the four quadrants is called a four quadrant converter. It is also called a dual or reversing converter. It provides control as well as regeneration in both directions of rotation, as shown in Fig. 3.69.

A dual converter can be realised by either an antiparallel connection of two 2 quadrant converters or a cross connection. Figures 3.70 (a) and 3.71 (a) show the two possible connections of 3-pulse controlled rectifiers. Depending upon the current direction, either converter I or converter II feeds the current to the dc motor.

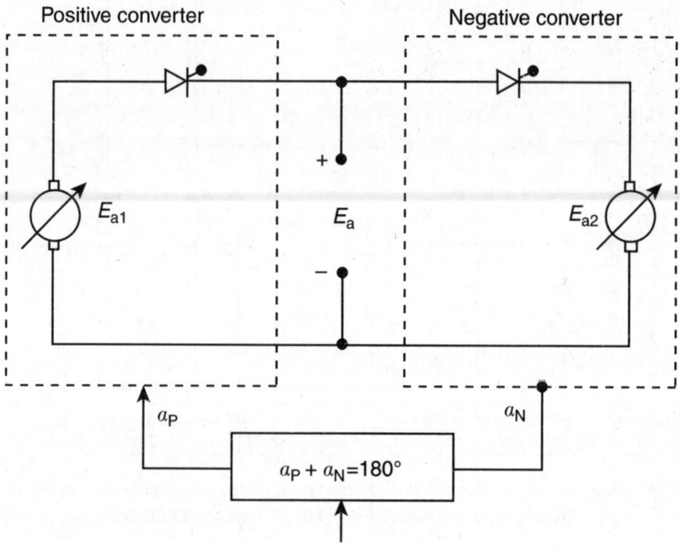

Fig. 3.69 *Principle of dual converter*

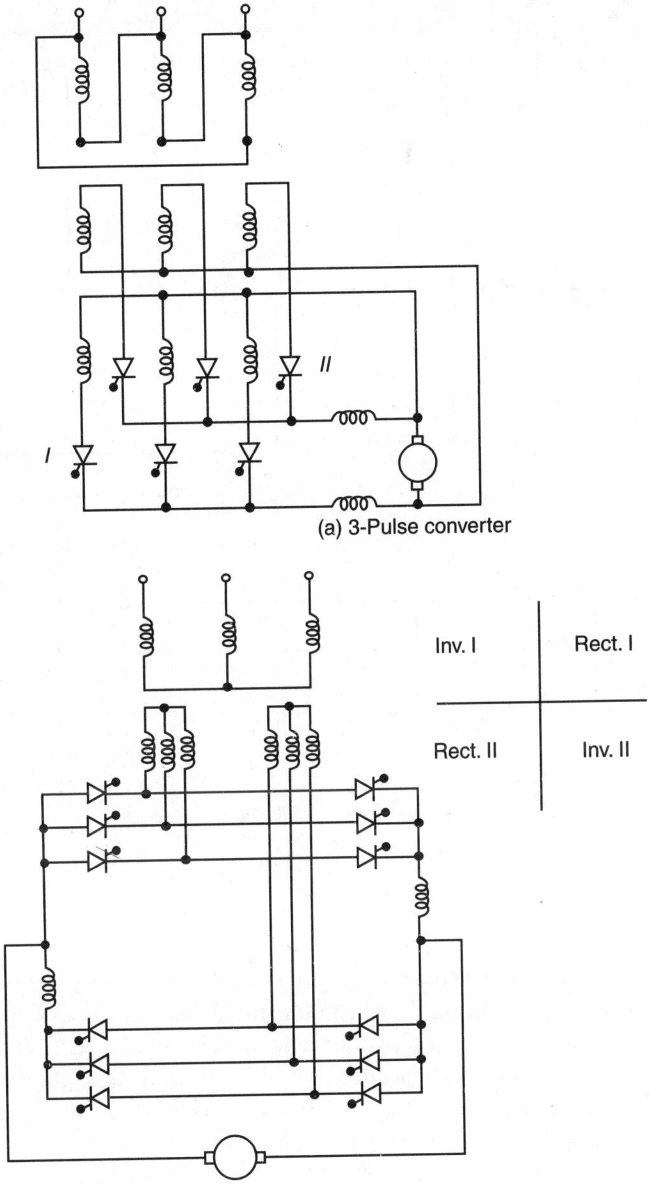

(a) 3-Pulse converter

(b) Bridge converters

Fig. 3.70 *Dual converters in cross connection*

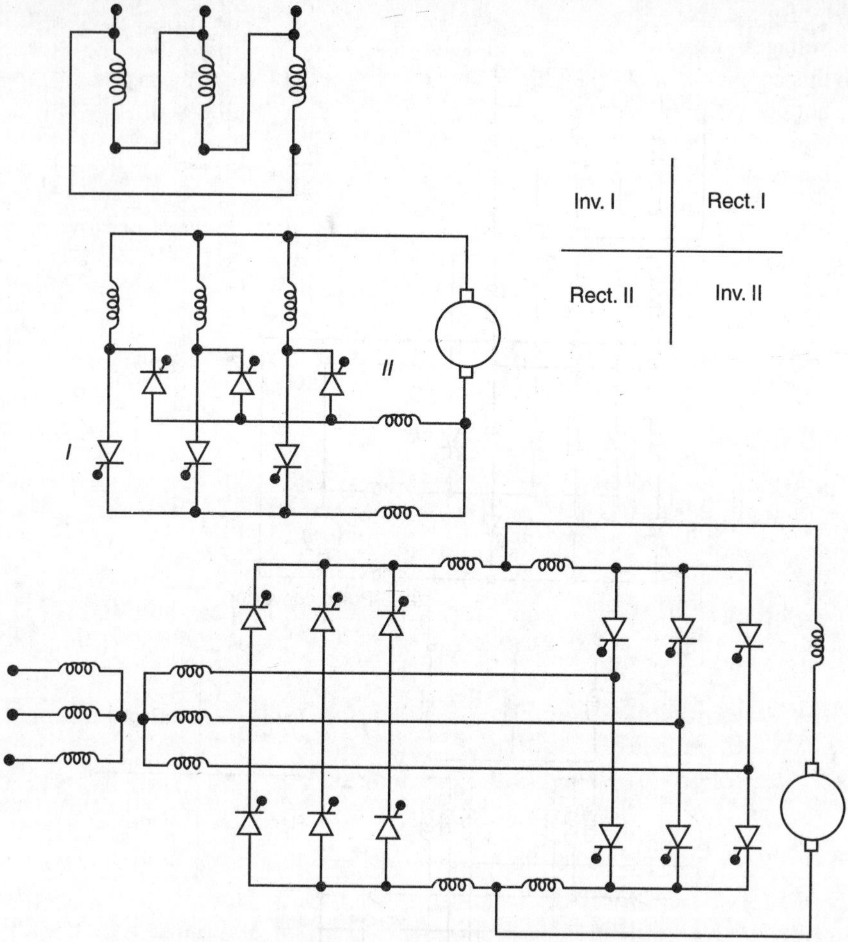

Fig. 3.71 *Dual converters in antiparallel connection*

In an antiparallel connection a single secondary of the transformer is sufficient. The two converters operate in opposite directions as far as the dc load is concerned. On the other hand, a cross connection has individual secondaries for the converters. An antiparallel connection is advantageous with 3 pulse converters, whereas a cross connection is better with 6 pulse converters. Figure 3.70(b) shows 6 pulse bridge converters supplying a reversible dc motor.

The block diagram of an antiparallel connection for a dual converter is depicted in Fig. 3.71(b). The firing angle of the two converters must be such that the average values of the voltages of the two converters are the same at the dc terminals, in both magnitude and polarity. Because of the nature of the connection this condition is possible only if one converter is a rectifier and the other an inverter. Thus, if a_P is the firing angle of the converter operating as a rectifier (say positive group), the firing angle of the converter operating as an inverter (say negative group) $a_N = 180 - a_P$.

By controlling the firing angles of the converters, the average values of the voltages of the converters are made equal and of opposite polarity. Even though this condition is satisfied, the instantaneous values of the dc output voltages are not equal. The difference of these voltages, also called loop voltages, causes a circulating current. The circulating current is limited by the smoothing inductance. The path of the circulating current is shown in Fig. 3.72, for different fir-

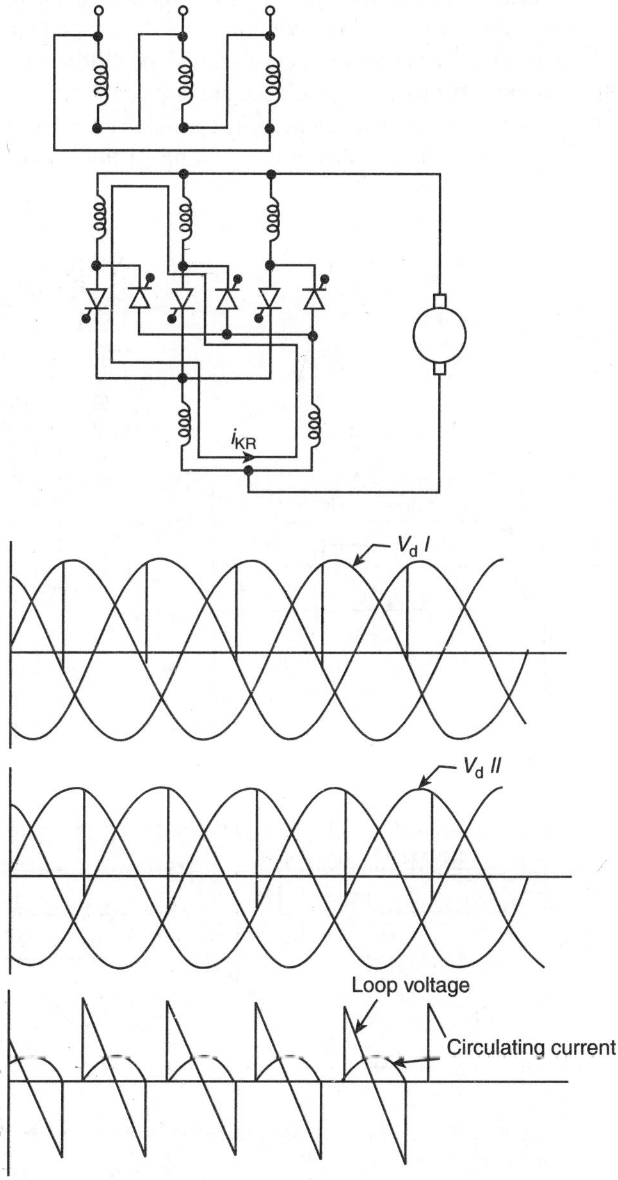

Fig. 3.72 *Loop voltage and circulating current of dual converters*

ing angles. One can see that the current is continuous for $\alpha < 60°$ and becomes discontinuous for $\alpha > 60°$. This is an additional loading on the thyristors and the transformer. However, it offers an advantage in that the circulating current does not permit the converter to go into discontinuous conduction and thus the associated difficulties are not present. The characteristics of the converter correspond to continuous conduction. The circulating current can be controlled. The control circuit adjusts the firing angles so that their sum is slightly less than 180°. This causes a difference in the average values, resulting in a dc component of the circulating current. This may reach unacceptable values if the difference of voltages is large and the limiting resistance and reactance are less.

Bridge circuits in an antiparallel connection forming a dual converter are shown in Fig. 3.71(b). This requires four current limiting reactors of large size and

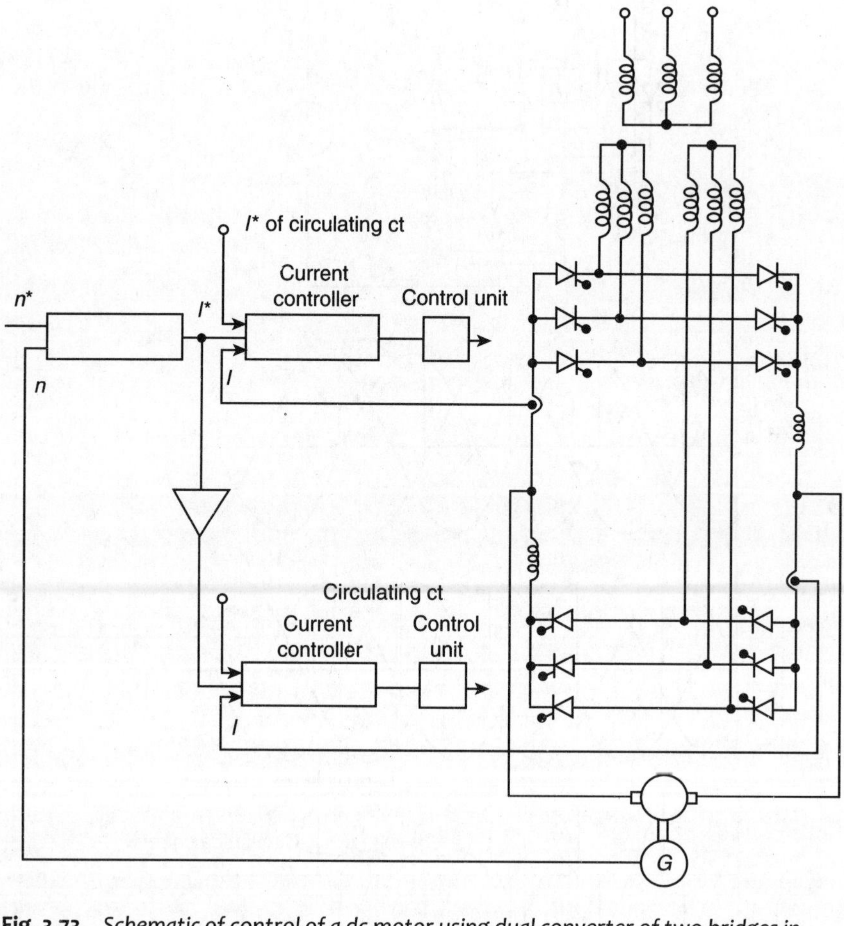

Fig. 3.73 *Schematic of control of a dc motor using dual converter of two bridges in cross connection*

a circulating current having three pulses is established. This has a disadvantage with regard to its control circuit and hence is rarely used.

The cross connected converter in Fig. 3.70 (b) shows better behaviour. The circulating current has a sixth harmonic ripple. The average value of the circulating current and the smoothing inductance are smaller than in the case of three pulse circuits. The schematic of the control of a dc motor using cross connected bridges is shown in Fig. 3.73.

A dual converter can also be operated in a non-circulating current mode. The converter shown in Fig. 3.74 is made up of bridge converters in an antiparallel connection. The control logic should be such that at any instant only one converter conducts and the firing pulses to the other are blocked. The control circuit needs to be more sophisticated. The change over from one converter to the other takes place in the following way:

i. The sign change is provided at the output of the speed controller.
ii. The speed controller is blocked by limiting the output voltage. The current controller is separated from the control circuit. The firing pulses are retarded to the inverter limit for the rapid building up of current.
iii. The zero crossing of the current is detected and the pulses of the conducting converter are blocked.
iv. The control circuit is again connected to the current controller. Control pulses to the converter and limiting of speed controller will be for the new direction of rotation.

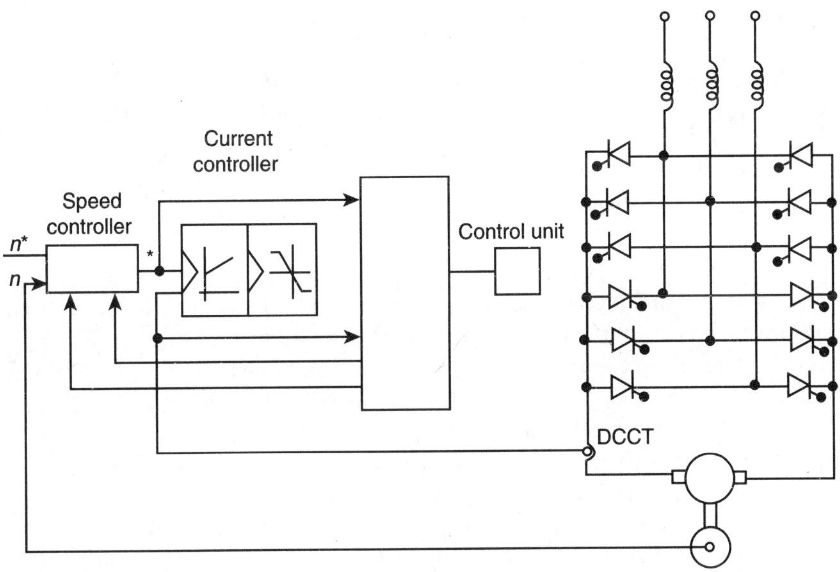

Fig. 3.74 *Dual converter in non-circulating current mode*

There should be a time lag of 3 ms for change over to occur, because simultaneous conduction of the converters must be avoided. A single control circuit and one smoothing inductance are sufficient. The latter can be small in size, if the armature inductance is large. A transformer can be dispensed with if the supply voltage is employed directly. The control circuit needs further refinement in case discontinuous conduction takes place.

3.3 DC CHOPPERS

DC Choppers are mainly dc to dc single stage conversion devices which provide a variable voltage on the load side when fed from a constant dc voltage source. The commutation of the current from the thyristors cannot be achieved by means of supply voltage. The necessary reactive power for the converter must be provided by means of energy storage elements in the circuit itself. A continuously variable voltage is available at the output terminals for feeding dc motors. This is more effective method than resistance control because of the absence of losses. The choppers can be used for two and four quadrant operation of dc motors.

3.3.1 Principles of dc Chopper

The basic circuit of a dc chopper is shown in Fig. 3.75. The main thyristor T_1 is turned ON and OFF periodically, so that the supply voltage V_d is available at the output as a pulse train. By changing the ON period of the thyristor the average voltage of the load can be varied. The switching of the thyristor is accomplished by means of a firing pulse from a control circuit. The turn off of the thyristor at the desired instant is achieved by a series circuit of an auxiliary thyristor and a capacitor connected across the main thyristor. A firing pulse is provided by the same control circuit to the auxiliary thyristor. This goes into conduction, applying the capacitor voltage to the main thyristor in the reverse direction.

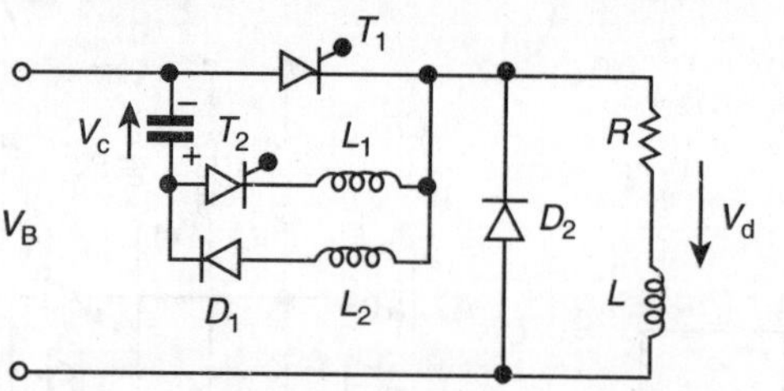

Fig. 3.75 *Basic circuit of a dc chopper*

The value of the capacitor is so chosen that the main thyristor has a negative voltage, and remains current free for a time greater than the turn off time (t_q) of the thyristor, so that the thyristor T_1 regains its positive blocking capability and a satisfactory commutation of the load current to the auxiliary thyristor takes place. The capacitor gets charged further by means of this constant load current, as shown in Fig. 3.76(b). When it gets charged to V_d in the opposite direction the auxiliary thyristor ceases conduction. The current has now been transferred to the free wheeling diode (D_2) which is there in the circuit to provide an alternative path for the load current when the main thyristor is switched off. This is required when the load is inductive. The free wheeling diode conducts until the main thyristor is again fired after the completion of T_{off}. Another circuit, comprising a diode and an inductance is placed in parallel to the auxiliary thyristor. This forms a resonating circuit with the commutating capacitor of the polarity needed for commutation of the main thyristor. When T_1 is fired in its sequence a local circuit is formed by the commutating capacitor, main thyristor T_1, the inductance L_2 and D_1. The capacitor discharges through the circuit. After half the cycle of oscillation of the circuit the capacitor is charged to V_d (in ideal conditions) to a polarity required for the next commutation. The diode blocks further discharge of the capacitor, thereby trapping the charge on the capacitor for the next commutation. The voltage and current waveforms are depicted in Fig. 3.76. To avoid the high rates of change of current in the thyristors, additional inductances are provided. Sometimes the line inductances may suffice in protecting the main thyristor from *di*/*dt*.

During the commutation the time taken by the commutating capacitor to reach zero voltage after the auxiliary thyristor is fired, is

$$t_s = \frac{V_d C}{I_d} \tag{3.86}$$

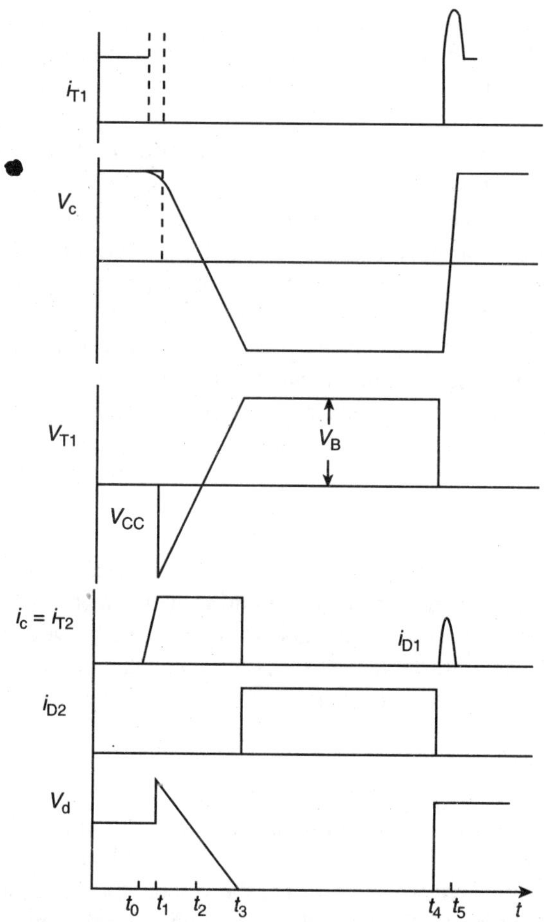

Fig. 3.76 *Wave forms of voltage and current of a chopper circuit*

This is the circuit turn off time and the negative voltage appears across the main thyristor for t_s. For satisfactory commutation the thyristor must acquire its positive blocking capability. In other words

$$t_s \geq t_q \tag{3.87}$$

Using this equation, the minimum value of capacitance can be determined using the formula

$$C \geq t_q \frac{I_d}{V_d} \tag{3.88}$$

Thus the minimum Value of the capacitance for commutation is directly proportional to the load current and the turn off time of the thyristor. The capacitance depends directly upon the source voltage. To limit the size of the capacitor, fast thyristors with a small turn off time (t_q) are used. There is a chance of failure of commutation if the capacitor is charged to a voltage lower than $V_{d'}$ for which it has already been designed. Fluctuations in V_d may also cause difficulties in thyristor commutation. The load current should not be below the value of I_d for which the capacitance is designed.

3.3.2 Effects of Inductances of the Circuit on the Performance of the Chopper

The preceding discussion of the chopper assumes ideal conditions and does not consider the effects of inductances in series with the thyristors. With choppers of large power, the leads connecting the chopper to the supply have a definite amount of inductance (L_e). Also, reactors are placed in series with the thyristors, to protect them from di/dt. These inductances affect the performance of the chopper and sometimes cause a failure of commutation. The energy stored in these inductances is transferred to the capacitor, thus increasing its voltage.

The input inductance has the disadvantage of increasing the voltage across the commutating capacitance. If the inductance is very large this voltage may attain very high undesirable values. Therefore a smoothing capacitor is required at the input. The capacitor and input inductance must resonate at a frequency that is l/3rd the chopper frequency. When a chopper is used for traction purposes, the input inductance varies and its minimum value must be considered. This frequency requirement is satisfied by an extra inductance between the line and the chopper. The value of this inductance is chosen such that the circuit formed by the capacitor, this and line inductance resonates at a frequency less than 1/3 chopper frequency.

The increased charge makes the corresponding energy flow to the source via the free wheeling diode D_2 and D_1. The inductance L_K and commuting capacitance form a resonant circuit, and a sinusoidal current is introduced which makes the voltage across the capacitor less than V_d (Fig. 3.77). When the thyristor T_1 is fired after the successful free wheeling of D_2, the voltage to which the capacitor gets recharged is less than V_d. Consequently the circuit turn off timc t_s decreases. The

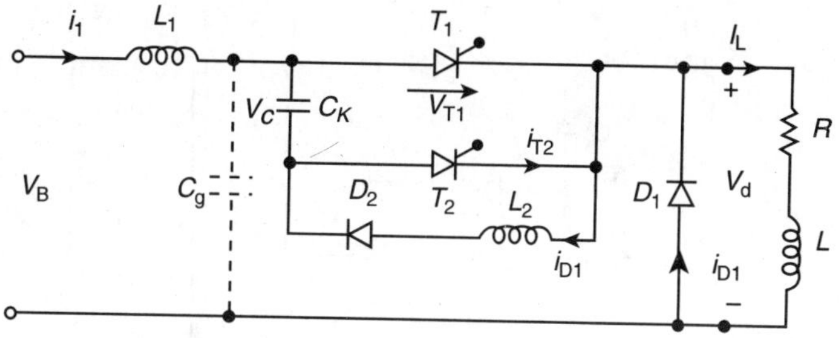

Fig. 3.77 *Chopper circuit with input inductance*

effect is pronounced at large load currents. If t_s becomes less than t_q, the commutation failure occurs. The voltage and current waveforms as affected by the inductances are shown in Fig. 3.78.

Commutation failure may be avoided by replacing the diode D_1 with a thyristor T_3. This introduces no additional complications in the control, since it is fired along with T_1. The commutating capability is clearly improved because the

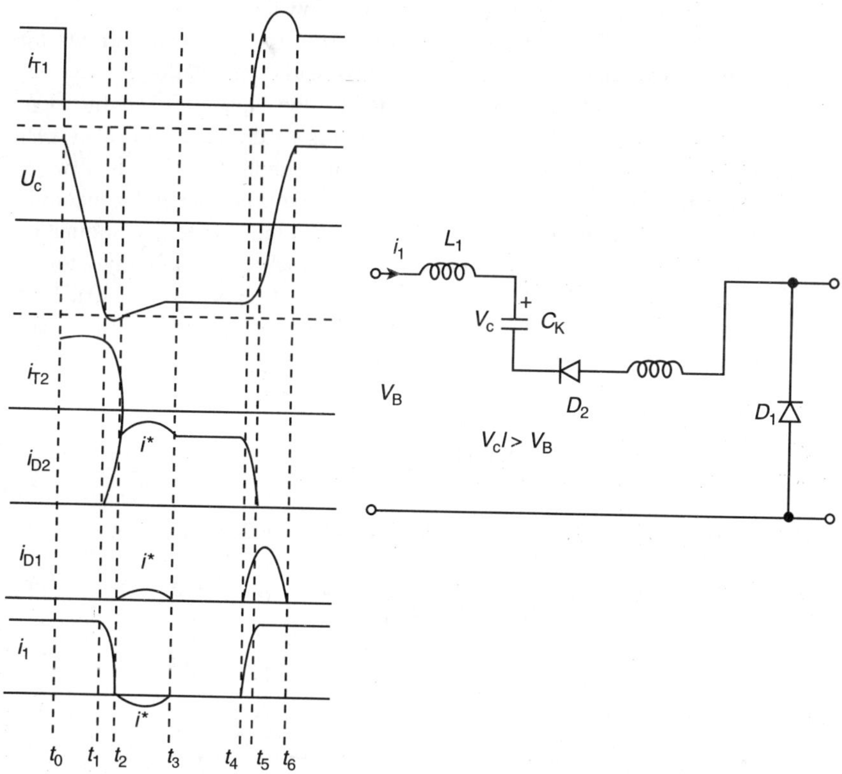

Fig. 3.78 *Voltage and current waveforms of chopper having input inductance*

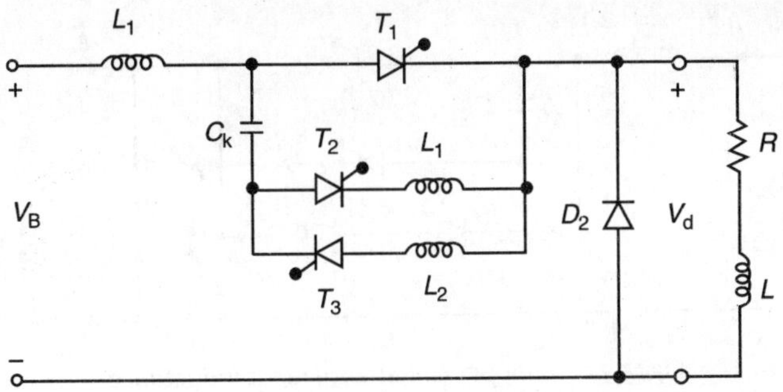

Fig. 3.79 *Modified chopper circuit taking the input inductance into consideration*

capacitor voltage is larger at larger currents. The introduction of T_3 (Fig. 3.79) and its firing along with T_1 prevents the flow of extra current, which is responsible for the reduction of capacitor voltage. The voltage of the capacitor is maintained at V_d.

The duration of the charging process of the condenser is dependent on the load current. Because of this, a minimum T_{ON} has to be assured and the time ratio T_{ON}/T (a) cannot be varied up to zero. There is a lower limit for a (i.e. T_{ON}/T), which increases as the load current decreases. At very light load currents the capacitor may not be charged to sufficient voltage and commutation failure may occur. Therefore sufficient T_{ON} must be provided for the capacitor to charge to V_d. As I_d approaches zero the charging and discharging of the capacitor takes very long time control of the chopper will not be of high quality. This difficulty can be avoided by making the commutation and associated processes independent of load current. To achieve this an additional circuit containing a series combination of a diode D_4 and inductance L_4 is connected in parallel to the main thyristor, as shown in Fig. 3.80. The charging of the capacitor is independent of load current. When the auxiliary thyristor is fired, the current gets transferred to T_2. The load current flows through the capacitor and discharges it. The capacitor forms a

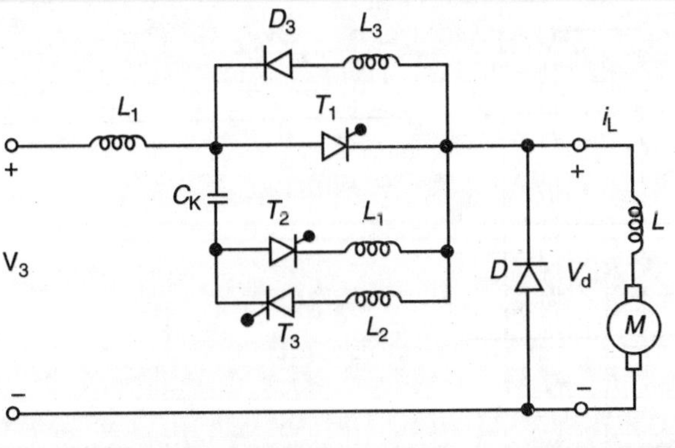

Fig. 3.80 *Modified circuit having capacitor charging independent of load current*

resonant circuit with $L_3 + L_4$ and is quickly charged to a higher voltage. T_2, therefore, conducts both the load current and this charging current. The charging current flows via D_4 and L_4, and the charging becomes independent of load current. These conditions are shown in Fig. 3.81. By a proper choice of inductance in the resonating circuit, the resonant frequency can be varied. The minimum value of T_{ON} is decided by the resonant frequency. It must be half the period of this frequency. The lower limit of T_{ON} can be made as small as possible by selecting the inductance properly.

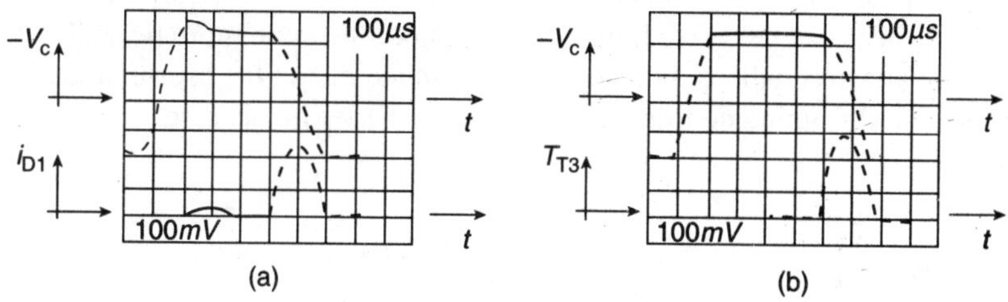

Fig. 3.81 *(a) Capacitor voltage and diode D1 current (b) Capacitor voltage and thyristor T3 current*

3.3.3 Performance of Chopper

Assuming steady-state conditions of the chopper, we have during T_{ON}

$$L\frac{di}{dt} + iR = V_d \tag{3.89}$$

The solution of Eq. 3.89 is

$$i_d = \frac{V_d}{R} + \left(I_0 - \frac{V_d}{R}\right)e^{-\frac{R}{L}t} \tag{3.90}$$

where I_0 is the current at the starting of T_{ON}. At $T_{ON} = aT$, the current is

$$I_a = \frac{V_d}{R} + \left(I_0 - \frac{V_d}{R}\right)e^{-\frac{R}{L}aT} \tag{3.91}$$

when the thyristor is OFF, free wheeling takes place and the operation is described by

$$L\frac{di_D}{dt} + Ri_D = 0 \tag{3.92}$$

which has the solution

$$i_D = I_a e^{\frac{-Rt}{L}} \tag{3.93}$$

Again taking $t = 0$ at the start of T_{OFF}, at the end of $T_{OFF} = (1 - a)T$ the current is

$$I_0 - I_a e^{-R(1-a)T/L} \tag{3.94}$$

Using these equations

$$I_0 = \frac{V_d}{R}\left(\frac{e^{-R(1-a)T/L}e^{-(R/L)T}}{1-e^{-(R/L)T}}\right) \tag{3.95}$$

The value of I_a can also be determined. The difference $I_a - I_0$ gives the peak to peak amplitude of the ripple superimposing the average dc current.

The amplitude of the ripple depends upon the time ratio a. Figure 3.82 depicts the variation of load current for different values of a. It can be seen from the figure that the ripple amplitude is maximum at $a = 0.5$ and decreases for both $a > 0.5$ and $a < 0.5$. These can be derived for d.c. series motor assuming it to be an R-L load using the above equations. For a separately excited motor, the back emf has to be considered. Assuming no losses in the motor, $E = aV_d$. In this case,

$$L\frac{di}{dt} + aV_d = V_d \tag{3.96}$$

solving which we get

$$i = I_0 + \frac{V_d}{L}(1-a)t \tag{3.97}$$

The current at the end of $T_{ON} = aT$ is

$$I_a = I_0 + \frac{V_d}{L}(1-a)aT \tag{3.98}$$

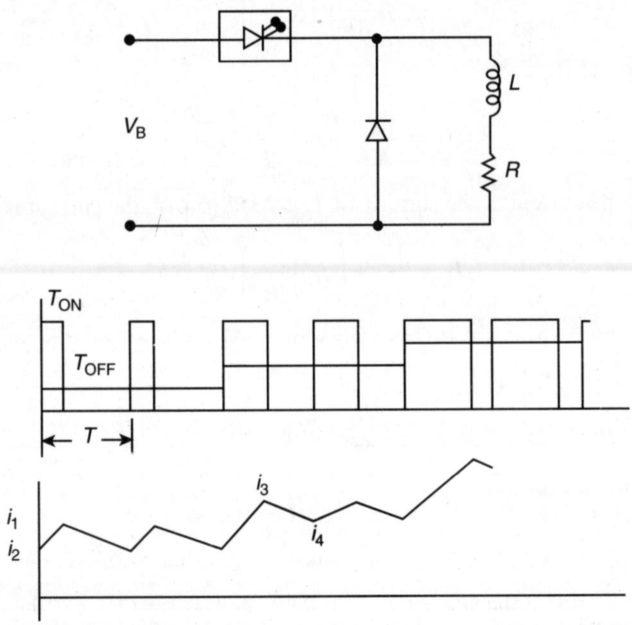

Fig. 3.82 *Constant frequency, variable T_{ON} control of a dc chopper*

from which

$$I_a - I_0 = \frac{V_d}{L}(1-a)aT \tag{3.99}$$

The solution for i shows a linear variation of current. The ripple superimposing the average dc current becomes triangular. The peak value of the ripple is

$$\frac{\Delta I}{2} = \frac{V_d}{2L}(1-a)aT \tag{3.100}$$

The rms value of the current

$$= \frac{V_d}{2\sqrt{3R}}(1-a)aT \tag{3.101}$$

From Eq. 3.100, we can see that the ripple has a maximum amplitude at $a = 0.5$. The ripple makes the load current discontinuous at very low values of load current. This lower limit of load current is influenced by the type of load. For *R-L* loads one can go down to very low values of current if the time constant is sufficiently large. For the back emf loads discontinuous operation occurs early. The lower limit of load current is high in this case.

The ripple in the load current can basically be reduced by increasing the chopper frequency and by introducing an extra inductance in the load circuit.

3.3.4 Methods of Controlling a Chopper

The basic principle of control of a chopper is the effective change of the value of the time ratio. This is done in two ways:

i. Time ratio control
ii. Current limit control

In the former, the ratio T_{ON}/T is varied. This can be achieved by varying T_{ON}, keeping the chopper period T constant, which is called pulse width control or constant frequency control. The waveforms of voltage and current for this type of control are shown in Fig. 3.83. The variation of the time ratio can also be achieved

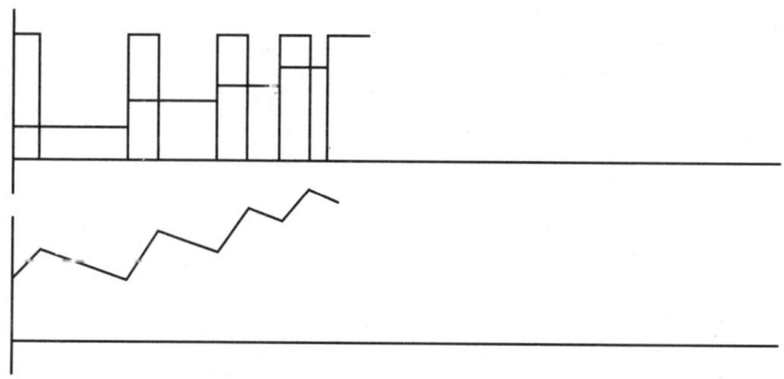

Fig. 3.83 *Control of a dc chopper*

by keeping T_{ON} constant and varying the chopper period T, which is called pulse frequency control or variable frequency control. The waveforms of voltage and current for this method are depicted in Fig. 3.83. From the waveforms it is clear that the maximum and minimum values of load current are decided by the ratio of T_{ON}/T. The variable frequency control is more prone to discontinuous conduction than constant frequency control.

In current limit control (Fig. 3.84), the load current is allowed to vary between two given (upper and lower) limits. The ON and OFF times of the chopper adjust automatically. When the current increases beyond the upper limit the chopper is turned off. The load current free wheels and starts to decrease. When it falls below the lower limit the chopper is turned ON. The current starts increasing in the load. The load current and voltage waveforms are shown in Fig. 3.84. The amplitude of the ripple can be controlled in this case by assuming proper limits of current. The lower the ripple current, the higher the chopper frequency. By this switching losses increase. Discontinuous conduction can also be avoided in this case, which makes this method superior to time ratio control in this regard.

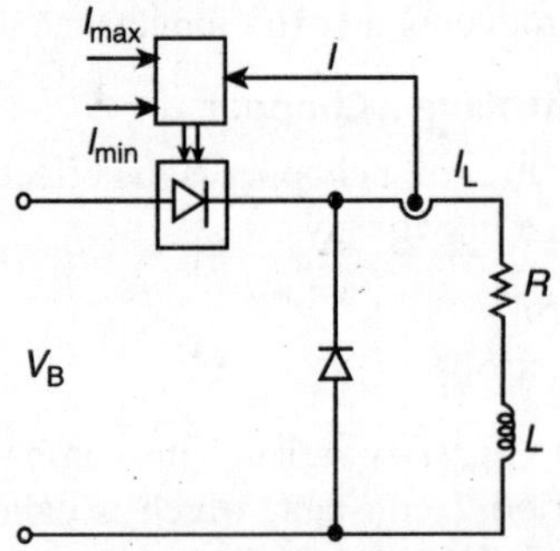

Current limit control of a dc chopper.

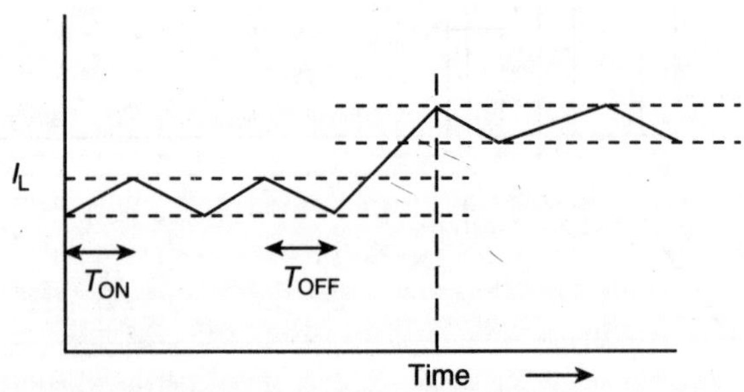

Fig. 3.84 *Current limit control of dc chopper*

3.3.5 Multi Quadrant Choppers

Regenerative Chopper Sometimes the energy of the load may have to be fed to the supply system. A Chopper in this mode is called regenerative. The chopper working as a switch is connected across the load, which is normally a dc machine and the diode is connected in the line. The dc machine works as a generator during this mode. Thus a regenerative chopper is obtained by changing the positions of the diode and thyristor switch of a normal chopper, as shown in Fig. 3.85. As long as the thyristor is ON, the generator and L_d are short circuited, the current in the load increases and energy is stored in the inductance. When the chopper is switched off a large voltage occurs across the load terminals. This voltage is greater than V_d and the energy stored in the inductance is fed to the supply system. When the voltage of the load falls to V_d, the diode in the line blocks the current flow, preventing any short circuit of the supply when the thyristor is switched on. Thus the complete energy of the load can be supplied to the source. Very effective braking of the motor is possible up to extremely small speeds. Regenerative braking is achieved here by changing the direction of current flow. A remote control may be arranged by a suitable logic to switch the normal chopper to a regenerative chopper and vice versa. A small delay is required, for transition from one mode to the other.

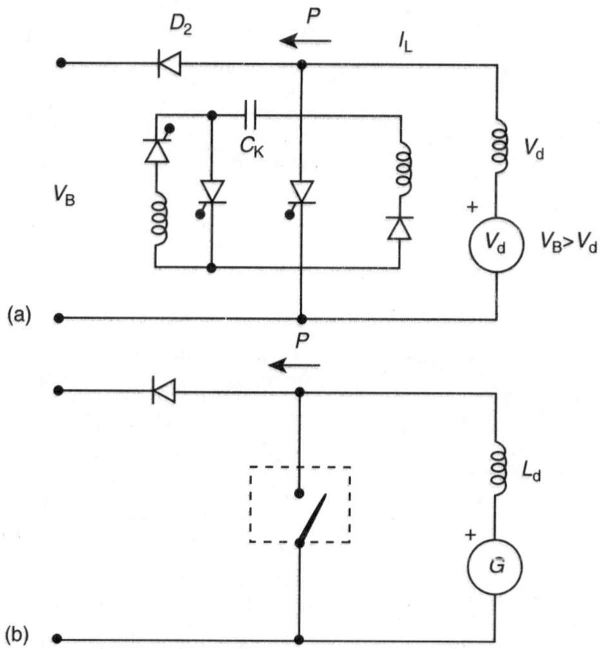

Fig. 3.85 *Regenerative chopper (a) Basic circuit, (b) Equivalent circuit*

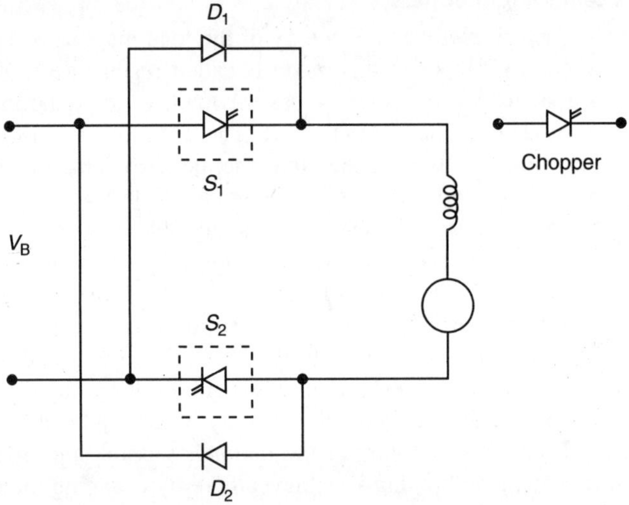

Fig. 3.86 *Two quadrant dc chopper*

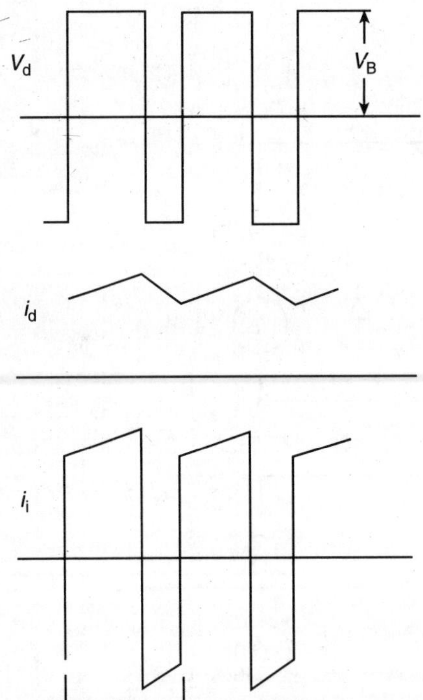

Fig. 3.87 *Current and voltage waveforms of 2-quadrant chopper*

Two Quadrant Chopper Sometimes a chopper may be required to provide a two quadrant operation by retaining the current direction in both motoring and braking modes. Such a two quadrant converter is shown in Fig. 3.86, and its waveforms in Fig. 3.87. The chopper permits a change in the polarity of terminal voltage keeping the direction of current constant. In Fig. 3.86 S_1 and S_2 are the choppers working as electronic switches and D_1 and D_2 are the feedback diodes. Assuming continuous conduction, both S_1 and S_2 operate simultaneously.

When they are ON, the load current I_d increases, with the rate of rise depending upon the back emf and torque of the machine. When they are OFF, the load current is fed to the supply through the diodes D_1 and D_2. The current decreases in this stage. From Fig. 3.87, it is clear that the load voltage is effectively negative when the diodes conduct. The average value of the output voltage of the chopper

$$V_{da} = \frac{1}{T}\left(\int_{0}^{T_{ON}} V_d\,dt - \int_{T_{ON}}^{T} V_d\,dt\right) \tag{3.102}$$

$$V_{da} = V_d(2a - 1) \text{ where } a = \frac{T_{ON}}{T} \tag{3.103}$$

V_{da} is zero for $a = 0.5$, the output voltage is an ac waveform with average voltage equal to zero. For $a > 0.5$ the average value of the dc voltage is negative. The chopper works in the regeneration mode. The power flows from the load to the supply and the dc machine operates as a generator. For control ratios $a < 0.5$, the operation is similar to the inversion of a line commutated converter ($a > 90°$). For $a > 0.5$ the chopper has normal operation. The power flows from the supply to the load. This operation of the chopper is similar to rectification of a line commutated converter ($a < 90°$). For, $a = 0.5$ the load voltage is zero. The control characteristic of a two quadrant chopper is shown in Fig. 3.87.

Four Quadrant Chopper A chopper circuit for four quadrant operation is shown in Fig. 3.88(a). This is obtained by connecting two 2-quadrant choppers antiparallel to each other. The circuit has characteristics similar to a dual converter. The load side voltage and current waveforms have both polarities. The dc machine connected to this chopper can be accelerated, braked and controlled so as to have smooth speed variation in both directions of rotation. This four quadrant chopper is a starting point for the development of a force commutated inverter, to obtain a variable frequency voltage at the load terminals from a constant dc supply. The sketch of the output voltage of the circuit operating as an inverter is given in Fig. 3.88(b).

3.3.6 Selection of Ratings of Devices of a Chopper

The waveforms of currents and voltages are useful in deciding the ratings of devices. The main thyristor carries the load current. Superimposed on this there is a current pulse during the recharging process of the capacitor, through the resonating

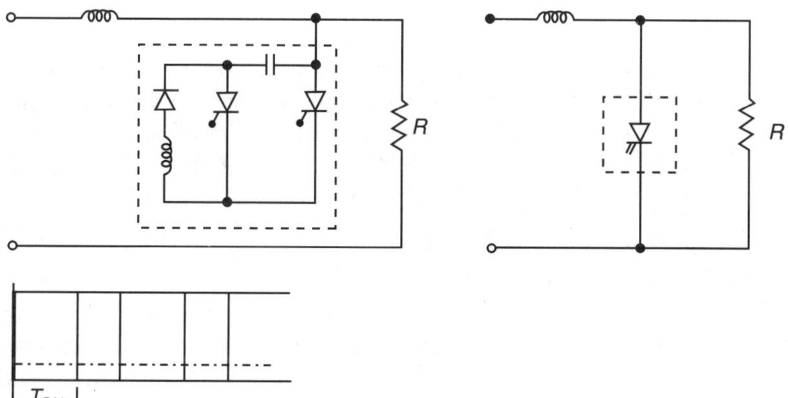

Fig. 3.88(a) *Four quadrant chopper*

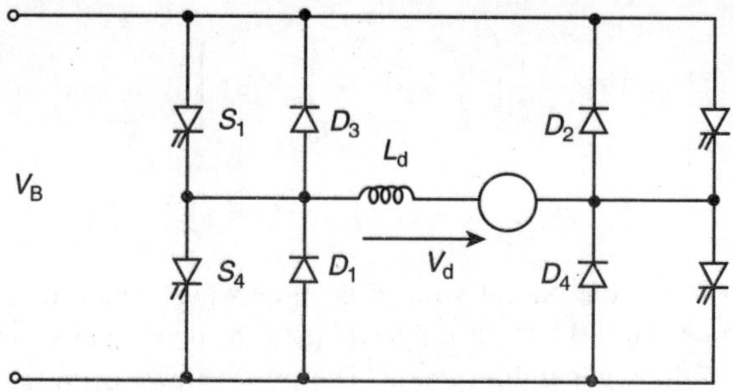

Fig. 3.88(b) *The output voltage waveform of a four quadrant chopper as a force commutated inverter*

circuit. Therefore the peak value of the main thyristor current is roughly thrice the load current. The maximum conduction time of the current of the main thyristor occurs when $a = 1$. Based on these facts, the current rating of the main thyristor can be decided. Besides, it must be an inverter grade thyristor with a small turn off time, which helps while using a small capacitor. The free wheeling diode conducts load current. Its period of conduction is a maximum and is equal to the chopper period when the time ratio $a = 0$. It must also be a fast diode, since normal diodes cause increased switching losses. The auxiliary thyristor conducts only for a short period which is independent of chopper frequency (80 to 200 μs). The peak value of the auxiliary thyristor current is twice the load current. A reactor may be placed in series with the thyristor for protection against high values of di/dt. In case the line inductance is not sufficient for this purpose it should be supplemented with an additional inductance. The thyristor T_3 (replacing the diode D_2) need not be of inverter grade. Its conduction time is 70 to 100 μs and it carries a current of peak value equal to the main thyristor current minus the load current. The diode D_4 conducts only pulses of current, for the same period as the auxiliary thyristors. The devices T_2, T_3, and T_4 are loaded for short intervals of time.

3.3.7 General Application of Choppers

Choppers are used in drives requiring loss free, efficient speed control with the possibility of regenerative braking. They can be used with reversible drives. Chopper fed dc motors are used in traction. They enjoy smooth starting, acceleration and braking. They can also be used to obtain a variable resistance from a fixed resistance, which is called a chopper controlled resistance. We shall now discuss the principles of chopper controlled resistance.

3.3.8 Chopper Controlled Resistance

By interrupting the flow of current through a resistance R in a periodic fashion using a switch, the value of resistance can be effectively varied. The interruption of the current can be accomplished by means of a dc chopper connected across the resistance, working as a switching device. When the chopper is on, the resistance is short circuited and the current is diverted through the chopper. When it is OFF, the supply voltage is connected to the resistance and the current flows through the resistance. If the chopper is always ON the effective resistance is zero. The effective resistance is R if the chopper is always OFF. Thus by controlling the ON/OFF ratio of the chopper, the effective resistance can be controlled in the range 0-R. Under ideal conditions the effective resistance can be shown to be

$$R^* = R\frac{T_{\text{OFF}}}{T_{\text{ON}} + T_{\text{OFF}}}$$

The chopper controlling the resistance is depicted in Fig. 3.89(a). The figure also gives the current waveform through the resistance. The voltage is in the form of pulses and the resistance is also called a pulse controlled resistance. The smoothing inductance L is used to reduce the ripple in the load current.

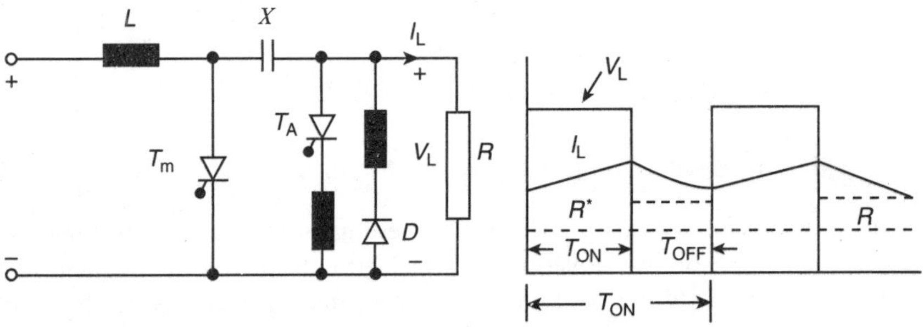

Fig. 3.89(a) *Chopper controlled resistance parallel connection R* = (0 to R)*

Sometimes a correction for the simple expression above may be required if the R-C time constant is not negligible compared to the chopper period T. In this case the capacitor current flowing through the resistance alters its effective value. Taking this into consideration

$$R^* = R\frac{T_{\text{OFF}}}{T} - \frac{RC}{T}\left(1 - e^{-T_{\text{OFF}}/RC}\right)\left(1 + \frac{V_c}{L}\right)$$

The resistance can be varied from 0 to ∞ by a simple modification in the circuit, as shown in Fig. 3.89(b).

A condenser is connected in series with the resistance and an R-C combination is placed across the chopper. When the chopper is always OFF, the capacitor does not allow any current through the resistance and the value of effective resistance is ∞. However, there is now a danger of high voltage across the main thyristor.

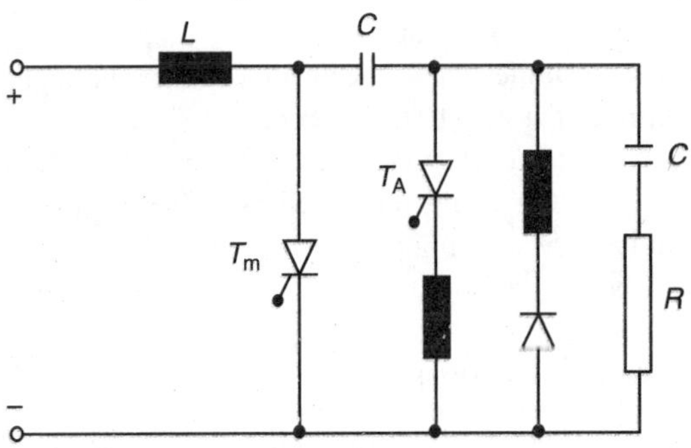

Fig. 3.89(b) *Modification to vary the range of resistance variation R* = (0 to ∞)*

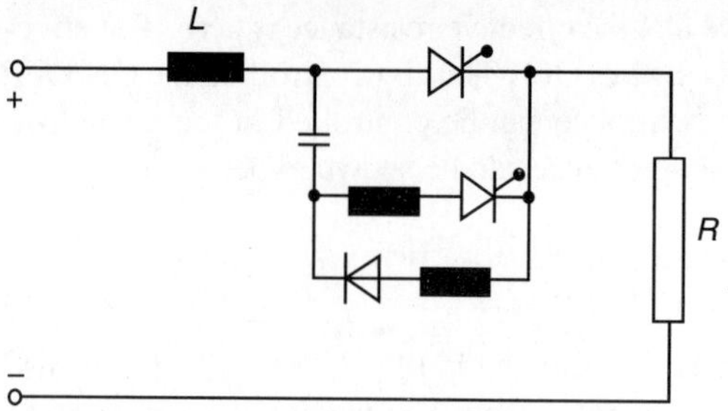

Fig. 3.89(c) *Series connection of chopper with resistance R* = (R to ∞)*

The chopper can also be connected in series with the resistance, as shown in Fig. 3.89c. When the chopper is ON the resistance is R and when it is OFF it is infinity. By varying T_{ON}/T_{OFF}, the resistance can be varied from R to infinity. In this case

$$R^* = \left(\frac{R}{\frac{T_{ON}}{T}}\right)$$

T_{ON}/T can be varied from 1 to 0 and R^* varies from R to ∞.

3.4 INVERTERS

Inverters are static power converters for converting dc to ac. By controlling the conducting periods of the thyristors it is possible to obtain variable frequency at the output terminals of the inverter. This variable frequency supply can be used to feed an ac motor, so as to control its speed. The commutation of the inverters requires reactive power, which a dc supply cannot provide. If the ac side of the inverter is also not in a position to provide this reactive power, special circuit elements which do so must be made available in the inverter circuit. The inverter in such a case is called a force commutated inverter. Sometimes the inverter can be commutated naturally, if the load is able to provide the necessary reactive power. This happens if the inverter feeds an over excited synchronous motor of a three phase system at constant frequency. However, an inverter feeding a three phase induction motor must necessarily employ force commutation as the latter cannot provide the requisite reactive power.

A force commutated inverter is equipped with capacitors and auxiliary thyristors, chosen such that satisfactory commutation takes place. The commutation can be a single stage one, using a counter voltage, or a two stage one providing an auxiliary passage to the current.

We know that a thyristor can be switched on and brought to a conducting state by providing firing pulses when it has a positive voltage. It does not cease to conduct, if the firing pulse is removed, but when the current through the thyristor falls below the holding value and stays at this value for a time greater than the turn OFF time of the thyristor. A commutation circuit must be designed to satisfy the following conditions:

i. The current through the conducting thyristor must go to zero.
ii. At the end of conduction the thyristor must continue to have a negative voltage (to maintain this zero current) for a definite amount of time,

called the turn off time, so that the thyristor regains its positive blocking capability.

iii. The change over of the current to the next thyristor.

The condition for the successful commutation of a thyristor is that a negative voltage must exist across the thyristor for a definite amount of time, after which the thyristor attains its forward blocking capability and a positive voltage can be applied across it. Normally it is sufficient that the time of application of the negative voltage be 25% greater than the turn off time of the thyristor.

The turnoff time of the thyristors considerably influences the design and choice of commutating circuit elements. The commutating capacitance is directly proportional to the turn off time. The latter also influences, in direct proportion the charge required to turn off the thyristor. If the capacitance is charged via the load, the load voltage cannot go below a certain value. The turn off time also limits the upper frequency of the inverter. Therefore it is desirable to use thyristors with the minimum possible turn off time for force commutated inverters.

A negative voltage to the outgoing thyristor of an inverter is normally applied by means of a capacitor via a commutating or auxiliary thyristor. For satisfactory commutation this capacitor must be charged to a sufficient voltage with correct polarity. The commutation circuit must have a provision to charge the capacitor. Basically the capacitor can be charged using

i. the input voltage of the inverter.
ii. the load current of the inverter.
iii. a separate voltage source.

Sometimes a combination of these three methods can be employed. Figure 3.90 shows the three possibilities. The first case is illustrated in Fig. 3.90(a) and is suitable if the input voltage is constant. The second case is depicted in Fig. 3.90(b), in which the capacitor charging is independent of input voltage, but on the load current. This circuit is not suitable for small currents, but has the advantage at higher currents of practically constant turn off time. The use of a separate

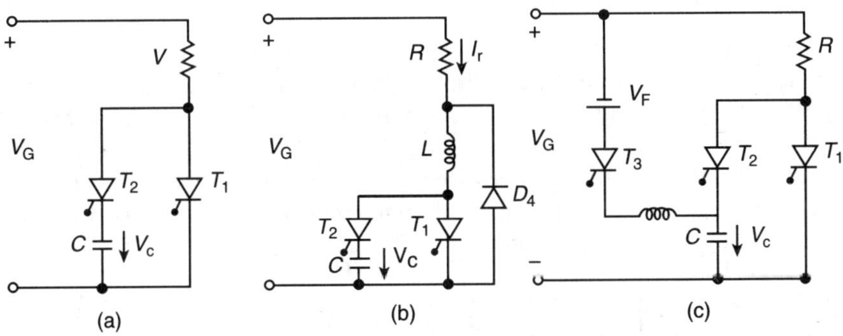

Fig. 3.90 *Methods of charging inverter capacitor (a) From the input voltage (b) Load dependent charging (c) From a separate voltage source*

voltage source to charge the capacitor is illustrated in Fig. 3.90(c). This method is suitable if there is a possibility of the input voltage falling below a certain value, or even disappearing.

To get an inverter which has an output voltage independent of the load current and which can provide the reactive current for the load, additional circuit elements are required. Sometimes diodes are connected antiparallel to the thyristors. These diodes enable one to use an inductive load on the inverter. They operate in the free wheeling as well as feed back modes. Commutating inductances are provided to prevent short circuits when both the incoming and outgoing thyristors conduct simultaneously. They also prevent the discharge of the commutating capacitance when commutation takes place through feed back diodes. Sometimes diodes are used in series with the thyristors to avoid the discharge of the commutating capacitance if the load contains a back emf or if there are sudden changes in the load. The commutating capacitor is connected between the diodes and the thyristors.

3.4.1 Three Phase Inverters

The variable frequency required for the speed control of three phase ac motors is obtained from a three phase inverter. To avoid magnetic saturation and to obtain constant flux conditions in the machine, the voltage fed to the motor must also be varied. Therefore an inverter feeding a three phase motor must be capable of providing a variable voltage, variable frequency supply. The required voltage control can be obtained either external to the inverter or within it (Fig. 3.91). In the former, the input voltage to the inverter is variable, whereas in the latter it is constant and the required variable voltage at the output terminals is obtained by controlling the inverter.

The dc voltage to the inverter is normally obtained by rectifying a 50 Hz supply using a bridge rectifier. The rectifier and inverter are interconnected by means of energy storing elements. These provide a kind of decoupling between the rectifier and the inverter. Such converters are called dc link converters. They provide a variable voltage, variable frequency supply from constant voltage, constant frequency mains, and are two stage conversion devices. The inductance in the dc link smoothens the current while the capacitor maintains the voltage constant.

Since the voltage is a controlled quantity, these are called voltage source inverters. When the voltage control is done external to the inverter, the line side rectifier must be a phase controlled one. By varying the firing angle the output voltage of the rectifier, and hence the input voltage to the inverter, can be varied. These are called square wave or variable voltage inverters. If the voltage control is available in the inverter itself, the input voltage of the inverter is constant and a simple diode rectifier suffices on the line side.

The inverter uses PWM for voltage control and hence is called a PWM inverter or constant voltage inverter (Fig. 3.93). In these inverters the voltage is maintained constant at a controlled value, irrespective of the load events. The capacitance across the inverter maintains the constant voltage.

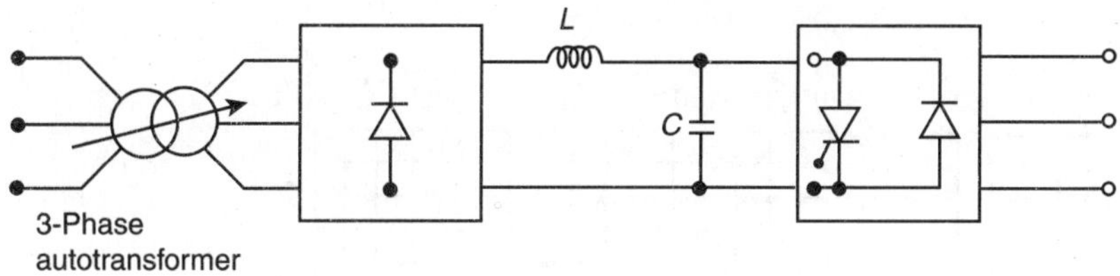

(a) Voltage control using an autotransformer at the converter input.

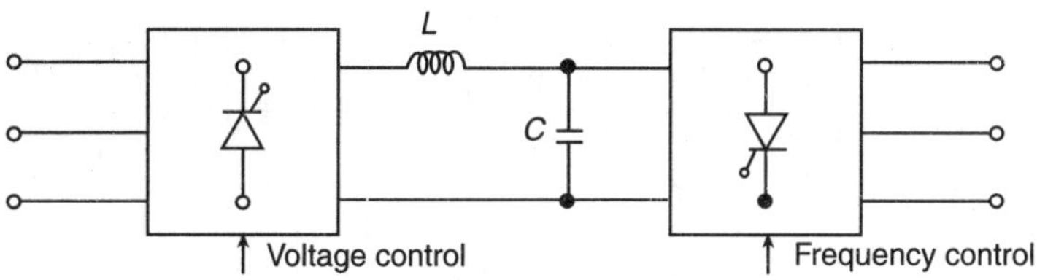

(b) Voltage control using phase controlled rectifier.

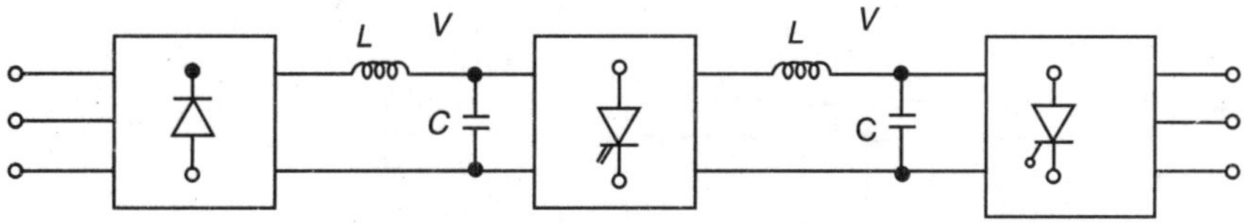

(c) Voltage control using dc chopper.

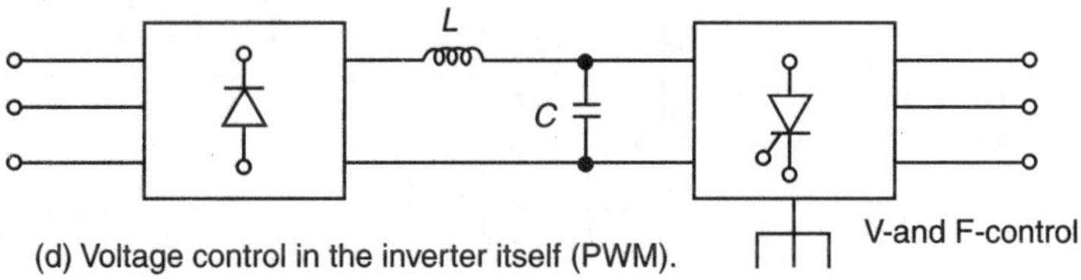

(d) Voltage control in the inverter itself (PWM).

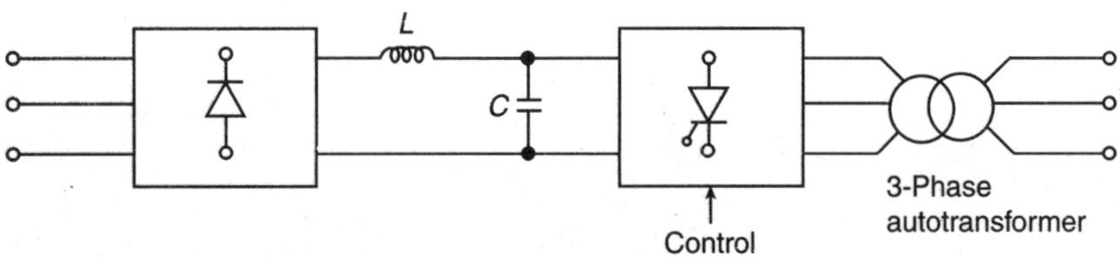

(e) Voltage control using autotransformer at the output.

Fig. 3.91 *Methods of voltage control of inverters*

Features of Variable Voltage Inverters The inverter has an impressed dc voltage. The output voltage of the inverter is decided by the firing and duration of the thyristors. The conduction of the thyristors can be either 180° or 120°, depending upon the control employed. In the case of 180° conduction three thyristors conduct at any instant, two in one half and the third in the other half. The motor phase

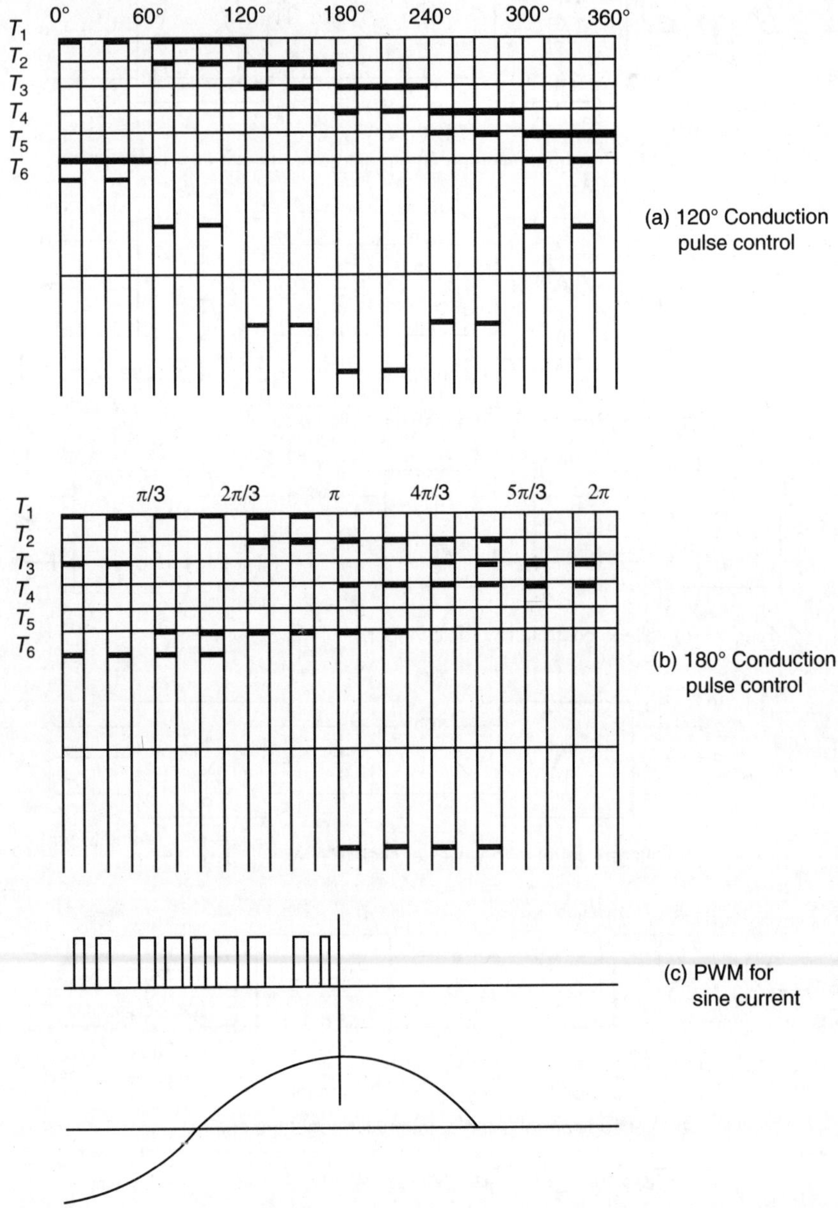

Fig. 3.92 *Principles of PWM*

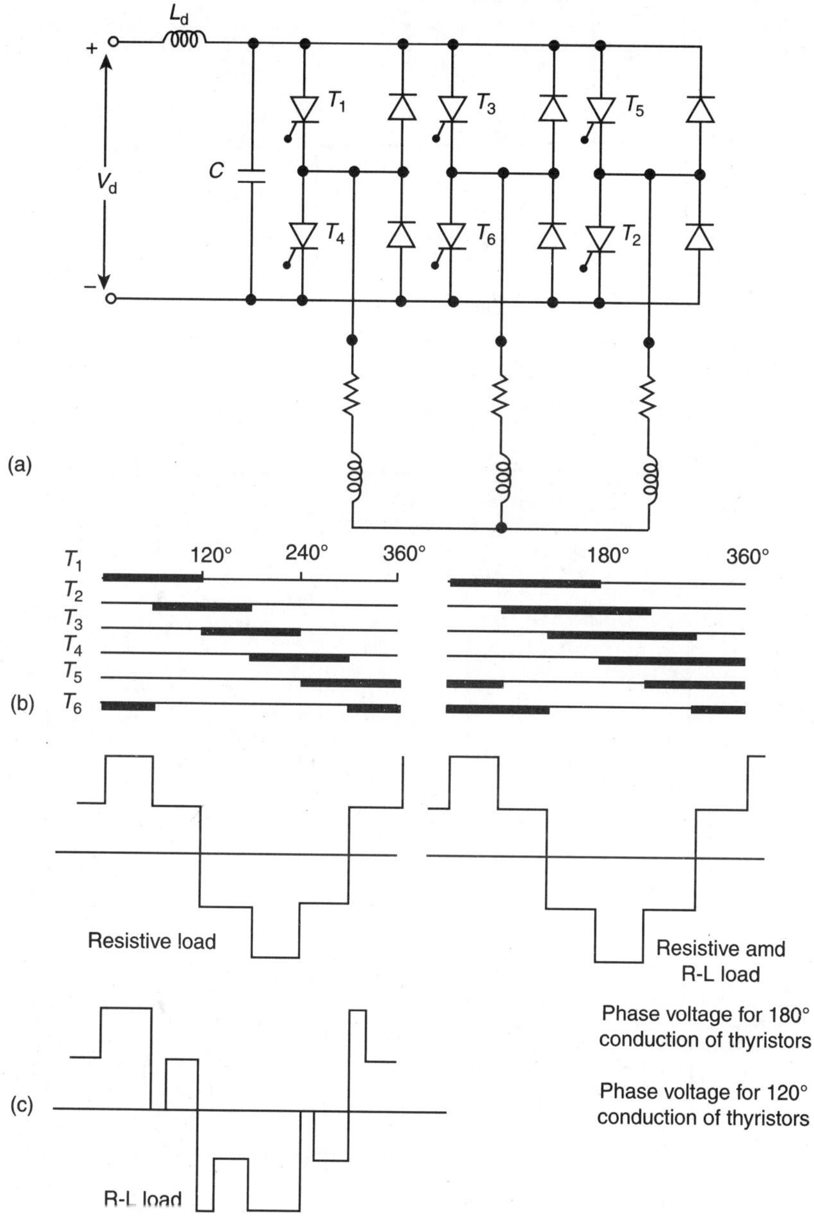

Fig. 3.93 *Voltage source inverter feeding R-L load, (a) Schematic diagram, (b) Sequence of thyristor conduction, (c) Phase voltages of load*

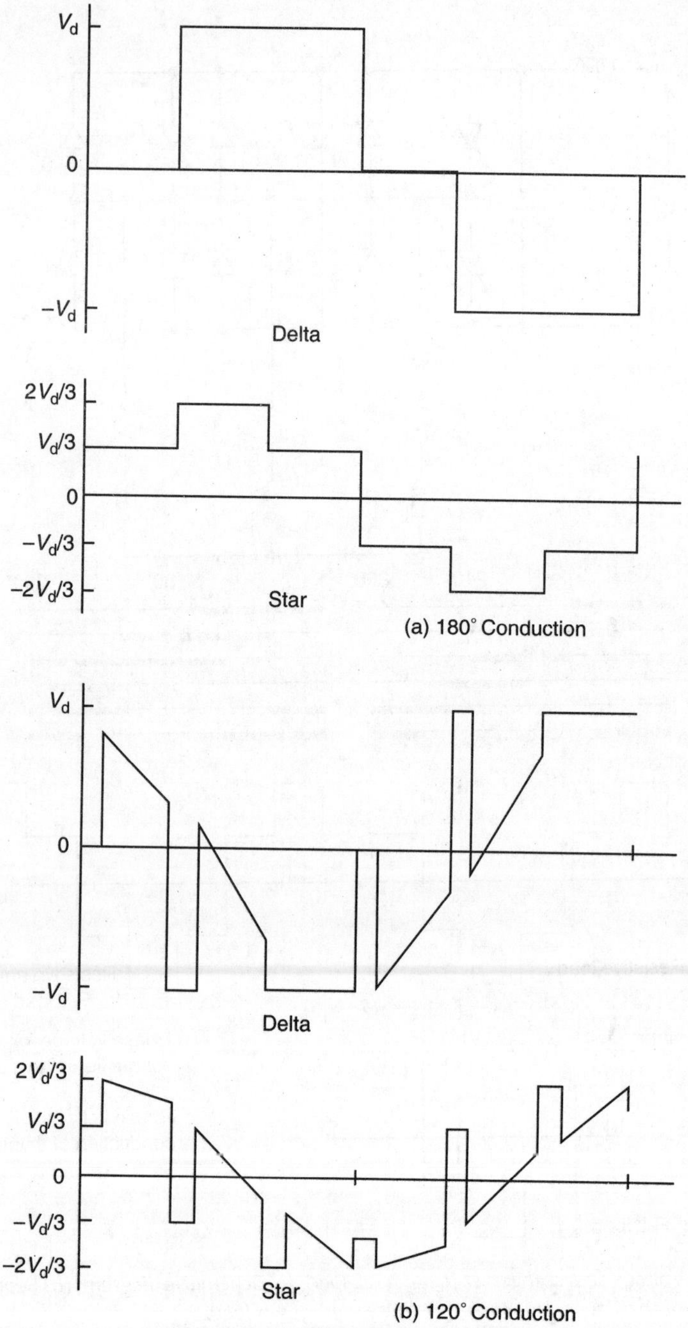

Fig. 3.94 *Voltage waveforms of VSI*

voltage can be determined from the known potentials of the output terminals. The output voltage is in the form of a square or stepped wave and is independent of load. In case of 120° conduction only two thyristors conduct at any instant, one in the top half and the other in the bottom half. The control is simple compared to the 180° case, but the output voltage is dependent on the load. The voltage wave forms of these inverters are shown in Fig. 3.94. Due to the extensive use the 180° of the 180° case, a variable voltage inverter is also called a square wave inverter.

At very low speeds there are commutation problems, as the input voltage may not be sufficient for commutating the thyristors. This imposes a lower limit on the frequency.

A multimotor drive is possible using this type of inverter. The commutation is load independent, and the converter and load need not be matched. The converter represents a source and the motor can be plugged on.

Output frequencies up to 1500 Hz are possible using this converter which makes it very suitable for high speeds. Transistorised converters of rating 10 KVA are available up to 6000 Hz.

The converters are built up to a rating of 200 KVA. Due to commutation problems at low speeds, the lowest operating frequency is 10 Hz. The speed control range is 1:20. The inverter is not suitable for acceleration of the motor on load and for sudden load changes. The dynamic behaviour is poor at low speeds and good at high speeds.

Regeneration requires an additional phase controlled converter at the line terminals, as shown in Fig. 3.95.

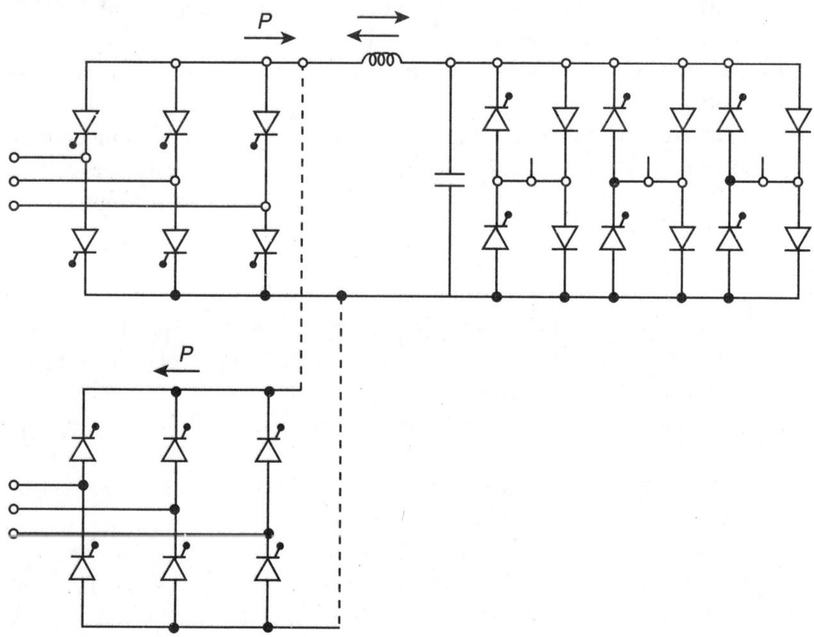

Fig. 3.95 *VSI for regeneration*

The output voltage has harmonic components which depend on the load. The load current is non-sinusoidal.

When an induction motor is fed from this type of inverter the harmonic content of the motor current is decided by the motor leakage reactance. The leakage reactance also influences the peak current, which in turn influences the choice of the inverter thyristors. The higher the leakage reactance, the smaller is the harmonic content and the peak value of the motor current. Both the harmonic losses and the torque pulsations are influenced by the leakage reactance. Open loop control is possible but there may be stability problems at low speeds. The line power factor is poor due to phase control.

Inverters with Constant Link Voltage In the case of these inverters, the voltage impressed is constant and voltage control is obtained in the inverter itself using the principle of PWM. The principle is illustrated in Fig. 3.93, with the resulting voltages shown. The problem of commutation at low speeds is avoided, which makes it possible to extend the speed control up to zero speed. The voltage waveform is not a square wave but is pulsed, depending upon the type of modulation employed. The inverter has a very good dynamic response.

The specific features of this type of inverter can be summarised as follows:

i. It has a constant dc link voltage and uses the PWM principle for voltage control.
ii. The output voltage waveform is improved with respect to the harmonic content, which is reduced. Therefore torque pulsations are not a problem, even at low speeds.
iii. The parallel operation of many inverters on the same bus system is possible.
iv. Uninterrupted operation using a buffer battery is possible for long periods.
v. When a diode rectifier is used the power factor on the ac side is excellent.
vi. The power and control circuits are complicated, when compared to square wave inverters.
vii. Four quadrant operation is possible. During braking a battery or another converter with phase control is used on the line side.
viii. Both single and multimotor operations are possible. Speed reversal is very smooth and it can be achieved with full torque capability of the motor. The dynamic behaviour is fast.
ix. Inverters are built up to 150 Hz and a rating of 450 kVA. The speed range is $1:\infty$.
x. The inverter and load need not be matched. The inverter can be considered as a source and the load plugged on.
xi. The harmonic content of the load current is further reduced by the load inductance; by leakage reactance in the case of a motor load. The peak current capability of the inverter can be smaller than that of square wave inverters, which decreases further if the leakage reactance is large. The filter size can be smaller than that of square wave inverters.

xii. Open loop operation is possible. The inverter has a fast response and very good dynamic behaviour.

3.4.2 Current Source Inverters for Feeding Three Phase Motors

In these type of inverters, the controlled quantity is the current in the dc link. The current from the dc source remains constant at the controlled value, irrespective of the load and events in the inverter. The voltage across the load adjusts itself. The link current is maintained constant by means of a large link inductance. The capacitance in the dc link can be dispensed with.

The dc link current is made to flow through the phases of the load alternately by controlling the inverter. These inverters are classified depending upon the commutation, and are shown in Fig. 3.96. The control of the link current is achieved by means of a phase controlled rectifier on the line side. As the current is a controlled quantity, feed back diodes are not required. The current source inverter shown in Fig. 3.96(a) employs individual commutation of phases. Auxiliary thyristors are used for commutation. The inverter shown in Fig. 3.96(b) employs sequential commutation and is rapidly gaining popularity. The diodes D1–D5 are used to prevent the discharging of capacitors through the load. They thus trap the charge on the capacitors. The inverter shown in Fig. 3.96(c) is a special case of an auto sequential commutated inverter.

The special features of current source inverters can be summarised as follows:

i. These inverters have load dependent commutation. As the load circuit, elements form a part of the commutation circuit, the inverter and the load (motor) must be matched with each other.

ii. The inverter has a very simple configuration due to the absence of free wheeling diodes.

iii. These are suitable for single motor operation.

iv. Since the dc link contains only inductance two/four quadrant operation is straightforward. No additional converter is required on the line side.

v. A phase controlled rectifier is invariably required on the line side. The variable dc link voltage is converted to constant current by means of a high inductance.

vi. Inverters employ forced commutation to give variable frequency currents.

vii. The currents are non-sinusoidal.

viii. Converter grade thyristors are sufficient and thyristor utilisation is good. The capacitance used for commutation has a value which is a compromise between the voltage spikes and the highest operating frequency of the inverter.

ix. The inverter can recover easily from commutation failure. In case of any fault, the link inductance prevents a fast rise of fault current so that by the time it reaches a large value, the fault can be suppressed.

x. There are stability problems at light loads and high speeds. A minimum current is required for satisfactory commutation.

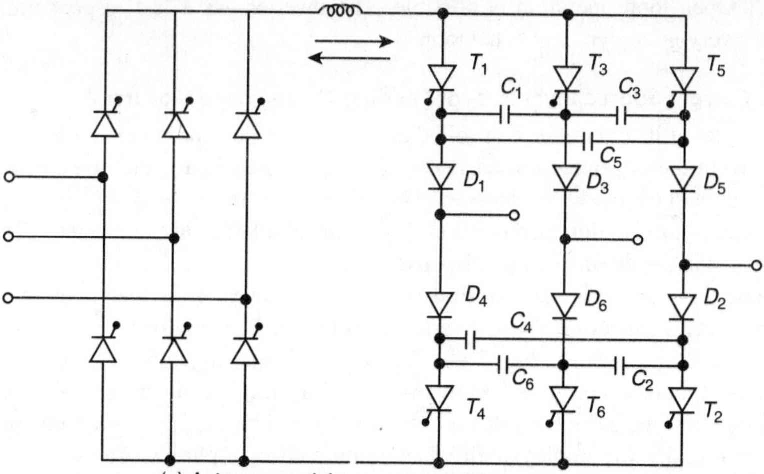

(a) Auto sequential commutated CSI.

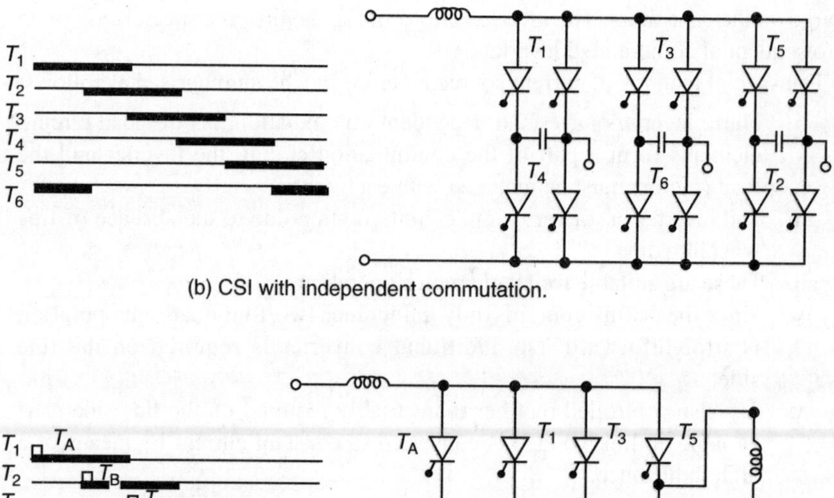

(b) CSI with independent commutation.

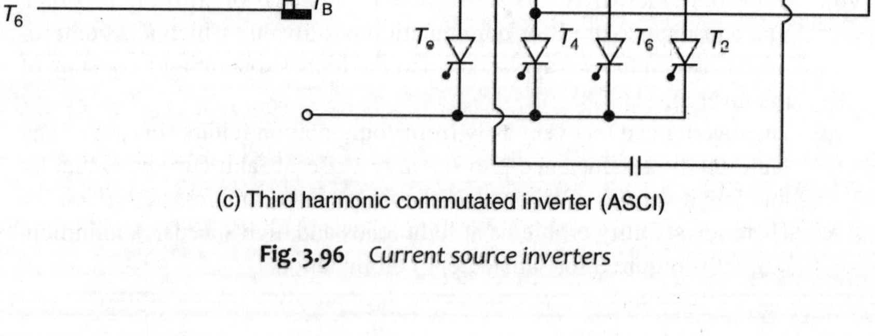

(c) Third harmonic commutated inverter (ASCI)

Fig. 3.96 *Current source inverters*

xi. Open loop operation is not possible, while operation over a wide range of frequencies is possible. Inverters have a sluggish dynamic response.

xii. The power factor is poor due to phase control on the line side. When a three phase motor is used as a load on the inverter, the leakage, reactance of the motor influences the harmonic voltage. It also causes voltage spikes during commutation. Being a parameter of the commutation circuit, it determines the time of commutation which limits the upper frequency. Therefore a motor must have low leakage reactance to have reduced harmonic voltages, small voltage spikes, and an increased range of speed control. The spikes in voltage influence the choice of the thyristors and affect motor insulation.

3.4.3 Comparison of Voltage Source and Current Source Inverters

The two types of dc link converter systems available for the speed control of ac motors are voltage source converters and current source converters. The dc link voltage is controlled in the former and the dc link current in the latter.

Both systems comprise a rectifier, which is usually a line commutated static converter, an intermediate circuit with energy storage and a force commutated inverter giving a variable frequency output.

In the case of a VSI the energy is stored in the capacitance. The commutation of the inverter takes place independently of the load, and the inverter can be considered as an alternating voltage source with variable frequency and amplitude.

In the current source inverter an inductor stores the intermediate circuit energy. The load becomes a part of the commutation circuit. The inverter is a source of alternating current of variable frequency.

The dc supply of voltage source inverters has a low impedance at all frequencies. The current in the load depends upon the load impedance.

The dc supply of current source inverters has a high impedance due to the link inductance holding the current constant. The output current is decided by the operation of the inverter and the voltage by the load impedance.

Voltage source inverters are suitable for loads of high impedance while current source inverters are suitable for loads of low impedance and unity power factor.

Voltage source inverters employ inverter grade thyristors, whereas CSI use converter grade ones.

Regeneration requires an additional phase controlled converter in a VSI, while it is straightforward in a CSI.

Voltage source inverters allow multimotor operation, unlike CSIs.

Thyristor utilisation is very good in current source inverters, whereas it is poor in voltage source inverter.

3.4.4 Voltage Control of Inverters

It has already been mentioned that inverters providing a variable frequency supply to three phase motors should be capable of providing a variable voltage. This is required to avoid saturation and ensure operation at constant flux density. The voltage control can be affected either external to the inverter or within it.

The voltage control external to the inverter can be done in two ways.

i. by varying the dc link voltage
ii. by varying the ac voltage at the output using a variable ratio transformer.

i. The variation of dc link voltage can be achieved in many ways. It has the advantage that the output voltage waveform is maintained over a wide range of frequencies. But at very low frequencies, the dc link voltage may be too low to commutate the inverter. This limits the lowest operating frequency and hence the frequency range. The dynamic response is also poor.

A variable dc supply can be obtained by using a phase controlled rectifier on the line side. A closed loop control varies the firing angle depending upon the frequency. The function generator (Fig. 3.97) gives a relation between the stator frequency and applied voltage to the stator for constant air gap flux or given flux conditions in the motor. The output of the function generator is voltage for a given value of f_s. This voltage is compared with the measured value of voltage and the error so obtained is used to change the firing angle of the converter on the line side. The frequency is obtained by controlling the firing and conduction of the thyristors of the machine side converter. A current loop is also employed to limit the current to safe values during dynamic operation of the system.

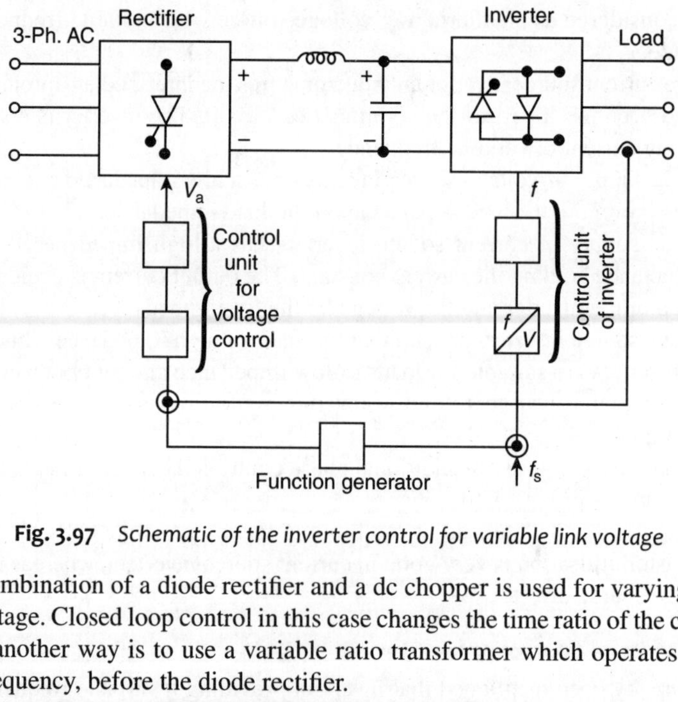

Fig. 3.97 *Schematic of the inverter control for variable link voltage*

A combination of a diode rectifier and a dc chopper is used for varying the dc link voltage. Closed loop control in this case changes the time ratio of the chopper.

Yet another way is to use a variable ratio transformer which operates at constant frequency, before the diode rectifier.

These methods are shown in Fig. 3.97.

ii. The voltage control of the inverter can be affected by means of a variable ratio transformer interposed between the motor and inverter. The method is very simple. Even in this case the waveforms of output voltage remain the same over a wide frequency range. The line side converter can be a simple diode rectifier. This provides a good pf. The main disadvantage of this method is that the transformer has to be designed for low frequencies and its size is large. The system also has an extremely poor dynamic response.

Voltage Control within the Inverter The dc link voltage is constant and the inverter is controlled to provide both variable voltage and variable frequency. As the link voltage is constant a simple diode rectifier may be employed on the line side. Variable voltage variable frequency supply to the motor is obtained within the inverter itself using suitable control based on the principles of PWM or PSM (phase shift modulation). The block diagram of control of the constant voltage inverter is shown in Fig. 3.98(a). The voltage is sensed and compared with the output of the function generator. The error is used to change the amplitude of the reference wave in order to obtain the desired value of voltage. The frequency of the reference wave is changed in order to get the desired frequency. As the inverter is supplied at constant voltage commutation problem at low frequencies disappears. The operating frequency extends to zero and the system has a very good dynamic response. The voltage waveform is not the same at all frequencies. At low frequencies the harmonic content may increase.

PWM techniques are very widely used. They sometimes provide harmonic elimination also. For this purpose, sinusoidal modulation is used. The principles of PWM are illustrated in Fig. 3.98.

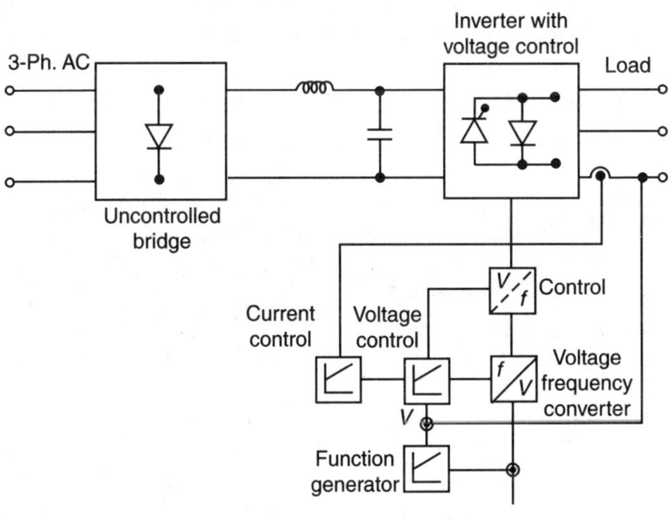

Fig. 3.98(a) *Voltage control within the inverter*

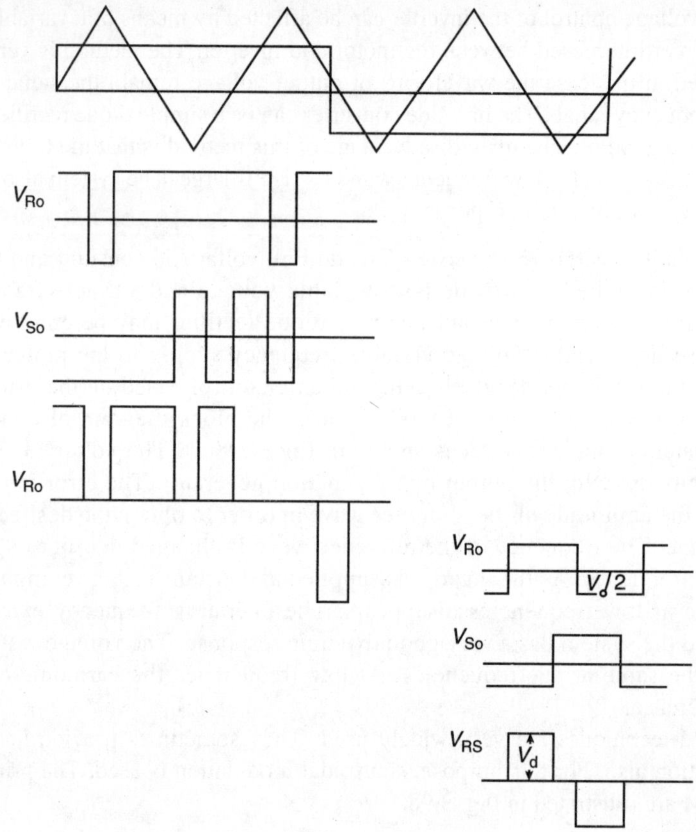

Fig. 3.98(b) *Representation of three phase inverter. (i) Voltage waveform for 180° (ii) voltage waveform for 120° for motor load*

The PSM technique for voltage control is illustrated in Fig. 3.99. This method has some disadvantages. The voltage waveform has a high harmonic content at lower frequencies and the utilisation of the thyristors is very poor. The technique involves two inverters, making it costly for low power applications.

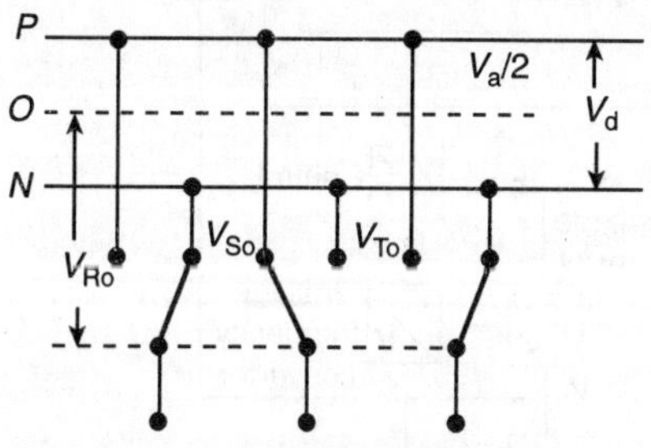

Fig. 3.98(c) *PWM principle using rectangular reference and triangular modulation*

3.4.5 Harmonic Neutralisation

The output voltage waveform of an inverter is non-sinusoidal. It contains a rich harmonic content. The harmonics cause additional losses and torque pulsations if a three phase motor is used as a load. These torque pulsations pose a problem at low speeds. Therefore it

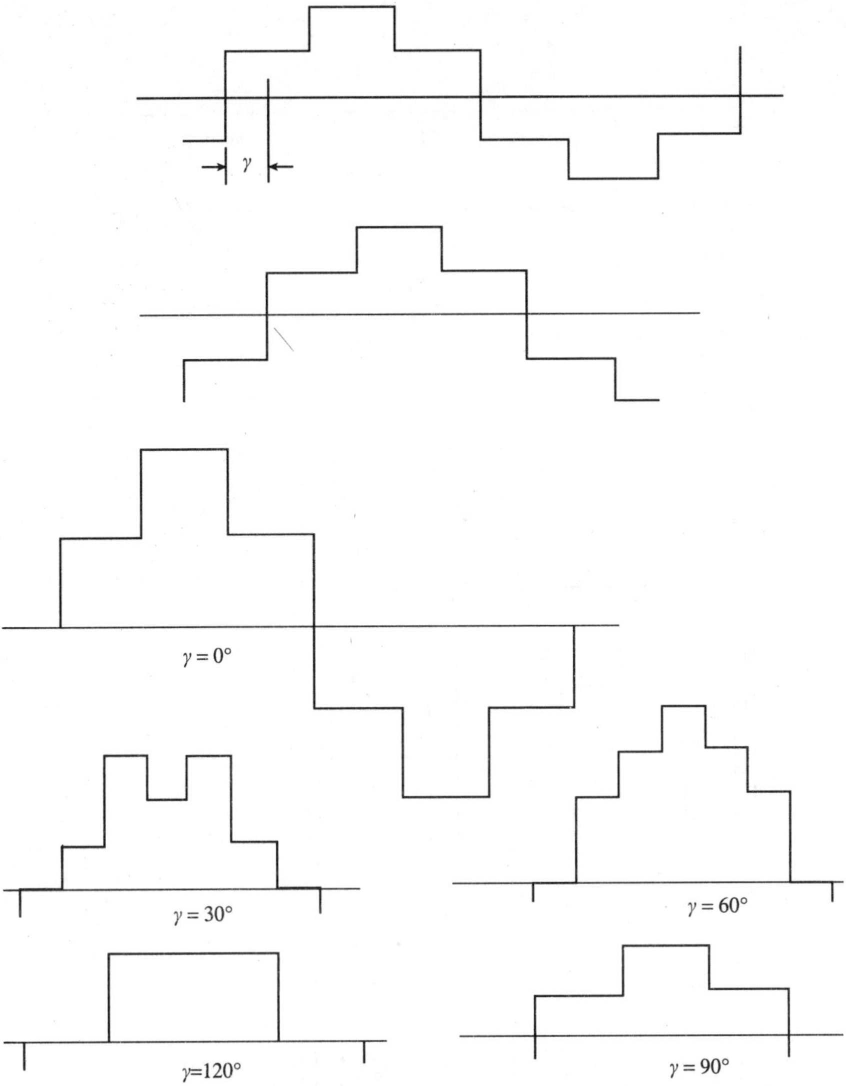

Fig. 3.99 *Phase shift modulation*

is necessary to improve the voltage waveform of an inverter and minimise the harmonic content. The following methods are employed to achieve this:

i. Reduction or even elimination of lower order harmonics by means of a switching process in the inverter. This method is called selective harmonic reduction and reduces the fifth and seventh harmonics to a minimum. The filter size decreases. The voltage waveform is shown in Fig. 3.100.

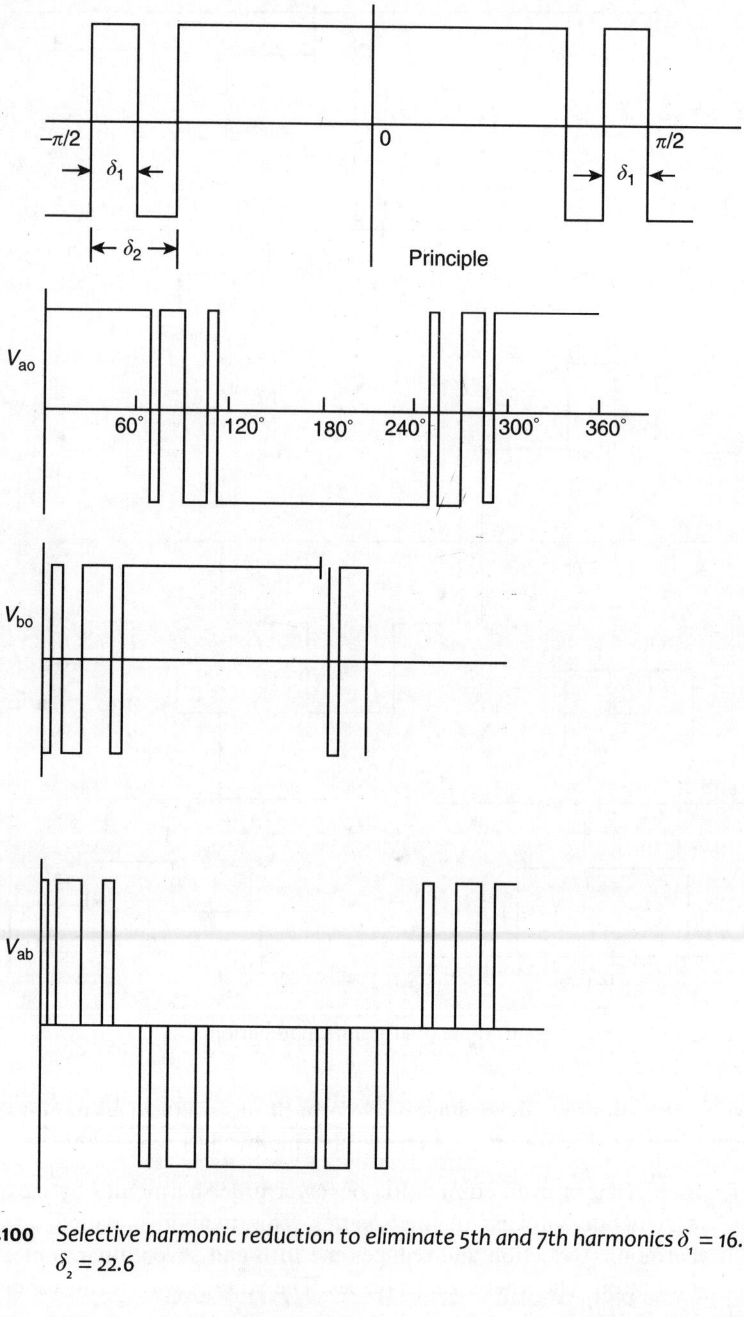

Fig. 3.100 *Selective harmonic reduction to eliminate 5th and 7th harmonics* $\delta_1 = 16.6$, $\delta_2 = 22.6$

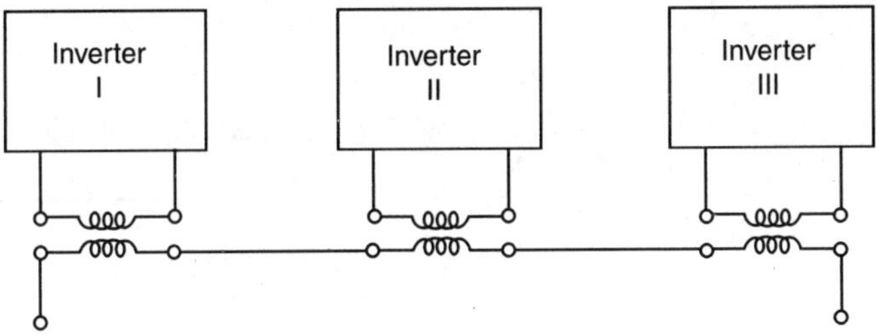

Fig. 3.101 *Series addition of inverter output voltages to form a stepped voltage waveform*

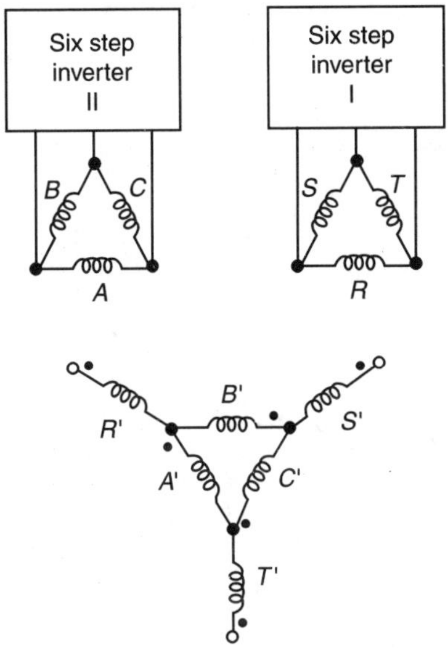

Fig. 3.102 *Synthesis of a three phase, 12 step wave*

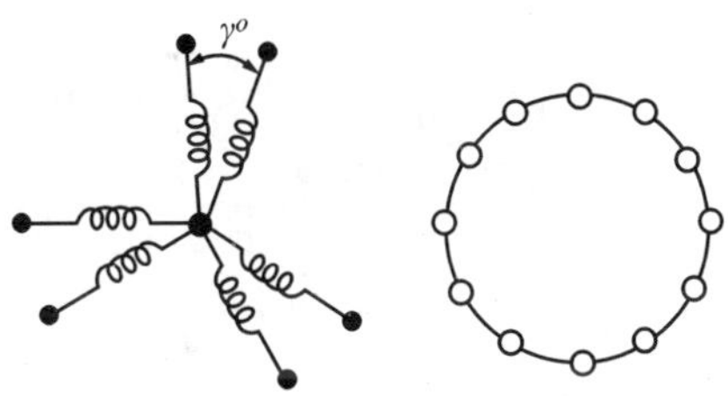

Fig. 3.103 *Six phase motor. Two sets of three phase windings having a displacement of γ^o*

ii. The number of pulses of the output wave can be increased to effectively reduce the harmonics. Several methods employed to increase the pulse number are shown in Figs 3.101, 3.102 and 3.103. A three phase output may be obtained using three single phase bridge converters as shown in Fig. 3.101. The resulting waveform has reduced harmonic content. The synthesis of the three phase waveform from two six-step inverters is illustrated in Fig. 3.102. The resultant twelve step waveform has a reduced harmonic content. The harmonics which are multiples of three are eliminated. The interconnection of the outputs of six-step inverters to get a twelve stepped waveform is shown in Fig. 3.103. Multistepped waveform with reduced harmonic content can be obtained using multiple phase shifted inverters. It is possible to get 12, 18, 24, … stepped waveform using suitable number of inverters and phase shift.

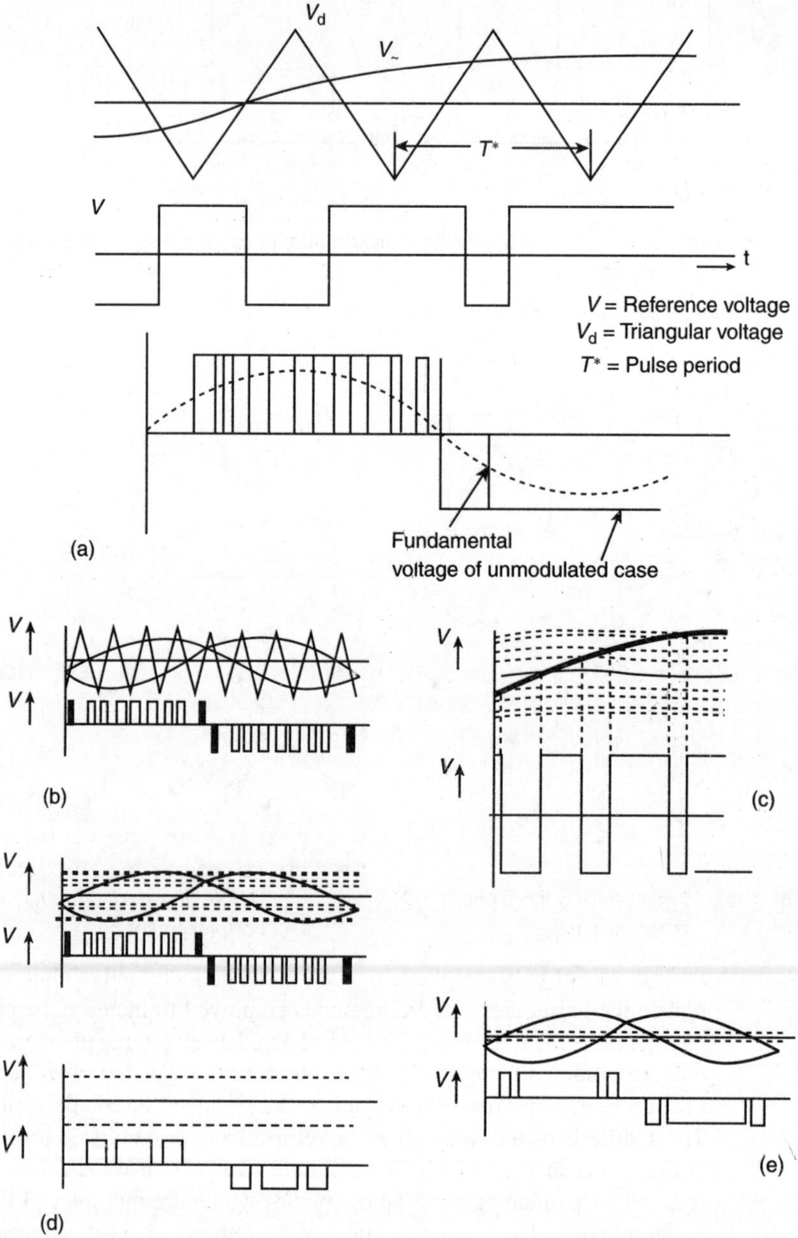

Fig. 3.104 *(a) Pulse width modulation methods. (b) Line voltage with synchronous modulation. (c) Principle of PWM using d.c. (d) Control range and output voltage for two different control (e) Extension of control range*

iii. Increasing the number of phases minimises the harmonic content. A twelve pulse inverter provides an output voltage free from fifth and seventh harmonics. The principle of the method and the waveforms are given in Fig. 3.103. Using this method, harmonics of order 5, 7, 17, 19 can be eliminated without affecting harmonics order of 11, 13, 23, 25, etc.

iv. Several harmonics can be eliminated by combining the phase displaced outputs of two inverters. This method has poor inverter utilisation and is not used.

v. Multiple pulse width control — By increasing the switching of the inverter voltage, waveforms having several pulses can be obtained, in which the content of lower order harmonics is minimised. The voltage waveforms are shown in Fig. 3.104.

vi. Pulse width modulation is also used to minimise the lower order harmonics. Using sinusoidal modulation it is possible to get an output voltage waveform which has a very low harmonic content, as shown in Fig. 3.104. In this process, the pulse duration is sinusoidally modulated. The modulation process and output voltage are shown in Fig. 3.104.

3.5 CYCLOCONVERTERS

A variable frequency supply for feeding a three phase motor can be obtained from a cycloconverter which operates on a 50 Hz supply and provides a variable frequency supply at the output (Fig. 3.105). These are single stage frequency conversion devices which make use of both line commutation as well as forced commutation.

The phase control of line commutated cycloconverters provides a variable voltage. The firing angle is varied over the period of the input voltage. The output voltage is made up of segments of the input voltage. The voltage control is possible in the inverter itself. Because of the phase control, the power factor is poor. A cycloconverter can be used to interconnect two three phase systems (Fig. 3.106). The firing angle is so varied that the output voltage has a small distortion and is nearly sinusoidal. The load inductance further reduces the harmonic content in the current. The utilisation of the thyristors is poor at low frequencies. The cycloconverter allows power flow in both directions at any power factor of the load. The converter has a complex control system, and the output frequency can be varied from 0-1/2 input frequency. They are normally developed for 400 MHz, so that the output frequency is in the range 0–200 Hz.

The disadvantages of a poor power factor and a limited frequency range can be eliminated if forced commutation is employed. In this method, each converter has antiparallel three phase bridges provided with a common control circuit. The circuit details are given in Fig. 3.108.

The main features of cycloconverters can be summarised as follows:

i. Voltage control can be achieved in the converter itself.

ii. A cycloconverter functioning by means of line commutation has poor p.f. and limitation of output frequency. Forced commutation can be

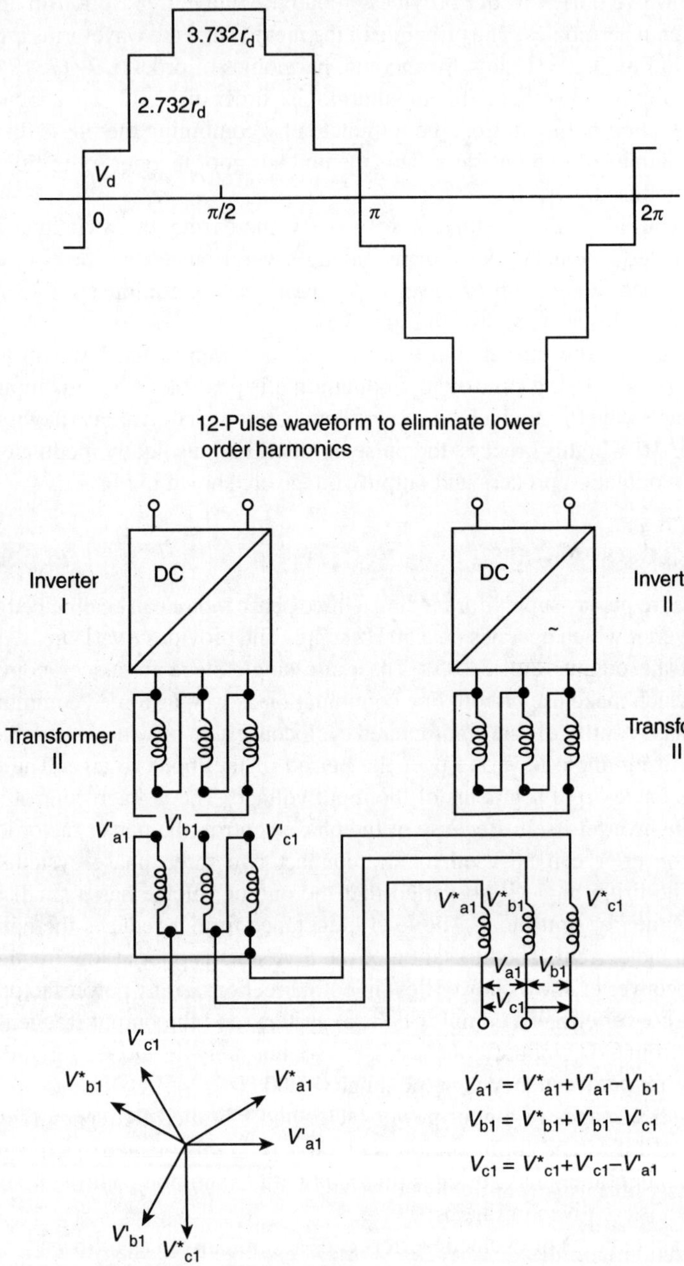

Fig. 3.105 *A method for synthesising twelve-pulse waveform*

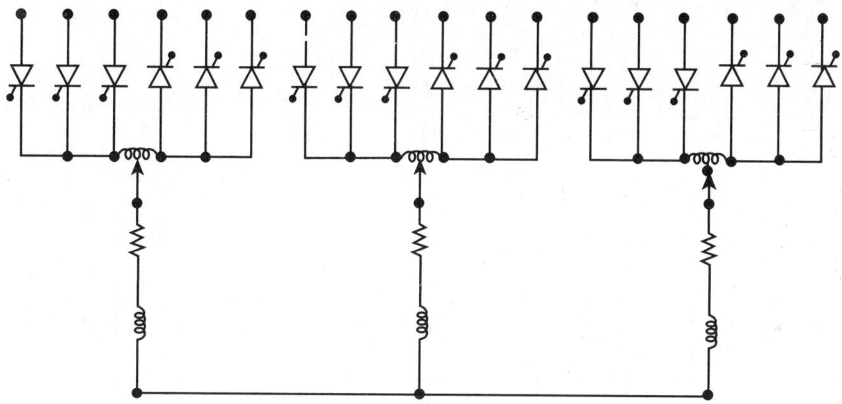

Fig. 3.106 *Three phase cycloconverter using mid-point three pulse converters*

employed to improve the p.f. and working range of frequencies. However this introduces switching losses.

iii. It is capable of power transfer in both directions between the source and the load. It can feed power to a load at any p.f. Regeneration is inherent over the complete frequency range.

iv. It delivers high quality sinusoidal output waveforms at all operating frequencies.

v. It requires many thyristors, which offsets the advantage of line commutation. However, no shut down is required if a thyristor fails. Output can be made available without any interruption, albeit with a slightly distorted waveform.

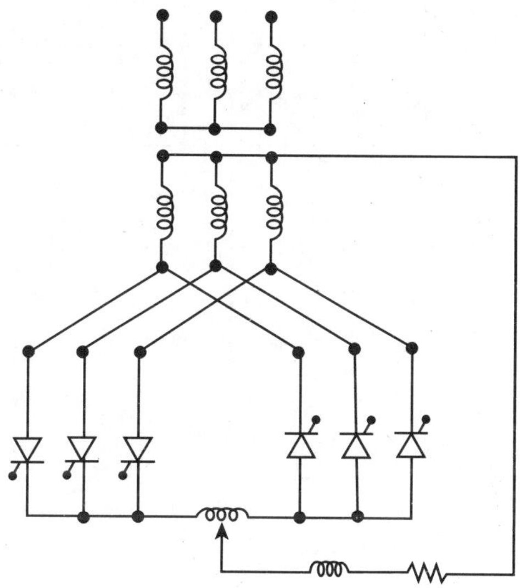

Fig. 3.107(a) *Single phase cycloconverter using three phase midpoint converter*

The above discussion of cycloconverter shows that an a.c. motor (synchronous or induction), when fed from a cycloconverters will have a very smooth low speed operation with least torque ripple. Four quadrant operation of the motor is straight forward as the cycloconverter allows power flow in either direction. Thus the cycloconverter is very attractive for feeding low speed, large power reversible ac motor drives.

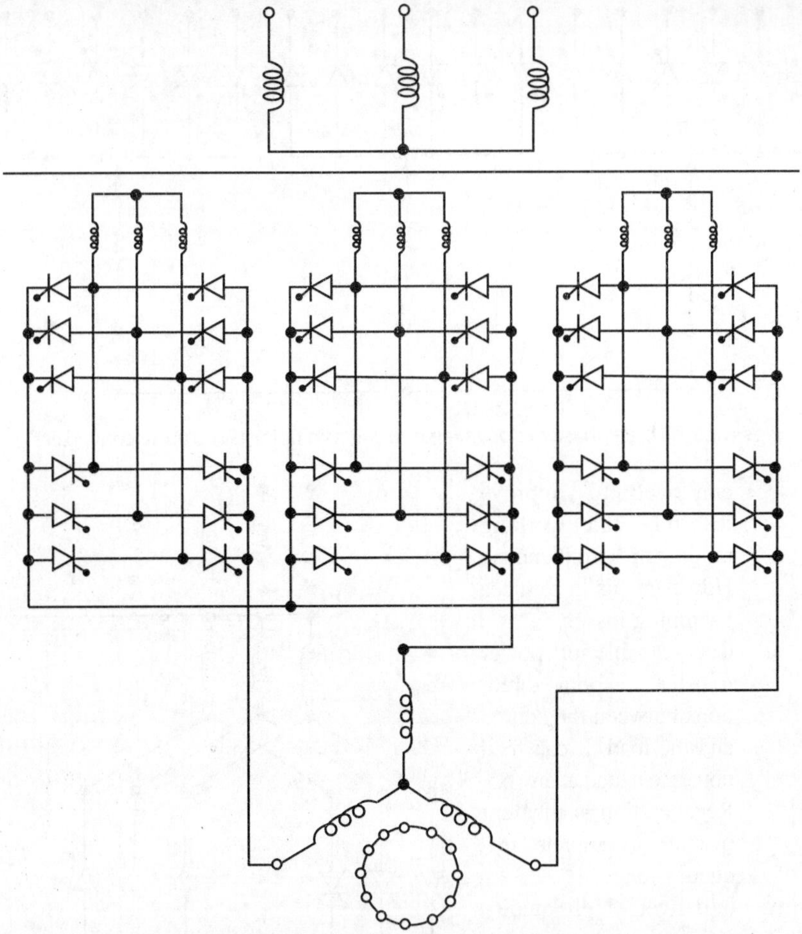

Fig. 3.107(b) *Three phase cycloconverter using antiparallel bridges*

3.6 AC VOLTAGE CONTROLLER

In earlier sections of this chapter, a detailed discussion of power conversion equipment which convert ac to dc and vice versa has been given. dc choppers do not actually perform the power conversion, but take part in controlling the current and voltage in the load from a constant dc voltage mains. AC voltage controllers also belong to this class of equipment to regulate the flow of current in the load. The load voltage can be varied in a stepless and smooth fashion and the load current can be controlled using this variable voltage. However, the frequency of the output is the same as that of the input. The source has a fixed frequency and fixed voltage. Introducing an ac voltage controller provides a variable voltage across the load at the same frequency as the source.

There are several areas, both in industry and household, where a variable voltage is required. Examples are control of lighting, heating, induction heating and

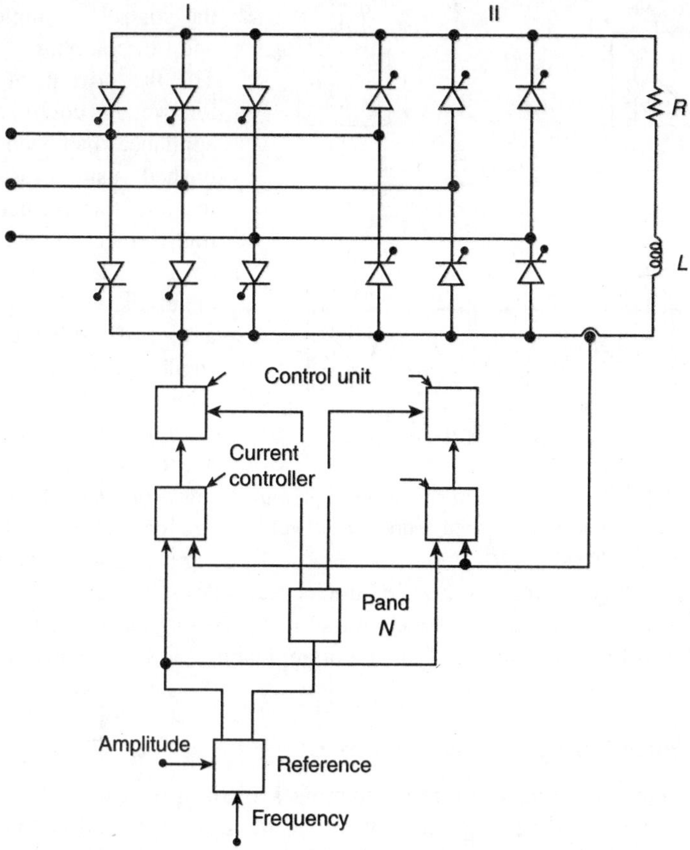

Fig. 3.108(a) *Basic circuit and control of a cycloconverter using antiparallel bridges*

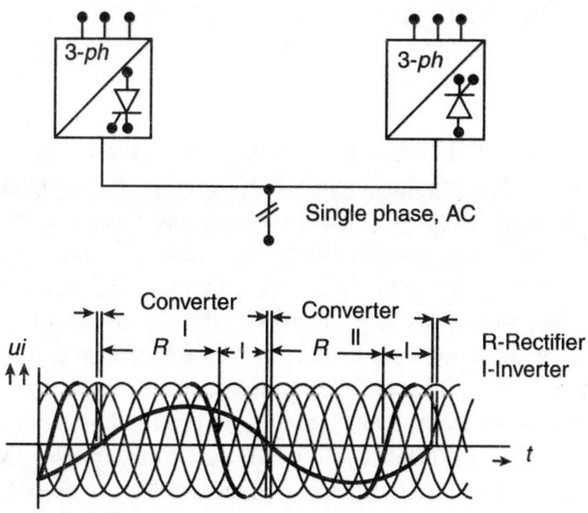

Fig. 3.108(b) *Principle of a cycloconverter and output voltage waveforms*

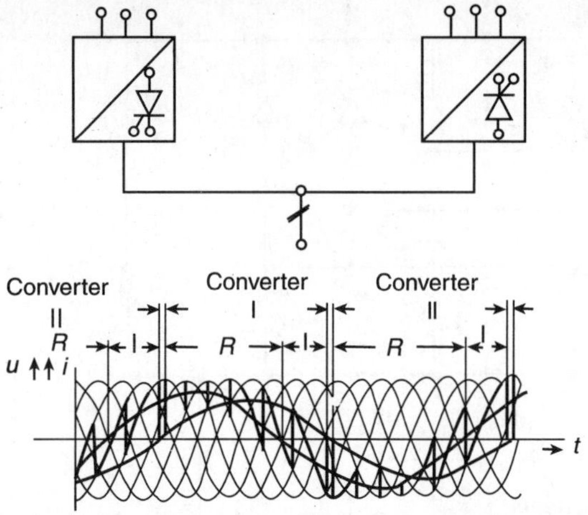

Fig. 3.108(c) *Principle of a cycloconverter with sinusoidal output voltage and output voltage waveforms*

the control of single phase and three phase motors. Till the advent of thyristors, voltage control for such appliances had been accomplished using resistances, reactors, transformers, autotransformer, potential dividers, etc. The development of thyristors has provided a way of getting a smooth stepless control of voltage without any delay. The class of converters used for voltage control and employing thyristors are called ac voltage controllers, ac voltage regulators or ac choppers.

These voltage controllers are used in electrical drives to control the speed of a single phase or three phase induction motors by varying the applied voltage of the motor. A brief description of the important features of these regulators is now given.

3.6.1 Single Phase AC Voltage Controller

A single phase ac voltage controller comprises a pair of back to back connected thyristors interposed between the source and the load. The load can be a pure resistance, pure inductance, or a combination of both. A single phase ac motor is also fed from the regulator. This kind of load consists of a back emf, besides a resistance and reactance. A control unit sends the firing pulses to the thyristors during the respective half cycles at the desired instant. The thyristors are symmetrically triggered if the firing instant is the same in both the half cycles, taking the zero of the half cycles as a reference. A simple single phase ac voltage controller is depicted in Fig. 3.109. The thyristor T_1 conducts during the positive half cycle and thyristor T_2 during the negative half cycle a is the firing angle. Up to the firing instant the supply voltage is across the thyristors in the corresponding half cycles and after this instant ($\omega t \geq a$) the source voltage is transferred to the load. Therefore, by changing the value of a the load voltage can be controlled both in the positive and negative half cycles. However, the waveforms of current and load voltage depend upon the load parameters and its impedance angle. To determine the average and effective values of these, their waveforms must be determined.

When the load is a pure resistance the current in the load is in phase with the voltage. At the end of each half cycle when the voltage passes through zero so does the current and the thyristor ceases conduction. Hence both the load voltage and the load current are known.

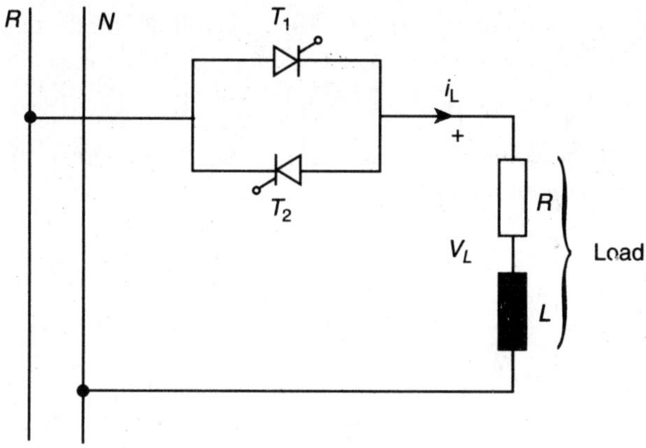

Fig. 3.109 *Antiparallel thyristors as ac voltage controller*

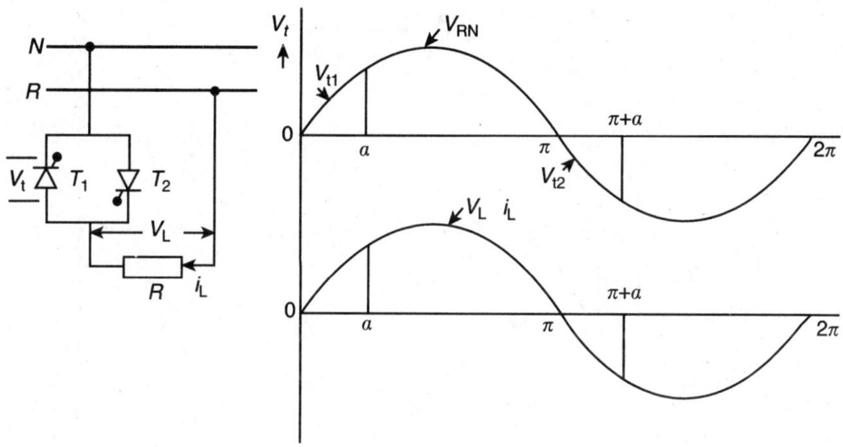

Fig. 3.110 *A single phase ac voltage controller feeding pure resistance. Voltage and current waveforms*

The thyristor voltage, load voltage and load current waveforms are depicted in Fig. 3.110 for a pure resistive load. The firing angle can be varied from 0 to 180°. The average and rms values of the voltage and current can be obtained over this range of firing angles.

If the supply voltage is given by

$$V_s = V_m \sin \omega t \tag{3.104}$$

the load voltage at any firing angle is

$$V_L = V_m \sin \omega t; \quad a \leq \omega t \leq 180° \tag{3.105}$$

The current at this instant jumps to $(V_m/R)\sin a$ and follows for the rest of the time.

$$i_L = \frac{V_m}{R}\sin \omega t \qquad a \le \omega t \le 180° \tag{3.106}$$

From the current and voltage waveforms of Fig. 3.110 the average and rms values can be obtained in the usual way as

$$V_{av\alpha} = \frac{V_m}{\pi}(1+\cos a) \tag{3.107}$$

$$V_a = V_m\sqrt{\frac{1}{\pi}\left(\frac{\pi - a}{2} + \frac{1}{4}\sin 2a\right)} \tag{3.108}$$

respectively.

Referring these values to the average and rms values at $a = 0$, we get

$$\frac{V_{av}a}{V_{avo}} = \frac{1+\cos a}{2} \tag{3.109}$$

$$\text{and } \frac{V_a}{V_o} = \sqrt{\left(1 - \frac{2a - \sin 2a}{2\pi}\right)} \tag{3.110}$$

The ratio V_a/V_o as a function of a is depicted in Fig. 3.111 and is called control characteristic.

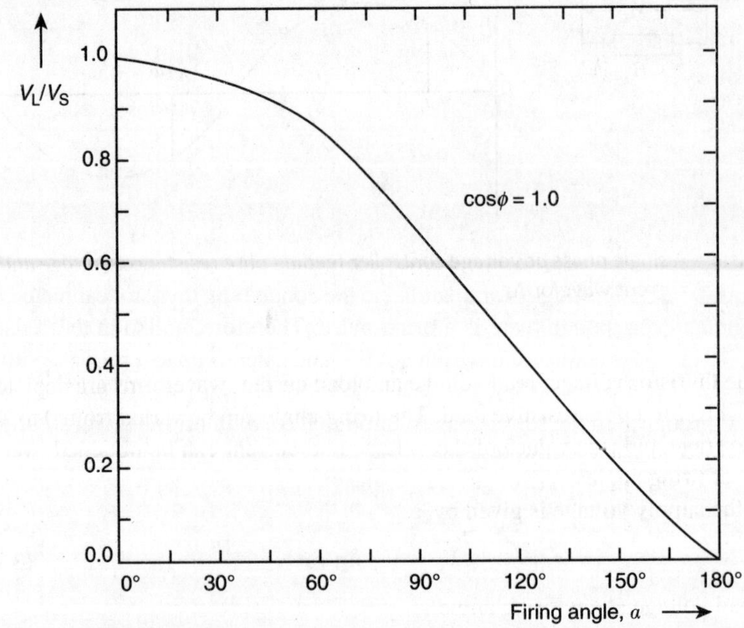

Fig. 3.111 *Control characteristic of single phase controller feeding pure resistive load*

The current waveform is similar to the voltage waveform the average and rms values of the currents can be obtained by dividing the corresponding voltage by the load resistance R.
Therefore, we have

$$\frac{I_{\text{Lava}}}{I_{\text{Lavo}}} = \frac{1 + \cos\alpha}{2} \tag{3.111}$$

$$\text{and } \frac{I_\alpha}{I_o} = \sqrt{\left(1 - \frac{2\alpha - \sin 2\alpha}{2\pi}\right)} \tag{3.112}$$

The power output at any firing angle α, referred to the power in the uncontrolled case ($\alpha = 0$), is

$$\frac{P_\alpha}{P_o} = 1 - \frac{2\alpha - \sin 2\alpha}{2\pi} \tag{3.113}$$

When the load is a pure inductance, the load current lags the voltage by 90°. The equation governing a pure inductance is (referring to Fig. 3.112).

$$L\frac{di_L}{dt} = V_L \tag{3.114}$$

$$\text{and} \quad i_L = \frac{1}{L}\int_{\hat{a}}^{\hat{a}} V_L dt \tag{3.115}$$

Again the load voltage is

$$V_L = V_m \sin \omega t \alpha < \omega t < (2\pi - \alpha) \cdot$$
$$(2\pi - \alpha) = \beta \tag{3.116}$$

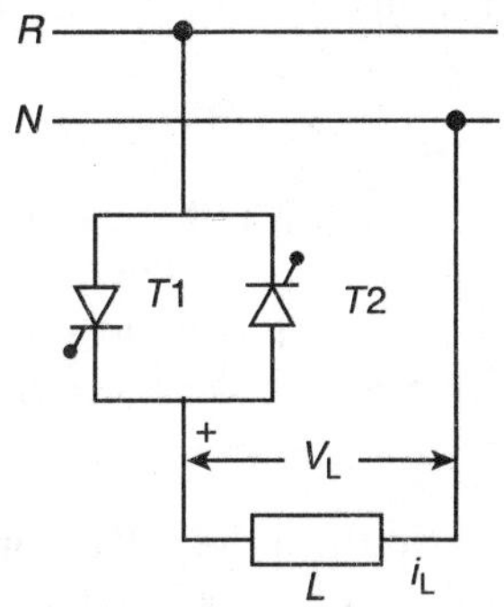

Fig. 3.112 *Single phase ac voltage controller feeding pure inductance*

When the voltage becomes zero at π the thyristor does not cease conduction but continues to conduct load current till the load current naturally goes to zero. Therefore the load voltage contains a portion of the negative half cycle also. The other thyristor, which is antiparallel to the conducting thyristor cannot be made to conduct even though there is a firing pulse. Therefore, unlike in the resistance case, the effective control starts from $\alpha = 90°$ and extends up to $\alpha = 180°$. With pure inductive loads control is not possible for angles less than 90°. When the voltage passes through zero the current passes through its maximum value. Substituting for V_L in Eq. 3.115 and performing the integration between the limits α and ωt we get

$$i_L = \frac{V_m}{\omega L}[\cos\alpha - \cos\omega t] \tag{3.117}$$

The maximum value of current occurs at $\omega t = \pi$ and is given by

$$I_{\text{Lm}\alpha} = \frac{V_m}{\omega L}(\cos\alpha + 1) \tag{3.118}$$

For $a > 90°$, $\cos a$ is negative and $I_{Lma} < V_m/\omega L$. This also shows that $a < \pi/2$ is not possible, as it makes

$$I_{Lma} > \frac{V_m}{\omega L}$$

which is not a practical case. The peak amplitude decreases as a increases and is zero at $a = \pi$.

The current which attains the peak value given by Eq. 3.118 at a given a falls to zero at $\omega t = (2\pi - a)$. The current wave is symmetrical about zero voltage. The current and load voltage waveforms are given in Fig. 3.113, for various values of a. The thyristor conducts for a period ranging from a to $2\pi - a$, given by $\delta = (2\pi - 2a)$. The time of conduction

$$t_c = \frac{2}{\omega}(\pi - a) \tag{3.119}$$

The time of conduction for $a \pm \frac{\pi}{2}$ (case of full control) is maximum, and is given by

$$t_c = \frac{\pi}{\omega} \tag{3.120}$$

From Eq. 3.119 it can be concluded that there is no control for a less than $\pi/2$. If $a < \pi/2$ the time of conduction exceeds π/ω which is not practically possible. Further, as a increases the time of conduction decreases and finally reaches zero when $a = \pi$. The range of firing angles for a pure inductive load is $\frac{\pi}{2} \le a \le \pi$.

Now that the voltage and current waveforms are known, the average and rms values of the load voltage and load current can be determined. The average load voltage

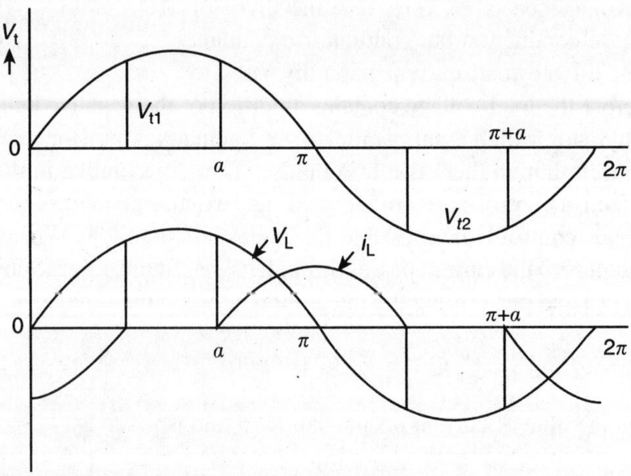

Fig. 3.113 *Load current and voltage waveforms*

$$V_{\text{Lav}a} = \frac{V_{\text{m}}}{\pi}(\cos a - \cos \pi) \qquad \frac{\pi}{2} < a < \pi \tag{3.121}$$

or
$$\frac{V_{\text{Lav}a}}{V_{\text{Lav}\,\pi/2}} = \frac{(1+\cos a)}{2} \quad \frac{\pi}{2} < a < \pi \tag{3.122}$$

The rms value of the load voltage

$$\frac{V_a}{V_{\pi/2}} = \sqrt{\left(2\left(1 - \frac{a}{\pi} + \frac{\sin 2a}{2\pi}\right)\right)} \tag{3.123}$$

The average value of the load current

$$I_{\text{Lav}a} = I_{\text{Lav}\,\pi/2}[(\pi - a)\cos a + \sin a] \tag{3.124}$$

or
$$\frac{I_{\text{Lav}a}}{I_{\text{Lav}\,\pi/2}} = (\pi - a)\cos a + \sin a$$

The rms value of the load current

$$\frac{I_a}{I_{\pi/2}} = \sqrt{\frac{4}{\pi}\left((\pi - a)\left(\cos^2 a + \frac{1}{2}\right) + \frac{3}{2}\sin a \cos a\right)} \tag{3.125}$$

The control characteristic ($V_a/V_{\pi/2}$) as a function of a is shown in Fig. 3.114. From the characteristic it is clear that the effective control starts at $a = \pi/2$. For $a < \pi/2$, $V_a > V_\pi/2$, it is not possible.

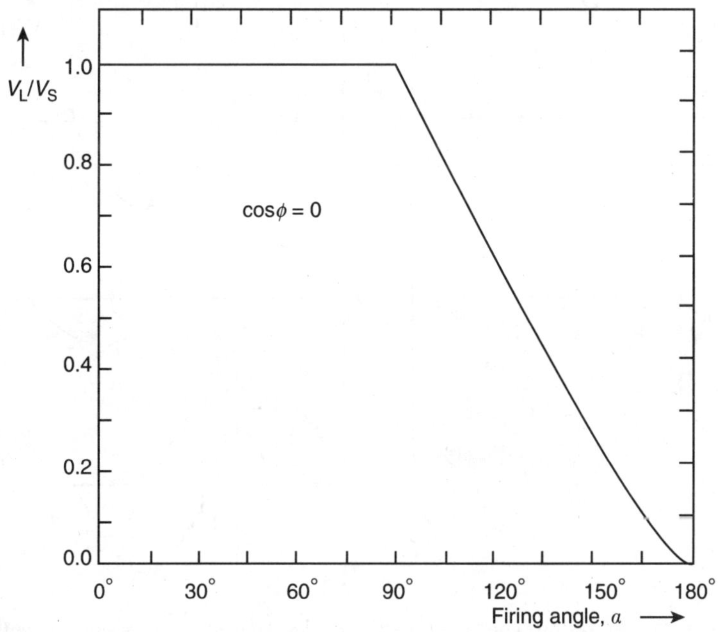

Fig. 3.114 *Control characteristic of single phase controller feeding pure inductive load*

When the load comprises both a resistance and an inductance in series, as shown in Fig. 3.115, the waveforms of voltage and current differ. The equation describing the load is

$$Ri_L + L\frac{di_L}{dt} = V_L, \qquad a > \omega t > (a + \delta) \tag{3.126}$$

Again, V_L follows the supply voltage waveform after the thyristor is triggered. The conduction time and effective range of control angles are decided by the ratio L/R of the load. Now, if a is the firing angle for $\omega t \geq a$

$$Ri_L + L\frac{di_L}{dt} = V_m \sin \omega t \tag{3.127}$$

Here too the current i_L does not fall to zero at the instant the voltage becomes zero. Therefore the conduction of the thyristor continues for some time (after zero voltage) during the negative half cycle also. During this period the antiparallel connected thyristor does not conduct, even though there is a firing pulse. The effective control (by the same reasoning as before for pure inductive loads) starts from $a = \theta$, the impedance angle of the load. Equation (3.128) is solved for a general solution for i_L, given by

$$i_L = \frac{V_m}{\sqrt{R^2 + \omega^2 L^2}} \sin(\omega t - \theta) + C_1 e^{-Rt/L} \tag{3.128}$$

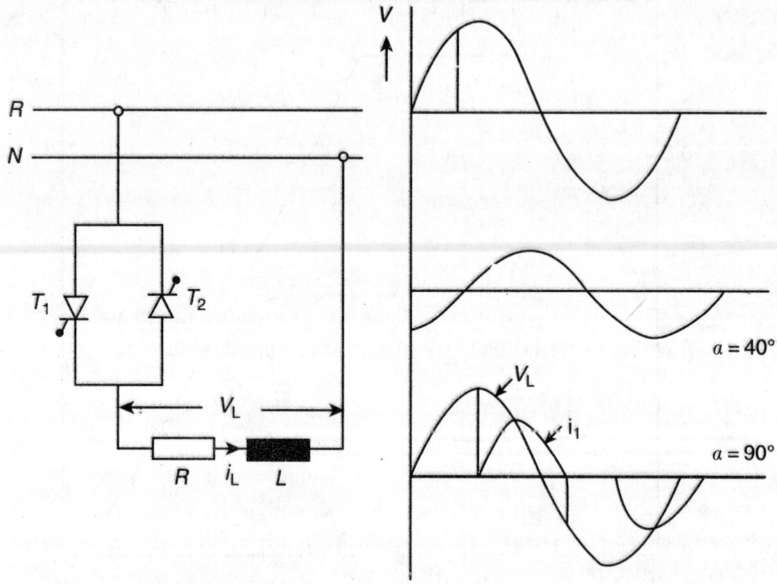

Fig. 3.115 *AC voltage controller feeding R-L load $\theta = 40°$ voltage and current waveforms*

The constant C_1 can be determined from the initial conditions, i.e., $\omega t = a$, $i_L = 0$

$$\text{Therefore } C_1 = -\frac{V_m}{\sqrt{R^2 + \omega^2 L^2}} \sin(a - \theta) e^{\frac{a}{\omega}\frac{R}{L}} \tag{3.129}$$

Using this in Eq. 3.128 we have

$$i_L = \frac{V_m}{\sqrt{R^2 + \omega^2 L^2}} [\sin(\omega t - \theta) - \sin(a - \theta) e^{-\left(t + \frac{a}{\omega}\right) R/L} \tag{3.130}$$

The lower limit of integration in determining the average value is a. The upper limit δ is the value of ωt where i_L = zero again. This can be found from Eq. 3.130 as

$$\beta = a + \frac{\omega L}{R} \log_e \frac{\sin(a - \theta)}{\sin(\beta - \theta)} \tag{3.131}$$

The time of conduction is

$$t_c = \frac{L}{R} \log \frac{\sin(a - \theta)}{\sin(\beta - \theta)} \tag{3.132}$$

Alternatively,

$$\sin(\beta - \theta) = \sin(a - \theta) e^{-\beta R / L\omega}$$

But $\frac{\omega L}{R} = \tan\theta$ and $\beta = a + \omega t_c$. Substituting we get

$$\tan(a - \theta) = \frac{\sin \omega t_c}{e^{-\omega t_L / \tan\theta} - \cos(\omega t_c)} \tag{3.133}$$

using which, t_c can be determined. If $a < \theta$, $t_c < \pi/\omega$ which is not possible. Therefore the minimum value of $a = \theta$. If $a = \theta$, the conduction is maximum, i.e.

$$t_c = \frac{\pi}{\omega} \tag{3.134}$$

Knowing the value of a and ($\omega t_c = \delta$) β, the waveforms are found and the average and rms values can be determined by integration, for example

$$\frac{V_a}{V_\theta} = \sqrt{\frac{(\sin 2a - \sin 2\beta) + 2(\beta - a)}{2\pi}} \tag{3.135}$$

The average value of the current is

$$I_{Lav} = \frac{V_m}{\pi\sqrt{(R^2 + \omega^2 L^2)}} [\cos(a - \theta) - \cos(\beta - \theta) + \tan\theta \sin(a - \theta) (e^{-\omega t_c / \tan\theta} - 1)] \tag{3.136}$$

and the average value of the thyristor current is

$$I_{\text{Tav}} = \frac{I_{\text{Lav}}}{2} \tag{3.137}$$

The average value of the voltage can be determined as

$$V_{\text{Lav}a} = \frac{V_{\text{m}}}{\pi}[-\cos(a + \omega t_{\text{c}}) + \cos a]$$

or

$$\frac{V_{\text{Lav}a}}{V_{\text{Lav}}} = \frac{1}{2}[\cos a - \cos(a + \omega t_{\text{c}})] \tag{3.138}$$

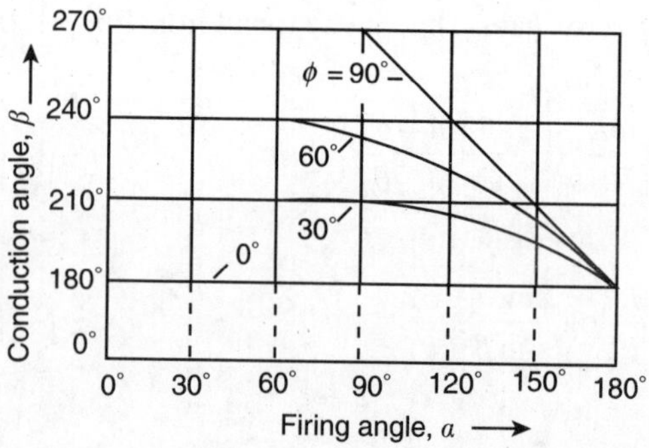

Fig. 3.116 *Control characteristic of a single phase voltage controller feeding R-L load*

The rms value of the current can be determined using the normal procedure of integration between the limits a and $(a + \omega t_c)$. The relationship between a and ωt_c for different impedance angles θ, is graphically represented in Fig. 3.116.

Power Factor and Reactive Power

The current waveform (flowing through the load) can be resolved into a Fourier series given by

$$i_{\text{L}}(t) = a_o + \sum a_v \cos v\omega t + \sum b_v \sin v\omega t \tag{3.139}$$

where a_o, a_v and b_v can be determined easily, knowing the waveform for $i_L(t)$. As the thyristors are symmetrically triggered, the current $i_L(t)$ is periodic and hence the value of $a_o = 0$. The load current

$$i_{\text{L}}(t) = \sqrt{2}I \sin(\omega t + \phi_1) + \sum_{v \to 2} \sqrt{2}I_v \sin(v\omega t + \phi_v) \tag{3.140}$$

where $\phi_1 = \arctan \dfrac{a_1}{b_1}$ and $\phi_v = \arctan \dfrac{a_v}{b_v}$

$$\sqrt{2}I_1 = \sqrt{(a_1^2 + b_1^2)} \text{ and } \sqrt{2}I_v = \sqrt{(a_v^2 + b_v^2)}$$

v is the order of the harmonic, ≥ 2.

The rms value of the load current

$$I = \sqrt{I_1^2 + \sum_{v \to 2} I_v^2} = \sqrt{I_1^2 + I_{\text{h}}^2} \tag{3.141}$$

As for the phase controlled rectifier, we define

$$\frac{I_1}{I} = g_i; \qquad \frac{I_{\text{h}}}{I} = \lambda_i \tag{3.142}$$

where g_i = distortion factor

λ_i = harmonic factor

$$g_i^2 + \lambda_i^2 = 1 \tag{3.143}$$

ϕ_1 is the phase difference between the line voltage and the fundamental of the current, given by

$$\phi_1 = \arc\tan \frac{a_1}{b_1} = \arc\tan \frac{\sin^2 a}{\pi - a + \sin a \cos a} \tag{3.144}$$

$\cos \phi_1$ is called the fundamental displacement factor and is depicted graphically as a function of a in Fig. 3.117.

The Total Power Factor The active power is contributed to only by the fundamental whereas the apparent power has both fundamental and harmonic content. Therefore the power factor

$$\text{P.F.} = \frac{P}{S} = \frac{V_1 I_1 \cos \phi_1}{V_1 I} = \frac{I_1}{I} \cos \phi_1 \tag{3.145}$$

$$\text{P.F.} = g_i \cos \phi_1$$

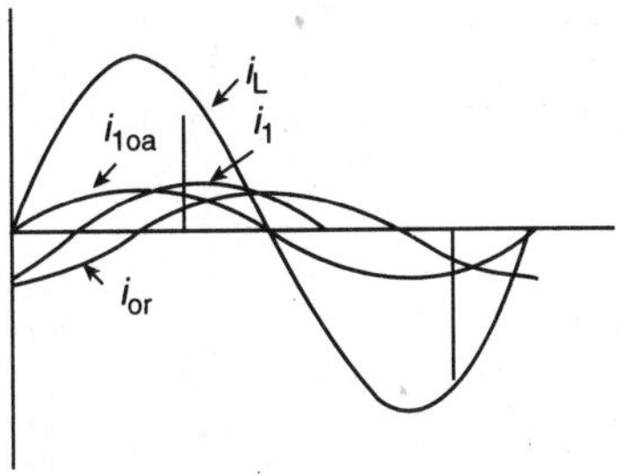

Fig. 3.117 *Fundamental displacement factor as function of a*

The total reactive power is Q. The fundamental reactive power $V_1 I_1 \sin \phi_1 = Q_1$

The reactive power due to harmonics $= \sqrt{Q^2 - Q_1^2} = Q_h$, called the harmonic reactive power.

3.6.2 Three Phase AC Voltage Controller

To control the current and voltage of three phase loads, three phase controllers are required. The single phase controller described previously can be introduced singly in each phase or line, to form a three phase controller. There exist a variety of connections for three phase controllers.

A three phase four wire controller is shown in Fig. 3.118. The load neutral and supply neutral are connected together. Each of the three controllers can be independently controlled to feed the load impedance. Each phase has the same relations as a single phase controller. The analysis is simple and straightforward since the system can be studied as if the loads here are supplied individually by single phase controllers. The neutral and line currents contain triplen harmonics along with other odd harmonics.

A similar connection, which can function as three groups of single phase controllers, is shown in Fig. 3.119. In this connection, three single phase converters supplying their loads are connected in delta. Each controller supplies its own load. Unlike the previous four wire star-connection, the triplen harmonics are absent here. The other odd harmonics are present.

There are certain types of connections which are difficult to analyse. One such circuit is a three phase, three wire star-connected controller, which is normally used when the source neutral cannot be loaded or is absent. The load neutral is

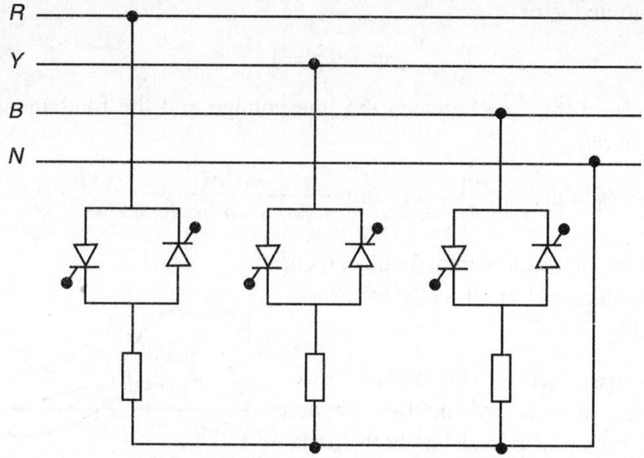

Fig. 3.118 *Three phase 4-wire voltage controller*

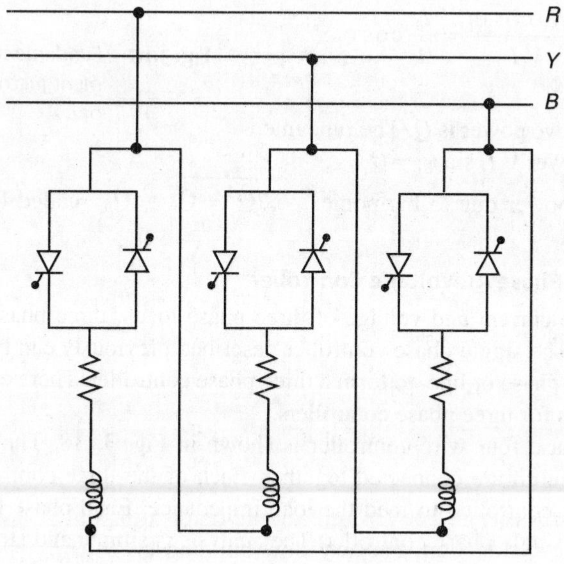

Fig. 3.119 *Delta-connected voltage controller*

isolated. The circuit is depicted in Fig. 3.120. The system is complicated and has to be studied and analysed as a three phase circuit.

Several other possible connections of three phase voltage controllers are depicted in Fig. 3.121. All of them have to be studied as three phase circuits.

The operation of a three phase controller is affected by both the load and the type of connections of the single phase controllers used to form the three phase unit. The analysis also differs for each configuration. It is difficult to summarise the features of each circuit.

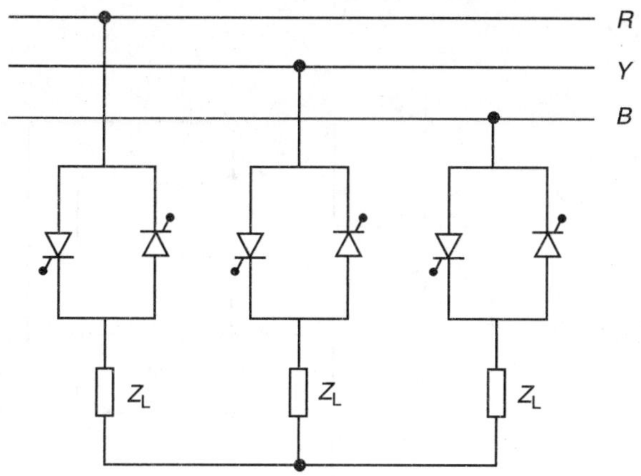

Fig. 3.120 *Symmetrical three phase, three wire voltage controller*

A three phase controller has symmetrical control if both the back to back connected thyristors have the same firing angle. It has asymmetrical control if the firing angles differ or if one of the thyristors is replaced by a diode, or if the controllers are placed in only two of the three lines.

We now discuss the features of a symmetrically controlled three phase, three wire, star-connected controller for both ohmic and inductive loads. The voltage and current waveforms and control characteristics are derived.

The schematic of a three phase, three wire voltage controller feeding a three phase, star-connected balanced resistance is shown in Fig. 3.122. Phase control of the thyristors is employed. The phase and line voltages of the three phase system are shown in Fig. 3.122. For a controller, the control pulse is of a long duration, equal to the conduction period of the thyristor. This is to make sure that the firing pulse is available at the gate whenever the thyristor is forward biased, so that the thyristor can go into conduction. It also ensures the firing of the thyristor whenever a forward current is expected. If, because of some circuit condition, the current goes to zero the thyristor turns off. A lengthy pulse can bring it into conduction. Further, slow building up of current in the load circuit when the thyristor is fired (to give maximum load voltage) may cause the thyristor to go to an off state if it is not fully turned on. In cases of inductive loads, the current zero occurs after the voltage zero and hence a trigger pulse must be present continuously, so that the thyristor turns on at the desired instant when the forward voltage occurs. However, when long gate pulses or a train of pulses are applied to the thyristor to ensure the conduction whenever the voltage is positive to the leakage currents during reverse bias of the thyristor must be acceptable. Further, the operation may be upset by the voltage sharing of the series-connected thyristors. The angle is reckoned from the zero of the voltage. The conduction period for the thyristor for ohmic load is 180° and the firing pulse occurs for 180°.

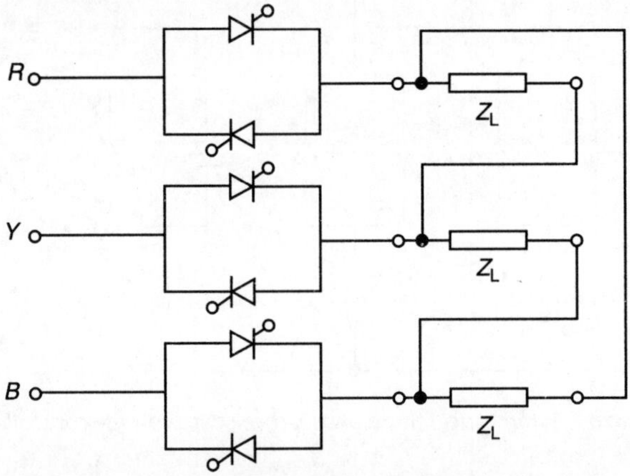

(a) 3-Wire controller feeding a delta connected load.

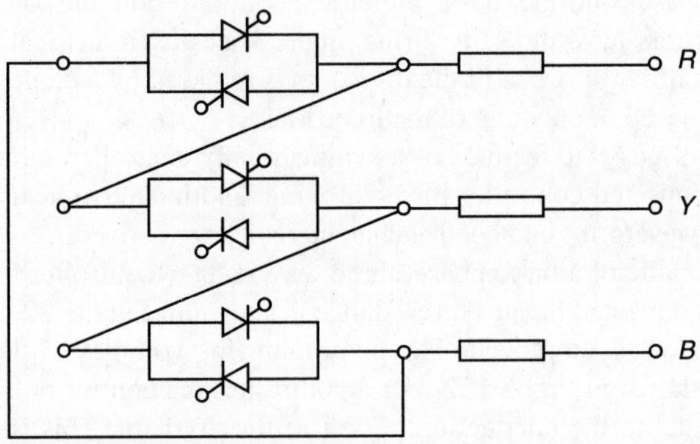

(b) Delta-controller.

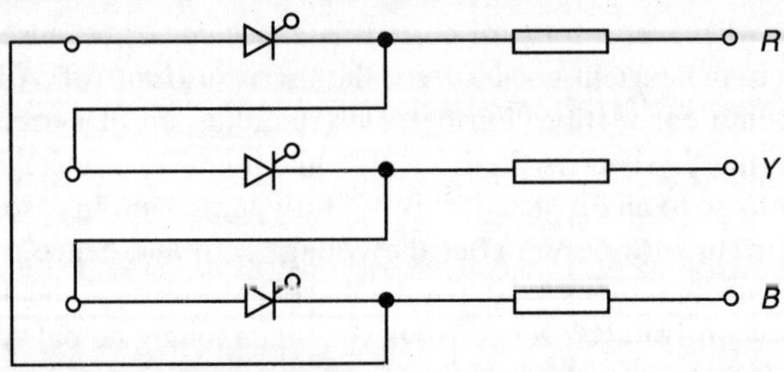

(c) Half wave delta controller.

Fig. 3.121(a) *Some practical symmetrical connections of three phase controller*

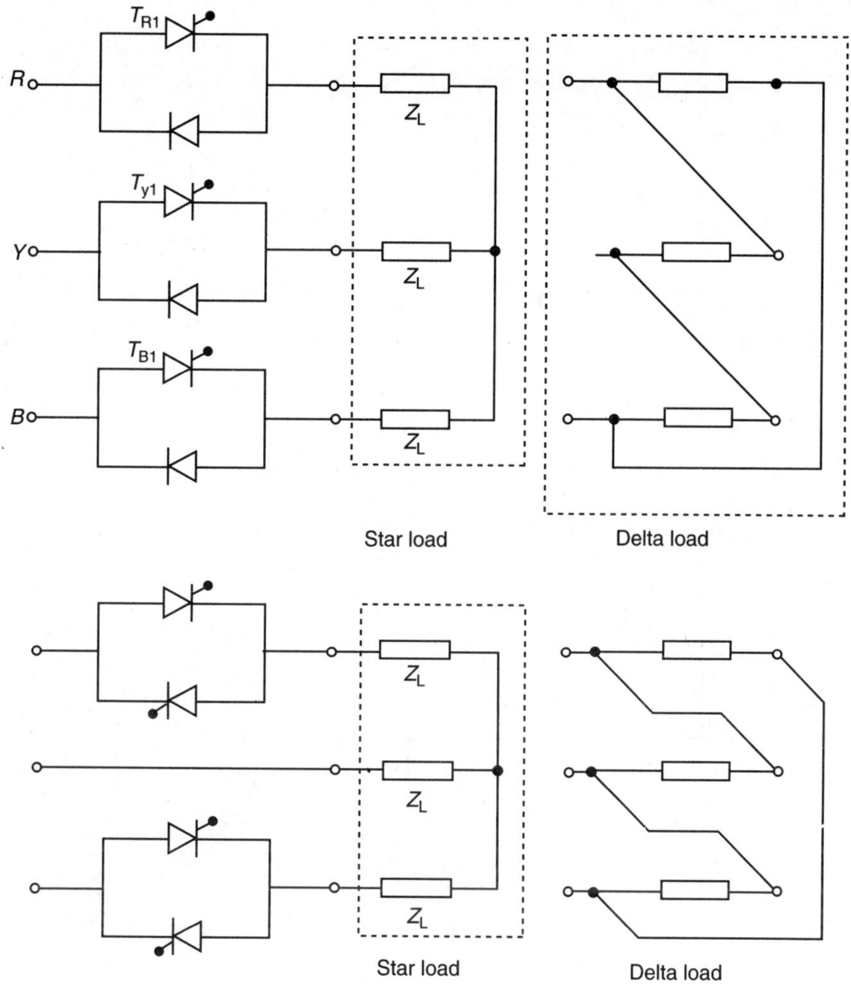

Fig. 3.121(b) *Typical asymmetrically controlled three phase controllers (i) thyristor diode circuit (ii) control in two lines*

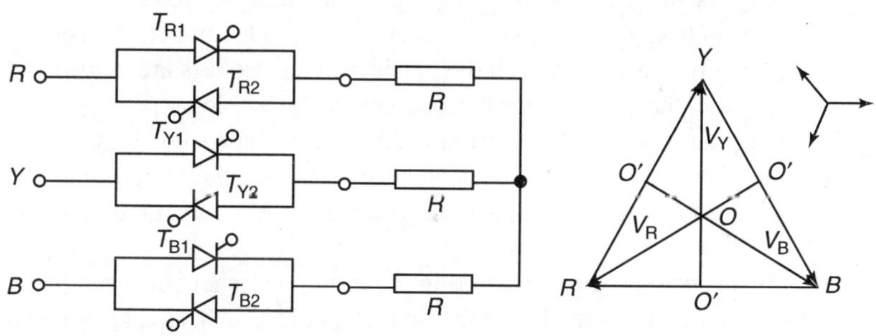

Fig. 3.122(a) *Three phase controller feeding a resistive load*

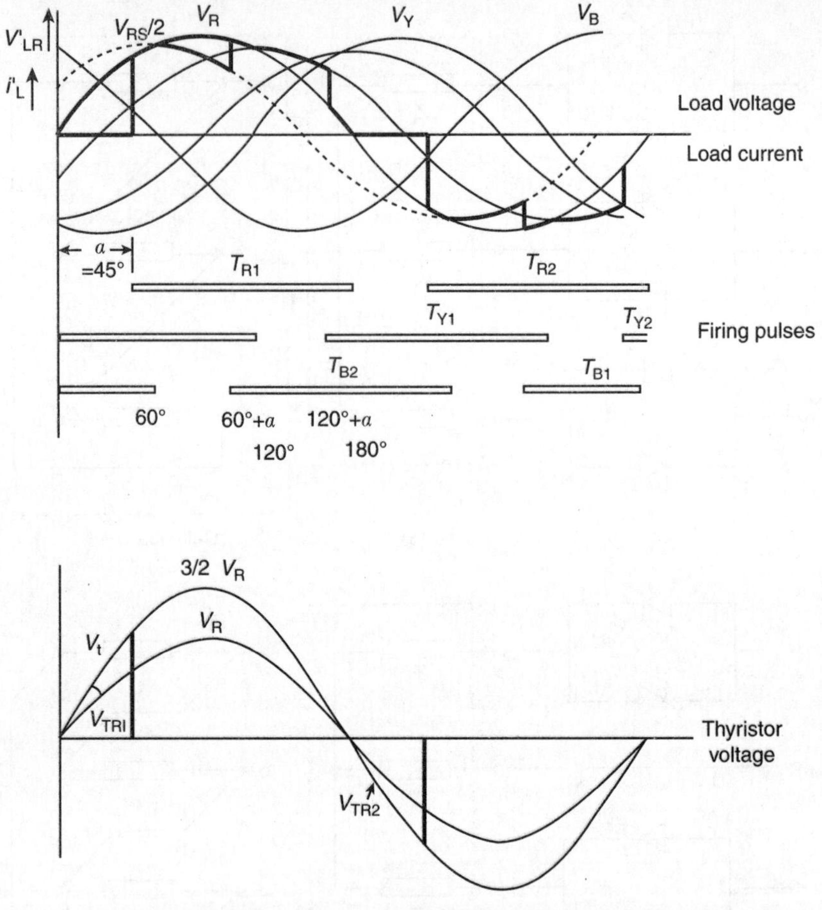

Fig. 3.122(b) *Voltage and current waveforms $a = 45°$*

The voltage across the load impedance can be determined as follows:

i. When a controller is in the non-conducting state the corresponding phase voltage of the load is zero assuming a star connected load.

ii. If two controllers conduct the voltage of the conducting phases is half the line voltage between which the conducting controllers are placed.

iii. If all the there controllers are conducting the load is effectively a three phase load supplied from a three phase balanced source. So, the load voltages can be determined using three phase circuit analysis. For balanced load the load phase voltage is same as source phase voltage.

iv. Only one converter receives firing pulse with no other converter being in the conducting state, no conduction takes place. All the phase voltages are zero.

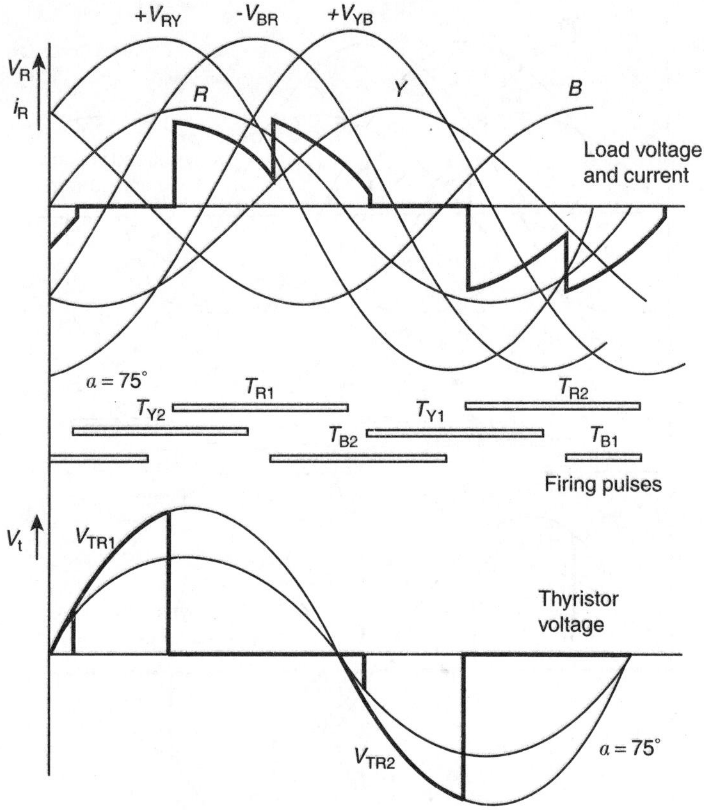

Fig. 3.122(c) *Waveforms for $\alpha = 75°$*

Owing to the dependence of conduction periods on the firing angle, and the overlapping of these periods, the control characteristics have to be determined for several ranges of α. The voltage waveform is determined and its effective value is calculated by integration.

For a three phase resistive load.

For firing angles in the range $0 < \alpha < 60°$

$$\frac{V_L}{V_s} = \sqrt{\left(1 - \frac{3\alpha}{2\pi} + \frac{3}{4\pi}\sin 2\alpha\right)} \tag{3.146}$$

For firing angles in the range $60° < \alpha < 90°$

$$\frac{V_L}{V_s} = \sqrt{\left(\frac{1}{2} + \frac{3}{4\pi}[\sin 2\alpha + \sin(2\alpha + 60°)]\right)} \tag{3.147}$$

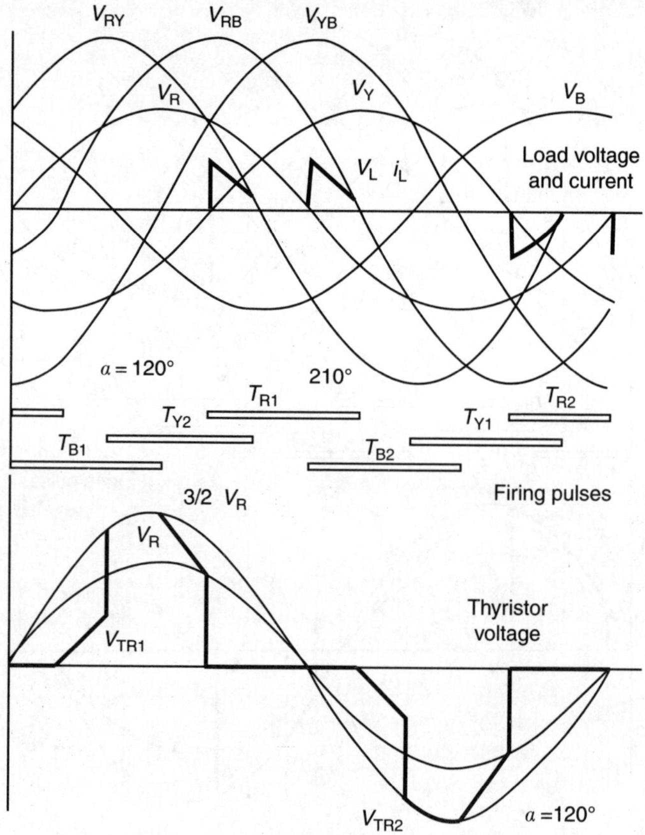

Fig. 3.122(d) *Waveforms for $\alpha = 120°$*

and finally in the range $90° < \alpha < 150°$

$$\frac{V_L}{V_s} = \sqrt{\left(\frac{5}{4} - \frac{3\alpha}{2\pi} + \frac{3}{2\pi}\sin(2\alpha + 60°)\right)} \tag{3.148}$$

The control characteristic showing $\frac{V_L}{V_s}$ as a function of the firing angle α is shown in Fig. 3.123.

When the voltage controller supplies a three phase inductive load the relations are different. It is known that effective control starts from $\alpha > 90°$. The thyristor conducts for a period of $2(\pi - \alpha)/\omega$ so that the conduction is between α and $(2\pi - \alpha)$. A three phase controller feeding star-connected inductances is shown in Fig. 3.124. The phase and line voltages of the source and load voltage for a typical firing angle are also shown. The effective value of the output voltage can be determined in the usual way. For the range of firing angles $90° \le \alpha \le 120°$

$$\frac{V_L}{V_S} = \sqrt{\left(\frac{5}{2} - \frac{3\alpha}{\pi} + \frac{3}{2\pi}\sin 2\alpha\right)} \tag{3.149}$$

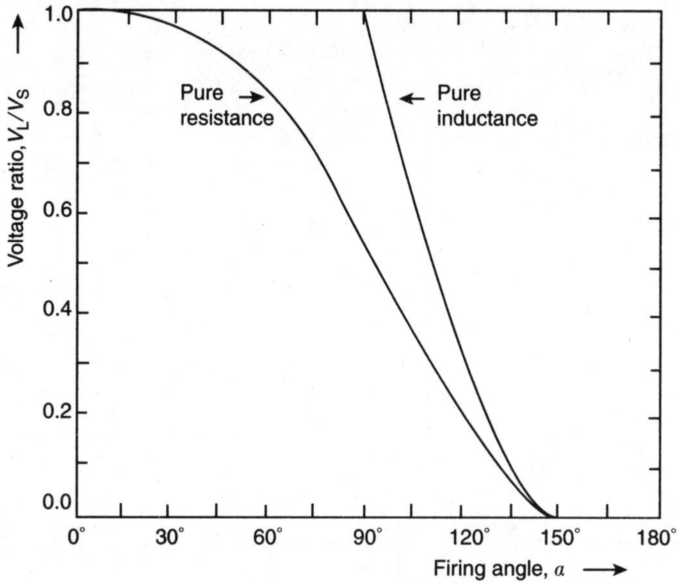

Fig. 3.123 *Control characteristics of a three phase voltage controller*

and in the range of firing angles $120° \le a \le 150°$

$$\frac{V_L}{V_S} = \sqrt{\frac{5}{2} - \frac{3a}{\pi} + \frac{3}{2\pi}\sin(2a + 60°)} \tag{3.150}$$

The control characteristic showing the variation of V_L/V_S as a function of a is depicted in Fig. 3.123.

The method of finding the load voltage waveform is now outlined for a pure resistance load in all the three ranges of firing angles described above. A value of $a = 45°$ is considered, which falls in the range of angles $0 < a < 60°$.

We refer to Fig. 3.122 where the line and phase voltages are given and the firing pulses indicated.

For $a = 45°$ (or in general $0 \le a \le 60°$) it can be easily observed from the firing pulse diagram that all the three controllers conduct, since they are forward biased. The thyristors T_{R1}, T_{Y2} and T_{B1} conduct. This conduction mode is depicted in Fig. 3.125(a). The load and source operate as a normal three phase system. The load neutral is at the same potential as the supply neutral. The phase voltages are decided by the supply system. This mode continues till the voltage V_B passes through zero. At this instant T_{B1} blocks and ceases to conduct. Only thyristors T_{R1} and T_{Y2} conduct. The currents are equal and opposite. This occurs in the range 60° to 60° + a. The line voltage V_{RY} is dropped across the two resistances and eventually the phase voltage is $V_{RY}/2$.

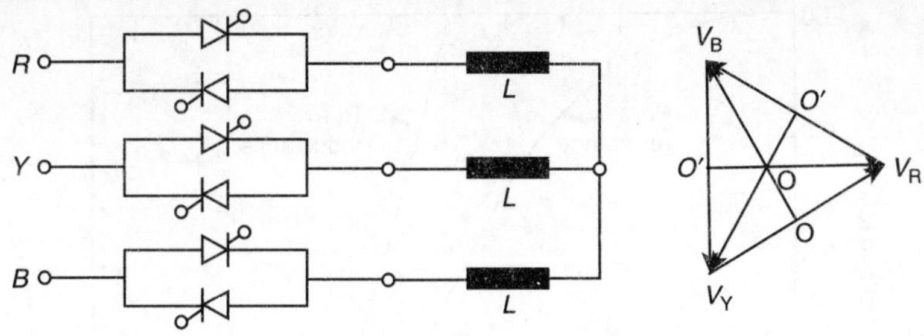

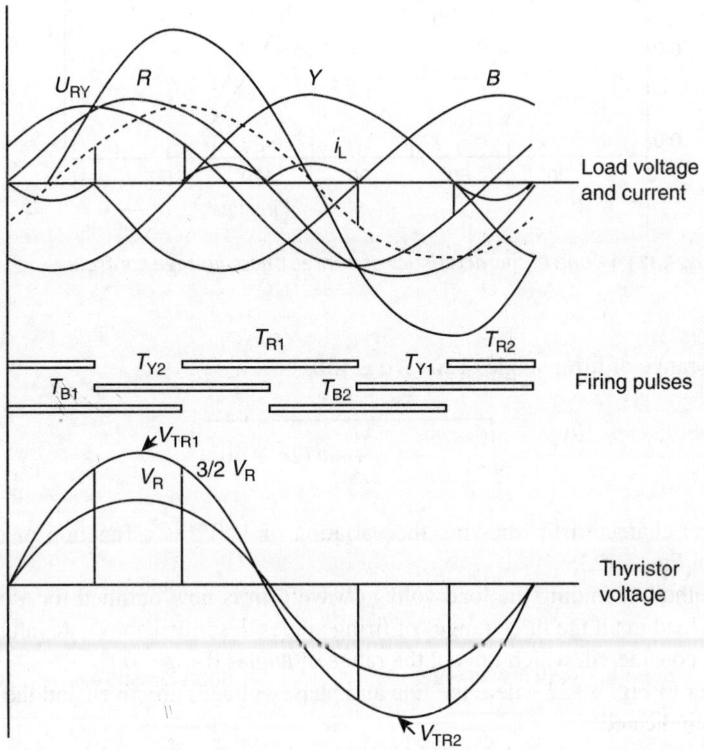

Fig. 3.124 *Waveforms of a three phase controller feeding pure inductive load*

In the range of angles $(60° + \alpha)$ to 120° the thyristor T_{B2} is forward biased and starts conducting with a firing pulse at the gate. Now again the controller is in a complete conduction state with all the three controllers in conduction. The thyristors now conducting are T_{R1}, T_{Y1} and T_{B2}. The load and source revert to a three phase system and the load voltages are specified by source voltages. At 120°, the phase voltage $V_Y = 0$ and the thyristor T_{Y2} stops conduction. The thyristors T_{R1} and T_{B2} conduct in the range 120 to $120° + \alpha$ making the load voltage equal to $\frac{1}{2} V_{BR}$.

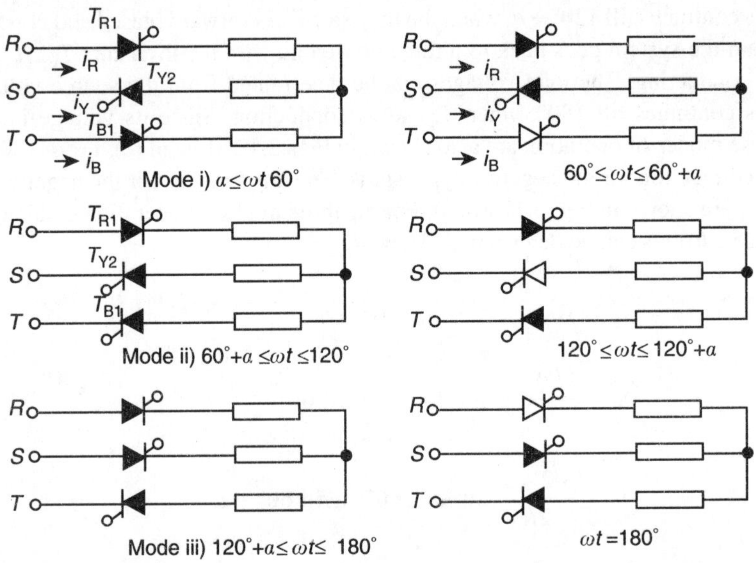

Fig. 3.125(a) *Modes of operation of three phase controller for a < 60° in the positive half cycle*

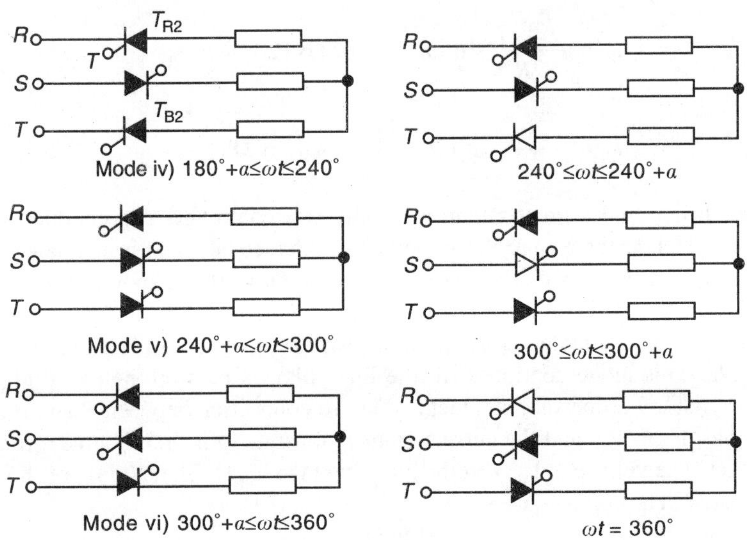

Fig. 3.125(b) *Modes of operation of three phase controller for a < 60° in the negative half cycle*

This continues till $120° + a$, when the thyristor T_{Y1} is forward biased and conducts. Again the system goes back to a three phase one with the thyristors T_{R1}, T_{Y1} and T_{B2} conducting. The load voltages can be determined from the source voltages. This continues till 180°, when T_{R1} stops conduction. Since its voltage is zero. These modes of operation are also shown in Figs 3.124(b) and (c). From $180° + a$ the conduction for the negative cycle starts. The three modes for the negative half cycle are shown in Fig. 3.125(d, c). For the three modes of operation detailed, the phase currents can be determined. They are:

Mode I

$$i_R = 0 \qquad 0 \text{ to } a$$

$$i_R = \frac{\sqrt{2}V_R}{R}\sin\omega t \qquad a \text{ and } 60° \tag{3.151}$$

$$i_R = \frac{\sqrt{2}V_{RS}}{2R}\sin\omega t \qquad 60° \text{ and } 60° + a$$

Mode II

$$i_R = \frac{\sqrt{2}V_R}{R}\sin\omega t \qquad 60° + a \text{ to } 120° \tag{3.152}$$

$$i_R = \frac{\sqrt{2}V_{RT}}{2R}\sin\omega t \qquad 120° \text{ to } 120° + a$$

$$i_R = \frac{\sqrt{2}V_R}{R}\sin\omega t \qquad 120° + a \text{ to } 180° \tag{3.153}$$

For firing angles greater than 60°, only two controllers can conduct. At no instant do all the three controllers conduct simultaneously. The system never operates as a three phase one. The modes for a typical $a > 60°$ are shown in Fig. 3.126. Up to a the load voltage is zero. At a the conduction of T_{R1} and T_{Y2} starts. The voltage V_{RY} is available across the two resistances in series. The phase voltage is $V_{RY}/2$. This mode continues till the line voltages pass through zero at 150°. The thyristor T_{B2} does not conduct and hence connection between R and B is not possible. $V_{RB} > V_{RY}$ and the current from Y transfers to B. In this mode, R and B conduct. The voltage is V_{RB} and the phase voltage $V_{RB}/2$. The modes are shown in Fig. 3.127. The currents are

Mode I

$$i_R = 0 \qquad 0 \text{ to } a$$

$$i_R = \frac{\sqrt{2}V_{RY}}{2R}\sin\omega t \qquad a \text{ to } 60° + a \tag{3.154}$$

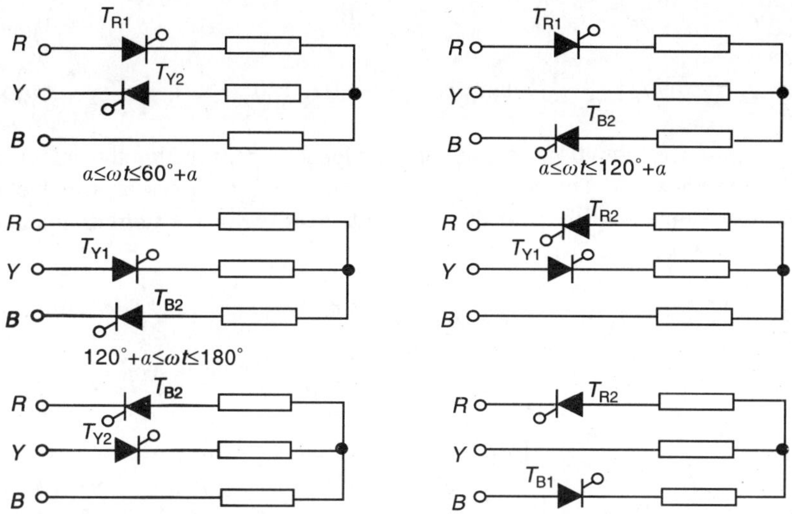

Fig. 3.126 *Modes of operation of three phase controller a > 60°*

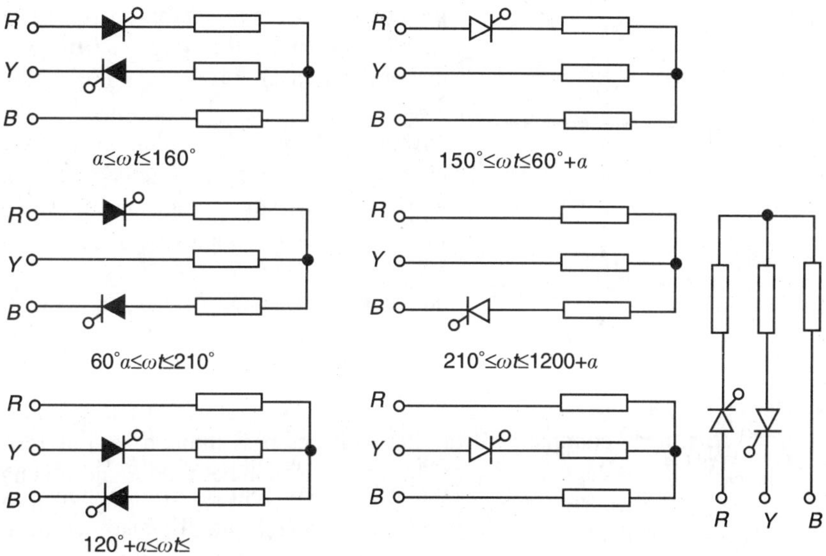

Fig. 3.127 *Modes of operation of three phase controller a > 90°*

Mode II

$$i_R = \frac{\sqrt{2}V_{RB}}{2}\sin\omega t \quad 60^\circ + s \text{ to } 120^\circ + a \tag{3.155}$$

For firing angles $a > 90°$, discontinuous conduction occurs within the half cycle. When the voltage V_{RS} passes through zero the load current ceases. Conduction does not resume until the next thyristor with positive voltage starts conducting. The currents are:

$$i_R = 0 \quad 30^\circ \text{ and } a$$

$$i_R = \frac{\sqrt{2}V_{RY}}{2R}\sin\omega t \quad a \text{ to } 150^\circ \tag{3.156}$$

$$i_R = 0 \quad 150^\circ \text{ to } 60^\circ + a \tag{3.157}$$

$$i_R = \frac{\sqrt{2}V_{RB}}{2}\sin\omega t \quad 60^\circ + a \text{ to } 210^\circ$$

The voltages and currents are shown in Fig. 3.128. For $a > 150°$, conduction is not possible. The current control characteristic showing the referred average and rms values as functions of a is given in Fig. 3.129.

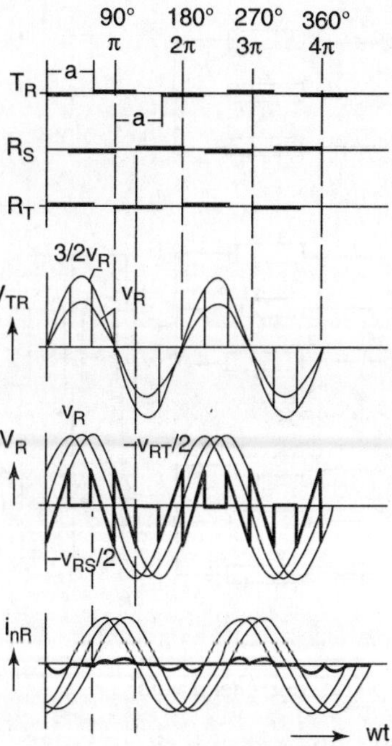

Fig. 3.128 *Voltage and current waveforms of a three phase voltage controller with inductive load*

Thyristor Voltages The voltage waveform of the thyristor must be known to decide its rating. Thyristor voltages are shown in Fig. 3.130 for the ranges of firing angles given above. The voltage of the thyristor between the instants of its zero current and the application of the firing pulse to make it conduct has to be considered to decide its rating. The voltage across the thyristor of R-phase is shown. For $a < 60°$, when one thyristor (T_R) is blocked the other two controllers conduct. The voltage between R and the load neutral is $\frac{3}{2}V_R$ (phasor diagram). This is applicable for a in the range of 60° to 90° also, and the voltages are shown accordingly. For $a > 90°$ discontinuous conduction occurs and the supply voltage during this period is across the thyristor. The thyristor voltage is shown taking this into consideration (Fig. 3.130).

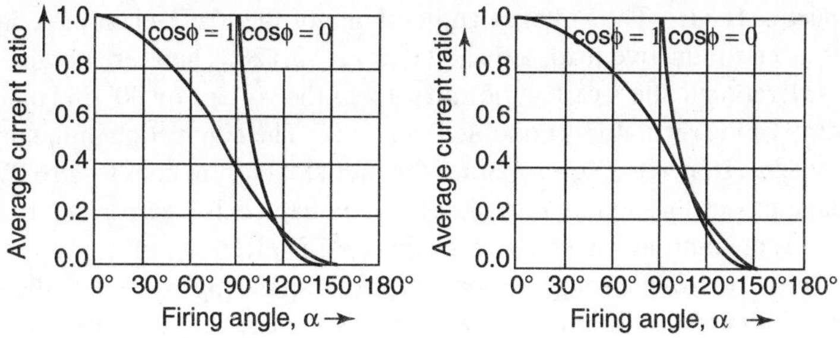

Fig. 3.129 *Current control characteristics of a three-phase controller*

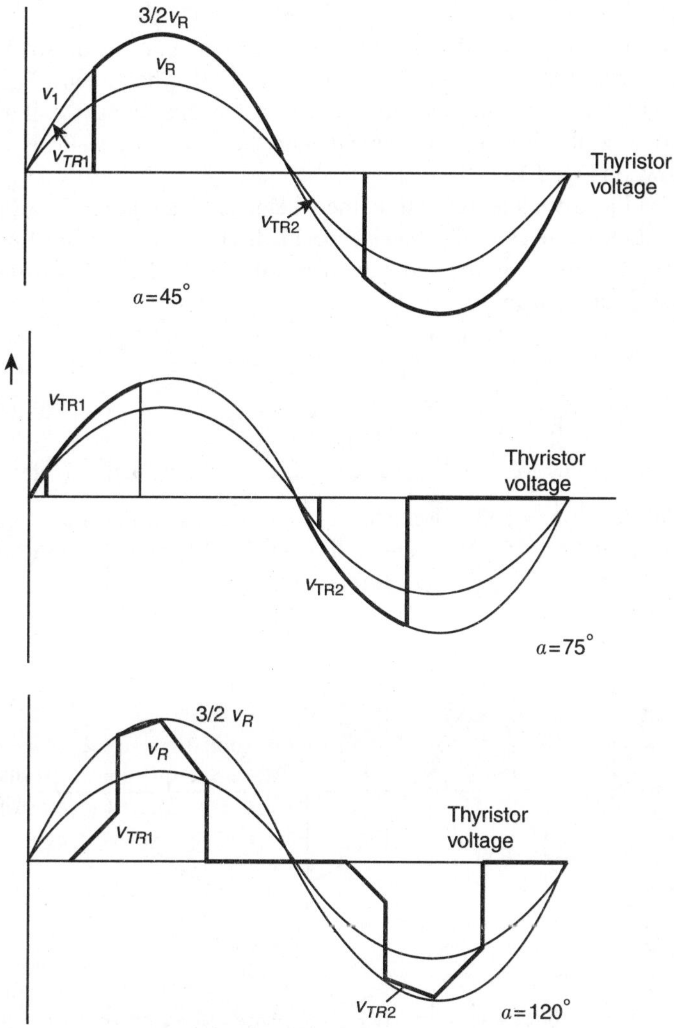

Fig. 3.130 *Thyristor voltage waveform of a three phase voltage controller*

Pure Inductive Loads The voltages and currents of the load, when the controller supplies a pure inductive load, are shown in Fig. 3.128. These are obtained with due consideration to the fact that the current lags the voltage by 90° and that effective control of the regulator is possible for $a > 90°$. The conduction of the thyristor is for $(2\pi - 2a)$, from a to $(2\pi - a)$. Since the thyristor current goes to zero 90° after the voltage passes through its zero, a negative voltage exists across the thyristor.

The current control characteristic for this case is given in Fig. 3.129.

The thyristor voltage for this type of load jumps to $3/2V_R$ when the thyristor is in a blocked condition. This voltage is the value between the line R and the load neutral. The load neutral has half the voltage of the line to which the conducting controllers are connected. The thyristor voltage for this loading is shown in Fig. 3.131.

A three phase controller also has a harmonic content in its load current. Reactive power is required both for control purposes as well as harmonics. The former is identified by the fundamental displacement factor. The fundamental displacement factor as a function of firing angle a, with the impedance angle as a parameter is shown in Fig. 3.132.

Based on the preceding discussion, the voltage and current ratings of the thyristor for a three phase controller can be specified. The selection of the thyristor is based on the maximum voltage and maximum value of the average or rms current. The thyristor currents are

$$I_{Tav} = \frac{I_{phase}}{2}, \quad I_{Trms} = \frac{I_{phase}}{\sqrt{2}} \tag{3.158}$$

$$\text{The voltage of the thyristor} = \frac{3}{2}\sqrt{2}V_R \tag{3.159}$$

The foregoing discussion of three phase voltage controllers makes it clear that

i. the topology of the circuit under study changes, depending upon the type of load and firing angle. The load voltages and current must be

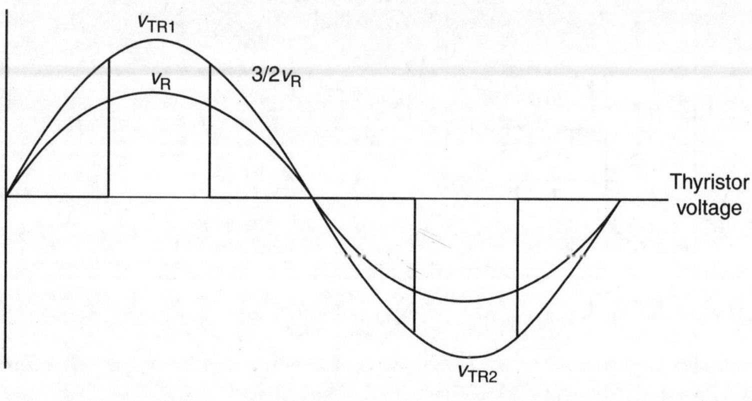

Fig. 3.131 *Thyrister voltage waveform of a three-phase voltage controller feeding pure inductive load*

independently studied and determined. No generalised analysis applicable to all circuits is possible.

ii. the analysis of a circuit can be easily done by identifying the modes of operation, so as to distinguish the controllers conducting and enable one to determine the load voltages.

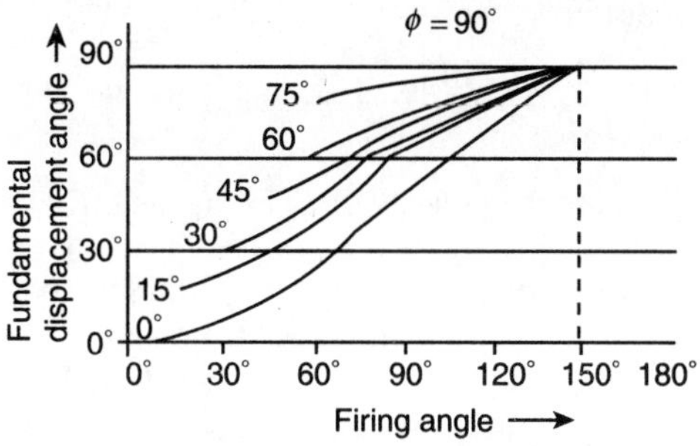

Fig. 3.132 *Fundamental displacement angle of a three phase controller*

iii. the load current waveform has rich harmonic content, calling for additional reactive power besides that required for control.

iv. the thyristor voltages and currents enable one to make a proper choice of the devices. The thyristors have the highest voltage of operation under low output conditions.

Worked Examples

3.1 *A midpoint two pulse converter feeds a load of resistance 2Ω in series with an infinite inductance and a back emf of 150 V. The converter is supplied by a converter transformer with a voltage of 400 V from the neutral to the line. If the firing angle is 30° determine the current in the load.*

Solution The terminal voltage of the converter

$$V_{dia} = \frac{2\sqrt{2}}{\pi} V_s \cos a = 311.76 \text{ V}$$

The average voltage across the inductance is zero. Therefore

$$I_d R_d = V_{dia} - E_b$$

$$I_d = \frac{311.76 - 150}{2} = 80.88 \text{ A}$$

$$\text{rms value of secondary current} = \text{rms value of thyristor current} = \frac{80.88}{\sqrt{2}} = 57.2 \text{ A}$$

rms value of primary current = 80.88 A

3.2 *In the previous example, the firing angle is increased simultaneously reversing the dc voltage in the load. Determine the firing angle of the converter if the current has to be maintained at 80.88 A.*

Solution

$$I_d = \frac{V_{dia} - E_b}{R_d}$$

Because the polarity of E_b is changed, $E_b = -150\ V$

$$80.88 = \frac{V_{dia} + 150}{2}$$

$$161.76 = V_{dia} + 150$$

$$V_{dia} = +11.76$$

$$\cos a = \frac{+11.76}{2\sqrt{2}/\pi\ 400}$$

$$a = 88.13^o$$

3.3 *In the previous example determine the value of α if the transformer has a leakage inductance of 2 mH. The supply frequency is 50 Hz. Determine the overlap.*

Solution

$$I_d = \frac{V_{da} - E_b}{R_d},\ E_b = -150\ V$$

$$V_{da} = V_{dia} - \frac{I_d X_c}{\pi}$$

$$= 0.9V_s \cos a - \frac{I_d X_c}{\pi}$$

$$X_c = 0.628$$

$$I_d \left[R_d + \frac{X_c}{\pi} \right] = 0.9V_s \cos a + 150$$

$$177.936 = 0.9V_s \cos a + 150$$

$$\cos a = \frac{177.936 - 150}{0.9V_s}$$

$$a = 85.549^o$$

$$\cos(a + u) = \cos a - \frac{I_d X_c}{\pi} \frac{2}{V_{dio}}$$

$$= 0.0776064 - 0.0898666$$

$$u = 5.1535^o$$

3.4 *A midpoint two pulse converter, shown in Fig. P 3.4 feeds an R-L load having a resistance of 12.5 Ω and infinite inductance. The fine reactance is 0.5 Ω. The voltage V_s = 200 V. Determine the average value of converter voltage $V_{d\alpha}$, average value of dc current I_d and overlap angle for α = 60°*

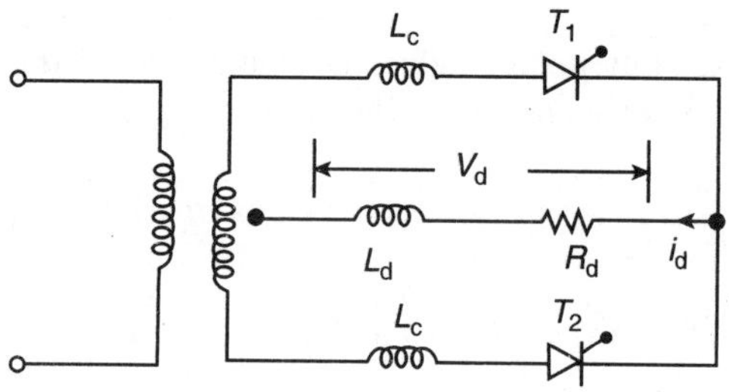

Fig. P3.4 *Two pulse mid point converter*

Solution The source inductance is responsible for the overlap causing a reduction in the terminal voltage of the converter. Therefore

$$V_{da} = \frac{V_{dio}}{2}\left[\cos a + \cos\left(a + u\right)\right]$$

$$\Delta V_{da} = \frac{V_{dio}}{2}\left[\cos a - \cos\left(a + u\right)\right] = \frac{I_d X_c}{\pi}$$

$$\text{Also } V_{da} = V_{dia} - \Delta V_{da} = V_{dia} - \frac{I_d X_c}{\pi}$$

The average value of voltage across the inductance is zero. Therefore

$$V_{da} = I_d R_d = V_{dia} - \frac{I_d X_c}{\pi}$$

leading after simplification to

$$I_d = \frac{V_{dia}}{\left(R_d + X_c/\pi\right)} = \frac{0.9V_s \cos a}{\left(12.5 + 0.5/\pi\right)} = \frac{90}{12.66}$$

$$= 7.11\text{A}$$

The average value of the dc voltage

$$V_{da} = I_d R_d = 7.11 \times 12.5 = 88.87 \text{ V}$$

The change in the terminal voltage

$$= 1.1321 \text{ V} = \frac{V_{dio}}{2}\left[\cos 60 - \cos\left(60 + u\right)\right]$$

$$\cos 60 - \cos(60 + u) = 0.0125785$$

$$u = 0.82876°$$

3.5 *A two pulse bridge rectifier feeds an R-L load having a resistance of 2. 5 Ω and infinite inductance, causing perfect smoothing. It is fed from a 220 V, 50 Hz supply having an inductance of 5 mH. The thyristors are ideal. Determine the average value of load current and overlap angle for firing angles of α = 0° and α = 30°.*

Solution Because of the source inductance there is an overlap causing a drop in the terminal voltage of the converter. Therefore

$$V_{da} = \frac{V_{dio}}{2}\left(\cos a + \cos\left(a + u\right)\right)$$

and $\Delta V_{da} = \frac{V_{dio}}{2}\left(\cos a - \cos\left(a + u\right)\right)$

It can be shown that

$$\Delta V_{da} = \frac{2I_d X_c}{\pi}$$

Therefore

$$V_{da} = V_{dia} - \frac{2I_d X_c}{\pi}$$

As the voltage across the load inductance is zero, we have

$$V_{da} = I_d R_d$$

Substituting, we get

$$I_d = \frac{V_{dia}}{\left(R_d + 2X_c/\pi\right)}$$

From the data of the problem

$$X_c = 2\pi fL_c = 314 \times 5 \times 10^{-3} = 1.57\Omega$$

$$R_d + \frac{2X_c}{\pi} = 2.5 + 1 = 3.5$$

$$V_{dia} = 0.9V_s \cos a = 0.9V_s \text{ for } a = 0$$

$$= 198 \text{ V}$$

$$I_d = \frac{198}{3.5} = 56.57 \text{ A}$$

Therefore $V_{da} = 141.43$ V

$$\Delta V_{da} = 56.57 \text{ V}$$

From equation for ΔV_{da} above for $a = 0$ we have

$$56.57 = \frac{198}{2}\left[1 - \cos u_o\right]$$
$$1 - \cos u_o = 0.5714$$
$$u_o = 64.62^\circ$$

For $a = 30^\circ$

$$V_{dia} = 0.9 \times 220 \times \cos 30^\circ = 171.47 \text{ V}$$

$$I_d = \frac{171.47}{3.5} = 48.99 \text{ A}$$

Therefore

$$\Delta V_{da} = 171.47 - 122.48 = 48.99$$
$$48.99 = \frac{V_{dio}}{2}\left[\cos 30 - \cos(30 + u)\right]$$
$$0.49488 = \cos 30 - \cos(30 + u)$$
$$= 0.866 - \cos(30 + u)$$
$$u = 38.22^\circ$$

3.6 *A two pulse, phase controlled, bridge converter operating from a 220 V, 50 Hz mains feeds a load comprising a resistance of 3 Ω, an infinite inductance and a dc source of voltage 200 V. The commutation inductance is 1 mH. For a firing angle of 120°, determine the average value of the load current and overlap angle.*

Solution Because of the source inductance there is an overlap, causing a drop in the converter voltage. Referring to Fig. P3.6.

$$I_d R_d = V_{da} - E_g$$

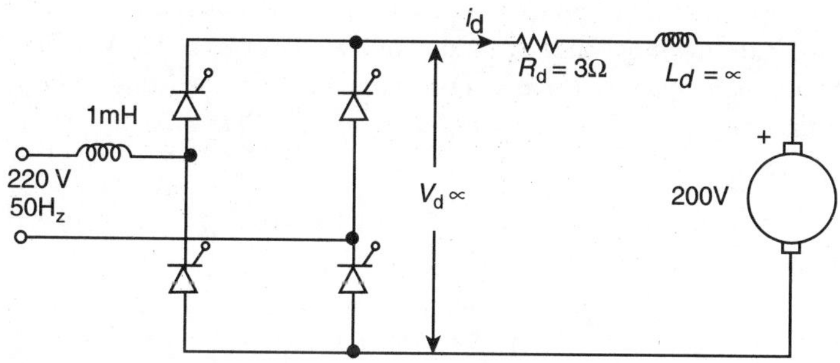

Fig. P3.6 *Two pulse bridge convener*

But $V_{da} = V_{dia} - \dfrac{2I_d X_c}{\pi}$

Substituting

$$I_d\left[R_d + \frac{2X_c}{\pi}\right] = V_{dia} - E_g$$

$$I_d = \frac{V_{dia} - E_g}{R_d + 2X_c/\pi}$$

$$V_{dia} = 0.9V_s \cos a = -99 \text{ V}$$

$$E_g = -200 \text{ V}\Omega, \qquad R_d = 3\Omega, \qquad L_c = 1\times10^{-3}\text{H}$$

Substituting

$$I_d = \frac{-99+200}{3+0.2} = \frac{101}{3.2} = 31.56 \text{ A}$$

$$V_{da} = I_d R_d + E_g = 31.56\times3 - 200$$

$$= -105.31$$

$$V_{da} = \frac{V_{dio}}{2}\left[\cos a + \cos\left(a+u\right)\right]$$

$$-102.26 = 99\left[-0.5 + \cos\left(a+u\right)\right]$$

$$= -49.5 + 99\cos\left(a+u\right)$$

$$\cos\left(a+u\right) = -0.5638$$

$$u = 4.316^\circ$$

3.7 *A three phase converter feeds a resistance load in series with a large inductance, causing perfect smoothing. The average load current is 50 A. The line to neutral voltage on the line side of the converter is 220 V at 50 Hz. The commutation inductance is 1.5 mH. Determine the average value of the dc voltage, and the overlap angle for a firing angle of 30°. Determine the rms value of thyristor current.*

Solution The commutation inductance causes an overlap because of which the terminal voltage decreases. When there is overlap, we have

$$V_{da} = V_{dia} - \frac{3I_d X_c}{2\pi} = 1.17V_s\cos a - \frac{3I_d X_c}{2\pi}$$

$$X_c = 314\times1.5\times10^{-3} = 0.471$$

Therefore

$$V_{da} = 1.17 \times 220 \times \cos 30 - \frac{3 \times 50 \times 0.471}{2\pi}$$

$$= 222.91 - 11.25 = 211.66 \text{ V}$$

Change in voltage = 11.25 V

$$11.25 = \frac{V_{dio}}{2}\left(\cos\alpha - \cos\left(a+u\right)\right)$$

$$= \frac{257.4}{2}\left(\cos 30 - \cos\left(30+u\right)\right)$$

$$0.0874125 = \cos 30 - \cos\left(30+u\right)$$

$$u = 8.87^{\circ}$$

3·8 *A three phase three pulse converter feeds an RLE load, having a resistance of 2.5 Ω, an inductance causing perfect smoothing and a negative voltage (voltage source) of 250 V. The converter is supplied from a three phase balanced supply at 50 Hz and line to neutral voltage of 150 V. Determine the mean value of load current at a firing angle of 120°. Assume the thyristors are ideal and commutation is instantaneous.*

Solution In the case of a converter feeding a back emf load in series with a resistance and inductance, the average current in the load is

$$I_d = \frac{V_{dia} - E_b}{R_d}$$

In the present case the load has a dc source $E_b = -250$ V.

$$V_{dia} = 1.17V_s \cos a = -87.75\text{V}$$

$$I_d = \frac{-87.75 + 250}{2.5} = 64.9 \text{ A}$$

The thyristor currents:

Average value = 21.63 A

rms value = 37.47 A

3·9 *If in Example 3.8 the line has an inductance of 3 mH, determine the firing angle to maintain an average current of 64.9 A in the load. What is the overlap angle?*

Solution Average value of load current = 64.9 A
Average value of load voltage = 162.25 V

When there is overlap

$$I_d = \frac{V_{dia} - E_b}{R_d + 3X_c / 2\pi}$$

Therefore $V_{dia} - E_b = I_d \left(R_d + \frac{3X_c}{2\pi} \right)$

$$= 64.9 \times 2.95$$

$$= 191.455^\circ$$

But $E_b = -250$ V

Therefore $V_{dia} = 191.455 - 250 = -58.545^\circ$

$$\cos a = \frac{-58.545}{1.17 \times 150} = -0.3336$$

$$a = 109.5^\circ$$

$$\cos(a + u) = \left(\cos a - \frac{3I_d X_c}{2\pi} \frac{2}{V_{dio}} \right)$$

$$= -0.3336 - 0.45 \times \frac{2}{0.17 \times 150} \times 64.9$$

$$a + u = 0.3179$$

$$u = 22.29^\circ$$

3.10 *A three phase fully controlled bridge rectifier is fed from a three phase balanced supply at 400 V and 50 Hz. The load consists of R = 15 Ω and a large smoothing inductance causing a perfect smoothing. Determine the average value of the load voltage, current and power dissipation for a firing angle of α = 75°. Assume the thyristor and supply to be ideal.*

Solution The average value of the load voltage of a three phase fully controlled bridge rectifier = 2.34 V_s cos a = 1.35 V_s cos a

$$V_{dia} = 1.35 \times 400 \times \cos 75^\circ = 139.76 \text{ V}$$

The average value of voltage across the inductor is zero. Therefore the average value of converter voltage is dropped across load resistance. Thus

$$V_{dia} = I_d R_d$$

$$I_d = \frac{V_{dia}}{R_d}$$

$$= \frac{139.76}{15} = 9.317\text{A}$$

The power dissipation $= V_{dia} I_d$

$$= 139.76 \times 9.317 = 1302.2 \text{ W}$$

3.11 *For the three phase bridge of problem 3.10 determine the fundamental displacement factor and power factor.*

Solution The fundamental displacement factor

$$\cos \phi_1 = \cos \alpha = 0.2588$$

The power factor $= g \cos \phi_1$

Where g is the distortion factor. For a three phase fully controlled bridge, it is $\frac{3}{\pi}$ for perfect smoothing.

Therefore the power factor is

$$= \frac{3}{\pi} \times \cos \phi_1 = \frac{3}{\pi} \cos \alpha = \frac{3}{\pi} \times 0.2588 = 0.2473$$

3.12 *For the three phase bridge of problem 3.11 determine the ratings of the thyristors.*

Solution *Voltage rating of the thyristor*

In a three phase fully controlled bridge rectifier the maximum voltage rating of a thyristor is the peak value of line voltage.

Therefore

$$V_{Th} = \sqrt{2} \times 400 = 565.69 \text{ V}$$

Using a factor of safety, 1000 V thyristors may be used.

Current ratings

The rms value of supply current

$$I_a = \sqrt{\frac{2}{3}} I_d = \sqrt{\frac{2}{3}} 9.317 = 7.607 \text{A}$$

This current flows only for half a cycle.

Therefore

$$I_{Th} = \frac{I_a}{\sqrt{2}} = \frac{7.607}{\sqrt{2}} = 5.379 \text{A}$$

or

$$I_{Th} = \frac{I_d}{\sqrt{3}} = 5.379 \text{A}$$

These are the ratings at $\alpha = 75°$

However if the firing angle ranges from 0 to 90°, the maximum current occurs at $\alpha = 0$

The maximum current rating of the thyristor

$$= \frac{5.379}{\cos a} = 20.783 \text{ A}$$

With a factor of safety, a thyristor of 25 A may be chosen.

3.13 *A three phase fully controlled bridge rectifier is supplied from a 415 V, 50 Hz supply having an inductance of 1.5 mH. The converter load consists of a resistance of 50 Ω and a large inductance, causing perfect smoothing. Calculate the average value of load current and voltage for firing angles of α = 0° and α = 60°. What are the overlap angles?*

Solution The source inductance of the converter is responsible for the overlap, which causes a drop in the voltage.

Therefore

$$V_{da} = \frac{V_{dio}}{2}[\cos a + \cos(a + u)]$$

$$\Delta V_{da} = \frac{V_{dio}}{2}[\cos a - \cos(a + u)]$$

$$\Delta V_{da} = V_{dia} - V_{da} = V_{dia} - \frac{6 I_d X_c}{\pi}$$

Also, the average value of the load voltage = $I_d R_d$. Therefore,

$$I_d \left(R_d + \frac{6X_c}{\pi} \right) = V_{dia}$$

leading to

$$I_d = \frac{V_{dia}}{R_d + 6X_c/\pi} = \frac{1.35 \times 415 \times 1}{50 + \frac{6}{\pi} \times 0.471}$$

$$= \frac{1.35 \times 415}{50 + 0.9} = 11.007 \text{ A}$$

The average value of converter voltage = 11.007 × 50 = 550.34 V

The ideal voltage = 1.35 × 415 = 560.25 V

Change in voltage = 9.9 V

$$9.9 - \frac{560.25}{2}[1 - \cos u_0]$$

$$1 - \cos u_o = \frac{9.9 \times 2}{560.22} = 0.3534$$

$$\cos u_o = 0.96466$$

$$u_o = 15.28^o$$

$$a = 60^o$$

$$I_d = \frac{1.35 \times 415 \times \cos 60^\circ}{50.9} = \frac{280.125}{50.9} = 5.5034 \text{ A}$$

The converter voltage = 5.5034 × 50

= 275.17 V

The inductive drop in terminal voltage = 4.953 V

$$\cos(\alpha) - \cos(\alpha + u) = \frac{4.953 \times 2}{560.25} = 0.0177$$

$$\cos(\alpha + u) = 0.5 - 0.0177 = 0.4823$$

$$u = 1.1631^\circ$$

3.14 *A dc chopper having an on time of 1.5 ms in a overall cycle time of 3 ms is supplied from a 200 V dc source. The load voltage waveform is in the form of rectangular pulses. Determine the average and rms values of the load voltage. Also determine the rms value of the fundamental current and ripple factor.*

Solution The waveform of the load voltage is shown in Fig. P 3.14. The control or time ratio of the chopper is

$$\gamma = \frac{1.5}{3.0} = 0.5$$

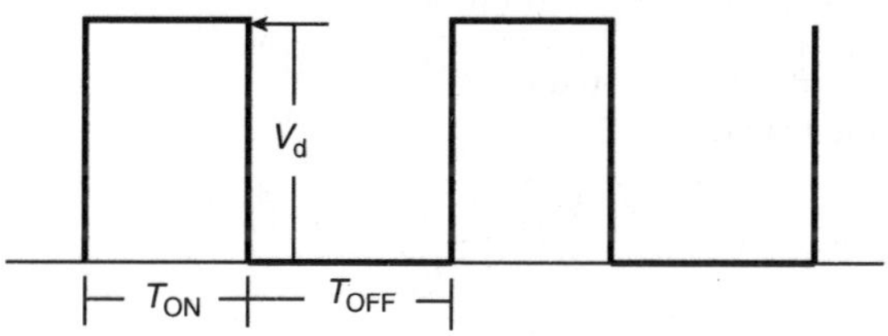

Fig. P3.14 *The load voltage waveform of a chopper*

The average value of the load voltage = 0.5 × 200 = 100 V. The rms value of the load voltage

$$= \sqrt{0.5} \times 200 = 141.42 \text{ V}$$

The fundamental of the load voltage

$$= \frac{V_s}{\pi} \sqrt{(1 - \cos 2\pi\, r)^2 + (\sin 2\pi\, r)^2} \quad 90.08 \text{ V}$$

$$\text{rms value} = 63.7 \text{ V}$$

$$\text{Ripple factor} = \frac{\sqrt{1-\gamma}}{\sqrt{\gamma}} = 1$$

3.15 *A dc chopper feeds an R-L load having a resistance of 1.5 Ω and an inductance of 3 mH. The ON time of the thyristor is 2 ms in an overall cycle time of 6 ms. Calculate the average value of the load voltage, the average current, and the maximum and minimum values of the current and power delivered to the load. The supply voltage of the chopper is 150 V.*

Solution The time ratio of the chopper $= \dfrac{2}{6} = \dfrac{1}{3}$

The time constant of the load $= \tau = \dfrac{3}{1.5} = 2$ ms

The angular frequency of chopper

$$\omega = \frac{2\pi}{T} = \frac{2\pi}{6} \times 10^3$$

The constant of the circuit, deciding the continuity of conduction is

$$= \frac{2\pi}{\omega\tau} = \frac{6}{1000} \times \frac{1000}{2} = 3$$

The conduction is continuous because $\dfrac{2\pi}{\omega\tau} >$ time ratio.

Average value of voltage $= \dfrac{1}{3} \times 150 = 50$ V

Average value of current $= \dfrac{50}{1.5} = 33.33$ A

Average power delivered = 1666.666 ω

The current in the load during T_{ON} is

$$i_1 = \frac{V}{R}(1 - e^{-t/\tau}) + I_o e^{-t/\tau}$$

at $t = T_{ON}$

$$I_{max} = \frac{150}{1.5}[1 - e^{-2/6}] + I_o e^{-t/\tau}$$
$$= 28.347 + I_o \times 0.717$$

During the OFF time of the chopper

$$i_2 = I_{max} e^{-t/\tau}$$

At thc cnd of T_{OFF} $i_2 - I_o$

$$I_o = I_{max}(e^{-4/6}) = 0.5134(28.347 + I_o \times 0.717)$$
$$= 14.554 + 0.3681 I_o$$
$$I_o = 23.032 \text{ A}$$
$$I_{max} = 28.347 + 23.032 \times 0.717$$
$$= 44.86 \text{ A}$$
$$I_{min} = 23.032 \text{ A}$$

3.16 *A single phase ac voltage regulator feeds an R-L load having an impedance angle of 60°. The angle of extinction is 215°. Determine the angle of firing. For this condition determine the fundamental displacement factor and the load voltage of regulator.*

Solution The current of a voltage regulator is given by

$$i = \frac{\sqrt{2}E}{Z}\sin(\omega t - \phi) - \frac{\sqrt{2}E}{Z}\sin(a - \phi)e^{-\left(\frac{\omega t - a}{\tan\phi}\right)}$$

At the angle of extinction $i = 0$

Therefore at $\omega t = x$

$$\sin(x - \phi) - \sin(a - \phi)e^{-\left(\frac{\omega t - a}{\tan\phi}\right)} = 0$$

$$x = 215^\circ \qquad \phi = 60^\circ$$

Substituting

$$0.42262 - \sin(a - \phi)e^{-\left(\frac{215\times\pi}{180} - a\right)/\sqrt{3}} = 0$$

$$0.42262 - \sin(a - \phi)e^{-\left(2.165 - \frac{a}{\sqrt{3}}\right)} = 0$$

$$0.42262 - \sin(a - \phi)e^{a/\sqrt{3}}e^{-2.165} = 0$$

$$a = 133^\circ\ 0.4187$$

$$a = 134^\circ\ 0.4255$$

By iteration $a = 134^\circ$

The conduction angle $\beta = 215 - 134 = 81^\circ$

Applying Fourier series expansion the fundamental coefficients are

$$a_1 = \frac{\sqrt{2}E}{2\pi}(\cos 2a - \cos 2x)$$

$$= \frac{\sqrt{2}E}{2\pi}[-0.0349 - 0.34202] = -\frac{\sqrt{2}E}{2\pi} \times 0.3769$$

$$b_1 = \frac{\sqrt{2}E}{2\pi}[2(x - a) - \sin 2x + \sin 2a]$$

$$= \frac{\sqrt{2}E}{2\pi}[2.826 - 0.9397 - 0.9994]$$

$$= \frac{\sqrt{2}E}{2\pi}[0.8869]$$

The fundamental displacement angle is

$$\tan^{-1}\frac{a_1}{b_1} = \frac{-0.3769}{0.8869}$$

$$\psi = 23.024^{\circ}$$

$$\cos\psi_1 = 0.920343$$

Problems

3.1 A single phase, half wave converter operating on a 220 V, 50 Hz supply charges a 72 V battery having an internal resistance of 0.5 Ω. Determine the value of the firing angle so that the charging current does not exceed 20 A. Determine the PIV of the thyristor.

3.2 A two pulse mid point converter is used to charge the battery of Problem 1. The converter in this case is fed from a converter transformer having a secondary voltage of 220 V between the line and centre tap. Determine the firing angle of the converter so that the charging current does not exceed 20 A. Determine also the PIV and average value of the thyristor current.

3.3 A 500 V, 3-phase, 50 Hz source supplies a 3-phase, 6 pulse bridge type converter, which feeds a dc load consisting of a back emf of 450 V and a resistance of 5 Ω. For firing angles of $\alpha = 30°$ and $\alpha = 60°$ determine the power supplied and the average thyristor current. Determine the input power factor and the reactive power supplied to the converter.

3.4 A 250 V, 50 Hz ac source supplies a single phase half wave converter supplying a load resistance of 10 Ω. The maximum current in the load has to be limited to 50 A. Calculate the desired control angle, the average thyristor current and the PIV of the thyristor.

3.5 A two-pulse midpoint converter with a voltage of 250 V between the centre tap and line is used to supply the load resistance of 10 Ω. Determine the control angle, the average thyristor current and the PIV of the thyristor if the load current must not exceed 15 A.

3.6 Compare the waveforms of the current in a dc load circuit (I_d), the thyristor input current (i_T) and the input current of a phase controlled two pulse midpoint converter for the two cases of loading (a) pure resistance ($L/R = O$), and (b) ideal smoothing. Compare also the effective values of line current, active power, power factor, and the output voltage as functions of the firing angle α.

3.7 A two pulse mid point converter is feeding a load containing a resistance of 10.5 Ω and very high inductance, which causes perfect smoothing on the dc side. The converter operates on a 220 V, 50 Hz supply. The reactance on the line side of the converter is 0.4 Ω. For a firing angle of 30° determine

(a) the average dc voltage
(b) the average dc current
(c) the overlap angle.

3.8 A two pulse mid point converter is feeding a load containing a resistance of 0.5 Ω and very high smoothing inductance. The converter is operating as an inverter and the load has a back emf of 200 V. The half winding on the secondary side of the transformer feeding the converter is 220 V and the frequency is 50 Hz. The commutation inductance (line side inductance) is 2 mH. Assume the thyristors to be ideal. Determine the firing angle of the converter to maintain a current of 200 A in the dc circuit.

3.9 A two pulse mid point converter has a free wheeling diode connected across the load, consisting of a resistance of 15 Ω and a large inductance giving ideal smoothing inductance. The converter is fed from a transformer having a half winding voltage on the secondary side of 220 V. The firing angle of the converter is 30°. Determine (a) the average value of the dc voltage, (b) the average value of the dc current, (c) the average value of the thyristor current and (d) the average value of the diode current.

3.10 A 2-pulse single phase bridge converter operating on a 220 V, 50 Hz supply feeds the following types of loads:

(a) $R = 10\ \Omega,\ L = 0,\ L_c = 0$
(b) $R = 10\ \Omega,\ L = \infty,\ L_c = 0$
(c) $R = 10\ \Omega,\ L = \infty,\ L_c = 2$ mH

Determine the average value of the dc voltage, dc current and the device currents for firing angle of $a = 30°$. Determine the overlap angle.

3.11 Determine the average value of the dc voltage and the dc current of a 2 pulse single phase bridge converter having the following details:

$$V_2 = \sqrt{2} \times 110 \sin \omega t,$$
$$\omega = 314 \text{ rad/sec.}$$

The converter transformer is ideal. The load consists of a resistance of 3 Ω, an inductance of very high value, causing perfect smoothing and a back emf of 60 V. The firing angle of the converter is 45°.

3.12 A 2 pulse phase controlled bridge converter operating from a 220 V, 50 Hz mains feeds a load comprising

$$R = 2\ \Omega,\ L_d = \infty \text{ and } E_b = 200 \text{ V}$$

The commutation inductance is 0.5 mH. For a firing angle of 120°, determine the average value of the load current and overlap angle.

3.13 A half controlled bridge converter has a firing angle of 75° and operates on 220 V, 50 Hz mains. It feeds a load having a resistance of 2 Ω and very large inductance sufficient to cause ideal smoothing. Determine the dc voltage and the average value of the dc current. Determine the rms and average values of the thyristor currents.

3.14 A 2-pulse phase controlled bridge converter has a free wheeling diode. It operates from a 220 V, 50 Hz mains to feed a load of 5 Ω resistance and very high value of inductance with perfect smoothing on the dc side. The commutating inductance can be neglected. Determine the average value of the dc voltage and the average value of the dc current for a firing angle of 60°. What are the rms and average values of thyristor and diode currents. Determine the fundamental displacement factor.

3.15 A 3-phase, 3 pulse converter feeds a purely resistive load of 5 Ω, the line voltage of the mains feeding the converter is 380 V at 50 Hz. The firing angle of the converter is 45°. Determine the mean values of output voltage and current. Determine also the average and rms values of the device currents.

3.16 A 3-phase, 3 pulse converter feeds a load having a resistance of 5 Ω in series with an inductance. Determine the value of the inductance in the load circuit to make the current just continuous.

3.17 A 3-phase, 3 pulse converter has a FWD connected across the load. The converter has an input voltage of

380 V at 50 Hz. The load comprises a resistance of 20 Ω in series with a very high value of smoothing inductance.

(a) Determine the average value of the dc voltage and dc current at a firing angle of $a = 60°$.

(b) If the inductance in the load is reduced to 80 mH and the FWD is removed, determine the firing angle at which the current becomes discontinuous.

3.18 A 3-phase, 3-pulse converter feeds a resistance load in series with a large inductance, causing perfect smoothing. The average load current is 50 A. The phase voltage on the ac side of the converter is 220 V at 50 Hz. The commutation inductance is 1.5 mH. Determine the average value of the dc voltage, and the overlap angle of a firing angle of 30°. Determine the rms value of thyristor current.

3.19 (a) For a P pulse uncontrolled rectifier in mid point connection show that the ideal output voltage is given by

$$V_{\mathrm{di}} = s\left(\frac{q}{\pi}\right)\sqrt{2}V_{\mathrm{s}}\sin\left(\frac{\pi}{q}\right)$$

where V_s is the voltage between the line and neutral on the ac side of the converter; s number of series connected converters; q number of commutations in a commutating group.

(b) Use the expression to determine the dc voltage of a 6-pulse mid point converter, 6-pulse bridge converter, 2-pulse mid point converter, 2-pulse bridge converter, and a 6-pulse converter with interphase transformer.

3.20 Show that under an ideal case of smoothing and instantaneous commutation

$$\frac{I_{v1}}{I_{11}} = \frac{1}{v}$$

On the line side of the converter. I_{vl} is the amplitude (rms value) of the v the harmonic and I_{11} that of the fundamental.

3.21 Discuss the effects of overlap on the harmonic content of the ac current on the line side. Show for the case of non-instantaneous commutation

$$\frac{I_{vl}}{I_{11}} = \frac{1}{v}\,\frac{\sin vu/2}{vu/2}$$

where μ is the overlap angle and v is the order of harmonic the fifth harmonic. Determine the ratios for the case of instantaneous and non-instantaneous commutations.

3.22 (a) What are the criteria for selecting the value of load inductance of a controlled rectifier?

(b) Show that for a P pulse converter, the inductance required is given by

$$L_{\mathrm{d}} = \frac{V_{\mathrm{di}}}{\omega I_{\mathrm{d}}}\sin a\left[1 - \frac{\pi}{P}\cot\frac{\pi}{P}\right]$$

Deduce the simplified expression for $P = 2$, 3 and 6.

3.23 A fully controlled 3-phase bridge rectifier is supplied by a 440 V, 50 Hz mains. The source has an inductance of 1 mH. The thyristors are ideal. Deduce a relationship between the average terminal voltage of the converter and firing angle a as well as load current. Draw the curve of voltage against firing angle for a constant load current of 50 A. Determine also the curve of voltage against load current for a given a of 30°.

3.24 A 3-phase bridge connected phase controlled rectifier feeds a load of 0.5 Ω in series with a very high value of inductance and a back emf of 190 V. The line to neutral voltage on the line side of the converter is 220 V. Assuming instantaneous commutation, determine the mean value of the rectified voltage, and the mean value of the direct current for $\alpha = 0$ and $\alpha = 60°$. Determine the rms value of the thyristor currents.

3.25 A 3-phase, 6-pulse bridge converter has the following particulars: Line to neutral voltage on the line side is 250 V at 50 Hz. The load circuit comprises a resistance of 2 Ω, a very large inductance making the current continuous and ripple free and a back emf of 400 V. The commutation inductance on the ac side, L_c = 2 mH. For a firing angle of $\alpha = 60°$, determine the average value of rectified voltage and the current in the thyristors.

3.26 If the converter of Problem 25 is in the inverting mode with a firing angle of $\alpha = 130°$, determine the average value of the load voltage and dc current and the power fed back to the supply. Consider $L_c = 0$ as well as 2 mH. Determine also the overlap angle.

3.27 A 3-phase half controlled bridge converter supplies a load comprising a resistance of 1.5 Ω in series with an inductance which is large enough to effect perfect smoothing. The line to neutral voltage on the line side is 200 V and the power delivered to the load is 15 kW. Determine the value of α. Find the mean values of thyristor and diode currents.

3.28 A 3-phase, 6-pulse bridge rectifier supplies a power of 400 kW to the dc load comprising an R-L-E. The maximum dc voltage is 500 V. Determine the inductance required in the load if the operation has to be continuous at 10% of the rated current and a firing angle of $\alpha = 90°$.

3.29 A dc chopper shown in Fig. P3.29 feeds an R-L load. The waveform of the load current for a definite operating condition is also shown in Fig. P3.29. The average value of the chopper output voltage is 150 V. Determine the values of R and L, neglecting commutation effects.

3.30 The basic chopper depicted in Fig. P3.20 feeds an R-L load. The load inductance is sufficiently large to make the load current constant and its variation during commutation negligible.

(a) Determine the minimum value of capacitance for commutation if the main thyristor T has a turn off time $t_q = 25\ \mu s$. The inductance L_1 can be neglected.

(b) What should be the value of commutating circuit inductance to allow an initial rise of current of 500 A.μs^{-1} in the auxiliary thyristors?

(c) What should be the value of L_2 if the recharging process is to complete in 100 μs?

(d) What is the maximum value of i_{T1} during the recharging process?

3.31 The following information is available about a basic chopper circuit: Input voltage = 150 V, Output voltage = 60 V, $E_b = 40$ V, $t_{on} = 0.5$ s, $T = 3$ ms. Determine the mode of operation of the chopper and the time constant of the load.

3.32 A 100 V dc chopper feeds an R-L load having $R = 5\ \Omega$ and $L = 40$ mH. A free wheeling diode is placed across the load. The load current varies between 10 and 12 A. Determine the time ratio of the chopper. What is the chopper frequency?

3.33 A chopper having a switching frequency of 250 Hz feeds a load having a pure inductance of 15 mH in series with a back emf. Determine the time ratio of the chopper to allow a current variation of 10 A in the load circuit. The input dc voltage to the chopper is 250 V.

3.34 A chopper operating from a 220 V dc source feeds an R-L load, having

$R = 5\ \Omega$ and L = 30 mH. The switching frequency is 200 Hz and the ON time of the chopper is 1.2 ms. Determine the limits of the load current.

3.35 Derive expressions for the variation of load current $i(t)$ of a dc chopper feeding an R-L load. Use the results to determine the dependence of current pulsation $\Delta i = i_{max} - t_{min}$ on the time ratio T_{ON}/T, with the time constant (T_L/T) of the load circuit as a parameter.

3.36 A dc chopper fed from 150 V feeds a load comprising $R = 0.2\ \Omega$, $L = 0.1$ mH and a back emf of 20 V. $T_{ON}/T = 0.33$ and the period is 3 ms.
(a) Determine the mode of operation of the chopper.
(b) Find the average values of output voltage and current.

Multiple-Choice Questions

3.1 A fully controlled line commutated converter operates as an inverter
(a) in the range of firing angles $0 \le a \le 90°$
(b) in the range of firing angles $90° \le a \le 180°$
(c) in the range of firing angles $90° \le a \le 180°$ with a suitable dc source in the load.
(d) When it supplies RLE load

3.2 The inverter limit of a line commutated inverter is due to
(a) Overlap alone
(b) Turn off time of the thyristor
(c) Turn on time of the thyristor
(d) Due to both overlap of the converter and turn off time of the thyristor

3.3 The overlap introduces
(a) additional reactive power requirement
(b) additional losses in the load
(c) additional harmonics in the load current
(d) a heavy short circuit current

3.4 In the thyristor power converters during discontinuous conduction
(a) the load current is zero even though the load voltage is present
(b) the load current and load voltage are both simultaneously zero
(c) the load current is present even though the load voltage is zero
(d) the current is ripple free

3.5 A free wheeling diode in a phase controlled rectifier
(a) enables the inverter operation
(b) smoothens the load current consequently the smoothing inductance required is small
(c) makes the converter draw additional reactive power
(d) impairs the line power factor

3.6 A phase controlled converter is designed for 220 V, 50 Hz supply. If this converter operates on 170 V, 50 Hz supply there is a possibility for
(a) commutation failure during rectification
(b) commutation failure during inversion
(c) increased voltage drop
(d) increased losses in the inverter

3.7 In a 2-pulse bridge converter with free wheeling diode, the width of the diode current pulse is
(a) π
(b) a
(c) $\pi - a$
(d) $\pi + a$

3.8 Mid point converters in general require input transformers. This input transformer is of special construction because
(a) harmonic currents are present in the load
(b) dc magnetisation of the core
(c) to give strength to the system
(d) to improve the converter behaviour

3.9 A voltage source inverter supplying an inductive load requires a gate pulse of long duration to turn the thyristor ON
(a) to ensure the initiation of conduction of the thyristor immediately when its voltage is positive
(b) because a wider pulse is required invariably to turn on the thyristor
(c) to reduce the thyristor losses
(d) because a wider pulse effectively provides an output with least harmonic content

3.10 The phase control employed for a cycloconverter imparts
(a) a very good power factor
(b) a very bad power factor
(c) undesirable harmonic behaviour
(d) poor voltage regulation

3.11 The conduction angle of an ac voltage controller depends on
(a) the variation in the input voltage
(b) the impedance angle of the load
(c) the supply frequency
(d) both supply frequency and voltage

3.12 A single phase ac voltage controller feeds a pure inductive load. Its control angle range
(a) $0 \le a \le 180°$
(b) $0 \le a \le 90°$
(c) $90° \le a \le 180°$
(d) $180° \le a \le 270°$

3.13 In a two quadrant chopper, the load voltage is varied from positive maximum to negative maximum by varying
(a) time ratio of the chopper from 0 to 1
(b) time ratio of the chopper from 1 to 0
(c) time ratio of the chopper from 0 to 0.5
(d) time ratio of the chopper from 0.5 to 0

3.14 A dual converter with circulating current mode
(a) has no possibility of discontinuous conduction
(b) requires a more sophisticated control
(c) has a very sluggish behaviour
(d) offers difficulties during transition from motor to generators

References

3.1 Bedford, B.D., and R.G. Hoft, *Principles of inverter circuits*, John Wiley and Sons, Inc., New York, 1964.

3.2 Dewan, S.B., and A. Straughen, *Power semiconductor circuits*, John Wiley and Sons Inc., New York, 1975.

3.3 Subrahmanyam, Vedam, *Thyristor control of electric drives*, Tata McGraw-Hill Publishing Co. Ltd., New Delhi, 1988.

3.4 Berde, M.S., *Thyristor engineering*, Khanna Publishers, New Delhi, 1988.

3.5 Sen, P.C., *Power electronics*, Tata McGraw-Hill Publishing Co. Ltd., New Delhi, 1987.

3.6 Dubey, G.K., S.R. Doradla, A. Joshi, and R.M.K. Sinha, *Thyristorized power controllers*, Wiley Eastern Ltd., 1986.

3.7 Heumann, K. and A.C. Stumpe, *Thyristoren*, B.G. Teubner, Stuttgart, 1969.

3.8 Bystron, K., *Leistungselektronik*, Carl Hanser Verlag, München, 1979.

3.9 Jager, R., *Leistungselektronik Grundlangen Und Anwendungen*, VDE Verlag GmBH, Berlin, 1977.

3.10 Thorborg, K., *Power electronics*, Prentice-Hall, International (UK) Ltd., 1988.

3.11 Lander, C.W., *Power electronics*, McGraw-Hill, London, 1981.

3.12 Seguier Guy, *Power electronic converters*, North Oxford Academic, 1986.

3.13 Patel, H.S. and R.G. Hoft, Generalised techniques of harmonic elimination and voltage control in thyristor converters, *IEEE transactions on industry applications*, Vol. IA9, pp.310–317, 1973 and Vol.IA-10, 1974, pp. 666–673.

3.14 Steigerwald, R.L. 'Characteristics of a current fed inverter with commutation applied through load neutral point', *IEEE transaction industry applications*, Vol. IA-15, pp. 538–553, 1979.

3.15 Subrahmanyam, Vedam, 'Analysis of commutation of a current source inverter feeding an induction motor using 2-axis variables', *IEE Proceedings*, Vol. 130, Pt.B, pp. 355–359, 1983.

3.16 Ward, E., 'Inverter suitable for operation over a wide range of frequency', *Proc. IEE,* 1974, Vol. 111, pp. 1423–1434.

3.17 Farrer, W., and J.D. Miskin, 'Quasi sine wave fully regenerative inverter', *Proc. IEE*, 1973, Vol. 120, pp. 969–976.

3.18 Bowes, S.R., 'New sinusoidal pulse width modulated inverter', *Proc. IEE*, Vol. 122, 1975.

3.19 Phillips, K.P., 'Current source converter for ac motor drives', *Conf. Record 1971*, IEEE IAS, pp. 385–392.

4 Control of Electric Motors

A separately excited dc motor is very versatile as a variable speed motor. Its speed can be varied by varying the applied voltage to the armature or field current. The speed control using the variation of armature voltage can be used for constant torque application in the speed range from zero to rated speed (base speed). Speeds above base speed are obtained by means of field weakening, the armature voltage being kept at the rated value. The speed control in this case is at constant power. In both cases the speed control is smooth and stepless. However, the dc motor has disadvantages as a variable speed drive, due to the presence of a mechanical commutator. The commutator actually limits the rating, highest speed of operation. It requires frequent maintenance and makes the dc motor unsuitable for application under certain environmental conditions.

With the advent of thyristors and thyristor power converters the variable voltage to the dc motor is obtained from static power converters. Phase controlled rectifiers provide variable dc voltage from constant voltage, constant frequency mains. The static apparatus is very efficient, compact and has a very good dynamic behaviour. It is very easy to provide a four quadrant drive with slight modifications in the converter.

A dc chopper can be used to obtain a variable voltage from a constant dc voltage. The average value of the output voltage can be varied by varying the time ratio of the chopper.

Besides the advantages, static converter fed dc motors have a performance which is very much different from the performance of the motor operating on pure dc. The motor current has ripple content which affects the commutation capability of the motor. Additional losses are present. There may be torque pulsations. When the phase control is used the line power factor is poor.

An ac motor is a constant speed motor. Its speed depends upon the supply frequency and number of poles. The mechanical commutator is not present and therefore an ac motor has several advantages compared to a dc motor. Besides robustness and simplicity of construction, e.g., cage type induction motor, an ac motor has advantages of less maintenance, increased power ratings, high speeds of operation, easily realisable explosion-proof construction, low inertia, and high power/weight ratio or increased power density. Because of these advantages, the ac motor finds application in reactor engineering where maintenance free operation is required, in air craft drives where high power density is required, and in the textile industry where dust and explosion-proof construction is required. Ac motor drives are built for high speeds in large power ratings and due to low inertia they have a fast response. The main disadvantage of these motors was that an efficient smooth speed control in a wide range was a problem.

The problem of speed control of ac motors has been very efficiently solved with the development of thyristor power converters which can provide variable frequency, variable voltage supply. Using these converters it has been possible to achieve an ac motor having dc motor characteristics. Developments are now oriented towards replacing the dc motor by means of inverter fed ac motors with improved dynamic response.

Induction motors operating on thyristor power converters have non-sinusoidal input waveform. These cause variations in the performance, such as additional harmonic losses, torque pulsations, etc. The additional harmonic losses are responsible for increased heating and temperature rise. Consequently the motor has to be derated, particularly at low speeds. Torque pulsations at low speeds are objectionable. Hence it is necessary to modify either the motor design or the inverter, for better performance.

The speed of a slip ring induction motor can be varied by using power converters in the rotor circuit. The rotor current waveform, which is non-sinusoidal, causes variations in the performance.

Synchronous motors are also becoming popular as variable speed drives using thyristor power converters. The variable frequency for the synchronous motors is given by cycloconverters, voltage fed inverters and current fed inverters. A synchronous motor has an advantage over an induction motor. When it is overexcited it operates at leading power factor. The machine voltages can be used for commutation of the inverter. Load commutation is thus possible and consequently the power circuit becomes simple. The inverter feeding a synchronous motor can have a control from the information of rotor position. All the six thyristors of the inverter are fired (once) in a sequence by the time the rotor moves by two pole pitches (i.e., 360° el.). The motor current completes one cycle. Thus the speed and frequency are synchronised using rotor position sensing. The control is called self control. The input frequency and rotor speed are tied together. This control imparts to synchronous motor the properties of a dc motor. The problem of stability disappears and the machine has a dynamic behaviour similar to that of a dc motor. A self controlled load commutated C.S.I. fed synchronous motor is now a competitor for both dc motor and induction motor.

The performance of a synchronous motor on variable frequency supply is of interest. This enables one to improve the inverter system or motor design for better performance.

4.1 INDUCTION MOTOR DRIVES

A converter fed induction motor has the following advantages over a line fed motor:

i. Smooth start up is guaranteed by variable frequency starting from a low value.
ii. Soft starting and acceleration at constant current and torque are possible.
iii. The network is no longer subjected to a high switching surge current as with the direct switch ON of cage induction motor, and as such, special starting equipment can be omitted even at high ratings.
iv. High moments of inertia can be accelerated without need to over dimension the motor.
v. The converter acts as a decoupling device. Therefore, feedback from the motor to the point of short circuit does not take place, when line short circuits occur. The short circuit rating, on the basis of which the switchgear has to be overdimensioned is therefore low, permitting a saving to be made.

As has already been stated, the induction motor speed can be controlled by supplying the stator a variable voltage, variable frequency supply using static frequency converters. Speed control is also possible by feeding the slip power to the supply system using converters in the rotor circuit. Basically one distinguishes two different methods of speed control.

i. Speed control by varying the slip frequency when the stator is fed from a constant voltage, constant frequency mains.
ii. Speed control of the motor using a variable frequency variable voltage, motor operating at constant rotor frequency.

   Speed control by variation of slip frequency is obtained by the following ways:

   (a) Stator voltage control using a three-phase voltage controller.
   (b) Rotor resistance control using a chopper controlled resistance in the rotor circuit.
   (c) Using a converter cascade in the rotor circuit to recover slip energy.
   (d) Using a cycloconverter in the rotor circuit.

4.1.1 Control of an Induction Motor by Stator Voltage Variation (Using a Three Phase Voltage Controller)

It is very well known that the torque of an induction motor varies directly in proportion to the square of the voltage. The torque of an induction motor is approximately given by

$$T_{\mathrm{d}} = \frac{m}{2\pi\, n_{\mathrm{s}}} \frac{V_{20}'^{2}}{\left(R_2'/s\right)^2 + x_2'^{2}} \frac{R_2'}{s} \tag{4.1}$$

and the torque speed curve is as shown in Fig. 4.1. The slip for maximum torque is given by

$$s_m = \frac{R_2'}{x_2'} \tag{4.2}$$

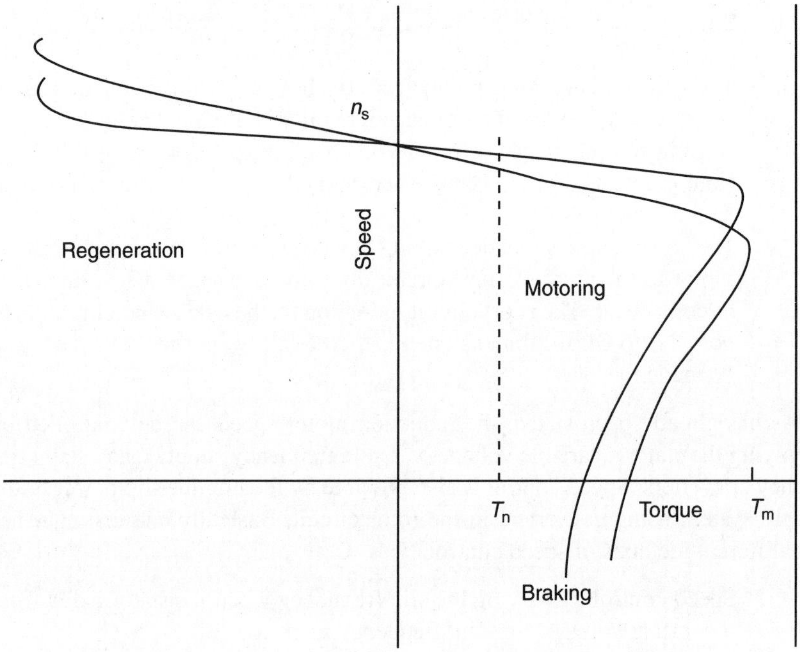

Fig. 4.1 *Speed-torque curve of a three phase induction motor*

which is independent of stator voltage. However, this can be varied by variation in rotor resistance. The value of maximum torque is given by

$$T_{dm} = \frac{m}{2\pi n_s} \cdot \frac{V_{20}'^2}{2x_2'} \tag{4.3}$$

This also changes as the square of the applied voltage. If the voltage is reduced to 80% the maximum torque falls to 64%. The variation in applied voltage is achieved by means of a voltage controller.

This method of controlling the speed of an induction motor is simple and economical. The stator voltage control is achieved by means of phase control of the antiparallel thyristors, connected as shown in Fig. 4.2. Figures 4.3(a) and (b) illustrate two practical connections of a voltage controller feeding an induction motor. In the connection of Fig. 4.2 the thyristors have to handle only phase current. The harmonic currents become higher. In Fig. 4.3 the harmonic penalty is rather more.

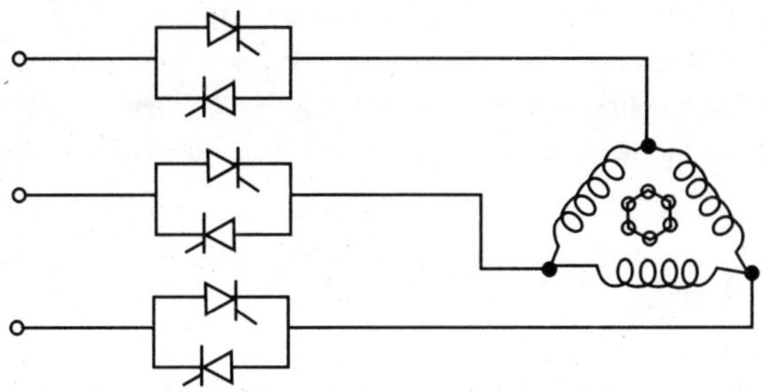

Fig. 4.2 *AC voltage controller using antiparallel thyristors for speed control of induction motor*

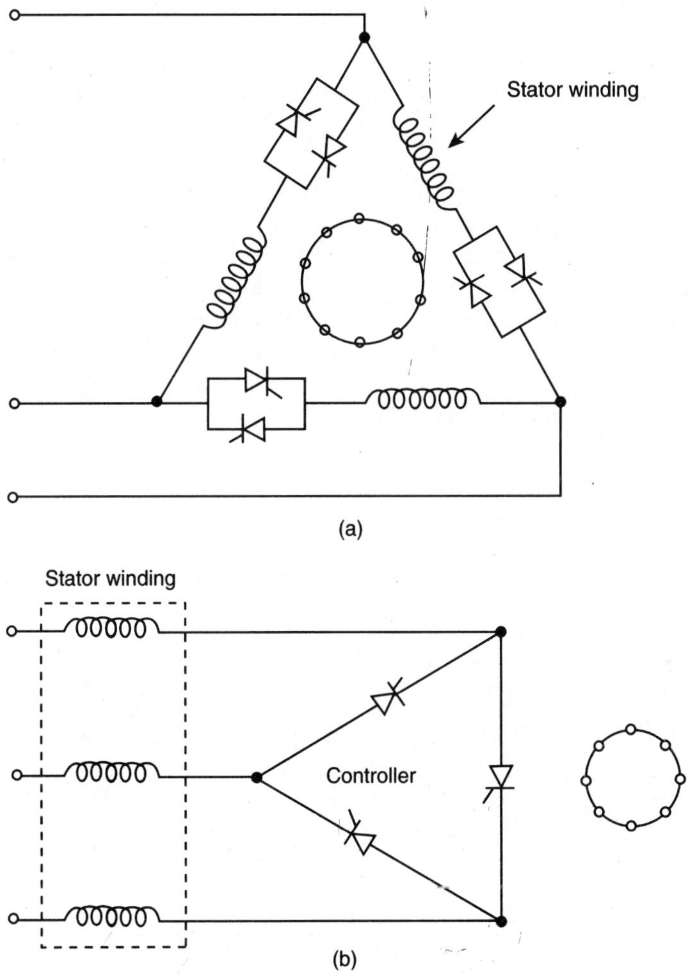

Fig. 4.3 *Practical connections of ac voltage controller (a) delta connected controller (b) neutral connected controller*

The stator voltage can be varied from zero to full value within the triggering angle range. The line side power factor is very poor because of harmonics and reactive power due to phase control.

When a cage induction motor is fed from a variable voltage supply for speed control the following observations may be made:

i. The torque speed curve beyond the maximum torque point has a negative shape. A stable operating point in this region is not possible for constant torque load.
ii. The voltage controller must be capable of withstanding high starting currents. The range of speed control is rather limited.
iii. The motor power factor is poor.

To obviate the above difficulties the induction motor must have a high resistance rotor. This makes the point of maximum torque shift towards $s = 1$, thereby reducing the unstable region of speed-torque curve. Due to increased rotor resistance the starting current decreases, the power factor improves and the range of speed control increases.

The method of speed control is therefore advantageous with a high resistance rotor. The speed-torque curves for this control are shown in Fig. 4.4(a). The current rating of the controller decides the possible torque at each speed. The limiting curve of the torque as a function of speed can be derived. Figure 4.4(b) shows this curve of limiting torque for design rating of the controller, which is three times the rated current of the motor. The current as a function of speed and stator voltage is depicted. During control, if the current value exceeds the limiting value, automatic current limit must be employed. By this the value of the firing angle gets adjusted until the permissible current flows.

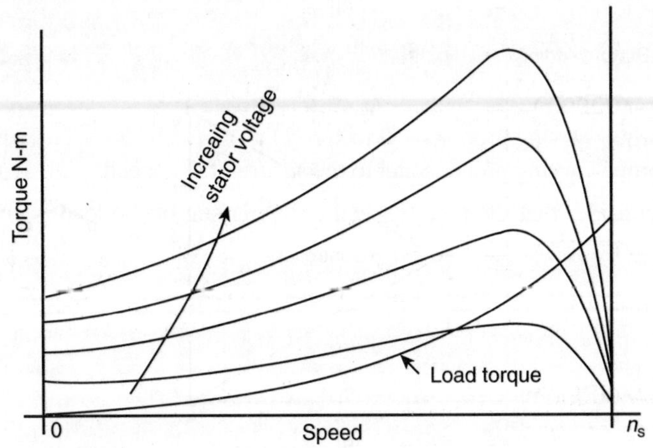

Fig. 4.4(a) *Speed-torque characteristics of an induction motor with variable stator voltage*

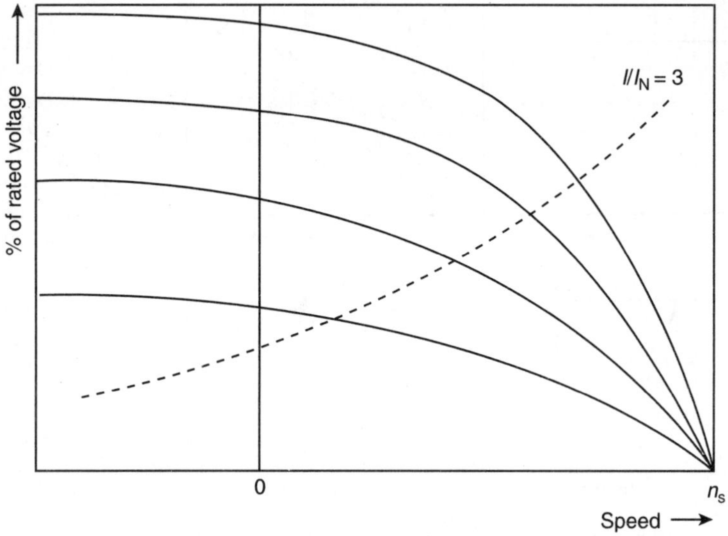

Fig. 4.4(b) *Typical limiting torque curve of an induction motor fed from a voltage controller*

Even though the method is simple and economical, the motor losses increase with increase in slip. The increase in the losses can be attributed to increase in the motor current due to drop in the air gap flux as well as to the high resistance of the rotor. The ratio r_1/r_2' can be taken to be representative in deciding these losses because increase in the resistance is instrumental in limiting the current drawn. The efficiency of the motor can be approximately given by

$$\eta = (1 - s)$$

where s is the slip of the motor.

The type of load driven by the motor influences the current drawn and losses of the motor as the slip varies. The normally occurring loads are

i. constant torque loads
ii. torque varying proportional to speed
iii. torque varying proportional to the square of the speed.

Let us consider that the torque speed characteristic of the load is given by a general equation

$$T_d = K_t n_r^x$$

If the value of $x = 0$, it amounts to constant torque load. $x > 0$ for variable torque loads (Fig. 4.5a)

The rotor copper losses

$$P_2 = sP_{D1}$$

where P_{D1} is the air gap power or power input to rotor.

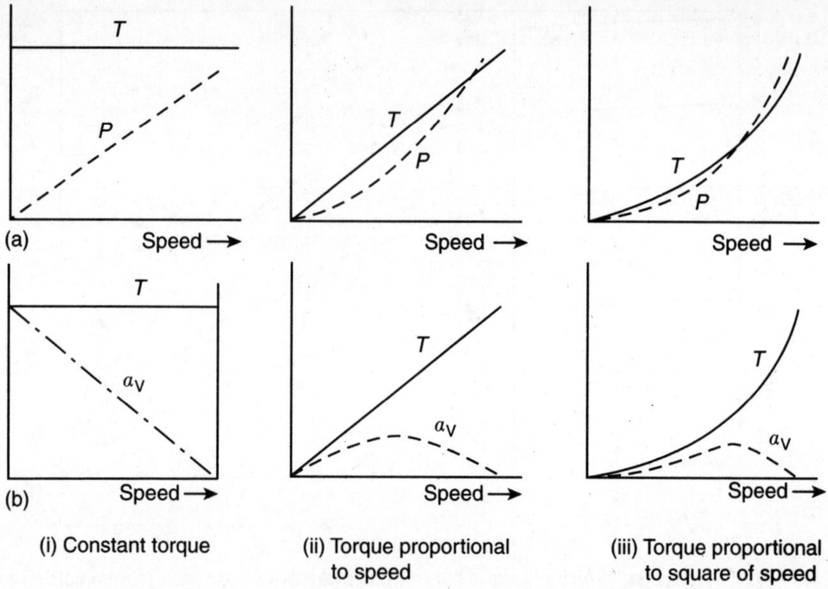

Fig. 4.5 *(a) Typical torque speed curves of normally occurring load (b) Variation of copper losses of a voltage controlled induction motor with speed*

But the slip of the motor

$$s = \left(1 - \frac{n_r}{n_s}\right)$$

and the power input to the rotor

$$P_{D1} = 2\pi n_s T_d = 2\pi n_s n_r^x \cdot k_t .$$

The maximum value of power transferred to rotor is

$$P_{D1\,max} = 2\pi n_s \cdot n_s^x \cdot k_t = k_t \cdot 2\pi\, n_s^{x+1}.$$

The copper losses when expressed as the ratio of $P_{D1\,max}$ we have

$$\frac{P_2}{P_{D1\,max}} = y = \left(1 - \frac{n_r}{n_s}\right)\left(\frac{n_r}{n_s}\right)^x$$

writing $\frac{n_r}{n_s} = s$ per unit speed we have

$$\frac{dy}{ds} = xs^{x-1} - (x+1)\, s^x = 0$$

for the value of s at which y is maximum. The value of

$$s_m = \frac{x}{x+1}$$

for maximum value of rotor copper loss. The maximum values of rotor copper losses are given by

$$y_m = \left(\frac{x}{x+1}\right)^x - \left(\frac{x}{x+1}\right)^{x+1}$$

As the value of x increases the value of y_m decreases. The maximum value of rotor current and per unit speed at which it occurs can also be determined. We have

$$T_d = \frac{3P}{\omega_s}\frac{I_2'^2 r_2'}{s} = k_1(1-s)^x \omega_s^x.$$

from which the rotor current

$$I_2' = k_1\sqrt{s^x(1-s)} = k_1\sqrt{s^x - s^{x+1}}$$

per unit speed for maximum value of rotor current is also

$$s_m = \frac{x}{x+1}$$

The maximum value of rotor current can be determined as

$$I_{2m}' = \sqrt{\frac{1}{x+1}\left(\frac{x}{x+1}\right)^x \frac{k_1}{3P}\omega_s^x}.$$

For special cases

$x = 0$	$y = 1.00$	$s = 0.0$
$x = 1$	$y = 0.25$	$s = 0.5$
$x = 2$	$y = 0.149$	$s = 0.67$

The variation of losses are shown in Fig. 4.5(b) for different types of loads. The constant torque loads are not favoured due to increase in the losses linearly with slip and becoming maximum at $s = 1.0$. This is obvious from the variation of flux as the voltage is varied for speed control. To maintain constant torque the motor draws heavy current resulting in a poor torque/ampere, poor efficiency and poor power factor at low speeds.

When the torque varies in direct proportion to speed the copper losses has a maximum value of 25% of rated power at a speed of 0.5 ω_s. For torques proportional to square of the speed, per unit speed at which copper losses are maximum is 2/3 and the maximum value of copper losses are 0.149 of rated power. The method of speed control can be advantageously employed for pump or blower type of loads where torque is proportional to square of the speed.

From the above discussion, this method of speed control is suitable only for the following cases:

i. For short time operations where the duration of speed control is defined.
ii. For speed control of blowers or pumps having parabolic or cubic variations of torque with speed. This is not suitable for constant torque loads due to increased losses and heating.
iii. For speed control of motor having poor efficiencies under normal operation.

The type of load (torque versus speed of the load) on the drive motor influences the losses in the motor. The non-sinusoidal input waveforms cause non-sinusoidal currents which increase the harmonic losses. Hence, the total losses increase particularly at low speeds and these losses cause a possible derating of the motor, or an over dimensioned motor must be used when this method is employed.

For blower type loads where $T \propto N^2$ the maximum value of current occurs at a speed of 2/3 base speed. This current depends upon the full load slip of the motor. For large slip motors (high resistance rotors) the ratio of maximum to rated current decreases. For constant torque loads the power losses increase with reduction in speed and reaches a maximum value at zero speed itself. For loads having $T \propto N$ the maximum occurs at 1/2 the base speed and the losses amount to 25% of the stator power.

The losses occurring in the motor are responsible for temperature rise of the motor. The losses must be kept within permissible value so that the motor operates always with permissible temperature rise. The different type of loads discussed have a tendency to increase the motor losses.

Further the waveform of the input voltage is distorted. The stator and rotor currents are non-sinusoidal with rich harmonic content. These harmonics cause additional losses. They can be taken as 50% of the rated copper losses.

The increase in the losses of the motor at large slips leads to a derating of the motor. A normal motor may be derated 5 to 6 times. If high resistance rotor is used the aerating factor decreases.

The total losses of the motor are

$$P_2 = f_v\left(1+\frac{r_1}{r_2'}\right)k_1 P_r \tag{4.4}$$

where f_v takes care of increase in losses due to distortion and is normally taken as 1.5.

P_r is the rating of the motor

K_1P_r is the total copper losses of the motor.

To maintain the permissible temperature rise, the permissible losses are

$$P_{vpv} = P_{Tvp}\cdot\frac{1-\eta}{\eta}a_v$$

where η is efficiency of the motor

a_v is the ratio of losses at minimum speed to the losses at rated speed.

P_{Tvp} rating of the motor.

The factor a_v depends upon the type of loading employed. For force cooled machines it is 1.0 and independent of speed. If the machine is self cooled it depends on speed. This value decreases as the lower limit of speed decreases. A typical characteristic for a two pole machine is shown in Fig. 4.5(c). The derating is also affected by the efficiency of the motor. As the efficiency decreases $\frac{1-\eta}{\eta}$ increases. For this type of speed control motors of large rotor resistance are used. They have poor efficiency. Consequently the derating of the motor decreases.

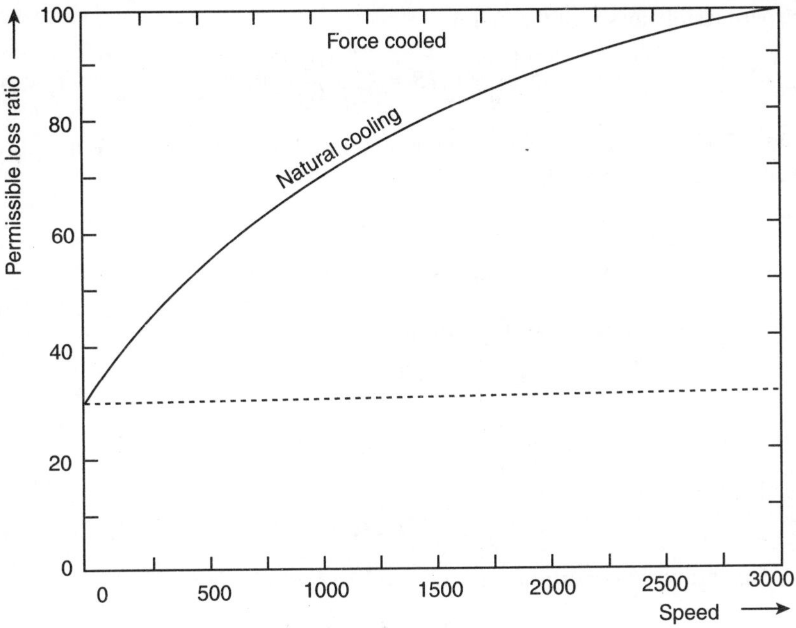

Fig. 4.5(c) *Permissible loss ratio as a function of speed*

The above discussion makes it clear that a given motor gets derated differently when it is driving different types of loads. In other words, the power required by the load is constant, the drive motor must be of different ratings for different types of torque-speed curves of the load. The following example illustrates this.

Example A 25 kW, 2-pole, 3-phase induction motor is fed from a three phase voltage controller for speed control. The motor has a stator to rotor resistance ratio of 0.7. The speed control is required from 3000 to 1400 rpm. Determine derating of this motor when it drives

i. constant torque loads
ii. loads having $T \propto N$
iii. loads having $T \propto N^2$

Assuming the efficiency of the motor to be 84%.

Solution

The copper losses of the motor

$$P_2 = 1.5\left(1+0.7\right)\frac{1400}{3000}P_{\mathrm{r}}$$

The permissible losses of the motor

$$= 25\frac{1-\eta}{\eta}0.75 = 25\times\frac{0.16}{0.84}0.75$$

where 0.75 is the ratio of the losses at minimum speed to the losses at rated speed. Equating the above we get

$$P_r = 3 \text{ kW}$$

Derating is $\frac{25}{3} = 8.33$ when it drives constant torque loads.

When the torque is proportional to speed

$$P_2 = 1.5\times1.7\times0.25P$$

Permissible losses $= 25\times\frac{0.16}{0.84}\times0.75$

Equating we get

$$P = 5.6 \text{ kW}.$$

when the torque is proportional to square of the speed

$$0.3825P = 3.57$$
$$P = 9.33 \text{ kW}$$

The rating of the motor can also be estimated if the load requires constant power. Form example if

$$P = 2 \text{ kW at a speed range of 3000 to 1400 rpm}$$

For constant torque loads.

$$\text{The rotor copper losses} = P_2 = 1.5\times1.7\times\frac{1400}{3000}\times2$$

assuming $r_1/r_2' = 0.7$ for the motor.
The permissible losses are

$$P_T\frac{1-\eta}{\eta}0.75 = P_T\frac{0.16}{0.84}0.75$$

assuming $\eta = 84\%$.

$$P_T = 1.5\times1.7\times\frac{1400}{3000}\times2\times\frac{0.84}{0.16}\times0.75 = 16.66 \text{ kW say 17 kW}$$

For $T \propto N$, $K_1 = 0.25$

$$P_2 = 1.5\times1.7\times0.25\times2$$
$$= P_T\frac{0.16}{0.84}\times0.75$$

$$P_T = 8.925 \text{ kW say 9 kW}$$

For $T \propto N^2$ $\quad K_1 = 0.15$

$$P_2 = 1.5 \times 1.7 \times 0.15 \times 2$$
$$= P_T \frac{0.16}{0.84} \times 0.75$$

$$P_T = 5.355 \text{ kW say } 5.5 \text{ kW}$$

As the motors are of different ratings, actual η of the motor must be considered. The derating factor is large for constant torque loads, as $x > 1$ the derating factor decreases.

Typical derating factors of an induction motor fed from a voltage controller are shown in Fig. 4.5(d) when the motor drives a fan or pump loads ($T \propto N^2$). The derating factor depends on the rated speed also.

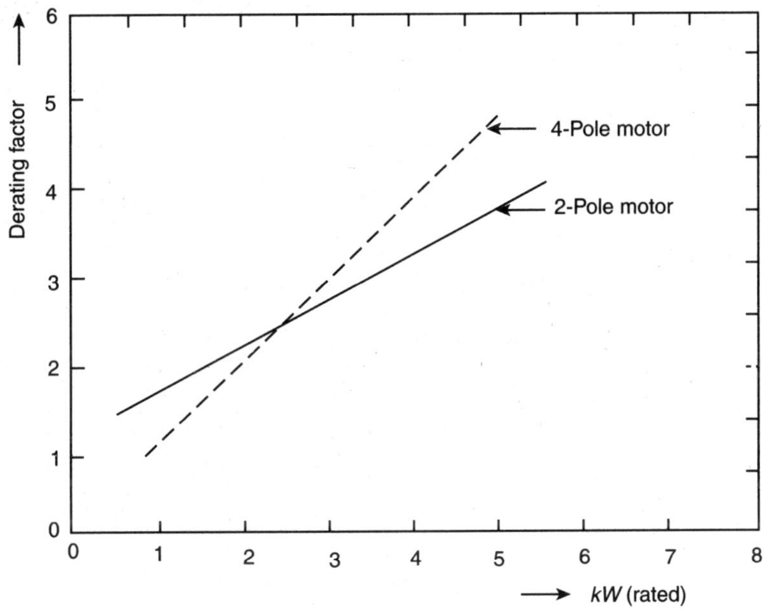

Fig. 4.5(d) *Derating factor of an electric motor operating on AC voltage controller*

Change of phase sequence changes the direction of the rotating magnetic field. The direction of rotation of the motor can be reversed by interchanging the connections to the two phases of the motor. Mechanical switches which are operated when the current is zero, may be employed. The method of speed reversal is shown in Fig. 4.6.

The drive is amenable for braking using the method of plugging. This is also called reverse current braking. The controller in this case must be designed to withstand the resulting currents which are greater than the starting current. If the current is limited, enough braking torque may not be present (Fig. 4.7). Regenerative braking is possible. In this case braking till zero speed is not possible. The method of dc dynamic braking is illustrated in Fig. 4.8. In this method the motor

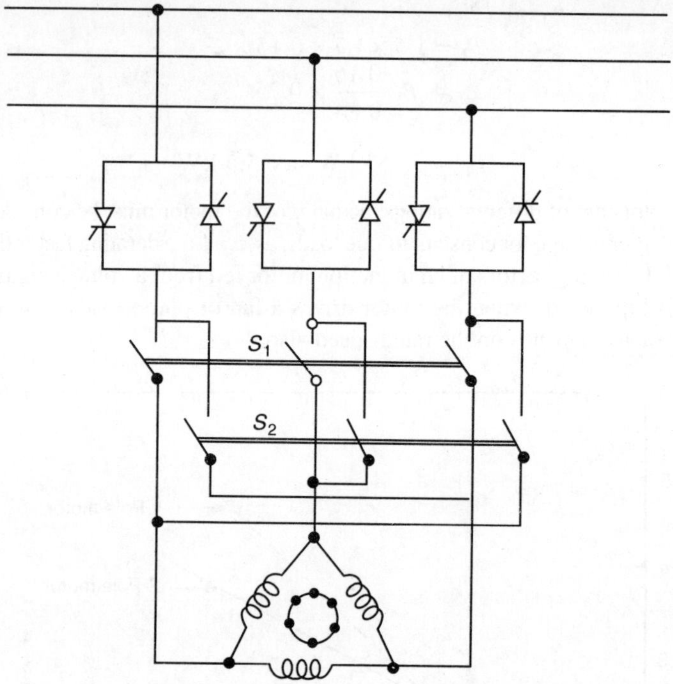

Fig. 4.6 *Speed reversal using voltage controller*

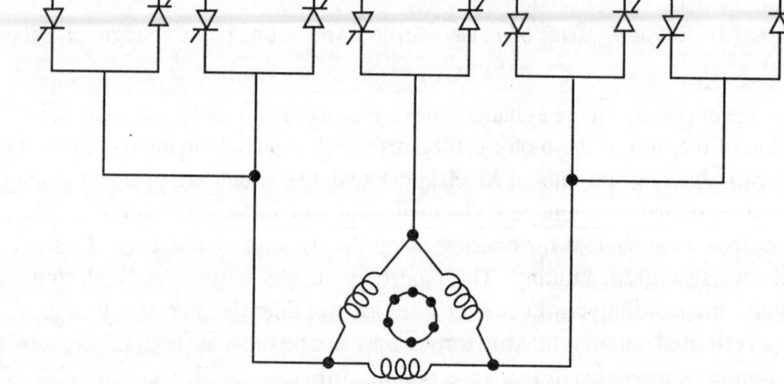

Fig. 4.7 *Braking of motor using plugging*

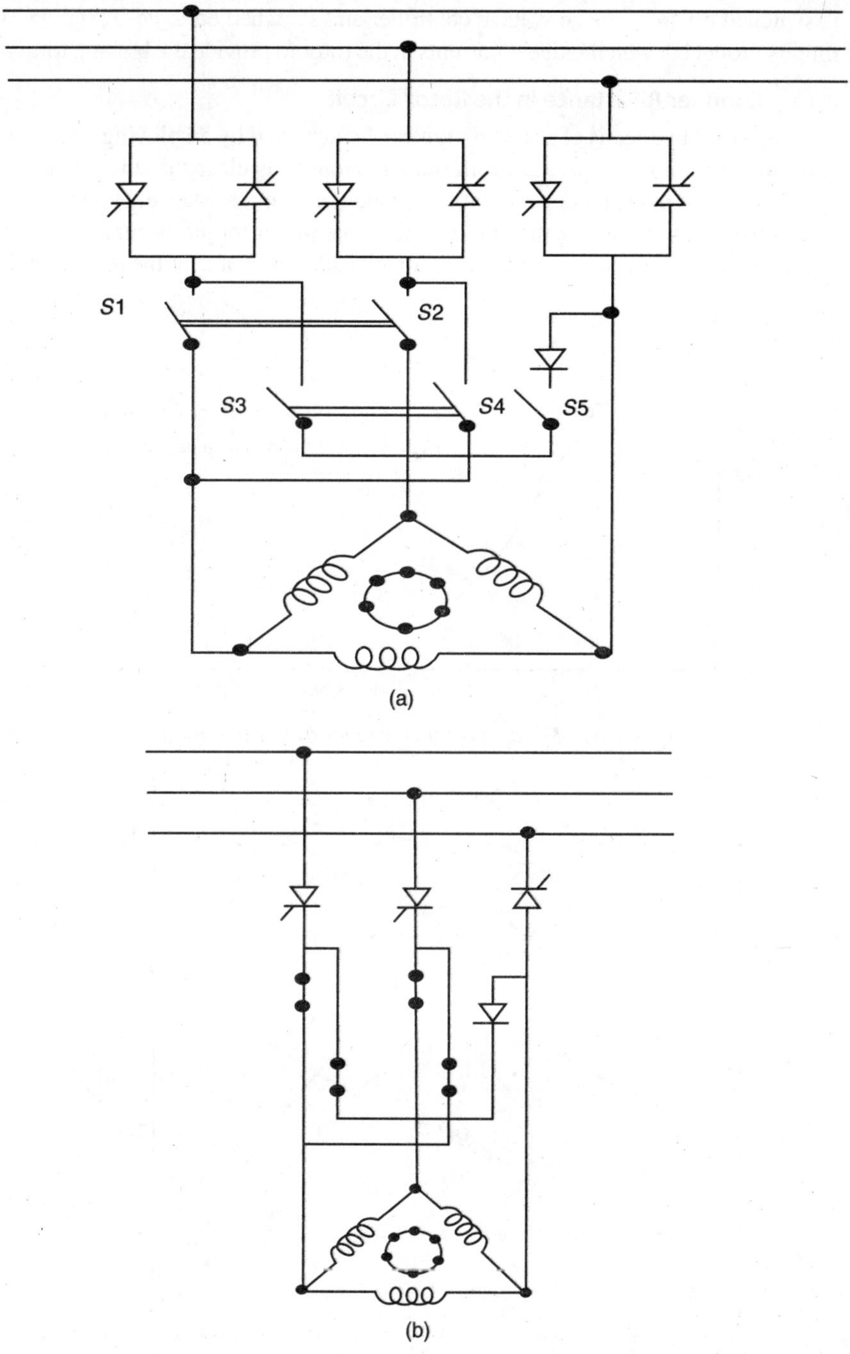

Fig. 4.8 *DC dynamic braking using AC voltage controller (a) basic circuit (b) under braking operation*

is switched off from the ac voltage controller and switched on to dc. A stationary mmf is produced, which induces currents in the rotor to provide the braking torque.

4.1.2 Chopper Resistance in the Rotor Circuit

Speed control by means of slip variation can be achieved by employing a variable resistance in the rotor circuit. From the equations it is clear that the maximum value of torque does not depend upon the value of rotor resistance. However, the rotor resistance influences the slip at which maximum torque occurs. A family of curves has been depicted in Fig. 4.9 for variable resistance in the rotor circuit.

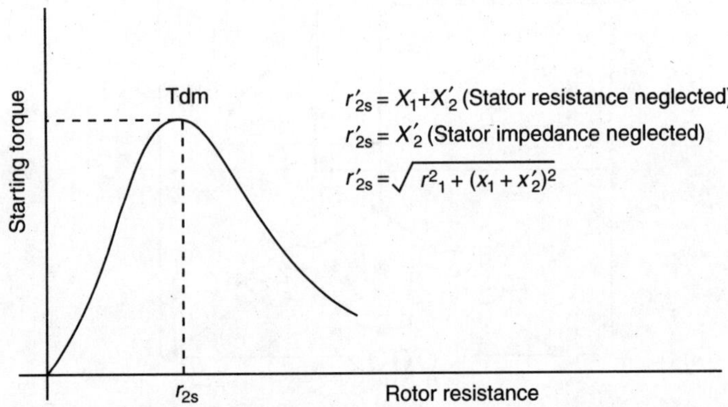

Fig. 4.9(a) *Effect of rotor resistance on starting torque*

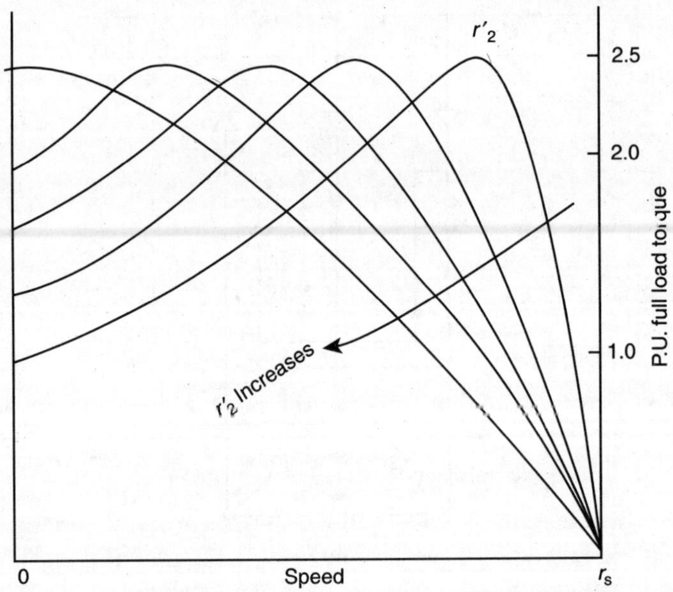

Fig. 4.9(b) *Effect of rotor resistance on speed torque curve*

For given load conditions it is clear that the speed can be varied. The slip of the motor at any given value of rotor resistance is given by

$$s' = s\,\frac{R_2' + R_{ex}'}{R_2'} \tag{4.5}$$

where R_2' is the rotor resistance and R_{ex}' is the resistance included.

The external resistance can be added very conveniently to the phases of a slip ring rotor. Conventional methods of variation of resistance by means of mechanical contactors are shown in Fig. 4.10.

With the development of thyristors it has become possible to use a chopper controlled resistance in the rotor circuit. The schematic diagram is shown in Fig. 4.11.

It has been discussed in Chapter 3 that a resistance connected across the output terminals of a chopper can be varied from 0 to R by varying the time ratio of the chopper. When the chopper is always OFF, the supply is always connected to the

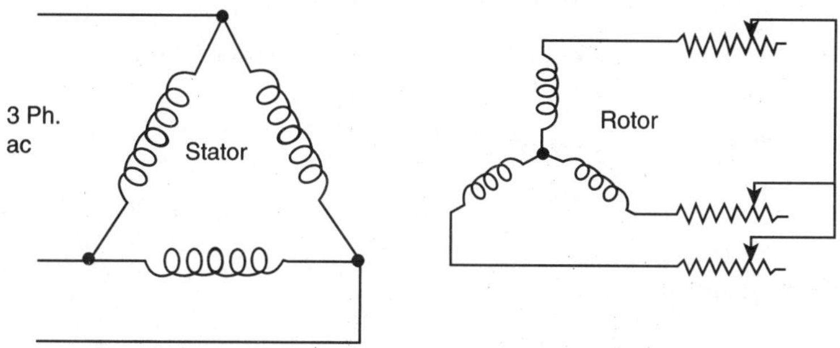

Fig. 4.10 *Conventional method of rotor resistance control*

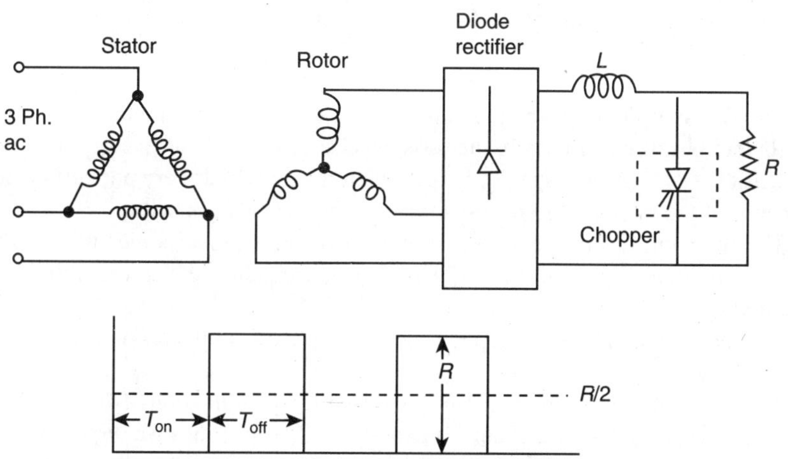

Fig. 4.11 *Chopper controlled resistance in the rotor circuit*

resistance R. The time ratio in this case is zero and the effective resistance connected is R. Similarly when the chopper is always ON, the resistance is short circuited. The time ratio in this case is unity and the effective resistance connected is 0. Hence by varying the time ratio from 0 to 1 the value of resistance can be varied from R to 0. The parameter Fig. 4.9 can be the time ratio instead of resistance.

The slip power of the rotor is rectified by a diode rectifier and is fed to the chopper controlled resistance (Fig. 4.11). The torque-speed curves can be drawn for different time ratios. For a time ratio of 1 we get normal characteristic of the motor. For a time ratio of 0 the characteristic corresponds to the one with complete resistance in the rotor circuit. A smoothing inductance L is used in the circuit to maintain the current at a constant value. Any short circuit in the chopper does not become effective due to L.

The value of R connected across the chopper is effective for all phases and its value can be related to the resistance to be connected in each phase if the conventional method has been used. Thus if R_{ex} is the resistance per phase we have

$$3(R_2' + R_{ex})I_L^2 = 3R_2' I_L^2 + RI_d^2 \tag{4.6}$$

$$\therefore R = 3 \cdot R_{ex} \left(\frac{I_L}{I_d}\right)^2 \tag{4.7}$$

But for a three phase bridge rectifier we have

$$\frac{I_L}{I_d} = \sqrt{\frac{2}{3}} \tag{4.8}$$

Using this relation in Eq. 4.7 we have

$$R = 2 \cdot R_{ex} \tag{4.9}$$

When once the value of R_{ex} is determined depending upon the range of speed control, the value of R can be determined. If the range of speed control is required down to zero speed, the value of R_{ex} and R can be decided based on the torque required at standstill.

The rating of the chopper thyristor decides the maximum rotor current of the motor. The speed control range is limited by the resistance. The method is very inefficient because of losses in the resistance. It is suitable for intermittent loads such as elevators. At low speeds, in particular, the motor has very poor efficiency. Because of the increased rotor resistance, the power factor is better.

The rotor current in this case is non-sinusoidal. The harmonics of the rotor current produce torque pulsations. These have a frequency which is six times the slip frequency.

The maximum torque developed is decided by the current carrying capacity of the chopper.

The range of speed control can be increased if a combination of stator voltage control and rotor resistance control is employed. Instead of using a high resistance rotor, a slip ring rotor with external rotor resistance can be used when stator voltage control is used for controlling the speed. The method is depicted in Fig. 4.12.

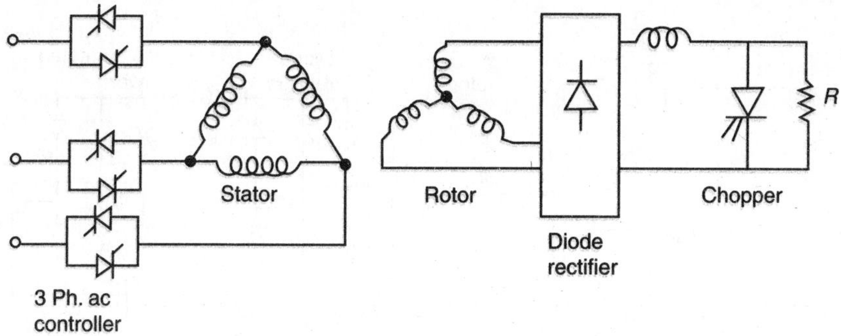

Fig. 4.12 *Combination of stator voltage control and rotor resistance control*

4.1.3 Speed Control Using Slip Energy Recovery Schemes

The foregoing discussion makes it very clear that the methods of voltage control and rotor resistance control have poor efficiency, particularly at low speeds, and find limited application. The slip power is wasted in the rotor resistance, either inherent in the rotor (high resistance rotors for voltage control) or connected in the rotor circuit. The power output of the motor is rather limited. For a given power output a machine which is overdimensioned must be used.

On the other hand if the slip power can be returned to the mains, the system can be made efficient and the capacity of the drive can be increased.

With the development of thyristors static power converters can be used in the conventional and Scherbius controls to recover the slip power to the supply for speed variation. The slip variation brings in speed control.

Figure 4.13 depicts a method of slip energy recovery. The rotary converter and motor generator (MG) set of the conventional system are replaced by a diode rectifier and phase controlled inverter. The combination is called a converter

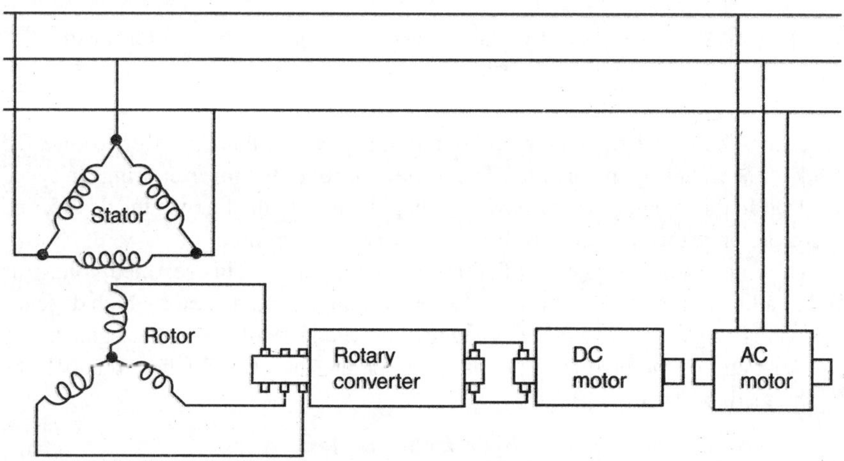

Fig. 4.13(a) *Conventional slip energy recovery scheme (Scherbius drive)*

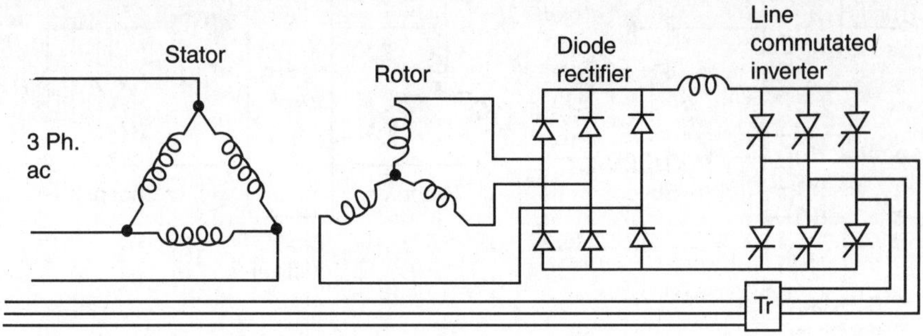

Fig. 4.13(b) *Slip energy recovery scheme using converter cascade in the rotor circuit. (Static Kramer system)*

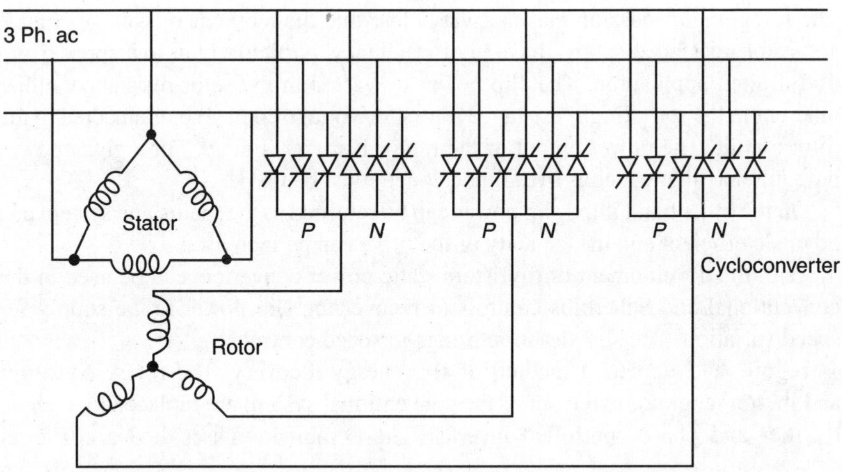

Fig. 4.13(c) *Slip energy recovery scheme using cycloconverter in the rotor circuit. (Static Scherbius system)*

cascade. The diode rectifier rectifies the slip power which is inverted and fed back to the mains by the inverter. The phase control of the inverter brings in speed control. In this method power flow is from the motor to the line via the dc link and therefore only speeds below synchronous speed are possible.

The rotor power is first rectified by a diode rectifier. This rectified voltage is fed to a line commutated inverter. The inverter applies a back emf which depends upon the firing angle. The output voltage of the diode rectifier must be balanced by the back emf of the inverter. The machine operates at a speed where this balance is obtained. The rotor current

$$I_2' = \frac{E_2' - E_4}{\sqrt{R_2'^2 + X_{\sigma 2}'^2}} = \frac{\bar{E}}{\sqrt{R_2'^2 + X_{2\sigma}'^2}} \tag{4.10}$$

To have a required value of current (I_2'), $\bar{E}$ must be constant. If E_4 varies, E_2' must also vary to keep $\bar{E}$ constant. The speed automatically varies to bring in balance.

As the slip power is returned to the mains the machine has fairly good efficiency. But due to the phase control of the inverter to effect the slip power recovery, the operating power factor is poor. A family of torque-speed curves are shown in Fig. 4.14. The firing angle changes the back emf which in turn shifts the speed-torque curve to the left. The machine always operates with nearly constant airgap flux, as decided by the fixed stator voltage and frequency. The torque developed is proportional to the dc link current. The motor has a speed-torque curve similar to that of a dc motor with separate excitation. The speed is reduced by increasing the back emf. The largest value of firing angle can only be 150° due to inverter limit of the inverter. Beyond this the firing angle cannot be increased.

The current capability of the converter cascade limits the maximum torque. The discontinuous conduction limit of the cascade decides the lowest torque. The dc link inductance connecting the diode rectifier and the inverter provides the constancy of the link current and decreases the ripple content. The ripple content is limited to 10%. The ripple content is due to the control of bridge. This causes a voltage regulation, due to which the speed of the motor is somewhat higher than the speed with continuous current.

The line power factor is poor due to the phase control of the inverter and associated reactive power for the converter. The induction motor draws a lagging current to establish the airgap flux. The power factor can be improved if the inverter

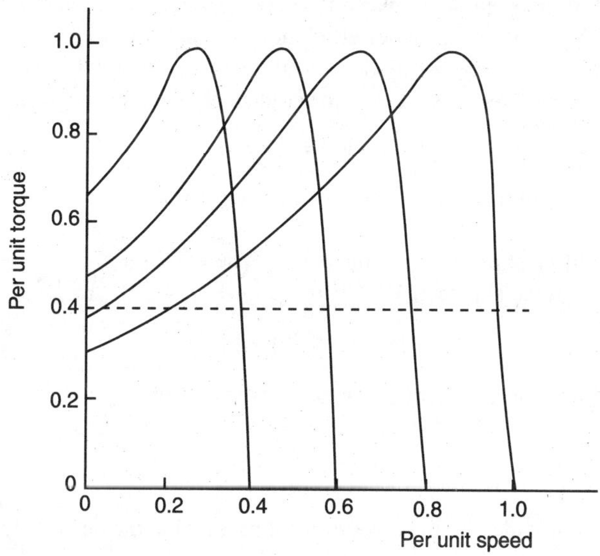

Fig. 4.14 *Speed-torque curves of subsynchronous converter cascade for no load slips of S_o = 0.4, 0.6, 0.8, 1.0*

can be operated near about 180° in the complete speed range. A transformer can be interposed in between the inverter and line. At the lowest speed the firing triangle is near 180° (taking the inverter limit into consideration) so that the transformer voltage matches the rotor voltage. Several methods are being developed to improve the power factor of slip energy recovery scheme employing converter cascade in the rotor.

The losses in the circuits cause a slight reduction in efficiency. This efficiency is further affected by additional losses due to the non-sinusoidal nature of the rotor current. Therefore the drive motor must be slightly overdimensioned. The motor used must have a rating 20% higher than the required power.

The lowest speed of the speed control range decides the design rating of the converter. The converter cascade handles only slip power. For limited speed ranges the rating can be low. Only if the speed range extends up to standstill must the cascade have full rating. When the lowest speed is other than zero, starting resistors are required to limit the starting current, as shown in Fig. 4.15. The changeover takes place from the starting resistors to the converter cascade very smoothly.

Non-sinusoidal rotor currents are responsible for torque pulsations which have a frequency six times the slip frequency. The harmonic currents are reflected to the ac line from the machine and inverter sides.

The converter cascade operates only at subsynchronous speeds due to the diode rectifier. Regeneration and speed reversal are not possible.

The design rating of the converter depends upon the speed control range required. The system is economical if this range is limited. Normally the machine cost is more. It finds application as a fan drive. Compared to the other methods of slip control this method has better efficiency but poorer power factor.

In a converter cascade the diodes and thyristors have a voltage rating decided by the lowest speed required. The rotor voltage at any slip s is given by

$$V_{2s} = sV_{20} \tag{4.11}$$

where V_{20} is the rotor voltage at standstill. If the speed range required is down to standstill the rotor voltage changes from 0 to V_{20}. If the speed range is from rotor, the corresponding slips are $s = 0$ and $s = s_m$. The maximum voltage between the slip rings is

$$s_m V_{20} \tag{4.12}$$

For a three-phase bridge rectifier the dc voltage output

$$V_{di} = \frac{3\sqrt{2}}{\pi} s_m V_{20} \tag{4.13}$$

The back emf provided by the inverter must be equal to the dc voltage of the rectifier. The inverter voltage

$$V_{dia} = V_{di} \cos \alpha \tag{4.14}$$

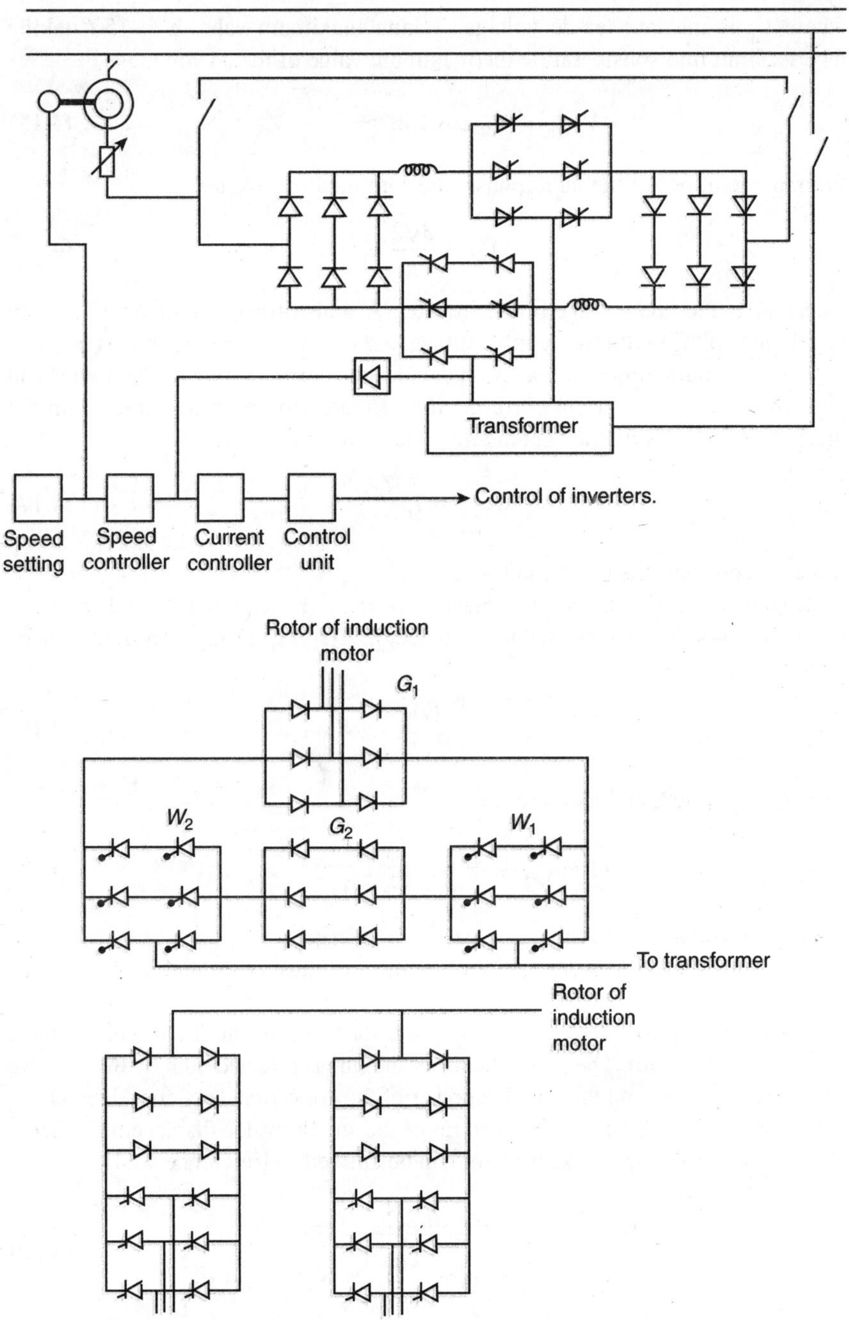

Fig. 4.15 *(a) Power circuit of interchangeable converter cascade (b) Series connection of converters for large voltage (large slips) (c) Parallel connection of converters for large currents (small slips)*

where V_{di} is the inverter dc voltage. Taking maximum value $a = 150°$ taking inverter limit into consideration the maximum value of the dc voltage is

$$V_{diam} = V_{di} \cos 150° = -\frac{\sqrt{3}}{2} V_{di} \tag{4.15}$$

From the relations of the three-phase line commutated inverter

$$V_{di} = \frac{3\sqrt{2}}{\pi} V_L \tag{4.16}$$

where V_L is the line voltage of the inverter. A transformer is sometimes used to match this voltage with the supply voltage and to have the desired speed range.

The maximum torque of operation decides the current rating of the diodes and thyristors. The rotor current corresponding to rated torque is available from the machine details. Using the data the maximum current

$$I_{max} = I_r \frac{T_{dm}}{T_{dr}} \tag{4.17}$$

where r represents the rated conditions.

Taking into consideration the reactive power for commutation and the form factor of rotor current by means of a factor $K(> 1)$ the rated current can be obtained as

$$I_r = \frac{P_r}{\sqrt{3}V_{10}} \frac{1}{\eta} \cdot K \tag{4.18}$$

The corresponding dc link current is

$$I_d = \sqrt{\frac{3}{2}} I_m = I_r \cdot \frac{T_{dm}}{T_{dr}} \frac{V_s}{V_{s_{min}}} \sqrt{\frac{3}{2}} \tag{4.19}$$

$V_s / V_{s_{min}}$ considers the possible fluctuation of voltage.

The converter comprising rectifier and inverter must withstand higher voltage at low speeds and higher current at small slips (high speeds).

The reactive power of this drive is more than that of the drive with variable resistance in the rotor. The power factor of the line is affected both by the reactive power of the motor and the reactive power of the converter. The latter depends on the speed range. The larger the speed range the smaller is the displacement factor. The fundamental displacement factor can be derived as [reference 4.24]

$$\cos \varphi_1 = \frac{(1-s)}{\sqrt{\left[(1-s)^2 + \frac{\sin \varphi_{1m}}{\cos \varphi_{1m}} + \sqrt{\frac{4}{3}} s_m^2 - s^2\right]}} \tag{4.20}$$

An interchangeable converter cascade can be used sometimes to handle higher voltages at low speeds and higher currents at high speeds. By this the design rating of the converter unit and the size of the reactor decrease.

Figure 4.15 depicts one such interchangeable converter cascade in which the converter comprises two rectifiers and two inverters. Depending on the situation they are connected either in parallel or in series. At a lower speed range where the voltage to be handled is more and current is small only one rectifier G_1 is connected to the rotor. The inverters W_1 and W_2 are connected in series via the rectifier G_2. In the upper speed range (voltage is small and current is high) the converters are connected as shown in Fig. 4.15(c). There exist two parallel paths of the converters, each being one half of the total bridge. A speed range of 1 : 3 can be obtained. There is an overlap region during transition of the connection. This actually restricts the frequency of switching ON and OFF of G_2 in operation around the point of interchange.

4.1.4 Braking and Four Quadrant Operation

Subsynchronous converter cascades have been used, till now, in applications requiring one quadrant operation. These can be employed for drives where at least one electrical braking is required. A four quadrant operation can also be made possible in these cascades, using suitable switching.

The braking is obtained by connecting the stator to the dc supply. The rectified output of the rotor can be used for the purpose. The reverse rotation is obtained by means of plugging using contactors. The drive is similar to a dc drive fed from a converter with armature current reversal.

A four quadrant subsynchronous converter cascade employing dc dynamic braking and speed reversal is shown in Fig. 4.16. The system operates as a normal converter cascade when the switches s_1, s_2, s_3 and s_6 are closed.

To have braking the machine is disconnected from the mains by opening s_3. The stator is connected to the rectified power supply from the rotor by closing s_5 and opening s_6. Braking torque equal to rated torque may be developed.

Contactor s_4 may be used to reverse the direction of rotation. The drive can operate in all four quadrants of the speed-torque plane.

The contactors must operate when the current is zero. Due to delay in operation of the contactors, sometimes mechanical braking is employed.

The drive operates free of loss as the energy during running and braking is fed back to the mains.

4.1.5 Static Scherbius Drive

The main limitation of the Kraemer drive discussed is the operation at sub-synchronous range due to diode rectifier. Super synchronous speeds can be achieved if the power is fed to the rotor from the ac mains. This can be made possible by replacing the converter cascade by a cycloconverter, as shown in Fig. 4.17. A cycloconverter allows power flow in either direction making the drive operate at both sub and super synchronous speeds. However, the converter becomes costly and has a complex control. The current in the rotor circuit is nearly sinusoidal. The torque pulsations and other reactions are minimal. The performance of the drive improves with respect to additional losses and torque pulsations. A smooth transition is possible from sub to super synchronous speeds

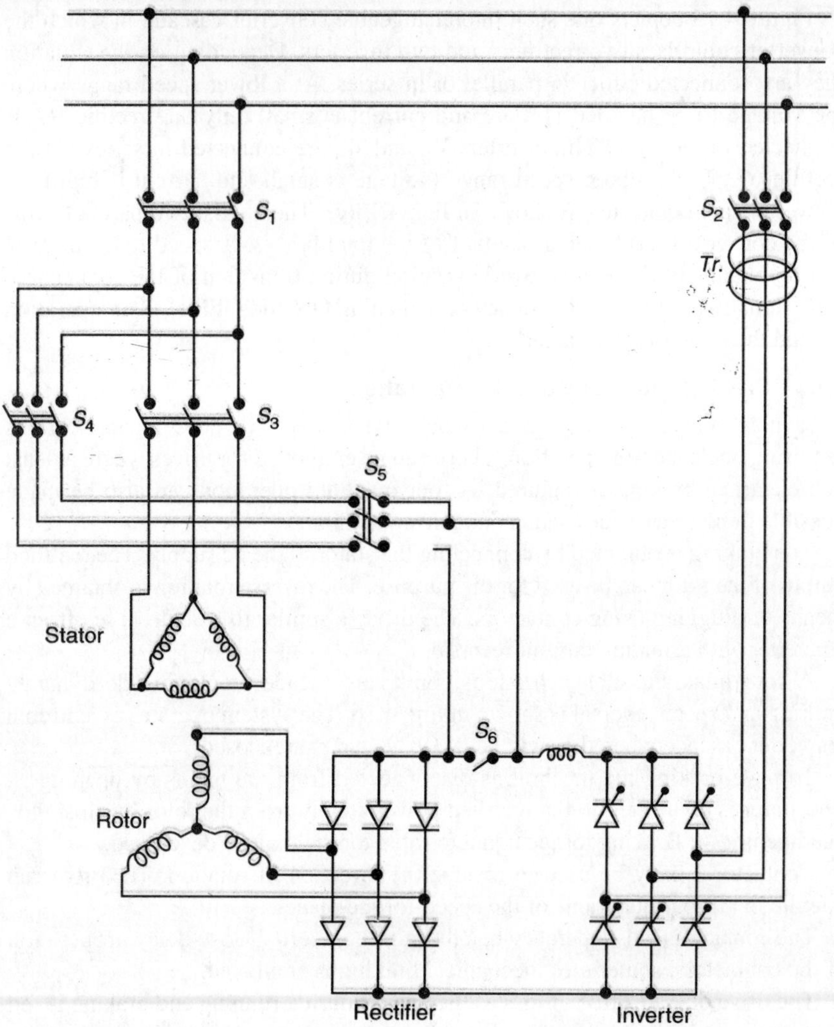

Fig. 4.16 *Four quadrant subsynchronous converter cascade with dc dynamic braking*

without any commutation problems. Speed reversal is not possible. A step up transformer may be interposed between the lines and the converter to reduce the voltage rating of the converter.

Static Scherbius drive, with power flow to and from the rotor, can also be realized by replacing the diode rectifier on the rotor side by means of a controlled bridge rectifier using thyristors, as shown in Fig. 4.18. This provides speeds below and above synchronous speed. The effects of non-sinusoidal rotor current are present. In addition the commutation of the bridge rectifier at very small slips (near synchronous speed) poses problems. Either forced commutation or artificial

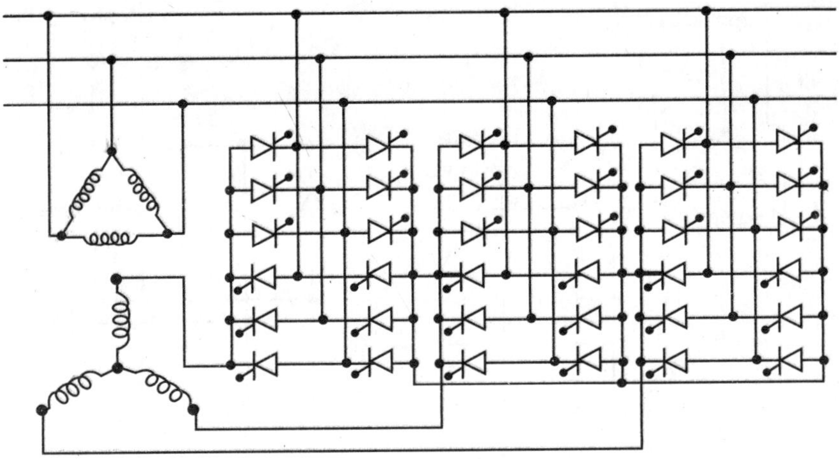

Fig. 4.17 *Static scherbius system using cychconverter in the rotor circuit*

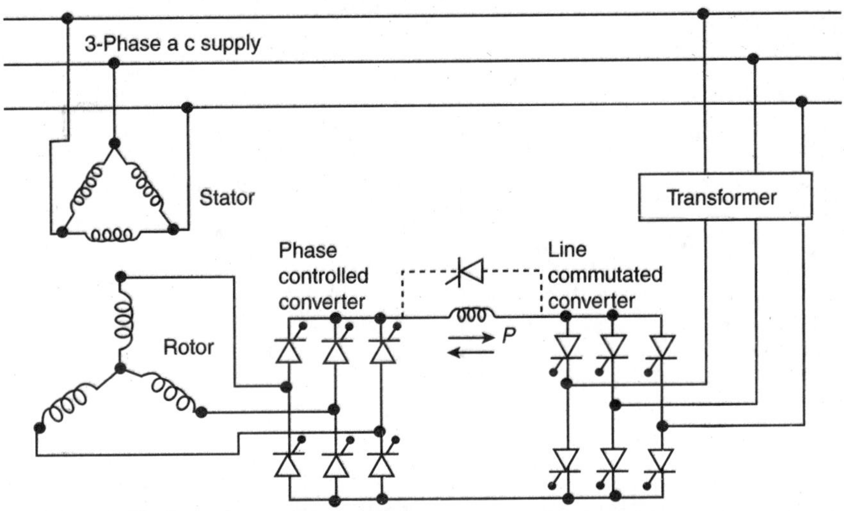

Fig. 4.18(a) *Sub- and super-synchronous converter cascade*

commutation of the thyristors is employed at and about synchronous speed. These methods are shown in Fig. 4.18. A current source inverter can be advantageously employed to obtain a Scherbius drive.

4.1.6 Speed Control of Induction Motor by Variable Frequency Supply

The speed of a squirrel cage induction motor can be controlled very effectively by varying the stator frequency. Further, the operation of the motor is economical and efficient, if it operates at very small slips. The speed of the motor is, therefore, varied by varying the supply frequency and maintaining the rotor frequency at the

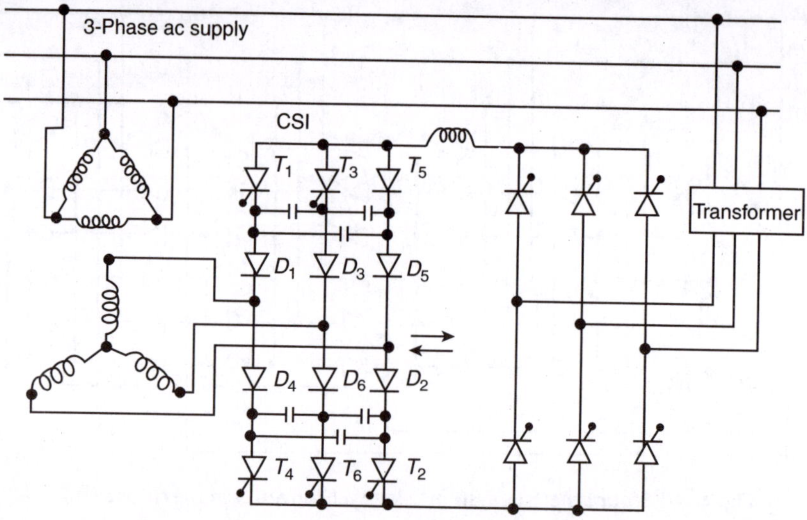

Fig. 4.18(b) *Use of CSI for super-synchronous operation*

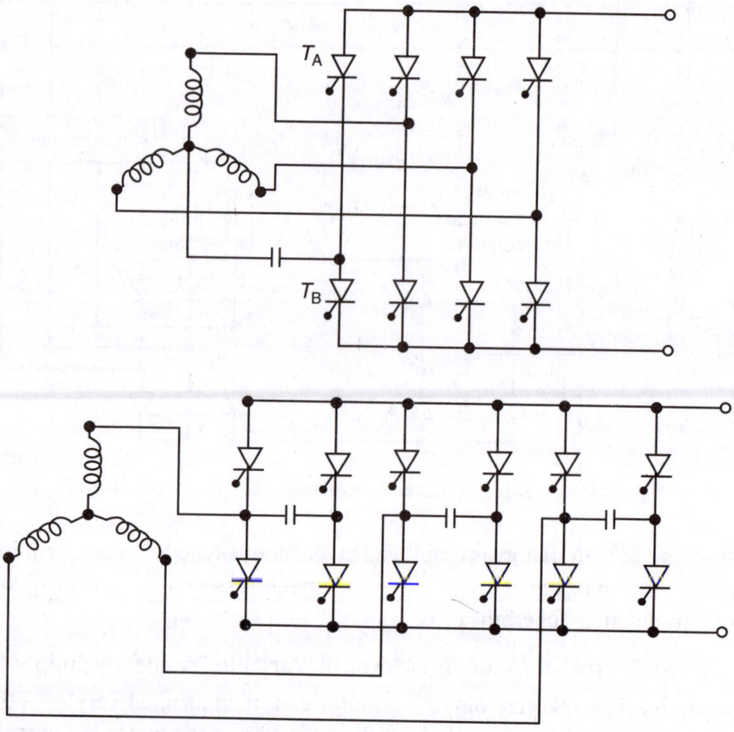

Fig. 4.18(c) *Alternate arrangements for the CSI for use with super synchronous converter cascade*

rated value or a value corresponding to the required torque on the linear portion of the torque-speed curve.

The control of a three-phase induction motor, particularly when the dynamic performance involved is more difficult than dc motors. The is due to

i. Relatively large internal resistance of the converter causes voltage fluctuations following load fluctuations because the capacitor cannot be ideally large.
ii. In a dc motor there is a decoupling between the flux producing magnetising current and torque producing armature current. They can be independently controlled. This is not the case with induction motors. For obtaining dc motor characteristics, the stator current must be separated into two components. The method is rather involved.
iii. An induction motor is very poorly damped compared to a dc motor.

For operation of the motor at small slips, it must be limited to the linear portion of the torque-speed curve. It can be shown that the torque developed is proportional to the square of the air gap flux, i.e.

$$T_d \propto \phi^2 \tag{4.21}$$

To retain the maximum torque capability at all operating points the flux must be maintained at its rated value. To achieve this supply voltage must also be varied simultaneously with the frequency. One simple and approximate method is to vary the voltage proportional to frequency such that

$$\frac{V_1}{f} = \text{constant}$$

Using the equivalent circuit of the motor (Fig. 4.19) it can be derived that

$$T_d = \frac{Pm_1}{2\pi}[V_1/f_1]^2 \frac{f_2 X_m^2 / R_2}{\left[R_1 + \dfrac{f_2}{f_1 R_2}(X_m^2 - X_{11}X_{22})\right]^2 + \left[X_{11} + \dfrac{f_2 R_1 X_{22}}{f_1 R_2}\right]^2} \tag{4.22}$$

where $X_{11} = X_1 + X_m$, $X_{22} = X_2 + X_m$

m_1 is number of stator phases.

f_1, f_2 are stator and rotor frequencies respectively

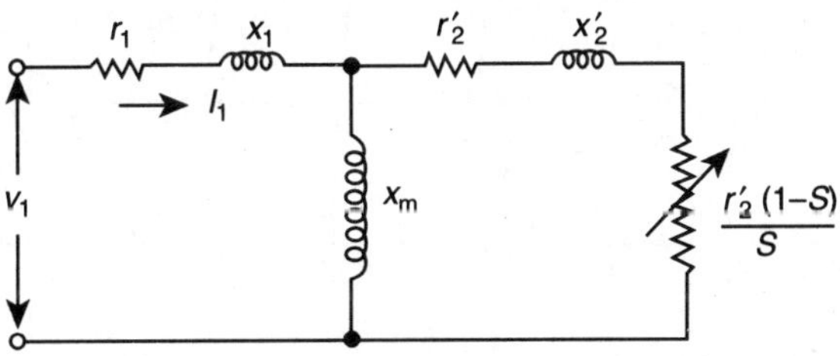

Fig. 4.19 *Exact equivalent circuit of a three phase induction motor (Core loss neglected)*

A family of curves showing the variation of torque with rotor frequency f_2 with the stator frequency f_1 as a parameter is shown in Fig. 4.20. From the expression as well as figure it is clear that at a given f_2 the developed torque decreases at small stator frequencies. The maximum torque developed also decreases. This is because at smaller stator frequencies, the stator resistance drop becomes a large

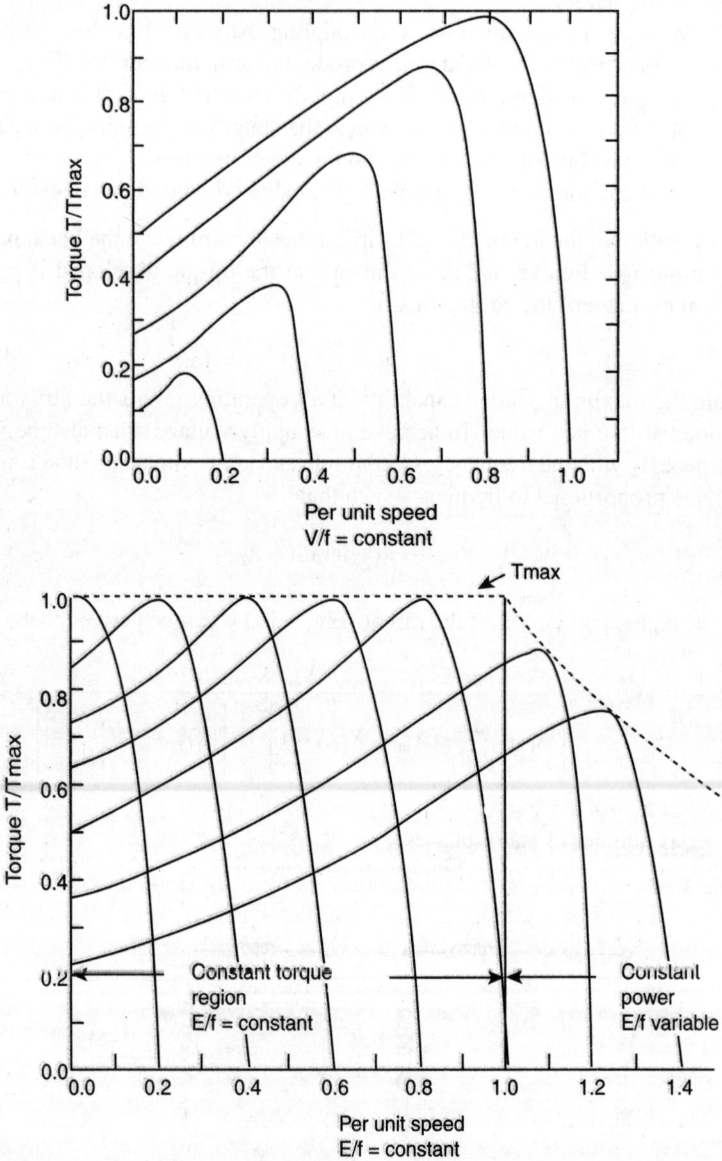

Fig. 4.20 *Torque speed curves of variable frequency induction motor*

part of the applied voltage. Consequently the air gap flux decreases, causing a depletion of torque developed at a given f_2, as well as maximum torque.

This control would be sufficient in some applications requiring variable torque, such as centrifugal pumps, compressors, and fans. In these, the torque varies as the square of the speed. Therefore at small speeds the required torque is also small and *V/f* control would be sufficient to drive these loads with no compensation required for resistance drop. This is true also for the case of the liquid being pumped with minimal solids.

However, in certain applications the pumps may require high starting torque. The torque developed with *V/f* constant may not be sufficient. In such cases the motor must be operated to provide its full torque capability at all frequencies. The motor must be supplied with an increased voltage to maintain the flux constant. Therefore the applied voltage to the induction motor has two components at low frequencies:

i. proportional to stator frequency
ii. to compensate for the resistance drop in the stator.

The second component depends on the load on the motor and hence on rotor frequency. A voltage boost proportional to slip frequency may be applied. This drive is the well known slip controlled drive.

The method of maintaining the flux constant by providing a voltage boost proportional to slip frequency is a kind of indirect flux control. This method of flux control is not desirable if very good dynamic behaviour is required. Direct flux control is required if better dynamic behaviour is required. This can be achieved by direct measurement of flux using search coils or calculation of the flux using machine parameters, voltages and currents. In the former the machine loses its robust nature whereas in the latter the errors are introduced due to the use of inaccurate parameters obtained from no-load tests. In all these methods the applied voltage V_1 is varied such that *E/f* is constant at all operating points, to maintain the air gap flux at its rated value. The inverter used to supply an induction motor must be capable of providing a variable voltage, variable frequency supply. The control used must be such that the air gap flux is constant at all operating points. The speed torque curves of the motor for this type of control are shown in Fig. 4.20(b).

4.1.7 Cycloconverter Fed Induction Motor Drive

Variable frequency supply to an induction motor may be obtained by using a cycloconverter. As has already been discussed, a cycloconverter is a single stage frequency conversion device which converts ac line frequency to a variable frequency. A three phase, three pulse cycloconverter feeding a three phase induction motor is shown in Fig. 4.21. A cycloconverter fed induction motor drive has the following features:

i. The voltage control is possible in the converter itself, so that the machine operates at its rated flux conditions.
ii. The cycloconverter operates by means of line commutation. No forced commutation is required as the necessary reactive power for commutation

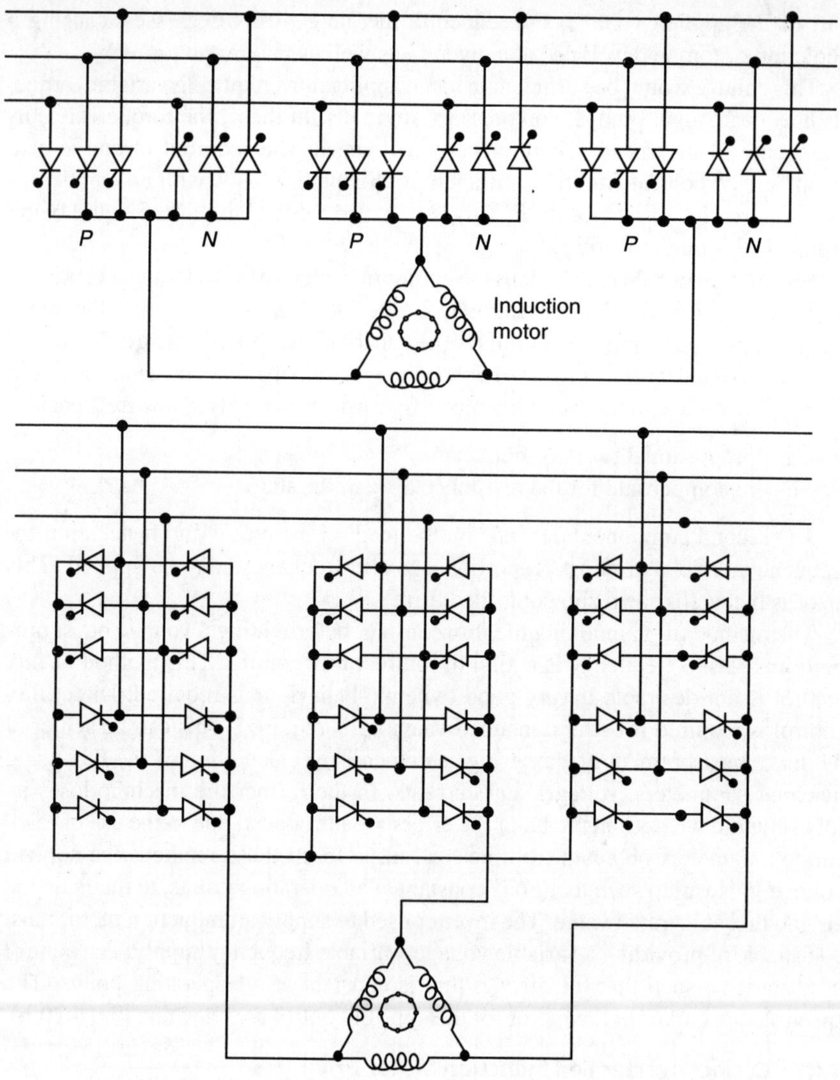

Fig. 4.21 *Cycloconverter feeding a three phase induction motor*

is provided by the line. Losses due to forced commutation can be eliminated. The converter operates at lagging powerfactor. The line power factor is very poor at light loads.

iii. The firing angles of the thyristors are modulated so that a mean sinewave voltage is synthesised from the input voltage waveform. Thus, a cycloconverter delivers a high quality sinusoidal waveform at all frequencies. When this voltage is impressed across the machine terminals a current which is almost sinusoidal will flow, as the machine inductance causes harmonic attenuation. The machine therefore does not have appreciable torque pulsations and harmonic losses and the operation is smooth and efficient.

iv. The cycloconverter is capable of power transfer in either direction between an ac source and motor load. It can feed power to a load of any p.f. Regeneration is inherent in the complete speed range. A four quadrant operation is simple and straight forward.

v. The output frequency of the cycloconverter is limited to 1/3 of input frequency. A speed control range of 0–33% of base speed is possible.

vi. It requires many thyristors, which offset the advantage of line commutation. However, if a thyristor fails no shut down is required. The output can be made available without any interruption, with a slightly distorted waveform.

The increased cost and complex control of a cycloconverter can be justified for large horsepower applications. The above features of a cycloconverter fed induction motor drive show that it provides a very smooth low speed operation with the least torque ripple. Thus it is very attractive for a low speed, large power, reversible drive. This drive finds application in gearless cement mill, or ball mill drives. The mill drive is directly connected to the motor which is supplied by a low frequency cycloconverter.

4.1.8 Voltage Source Inverter Fed Induction Motor Drive

Voltage source inverter is a kind of dc link converter, which is a two stage conversion device. A three phase supply is first rectified using a rectifier on the line side. The rectified dc is inverted to ac of desired frequency by an inverter on the load side, as shown in Fig. 4.22. When the load cannot provide the required reactive power for the inverter, the inverter has to necessarily be a force commutated one.

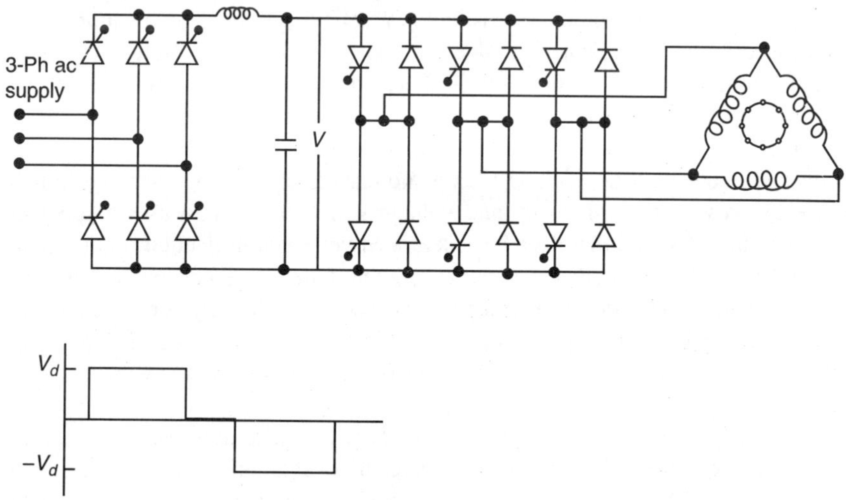

Fig. 4.22 *Square wave voltage source inverter feeding a three phase induction motor*

The inductance in the dc link circuit provides smoothing whereas the capacitance maintains the constancy of link voltage. The link voltage is a controlled quality. The instantaneous voltage at the machine terminals (at the output terminals of the inverter) is at all times directly proportional to the dc link voltage (dc supply which is of low internal impedance) and the machine current (output) is a function of load admittance. Therefore, the ideal load for a voltage source inverter, in view of harmonics in load current, should be highly inductive and have a low powerfactor.

By proper switching of the inverter thyristors the dc link voltage is impressed across the phases of the induction motor alternately. The voltage waveform at the output terminals depends purely on this switching and conduction of thyristors. If the thyristors conduct 180° in a period, the output voltage is in the form of a square wave. In this case the voltage control is obtained by means of a phase controlled rectifier on the line side. The dc link voltage is variable and the output voltage waveform remains the same at all loads and frequencies. The inverter is also called a square wave inverter, as the output voltage is a square wave.

These inverters have commutation problems at very low frequencies, as the dc link voltage available at these frequencies cannot charge the commutating capacitors sufficiently enough to commutate the thyristors. This puts a limit on the lower frequency of operation. To extend the frequency towards zero, special charging circuits must be used. The speed control range of the induction motor operating on a square wave inverter is 1:20.

The polarity of the dc link voltage cannot be changed. Hence during regeneration the current direction in the link circuit must be reversed. A separate phase controlled converter is required on the line side for regeneration, as shown in Fig. 4.22. Dynamic braking can be employed by means of resistors switched. The dynamic behaviour of the system is not very good at low frequencies.

Soft starting of the motor is possible. A machine operating on a variable frequency, variable voltage converter does not require additional starting equipment. The machine normally operates on the linear portion of the torque-speed curve and does not see locked rotor torque and current in its starting. The machine has identical characteristics from the synchronous point to the maximum torque point, at every frequency. The machine torque and current can be kept constant in this range by varying the frequency and voltage simultaneously to assure constant flux. The machine can therefore be accelerated very smoothly at constant torque and current to the required speed, by varying the stator frequency and keeping the rotor frequency constant. The acceleration is at constant flux slip controlled mode. The mode in which the slip is kept constant and flux is controlled (constant slip flux controlled mode) is also possible for starting purposes. These methods add to the efficiency of the drive.

When the slip is used as a controlled quantity to maintain the flux constant in the motor the drive is called slip controlled drive. By making the slip negative (i.e., decreasing the output frequency of the inverter) the machine may be made to operate as a generator and the energy of the rotating parts fed back to the mains by an additional line side converter or dissipated in a resistance for dynamic braking. By keeping the slip frequency constant (or controlling slip), braking at constant torque and current can be achieved. Thus, braking is also fast.

Since the voltage can be varied to maintain the constant flux, constant torque operation is possible up to rated frequency. Beyond the rated frequency, the voltage remains at its rated value and the machine operates in flux weakening mode. The motor gives constant output at all speeds and is called constant horse power mode. These modes are depicted in Fig. 4.23.

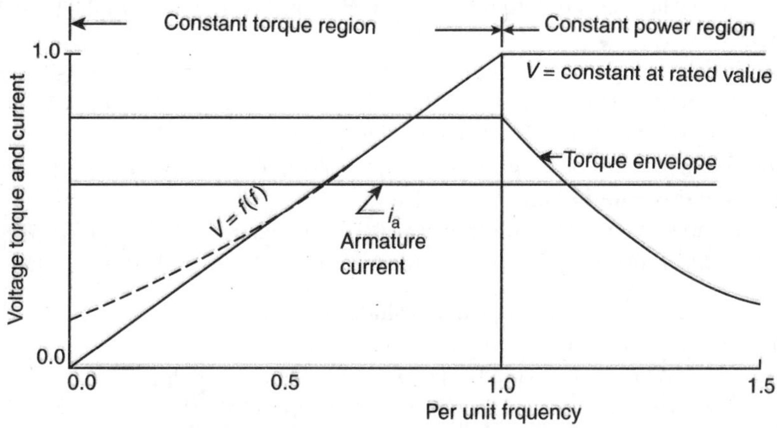

Fig. 4.23 *Constant and variable torque regions of variable frequency induction motor*

The motor receives square wave voltages. This voltage has harmonic components. The resulting armature current is non-sinusoidal, having peaks. These peaks actually decide the design rating of the inverter. The harmonics of the stator current cause additional losses and heating. The motor therefore requires a derating, or for a given horse power an overdimensioned motor must be used.

These harmonics are also responsible for torque pulsations. The reaction of the fifth and seventh harmonics with the fundamental gives rise to the sixth harmonic pulsations in the torque developed. For a given induction motor fed from a square wave inverter the harmonic content in the current tends to remain constant independent of input frequency, within the range of operating frequencies of the inverter.

The peak and harmonic currents of the line as well as stator current are influenced by the leakage reactance of the motor. Higher the leakage reactance smaller is the harmonic content and the peak value of the stator current. It is therefore necessary to choose an induction motor having a large leakage reactance for operation on a voltage source inverter. As the peak currents are less, the design rating of the inverter decreases.

Open loop control is possible, but may have stability problems at low speeds.

Multimotor operation is possible and commutation is load independent. The converter represents a source and the motor can be just plugged on. Therefore, no matching between converter and load is necessary.

As slip controlled drive, a VSI fed motor has the following additional features:

An indirect flux control can be achieved by slip control. As the steady state quantities are specified in the control, the dynamic behaviour may not be satisfactory.

Precise control of torque over a wide range of speeds is possible (sometimes down to standstill). It is a very efficient drive, having a very good efficiency and power factor, when the slip frequency is limited to the linear portion of the torque-speed curve.

No starting equipment is required and very fast acceleration is possible at constant torque and current. Regeneration or dynamic braking is also possible at constant torque and current. Four quadrant drive is possible.

Closed loop control of frequency provides a variable speed drive having characteristics of dc motor in Ward Leonard system.

The features of an induction motor fed from a square wave inverter can be summarized as follows:

i. The inverter has impressed dc voltage of variable amplitude.
ii. Advantageous for multimotor drive.
iii. Commutation is load independent. Converter and load need not be matched. The converter represents a source to which the motor can be just plugged on.
iv. At present converter output frequencies up to 1500 Hz are possible. This drive is very much suitable for (motors of) high speed operation. The drives are available up to rating of 200 KVA.
v. The lowest operating frequency as limited by commutation is about 5 Hz. Speed range is 1:20.
vi. Not suitable for acceleration on load and sudden load changes.
vii. Dynamic behaviour is fairly good at high speeds.
viii. Dynamic braking is possible. Regeneration (Four quadrant operation) requires an additional converter connected antiparallel to the line side one. Speed reversal is obtained by changing the phase sequence.
ix. The input voltage to the motor is nonsinusoidal. This results in additional losses, heating and torque pulsations.
x. Motor should have sufficiently large leakage inductance to limit the peak currents and decrease the harmonic content.
xi. Open loop control of the motor is possible, but may have stability problem at low speeds.
xii. Line power factor is poor due to phase control.
xiii. It can be operated as a slip controlled drive.
xiv. It finds application as a general purpose industrial drive for low to medium power.

4.1.9 Pulse Width Modulated Inverter Fed Induction Motor Drive

Voltage control in the square wave inverter has been external to the inverter, by means of a phase controlled rectifier on the line side. This posed some practical application problems on the drive by limiting the lowest operating frequency and introducing torque pulsations and harmonic heating. However, the harmonic effects can be decreased by using a motor of larger leakage reactance.

The problems can be reduced to a minimum if voltage control is obtained in the inverter itself. The inverter is supplied with constant dc voltage and the inverter is controlled so that the average amplitude of the output voltage is variable. By this, the operation of the inverter can be extended up to zero frequency, as the commutation is effective at all frequencies. By this control the output voltage is no longer a square wave but a pulsed wave, the average of which tends to be sinusoidal if sinusoidal modulation is used. The principle of modulation and output voltage waveform are shown in Fig. 4.24. The stator current also tends to be sinusoidal. Harmonic effects are reduced to a minimum and torque pulsations are minimal. The dynamic behaviour of the drive at low speeds is improved. As there is no limitation on the lowest frequency the speed range of the drive is 1:∞. The machine need not be specially designed to have large leakage reactance. Normal motors can be employed and the inverter output waveform causes the least harmonic effects. The filter size in the dc link can be smaller. The peak currents are also small and hence the converter design rating decreases. However the methods of modulation employed make the control circuit complex and costly. The frequent switching of the thyristors introduces switching losses and these inverters are built to an upper frequency of 150 Hz. A slip controlled drive is possible. Dynamic braking can be employed. The K.E. of rotating parts is dissipated in an external resistance. If a battery is used to supply the inverter, the regeneration is simple and straight forward. If the dc supply is obtained using a rectifier another phase controlled rectifier which is connected antiparallel to the previous one is required on the line side. It is suitable for both single and multimotor operation. Uninterrupted operation can be made possible when a buffer battery is used. During motoring a diode rectifier is sufficient on the line side and the line p.f. is better. Slip control can be had to keep the flux constant. Thus it has the merits of a slip controlled drive. Selected harmonic neutralisation principles can be used to eliminate lower order harmonics. Soft starting without any starting equipment is possible with acceleration and braking at constant torque.

The specific features of PWM inverter induction motor drive can be summarized as follows:

i. The inverter has constant dc link voltage and employs PWM principle for both voltage control and harmonic elimination.
ii. The output voltage waveform is improved, with reduced harmonic content. The amplitude of torque pulsations is minimal even at low speeds.
iii. Parallel operation of many inverters on the same dc bus.
iv. Uninterrupted operation is possible when a buffer battery is used.

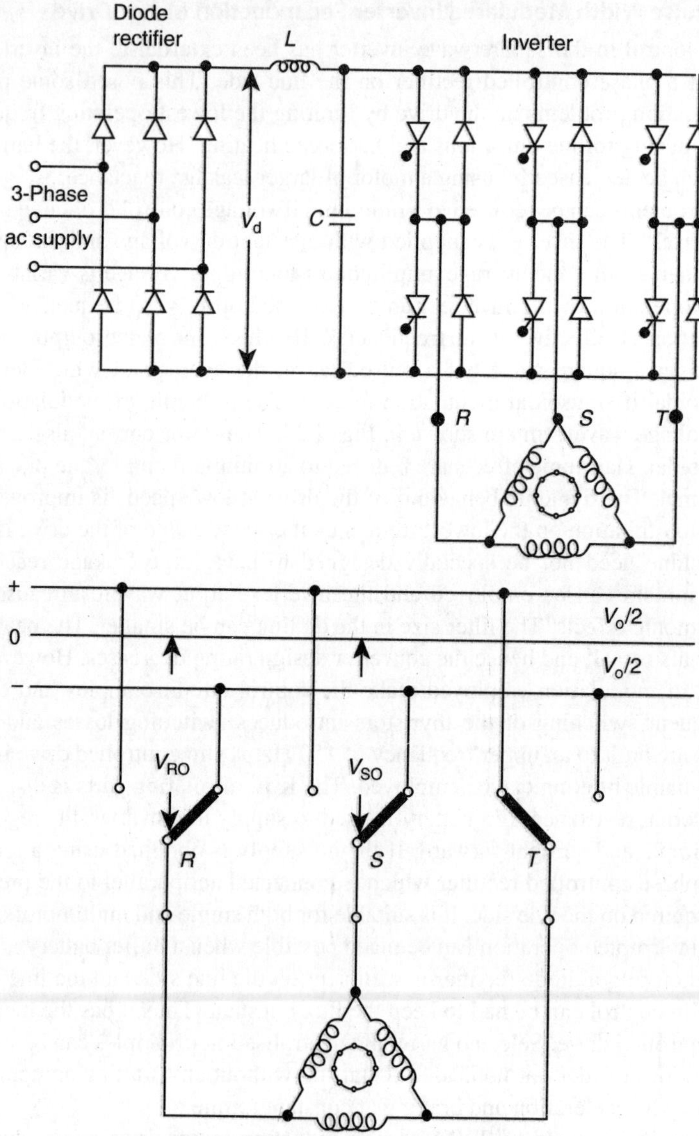

Fig. 4.24(a) *Constant voltage PWM inverter*

v. The power factor of the system is good as a diode rectifier can be employed on the line scale.

vi. The control is complicated.

vii. Four quadrant operation is possible. During regeneration a battery or another converter with phase control may be used. Dynamic braking can also be employed.

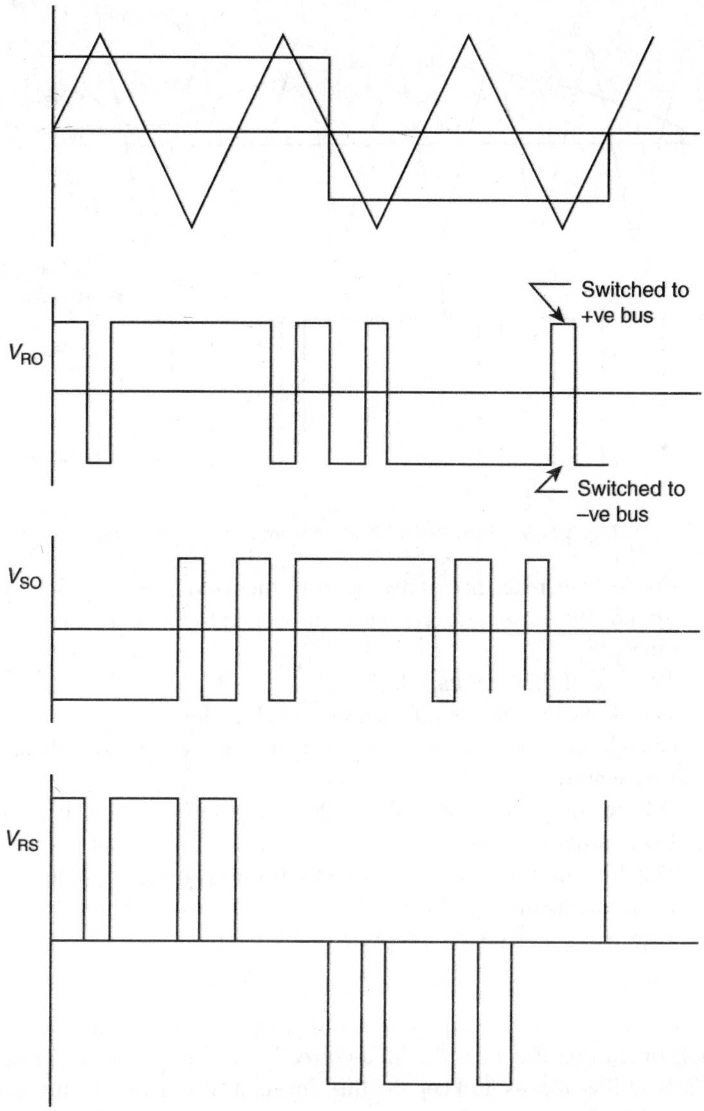

Fig. 4.24(b) *Principle of PWM and line voltage waveforms*

viii. Single and multimotor operations are possible. Smooth changeover of voltage and frequency values at zero crossing for speed reversal with full torque capability at standstill.

ix. Due to switching losses, the highest operating frequency is 150 Hz. Speed range 1: ∞.

x. The inverter and load need not be matched. The converter operates as a source to which the motor can be plugged.

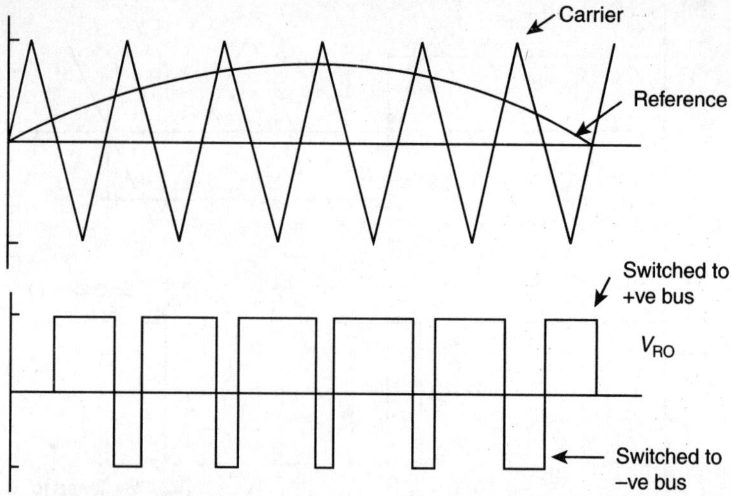

Fig. 4.24(c) *Sinusoidal PWM potential of R only shown*

xi. The leakage reactance of the motor further smoothens the motor current, effectively decreasing the harmonic content and peak values of stator current.

xii. The size of the filter also decreases.

xiii. The efficiency may be affected by switching losses.

xiv. Open loop operation is possible. The drive has a very good dynamic and transient response.

xv. Selected harmonic elimination techniques may be employed to eliminate lower order harmonics.

xvi. The drive finds application in low to medium power. Commercial drives are available up to 450 kVA.

xvii. Transistors may be used in place of thyristors.

4.1.10 Current Source Inverter Fed Induction Motor Drive

In a voltage source inverter fed induction motor the applied voltage to the stator is proportional to the frequency, with a correction for the stator resistance drop, particularly at low speeds, to keep the flux constant. It is a very well known fact that the current drawn by an induction motor does not depend on the stator frequency when the air gap flux is constant. There exists a fixed relationship between the slip frequency and stator current for rated flux in the air gap, as shown in Fig. 4.25. By controlling the slip of the motor, the stator current can be controlled. An indirect flux control is therefore possible. The control is simpler than voltage control. The curve between slip frequency and stator current can be calculated using the equivalent circuit. A PWM inverter can be controlled to provide the desired currents in the motor.

In a dc link converter, if the dc link current is controlled the inverter is called a current source inverter. The current in the dc link is kept constant by a high

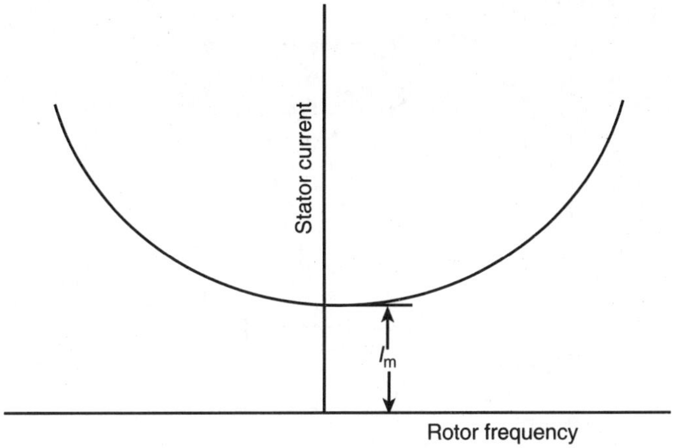

Fig. 4.25 *Stator current of an induction motor as a function of rotor frequency for constant flux*

inductance and the capacitance of the filter is dispensed with. The variable dc link voltage is converted to a current source by means of the inductance. The dc supply is of high impedance. As the link current is held constant, the output current waveform is determined by the operation of the inverter, while the output voltage is determined by the nature of the load impedance. A current source inverter is suitable for loads which present a low impedance to harmonic currents and have unity p.f.

A current source inverter has a very simple configuration. No feedback diodes are required. A phase controlled rectifier is used on the line side to provide a current control. As the dc link contains only the inductance, regeneration is possible by changing the polarity of voltages and maintaining the current direction. Hence, a four quadrant drive is simple and straightforward. It provides effective buffering of the inverter output from supply voltage variations. Direct control of stator current allows a precise closed loop control to be implemented with relative ease.

The commutation of the inverter is load dependent. The load parameters form a part of the commutation circuit. A matching is therefore required between the inverter and the motor. Multimotor operation is not possible. The inverter must necessarily be a force commutated one, as the induction motor cannot provide the reactive power for the inverter.

The constant dc link current is allowed to flow through the phases of the motor by control of the inverter, and therefore the motor current is a quasi-square wave. The motor voltage is almost sinusoidal with superimposed spikes, due to commutation. These voltage spikes decide the voltage rating of the thyristors and also affect the insulation of the motor. These spikes can be limited if the machine has small leakage reactance or if the commutating capacitors are large. A machine with smaller leakage reactance is suitable for current source operation to keep the voltage spikes and the harmonic losses to a minimum. The effect of torque pulsations decreases and the frequency of operation can be increased.

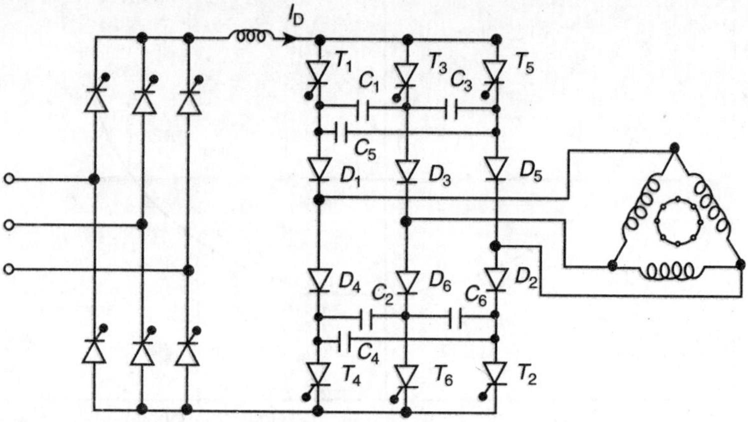

Fig. 4.26(a) *Auto sequential commutated CS inverter feeding a three phase induction motor*

The commutating capacitance is chosen to strike a compromise between voltage spikes and the highest operating frequency. The commutation requires a definite minimum current. The inverter has the ability to recover from commutation failure. The link inductance causes a slow rise of the fault current and by the time it reaches high values, the fault may be cleared.

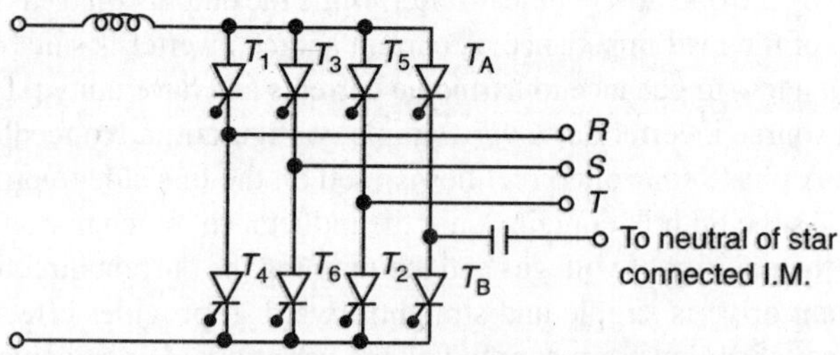

Fig. 4.26(b) *Third harmonic ASCSI*

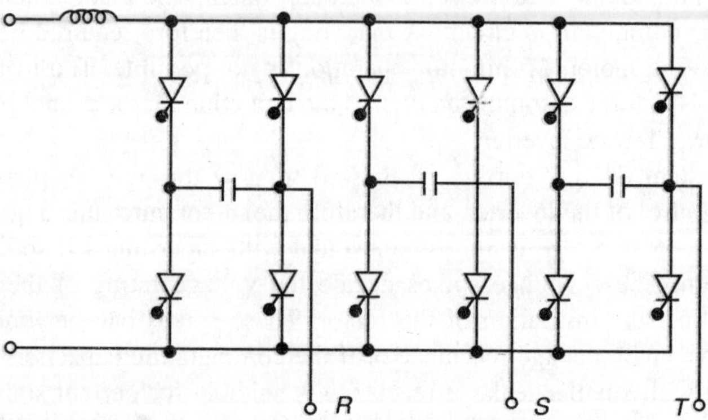

Fig. 4.26(c) *CSI employing individual commutation*

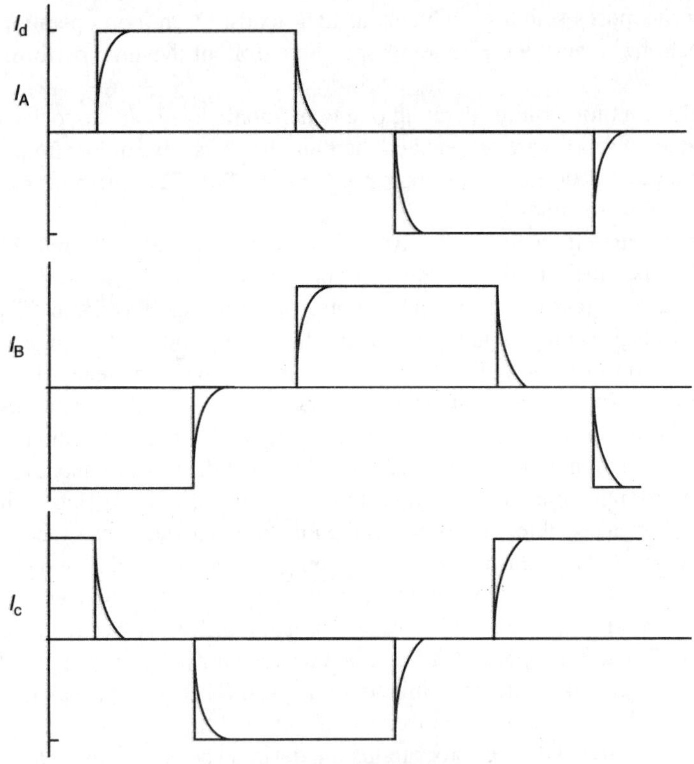

(a) Current waveforms

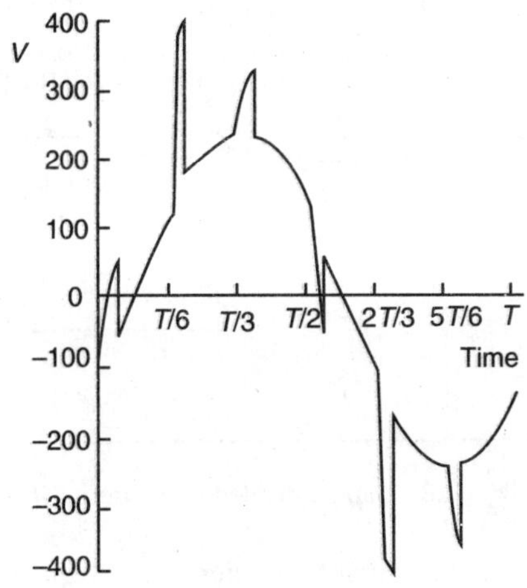

(b) Voltage waveform

Fig. 4.27 *Current and voltage waveforms of CSI fed induction motor*

The drive poses stability problems at light loads. Open loop operation is not possible. It has a very wide range of speed control but dynamic performance is poor.

The drive motor requires derating due to harmonic losses and associated heating. Torque pulsations are present and their amplitude is large at low frequency of operation, due to additional harmonics in the rotor flux. The line power factor is poor, due to phase control.

Up to the rated frequency, the drive is in a constant torque mode and above the rated frequency the drive is in a constant horse power mode.

The stator current of an induction motor operating on a variable frequency, variable voltage supply is independent of stator frequency if the airgap flux is maintained constant. However, it is a function of the rotor frequency. The torque developed is also a function of rotor frequency only. Using these features a slip controlled drive (Fig. 4.28) can be developed employing a current source inverter to feed an induction motor. The relationship between the rotor frequency and stator current for rated flux in the airgap is introduced in the control. Thus an indirect control of flux is possible. The output of the function generator gives the reference value of current. The measured current is compared with the reference value and the error is used to alter the firing angle of the phase controlled converter on the line side. The input to the function generator is the difference between the reference speed and actual speed and it can be considered as slip frequency which is added to the frequency corresponding to rotor speed. This gives the value of stator frequency and the machine side inverter is controlled to give this frequency. The control is operative until the rotor attains the desired speed with the required slip frequency. The slip controlled drive has the following advantages:

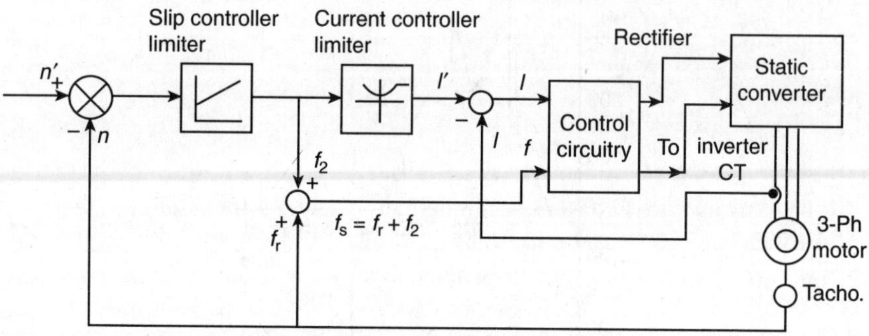

Fig. 4.28 *Slip controlled drive using CSI*

i. Controlled slip drive is highly efficient.
ii. Precise control of torque is possible over a wide speed range.
iii. The slip frequency can be any value up to the value corresponding to breakdown torque. The operation is at a very good power factor. The operation is very stable.

iv. The rotor can be accelerated at constant torque and current maintaining the rotor frequency at a suitable value. Fast acceleration.
v. As this leads to soft starting the motor does not see the blocked rotor currents and the associated voltage slips are not there.
vi. Special rotors with high starting torque are not necessary. Rotors with low resistance may be used so that the losses are limited.
vii. Regenerative braking can be incorporated. Braking at constant torque is possible.
viii. The drive has efficiency comparable to a thyristorized dc drive.

The added advantages of squirrel cage induction motors like high power to weight ratio, less maintenance, low inertia, no limitations on the power ranges and speed ranges make slip controlled drive a real competitor to dc motor drives.

Selected harmonic elimination methods or PWM principles can be employed to reduce the effects of torque pulsations, particularly at low speeds. In these methods control can be achieved by controlling the dc link current and PWM principle can then be used solely to control the harmonic content of the current waveform. This separation of the current and harmonic control functions allows the choice of PWM control strategy being directed solely towards improving the motor torque pulsations and reducing the harmonic losses.

The majority of PWM strategies for a current source inverter are based on selected harmonic elimination techniques. They are used to eliminate the lower order harmonics from the stator current and the methods result in the elimination of lower order torque ripples (Fig. 4.29).

In recent years PWM strategies have been specially developed to minimise the rotor speed ripple due to torque pulsation, to result in a drive having smooth rotation at low speeds (Fig. 4.30).

The general features of an induction motor on a current source inverter can be summarised as follows:

i. *Load dependent commutation:* As the load parameters form part of the commutation circuit, the inverter and motor must be matched.
ii. The inverter has a simple configuration. FWD's are absent.
iii. Only single motor operation
iv. The dc link contains only inductance. To maintain constant current this has to be very large. Two quadrant operation is straightforward
v. Invariably, a phase controlled rectifier is required on the line side. The variable dc link voltage is converted to a constant current source by means of high link inductance.
vi. The inverter is force commutated to give variable frequency currents to feed the motor.
vii. The value of the capacitance is a compromise between the voltage spikes and highest operating frequency. Larger the capacitance smaller the voltage spikes. The highest operating frequency is also limited.

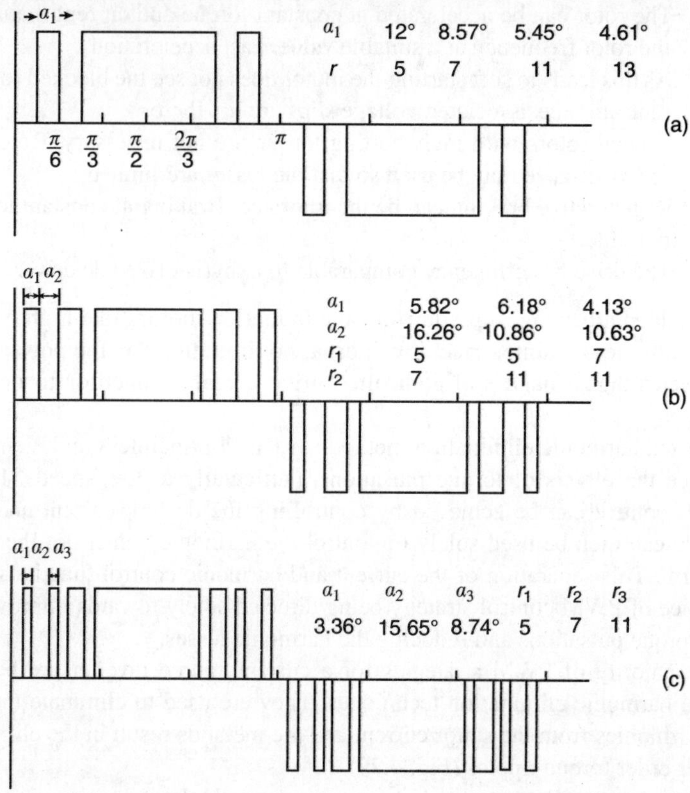

Fig. 4.29 *Pulsing the current of a CSI for selective harmonic elimination (a) to eliminate one harmonic (r) (b) to eliminate two harmonics (r_1, r_2) (c) to eliminate three harmonics (r_1, r_2, r_3)*

viii. The leakage reactance of the motor influences the harmonic voltages. It is also responsible for spikes of voltage during commutation. The leakage reactance being a parameter of the commutation circuit, determines the time of commutation, and consequently the upper operational frequency is limited. A motor must have smaller leakage reactance to have reduced harmonic voltages and spikes of voltage and to increase the range of speed control. The spikes influence the rating of the thyristor and affect the insulation. The motor size becomes larger if leakage reactance is small.

ix. Converter grade thyristors are sufficient. Thyristor utilisation is good

x. The inverter recovers from commutation failure. The link inductance causes a slow rise of fault current and by the time it reaches high value it can be suppressed.

xi. There is a stability problem at light load. A minimum current should be there for commutation.

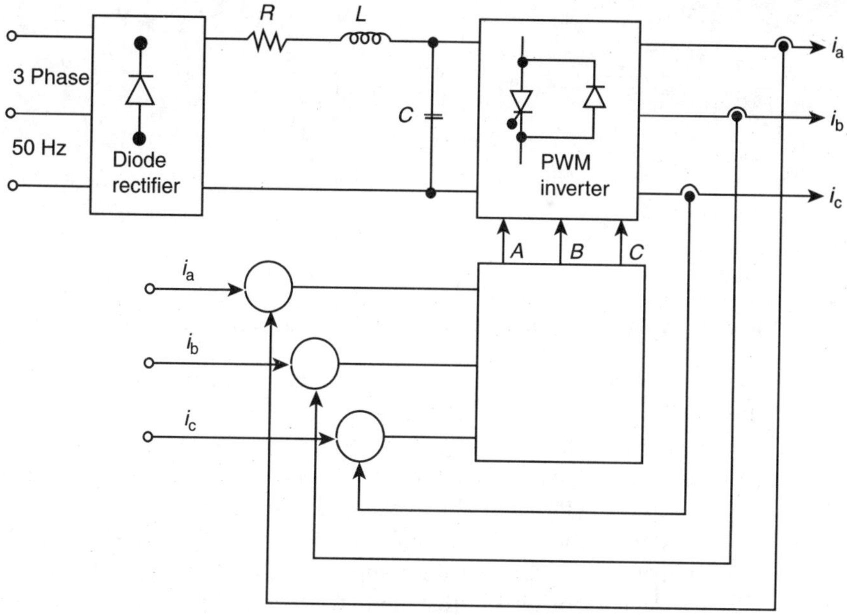

Fig. 4.30 *PWM inverter for current control*

xii. Open loop operation is not possible. Dynamic response is sluggish
xiii. The line p.f. is poor due to phase control.
xiv. Finds application as medium to high power drive
xv. Torque pulsations cause speed oscillations at very low speeds. PWM strategies are being employed to eliminate the speed oscillations and make the running smooth.
xvi. Both constant torque and constant horse power operations are possible.

Field Oriented Control In the slip controlled drives using VSI or CSI discussed in the foregoing sections, the stator voltage or stator current is controlled using slip frequency. They are controlled in magnitude only. The stator current control does not take care of its phase position with respect to flux. The control does not provide satisfactory dynamic behaviour. There exists an oscillatory response to changes in the rotor frequency. The dynamic response may be improved using the principle of field orientation where the stator current is controlled both in magnitude and phase position with respect to flux.

In induction motor drives the stator current has the function of providing flux as well as torque. The induction motor will have an operation similar to that of a dc motor if the stator current components, namely flux producing and torque producing are separately controlled (Fig. 4.31). This is actually the case in a dc motor where the torque depends on armature current and the flux on field current. There is an inherent decoupling between them but for the effects of armature reaction. These effects can be eliminated by means of compensating windings and there is

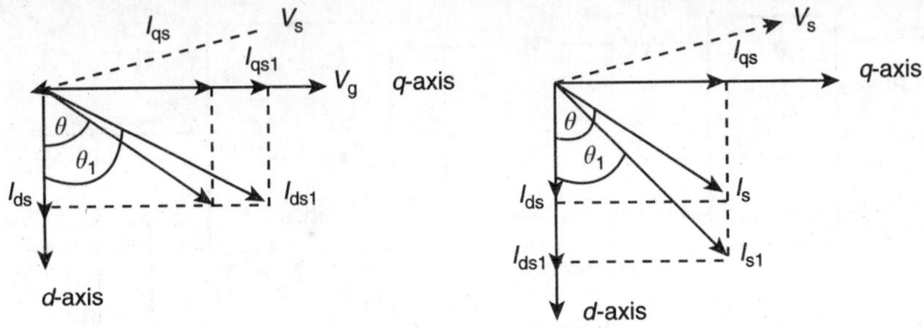

Fig. 4.31 *Principle of field orientation to control torque and flux individually*

perfect decoupling in a separately excited dc motor. This kind of decoupling is being attempted in the control of induction motors. The principle is called field orientation or vector control. This control improves the dynamic performance of the drive at all speeds. The stator current is decomposed into two components one along the *d*-axis and the other along the *q*-axis. The reference axes have been chosen such that the rotor flux is available completely along the direct axis. Its quadrature component is zero. Thus the component of current along the *d*-axis is the flux producing component and that along the *q*-axis is the torque producing component. By varying these components independently we can have independent flux control and torque control. These are depicted in Fig. 4.31(a) and (b). In the former the variation of flux is shown by varying direct axis component of current whereas in the latter the variation of *q*-axis component of stator current is depicted. Therefore the control of stator current amounts not only the variation of its magnitude but also its phase angle. The method is therefore called vector control.

Vector control techniques employed to keep the air gap flux constant impart poor dynamic characteristics to the drive, as the torque follows the slip frequency with a delay. Techniques have been developed to keep the rotor flux constant. In these methods the torque follows without any delay, thereby improving the dynamic behaviour. Such a high quality dynamic behaviour is required for induction motors used as actuators. Figure 4.32 depicts the implementation of the principle of field orientation. The actual value of rotor flux is compared with the reference value and the error so obtained is used to control the direct axis component of current. Normally this component is maintained constant so that the rotor flux is constant. The quadrature axis component is controlled using the error signal obtained from the comparison of actual torque and reference torque. The components are in the synchronously rotating frame. Therefore the decoupling between flux producing and torque producing components of the armature current necessitates a reference coordinate system, and transformation of the quantities to this reference frame and finally to the stator frame.

In this control, which provides a very good dynamic behaviour without a torque transient, secondary flux is required, which can be made available in two ways:

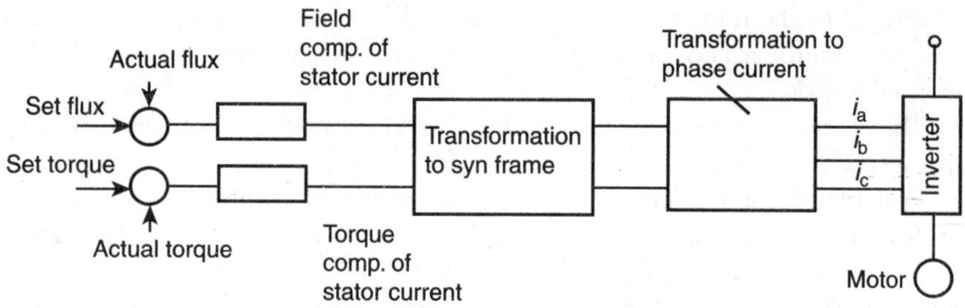

Fig. 4.32 *Implementation of field oriented control*

i. direct measurement using flux sensing coils.
ii. indirect estimation of the flux using a machine model, using easily measurable terminal quantities, such as voltages and currents.

The first method using direct measurement gives good results and is probably the most accurate control method available. The measurement is done by means of search coils, Hall probes or any other flux measuring techniques. The measured flux is used to effect the required decoupling between the torque producing and flux producing components of the stator current. The method is essentially insensitive to parameter variations. However, the cage motor loses its robustness and simplicity of construction.

If one tries to retain the robustness and simplicity of the motor, the flux is obtained using the second method. The rotor flux is estimated from the stator voltage vector, current vector and rotor speed. This estimated flux is fed to the torque controller. This approach is sensitive to errors in parameters. The rotor resistance, leakage reactance and other parameters must be accurately determined to achieve a performance equivalent to direct measurement. Unfortunately, the parameters of the motor used in the calculation are determined from no-load tests and do not represent the values of the parameters actually present at the operating point. Further, these parameters vary widely with saturation, temperature, frequency and current amplitude. The secondary flux level may be changed by the parameter variation. These variations in parameters result in erroneous flux control, which deteriorates the dynamic performance.

To avoid the errors due to variation of values of parameters (mainly rotor resistance), either due to incorrect estimation or due to operating conditions of the motor, automatic parameter identification or adaptation has been employed. The methods identify the changes in the performance due to the variation of parameters and correct the parameters accordingly. The error between the estimated value of flux and desired flux in the motor is made use of to correct the most influential parameter, which is the rotor resistance or rotor time constant, so that the machine model gives the required value of flux without any error.

Another method discussed recently is an on-line technique for establishing the exact value of rotor resistance of the induction motor. Identification is achieved by injecting a negative sequence current and detecting the negative sequence voltage.

The value of the rotor resistance is calculated using the information. The field oriented controller corrects the value of rotor resistance without the need for a thermal sensor.

The methods of state observer feed back are also employed for parameter identification.

With the advent of microprocessors (μps) and microcomputers it is now-a-days possible to solve the problems connected with drives effectively with little complexity in the control circuitry. More sophisticated controls in the drive system are possible. The flux vector can be very easily determined. To achieve the matching of the model of the motor and the identification of rotor parameters by a correlation procedure, a μp can be very effectively implemented without any additional measurements. The μps also facilitate the implementation of sophisticated algorithms for generating firing sequences of the inverter. Also, the elaborate process of field oriented control of the overall drive system has become economically feasible, as the expensive hardware used so far can now be substituted by software.

The field orientation can be very easily implemented with CSI feeding an induction motor. As the PWM inverter has good dynamic behaviour, these are used with a current control on the output side. Figure 4.33 depicts the principle of current control using PWM inverter. This has the advantages of both VSI and CSI.

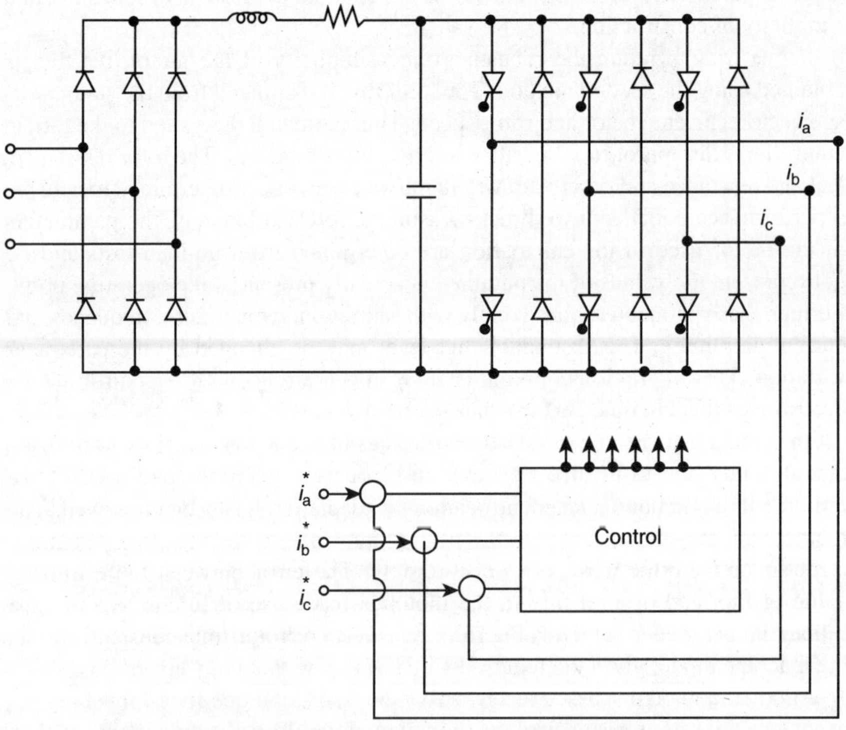

Fig. 4.33 *Current control using PWM inverter*

Induction Motors in the Flux Weakening Mode In the case of dc motors speeds above base speed are obtained by decreasing the field current at constant rated armature voltage. The torque developed decreases. A constant power mode can be realised in this speed range and is called flux weakening mode.

A similar behaviour can be observed in the case of induction motors. The voltage reaches its rated value for rated frequency. For frequencies above this value the inverter voltage is kept constant. The speed of the motor increases in proportion to the frequency. Due to the increase in the frequency, the air gap flux decreases. The torque at a given rotor frequency is inversely proportional to the square of the stator frequency and the power developed is not constant. The dynamic behaviour under weakened flux conditions is very poor. This can be improved by varying slip frequency for maximum torque in proportion to frequency.

A motor having current feeding has a good dynamic behaviour when there is reserve voltage at the inverter terminals. When once a certain value of upper frequency is reached, the back emf of the motor is equal to the applied voltage and there is no reserve voltage for current control. The actual value of stator current and rotor flux deviate from the desired values and the drive has a poor dynamic behaviour. Improvement of dynamic performance of the motor in the field weakening mode is a problem of interest.

This can be done by controlling the amplitude of rotor flux and hence the back emf, so that sufficient voltage reserve is available in the speed range above base speed. Here also the direct and indirect methods of flux control can be used.

4.2 SYNCHRONOUS MOTOR DRIVES

4.2.1 General Considerations

Variable frequency drives employing synchronous motors are receiving considerable interest, and are even becoming competitors to both induction motors and dc motors. These are popular as drives for very high power applications, such as pumps, fans, conveyors, etc. The stator of the synchronous motor is supplied from a thyristor power converter capable of providing a variable frequency supply. The rotor, depending upon the situation, may be constructed with slip rings, where it conforms to a conventional rotor. It is supplied with dc through slip rings. Sometimes rotor may also be free from sliding contacts (slip rings), in which case the rotor is fed from a rectifier rotating with rotor.

An induction motor operates at lagging power factor and hence the converter supplying the same must invariably be a force commutated one. A synchronous motor, on the other hand, can be operated at any power factor by controlling the field current. For operation at unity power factor and lagging power factors the inverter must be force commutated. However if the power factor is near unity the inverter size is reduced, besides reducing armature copper loss. For operation at leading power factors the motor must be overexcited. The motor is in a position to supply the reactive power required by the inverter. In other words, the machine voltages can be used for commutation. The inverter in this case is load commutated. The load commutation is possible with cycloconverters and current source inverters but not with voltage source inverters.

A machine is said to be self controlled if it gets its variable frequency from an inverter whose thyristors are fired in a sequence, using the information of rotor position or stator voltages. In the former, a rotor position sensor is employed, which measures the rotor position with respect to the stator and sends pulses to the thyristots. Thus the frequency of the inverter output is decided by the rotor speed. This kind of control makes it possible to control the angle between the stator and rotor mmf or load angle. The machine behaviour is decided by the torque angle, and voltage/current. Such a machine can be looked upon as a dc motor having its commutator replaced by a converter connected to stator. The rotor may be a conventional one supplied by dc power through slip rings, or it may be of special construction to avoid sliding contacts. In this case it is fed from a rectifier rotating with the rotor. The self controlled motor has properties of a dc motor both under steady-state and dynamic conditions and therefore is called commutator less motor (CLM). These machines have better stability behaviour. They do not fall out of step and do not have oscillatory behaviour, as in normal synchronous motors.

Alternatively, firing signals can be derived from the phase position of stator voltages in which case the mechanical rotor position sensor can be dispensed with. The synchronous machine with the inverter can be considered to be similar to a line commutated converter where the firing signals are synchronised with the line voltage. In this case the firing pulses of the inverter are synchronised with the machine voltages. The natural commutation using machine voltages takes place. Thus the machine voltages are used for both control as well as commutation. The frequency of the inverter is the same as that of the voltages.

The dynamic behaviour of the motor depends upon the type of control used. The machine however retains the properties of a dc motor in both the types of control.

An overexcited synchronous motor will be able to provide the necessary reactive power required for commutation of thyristors. The inverter commutation circuit has a very simple configuration as the capacitors, diodes and auxiliary thyristors are absent. The converter is economical with better utilisation of the thyristors. However, the machine commutation has a disadvantage. At very low speeds (up to about 10% of base speed) the machine voltages are insufficient to provide satisfactory commutation. Special methods are required to run the motor at these speeds, beyond which machine commutation can be used. Commutation assistance is required only at low frequencies and hence the commutation circuit required will be relatively simple and cheap. Artificial commutation of conducting thyristors is possible by means of a controlled thyristor across the link inductance. The motor is underutilised as it is operating at leading power factor.

A self controlled synchronous motor is, therefore, a substitute for a dc motor drive and finds application where a dc motor is objectionable due to its mechanical commutator, which limits the speed range and power output.

4.2.2 Control of Synchronous Motors on Variable Frequency Supply

As has already been discussed, the speed of a synchronous motor can be varied by supply frequency variation similar to an induction motor. The applied voltage must be varied in this case also to have the maximum torque capability and to avoid saturation at all frequencies. A synchronous motor can receive its variable voltage, variable frequency supply from a cycloconverter or a dc link converter. The dc link converter can have a voltage source inverter to feed impressed voltages or a current source inverter to feed impressed currents. Further, the voltage source inverter can be a square wave inverter with variable link voltage or a PWM inverter with constant link voltage. These types are depicted in Fig. 4.34. In all these cases self control can be employed with the motor in the CLM mode which imparts a very good steady-state and dynamic behaviour similar to a dc motor. But machine commutation may not be always possible.

A separate control of the inverter feeding a synchronous motor can also be used. In this case the speed of the motor is determined from an external frequency obtained from a crystal oscillator (Fig. 4.35). The performance of the motor is similar to a conventional synchronous motor with problems of hunting for sudden changes of load. Multimotor operation is possible here. This is seldom used with inverter feeding.

Principle of Self Control Self controlled mode of a synchronous motor facilitates unique positioning of stator voltage vector and rotor flux vector. This is similar to a dc motor where the armature mmf and field mmf are in space quadrature by proper positioning of the brushes in the neutral zone. The principle of self control can be explained for a three phase motor, referring to Fig. 4.36. The three windings of a three phase synchronous motor are 120° away from each other in space and are connected in star. These windings are connected to a dc supply through a thyristor inverter having thyristors $T1$ and $T6$. The rotor position decides the thyristors to be fired in the sequence. Depending upon the thyristors conducting, the currents in the windings flow. These produce an mmf fixed in space till the other thyristor in the sequence is fired and is ahead of the rotor mmf. The stator mmf changes its position when the next thyristor is fired. Thus a dc link current flows through the phases of the motor alternately as the inverter is controlled. These currents effectively produce an mmf which rotates by two pole pitches by the time all the thyristors are fired once in the sequence. The rotor carrying a dc excitation produces a field mmf which rotates with the rotor. The angle between the stator mmf and rotor mmf decreases owing to the rotation. When once the angle takes the value of 60°, the next thyristor in the sequence is fired, so that the stator mmf jumps by 60°, making the angle between the mmfs 120°. Thus at the start of conduction of a pair of thyristors stator windings, the mmfs have an angle of 120° which slowly decreases as the field rotates. When the angle reaches 60° the next two windings conduct as the next thyristor is fired to bring the angle to a value of 120°. Thus the angle between the mmfs changes from 120° to 60°.

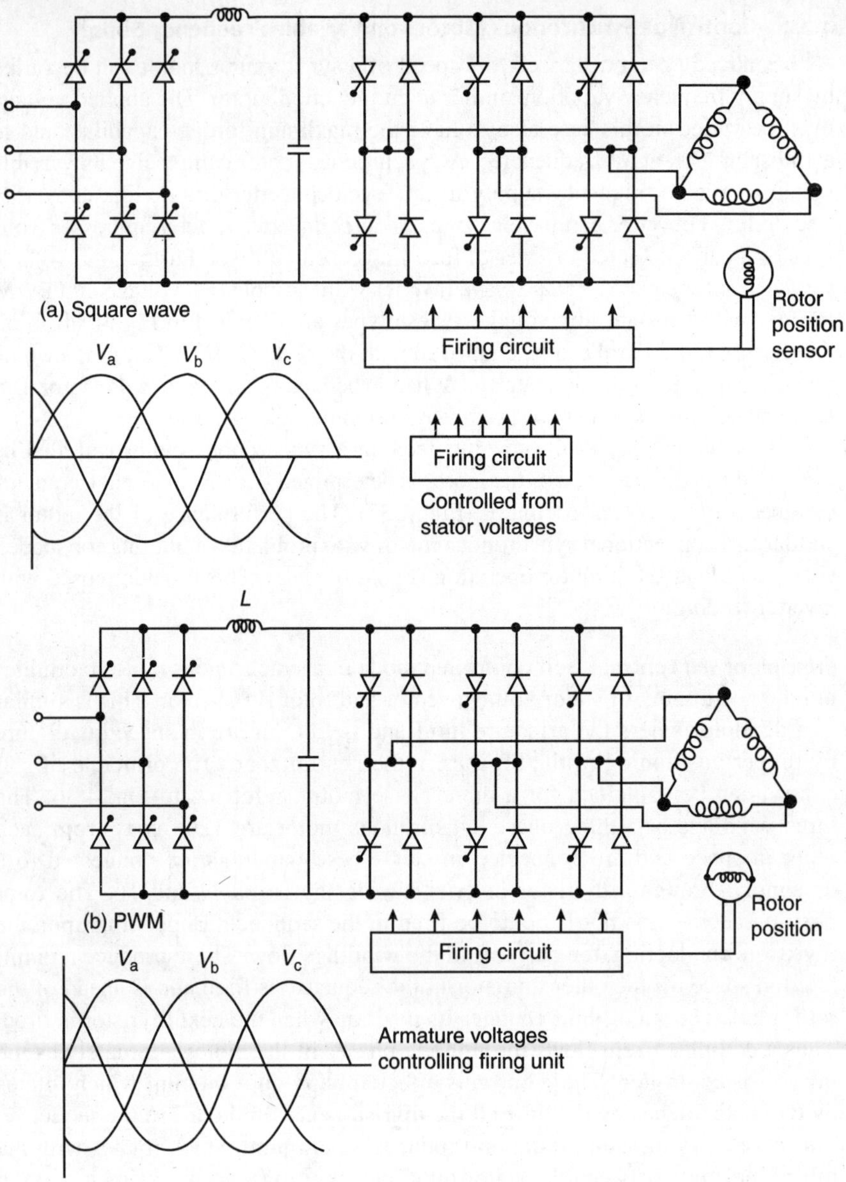

Fig. 4.34 *Self control of synchronous motor fed from VSI*

Referring to Fig. 4.37 at the instant the phases *S* and *R* start conduction due to conduction of *T*1 and *T*6, the angle between the mmfs is 120°. Due to rotation it slowly decreases and at the instant t_2 when the angle is 60°, *T*2 is fired. The commutation takes place during which *T*1 stops conduction and *T*2 takes over. The stator mmf has shifted to a new position by 60° making the angle again 120°.

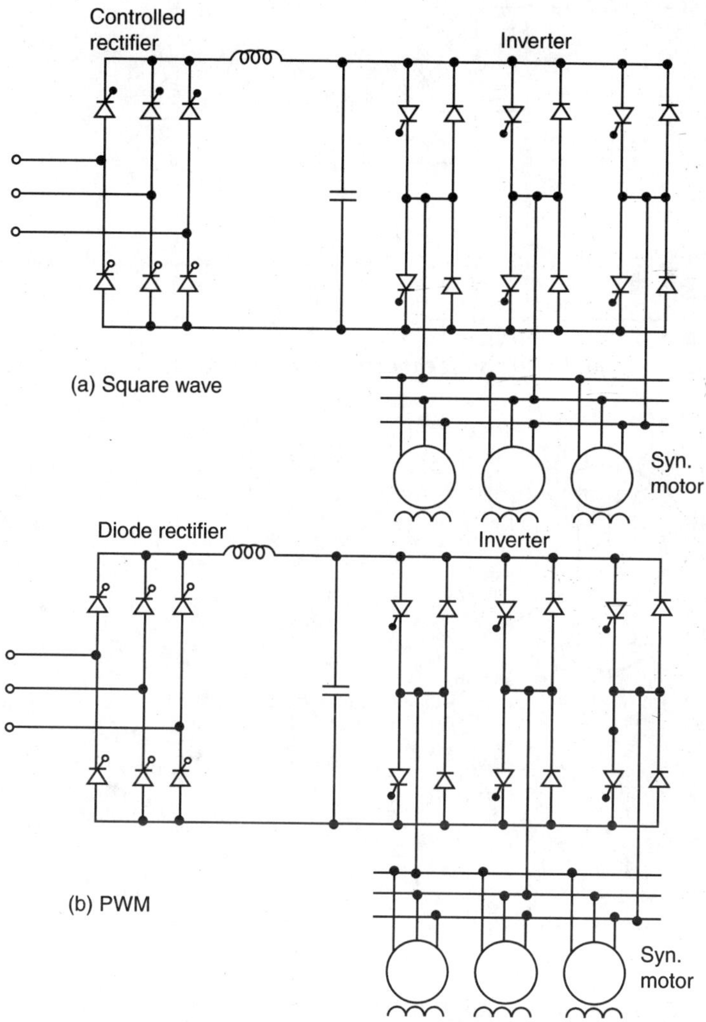

Fig. 4.35 *Separately controlled synchronous motor fed from VSI*

Similar operations repeat at firing of each thyristor (Fig. 4.37). Following this the torque developed by the motor is not a constant one but has pulsations varying between a maximum and 0.86 of maximum values (Fig. 4.38). These pulsations can be reduced to a minimum by increasing the pulse number of the inverter. In the case of dc motors the pulsations are almost absent because of a large number of commutator segments which correspond to pulses of the inverter. The brushes on the commutator decide the commutation instants. By virtue of their lying in the neutral zone the angle between the mmfs is maintained at 90°. In the case of a self controlled motor the commutation instants are decided by rotor position or synchronisation of machine voltages, as explained previously. The inverter here can be considered as a six segment commutator.

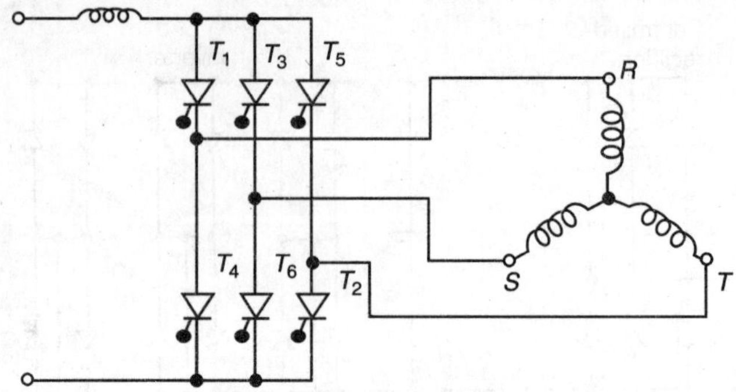

Fig. 4.36 *Control of thyristors either from rotor position sensor or voltages*

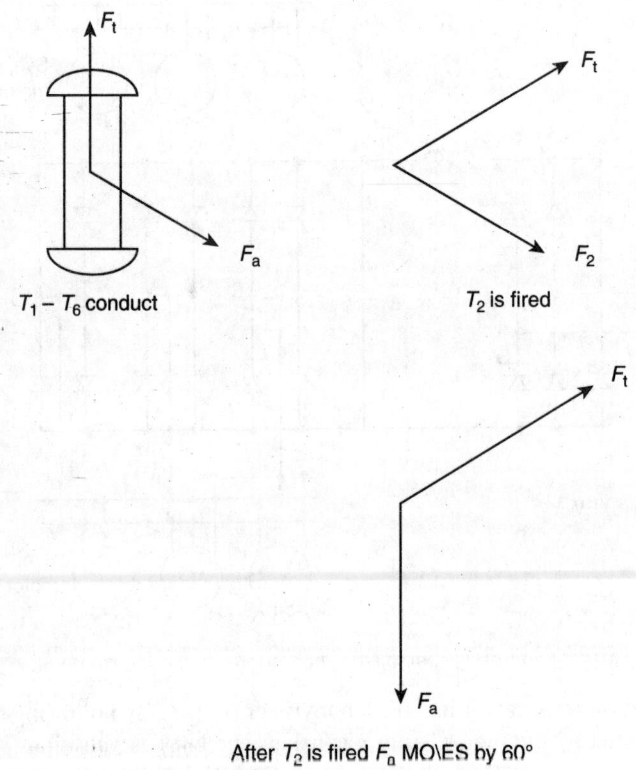

Fig. 4.37

In self controlled mode the power factor can be maintained at the desired value by using a closed loop control. A constant air gap flux can be achieved by independent field control. Field weakening exists, similar to the case of an induction motor, when the inverter voltage and field current have reached their rated values.

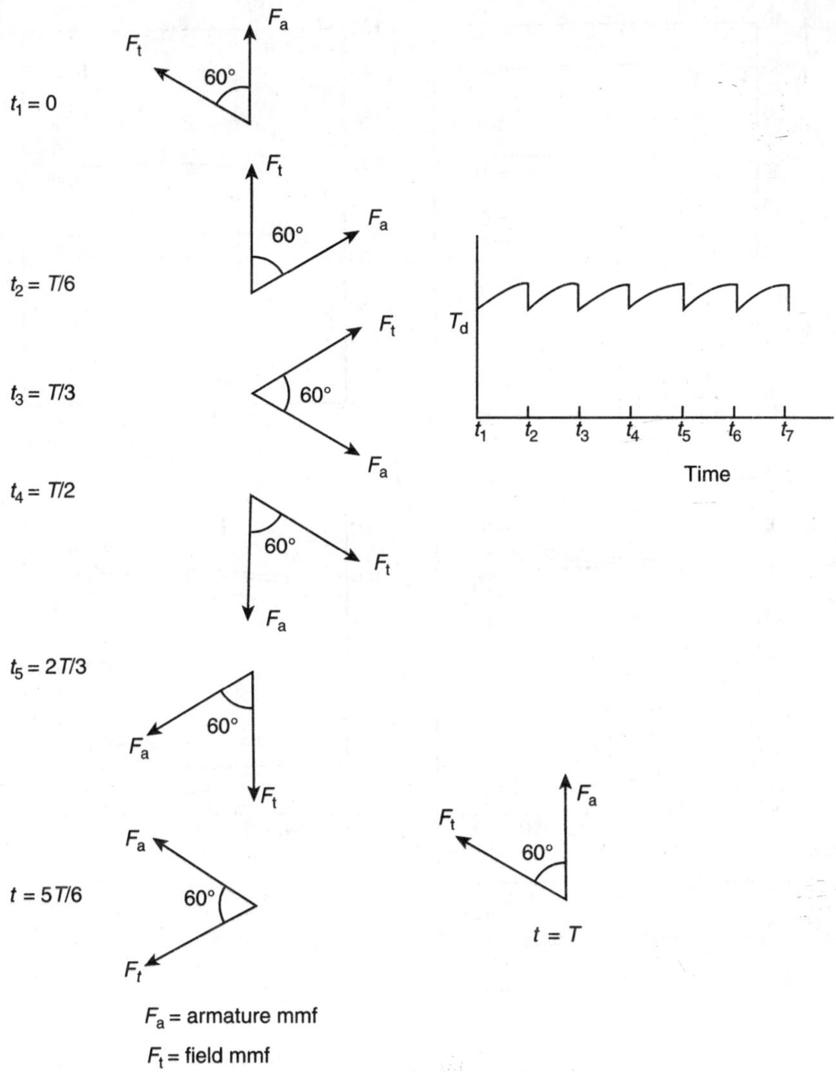

Fig. 4.38 *Pertaining to pulsating torque behaviour of an inverter fed synchronous motor*

In a synchronous motor on a variable frequency supply, several modes of control exist. Both terminal voltage and excitation voltage can be varied proportional to frequency, i.e., V/f and E/f are constant. The value of E/f is constant if field current is held at a constant value. If the applied voltage is varied proportional to frequency and the effect of stator resistance is neglected, constant flux operation results in the basic speed range. However, at low frequencies the stator resistance drop becomes comparable to the applied voltage and hence a correction

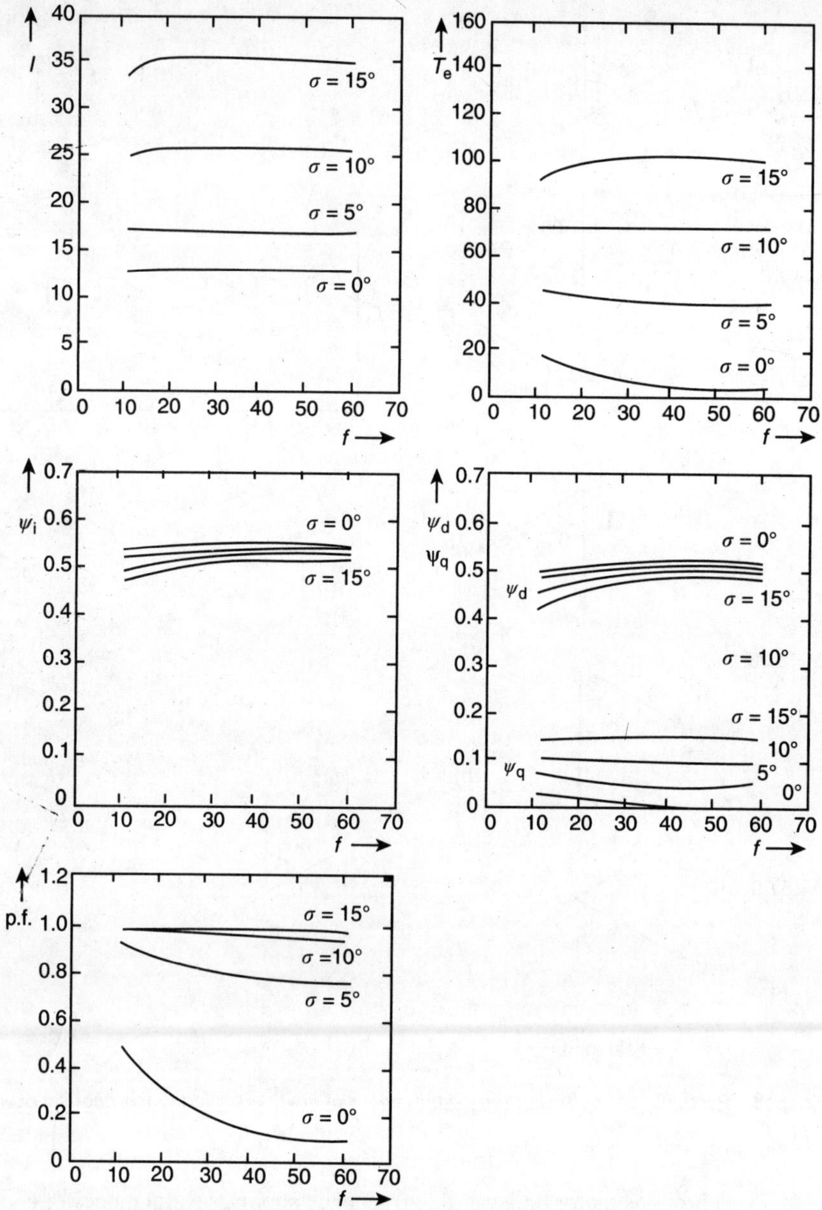

Fig. 4.39 *Characteristics of a variable frequency synchronous motor (V/f and E/f constant)*

is required to compensate for the resistance drop. The motor has a speed-torque curve similar to that of a separately excited dc motor. Also, when the effects of stator resistance are compensated, it can be easily seen that the stator current is independent of supply frequency (Fig. 4.39).

A control of the motor with constant V and E/f on the other hand, imparts a series characteristic to the motor which makes it suitable for traction purposes. The problems of saturation have to be taken care of at low frequencies (Fig. 4.40).

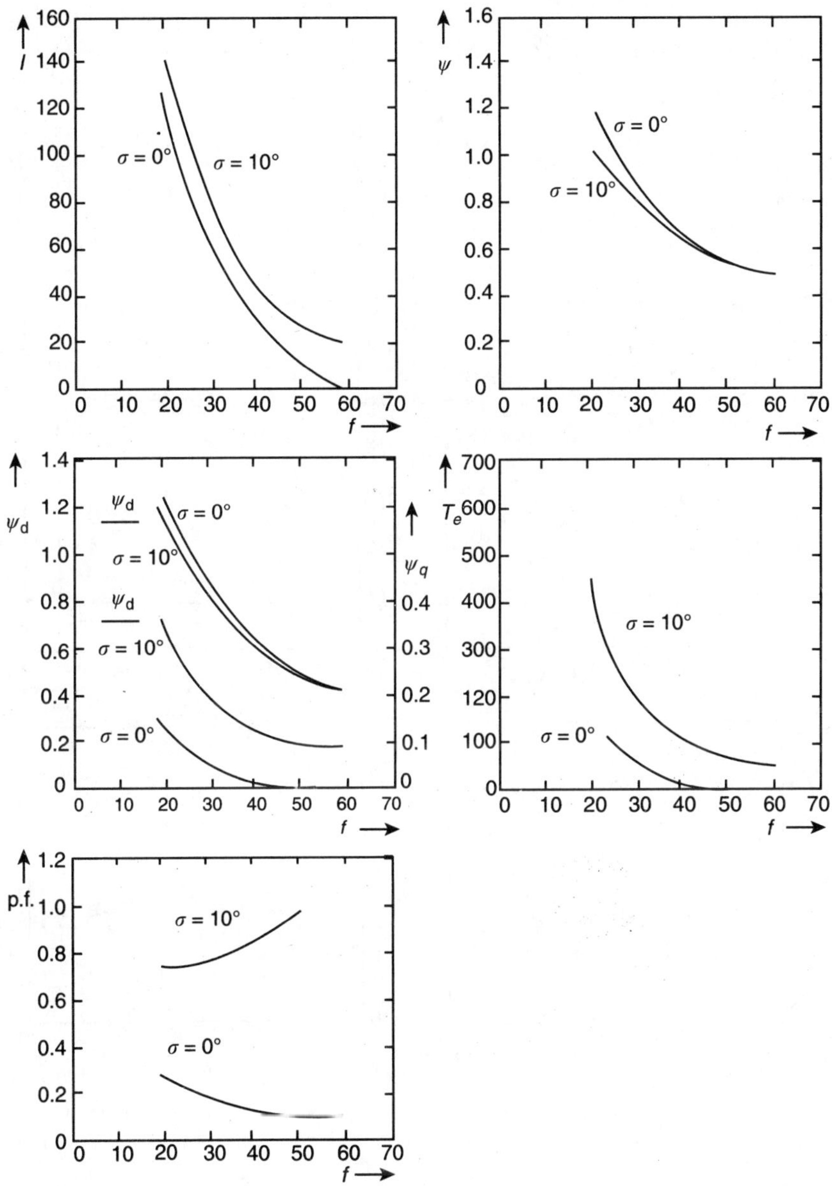

Fig. 4.40(a) *Characteristics of variable frequency synchronous motor (V and E/f constant)*

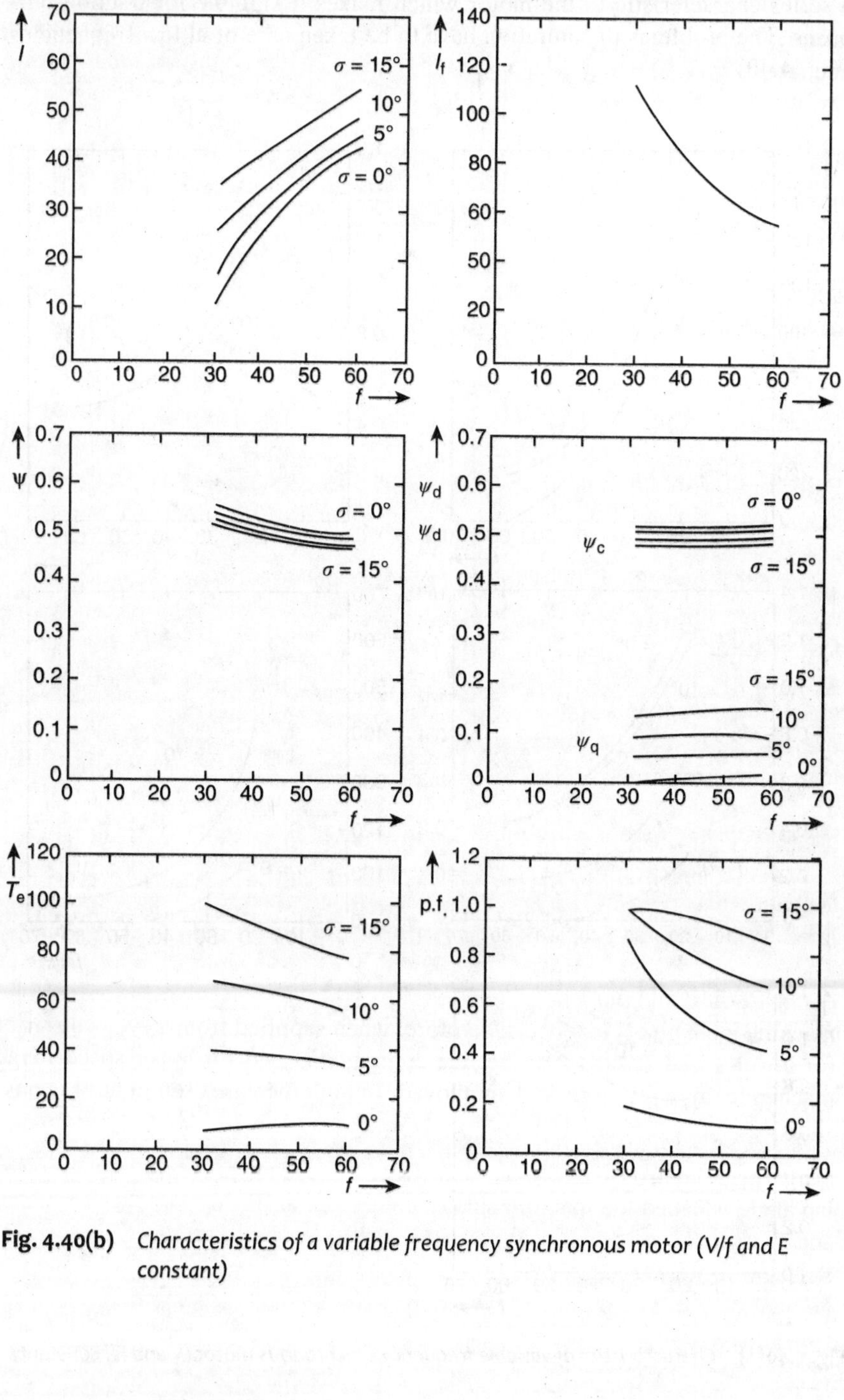

Fig. 4.40(b) *Characteristics of a variable frequency synchronous motor (V/f and E constant)*

When fed from a VSI the load commutation is not possible. As the power factor can be controlled by field excitation, UPF operation of the motor reduces the size of the inverter and its commutation circuit. The armature copper losses are also reduced.

The above controls can be realised with PWM inverters also, in which the voltage control is achieved in the inverter itself. By this the speed range is extended up to zero; the harmonic content of the stator current decreases. Consequently, the torque pulsations are also decreased.

An examination of the control to keep V/f and E/f constant shows that the armature current is independent of frequency when the effects of resistance have been corrected, particularly at low frequencies. This leads to a simple and straightforward control of stator current to maintain constant flux in the motor. This makes the operation of a synchronous motor for a current source inverter feasible. Self controlled CLM operation and load commutation are possible. As has already been stated, commutation assistance is required at low speeds. Forced commutation may be employed at low frequencies. The circuit becomes simple and less expensive if the p.f. is maintained at unity. Field regulation may also be employed to keep the air gap flux constant. Current control is advantageous as it assures reserve voltage. However when the frequency reaches the rated value, this reserve is not present and the field weakening starts. A motor for CSI operation should invariably have damper windings. The subtransient (leakage) reactance is effective during commutation. This must be as small as possible to limit voltage spikes. The dampers actually decrease this reactance. The overlap decreases. The control sometimes requires force commutation in which case p.f. can be controlled to unity in view of the advantages discussed already. The field mmf required is however greater than that of an induction motor due to armature reaction.

A cycloconverter can also be used to provide variable frequency supply to a synchronous motor. This can be either line commutated or machine commutated. When line commutated there exists a limitation on the output frequency. The speed variation can be in the range 0–1/3 of base speed. When machine commutation is employed the speed range can be 10%–100% base speed. The machine must be overexcited. Therefore, when supplied from a cycloconverter for lower speed range it can be line commutated and for upper speed range machine commutation may be employed. In both the cases self control is possible.

Margin Angle Control The difference between the lead angle of firing and the overlap angle is called the margin angle of commutation. If this angle is very small and is less than the turn off angle of the thyristor, commutation failure occurs. Safe commutation is assured if this angle has a minimum value equal to the turn off angle of the thyristor. The type of sensing influences the margin angle. When voltage sensing is used the firing angle of the inverter thyristors may be constant.

The angle of overlap increases as the dc link current increases. Therefore at large values of link current the margin angle decreases and this may cause failure of commutation at a particular current. To avoid this the load angle of firing has to be simultaneously increased. When this is done certain disadvantages, such as increased torque pulsations at low speeds, poor power factor on line side, and poor efficiency would occur.

To have better performance with respect to the above quantities, the firing angle is controlled such that the margin angle remains constant at the minimum value required and ensures safe commutation at all operating points. The control can be employed with self controlled synchronous motors using either rotor position sensing or voltage sensing using closed loop control, to achieve this. The margin angle kept constant at steady-state may not remain constant under dynamic conditions.

The advantages of margin angle control can be summarized:

i. Commutation failure is prevented.
ii. The maximum power output can be increased by simultaneous control of field current to compensate for armature reaction. This improves the overload capacity of the motor.
iii. There is improvement in power factor.
iv. The torque pulsations under light load conditions are reduced. The ripple content of the dc link current also decreases.

However, a disadvantage of the method, is a limited speed range with a limit on the upper speed.

To increase the upper speed limit, margin time control is employed. This improves the performance of the drive.

4.2.3 Voltage Source Inverter Fed Synchronous Motor Drive

An inverter fed synchronous motor has been very popular as a converter motor in which the synchronous motor is fed from a CSI having load commutation. Of late more attention is being paid towards understanding the behaviour of synchronous motors fed from a VSI. These drives can also be developed to have self control, using a rotor position sensor or phase control methods. It has been reported in the literature that these drives might impose fewer problems both on machine as well as on the system design. A normal VSI with 180° conduction of thyristors requires forced commutation and load commutation is not possible.

A typical power circuit of a voltage source inverter is shown in Fig. 4.41. Three combinations are possible, to provide a variable voltage variable frequency supply to a synchronous motor (Fig. 4.42). The voltage control can be obtained external to the inverter using a phase controlled rectifier. The link voltage is variable. This has the disadvantage that commutation is difficult at very low speeds. As the output voltage is a square wave the inverter is called variable voltage inverter

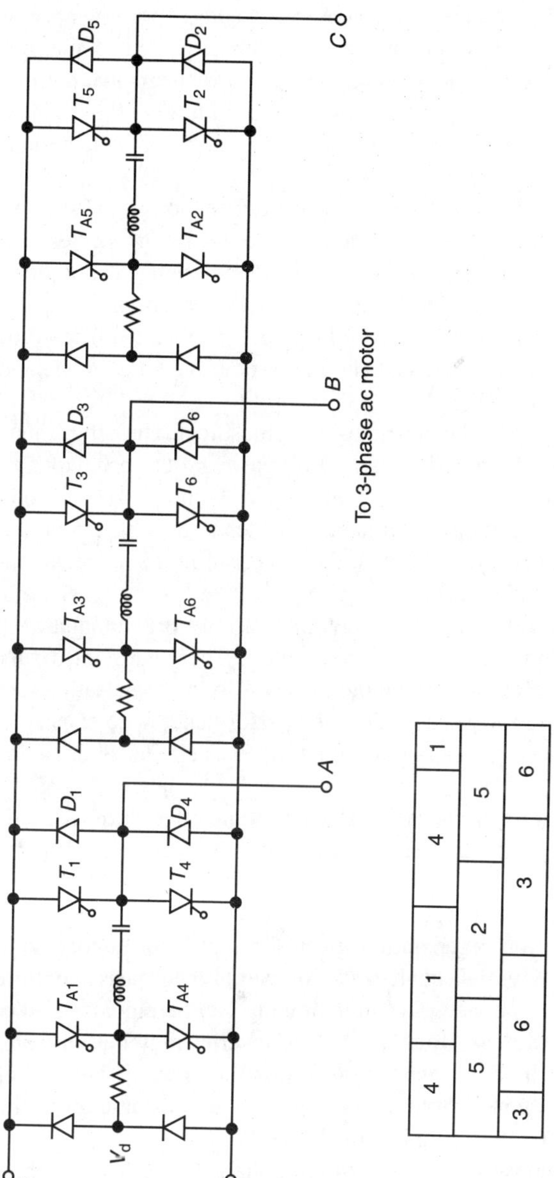

Fig. 4.41 *Power circuit of a VSI*

or square wave inverter. The second alternative is to have voltage control in the inverter itself, using principles of PWM or PSM. The inverter is fed from a constant link voltage. A diode rectifier would be sufficient on the line side. This does not have difficulties of commutation at low speeds. Very low speeds up to zero can be obtained. The third alternative is to interpose a dc chopper in between the rectifier and the inverter. The system may appear cumbersome at first sight, but it has advantages. Three simple converters are used to give the desired result. It is possible to reduce the size of link inductance by having a synchronous control of the chopper.

A voltage source inverter feeding a synchronous motor can have either separate control or self control. In the former the speed of the motor is determined by external frequency from a crystal oscillator. Open loop control is possible. The motor has instability problems and hunting, similar to a conventional motor. In the latter the inverter is controlled by means of firing pulses obtained from a rotor position sensor or induced voltage sensor. The motor is in the CLM mode and has better stability characteristic (Fig. 4.43).

The output voltage of the inverter is non-sinusoidal. The behaviour of the motor supplied from the inverter is entirely different from the behaviour of the motor operating on a conventional sinusoidal supply. A knowledge of the behaviour is essential. The steady-state performance enables one to have a proper choice of the thyristors, and also to determine the effects of non-sinusoidai waveforms on torque developed and machine losses.

The stator current drawn by the motor when fed from the square wave inverter has sharp peaks and is rich in harmonic content. These harmonics can cause additional losses and heating of the motor. They also cause pulsating torques which are objectionable at low speeds. Thus the performance with respect to additional heating due to harmonics, and pulsating torques is similar to that of an induction motor.

When a PWM inverter is used, these harmonic effects are reduced. The stator currents are less peaky and have reduced harmonic content. Accordingly additional losses due to harmonics, consequent motor heating and torque pulsations are decreased. These effects become minimal.

The discussion on regeneration given for induction motors holds good for these cases also. With the square wave inverter another phase controlled rectifier is required on the line side. Dynamic braking can be employed. When a PWM inverter is used, two cases may arise. The inverter may be fed from a constant dc source in which case regeneration is straight forward. The dc supply to the inverter may be obtained from a diode rectifier. In this case an additional phase controlled converter is required on the line side.

A square wave inverter drive must have a phase controlled converter on the line side. Due to phase control the line power factor is very poor. A diode rectifier is sufficient in the case of PWM inverter. The line p.f. improves to unity. In either case the machine p.f. can be improved by field control. With a view to minimizing

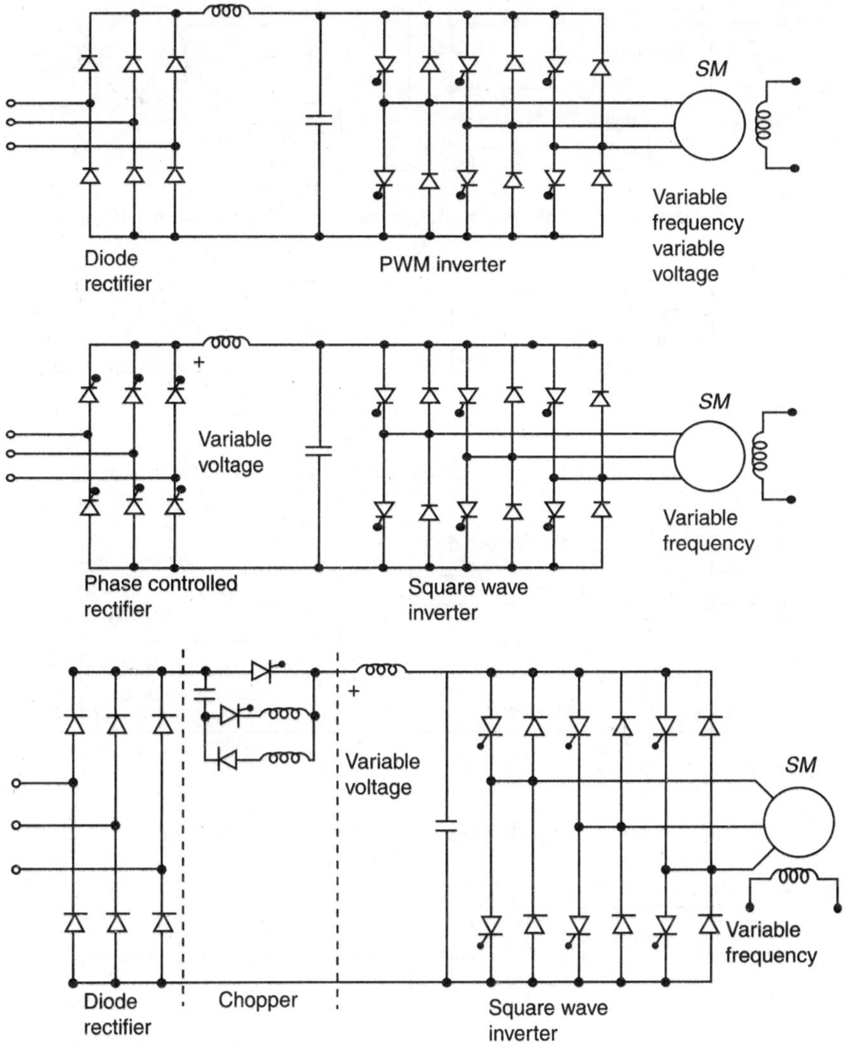

Fig. 4.42 *Possible combinations of voltage source dc link converters to obtain a variable voltage variable frequency supply to feed a synchronous motor*

the inverter size as well as losses in the inverter and motor, it is advantageous to operate the motor at UPF.

A VSI drive provides reasonably good efficiency. Converter cost is high and multimotor operation is possible. Open loop (separate) control may pose stability problems at low speeds. CLM mode is very stable. PWM drive has a better dynamic response than a square wave drive. This finds application as a general purpose industrial drive for low and medium powers.

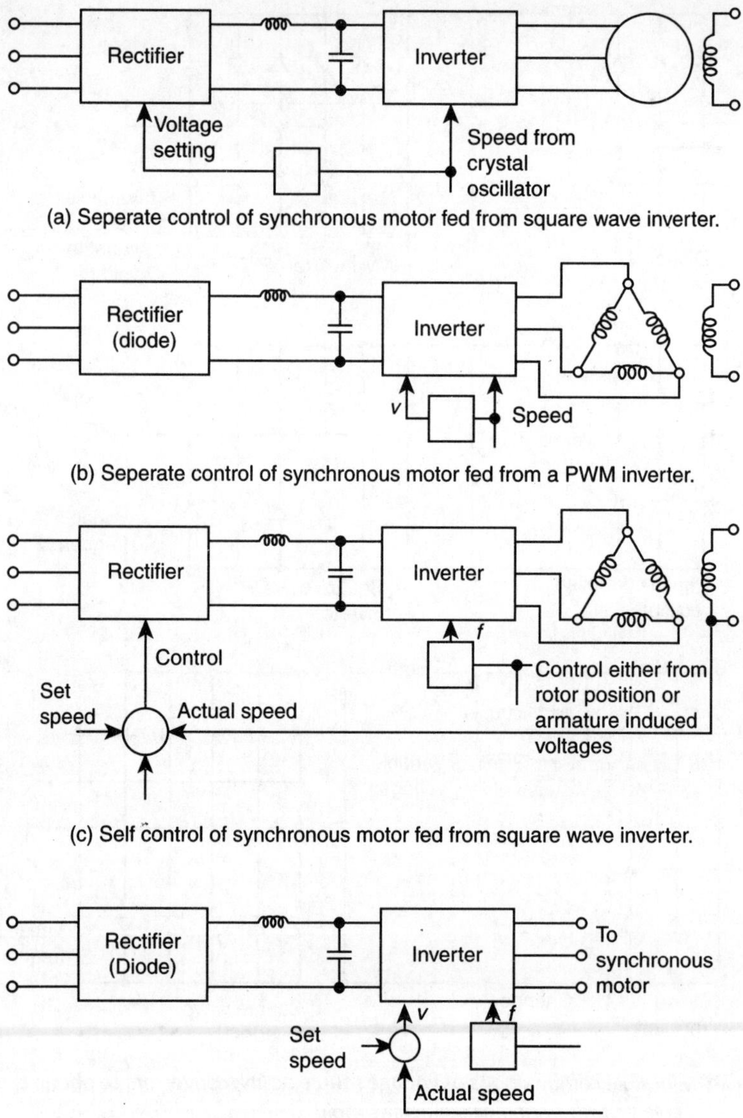

(a) Seperate control of synchronous motor fed from square wave inverter.

(b) Seperate control of synchronous motor fed from a PWM inverter.

(c) Self control of synchronous motor fed from square wave inverter.

(d) Self control of synchronous motor fed from square a PWM inverter.

Fig. 4.43 *Principles of separate and self control*

General Features of VSI Fed Synchronous Motors—a Summary

i. When fed from a square wave inverter the drives have a speed range from medium to high speed. This is because of low voltages at small frequencies, which are insufficient for commutation. When fed from a PWM inverter a wide range of speeds, down to zero speed is possible. Inverter voltage is constant at all frequencies.

ii. Dynamic braking is possible. Regeneration reauires another phase controlled converter on the line side. However, if in PWM case, the inverter is s'ıpplied from a constant dc source the regeneration is straightforward.

iii. A square stepped voltage wave is applied to the motor which results in a current having rich harmonic content. This causes additional heating and torque pulsations. The effects are more at low speeds. PWM principles employed result in nearly sinusoidal voltage applied to the motor. The harmonic content of stator current decreases. The additional losses, heating, and torque pulsations are minimal.

iv. The phase control on the line side of a square wave inverter causes a poor line p.f. With PWM inverter a diode rectifier may be sufficient for motor operation and the line p.f. is near unity. Machine p.f. can be improved by field control. It is operated at UPF in view of better performance. It is usually advantageous to operate the motor at UPF as this reduces the armature current to a minimum for a given output. Consequently motor armature copper losses and inverter losses are reduced. The VA rating of the inverter is also small.

4.2.4 Synchronous Motor Fed from a Cycloconverter

DC link converter is a two stage conversion device which provides a variable voltage, variable frequency supply. Variable voltage, variable frequency supply can be obtained from a cycloconverter which is a single stage conversion equipment. The power circuit of a cycloconverter feeding a synchronous motor is shown in Fig. 4.44. This has several differences compared to a dc link converter.

The line voltages are made use of to commutate the thyristors of a cycloconverter. The output frequency can be varied from 0–1/3 of the input frequency. The range of speed control is therefore limited, extending from 0–1/3 base speed.

Cycloconverters are inherently capable of power transfer in both directions. Four quadrant operation is simple.

A cycloconverter in the above speed range gives a high quality sinusoidal output voltage. The resulting currents are also nearly sinusoidal. The harmonic content of the current is small. Consequent effects of harmonic current, such as losses, heating and torque pulsations are minimal.

The line power factor is somewhat better because the machine power factor can be made unity.

A cycloconverter requires a large number of thyristors and its control circuitry is complex. Converter grade thyristors are sufficient but the cost of the converter is high.

The efficiency is good and the drive has a good dynamic behaviour. The operation in CLM mode is popular.

A cycloconverter drive is attractive for low speed operation and is frequently employed in large, low speed reversing mills requiring rapid acceleration and

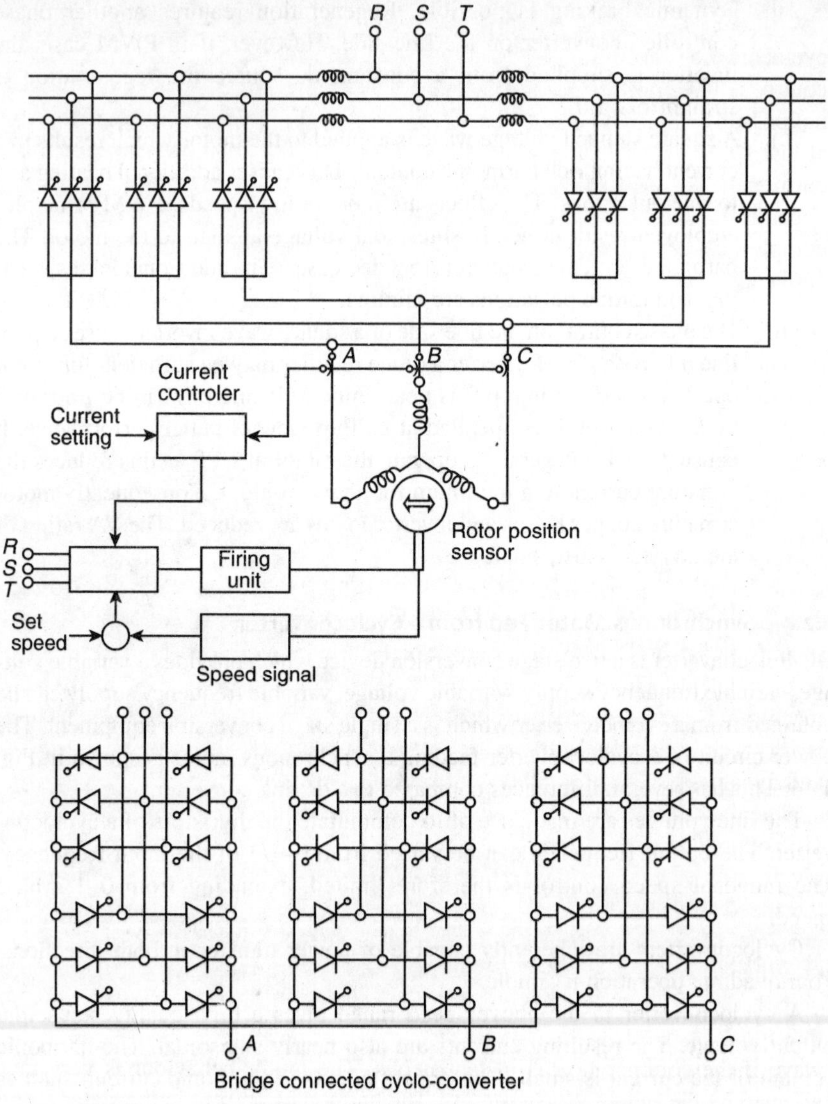

Fig. 4.44 *Cycloconverter feeding a three phase synchronous motor. Self control (CLM mode). Cycloconverter can have either line commutation or machine commutation*

deceleration. Typical applications are large gearless drives, e.g., drives for reversing mills, mine hoists, etc.

A cycloconverter can also be commutated using the load voltages if the load is capable of providing the necessary reactive power for the inverter. An over-excited synchronous motor can provide the necessary reactive power. Hence a cycloconverter feeding such a motor can be load commutated. The range of speed control is from medium to base speed. At very low speeds load commutation is not possible. The speed range can be extended to zero if line commutation is used at low speeds. Four quadrant operation is simple. The problems associated with harmonics are minimal due to high quality of the output. The line power factor depends on the angle of firing and is poor. The cost of the converter is high with complex control. Its efficiency is good and the drive has a fast response. It finds application in high power pump and blower type drives.

4.2.5 Current Source Inverter Fed Synchronous Motor Drive

A synchronous motor draws a stator current which is independent of stator frequency when *V/f* and *E/f* are maintained constant and armature resistance is neglected. The motor also develops constant torque. The flux also remains constant. Therefore, by controlling the stator current of a synchronous motor we can have flux control as well as torque control. As has been discussed in the case of the induction motor, current control is simple and straightforward. A synchronous motor is fed from a current source inverter. A synchronous motor can have either separate control or self control. Due to stable operation self control is normally employed, by using either rotor position sensing or induced voltage sensing. The motor operates in CLM mode. When fed from a CSI the synchronous motor can be operated at leading power factor so that the inverter can be commutated using machine voltages. A load commutated, CSI fed self controlled synchronous motor is very well known as a converter motor. It has very good stability characteristics and dynamic behaviour similar to a dc motor.

Due to machine commutation the working speed range starts typically above 10% of base speed and extends up to base speed. By using (assisted) forced commutation the lower speed limit can be extended to zero. During the operation in the speed range from 0 to 10% of base speed (above which load commutation is possible) the machine can be operated at UPF.

When fed from a CSI, the synchronous motor is supplied with currents of variable frequency and variable amplitude. The dc link current is allowed to flow through the phases of the motor alternately. The motor currents are quasi-square wave if the commutation is instantaneous. The motor behaviour is very much affected by the square wave currents. The harmonics present in the stator current cause additional losses and heating. They also cause torque pulsations, which are objectionable at low speeds.

A current source inverter is inherently capable of regeneration. No additional converter is required, and four quadrant operation is simple and straightforward.

Due to overexcitation the machine power factor is leading. The motor is utilised less. The phase control on the line side converter for current control in the dc link causes the power factor to become poor at retarded angles of firing. The cost of the inverter is medium, due to absence of commutation circuit. The drive has moderately good efficiency and is popular as CLM in medium to high power range.

Voltage spikes during commutation occur in the terminal voltage. These depend on the subtransient leakage reactance and affect the insulation of the motor also. The motor must have damper windings to limit the voltage spikes. Application of this type of drive is in gas turbine starting, pumped hydroturbine starting, pump and blower drives, etc.

4.2.6 Current Source Inverter with Forced Commutation

As has already been discussed, the disadvantages of machine commutation are (a) limitation on the speed range (b) the machine size is large and due to overexcitation it is underutilised.

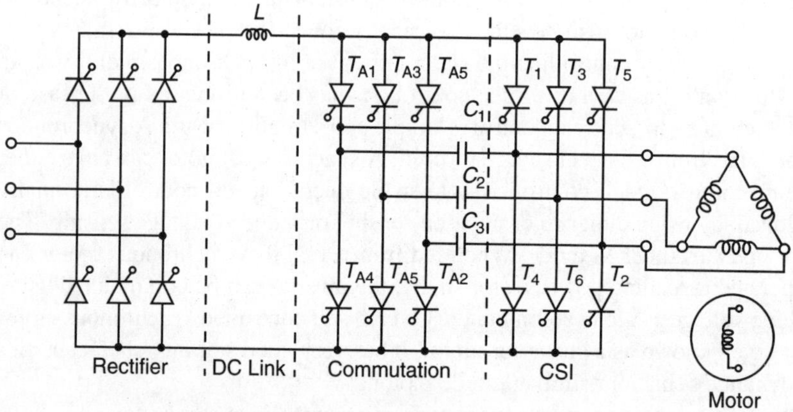

CSI with individual commutation for feeding synchronous motor.

(i) Used for providing commutation at low speeds

(ii) Synchronous motor may be operated at UPF

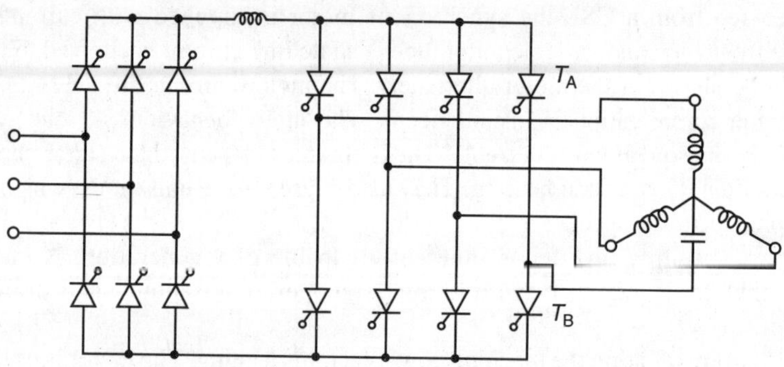

Third harmonic ASC CSI for use with synchronous motor at low speeds for starting.

Fig. 4.45(a) *Alternative CSIs for feeding synchronous motor. These are also used to provide commutation assistance at low speeds until the machine commutation is effective*

Sometimes it may be required to provide forced commutation for the inverter. Obviously the speed range can be extended from zero to base speed. The discussion regarding regeneration, harmonics and torque pulsations hold good for this case also. The line power factor is poor. However the machine is operated at UPF to obtain the advantages already discussed. The cost of the inverter increases due to forced commutation. The efficiency of the drive is good and it is popular as a drive in the low to medium power ranges in CLM mode. The drive cannot be operated in open loop. Stability improves in CLM mode. The problem of voltage spikes is present here also and commutation is simple.

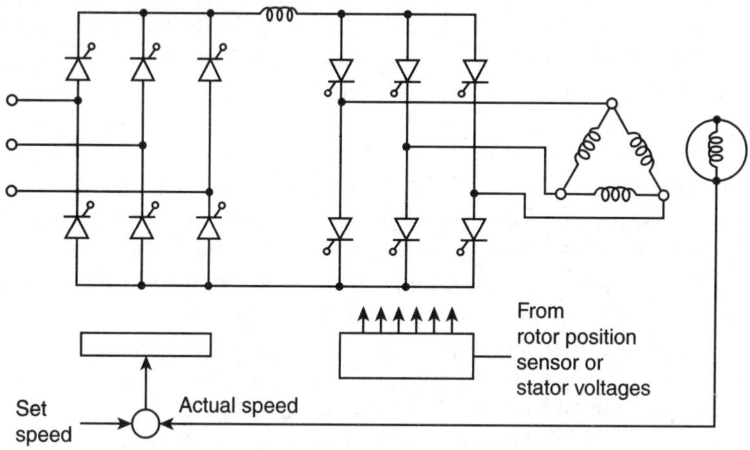

Fig. 4.45(b) *Self controlled synchronous motor utilising load commutation over exciting the field*

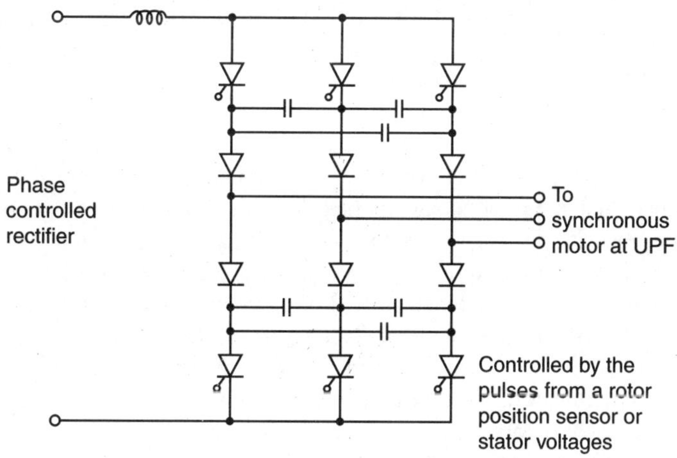

Fig. 4.45(c) *CSI for self controlled synchronous motor using forced commutation at UPF*

4.2.7 Starting Schemes for LCI Fed Synchronous Motors

It has been pointed out that a synchronous motor, even though complicated in its construction, has an advantage that it can provide the necessary reactive power required by the inverter, except at low speed. Among all the drives possible with synchronous motor, LCI fed synchronous motor drive is popular in CLM mode, and is known as converter motor. At low speeds, i.e., speeds below 10% of base speed, commutation should be assisted. Several methods are employed for starting and bringing the motor to a speed where load commutation can take over.

i. Forced commutation can be employed at very low speeds. The commutation circuit can be switched off when the speed is reached at which machine voltages are capable of commutating the inverter. As forced commutation is required only at low speeds, the size of the commutation circuit is relatively small. A CSI using individual commutation is very suitable and is shown in Fig. 4.46(a).

ii. Forced commutation at low speeds can be obtained by means of a fourth leg of the inverter containing auxiliary thyristors (Fig. 4.46(b)). A commutating capacitance is connected across the start point and common point of the two thyristors. At low speeds the voltage of capacitance is used to quench the thyristors. When the motor attains the speed at which machine commutation can take place the fourth leg is cut off. The inverter is called third harmonic commutated inverter and its operation has been discussed in detail in Chapter 3.

iii. DC Link current interruption can be employed to achieve the operation at low speeds. This is a kind of artificial quenching. The dc link current is interrupted at the instant of commutation and at the same time the line side converter is controlled so that it goes to inversion from rectification. The rotor position sensor sends information to the control unit of the machine side converter to block the firing pulses to the outgoing thyristor and provide them to the incoming one. The dc link voltage has changed its polarity due to transition of line side converter. Consequently the dc link current decays to zero and is maintained at zero value for a time greater than the turn off time of the thyristor. After this dead zone the line side converter is again made a rectifier. The dc link current builds up and flows through the machine phases via the new thyristor. There is a similar sequence if operations take place in the other commutations.

The interreption of link current to zero at the instant of commutation and its building up to the reference value after commutation are delayed by the link inductance. When the inductance is large and frequency is high the current may not even reach its reference value by the time the next commutation starts. To make the current variation faster, a thyristor is placed across the link inductance. Only machine and line inductance will be effective during variation of current. This thyristor is fired just at the instant when the zero current should exist. At the end of commutation, when the link voltage changes its sign it automatically ceases conduction. The link current flows through the inductance and the new phases via the thyristors.

The schematic diagram of connections of this method are shown in Fig. 4.46(c). The line side converter receives its firing pulses from CUI, whereas CU II

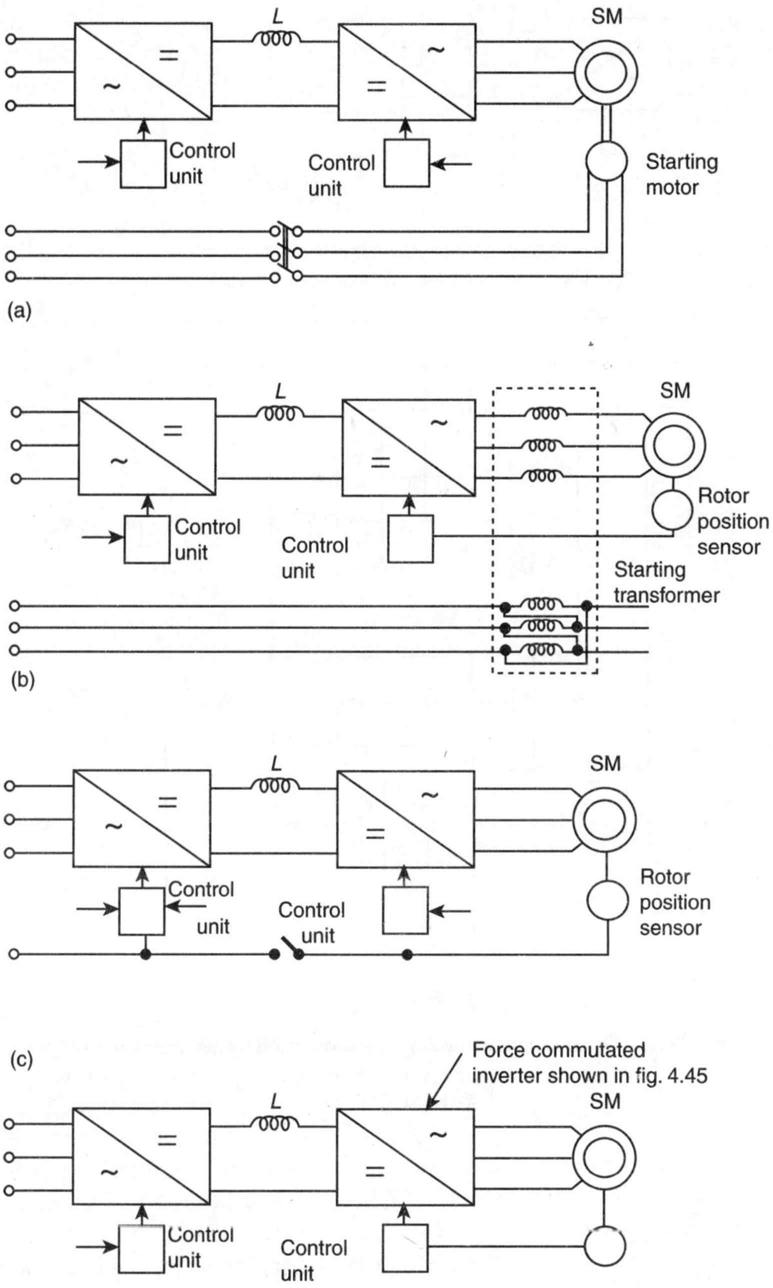

Fig. 4.46 *Starting schemes for a LCI fed synchronous motor*

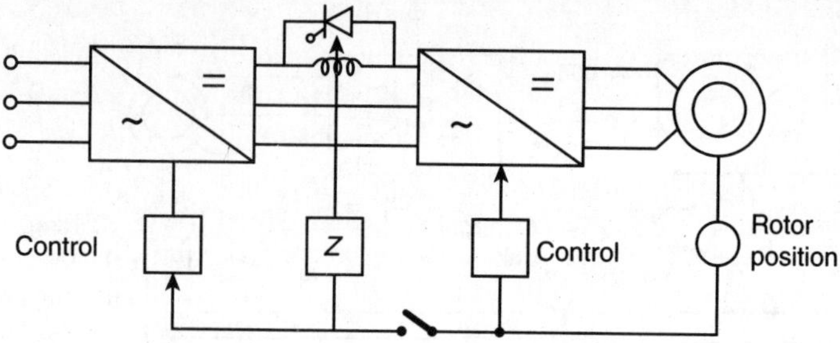

(d) Use of a controlled thyristor across dc link inductor.

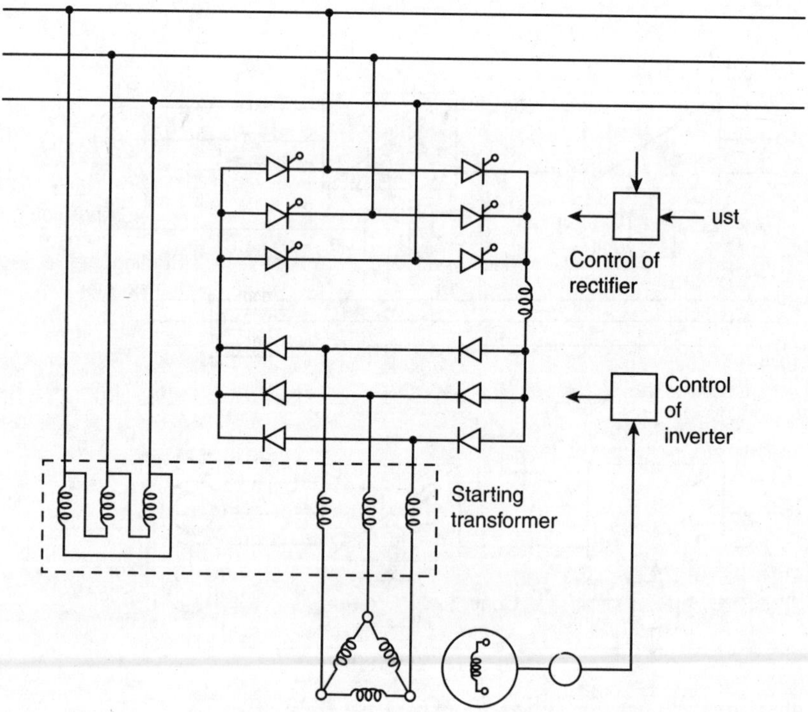

(e) The commutation voltages are provided by the transformer.

Fig. 4.46 *(Contd.)*

provides firing pulses to the machine side inverter. The current controller provides reference current. The thyristors *T*21 and *T*26 are fired in the inverter. The machine current builds up to the reference value. The rotor starts to move and after it has moved by the set amount the rotor position sensor sends information to block the firing pulses to *T*21 and provide them to *T*22. This is to advance the stator mmf by 60°. At the same time the line side converter starts working as an inverter. The link voltage changes its polarity. The link current is interrupted. The decay of current to zero is made faster by short circuiting the link inductance. The current through the outgoing thyristor is zero and it is maintained for a time greater than turn off time to ensure its forward blocking capability. After this time, the line side converter is changed to rectification. The link voltage polarity changes. The current builds up to reference value. The thyristor across the inductance ceases to conduct. The thyristors *T*22 and *T*26 conduct till the next commutation in the sequence starts, i.e., after a definite interval when the rotor has moved by 60°, *T*24 is fired and a similar process takes place to commutate *T*26 to *T*24.

iv. *Use of an additional transformer* A method suggested by E. Kubler, making use of an additional inductive ac voltage for commutation of current from one thyristor to the other can be used. The details are given in Fig. 4.46(d). The process is explained assuming that the thyristors *T*21 and *T*26 conduct to start with. When the rotor rotates by the desired angle the pulse to *T*21 is removed and the thyristor *T*22 is fired. When there is no induced voltage, the current divides between *T*21 and *T*22. This induced voltage is introduced by means of a transformer. A voltage is induced in the current carrying winding of the auxiliary transformer. This voltage is short-circuited due to the conduction of *T*21 and *T*22 and acts as a commutation voltage. This decreases the current to zero in *T*21 and increases current to dc link value in *T*22. The thyristor *T*21 is quenched and the current commutates over to *T*22.

In the next commutation the thyristor *T*24 in the other half of the bridge is fired and *T*26 is commutated to *T*24. In this way in a sequence commutations take place to accelerate the rotor. When the desired speed is reached the machine commutation takes place. The primary of the transformer is disconnected from the line and secondary is short-circuited.

For satisfactory commutation the machine frequency must be smaller than the transformer frequency. This condition is satisfied because the commutation assistance is required up to 10% of rated frequency. The rating of transformer need be only 10% of machine rating.

v. Sometimes when the power is small an auxiliary motor can be used to run up the synchronous motor to the desired speed.

Among the methods discussed above the methods using forced commutation are universally applicable ones. They are suitable for high speed motors having high starting torque. They make use of the current handling capacity of inverter thyristors and provide symmetrical currents during starting.

The other methods can be employed when these qualities are not of very much significance.

4.3 DC DRIVES

Among the electric motors, the dc motors are very versatile in that they provide a smooth speed control over a wide range. They have been very widely used in the industry as variable speed drives. The main disadvantage of these motors is the presence of a mechanical commutator which limits the maximum power rating and the speed. With the advent of thyristors the static Ward Leonard systems have been developed for one quadrant, two quadrant and four quadrant operations. The following sections deal with the aspects of dc drives fed from thyristor power converters.

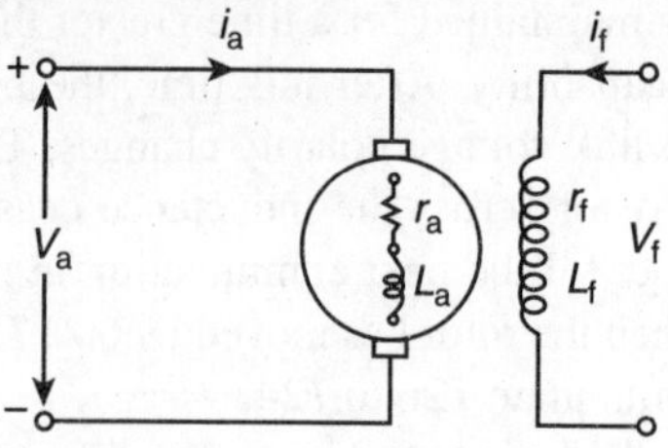

(a) Seperately excited dc motor

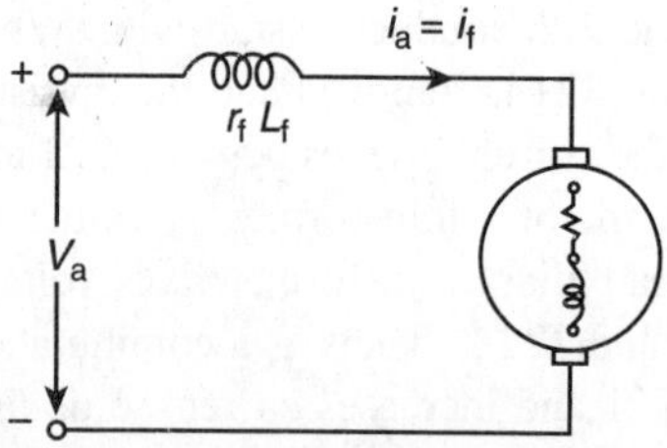

(b) Series excited dc motor

Fig. 4.47 *Circuit configurations of dc motor*

4.3.1 Basic Machine Equations

When dc voltage is applied to the armature of a dc motor with its field excited by dc, a torque is developed and the armature rotates. It accelerates to a speed at which the emf induced in the armature conductors balances the applied voltage and the following equation is satisfied.

$$V_a = I_a r_a + E_b \tag{4.23}$$

E_b is the induced emf in the armature by virtue of rotation in the uniform magnetic field at constant speed N. The nature of this emf is to oppose the terminal voltage and therefore it is also called back emf. The circuit conditions are shown in Fig. 4.47. Figure 4.47(a) depicts a separately excited motor where the excitation is provided by a current independent of I_a. Figure 4.47(b) depicts a series motor in which armature and field currents are the same. The generated voltage is given by

$$E_b = \frac{\phi Z N}{60} \cdot \frac{2P}{2a} V \tag{4.24}$$

where ϕ is flux per pole
Z is number of armature conductors
N is speed in rpm
$2P$ is number of poles
$2a$ is number of parallel paths in the armature

For a given machine

$$E_b = K_e \phi n = k_a \phi \omega \tag{4.25}$$

where $\omega = 2\,\pi\,n$

$$k_e = \frac{z}{za} \cdot 2p \text{ and } k_a = \frac{k_e}{2\pi}$$

The torque developed by the armature is given by

$$T_d = \frac{1}{2\pi}(2p\phi)\frac{I_a}{2_a} \cdot Z\ \mathrm{Nw}\ m = K_a \phi I_a \tag{4.26}$$

K_a is called armature constant of the motor.

Using Eqs 4.23 and 4.25 we have

$$n = \frac{V_a - I_a r_a}{K_e \phi}$$

In a separately excited motor ϕ can be assumed to be constant at a given field current when armature reaction is neglected or compensated.

From eq. 4.27 it is clear that the speed of a dc motor can be varied by changing the value of $(V_a - I_a r_a)$.

This is normally is achieved in two ways:

i. By varying the value of applied voltage to the armature, as is done in Ward Leonard system (Fig. 4.48).
ii. By inserting an extra resistance in series with the armature (Fig. 4.49).

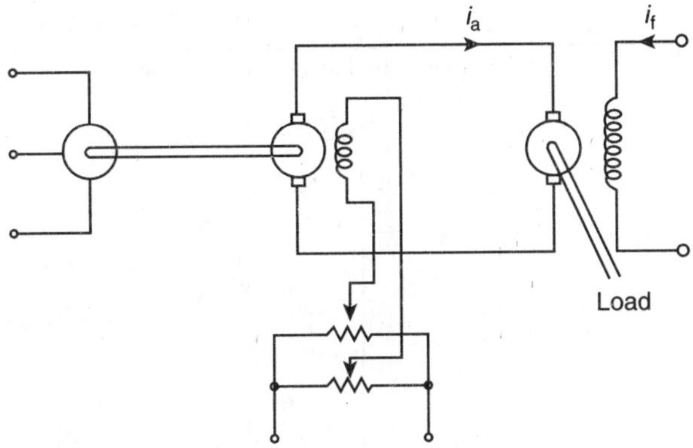

Fig. 4.48 *Conventional Ward Leonard control*

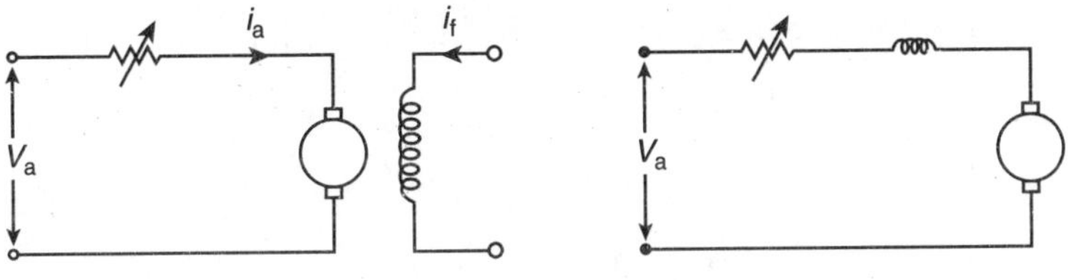

Fig. 4.49(a) *Armature resistance control*

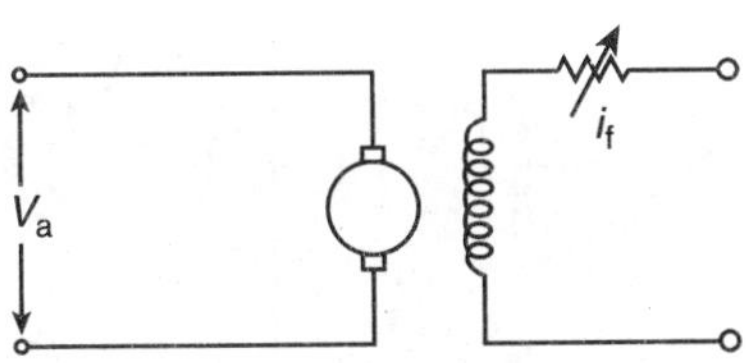

Fig. 4.49(b) *Field control*

In the former, a variable voltage supply is required. The operation is loss free and efficient. The torque versus speed characteristic is shown in Fig. 4.50. In the latter, extra resistance inserted involves extra losses and the motor becomes less efficient. The speed-torque characteristic is shown in Fig. 4.51. In both the cases

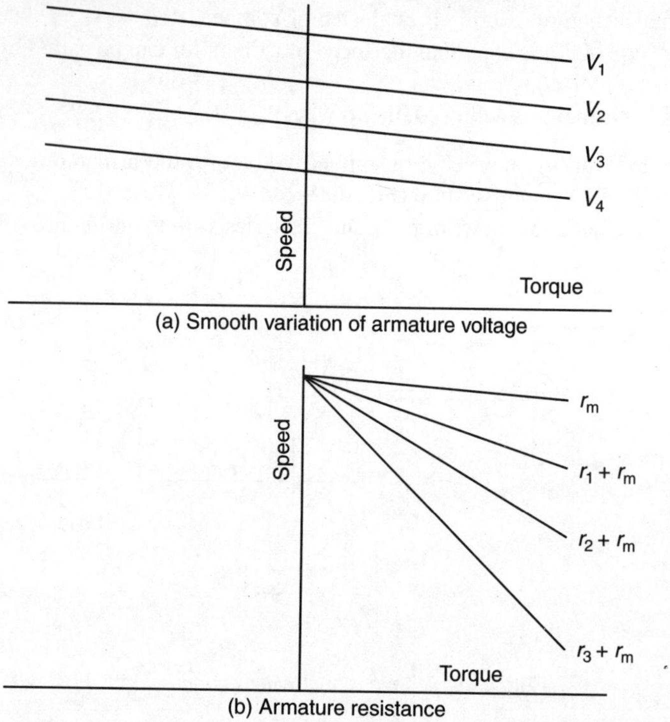

Fig. 4.50 *Torque-speed curves of a separately excited dc motor*

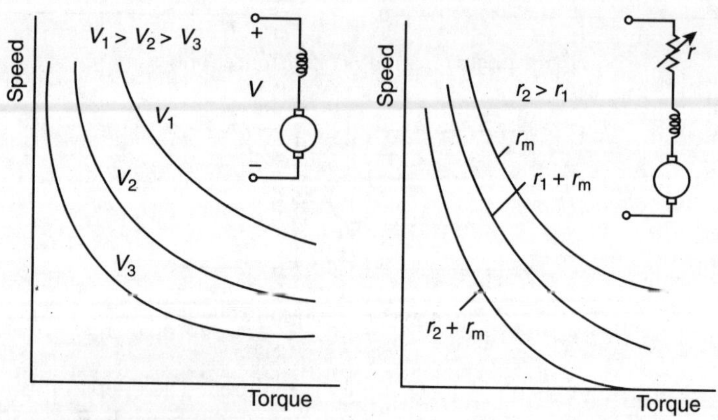

Fig.4.51(a) *Torque-speed curves of a series motor for different armature resistances or variable armature voltage*

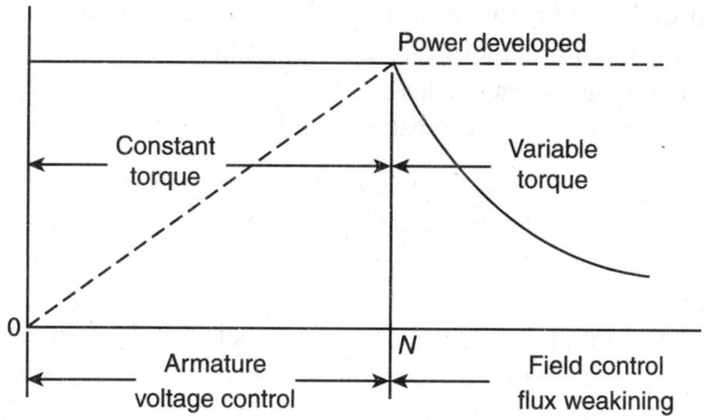

Fig. 4.51(b) *Modes of operation of separately excited dc motor*

very small speeds up to zero speed are possible. The motor operates at constant torque from zero to base speed.

The speed of a dc motor is also varied by variation of field flux. Speeds above base speeds are possible in this method of control. This method is suitable for variable torque loads. The operation of a dc motor with variable speed is depicted in Fig. 4.52.

Speed control by means of a variable voltage is very efficient. This variable voltage is obtained from a generator. The method is known as the Ward Leonard method. It is very versatile and has the following features:

i. The motor can be accelerated at constant torque (constant armature current) by suitably adjusting the field of the motor.
ii. The regenerative braking of the motor is possible and the motor can be brought to rest very fast.

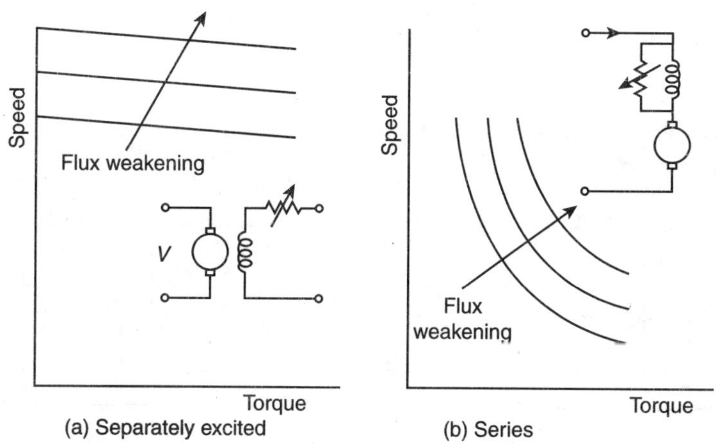

Fig. 4.52 *Speed torque curves of dc motor with flux variation*

iii. Four quadrant operation of the motor is straightforward.
iv. The capital cost is more, as an extra M G set is required.
v. The overall efficiency is less.
vi. The equipment requires space.

The advent of thyristors and the development of power converters using thyristors has made the speed control of electric motors easier and more straightforward. Phase controlled rectifiers provide a variable dc voltage to the armature of a dc motor (Fig. 4.53). These converters can also operate as inverters, in which case regenerative braking of the dc motor is possible. By suitable connection of these converters, a reversible drive allowing motoring and regeneration in both the directions of rotation is possible. These converters are more flexible, have a faster response and occupy less space. A static Ward Leonard system can be made possible using these converters. However, there are certain drawbacks, such as poor power factor on the ac side due to lagging current, non-sinusoidal input current having rich harmonic content, etc. The load voltage is superimposed by a ripple content. The load current also has a ripple making the ratios of peak to average current and rms to average current greater than one. A large inductance is required on the load side to smoothen the current. Sometimes the load current is discontinuous, which reduces the performance. The ripple content affects the motor heating and commutation. Therefore it is necessary to develop the rectifier which provides a supply to the motor causing only very little variations in the

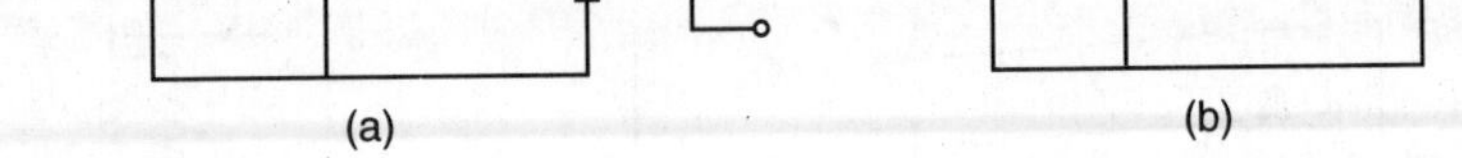

Fig. 4.53 *Phase controlled rectifier feeding a dc motor (a) Separately excited (b) Series motor*

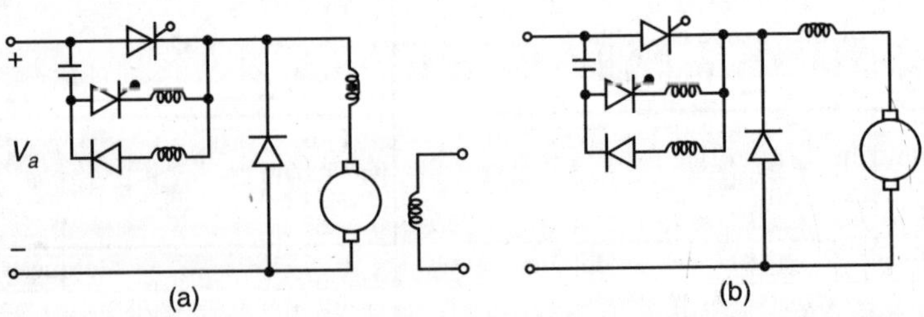

Fig. 4.54 *DC chopper feeding dc motor (a) Separately excited (b) Series motor*

performance as compared to a normal dc supply. Sometimes the dc motor design is modified so that it can be used on any converter.

A dc motor on normal dc supply requires a starting resistor to limit the starting current. When fed from a thyristor converter the starting resistance can be dispensed with and soft starting is possible. The motor can be accelerated at constant torque.

The dc chopper can be interposed between dc mains of fixed voltage and the dc motor provide a variable voltage to the motor to control the speed. Regeneration is possible in this case also. A static Ward Leonard scheme is possible for two and four quadrant operation. The output voltage of the chopper is in the form of pulses. The time ratio of the chopper can be controlled to vary the average voltage. The output current varies exponentially during T_{ON} and T_{OFF} of the chopper. The output current varies between two limits. There is a possibility of discontinuous conduction if the ripple is more and the load current is small. The chopper can also be controlled with specified current limits, the method being known as current limit control. This fixes the chopper frequency. The ripple content can be limited by proper choice of the limits. The current limit control is less prone to discontinuous conduction. The increased chopper frequency introduces losses. Soft starting and acceleration are possible. Since the battery is supplying the power the problems of harmonic content and power factor are absent.

The performance of a dc motor when operating on phase controlled converters or dc choppers differs very much from the performance when operating on a normal dc supply. The ripple content of the load current affects the motor performance, whereas the harmonics and poor power factor affect the line performance. An understanding of the behaviour helps in improving either the converter or motor design, to achieve better performance.

4.3.2 Performance of DC Motors Operating on Phase Controlled Converters

The nature of the output voltage and output current of a phase controlled converter impains the performance of the dc motor operating on these converters. The output of the converter is in the form of pulses. The average voltage is superimposed by an ac ripple which results in a deterioration of the operation. The motor current also has ripple content. The torque produced is given by the average current whereas the rms current influences the copper loss and heating. The rms to average value of the motor current is greater than one. So, for a given armature heating the torque capability of the motor decreases. The ratio of peak value to average value of the current is also greater than unity. The peak value affects the commutating capability of the motor. The ripple content also affects the commutation of the motor. The motor current is not always continuous; the ripple amplitude makes it discontinuous. Under this condition both load voltage and load current are zero. There is loss of torque and the performance deteriorates because of poor speed regulation. Discontinuous conduction also affects the commutation. The performance improvement can be obtained by a proper choice of the converter, e.g., using a converter with increased pulses. Sometimes the motor design can be altered.

Discontinuous Armature Current and its Effects on the Performance of the Motor When a dc motor is fed from a phase controlled converter the current in the armature may flow in discrete pulses or it may flow continuously with an average value superimposed on by a ripple. The former is called discontinuous conduction and the latter is called continuous conduction. In the case of discontinuous conduction the load voltage is back emf and load current is zero. The converter output voltage (average value) is also more than what would occur if the conduction were continuous for the same firing angle. The control characteristics are valid only for continuous conduction. The voltage waveform must be known to determine the average dc voltage when there is discontinuous conduction. The discontinuous conduction has the following effects on the motor performance:

i. The motor has a large speed drop as it is loaded. Speed falls very fast and speed regulation is very poor. As the current flows in pulses there is a loss of torque. The determination of the speed-torque characteristic is involved. For impact loads there is a substantial drop in the speed. An additional inductance in the load circuit improves the performance.
ii. The range of speed control is limited and speed oscillations are present.
iii. The ratios of peak to average and rms to average currents become more. The former deteriorates the commutating capability whereas the latter results in increased heating of the motor. For rated armature current to develop rated torque, the motor gets overheated; for normal heating there is loss of torque.
iv. It affects the commutation by increasing bar to bar voltage of the commutator.
v. The dynamic response of the motor is very poor.

A sufficient value of inductance in the load circuit improves the performance of the motor.

4.3.3 Commutation of the Motor

Commutation and commutator wear and tear of the dc motor must be given due consideration when the motor operates on solid state converters. Spark-less commutation must be aimed at. The reactance voltage and dynamically induced voltage in the coil undergoing commutation affect the rate of change of current in the coil. Interpoles are provided to establish a flux to compensate for the above voltages and to achieve sparkless commutation. The commutation is sparky when the field is weak, the load current is high and the speed is large. The dc motor must be used on power converters with a caution because commutation problems occur. These problems are due to

i. ripple content of the armature current.
ii. discontinuous conduction.

The commutation of a dc motor fed from a normal dc supply and rectified power supply is shown in Fig. 4.55.

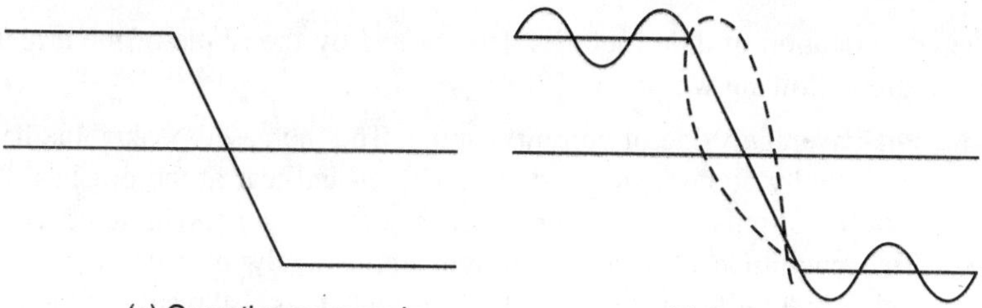

(a) Operation on pure dc (b) Operation, rectifier

Commutation of current in a dc motor.

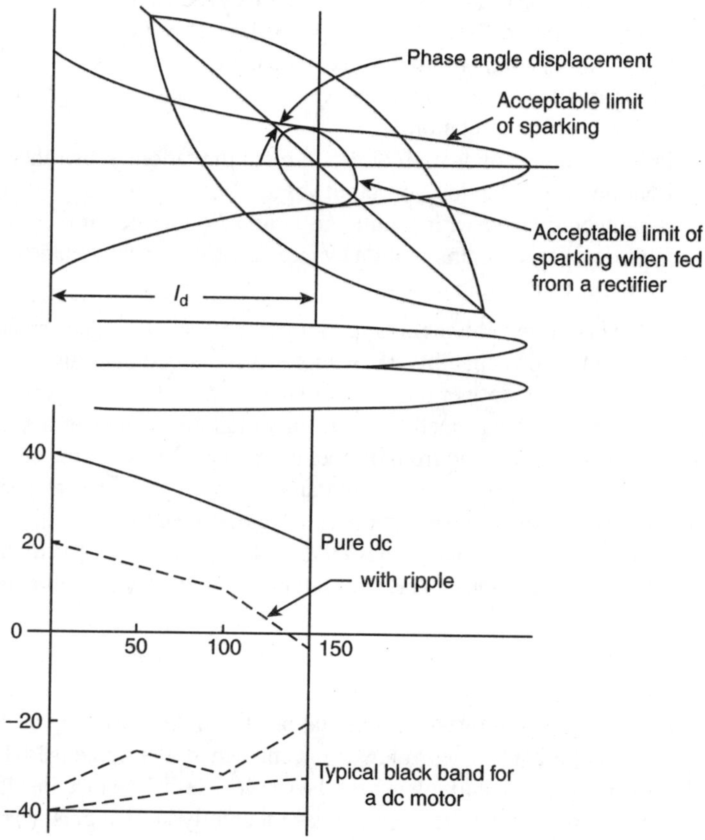

(c) Pertaining to black band

Fig. 4.55 *Commutation of a dc motor fed from a rectifier*

The commutation of a dc motor is deteriorated by the ripple of the armature current in the following ways:

i. Peak/average value of current is more. The increased peak value has to be actually commutated and causes wear and tear at the brushes if the number of pulses occurring per brush is more than one in number.
ii. The magnitude of armature current immediately after commutation is rather indefinite due to slope of the current. This also results in indefinite value of time when the commutation bar leaves the brush.
iii. The ripples produce eddy currents and associated fluxes in the interpoles. These additional fluxes react with the flux produced by average current. There is a phase shift in the interpoles flux. This modified flux is not in a position to compensate the reactance voltage.
iv. The armature current ripple increases the reactance voltage of the coil undergoing commutation.
v. Discontinuous conduction affects the commutation. When there is discontinuous conduction the output voltage of the converter is greater than what would occur with continuous conduction at the same firing angle. This deteriorates commutation by increasing bar to bar voltage and peak/average current ratio.

The machine is more vulnerable to sparking if the peak/average current ratio is more. The commutation capability of the motor is measured in terms of the width of the blackband. The narrower the band, greater the possibility of sparking at the brushes. This band is determined by a test in which the current in the interpole winding is bucked and boosted from its operating point until sparking occurs. The limits of this current represent blackband and can be represented as per unit of rated current. The wider this band, more satisfactory the commutation. The band width decreases when the armature has increased ripple and speed is high.

Satisfactory commutation can be achieved in a dc motor by the following modifications:

i. The peak value of the armature current can be controlled by increasing the pulse number of the converter. Additional inductance in the armature circuit helps improve the commutation. The inductance may be designed such that it contributes at lower currents where the ripple is high. It gets saturated at high currents where the ripple is less. The machine itself can be designed so that its armature has inherently large inductance. This is done by making the field weak, armature large and air gap small.
ii. The commutating capability of the motor can be improved by laminating the frame as well as the interpole body. These reduce the effective phase shift in the interpole flux due to eddy fluxes caused by the ripple content. With this type of motor a converter with less number of pulses can also be used.
iii. Discontinuous conduction is also reduced to a minimum by the additional inductance. The choice of inductance is made to avoid discontinuous conduction rather than smooth the output ripple.

Speed-Torque Characteristic A dc motor which is fed from a converter shows a very poor speed regulation compared to the operation on a normal dc supply. This is more pronounced at low speeds and low torques. The average current is representative of the torque developed. As has already been explained, with converter supplies average current is small for a given armature heating. Discontinuous conduction affects speed regulation. This can be improved by means of a FWD or a semi converter.

Line Side Performance The converter operation affects the line performance also. As the converter firing angle is retarded to get low speeds the power factor becomes poor. This is because of the reactive power requirement of the converter and also the harmonic content of the line current. The line power factor can be improved by use of semi converters and converters with FWD. An inductance on the load side also improves the harmonic content on the line side. The drive is not capable of regeneration.

4.3.4 Additional Losses and Low Speed Operation

The distorted armature current of the d.c. motor fed from phase controlled converter causes additional losses in the armature conductors. These losses are due to the increased RMS value of the armature current and also the possible skin effect of the armature conductors. The skin effect causes an increase in the effective resistance of the armature. The additional losses may be present in the interpole winding and compensating winding. The interpoles may also have additional iron losses.

However, the increase in the losses is significant only in respect of temperature rise but not efficiency. The increase in temperature rise may lead to a derating of the motor. The increase in the firing angle of the converter increases the ripple factor of current waveform. So, as the value of a increases these additional losses may be more and resulting derating is also more. Further the motor speed decreases as a increases. At low speeds the natural ventilation is poor. If there is no provision for additional ventilation, this will also cause a further derating of the motor. Therefore a d.c. motor operating on phase controlled converters cannot be fully loaded, particularly at low speeds.

4.3.5 Phase Controlled Converter Fed DC Drives

A detailed discussion has been given how the converter operation affects the performance. There are several types of converters which can be used for feeding dc motors. These were described in Chapter 3 in detail. To arrive at a suitable converter-motor system an evaluation of the performance of a dc motor operating on different types of converter may be required. A systematic description of these drives is given below. The parameters of importance are as follows:

i. on the motor side:
   (a) average current of the motor representative of torque developed
   (b) rms value of the current which represents the motor heating and losses

   (c) peak value of current which represents commutating capability of the motor
   (d) nature of current whether continuous or discontinuous
   (e) torque-speed characteristic, associated speed regulation, etc

ii. on the line side:
   (a) fundamental displacement factor
   (b) harmonic factor
   (c) total power factor of the system.

Single Phase Drives In these drives phase controlled converters operating from a single phase are used to drive the motor. These are used for low and medium power applications. These have inherently poor speed regulation with open loop control. This can be improved with closed loop operation.

These drives can be half wave or full wave drives. A full wave drive can be supplied from a fully controlled or half controlled converter. The former is capable of two quadrant operation whereas only one quadrant operation is possible with the latter, as it contains diodes in several positions. A dual converter can be obtained by a connection of two two-quadrant converters. Table 1 summarises these drives.

Single Phase Separately Excited dc Motor Drives The circuit arrangement of a separately existed dc motor drive fed from a single phase controlled converter is shown in Fig. 4.56. The block 1 can be any one of the single phase converters.

Half Wave Drives The dc motor is supplied from a single phase half wave circuit (Fig. 4.56(a)). Low cost and simplicity are the advantages of this drive. This can provide only one quadrant operation. Regeneration is not possible. The conduction angle of the thyristor is very small, resulting in a very low average current. The torque developed is very small resulting in a loss of torque at the rated rms current. The rms to average current ratio is also more. The motor current is always discontinuous. The ripple frequency is equal to supply frequency. The current and voltage wave forms are shown in Fig. 4.57.

The freewheeling diode across the load improves the performance. The speed regulation is very poor. At low speeds the motor receives power in pulses and the motor may chug when the load is high. The speed oscillation is quite high. The supply transformer has premagnetisation due to dc component of the load current. The application of this drive is limited to low powers.

Full Wave Drives The dc motor is supplied by a full wave converter. A full wave drive using a fully controlled converter is shown in Fig. 4.56(b). The wave forms of voltage and current in the load are shown in Fig. 4.58. Having thyristors in all positions, the firing of the thyristors allows negative average voltages, making the power flow from load to the supply. The motor can be braked effectively using regenerative braking. This can be made possible at constant current. As the load current flows both during positive and negative half cycles, the average value of

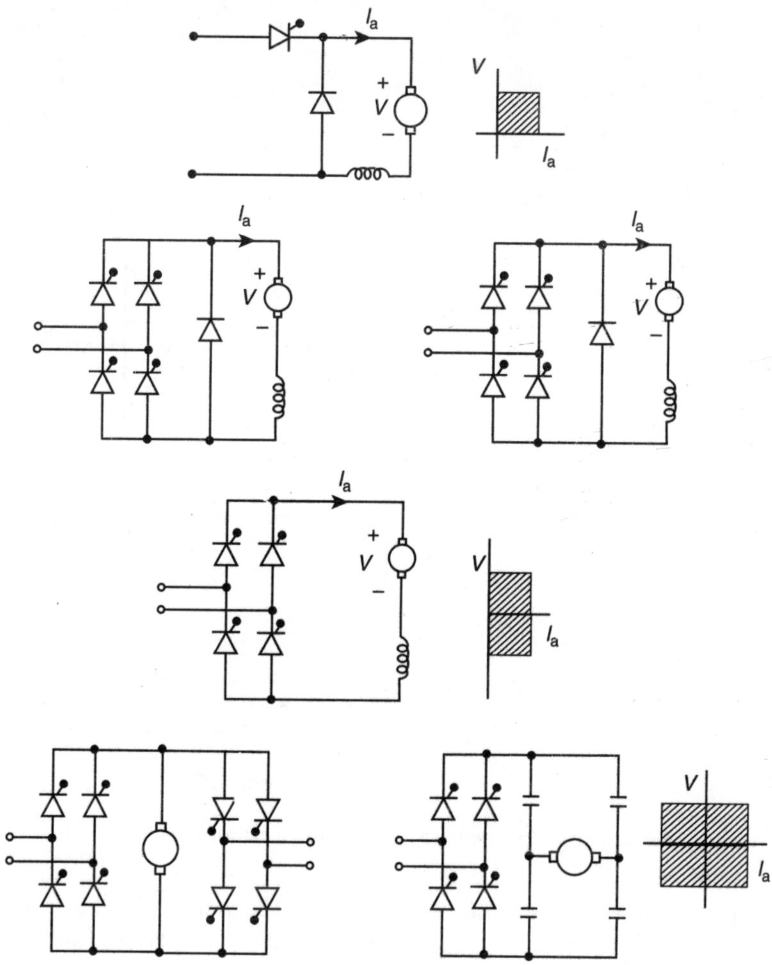

Fig. 4.56 *Summary of single phase drives using separately excited motor*

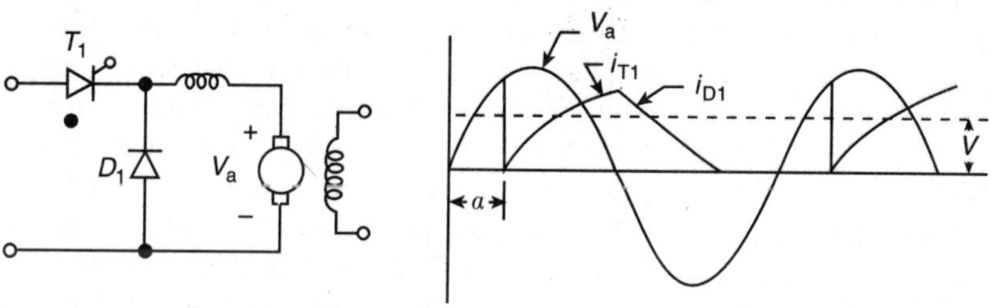

Fig. 4.57 *Half wave drive and its voltage and current waveforms*

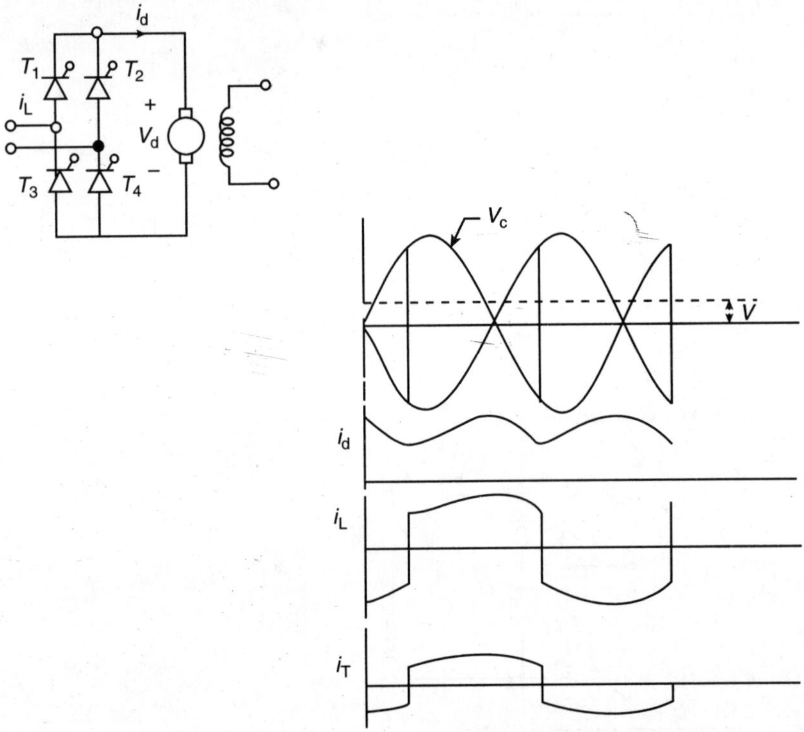

(a) Rectilication

Fig. 4.58(a) *Single phase full wave drive in motoring mode*

the current is more than that in the half wave drive. The torque capability increases for a given armature heating. The ratios of peak to average and rms to average are better here. The speed oscillation is less. The speed regulation improves because of increased conduction of current in the load. The number of pulses are two and pulse frequency in the load is $2f$. The ripple amplitude is smaller in this case. The speed oscillations are decreased. The discontinuous conduction is present and the wave forms for this case are shown in Fig. 4.59. This affects the speed regulation. An additional inductance in the armature improves the performance. It reduces the ripple content, reduces the possibility of discontinuous conduction, improves speed regulation, etc. This inductance also affects the performance of the drive on the line side. The harmonic content of the line current is more at lower values of inductance because of possible discontinuous load current. As the inductance increases the harmonic factor decreases. The peak value of current decreases with additional inductance. This improves the commutating capability. The losses and heating of the motor are less as the rms value of current decreases. The torque capability of the motor increases. As has already been stated the motor may be designed with a high value of armature inductance. Otherwise extra inductance

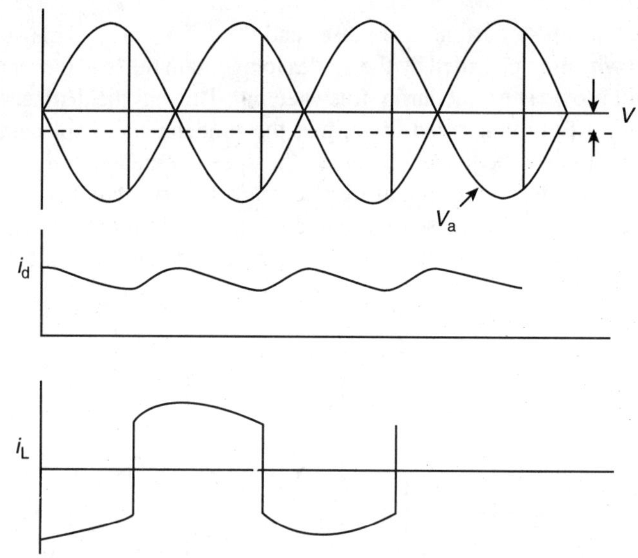

Fig. 4.58(b) *Full wave single phase drive during regeneration*

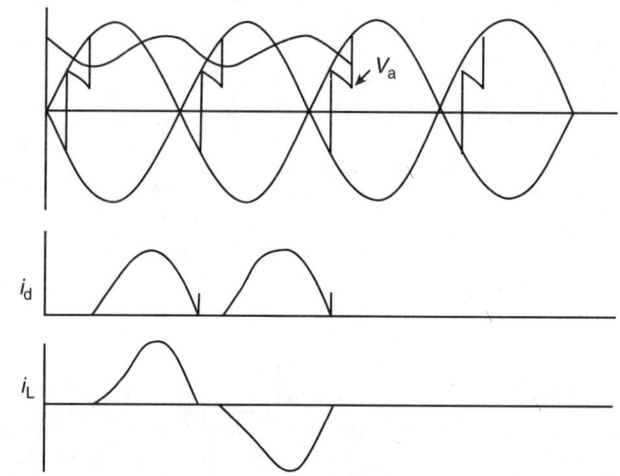

Fig. 4.59 *Voltage and current waveforms in discontinuous mode*

occupies space. Inductance has no effect on the fundamental displacement factor. There can be slight improvement in the power factor as the harmonic factor is affected by this inductance. The serious drawback with a fully controlled converter is poor power factor.

A full wave drive fed from a semi converter is depicted in Fig. 4.56(c). The wave forms of voltage and current are shown in Fig, 4.60. A reference to Fig. 4.60

shows that the supply current is in pulses and flows for $(\pi - \alpha)$ This is due to the natural freewheeling provided by the diodes to maintain the load current. The supply does not provide current during freewheeling. This has the distinct advantage, compared to a fully controlled drive, that the fundamental displacement factor of the line current is better. However, as the speed decreases, i.e., as the firing angle is delayed, the fundamental displacement factor decreases. A semi converter drive shows an increased harmonic content at low speeds. This is again due to the dependence of the width of the current pulse on firing angle. The peak

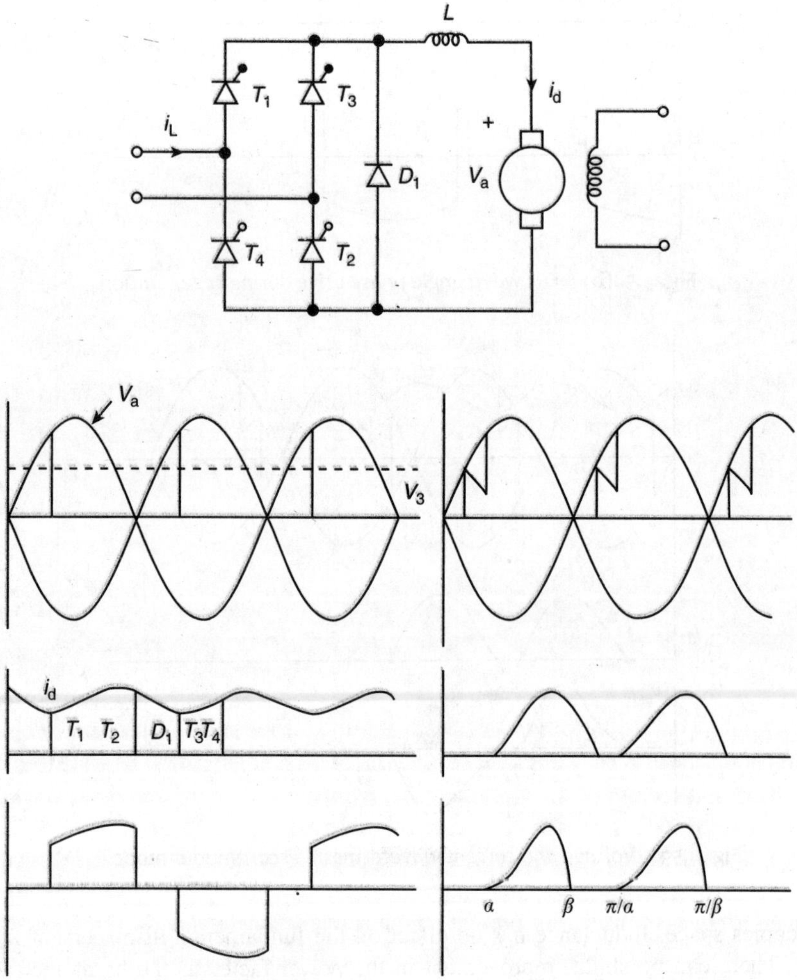

(a) Continuous conduction

(b) Discontinuous conduction

Fig. 4.60 *Voltage and current waveforms of a semiconverter circuit* value is less due to freewheeling. This improves the commutating capability of the motor, particularly at low speeds and light loads. The rms current is also less in a semi converter, which reduces motor heating compared to a fully controlled one. The heating of the motor is 44% less. Discontinuous conduction is present at light loads and low speeds and this causes a speed regulation poorer than the full converter. The freewheeling does not allow negative excursions of the load voltage and therefore negative average voltage is not possible. Hence, a semi converter is used with one quadrant drive where no regeneration is required. The advantages of improved power factor, better commutating capability and low cost due to diodes makes a semi converter drive applicable to all cases where regeneration is not required.

The saving of reactive power and hence improvement in the fundamental displacement factor can be achieved by a full converter with a freewheeling diode. The schematic diagram of the converter feeding a dc motor is shown in Fig. 4.56(c). The diode provides an alternative path for the load current and provides the same effects as in a semi converter. The freewheeling diode is effective in the complete speed range. The diode does not allow negative voltages of the load and hence there is no regeneration.

It is possible to have the regeneration mode besides the advantages mentioned above during rectification by having thyristors with optional freewheeling. Two thyristors ($T2$, $T4$ for symmetrical connection; $T3$, $T4$ for unsymmetrical connection) of a fully controlled converter are fired at $\alpha = 0$ so that they act as diodes to provide freewheeling during rectification (Fig. 4.61). During inversion they are operated as normal thyristors. The freewheeling diode is also replaced by a thyristor which is fired at $\alpha = 0$ when freewheeling is required and is blocked when it is not required. This optional freewheeling provides all the advantages of one quadrant converters as well as making inversion possible. However it may be noted that these advantages are not present during inversion.

Single Phase dc Series Motor Drives In the series motor the field is connected in series with the armature. The field current is the same as the armature current. The schematic of a series motor fed from a phase controlled converter is shown in Fig. 4.62. Series motors are capable of high starting torque and constant power operation at all speeds. They are used in traction, cranes, hoists, etc. The waveforms of voltage and current of a series motor fed from a thyristor converter are shown in Fig. 4.63. Both continuous and discontinuous conduction are considered.

It is worth noting the differences in the operation of series motors and separately excited motors while operating on power converters. In a separately excited motor there is always a back emf present which actually accelerates the decay of armature current, thereby making discontinuous conduction occur, particularly at low currents. Discontinuous conduction occurs over a wide range of operating conditions, on the other hand, in series motors; the back emf being proportional to I_a, does not contribute to discontinuous conduction very much. As the motor current

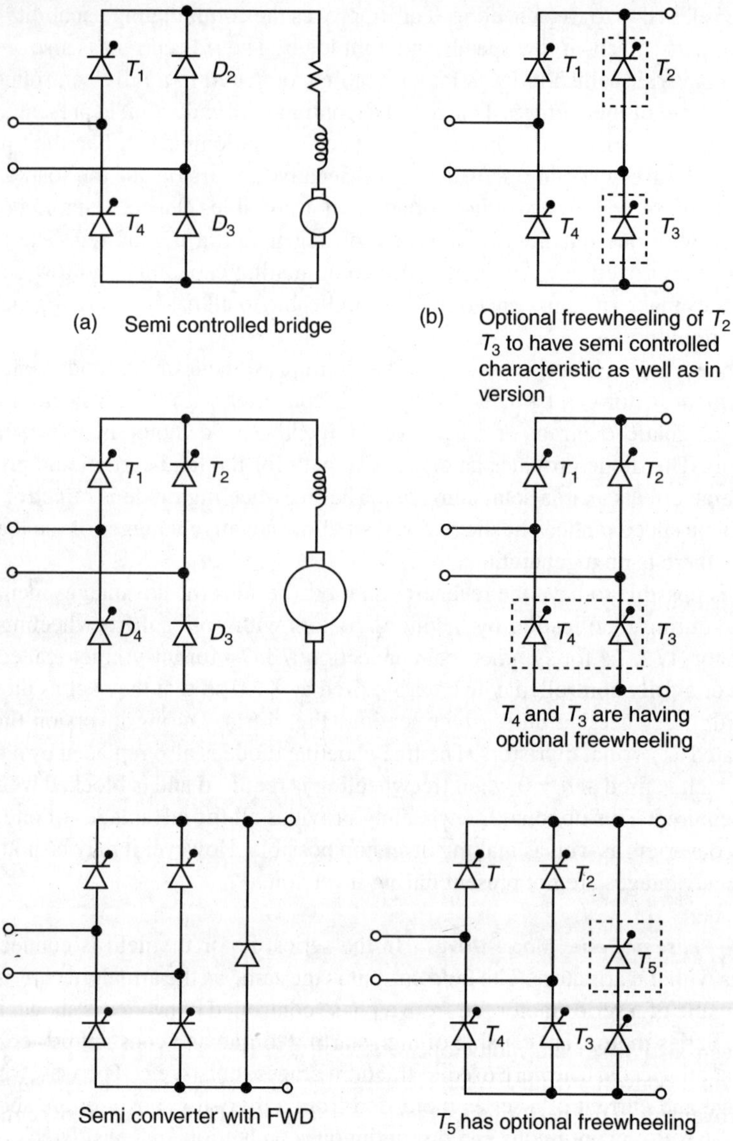

Fig. 4.61 *Optional freewheeling in semi converters and converters with FWD*

decreases it tends to become continuous, unlike in the case of a separately excited motor. Therefore the current of a series motor is continuous over a wide range of speeds. It may become discontinuous at high speeds and low currents of the motor. In this mode the motor terminal voltage is the same as back emf due to residual magnetism. The ratios of peak/average and rms/average currents of a series motor

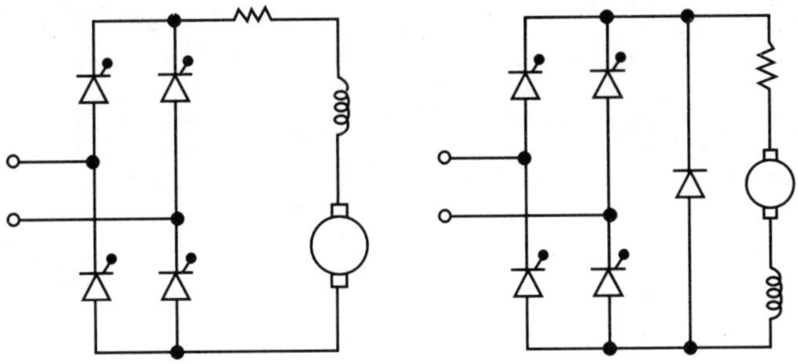

Fig. 4.62 *Series motor operating on full and converters*

are smaller than those of a separately excited motor.

Another point of difference, which is also a reason for better performance of a series motor compared to a separately excited one, is the smoothing effect provided by field inductance, which effectively increases the circuit inductance. If a further improvement is required in the performance, a small inductance is sufficient.

A comparison of series motor operation on full and semi converters shows that the current is continuous in almost the entire region of operation when operating on semi converters. The addition of a small inductance makes the current continuous over the complete region. This can be seen also when the motor operates on a converter with FWD. The improvement in the performance can also be observed with respect to p.f., and ratios of peak to average and rms to average currents. This is due to the free wheeling of the current through the diodes of a semi converter or through the FWD of the converter with FWD.

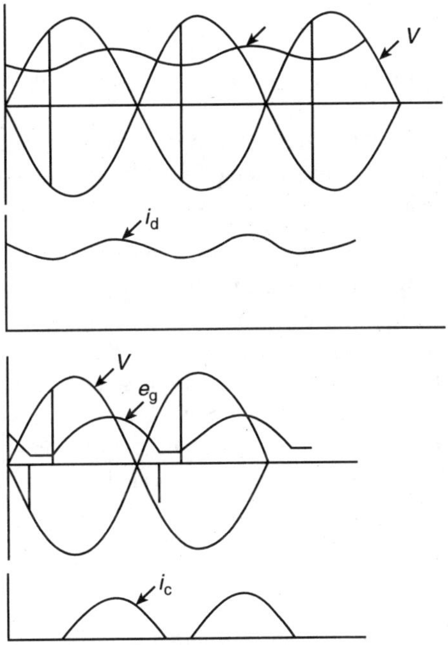

Fig. 4.63(a) *Voltage and current waveforms of full converter for continuous and discontinuous operation*

A series motor has a torque-speed characteristic making it suitable for constant power operation. However, when it is fed from a converter it does not exhibit constant power operation at all speeds for a given firing angle. To achieve constant power at all speeds, adjustment of firing angle is required. A comparison of constant power operation of a series motor operating on full and semi converters

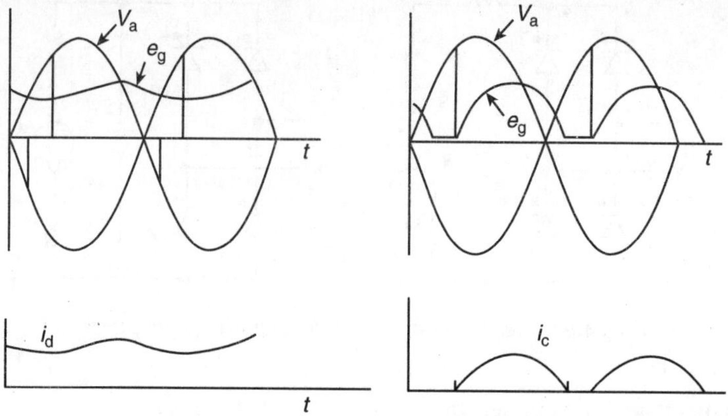

Fig. 4.63(b) *Voltage and current waveforms of semi converter feeding a dc motor*

show that it has better operation on semi converters with respect to power factor, discontinuous conduction, etc. Similar improvement in performance is possible using a full converter with a freewheeling diode.

Three-Phase Drives Single phase drives discussed in the previous section are employed for low and medium powers. When the power of the drive is very large the three-phase converter is preferred for supplying the load. As has been described in Chapter 3, increase in the number of pulses of the output voltage improves the performance. A three-phase converter can be used to give three or six pulses in the output. The output voltage has less ripple when the number of pulses increase. Therefore three-phase converters produce an output voltage with reduced harmonic content. This has an effect on the peak/average and rms/average ratios which effectively decrease. The filtering requirements are therefore less in the output circuit. The speed regulation of the drive improves and speed oscillations of the motor are small. In the light of the above, the motor performance will also be better in three-phase drives.

Three-phase drives can also be classified as

i. half wave drives
ii. full wave drives
iii. reversible drives

These are listed in Table 2.

Three-Phase Half Wave Drive A simple three-phase mp converter drive is shown in Fig. 4.64(a). The output ripple has a frequency of 3f. This is suitable for two quadrant operation. Among the three-phase drives this has higher peak/average and rms/average current ratios. These result in losses of motor and also affect the commutation capability of the motor. They are troublesome, particularly at low speeds, as the % ripple at low speeds is more. The self ventilation of the motor

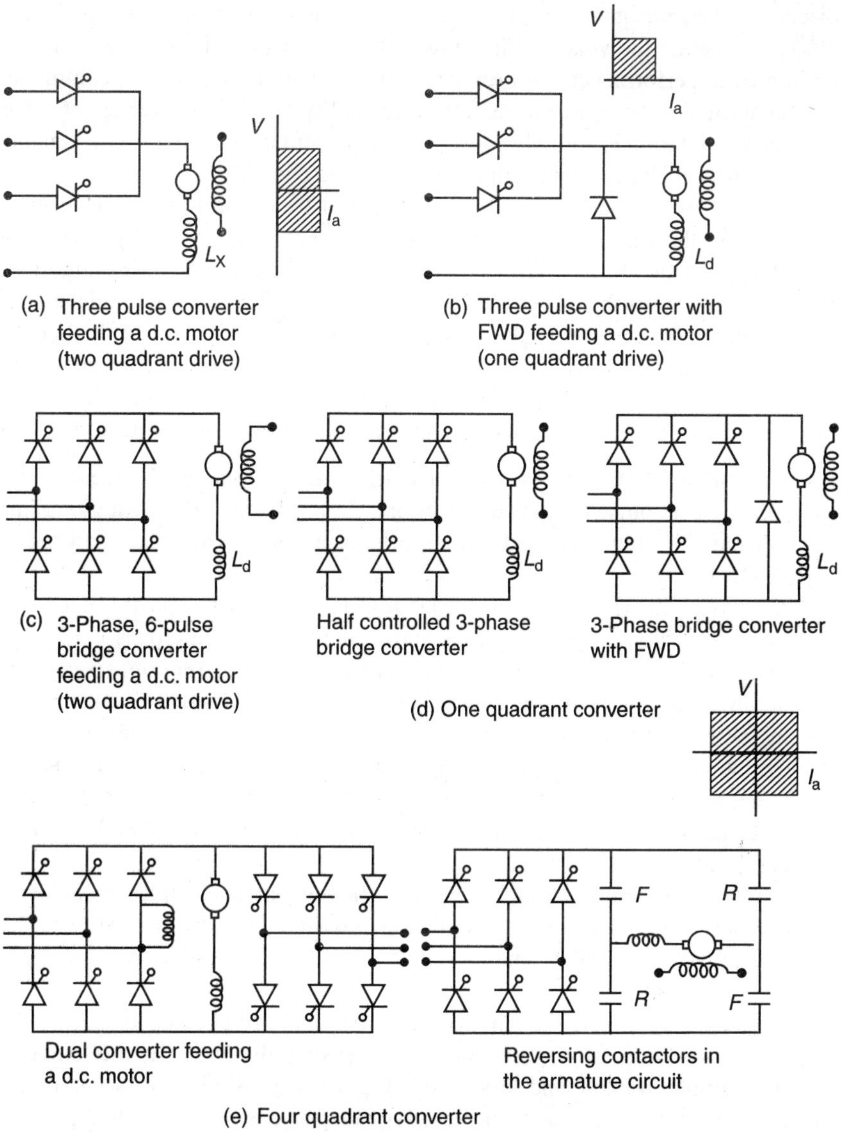

Fig. 4.64 *Summary of three-phase drives*

also decreases. This may necessitate forced ventilation at low speeds. Further the dc component of the load current premagnetises the transformer core and a big transformer is required. To avoid this dc magnetisation, special transformer winding connections may be required.

The performance of the motor may be improved by an additional inductance in the load circuit or by suitably modifying the dc motor, as has already been discussed, i.e., designing the motor with an inherently large armature inductance, laminating the yoke and interpoles. The motor in such a case would have an improved performance with respect to discontinuous conduction, commutating capability, dynamic response, torque capability for a given heating, etc. The regenerative braking is possible. However, the power factor is poor and the quality of p.f. is affected by the delay angle and harmonic content.

The drive performance can also be improved by connecting a freewheeling diode across the supply. The connections are shown in Fig. 4.64(b). This drive is preferred to the above if no regeneration is required. The freewheeling diode effectively improves the line power factor, besides providing the beneficial effects of decreasing the ratios of peak/average and rms/average currents of the load by providing a kind of smoothening of the load voltage. The armature heating is therefore less. The amount of inductance required in the load for filtering the harmonics is less. The load current is less prone to discontinuous conduction.

Full Wave Drive A three-phase full wave drive may be obtained by feeding a dc motor from a six pulse bridge converter or a six pulse converter with interphase transformer. The latter is used for large load currents. These are depicted in Fig. 4.64(c).

The increase of the pulses in the load is advantageous, as it effectively decreases the ratios of peak/average and rms/average currents. This improves the performance of the motor, and the filtering requirements are less. The current is continuous in almost all the operating points because of the faster rate of switching of the thyristors. The line side performance also improves because of less harmonic content in the line current. A slight improvement in the total power factor may be there. These drives have better speed regulation than single phase drives. A full converter allows negative average voltage at the load terminals thereby making regeneration possible.

To improve the line power factor, mainly the fundamental displacement factor, a semi converter bridge circuit is used (Fig. 4.64(d)). Half the number of thyristors are replaced by diodes. These provide natural freewheeling for firing angles greater than 30°; up to firing angle = 30° the converter output has six pulses. Because of a reduction in the output pulses, the output ripple increases. The peak to peak value of current is more than that in the six pulse case. But the semi converter is simple, reliable and less costly. There is a possibility of discontinuous conduction at light loads due to increased ripple, particularly at large firing angles. The fundamental displacement factor is better than that of a full converter. This improvement is slightly augmented by the harmonic content on the line side. The effect in the improvement in the total p.f. is not very much. The regeneration is not possible as the diodes do not allow negative voltage.

The semi converter bridge has only three pulses in the larger portion of the operating range. Even though this converter has better performance with regard to p.f. the peak to peak current increases and this is 3.11 times that of a six pulse converter(Fig. 4.65). The motor experiences more heating and the commutation tends to be sparky for a given torque.

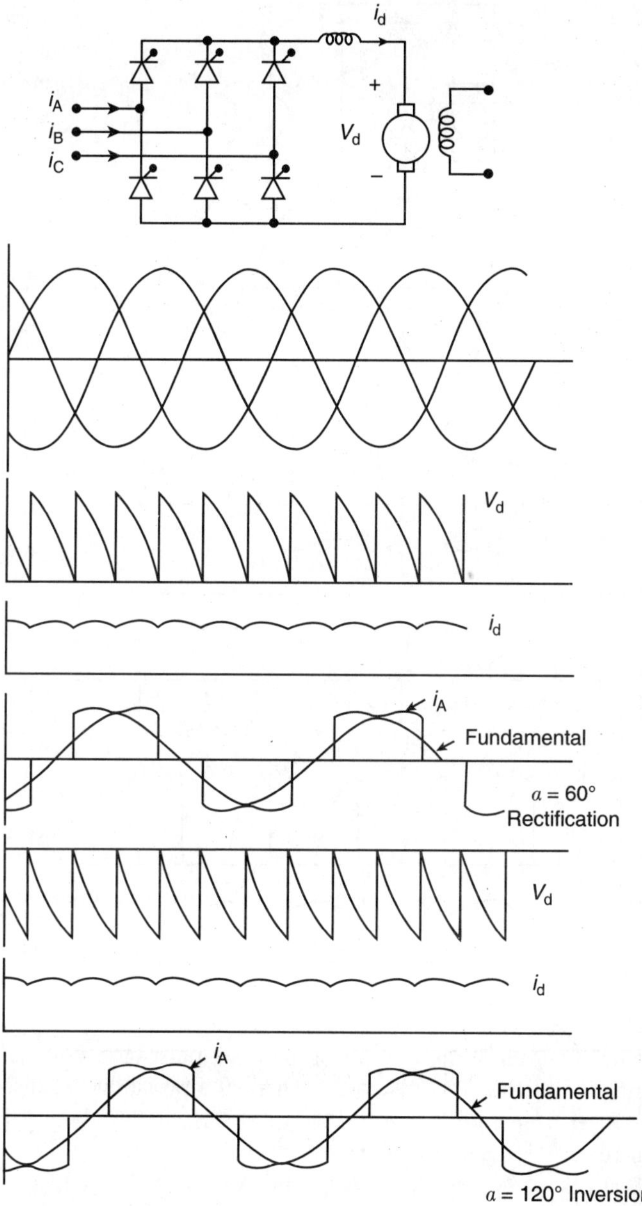

Fig. 4.65(a) *Voltage and current waveforms of a fully controlled three phase bridge feeding a dc motor*

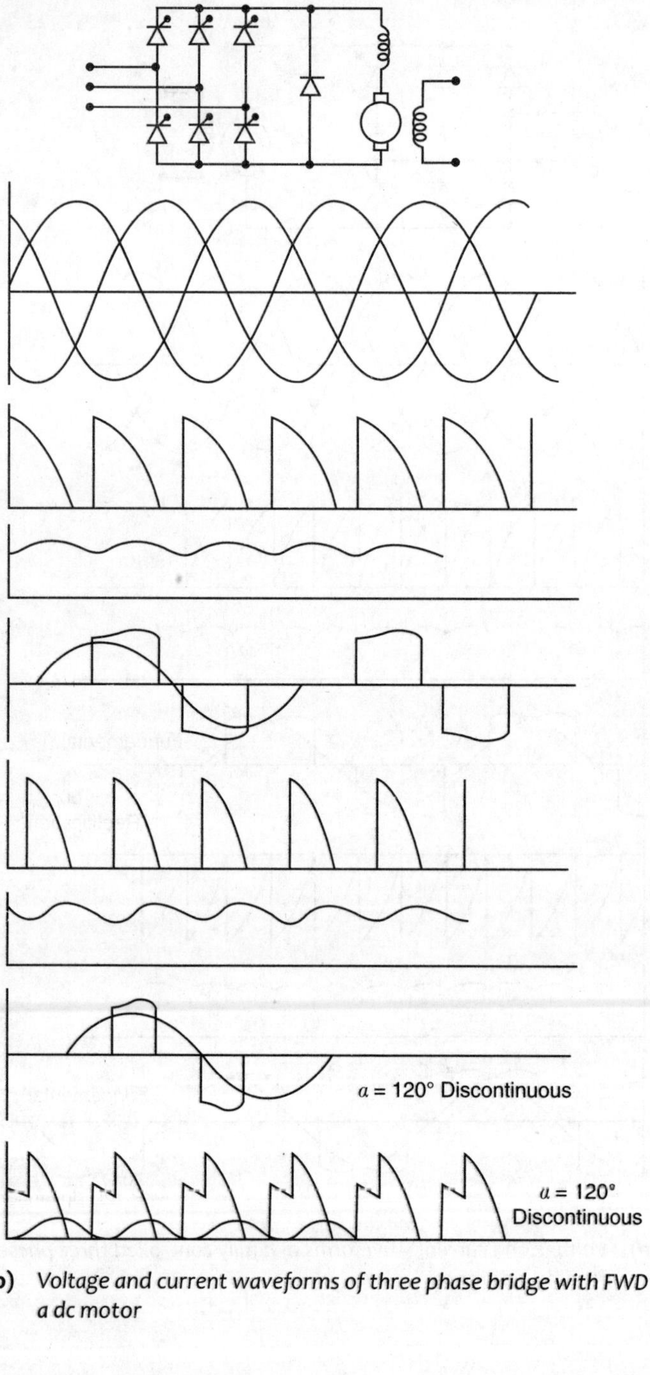

Fig. 4.65(b) *Voltage and current waveforms of three phase bridge with FWD feeding a dc motor*

As the converter is simple, reliable and less costly, it seems to be worthwhile considering the modification of the motor design or the ways of reducing the peak to peak current. An inductance in the armature circuit is effective in reducing peak to peak current, as it smoothens the ripple content of armature current. The additional inductance is bulky, increases the cost, occupies space and results in additional losses.

Because of space considerations, it is some times imperative to dispense with additional inductance. In such case the motor itself is modified in its design. (These modifications have already been indicated). These are laminating the yoke, interpoles, increasing the number of commutator segments and designing the armature with sufficiently large inductance. The laminated structures improve dynamic response also.

One quadrant operation can also be obtained by a freewheeling diode across the load. This also provides the advantages of a semi converter, but in a limited range of firing angles. This is because the diode participates in conduction only after $\alpha = 60°$. Up to this angle the advantages of freewheeling are not there. Therefore up to $\alpha = 60°$ it is six pulse converter. This circuit must be used with caution.

A Comparison of Three-Phase and One-Phase Drives The above discussions show that three phase drives have better operation compared to one phase drives. The drive motor has better commutating capability, torque capability, etc. Filtering requirements are less, speed regulation is better and line harmonics are reduced. A short comparison of three phase and single phase drive can be summarised as follows:

i. Three phase drives are used for large power ratings.
ii. The increased pulse frequency of the output ripple reduces the filter requirements as the peak to peak ripple content decreases. The amount of inductance required for smoothing is also less.
iii. Peak/average and rms/average current ratios in the load are smaller in three-phase drives. This tends to improve the performance of the drive motor. The outlay of smoothing inductance is also less. The heating of the motor for a given torque decreases compared to single phase drives.
iv. The commutating capability of three-phase drives is better than that of one phase drives. This is because three-phase drives are less prone to discontinuous conduction.
v. The line power factor is poorer in a full converter as the firing angle is delayed. This is because of a deterioration in the fundamental displacement factor as well as harmonic factor of the line current. Semi converters and converters with FWD improve the fundamental displacement factor. This improvement is possible in three phase drives only in a certain range of firing angles. On the other hand single phase drives offer this improvement in the complete range of firing angles. Also, in three phase drives when freewheeling comes into existence the number

of pulses reduces to three and the performance is poorer than full converter. Therefore the performance improvement in single phase drives is considerable compared to three phase drives.

4.3.6 Methods to Improve the Power Factor

We have seen that a phase controlled converter requires reactive power for control and commutation. The harmonics do not contribute to the active power loading but they contribute to the reactive loading of the line. Because of these the line power factor is poor. A semi converter has been found to improve the reactive power requirement which improves the line power factor. But it has a disadvantage that it cannot be used if regenerative braking is required. Optional freewheeling is a solution for this, but the control system becomes complex.

Several forced commutation methods are available to improve the performance of dc motors operating on phase controlled converters. These can be broadly divided into single firing and multiple firing schemes. In single firing schemes a thyristor is triggered once every half cycle and it is turned off using forced commutation. This means that the angle of quenching of the thyristor is also varied. These single firing schemes are extinction angle control (EAC) and symmetrical angle control (SAC) (Fig. 4.66). In normal phase angle control using natural

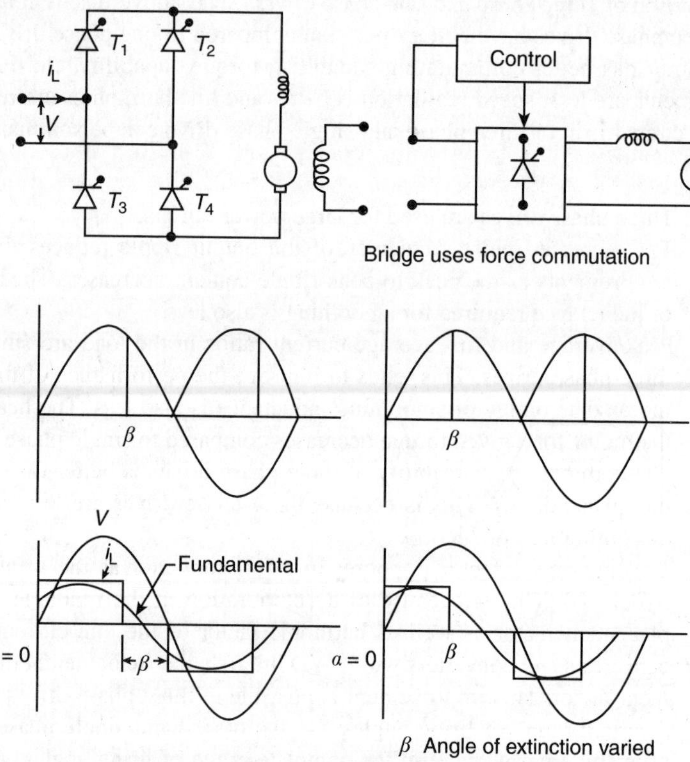

Fig. 4.66 *Principle of EAC voltage and current waveforms*

commutation, the turn off of the thyristor takes place naturally. The drive performance has been discussed in detail. In EAC a thyristor is fired at its natural firing instant and its quenching angle β is varied. The commutation of the thyristor is achieved by one of the forced commutation methods. In the symmetrical angle control, if a is the firing angle, the angle of quenching $\beta = 180 - a$. Thus both a and β are varied.

The waveforms of voltage and current are shown in Fig. 4.67 for all the cases. PAC waveforms are also included for comparison. The waveforms of PAC show deteriorating fundamental displacement factor as a is varied. In the EAC the angle of quenching is varied to vary the applied voltage to the motor. Fig. 4.66 shows that the fundamental of the input current pulse leads the applied voltage. Besides improving the power factor, EAC provides the effects of a leading power factor. In SAC both angle of firing as well as quenching are varied such that a symmetrical current pulse is obtained. The width of the current pulse changes with the control. The fundamental of this current pulse changes in amplitude but is fixed with respect to the voltage wave as shown in Fig. 4.67. The fundamental displacement factor is always unity and is independent of a. This improves the power factor.

In all the single firing schemes there is a possibility of the load current becoming discontinuous. This happens particularly at light loads and when the armature inductance is low. Discontinuous conduction affects both the static and dynamic behaviour of the motor and the drive performance is poor. The peak armature currents cannot be controlled and hence commutating capability of the motor gets impaired. Harmonic losses are more and the motor heating limits the torque capability of the motor. The ripple content causes torque pulsations. It also makes the armature current discontinuous. The peak/average and rms/average current ratios can be reduced by a proper additional inductance in the armature circuit. This

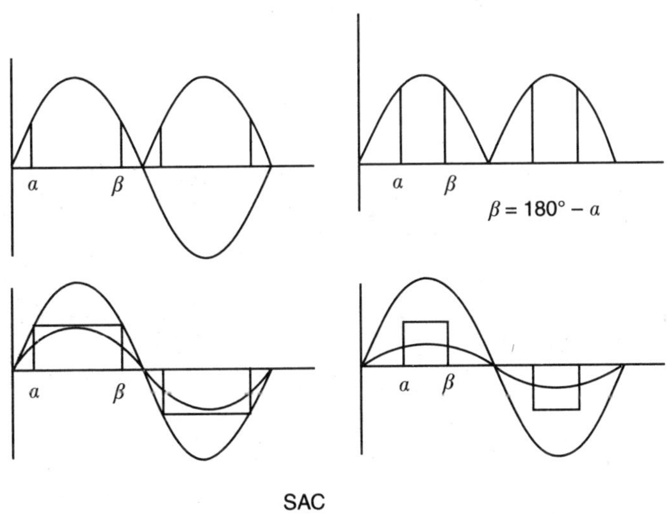

Fig. 4.67 *Principle of SAC voltage and current waveforms*

adds to the bulk of the system, makes the transient response sluggish and introduces additional losses.

In the multiple firing schemes a thyristor is triggered ON and OFF several times in a half cycle. Obviously the schemes employ forced commutation. These are again current limit control (CLC) and the time ratio control (TRC) (Fig. 4.68). In the former the converter thyristors are turned ON and OFF such that the motor current varies between two limits. When the motor current reaches the upper limit the thyristor is turned off. The motor current freewheels and decays. When it decays to the lower limit the thyristor is again fired. The waveforms of voltage and current are shown in Fig. 4.69. The difference between the two limits determines the switching frequency and ripple content in the armature current. If the difference

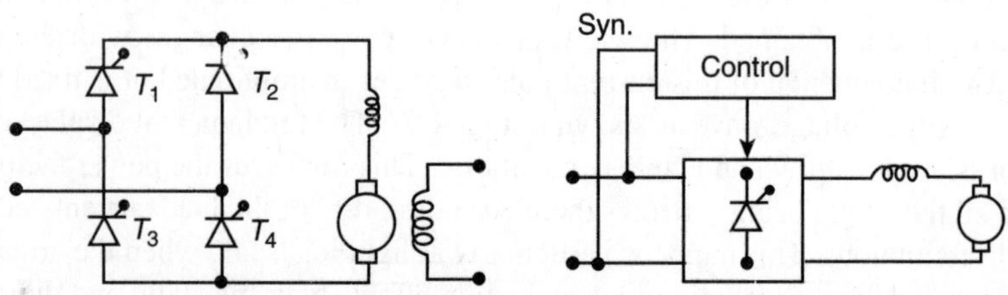

Fig. 4.68 *Multiple firing scheme to improve powerfactor*

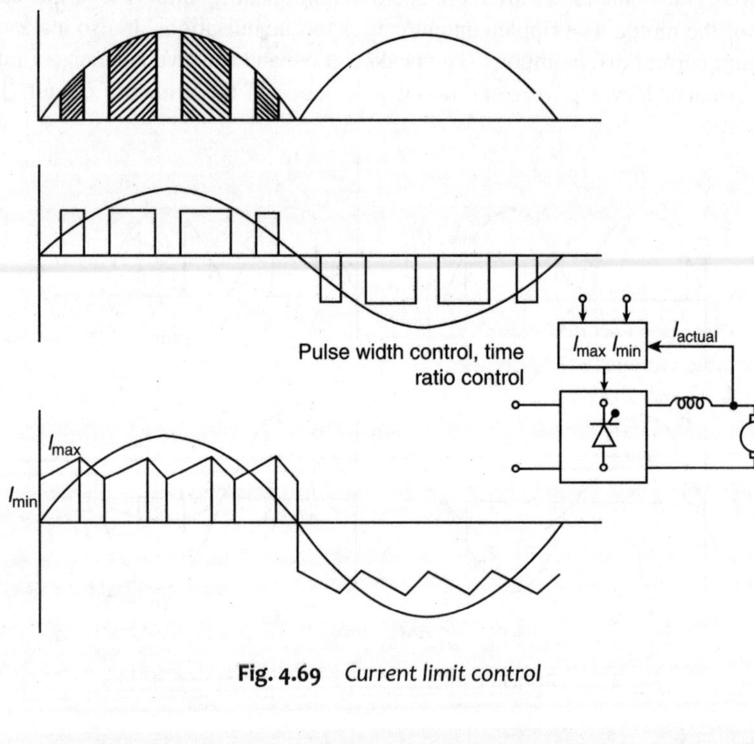

Fig. 4.69 *Current limit control*

is small the ripple content is also small, but the switching frequency is large. This introduces switching losses. A compromise has to be made. The peak current can be controlled thereby improving the commutating capability of the motor. The discontinuous conduction can be eliminated. From the figure it is clear that the fundamental displacement factor is unity. As the harmonic factor is also improved there is a net improvement in the power factor. The inductance in the load circuit for smoothing purposes is rather small and this improves the dynamic response.

In the time ratio control the T_{ON} and T_{OFF} of the thyristor are of given value. These are variable for variation in the output voltage. During T_{ON} the voltage is applied to the motor and the motor current increases. During T_{OFF} the motor is switched off and the current decays. The waveforms are shown in Fig. 4.70. The switching frequency is predetermined. The ripple content is reduced by the large number of switchings per cycle. This increases switching losses. The peak current is not under control. There is possibility of discontinuous conduction. The ripple content may be more. The performance of the motor is slightly inferior compared to CLC, but it is improved in comparison with single firing schemes. In this scheme also, the fundamental displacement factor is unity. The improvement in the total power factor depends on harmonic content.

The performance of a dc drive on all these schemes is compared in Table 4.1.

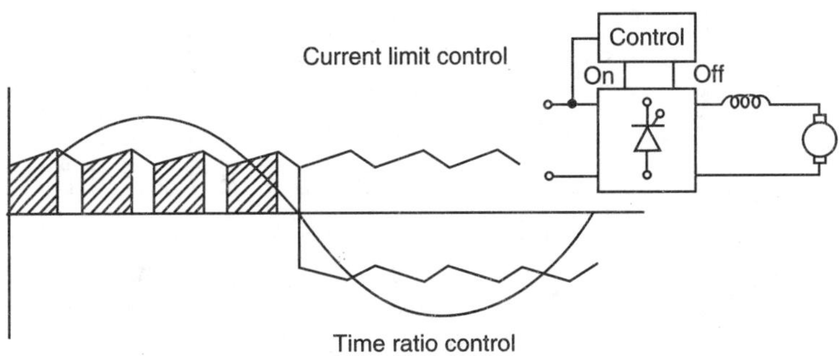

Fig. 4.70 *Time ratio control*

4.3.7 Chopper Fed dc Drives

The variable voltage to the armature of a dc motor for speed control can be obtained from a dc chopper which is a single stage dc to dc conversion device. The voltage variation at the load terminals can be obtained by using either current limit control or time ratio control. In the former, as has already been discussed, the chopper is controlled such that the load current has a variation between two limits. When the current reaches the upper limit the chopper is turned off to disconnect the motor from the supply. The load current freewheels through freewheeling diode and decays. When it falls to the lower limit the chopper is turned on, connecting the motor to the supply. An average current is always maintained. When the chopper is controlled by TRC the ratio of T_{ON}/T_{OFF} of the chopper is changed. In this

Table 4.1 *Comparison of schemes for power factor improvement*

Basis of comparison	*Significance*	*Single firing schemes*	*Multiple firing schemes*
1. Torque-speed characteristic	Signifies the range of speed control, constant torque or constant power operation over the range.	In all the schemes the curves are similar and are drooping in nature. Discontinuous current at several control angles makes the range of speed control limited.	The torque-speed characteristics of TRC scheme are similar to those of single firing schemes with a difference that the possibility of discontinuous conduction is reduced. Consequently the range of speed control increases. In CLC the torque-speed characteristic is flat at lower limits of armature current and thus the drive has a wide range of speed control at constant torque. In series motors as the limit of current rises increasing the average motor current range of speed control becomes small due to saturation.
2. Fundamental displacement factor.	Influences the total power factor.	PAC has a very poor displacement factor. EAC has a leading factor and SAC has unity displacement factor.	TRC and CLC have almost unity displacement factor.
3. Power factor	Decides the kVA demand from the supply. A drive is designed to have a good power factor.	PAC has poor power factor whereas EAC and SAC have better power factors. The power factor of all schemes improves with speed and power output.	TRC and CLC provide essentially the same p.f. TRC has a better p.f. than CLC at low power output.

4. Harmonic content or ripple content	Responsible for pulsating torques and harmonic losses. These increase the RMS value of current and cause additional heating of the motor.	PAC has little harmonic distortion. In general harmonic distortion is very small in single firing schemes. Lower order harmonics predominate. Because of the symmetry of waveform even harmonics are not present.	These have more harmonic distortion. Both lower and higher order harmonics are present. Higher order harmonics can be filtered, even harmonics are not present in this case also.
5. Peak armature current	Determines the commutating capability of the drive.	Peak armature current cannot be controlled in these schemes. The commutating capability is impaired. The peak armature current is essentially the same in all schemes.	In CLC peak armature current is controlled and hence the motor has good commutation capability. TRC has a peak current which is smaller than the single firing schemes. Its peak current cannot also be controlled. Between CLC and TRC, CLC is better as far as peak current and associated commutation capability is concerned.
6. Discontinuous conduction	Affects the dynamic and static behaviour of the motor.	These exhibit discontinuous conduction. Continuous conduction can be achieved by an additional inductance in the motor circuit which adds to the bulk of the system, increases losses and imparts sluggish transient behaviour.	The discontinuous conduction can be reduced to a minimum by proper control in both CLC and TRC. The drives employing these schemes have a wide range of speed control with very good dynamic and static behaviour.

case the operation is at fixed frequency if $(T_{ON} + T_{OFF})$ is kept constant. T_{ON} only is varied to obtain voltage control. The operation will be at variable frequency with T_{ON} kept constant and $(T_{ON} + T_{OFF})$ varied. But owing to several advantages of simplicity, a fixed frequency TRC is normally used. Chopper circuits are used to control both separately excited and series motors.

Chopper circuits have several advantages over phase controlled converters:

i. Ripple content in the output is small. Peak/average and rms/average current ratios are small. This improves the commutation and decreases the harmonic heating of the motor. The pulsating torques are also less.
ii. The chopper is supplied from a constant dc voltage using batteries. The problem of power factor does not occur at all. The conventional phase control method suffers from a poorer power factor as the angle is delayed. This means that the cur ent drawn by the chopper is smaller than in a ac/dc phase controlled converter.
iii. The circuit is simpe and can be modified to provide regeneration.
iv. The control circuit is simple.

However, because of the forced commutation employed, the chopper may be costlier than a phase controlled converter.

Chopper Drives Using TRC

Separately Excited DC Motor

A separately excited dc motor fed from a dc chopper employing TRC strategy is shown in Fig. 4.71. The variation of time ratio at constant chopper frequency provides a variable voltage. The load current can be determined using the relevant equations. During ON period the current in the load grows and in the OFF period the current decays. The current waveform is shown in Fig. 4.72. Due to back emf present in the circuit the variation of currents is almost linear. A detailed study of the drive on TRC strategy shows that the motor has the following features:

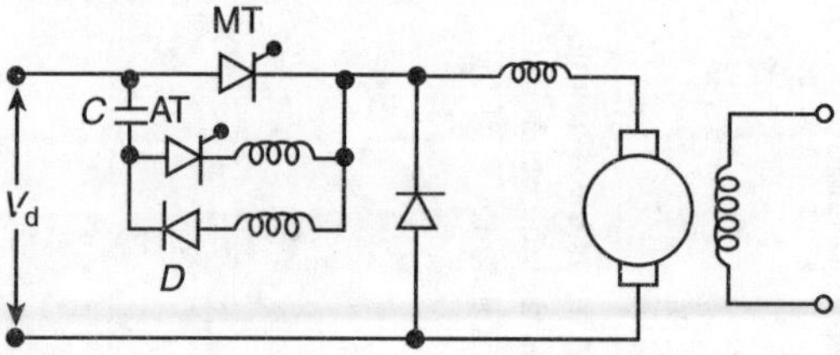

Fig. 4.71 *Separately excited DCM fed from chopper*

A combination of constant frequency and variable TRC strategies gives a better performance. The question of power factor does not occur when the chopper is

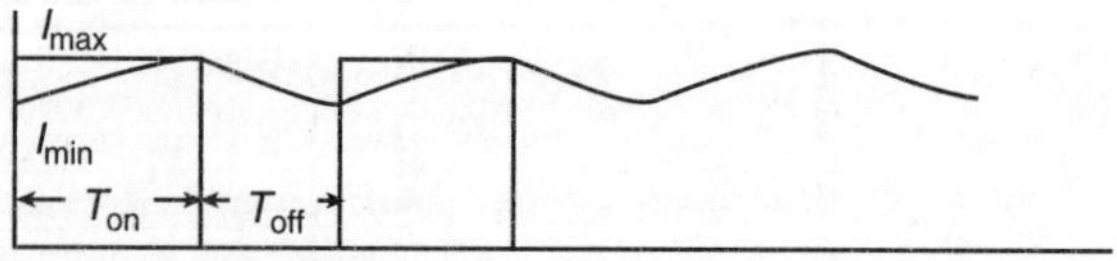

Fig. 4.72 *Voltage and current waveforms for timing ratio control (continuous conduction)*

fed from a constant voltage source such as batteries. Sometimes the dc voltage is obtained from a diode rectifier. In this case the power factor on the ac side is better than that in phase control. But regeneration is not possible in this case, whereas it is possible when fed from batteries.

At very low frequencies of operation and smaller time ratios the load current may become discontinuous. The output ripple depends on TR. At low currents the ripple becomes responsible for discontinuous operation. At very light loads if the average current is smaller than 1/2 (peak to peak) of the ripple amplitude, the load current becomes discontinuous, as shown in Fig. 4.73. Discontinuous conduction occurs if the freewheeling period is larger. In other words, ON time is less than a critical value. As far as the discontinuous conduction is concerned, choppers are similar to ac/dc converters. The peak current cannot be controlled. The limits between which the current varies depends upon TR. The peak current and discontinuous conduction affect the commutation capability of the motor and its speed control range. However the operation of the motor on chopper is better than that on ac/dc converters. In the case of a dc motor fed from a chopper the discontinuous conduction can be eliminated by increasing chopper frequency or by adding extra inductance in the armature circuit. The amount of inductance is rather small.

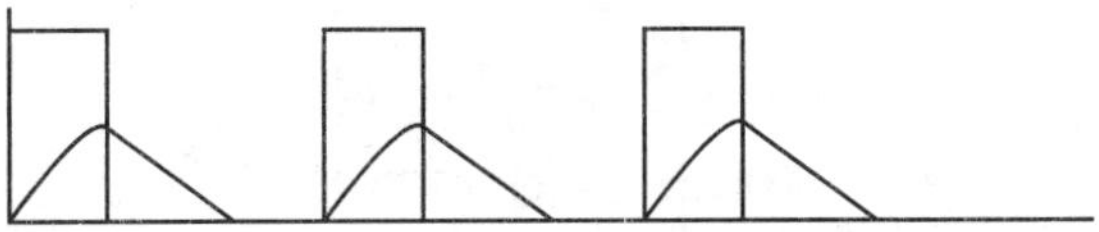

Fig. 4.73 *Voltage and current waveforms for time ratio control*

The torque-speed characteristic is drooping in nature. The speed regulation of the motor is better with chopper control than with the phase angle control. The speed-torque characteristic is essentially the same at all time ratios, (Fig. 4.74).

The ratios of peak/average and rms/average current depend upon TR and hence speed. Therefore the chopper control shows improved performance at high speed. The performance can be improved by an inductance in the armature circuit. The peak current can be significantly reduced so that the commutating capability increases. The heating of the motor will be less. Multi quadrant operation is possible.

Series Motor Series motors produce high starting torque. They are capable of constant power operation at all speeds. They are therefore suitable for traction purposes. Choppers are used to control these motors. Both the control strategies of TRC and CLC are used. The schematic of a series motor fed from a chopper operating with fixed frequency variable TR is shown in Fig. 4.75.

As in the case of separately excited dc motor, the current drawn by the motor is smaller than that of PAC. When a battery is used to supply the chopper, regeneration is possible and power factor does not pose a problem. However when the

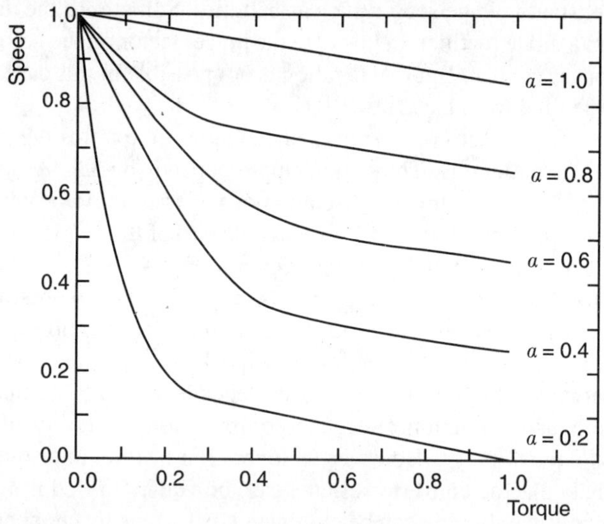

Fig. 4.74 *Speed-torque curves of a dc motor fed from a chopper having TRC*

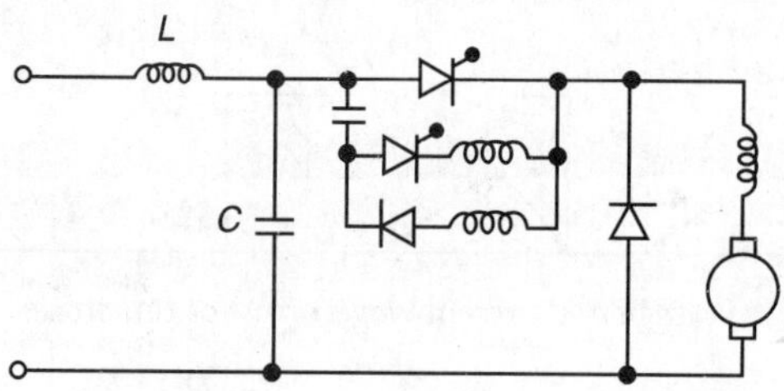

Fig. 4.75 *Series motor fed from a dc chopper*

dc voltage is provided by means of diode rectifier, regeneration is not possible and the power factor is better than that of a phase controlled rectifier.

The ratios of peak/average and rms/average are functions of the time ratio. As compared to phase controlled converters, these are considerably smaller in chopper control. Chopper control improves the performance, the improvement being more at high chopping frequencies.

The discontinuous conduction takes place at very small current ratios. As has already been discussed previously, the series motor is less prone to discontinuous conduction than a separately excited motor. This is because the back emf in a series motor is a function of armature current. Also the speed control range of a chopper fed motor is better than that of the motor fed from a phase controlled rectifier. The discontinuous conduction can be eliminated by means of a proper armature inductance. The ripple content gets reduced, which decreases the possibility of discontinuous conduction. High chopper frequency helps in eliminating the discontinuous conduction in the load at small time ratios.

Typical torque-speed characteristics of a series motor for different time ratios are shown in Fig. 4.76. The speed control range is affected only at small time ratios. As the time ratio increases the characteristic moves towards the left and the range of speed control also increases.

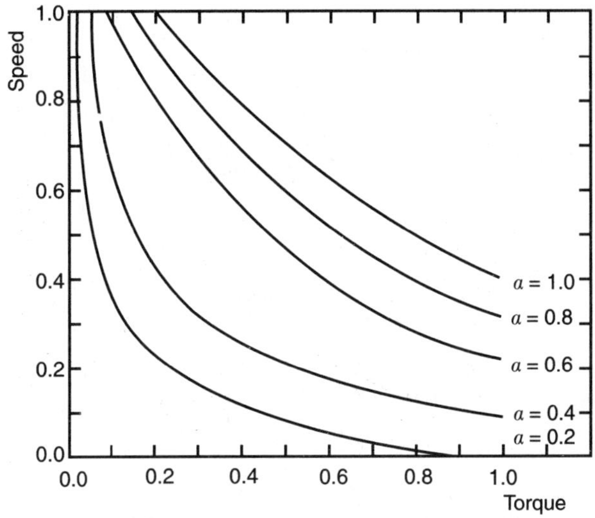

Fig. 4.76 *Speed-torque curves of o chopper fed dc series motor*

The armature current pulsations may cause armature heating. However, this is smaller compared to the operation of motors with phase control.

As the speed increases the average current drawn decreases. A typical current versus speed characteristic is shown in Fig. 4.77. This can be expected in a series motor as the field current is the same as the armature current. At small time ratios and low armature current there is a possibility for discontinuous conduction. So high speeds may not be possible at small time ratios, as is evident from Fig. 4.76.

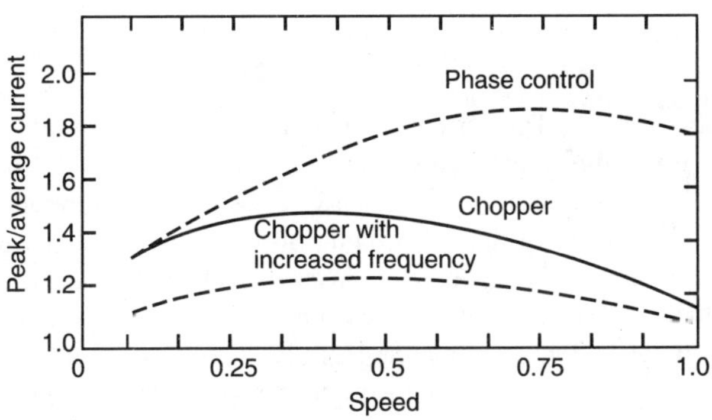

Fig. 4.77(a) *Typical current vs speed characteristics of a series motor*

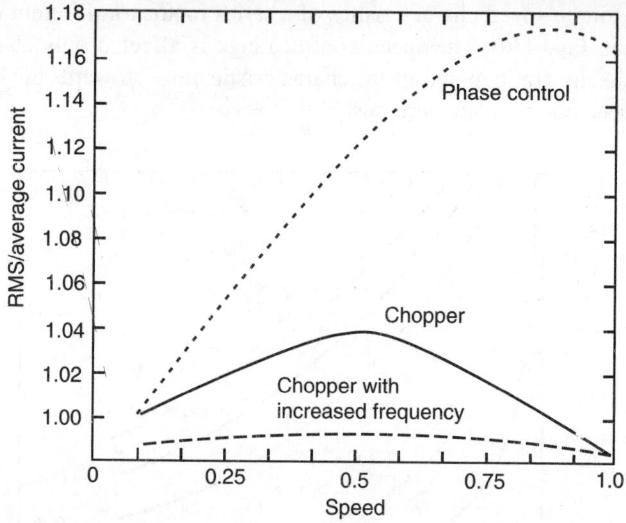

Fig. 4.77(b) *Typical current vs speed characteristics of a series motor*

Because of saturation very low speeds are not possible. As the time ratio increases the lower limit on the speed also increases. At smaller time ratios higher speeds are not possible and at larger time ratios lower speeds are not possible.

Regeneration is possible using a two quadrant chopper circuit. A stable series generator operation is possible using a chopper circuit.

The peak currents also depend upon the time ratio and affect the commutating capability. The commutating capability of a chopper fed motor is better than that of a motor fed from a phase controlled converter.

The above discussion shows that the motor performance is better at high chopper frequencies. But when the time of commutation becomes comparable to the ON time of the chopper, it introduces a voltage regulation. The switching losses are more.

A chopper fed series motor is preferable to a series motor fed from a phase controlled converter because of improved performance.

Chopper Drives with CLC The current limit control used with phase controlled rectifiers can be used with choppers also. A schematic of a motor fed from a chopper employing CLC is shown in Fig. 4.78. CLC gives better performance compared to TRC. This is because of the defined current limits. The peak current is under control. The commutation capability increases. The ripple content is controlled by making the difference between the current limits small. The motor heating is less, but the chopper frequency increases, introducing losses and voltage regulation due to commutation. The speed-torque characteristic is improved. At high currents the lowest speed is limited due to saturation. At lower currents there is the possibility of discontinuous conduction only if the ripple is high. By proper

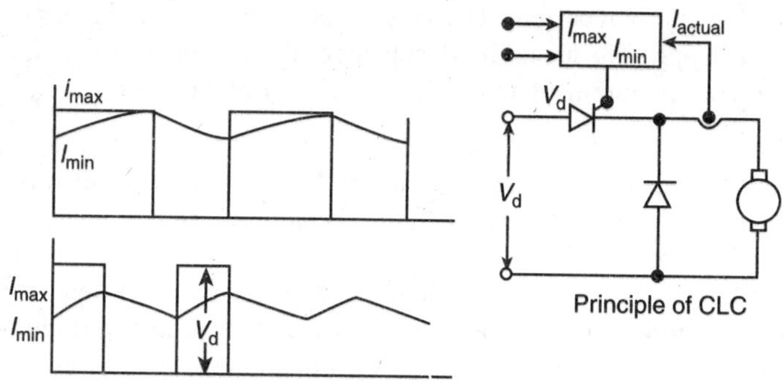

Current limit control

Fig. 4.78 *Current limit control in chopper drives*

control we can completely eliminate the discontinuous conduction. These advantages are available with both series and separately excited dc motors.

4.3.8 Reversible Drives

When speea control in both forward and reverse directions is required a reversible drive is used. In several applications the speed reversal may be required very frequently in which case a regenerative reversal may be advantageous. The drive is very efficient as the kinetic energy of rotating parts is returned to the supply.

The conventional Ward Leonard control (Fig. 4.79) is extensively used where smooth speed control in either direction is required. The drive is capable of operation in all the four quadrants. A regenerative speed reversal is possible by adjusting the field of variable voltage generator. The acceleration and deceleration can be

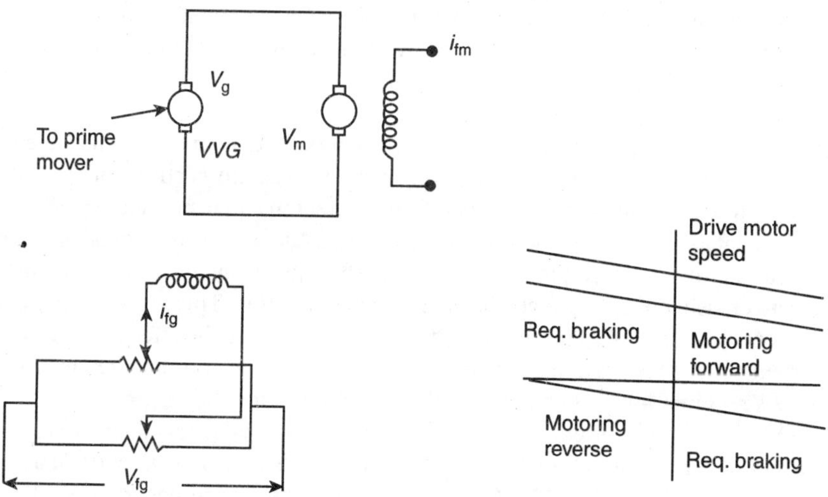

Fig. 4.79 *Conventional Ward Leonard control of a dc motor*

controlled at constant current. The kinetic energy of the rotating parts can be fed back to the supply. The major disadvantages of this system are (a) high initial cost, (b) large space requirement and (c) low efficiency of the drive.

Reversible Drives Using Phase Controlled Converters The advent of thyristor power converters has made static speed control of dc motors very popular. A two quadrant converter is used where regeneration is required. Due to unidirectional current carrying property of thyristors the two quadrant operation is achieved by changing the polarity of voltage, retaining the direction of current. However for reversing the direction of rotation and achieving the speed control in the reverse direction, one should be able to reverse the direction of armature current with the voltage variation from a positive maximum to negative maximum. A static ward Leonard scheme is possible with the help of static power converters having the features of regenerative speed reversal.

The speed of a dc motor can be reversed by changing

i. the direction of armature current.
ii. the direction of field current.

The reversal of current can be achieved by using either mechanical contactors (change over switches) in conjunction with a two quadrant converter, or a dual converter, one feeding the motor in the forward direction and the other in the reverse direction. The possible ways of obtaining reversible drives are given in Fig. 4.80. Each of the drive system has its own merits and demerits.

The field current reversal to obtain a reversible drive is cheaper and simpler compared to armature current reversal. This is because the power level to be handled by the converter is small. At weak field currents the commutation may pose problems. The time constant in the speed reversal is large for field current reversal. Field forcing may be employed to make the current reversal fast. The response of the drive is faster with armature current reversal than with field current reversal, even when field forcing is employed. Also, when compared on the basis of control circuits, the field current reversal is complex.

The reversal of the current may be achieved by mechanical change over switches or dual converters. The use of contactors is economical whereas the use of dual converters is costly. When frequent reversals are needed, there will be wear and tear on the contactors. Therefore frequent maintenance and replacement of the contactors may be required. In such cases a dual converter may be preferred so that the maintenance problem is minimised. The change over must take place at zero crossing of the current when the switches are used. This gives a long life to the contactors. Using a current limit the speed reversal may be effected at constant current so that the drive has a very fast response. However a definite amount of time is required for the opening and closing of contactors.

When dual converters are employed, they may be operated in circulating current or circulating current-free mode. Current reversal is fast, particularly in circulating current mode. The changeover from motoring to generator is almost instantaneous in the case of the circulating current mode, as this mode makes one

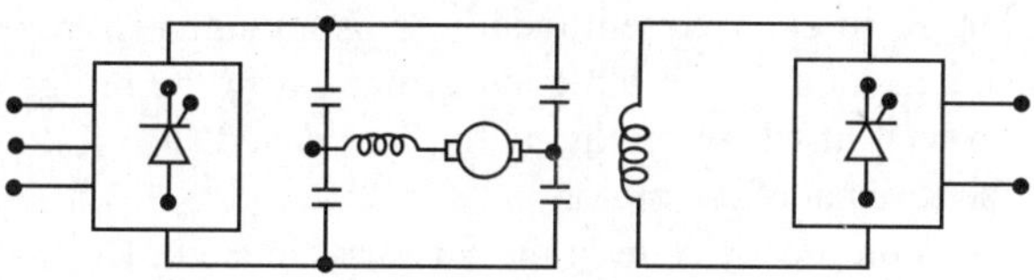

Reversible drive using contactors to reverse armature current.

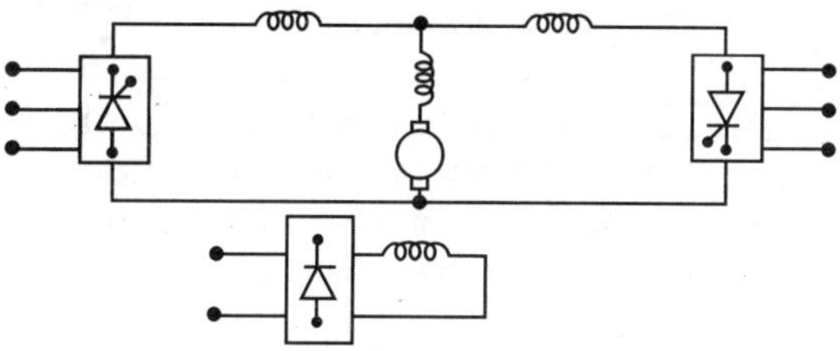

Reversible drive using a dual converter in circulating current mode to reverse the armature current.

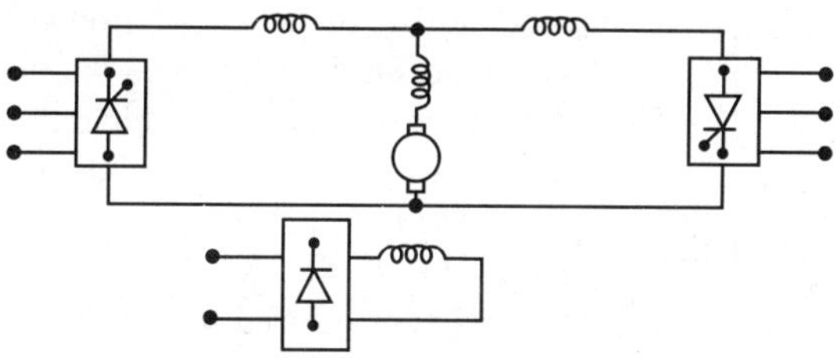

Reversible drive using a dual converter in non-circulating current mode in the armature.

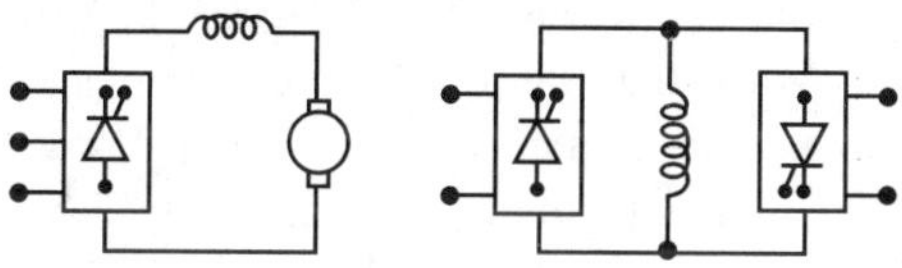

Reversible drive using dual converter in the field.

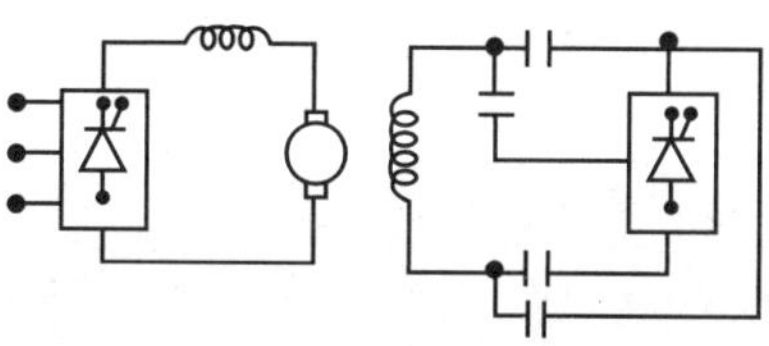

Reversible drive using contactors in the field.

Fig. 4.80 *A summary of reversible drives*

converter operate as a rectifier and the other as an inverter. In the case of circulating current-free mode a time lag of 20 ms may be required for change over, once the zero is detected.

Reversible Drive Using Armature Current Reversal A schematic of a reversible drive using contactors to reverse the armature current for reversing the rotation is shown in Fig. 4.80. The contactors F_1 and F_2 are provided for rotation in the forward direction, whereas R_1 and R_2 in the reverse direction. This arrangement reverses the current in the armature retaining the direction of current in the bridge thyristors. During speed reversal, the load on the motor is removed. The following stages are involved in the reversal of the motor.

Stage 1: The motor is in the steady state in the forward direction. During this stage F_1 and F_2 are in the closed position. The armature current flows in the armature from A to B. At the instant of providing the signal to reverse, the motor is unloaded. A small no-load current flows through the motor.

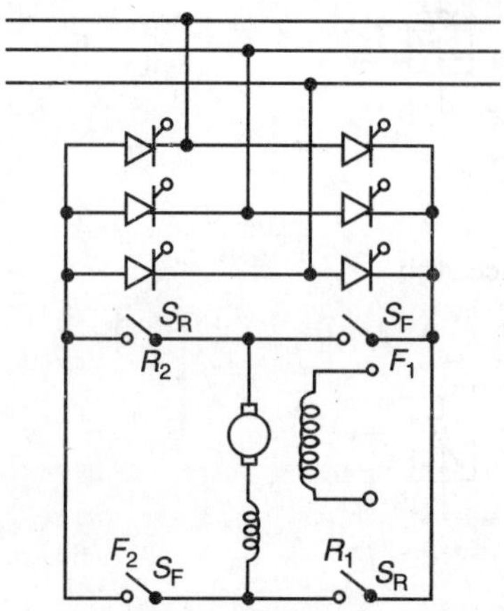

Fig. 4.81 *Three-phase four quadrant drive using armature reversal*

Stage 2: The speed reversal takes place in this stage. The contactors must be operated at zero current, for an longer life. The firing angle is retarded such that the applied voltage is less than the induced voltage. This makes the armature current zero. The contactors F_1 and F_2 are opened using a zero current detector. The switches R_1 and R_2 are closed. Opening and closing of switches introduce a kind of time lag. During this time the armature of the motor is not energised and the motor starts coasting. Speed may be assumed 10 be constant due to large inertia.

After closing the switches R_1 and R_2 the firing angle is advanced so that the armature current flows in the opposite direction, i.e. in the direction of induced voltage. The motor comes to rest. The firing angle is adjusted such that the retardation takes place at constant current. The motor operates as a regenerative brake during this time. All the kinetic energy is returned to the mains by the time the motor approaches zero speed.

The firing angle is further advanced, the converter operates as a rectifier and the machine accelerates in the reverse direction. The motor attains rated speed in the reverse direction. Finally the current drops to no load value.

Stage 3: The load is reapplied to the motor. The motor takes a current to drive the load. There is a small drop in the speed.

Reversible Drive Using Dual Converter It is possible to reverse both voltage and current of the load using a dual converter. The current reversal is very fast in this case as the system is static without any mechanical parts. A dual converter can be operated in either circulating current mode or circulating current-free mode. In the former there will not be any discontinuous conduction and control becomes very simple. About 15 to 25% of full load current is allowed as circulating current. This is limited to a chosen value by means of reactors. The change over from converter I to converter II is almost instantaneous. On the other hand, control of the converter in circulating current free mode, becomes complex. This is because change over from positive converter to negative must take place when the current is zero. The zero detection circuit should be effective during discontinuous conduction also.

A scheme for speed reversal using a dual converter is shown in Fig. 4.82. The following stages are involved in the speed reversal. The circulating current mode is assumed.

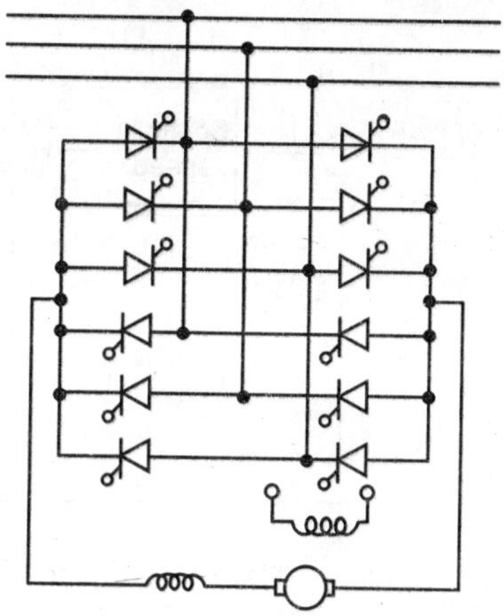

Fig. 4.82(a) *Three phase four quadrant drive using dual converter*

Stage 1: The motor is under steady state in the forward direction. The converter I operates as a rectifier and converter II as inverter. The motor is unloaded.

Stage 2: This is the speed reversal stage. It is initiated with a command to reverse the speed. The firing angles are adjusted such that the converter voltage changes and the current is maintained at the limit. Converter II takes over the conduction and the power goes to the supply, as the voltage induced and I_a are in the same direction. As the kinetic energy is supplied to the line the motor retards to zero speed.

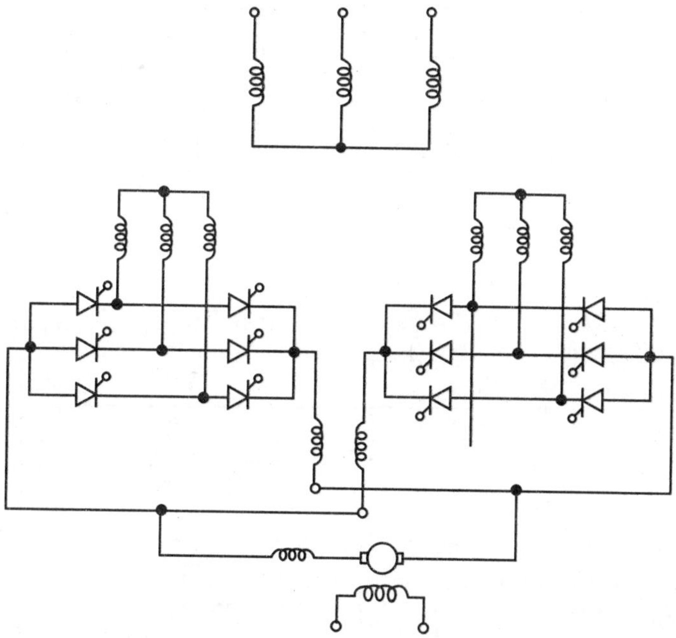

Fig. 4.82(b) *Dual converter in circulating current mode to feed a dc motor*

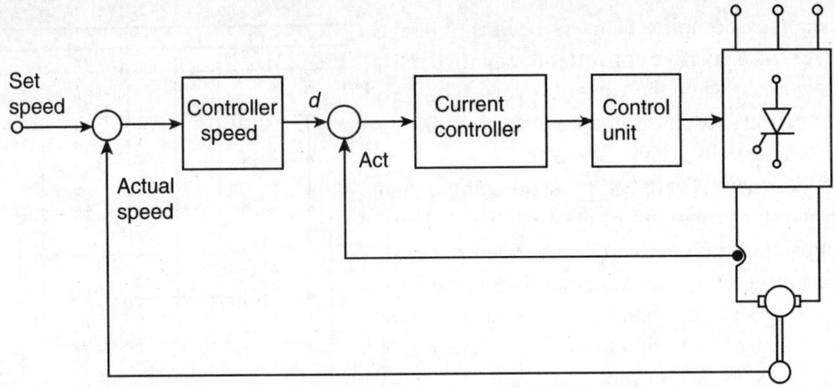

Fig. 4.82(c) *Block diagram of speed control of dc motor*

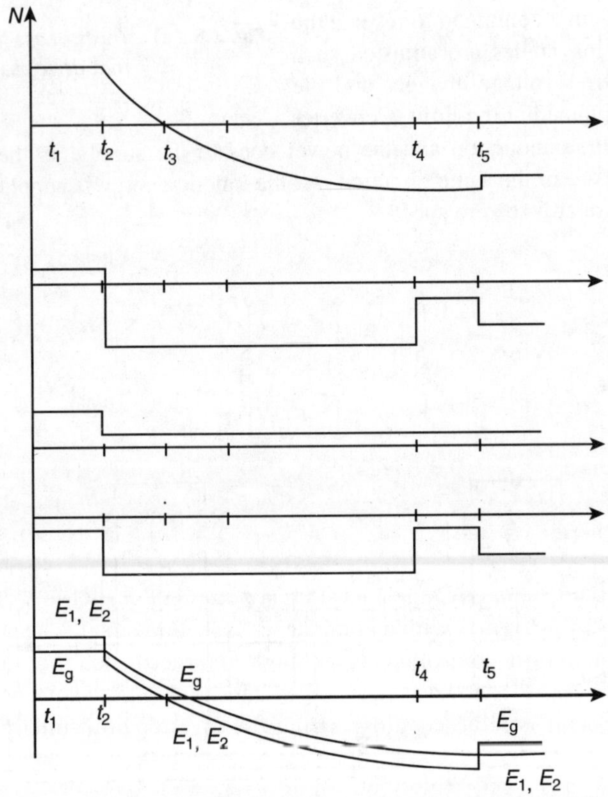

t_2 – Speed reversal command, t_2 t_3 – Motor decelarates regeneratively to zero speed

t_3 t_4 – Motor accelerates in the reverse direction t_5 – Load is reapplied

Fig. 4.82(d) *Variation of speed, current and voltage during speed reversal*

From this point onwards converter II operates as a rectifier and converter I as inverter. As the firing angle is changed the motor accelerates in the reverse direction to rated speeds.

Stage 3: The load is reapplied to the motor in the reverse direction. The waveforms of current, voltage and speed are shown in Fig. 4.82.

Reversible Drive Using Field Reversal A reversible drive using field reversal can be realised as shown in Figs. 4.83 and 4.84. In Fig. 4.83, the armature is fed from a two quadrant converter with change over mechanical contactors in the field, which is also fed from a two quadrant converter.

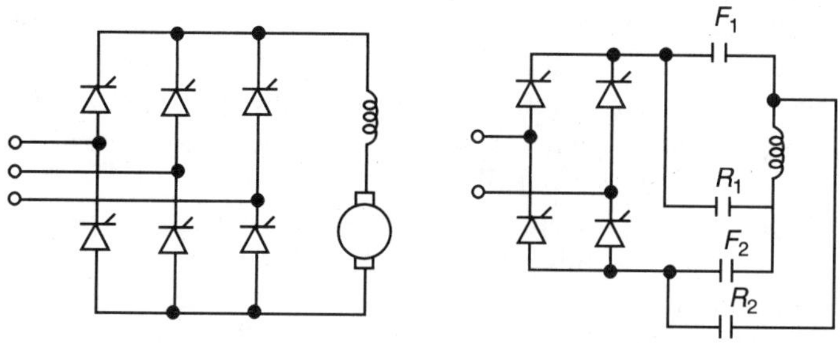

Fig. 4.83 *Reversible drive using contactors in the field*

In the process of reversal of speed, the field current is reversed very fast using field forcing. Speed reversal using the above scheme is achieved in the following stages. F_1 and F_2 are for the forward direction and R_1 and R_2 are for reverse direction.

Stage 1: The motor is under a steady state, running at the rated speed in the forward direction. The converter in the field circuit is fired at a firing angle such that about 30% of the maximum voltage is the rated voltage of the field and it is applied during running operation. For example $a_f = 70°$ and $V_f = V_{fm} \cos 70 = 0.34\ V_{fm}$, where V_{fm} is the voltage of the converter at zero firing angle.

Stage 2: This stage starts with a command of speed reversal. The converter feeding the armature is controlled (firing angle retarded) such that the armature current is zero (applied voltage E). The firing angle of the converter feeding the field circuit is retarded fully, taking the inverter limit into consideration. At this instant the converter voltage reverses and its magnitude would be 95% V_{fm}.

e.g., a is retarded to 170° and

$$V_f = V_{fm} \cos 170° = -0.98\ V_{fm}$$

This voltage is three times the rated voltage in the reverse direction. The field current decreases very fast. The magnetic energy stored in the field is also fed back to the mains. During this time the motor back emf E also decreases.

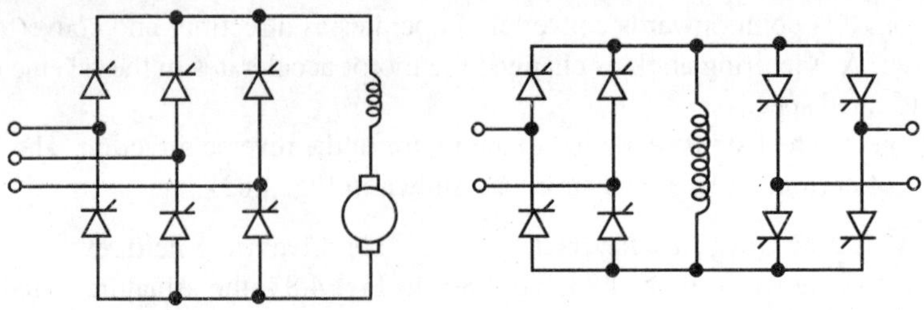

Fig. 4.84 *Reversible drive using a dual converter in the field*

When the field current reaches zero value, the contactors F_1 and F_2 are open and R_1 and R_2 closed. The field current flows in the opposite direction and is built up by increased voltage ($a = 10$)(.95 V_{fm}). As the field current builds up in the negative direction the motor retards regeneratively under constant armature current. The firing angle of the converter is changed such that normal voltage is applied to the field.

The motor speed reaches zero and it accelerates to the rated speed under constant armature current. At this point a load is applied to the motor.

When a dual converter is used in the field circuit for field reversal, it is preferable to operate it in the circulating current-free mode because the field current must go to zero. When a dual converter is used, the stages of speed reversal are almost the same as above. Converter I allows the current for forward rotation and converter II for reverse rotation. The only difference is, when once the field current has become zero the pulses to converter I are blocked and pulses to converter II are given. This requires a delay time similar to the one required by the contactors. Later, building up of current in the field and regenerative braking of the motor are similar.

These schemes have complex control circuitry. Even with field forcing, the response (time) of these schemes is slower than that of the schemes employing armature current reversal. The total speed reversal depends on the mechanical time constant of the drive.

Reversible Drives Using Choppers Using two quadrant choppers a reversible drive can be achieved, because this combination allows reversal of both current and voltage of the motor terminals. We can get a regenerative reversible drive. The motor can operate in all four quadrants. A scheme of a reversible drive using choppers is shown in Fig. 4.85(a). The modes of operation are illustrated in Fig. 4.85(b). The two choppers are connected in antiparallel and supply the load. The motor operation in both the directions is achieved by operating the choppers 1 to 4 to provide voltage and current of desired polarity.

Control of choppers S_1 S_2 with their diodes D_1 D_2 provides operation in I and IV quadrants with positive current and V_a reversible by varying T_{ON}/T_{OFF} ratio. S_3 S_4 with diodes D_3 D_4 comprise another chopper providing operation in II and

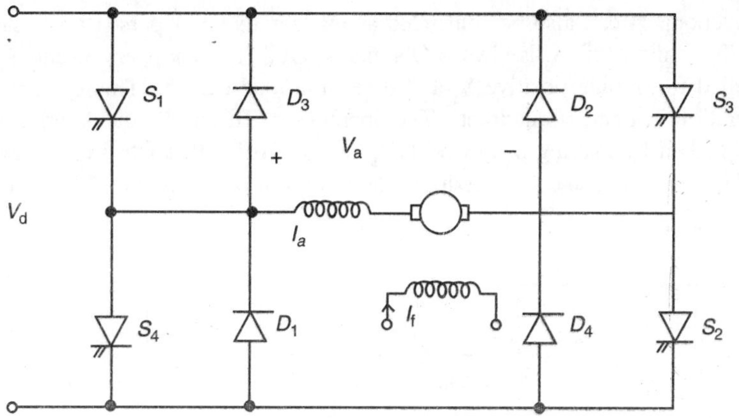

Fig. 4.85(a) *Chopper fed four quadrant dc drive*

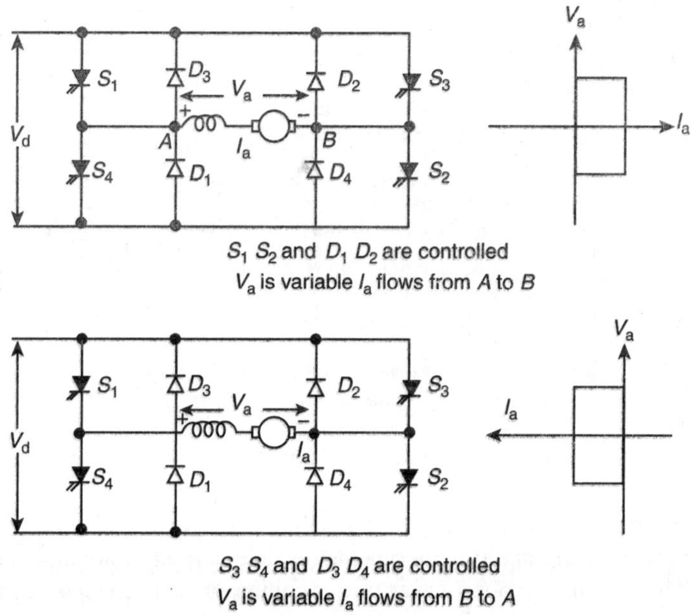

Fig. 4.85(b) *Modes of operation of a four quadrant chopper (antiparallel connection of two quadrant choppers)*

III quadrants with negative current and reversible V_a. Thus the load voltage and current can take up both possible directions. The circuit has similar behaviour of a dual convenor made up of two antiparallel connected fully controlled line commutated converters.

The choppers can also be controlled as regenerative choppers (step up choppers). To achieve this S_2 is always ON and S_3 OFF. The choppers S_1 and S_4 are controlled to provide positive V_a and reversible load current. This corresponds to operation in I and II quadrants. The operation in III and IV quadrants can be accomplished by making S_4 always ON, and S_1 OFF with control on S_1 and S_2 to make V_a negative and I_a reversible. The modes of operation are shown in Fig. 4.85(c).

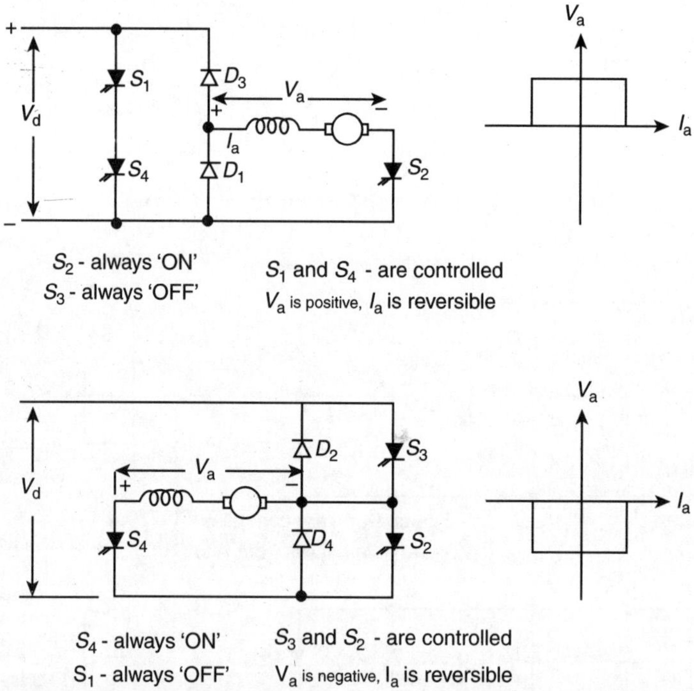

Fig. 4.85(c) *Modes of operation of four quadrant chopper (regenerative choppers)*

Four quadrant operation can also be obtained with a two quadrant chopper supplying the armature and using a scheme for field current or armature current reversal. This scheme involves fewer power components. The response is very sluggish, as has already been discussed.

Closed loop control systems are being developed to obtain a precise control of speed with regeneration in both directions for thyristor drives.

A comparison of a reversible drive using a dual converter and a four quadrant chopper is rather interesting. A four quadrant chopper is costlier than a dual converter. This is because of commutation circuits required by each chopper. A dual converter has a simpler control, if circulating current is allowed. A detailed investigation may be required to compare the performance of the drives under dynamic and steady state conditions.

4.4 PERMANENT MAGNET SYNCHRONOUS MOTOR

A conventional synchronous motor is a doubly fed machine. The stator is provided with a 3 phase balanced source of ac voltages or currents. The rotor contains a field winding excited from a dc source. For development of torque, it is necessary that both the stator and the rotor mmfs must have zero relative speed. The torque is possible only when the rotor rotates at the synchronous speed, i.e., the speed of the stator mmf. With the availability of permanent magnets, the field winding of the conventional motor is replaced by good-quality permanent magnets. This dispenses with the field winding which adds to the weight, cost and losses. These permanent magnet motors can be fed from a power electronic converter. The converter devices are fired to conduction, based on the rotor position. For every 60° of rotation, a device is fired in a given sequence. By the time the shaft completes a rotation of two pole pitches, the firing of all the devices is complete and one cycle of the stator excitation is complete. The control needs a position sensor (i.e., a shaft encoder). The shaft encoder sends firing pulses to the devices of the converter. This type of control is well known as self-control. Here, the frequency of the stator excitation is made to slave the rotor speed. This is similar to the action taking place in a dc motor. Any changes in the rotor speed automatically changes the frequency of the armature excitation. The conventional control is known as separate control where the frequency of the stator excitation is varied to bring in the speed variation. The speed is slaved to the frequency which causes hunting and stability problems. On the other hand, self-control using a rotor position sensor imparts to the motor a speed similar to that imparted by a mechanical commutator to the dc motor.

DC motors have been enjoying wider application in the industry as high-performance servomotors with all their other weaknesses presented by mechanical commutator. Obviously, a PM synchronous motor has an added advantage of low inertia, due to the absence of field winding, compared to a conventional synchronous motor, and due to absence of field winding and commutator when compared to a dc motor.

A synchronous motor (conventional or PM) when operated in self-control loses its identity as far as its behaviour is concerned. It possesses an entirely different behaviour with several advantageous features, mainly having behaviour similar to that of a separately excited dc motor. The synchronous motor is called a Commutator Less Motor (CLM).

The PMSM is becoming a serious competitor to dc as well as induction motors for high performance applications. Its high efficiency, small size, low inertia, large power density and larger torque/ampere puts the motor in the forefront for high performance servo applications. Other industrial applications are in the area of robotics and aerospace actuators. The PMSM has also the following features:

- High speed operation
- Least maintenance
- Absence of limitations due to mechanical commutator in respect to higher voltage, higher current ratings and higher speed capabilities.

4.4.1 Mathematical Modelling of a PMSM

With the advent of power converters which can feed the stator of a PMSM and also powerful magnets, there is a constant attention paid to this motor as a servo drive in closed loop control. Different current control methods are evolved in order to make the best utilisation of the magnets. Before the availability of these powerful magnets, there was fear against demagnetization of the magnets. The only control that was prevalent was the current control having no direct demagnetising. Therefore, a surface-mounted rotor can be considered as uniform air gap motor or round rotor motor. It has no saliency. The IPM motor, on the other hand, has its permanent magnet buried into the rotor iron. Again considering the permeability of the magnet equal to that of the air gap, the direct axis inductance is less than the quadrature axis inductance, where the path of the flux is complete in iron. The modern permanent magnets made are of rare-earth magnetic components. The powerful magnets available at present have made researchers think about developing controls having a direct demagnetising effect, causing demagnetisation. Also, the drive technology is heading towards dispensing with flux sensors and speed sensors. The estimated speed and flux are used in different controls. For the estimation of flux and speed, a mathematical model is required. Modern control techniques look forward for adaptive observers. Model of a PMSM can be obtained by making simple modifications to the model of a wound-field synchronous motor. A few of the differences are the following:

- Wound field synchronous motor has damper cage. In a PM motor the magnets are embedded in the rotor structure or they are pasted to the surface of the rotor. The former is equivalent to salient pole type whereas the latter to the round rotor type.
- The synchronous PM motor having permanent magnets pasted to the surface of the rotor is considered equivalent to a uniform air gap machine in which $L_d = L_q = L_s$. So the models of SPM and IPM motors can be deduced suitably.
- The permanent magnet has permeability equal to that of an air gap. Therefore, the PM embedded in the rotor causes a reluctance equal to the air gap.
- On the other hand, the volume between the magnets is the quadrature axis made up of magnetic material. Unlike the case of a conventional synchronous motor, the quadrature axis inductance L_q of a PM motor will be greater than the direct axis inductance L_d.
- Also, as there are no damper windings; the damper currents, damper inductances and damper resistances are zero. Also, since there is no field winding, the field current is zero. Substituting zeros for the damper and field winding parameters in the model of a wound-field synchronous motor, we get the model of a PM motor.

The stator of a PMSM is similar to that of a conventional wound-field synchronous motor. It has a distributed short pitched winding. The three phases are λ or Δ connected. The exciting coils on the magnetic poles are replaced by permanent magnets which are fixed on the rotor surface or placed well inside the rotor structure. The former is known as Surface Mounted Motor (SPM), and the latter is known as Interior Magnet (IPM) motor. The magnet can be considered as equivalent to its air gap as far as its permeability is concerned, i.e., its relative permeability $\mu_r = 1$ and $\mu = \mu_r \mu_o$. The materials used are of high quality. These possess high resistivity and hence the induced currents in the rotor are negligible. Between the PM synchronous motor and the conventional wound-field motor, there is no difference between the back emf induced in the stator winding by permanent magnet flux and the flux produced by an excited coil on the pole body. The mathematical model of a PMSM will be almost similar to that of a wound-field synchronous motor. The following assumptions are made while deriving the model of a PMSM.

- Neglect of saturation of parts of magnetic circuit. However, the changes in the motor parameters due to saturation may be considered wherever applicable.
- The induced emf is sinusoidal.
- Hysteresis and eddy current losses are neglected. The transients in the field flux are not present.
- There are no damper windings on the rotor.
- The system is balanced.

With the above assumptions, the two-axis model of the 3 phase stator in rotor (synchronous rotating) reference frame is

$$V_d = r_s i_d + \frac{d\psi_d}{dt} - \omega_s \psi_q \tag{4.27}$$

$$V_q = r_s i_q + \frac{d\psi_q}{dt} - \omega_s \lambda_d \tag{4.28}$$

where ψ_q is the quadrature axis flux linkages given by

$$\psi_q = L_q i_q \tag{4.29}$$

And ψ_d is the direct axis flux linkages given by

$$\psi_d = L_d i_d + \psi_{af} \tag{4.30}$$

where ψ_{af} is the flux in the air gap due to the magnet and it is no-load flux in the air gap.

In the above equations which describe the mathematical model of PM synchronous motor,

- v_d and v_q are the direct and quadrature axis stator voltages
- i_d and i_q are the direct and quadrature axis stator currents
- L_d and L_q are the direct and quadrature axis synchronous inductances
- r_s and ω_s are the stator resistance per phase and angular velocity of stator mmf.

The electromagnetic torque developed is

$$T_d = \frac{3}{2} P \left[\psi_{af} i_q + \left(L_{ds} - L_{qs} \right) i_d i_q \right] \tag{4.31}$$

P being the number of pole pairs of the motor.

The electromechanical dynamic torque equation of the motor is

$$T_d = T_L + J\frac{d\omega_r}{dt} + B\omega_r \tag{4.32}$$

where T_L is the load torque
J is the moment of inertia
ω_r is the rotor angular velocity
B is the damping torque coefficient

The mechanical angular velocity of the rotor is related to the electrical angular velocity by

$$\omega_s = P\,\omega_r \tag{4.33}$$

The machine model is nonlinear due to the presence of the products showing cross coupling of the state variables such as speed, i_d and i_q.

The equations 4.27 to 4.32 may be written to represent the state space model of PMSM for dynamic simulation purposes as

$$\begin{aligned} pi_{ds} &= \frac{1}{L_{ds}}\left\{v_{ds} - r_s i_{ds} + \omega_s L_{qs} i_{qs}\right\} \\ pi_{qs} &= \frac{1}{L_{qs}}\left\{v_{qs} - r_s i_{qs} + \omega_s L_{ds} i_{ds} - \omega_s \psi_{af}\right\} \\ p\omega_r &= \frac{1}{J}\left\{T_d - T_L - B\omega_r\right\} \end{aligned} \tag{4.34}$$

The well-known Park's transformations are applicable here to determine the equivalent two-axis components of 3 phase stator currents. The two axis components of current in the rotor reference frame are Fig.

$$\begin{pmatrix} i_{ds} \\ i_{qs} \\ i_o \end{pmatrix} = \frac{2}{3}\begin{pmatrix} \cos\theta_s & \cos\left(\theta_s - \frac{2\pi}{3}\right) & \cos\left(\theta_s - \frac{4\pi}{3}\right) \\ \sin\theta_s & \sin\left(\theta_s - \frac{2\pi}{3}\right) & \sin\left(\theta_s - \frac{4\pi}{3}\right) \\ \frac{1}{2} & \frac{1}{2} & \frac{1}{2} \end{pmatrix}\begin{pmatrix} i_a \\ i_b \\ i_c \end{pmatrix} \tag{4.35}$$

Suffixes *a, b, c* represent the three phases. Inversely, the phase currents are determined from the two-axis currents using

$$\begin{pmatrix} i_a \\ i_b \\ i_c \end{pmatrix} = \begin{pmatrix} \cos\theta_s & -\sin\theta_s & 1 \\ \cos\left(\theta_s - \frac{2\pi}{3}\right) & -\sin\left(\theta_s - \frac{2\pi}{3}\right) & 1 \\ \cos\left(\theta_s - \frac{4\pi}{3}\right) & -\sin\left(\theta_s - \frac{4\pi}{3}\right) & 1 \end{pmatrix}\begin{pmatrix} i_{ds} \\ i_{qs} \\ i_o \end{pmatrix} \tag{4.36}$$

The above transformations apply equally well to the flux linkages. They apply to voltages by definition. The main criterion underlying the transformation is the power invariance.

The power input to the actual machine is

$$P_i = v_a i_a + v_b i_b + v_c i_c \tag{4.37}$$

in phase variables and the power input to the equivalent two-axis model is

$$P_i = \frac{3}{2}\left[v_d i_d + v_q i_q \right] \tag{4.38}$$

Due to balance, structures of stator and rotor, and thereby the resulting uniform air gap of the induction motor, there exist three different reference frames—stator reference frame, synchronously rotating reference frame and rotor reference frame. In the case of a PMSM, which is only a particular case of conventional wound-field motor, having saliency in the rotor structure does not result in the uniform air gap. The rotor reference frame is convenient here. Also, because the angular velocity of the stator mmf is the same as that of the rotor, the synchronous rotating reference and the rotor reference frame are one and the same. However, in SPM machines, where the air gap is uniform, we can have a stator reference frame by substituting zero for the speed of the reference frame ($\omega_s = 0$).

4.4.2 Current Control Methods for Permanent Magnet Motors

Permanent magnet motors are becoming very popular in a wide variety of industrial applications. They are very attractive for servo applications, and robot drives due to their low inertia of the rotor. Their attractive power density and efficiency characteristics make them compete with the conventional ac motors. Their fast response and compactness are the most important factor responsible for generating interest. The well-established developments and findings in more powerful and cost-effective permanent magnet materials are serving to accelerate this interest further.

A very well known and widely applied control method for PMSM is to make the direct axis component of the stator current zero, i.e., $i_d = 0$, and the direct-axis component of stator current becomes nonexistent. The total stator current is made up of only the torque producing a quadrature component. This method avoids the direct demagnetisation of the permanent magnets. With recent development in the material technology, the magnetic materials with large coercive forces are made available and these do not limit the control to the above method. Several control methods improve the performance and ability of a PM motor. The developments in the rotor configurations and permanent magnet geometries have given rise to two types of motors. These are Surface Mounted Magnet motors (SPM) and Interior Magnet (IPM) motors. The parameters like reactances and resistances in the two axes of the motor are greatly influenced by the configurations.

The versatility of these control methods may be evaluated by the following criteria or the coefficients of the motor and power electronic converter feeding the motor.

Voltage Ratio K This ratio can be defined as the no-load terminal voltage to the load terminal voltage. The ratio is closely related to inverter capacity. For machines having larger K, an inverter of larger capacity is required. The inverter should be capable of providing larger voltages as the motor is mechanically loaded.

The Demagnetising Coefficient ξ This coefficient is defined as the ratio of direct-axis armature reaction flux linkages to the permanent magnet flux linkages. This ratio is very closely related to the air gap flux under loaded conditions of the motor. The increasing value of ξ indicates the decreasing value of the air gap flux. This, in turn, is associated with reduced developed torque and reduced output power.

The Power Factor cos ϕ and the Torque Angle δ The torque angle δ is the space angle between the induced voltage ($\omega\psi_{af}$) and the applied voltage V_1. The parameters cos ϕ, tan δ effectively decide the performance characteristics and inverter capacity.

Referring to phasor diagram of the motor, the power factor $\cos\phi = \cos(\delta - \beta_o)$ and

$$\tan\delta = \frac{rI_a \sin\beta_o + \omega L_q I_a \cos\beta_o}{rI_a \cos\beta_o - \omega L_d I_a \sin\beta_o + \omega\psi_{af}} \tag{4.39}$$

The basic variables, the direct and quadrature axis synchronous inductances L_d, L_q and the magnet flux linkages ψ_a are included in the parameters and equations. The armature current, torque developed and the resistance and reactance parameters are normalised as

$$I_a^n = L_d I_a; \qquad T^N = \frac{T}{p\psi_a^2 / L_d}; \qquad r^n = \frac{R}{\omega L_d}$$

With this, the normalised value of armature current has a range of values from 0.2 to 0.4. The operation of the motor is confined normally to high speeds. The resistance drop can be neglected compared to ωL_d at these speeds and frequencies. The normalised basic equations are

$$T^N = -I_a^N \left[\cos\beta_o + \frac{1}{2}(\rho - 1) I_a^N \sin\beta_o \right]$$

$$K = \left\{ \left(1 - I_a^N \sin\beta_o\right)^2 + \left(\rho I_a^N \cos\beta_o\right)^2 \right\}^{\frac{1}{2}}$$

$$\xi = I_a^N \sin\beta_o$$

$$\tan\delta = \frac{\rho I_a^N \sin\beta_o}{1 - I_a^N \sin\beta_o} \tag{4.40}$$

In the above equations, ρ is the saliency coefficient and is the ratio of quadrature axis synchronous inductance to the direct axis synchronous inductance

$$\rho = \frac{L_q}{L_d}$$

The relative permeability of the permanent magnet is near unity. The magnet behaves like an air gap (or equivalent to a non-magnetic material). The machine direct and quadrature axis inductances depend on the rotor configurations and the geometry of the permanent magnet. The surface magnet motor (SPM) exhibits negligible saliency and hence $\rho = 1$. For the Interior Magnet Motor indicating that the quadrature axis inductance L_q exceeds the direct axis inductance L_d.

Control methods of these motors can be different for specific values of β_o, the angle between induced emf phasor E and the armature current phasor, I_a. Control of the value of β_o imparts to the motor effectively an altogether different performance.

Present-day permanent magnets can tolerate demagnetisation. Due to large coercive force they need for complete demagnetisation, two extra control methods can be considered. In one control method the power factor cos ϕ can be maintained constant, may be in particular at unity. In the other method, the flux linkages may be maintained constant. Both of these are effected by controlling the two axis components I_d, I_q or by controlling the phase position of the current with respect to the applied voltage.

4.5 CLASSIFICATION OF PERMANENT MAGNET SYNCHRONOUS MOTORS

The permanent magnet synchronous motors can be broadly classified into two major heads:

i. Interior magnet motors
ii. Exterior magnet motors

There are some similarities between these two types. Both the types of motors have permanent magnets in the rotor to produce the desired magnetic field. The stators of these motors are wire wound. There are three phases in the stator. The three phase winding is wound for the same number of poles as in the rotor. The stator is excited by a three phase balanced system of currents. This three phase system of currents produces a space distributed magnetomotive force rotating in electromagnetic torque is developed when the rotor speed is same as that of the stator mmf (flux). However, the waveforms of the exciting currents of the two motors are different. The waveforms of the back emfs induced in the stator phases of the two motors also differ. The interior magnet motors have a sinusoidal back emf whereas the exterior magnet ones have a trapezoidal back emf. Obviously, the performance characteristics of the two types of machines differ. They differ also in the operating requirements.

Another aspect of difference between the two motors lies in their construction. In the interior magnet motor, as the name indicates, the magnets are well buried in the rotor. On the other hand, in the other type, the magnets are pasted to the surface of the rotor. Based on the construction they are called Interior Permanent Magnet (IPM) motor and Surface mounted Permanent Magnet (SPM) motor respectively. In the SPM motor, the magnet may occupy 120 degrees to 180 degrees of the pole pitch. This may influence the ripple torque of the motor.

The permeability of the magnet is considered to be nearly equal to that of air. In the interior-type motors, permanence along the direct axis is air and is smaller than that along the quadrature axis where it is completely iron of the rotor core. Therefore, the armature reaction effect is weaker along the direct axis than the armature reaction along the quadrature axis. The direct axis synchronous reactance (x_d) is less than the quadrature axis synchronous reactance (x_q). The construction of IPM motor is similar to that of salient pole construction of wound field motor. In an SPM motor, the magnet on the surface of the motor is equivalent to the air gap. The motor is considered to be a cylindrical rotor construction of the conventional wound field motor. Therefore, in the SPM motor the synchronous reactance along the direct and the quadrature axes is the same. As the permanent magnets are embedded in the rotor core, the IPM motor is mechanically stronger than the SPM motor. This feature gives the IPM motor an edge over the SPM motor in that it can be run at higher speeds than its counterpart, the SPM motor.

Both the types of permanent magnet motors are characterised by some unique advantages of very high efficiency, large power density and low inertia. High efficiency permits a totally enclosed design with surface cooling. The large unsurpassed power density is imparted by the use of rare-earth permanent magnets enabling high flux densities in the air gap. Lower inertia results in high torque-to-inertia ratio and improves the dynamic performance of the motor. These favourable properties make the permanent magnet motor suitable as an extremely compact, fast and rugged mechanical actuator. The se motors can be controlled to have very good torque-per-ampere characteristics. These motors do not have field windings. The absence of the field winding adds to the compactness and efficiency of the motor. This feature, on the other hand, warrants for a power electronic converter to feed the power to the stator for drive operation. This converter would be a costly burden if only constant speed operation is required. However, power conversion, being capable of providing variable frequency, variable voltage/current source, makes the control versatile in making the motor a high-quality variable speed motor. These permanent magnet motors are becoming strong competitors to both the dc and ac motor drives, particularly as a servo motor. The IPM motor is called a Permanent Magnet Synchronous Motor (PMSM) whereas the SPM motor as a Brushless DC Motor (BLDC). These are also known as sinusoidal motors and trapezoidal motors respectively, based on the waveform of the excitation.

4.5.1 Control Issues of PMAC Motors

The differences in the construction of both the categories of the permanent magnet motors are in the nature of the ac waveforms they are fed with, get reflected in their torque production characteristics, drive or motion control characteristics, etc. Some of these differences are obvious as they occur due to constructional differences and some due to the difference in the excitation waveforms. These waveforms of excitation which are basically different, and typical constructional differences are though the differences in the excitation waveforms are very much obvious, a close observation of these would reveal some similarities. The six-step waveform from a six pulse bridge converter contains a predominant fundamental component. The trapezoidal excitation of a BLDC motor is similar to a six-step waveform and has a predominant fundamental component. The sinusoidal excitation to the PMSM (IPM) can be obtained by PWM control of the six-step inverter. On the other hand, the constructional differences impart different characteristics to the motor. The differences may require somewhat detailed discussion. The interior permanent magnet (IPM) motor has a salient pole structure; whereas the surface mounted magnet (SPM) motor has a cylindrical rotor structure. These features bring in the differences in the performance characteristics like torque per ampere, power factor, etc. These characteristic differences also affect the capabilities of the inverter feeding the motor.

A detailed discussion of the characteristic differences and similarities between the two motors follows.

4.5.2 Converter Configuration and Application to BJTs, MOSFETs and IGBTs

When supplied from the utility sources, the PMAC machine requires a power converter to transform a fixed frequency, fixed voltage source to a variable frequency variable voltage/current source to be fed to the motor. This power electronic converter is the conventional dc link converter which performs the conversion in two stages. These are the most preferred ones to cycloconverters which are single stage power converters. The dc link power converters have flexibility as far as the output voltage and frequency are concerned. Further, these have several advantages over the single stage cycloconverter type when the frequency conversion in the complete base frequency range is required. Further, dc link converters are available as dc link voltage source converters, dc link PWM converters and dc link current source converters. The permanent magnet motors, both sinusoidal and trapezoidal ones, can be operated on these converters. A dc link voltage source converter transforms constant voltage constant frequency mains to variable frequency variable voltage source. The output of the converter is a six-stepped quasi square wave voltage. The utility source is converted to dc by means of a line-commutated phase controlled ac to dc converter. The current ripple and voltage ripple are filtered by inductance and capacitance respectively of an LC filter in the dc link circuit. The dc link voltage is inverted to ac voltage of desired frequency by the six-step McMurray inverter. The necessary voltage control is accomplished by controlling the line side converter in the first stage itself. The devices of this converter are phase controlled. The conduction period of the machine side converter is controlled to make the output have the desired frequency. Thus, the applied voltage to the motor at the input has its magnitude and frequency such that it establishes the required rated flux in the base speed region.

Using PWM schemes (or principles) to the dc to ac inverter feeding the motor, the voltage control can be accomplished in the inverter itself. The sinusoidal PWM gives and output voltage approaching a sinusoid with the least harmonic component. The advantage of a PWM inversion is that both the voltage and frequency control can be accomplished simultaneously followed by harmonic neutralisation. This also allows the use of a diode rectifier on the line side making the converter economical and the power factor unity on the line side. For small power applications, the self-quenching devices like BJTs, MOSFETs, and IGBTs with integrated feedback diodes can be used as the devices in a line-side rectifier.

Further, PWM principles can be extended to control the voltage source inverter to provide a sinusoidal current to the motor load. Current-controlled PWM techniques like hysteresis control; delta modulation, etc., can be employed.

The dc link current source inverter can be employed to feed PMAC machines. This is also a two-stage converter. The variable dc link voltage at the output terminals of the line side ac to dc phase controlled converter is converted to a constant current by a very large inductance. The capacitor in the dc link is dispensed with. The absence of the capacitor permits a straightforward voltage reversing capability in the dc link. This makes the regeneration also straightforward. In the case of a voltage source inverter, this is not possible. An additional inverse parallel converter is required on the line side for this purpose.

4.5.3 Control Structures for PMAC Motors

Even though the waveforms of excitation currents and constructional features of the two categories of PMAC motors are quite different, there are several important characteristics which are shared in common by the drives developed using these two types of motors. A majority of PMAC motor drives and their applications incorporate closed loop control. For applications with constant air gap flux, a simple *V/f* control can be employed if there is no demagnetisation of the magnet flux in order to achieve a high-quality torque control. If any demagnetisation takes place, it can be compensated by proper current control. Current-fed operation, where direct torque control is neither complex nor difficult, can also be incorporated in closed loop control. Current control techniques can be easily implemented with PMAC motors, where direct association of flux and torque is possible for such a control.

Voltage source inverters are more flexible compared to current source inverters. PWM techniques can be employed to control the current to have a desired waveform preferably a sinusoidal one. The salient features of a voltage source inverter may be summarized as follows:

i. Very fast dynamic performance compared to the conventional Current Source Inverter (CSI) with block current waveforms due to obvious reasons. One reason to mention this is the absence of large link inductance. The motor winding inductances are sufficient to act as link inductances.
ii. For small power levels self-quenching devices like BJTs, MOSFETs and IGBTs can be made use of as active power-switching devices.

iii. These have advantages over inverters having thyristors as the power switching devices which require special turn-off circuits.
iv. Absence of commutation problems at the low speed region. Speed control can go down to nearly zero speed.
v. As the above devices allow high frequency switching, the waveform of the current fed to the phase windings of the stator can be sinusoidal in nature. PWM principles are employed to control the wave shape of the current.
vi. The topology of the converter configuration will be the same for control of either sinusoidal or trapezoidal machines. There are, however, differences in the switching sequence of the converter to distinguish between the waveforms of the excitation current fed to the stator of the motor.

The conventional current source inverter configuration can be obtained by making simple modifications to the voltage source inverter topology. Shunt capacitances may be replaced by series inductances to effectively raise the input impedance of the inverter. This acts as the dc link inductance, which can effectively make the link current ripple free and constant. With a line-commutated phase controlled converter on the line side, the converter can be controlled to provide a variable voltage, which is converted to a constant, ripple free current source by dc link inductance. However, current source implementations that are reported in the literature for supplying the widely popular PMAC motors are associated, in general, with the combinations of various factors like cost, weight and dynamic response.

4.5.4 Advanced Control Techniques

Sensor Less Control One of the most attractive and state of the art areas of research and development involving the PM motors has been the development of the techniques for eliminating the speed and position sensors, which are necessary for the self-control of these motors. Using the measured terminal-values of current and voltage variables and suitable mathematical model of the motor position, speed (and, if necessary, flux) are estimated for the desired control of the motor. The results, which may be sometimes not reliable due to variation of the parameters of the motor due to external and internal disturbances like loading of the motor, winding temperature, magnetic saturation, etc., are corrected by proper identification of the parameters and their adaptation. For this adaptation of the parameters, well-known techniques like Kalman filter algorithm and Artificial Intelligence (AI) techniques like Artificial Neural Networks (ANN), fuzzy logic principles and the controllers obtained using suitable fusion of these techniques. The ongoing present-day research concentrates on the elimination of the mechanical speed or position sensor (shaft encoder) on the machine shaft without deteriorating the dynamic performance of the motor and control system of the drive. The speed and position sensorless drives have the following features.

- Reduced hardware complexity and lower cost
- Reduced size of the motor
- Elimination of the sensor cable

- Better noise immunity
- Increased reliability
- Reduced maintenance requirement

The mechanical sensor which is fixed, normally, on the extended shaft may affect the mechanical strength and vibration of the shaft. It adds to the bulk of the machine increasing the inertia of the drive system. It adds to the cost, making the system expensive. It prevents the use of the system in hard environments, such as high temperature and pressure conditions.

Between the two kinds of PM motors, the trapezoidal machines (BLDC motors) prove their suitability for estimation of the speed from the measurement of back emf. The feature of the motor is that it has at any instant of time, only two phases excited during a 60° interval. The back emf of the unexcited winding can be utilised for estimating the speed as well as the position signal. A variety of algorithms have been developed in the literature which specifically employ the measurement of the third phase (unexcited) winding emf to determine the commutation instants of the devices of the inverter. This eliminates the mechanical position sensor on the shaft. Some of these schemes have been successfully implemented and they are in commercial production.

4.5.5 Extended Kalman Filter for the State Estimation of PMSM

The Extended Kalman Filter (EKF) is an optimal recursion algorithm for state estimation of non-linear systems. Its features are as follows.

- Capability to process all the measurements regardless of their accuracy and precision
- Capability to provide a quick and accurate estimate of any state variable of interest
- Rapid convergence

The EKF approach is ideally suited for the estimation of the states of PMSM, which include the rotor position and speed of the rotor. Apparently, it is a viable and computationally efficient candidate for online estimation of speed and rotor position. This capability is imparted to the algorithm by

- A sufficiently well-laid-down mathematical model which describes the motor dynamics; the knowledge of the system and dynamics measuring equipment are easily available
- The system noise, both process and measurement, system disturbances and uncertainties of the system model can be described satisfactorily
- Any available information about the initial condition of the states is of interest
- Measurable motor terminal voltages and currents

All steps involved in the computation of this computationally intensive algorithm require matrix or vector operations. An efficient formulation of the algorithm is a necessity. The computational requirements include computation time per cycle and the required memory storage. Prior knowledge of these helps the choice of meaningful data, sampling rate and the required memory size of the system. The constraints on the computer would be speed (cycle execution time), calculation capability (instruction set), and the type of arithmetic used (floating point or fixed point).

The rotor position and speed can be independently estimated from the measured terminal voltages and currents given as inputs, the model and use of EKF algorithm.

A permanent magnet motor with surface mounted magnets, i.e., the SPM motor, which has the configuration of a cylindrical rotor with a uniform air gap can be modeled using the dynamic model in a static reference frame. The SPM motor can be assumed to have sinusoidal excitation. The state space model of the motor for the dynamic analysis has as its state variables, the two-axis components of stator current $i_{\alpha s}$ and $i_{\beta s}$, the rotor position θ_r, and rotor angular velocity ω_r. The state space model in matrix notation is given by

$$p\begin{bmatrix} i_{\alpha s} \\ i_{\beta s} \\ \omega_r \\ \theta_r \end{bmatrix} = \begin{bmatrix} -\dfrac{r_s}{L_s} & 0 & \lambda_{af}\sin\theta_r & 0 \\ 0 & -\dfrac{r_s}{L_s} & -\lambda_{af}\cos\theta_r & 0 \\ -\dfrac{3}{2}\lambda_{af}\sin\theta_r & \dfrac{3}{2}\lambda_{af}\cos\theta_r & -\dfrac{B}{J} & 0 \\ 0 & 0 & 1 & 0 \end{bmatrix} \begin{bmatrix} i_{\alpha s} \\ i_{\beta s} \\ \omega_r \\ \theta_r \end{bmatrix} + \frac{1}{L_s}\begin{bmatrix} v_{\alpha s} \\ v_{\beta s} \\ T_L \\ 0 \end{bmatrix} \tag{4.41}$$

The two-axis components of the stator voltages $v_{\alpha s}$, $v_{\beta s}$ and average load torque T_L are the control inputs. $v_{\alpha s}$, $v_{\beta s}$ can be determined using Park's transformations in static reference frame. Using Eqn. (4.41). The numerical solution of state space model provides the current components $i_{\alpha s}$, $i_{\beta s}$, ω_r and θ_r as state variables. The above state space model has cross-coupling terms making it nonlinear. From this, the two-axis stator current components can be calculated.

4.6 CYCLOCONVERTER-FED SYNCHRONOUS MOTOR

A cycloconverter is a single-stage static frequency converter capable of converting source frequency to a variable frequency. This, with a suitable transformer on the input side, may be interposed between the ac mains and ac motor. The converter transformer has three different secondaries, each for a phase of the motor. The frequency range of the cycloconverter output is 0 to 33% of the rated frequency. The upper frequency limit is in view of the distortion of the output. This converter can be used for speed control of both induction and synchronous motors. When used with a synchronous motor, it can operate both in the line-commutated and load-commutated modes.

Two anti-parallel connected three-phase bridge converters form a single phase cycloconverter converting a 3-phase input to a single-phase output of variable frequency. Three of such units are required to provide a 3-phase output. The number of devices of a single-phase converter are 12 and those of a 3-phase unit are 36. The application of the cycloconverter is cost justified only at high power levels. The output of a cycloconverter is fabricated from the segments of the input voltage and hence the output voltage waveform is nearly sinusoidal. Further, the motor inductance smoothens the motor current waveform. The effects of harmonics on the performance of the motor is least as the input voltage to the motor and motor current contain almost negligible harmonics. Very smooth low-speed operation is possible. The cycloconverter allows energy transfer in both the directions, thereby facilitating a straightforward four-quadrant operation of the motor. As the converter current can be adjusted to have any desirable phase difference with respect to phase voltage, it can feed power to loads of any power factor. The cycloconverter operates both in the circulating current mode and circulating current-free mode. In the former, both the converters operate simultaneously. In the latter, the conducting bridge is fired and the latter is not in operation and does not get firing pulses. The transfer between bridges takes place only when the polarity has to reverse.

The leakage inductance of the transformer forms the main part of commutation inductance, and it can be assumed to be concentrated in the secondary side for all purposes of commutation analysis. However, under line-commutated mode the power factor is very poor.

As has already been indicated, under line-commutated mode, the output voltage of the converter, which is made up of the segments of the input voltage, is nearly sinusoidal and the machine inductance causes further smoothing of motor current. Larger the inductance, better the smoothing action. During operation, subtransient inductance is very effective. The value of this is reduced by the presence of damper windings. In view of the smoothening of motor current, the inductance must be large and no damper windings need be there.

In contrast to the above, when the converter operates in the load-commutated mode it can work as a current source converter, in which case the load voltage (motor voltage) is decided by the operating conditions of the motor. The motor voltage becomes the converter voltage also. To reduce the voltage spikes during commutation, the commutation time for the subtransient reactance should be small. In such a case, the motor should have damper windings.

The synchronous motor is operated in the self-controlled mode, wherein the stator frequency is decided by the motor speed. This control is accomplished by means of a shaft encoder giving the rotor position information. By the time the rotor moves by two pole pitches, one cycle of frequency is completed. Thus, the frequency of the stator becomes a slave of the rotor speed [commutator action in the case of dc motor]. Obviously, the synchronous motor becomes free from hunting and instability problems. Therefore, a variable frequency synchronous motor in the self-controlled mode need not have damper windings. The starting of the motor does not pose any problem because static frequency controller-cycloconverter can be used to provide good starting with variable frequency.

The necessity of damper windings is decided by only the subtransient operation of the motor fed by VS or CS operation of the cycloconverter.

Another attractive feature of the synchronous motor is the possibility of load commutation of the inverter if the synchronous motor is over exerted. The cycloconverter feeding the motor offers another flexibility. At low frequencies where the machine voltages are too small to render load commutation, the line commutation of the converter can be made use of and when once the speed increases to a value where the machine voltages are sufficiently large for load commutation to take over, one can change over to load commutation. This increases the speed control range of the motor up to the rated speed and also above.

Self-control imparts to a synchronous motor the performance of a separately exited dc motor with an additional feature that the angle between the stator and rotor mmfs can be controlled. This feature is not present in a dc motor. The effects of zero sequence voltages of the converter output may be kept away by star connection of the motor with an isolated neutral.

The stator of a 3 phase synchronous motor is fed from current sources provided by the cycloconverter, whereas the field winding is fed from a current source obtained by means of a phase-controlled rectifier. The field current establishes a space distributed flux in the air which rotates along with the rotor structure. The rotor position information from a rotor position sensor is used to fire the thyristors of the cycloconverter to excite the stator with a frequency in synchronism with rotor speed. The stator currents produce an mmf rotating at the same speed as the rotor mmf, however with a definite space angle to produce the desired torque. The two mmfs produce a resultant mmf (air flux). The torque developed is proportional to the area of the triangle (Fig. 4.86). The torque developed T_d is, therefore, a function of the stator current, field current and the angle between them.

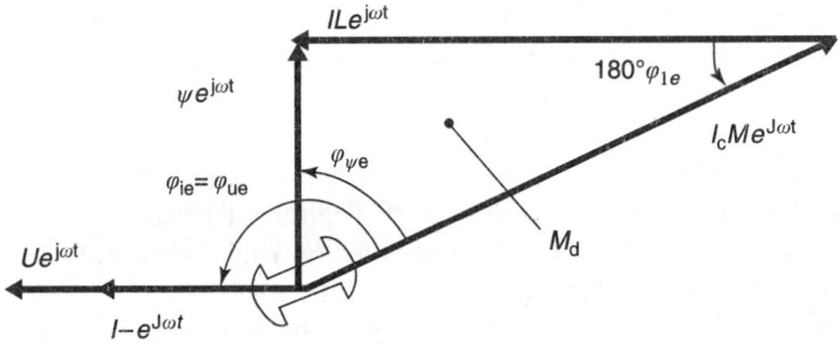

$D = 0°, -120°, +120°$:
a b c

$u_{a,b,c} = U\cos(\omega_t + \varphi_{ue} + D)$ $\qquad U = Ue^{j\varphi ue}$

$\psi_{a,b,c} = \psi\cos(\omega_t + \varphi_{\psi e} + D)$ $\qquad \psi = \psi e^{j\varphi\psi e}$

$i_{a,b,c} = I\cos(\omega_t + \varphi_{ie} + D)$ $\qquad I = Ie^{j\varphi ie}$

$i_e = I_e$ $\qquad I_e = I_e$

Fig. 4.86 *Phasor diagram*

The resultant stator flux linkages and the stator voltage can be represented by

$$\psi = \text{LI} + \text{MI}_f$$

$$v = j\omega\psi$$

neglecting the stator resistance.

The best magnetic utilisation of the motor is accomplished if only the motor operates at the rated flux. The current needed from the stator side will also be the least (in the converter as well as the motor). The stator current, field current and the angle between them are controlled such that

i. resultant air flux remains constant at all load conditions and speeds at rated value, and
ii. the stator current is purely active in nature, i.e., it is in phase with voltage.

Neglecting stator copper losses, the power developed = power input to the motor.

The torque developed = $\dfrac{p_d}{2\pi n_s} = T_d$.

The independent speed and torque control are the chief objectives. These are realised by maintaining the air flux as a constant at the rated value and controlling the stator current to be purely active.

The speed controller produces a set of current references based on the speed error at the input. These are stator current references as well as rotor current references. These reference currents are decided by the control requirement and forced into the motor for the desired torque, whereas independent frequency variation provides the speed. The stator current controls via cycloconverter impress the required stator mmf and the rotor current control impresses the dc excitation to the rotor via the rectifier in the field circuit. The cycloconverter is also controlled to provide the required phase difference between the stator and rotor mmfs. The stator current, rotor current and the phase difference between them (which produce torque) are so coordinated that the (stator) flux remains constant and the stator current is in phase with the voltage. This coordination may be accomplished using suitable function generators. Using the cycloconverter on the stator side, phase-controlled rectifier on the rotor side and self-control independent torque and speed control may be realised for a synchronous motor. High torque even at every low speeds including standstill conditions is possible.

Feeding of the motor with a stator current having desired phase shift with respect to rotor mmf is possible in the self-control of the motor and this control eliminates the hunting and instability present in conventional behaviour of synchronous motor and imparts to the synchronous motor the performance of a separately excited dc motor.

There are other sophisticated controls of a synchronous motor which more or less can be realised based on the above principles of control. Vector or field-oriented control uses the machine models for determining the control parameters such as flux. These models make use of the motor parameters. The control of the motor thus results in parameter-sensitive motor behaviour which may impair the concept of field orientation and the desired dynamic performance of the motor may not be realised. To improve the dynamic performance even at heavy peaks of load torques, the following may be ensured with additional loops:

—a more direct field orientation of the stator current
—a more direct stator flux control with least parameter-sensitive flux estimation

The rated flux operation can be realised by operating the motor with a stator voltage proportional to frequency (with compensation for stator resistance voltage at low frequencies) and constant field current. The voltage variation can be accomplished by varying the firing angle α from 0° to 90°, the latter being for standstill conditions. The value of α decreases as the frequency increases and becomes zero at the rated frequency. The power factor becomes closer to unity if the cycloconverter operates on the trapezoidal mode $\alpha \rightarrow 0$.

The cycloconverter feeding the motor can also make use of machine voltages for commutation by overexciting the motor. The performance features of a CSI fed synchronous motor with load commutation are applicable for cycloconverter-fed synchronous motor with load commutation.

Applications The cycloconverter-fed synchronous motors are best suited where the drive requirements are four-quadrant operation at high power levels in the low-speed range with a limit on the maximum speed.

- Gearless cement mills are the first application of these drive motors. The motor has a large number of poles with limited frequency of operation for driving the tube mill.
- Another typical application is for mine hoists with high power ratings.
- Reversing steel rolling mills with very high dynamic requirements for torque and regenerative speed reversal.
- Diesel-generator-fed cycloconverter synchronous motors with power ratings of about 20 MW are used in ships and ice breakers.
- These are also being used in diesel traction systems.

The cycloconverter-fed synchronous motor drive systems are fully developed mature systems. As such, they do not show any advantages from the high-speed self-turn off modern power semiconductor devices. They will remain a standard solution for high torque, low-speed reversible drives.

Worked Examples

4.1 *A single phase, single pulse controlled rectifier is fed from a 120 V, 60 Hz supply. This provides a variable voltage to the armature of a separately excited dc motor having an armature resistance of 10 Ω. Due to high inertia* *the motor speed is constant, providing a back emf of 50 V when the thyristor is triggered continuously. Neglecting the armature inductance determine the average value of current in the motor.*

Solution The thyristor starts conducting when the instantaneous voltage = 50 V. Therefore

$$a = \sin^{-1}\frac{50}{\sqrt{2}\times 120} = 17.14°$$

The current flows in the load from a to $180 - a$. The current

$$i = \frac{\sqrt{2}V\sin\omega t - E_b}{10}$$

$$I_{av} = \frac{1}{2\pi}\int_{a}^{180-a}\left(\frac{\sqrt{2}V\sin\omega t - E_b}{R}\right)d(\omega t)$$

$$= \frac{1}{2\pi R}(2\sqrt{2}V\cos a - E_b(\pi - 2a))$$

$$= \frac{1}{62.8}\left(2\sqrt{2}\times 120\cos 17.14 + 50(2.542)\right)$$

$$324.288 \qquad\qquad -127.1$$

$$= 2.89 \text{ A}$$

4.2 *A single pulse converter feeds a separately excited dc motor having an armature resistance of 1 Ω and inductance 50 mH. The supply voltage is 220 V at 50 Hz. The firing angle is 75° and the back emf is 130 V. Assuming the devices to be ideal determine the current waveform, its mean value and converted power. There is a FWD across the load.*

Solution If i is the current through the load from 75° to 180° the equation is

$$Ri + L\frac{di}{dt} + E_b = V$$

from 180° to $180 + \beta$ the equation is

$$Ri + L\frac{di}{dt} + E_b = 0$$

When the thyristor conducts the current i is given by

$$i = \frac{\sqrt{2}V}{z}\sin(\omega t - \phi) - \frac{E_b}{R_a} + Ae^{-t/\tau}$$

$$
\begin{aligned}
i(a) &= 0\\
z &= 1 + j314 \times 50 \times 10^{-3} = 1 + j15.7 = 15.732\\
\phi &= 86.36°, \tau = 50 \times 10^{-3} = 0.05 \text{ s}\\
\omega\tau &= 314 \times 0.05 = 15.7\\
i &= 19.774 \sin(\omega t - 86.36) - 130 + Ae^{-\omega t/15.7}\\
a &= \omega t = 75°\\
i &= 0 = 19.774 \sin(75 - 86.36) - 130 + A0.92\\
A &= 145.54\\
i &= 19.774 \sin(\omega t - 86.36) - 130 + 145.54e^{-\omega t/15.7}\\
\text{At } \omega t &= \pi\\
i &= 19.734 - 130 + 119.158 = 8.892.
\end{aligned}
$$

During freewheeling

$$
\begin{aligned}
i &= -\frac{E_b}{R} + A_1 e^{-t/\tau}\\
8.892 &= -\frac{E_b}{R} + A_1 = -130 + A_1\\
A_1 &= 138.89\\
i &= -\frac{E_b}{R} + 138.89e^{-t/\tau} = 0\\
e^{-t/\tau} &= \frac{130}{138.89}\\
-t/\tau &= -0.06615\\
\omega t &= 59.53°
\end{aligned}
$$

The current is

$$i = 19.774 \sin(\omega t - \phi) - 130 + 145e^{-\omega t/\omega\tau}$$

$$75° \text{ to } 180°$$

$$i = 138.89e^{-\omega t/\omega\tau} - 130 \quad 180° \text{ to } 239.53°$$

4.3 *A separately excited dc motor rated at 10 kW, 240 V, 1000 rpm is supplied from a fully controlled two pulse bridge converter. The converter is supplied at 250 V, 50 Hz supply. An extra inductance is connected in the load circuit to make the conduction continuous. Determine the speed, power factor and efficiency of operation for thyristor firing angles of 0 and 60° assuming the armature resistance of 0.40 Ω and an efficiency of 87% at rated conditions. Assume constant torque load.*

Solution

The input to the motor at rated conditions $= \dfrac{10 \times 10^3}{0.87} = 11.494 \times 10^3$ W

The supply current to the motor = 47.89 A

Neglecting field copper loss the armature current = 47.89 A

The back emf at rated conditions

$$= 240 - 47.89 \times 0.4 = 220.843 \text{ V}$$

At $\alpha = 0$ the converter voltage

$$= 0.9 \times 250 = 225 \text{ V}$$

As the load torque is constant the armature current is the same. Therefore back emf.

$$= 225 - 47.89 \times 0.4 = 200.844$$

$$\text{speed} = \frac{200.844}{220.843} \times 1000 = 909.4 \text{ rpm}$$

$$\text{Fundamental displacement factor} = 1$$

$$\text{distortion factor} = 0.9$$

$$\text{The power factor} = 0.9$$

$$\text{The input} = 240 \times 47.89 \times 0.9 \times 10^{-3} = 10.344 \text{ kW}$$

Assuming for constant torque the output varies in the direct proportion of speeds output at lower speed

$$= \frac{10 \times 909.4}{1000} = 9.094 \text{ kW}$$

$$\text{Efficiency} = 87.9\%$$

$$\text{At a firing angle } \alpha = 60°$$

$$\text{The converter voltage} = 0.9 \times 250 \times \cos 60 = 112.5 \text{ V}$$

$$\text{The back emf} = 112.5 - 47.89 \times 0.4 = 93.344 \text{ V}$$

$$\text{Speed} = \frac{1000 \times 93.344}{220.843} = 422.7 \text{ rpm}$$

$$\text{Output} = \frac{10 \times 422.7}{1000} = 4.227 \text{ kW}$$

$$\text{Input} = 112.5 \times 47.89$$

$$\text{Efficiency} = 78.46$$

4.4 *A single phase bridge converter fed from a 250 V, 50 Hz supply is used to control the speed of a separately excited dc motor having an armature resistance of 1.5 Ω and inductance 30 mH. The back emf at one speed of operation is 100 V and the converter control angle is 30°. Determine the average and rms values of the armature current, the power input to the motor and power factor of operation.*

Solution The average value of the converter voltage at $\alpha = 30°$

$$= 0.9 \times 250 \times \cos 30 = 194.85 \text{ V}$$

The average value of voltage across the inductance is zero.

$$\therefore \qquad I_d = \frac{194.85 - 100}{1.5} = 63.24 \text{ A}$$

If the inductance is sufficiently large the rms value of

$$I_d = 63.24 \text{ A}$$

Power delivered to the motor = 194.85 × 63.24 × 10^{-3} = 12.284 kW
Power factor = 0.9 cos 30° = 0.78

Because of the definite value of inductance the rms value of current must be found out from the wave form of the current.

$$i(t) = \frac{\sqrt{2}V}{z}\sin(\omega t - \phi) - \frac{E_b}{R} + Ae^{-t/\tau}$$

A can be found out from the condition

$i(a/\omega) = i\left(\dfrac{\pi + a}{\omega}\right)$ because the current is continuous. Using the numerical values $i(t)$ and its rms value can be determined.

4.5 *A separately excited dc motor is fed from a three-phase six pulse fully controlled bridge converter. The motor develops its full load torque at a rated speed of 1800 rpm taking a current of 60 A from a 400 V supply. Determine the rms value of supply voltage if the motor runs at its rated conditions for a = 0. What is the range of firing angles for a speed control of 1800 to 900 rpm. The armature resistance is 0.5 Ω. The supply and thyristors are ideal.*

Solution Average value of converter voltage

$$= 2.34\, V_s \cos a = 1.35 V_L \cos a$$

At a = 0 this value = 400 V.

$$\cos a = \frac{400}{2.34 V_s} = 1.$$

$$V_s = \frac{400}{2.34} = 171 \text{ V}, \quad V_L = 296 \text{ V}$$

Induced voltage = 400 – 60 × 0.5 = 370 V
E_2 at 900 rpm = 185 V
V_{dia} at 900 rpm – 185 + 30 = 215
$2.34 V_s \cos a$ = 215

$$\cos a = \frac{215}{2.34 \times 171}$$

$$a = 57.5°$$

4.6 *A three-phase six pulse bridge converter working on a 500 V 50 Hz supply feeds a dc motor having a rated voltage of 250 V. The motor is separately excited and draws an armature current of 181 A at 250 V and runs at 1500 rpm. The motor drives a load having a torque speed characteristic given by*

$$T_L = 0.64\,\omega^2$$

If the speed control of this motor is required from 1500 to 500 rpm determine the range of firing angles. If the firing angle α = 0 for operation at rated speed, determine the line voltage. What is the range of firing angles in this case? What is the advantage of the second use? Resistance of the armature = 0.1 Ω.

Solution The back emf at rated speed

$$= 250 - 181 \times 0.1 = 231.9 \text{ V}$$

The back emf at 500 rpm

$$= \frac{231.9}{3} = 77.3 \text{ V}$$

Firing angle for rated speed

$$\cos\alpha = \frac{250}{1.35 \times 500}$$

$$\alpha = 68.26°$$

The armature current at lower

$$\text{speed} = I_{a2} = \frac{I_{a1}}{9} = 20.11 \text{ A}$$

$$V_2 = 77.3 + 20.11 \times 0.1 = 79.31 \text{ V}$$

$$\cos\alpha = \frac{79.31}{1.35 \times 500} = \alpha = 83.25°$$

The second case can be achieved by a step down transformer.

At $\alpha = 0$

$$1.35 V_L = 250$$

$$V_L = \frac{250}{1.35} = 185.2 \text{ V}$$

Firing angle for second speed

$$\cos^{-1}\frac{79.31}{1.35 \times 185.2} = 71.505°$$

The reactive power requirement gets reduced. Power factor improves. However, the leakage inductance must be considered while determining the firing angles.

4.7(a) *A three phase, half controlled bridge rectifier fed from a 300 V, 60 Hz supply provides a variable voltage supply to the armature of a separately excited dc motor. The specifications of the motor are*

$R_a = 0.02\Omega \quad L_a = 0.002$ H

The constant of the motor 2.25 vs/rad
rated current is 500 A

Determine the firing angle a so that the motor runs at a rated speed of 1500 rpm.

Solution The speed of the motor = 1500 rpm

= 157 rad/s

The induced voltage or back emf of the motor = 353.25 V

The terminal voltage = 363.25 V

The converter voltage = $1.35 \times 300 \cos a$

$$a = \cos^{-1} \frac{363.25}{1.35 \times 300} = 26.25°$$

4.7(b) *If in the above problem the supply has an inductance of 1.0 mH, determine the speed at which the motor would run for a = 26.25°.*

Solution Supply reactance $X_c = 2\pi f L_c = 0.3768$

$$I_d = \frac{V_{dia} - E_b}{R_d + \frac{3X_c}{\pi}} = \frac{V_{dia} - E_b}{0.38}$$

$V_{dia} - E_b = 190 \qquad \therefore E_b = 173.25$

$363.25 - E_b = 190 \qquad$ speed is 49.04 % of the speed in the previous problem

4.8 *A 600 V dc series motor has a combined armature and field resistances of 0.16 Ω. The motor draws an armature current of 210 A while running at a speed of 600 rpm. The motor drives a load having a torque speed characteristic*

$$T_L = A\omega^2$$

Determine the constant A while running at a speed of 600 rpm.

This motor is to be controlled by a three phase fully controlled bridge converter. Determine the input voltage of the converter if the speed of 600 rpm is required at a = 0. Assume a linear magnetisation characteristic determine the range of firing angles for a speed control range 600-0 rpm.

Solution

$$T_d = \frac{E_b I_a}{2\pi n} = \frac{(600 - 0.16 \times 210)210}{2\pi \times 10}$$

$$= \frac{566.4 \times 210}{62.8} = 1894 \text{ Nm}$$

$$\omega = 2\pi \frac{N}{60} = 2\pi \times 10 = 62.8 \text{ rad/s}$$

$$T_L = A \times (62.8)^2 = 1894$$

$$A = \frac{1894}{(62.8)^2} = 0.4802$$

$$T_d = K_t I_a^2$$

Therefore

$$\frac{I_{a1}}{I_{a2}} = \frac{N_1}{N_2} \quad I_{ax} = I_{a1} \frac{N_X}{N_1}$$

$$\text{Also } \frac{E_1}{E_2} = \frac{\phi_1 N_1}{\phi_2 N_2} \quad E_X = \frac{I_{ax} N_X}{I_{a1} N_1} \cdot E_1$$

Speed	I_a	$I_a R_m$	E	V	*a*
500	175	28.0	393.33	421.33	45.47°
400	140	22.4	251.73	274.13	62.85
300	105	16.8	141.60	158.40	74.69
200	70	11.2	62.90	74.10	82.90
100	35	5.6	15.73	21.33	87.96
0					90.00

4.9 *A three phase dual converter feeds a 500 V, 60 A dc motor with separate excitation. The armature resistance is 1.5 Ω. The converter is fed from a 420 V, 50 Hz supply. Assuming a voltage drop of 20 V in the converter determine the firing angle and back emf for*

i. *motoring operation at full load current with motor terminal voltage of 450 V.*
ii. *regeneration operation at full load current with terminal voltage of 450 V.*
iii. *the motor is plugged at a terminal voltage of 400 V with a current limiting resistor of 5Ω.*

Solution

i. The average value of converter voltage

$$= 1.35 V_L \cos a = 450 + 20 = 470$$

$$\cos a = \frac{470}{1.35 \times 420}$$

$$a = 34.01°$$

Back emf = 450 – 60 × 1.5 = 360 V

ii. During regeneration the induced voltage = –450 – 90 = –540 V.
The converter voltage = –450 + 20 = –430 V

$$\cos a = \frac{-430}{1.35 \times 420}$$

$$a = 139.32°$$

iii. When the motor is plugged the motor voltage is –400 V, I_a = 50 A
Back emf = –400 – 90 = –490 V
The converter voltage = –400 + 20 + 5 × 60
= –80 V

$$\cos a = \frac{-80}{1.35 \times 420} \quad a = 98.52$$

4.10 *A dc chopper feeds a dc series motor. The supply voltage to the chopper is 500 V. The total current is found to vary between two current limits having a difference of 15 A. The time ratio of the chopper is 0.6 and its pulse frequency 80 cycles/s. Determine the armature inductance of the motor.*

Solution The average voltage of the chopper V_{av} = 500 × 0.6 = 300 V.
The voltage across inductance is ac with its peak value = 500 – 300 = 200 V

$$L\frac{\Delta I_a}{\Delta t} = 200 \text{ V}$$

$$\Delta t = T_{ON} = 0.6 \times T = \frac{0.6}{80}$$

$$= 7.5 \text{ ms}$$

$$L = 200 \times \frac{\Delta t}{\Delta I_a} = \frac{200}{15} \times 7.5$$

$$= 100 \text{ mH}$$

4.11 *A 240 V, 50 Hz, four pole star connected three phase induction motor has the following equivalent circuit parameters.*

$r_1 = 0.25\Omega$, $x_1 = 0.36\Omega$, $x_m = 17.3\ \Omega$
$r_2 = 0.60\Omega$, $x_2 = 0.36\Omega$

The speed of the motor is to be controlled by a three-phase voltage controller from 1400 to 600 rpm. The motor drives a fan load having characteristic

$$T_L = (1.4/10^3)\omega^2$$

Calculate the range of firing angles.

Solution Synchronous speed of the motor = 1500 rpm

$$\text{Slip at 1400 rpm } s_1 = \frac{100}{1500} = 0.067$$

$$\text{Slip at 600 rpm } s_2 = \frac{900}{1500} = 0.6$$

Rotor impedance $8.955 + j0.36 \quad 1 + j0.36$

Stator impedance $0.25 + j0.36 \quad 0.25 + j0.36$

$Z \quad 9.205 + j0.72 \quad 1.25 + j0.72$

Input impedance $j17.3\| \quad j17.3\|$

$Z_m\|Z \quad (9.205 + j0.72) \quad (1.25 + j0.72)$

$7.894\angle 31.53° \quad 1.382\angle 33.92°$

$$T_L = 1.4 \times 10^{-3}\,\omega^2$$

$$T_1 = 1.4 \times 10^{-3}\left(\frac{2\pi \times 1400}{60}\right)^2 = 30.06 \text{ Nm}$$

$$\frac{3}{2\pi n_s}\frac{I_{21}^2 r_2^1}{s_1}(1 - s_1) = 39.44$$

$$I_{21} = 13.25\ A \quad I_{22} = 18.17 \text{ A}$$

Base currents at these speeds are

$$I_{b1} = 17.55 \text{ A} \qquad I_{b2} = 100.27$$

Per unit rotor currents

$$I_{P1} = 0.755 \qquad I_{P2} = 0.1146$$

$$I_{11} = \frac{(r_1 + r_2'/s) + j(x_1 + x_2')}{jX_m} I_2'$$

$$= 15.47 \text{ A}$$

$$I_{12} = 18.97 \text{ A}$$

Per unit stator currents = 0.8815

0.1892

From the graph the firing angles are 32° and 115°.

4.12 *A 100 hp, 460 V, 60 Hz star connected squirrel cage induction motor has the following equivalent circuit parameters.*

$$r_1 = 0.06\ \Omega,\ r_2' = 0.35\ \Omega,\ x_1 + x_2' = 0.6\ \Omega,\ X_m = 8\Omega.$$

The motor drives a fan which requires 100 hp at a speed of 1000 rpm. Determine the firing angles required for a speed range of 200 to 1000 rpm.

Solution Obviously the motor must have six poles with a synchronous speed of 1200 rpm.

$$\text{Slip at 1000 rpm} = \frac{1200 - 1000}{1200} = 0.167$$

The input impedance of the motor at this slip

$$= \frac{j8(2.16 + j0.6)}{2.16 + j8.6} = 2.02\angle 29.56°$$

$$\text{The stator current} = \frac{460}{\sqrt{3}} \times \frac{1}{2.02} = 131.48 \text{ A}$$

$$\text{Slip at 200 rpm } \frac{1000}{1200} = 0.8333$$

$$\text{The input impedance } = \frac{j8(0.48 + j0.6)}{0.48 + j8.6} = 0.7133\angle 54.56$$

$$\frac{I_a^2}{131.48^2} = \frac{0.833(1-0.833)^2}{0.167(1-0.167)^2}$$

$$I_a = 58.86$$

Per unit value = 0.16

The firing angle is 115°

The range of firing angles 30 – 115°

4.13 *A three phase slip ring induction motor has a chopper controlled resistance in the rotor circuit for speed control. With the chopper completely ON always, the maximum torque occurs at a slip of 0.2. With the chopper completely OFF the maximum torque occurs at a slip of 1. Determine the value of resistance in the chopper.*

Solution Neglecting the stator impedance

$$\frac{s_{m2}}{s_{m1}} = \frac{r_2 + r_a}{r_2} = \frac{1}{0.2} = 5$$

$$r_a = 4r_2$$

If R is the resistance in the chopper the copper loss

$$= 3I_2^2 r_2 + R\left(\frac{T_{ON}}{T}\right)^2 I_d^2$$

But

$$\frac{I_2}{I_d} = \sqrt{2/3}$$

$$\text{Copper loss} = 3I_2^2 r_2 + R\left(\frac{T_{ON}}{T}\right)\frac{3}{2}I_2^2$$

$$= 3I_2^2(r_2 + r_{ad})$$

$$\therefore \quad r_{ad} = \frac{1}{2}\left(\frac{T_{ON}}{T}\right)R$$

$$r_{ad} = r_a$$

$$\text{if } \frac{T_{ON}}{T} = 1, r_a = \frac{R}{2}$$

$$\text{Substituting } \frac{R}{2} = 4r_2$$

$$R = 8r_2$$

A resistance $R = 8r_2$ must be placed so that the torque speed curve will have its slip for maximum torque at unity, if the chopper is always OFF.

4.14 *A 440 V, 50 kW, 50 Hz, three phase slip ring induction motor has the following equivalent circuit parameters*

$$r_1 = 0.07\ \Omega \qquad r_2 = 0.05\ \Omega$$
$$x_1 + x_2' = 0.2 \qquad \Omega X_m = 20\ \Omega$$

The speed of the motor at rated load is 1420 rpm. Determine the resistance required in the chopper circuit so that the speed can be controlled in the range 1420-1000 rpm at constant torque. Determine the TR for 1100 rpm.

Solution The torque is constant in the range of speeds. I_d is constant leading to the relation

$$\frac{r_2 + r_a}{r_2} = \frac{s_2}{s_1} = \frac{500}{80}$$
$$r_2 + r_a = 0.3125$$
$$r_a = 0.3075$$
$$R = 2r_a = 0.615$$

Time ratio for 1100 rpm:
slip rpm = 400 rpm

$$\frac{r_2 + r_a}{r_2} = \frac{500}{80} = 5$$
$$r_a = 4r_2 = 4 \times 0.05 = 0.2$$
$$0.2 = \frac{0.615}{2}\left(\frac{T_e}{T}\right)$$
$$\frac{T_e}{T} = 0.65 \quad \frac{T_{ON}}{T} = 0.35$$

4.15 *A slip ring induction motor incorporating slip energy recovery scheme for speed control drives a fan load having a characteristic of $T_L = KN^2$. The motor is rated at 440 V, 50 Hz, 100 kW. The speed is to be controlled from a rated value of 1420 to 750 rpm. The equivalent circuit parameters are*

$$r_1 = 0.052\ \Omega, \quad r_2 = 0.06\ \Omega,$$
$$X_m = 10\ \Omega, \quad x_1 + x_2' = 0.29\ \Omega \quad T_p/T_s = 1.2$$

Determine the firing angle range.

Solution The rated load is assumed to occur at $a = 90°$.

$$\text{slip} = \frac{60}{1500} = 0.04$$

The torque developed $s \dfrac{3}{2\pi n_s} \dfrac{I_2^2 r_2}{s} = KN^2$

$$\text{The Current } I_2' = \frac{V}{\left(\left(r_1 + \dfrac{r_2'}{s}\right)^2 (x_1 + x_2')^2\right)^{1/2}}$$

$$I_2' = 125.6 \text{ A}$$

$$\frac{60}{2\pi \times 1500}(125.6)^2 \times 1.5 = K(1420)^2$$

from which $K = 7.47 \times 10^{-5}$

$$T_{\text{rated}} = 150.72 \text{ Nm}$$

$$\text{Slope of the characteristic} = \frac{150.72}{60}$$

T at 750 rpm = 42.02
The slip rpm x such that

$$\frac{42.02}{x} = \frac{150.72}{60}$$

$$x = 16.73$$

No load speed 766.73 = 767
No load slip = 0.489

$$1.12\cos a = -0.489$$

$$a = \cos^{-1}\frac{-0.489}{1.12}$$

$$a = 115.88°$$

4.16 *A three phase, 400 V, 50 Hz, four pole, delta connected squirrel cage induction motor is fed from a six pulse bridge inverter supplied from a dc source, such that the fundamental of the motor voltage is the same as the motor rated voltage at 50 Hz. The equivalent circuit parameters of the motor are as follows: Stator impedance (1 + j2.6)Ω rotor impedance (0.5 + j2.4)Ω full load speed is 1425 rpm at 50 Hz, calculate the current wave of the motor assuming V/f is constant for (a) 50 Hz (b) 25 Hz and (c) 5 Hz.*

Solution The voltage waveform of a delta connected three phase motor fed from an inverter is as shown in Fig. p 4.16
The fundamental of this is

$$V_1 = \frac{4}{\pi} V_{dc}\sqrt{3} = 400$$

$$V_{dc} = \frac{400 \times \pi}{4\sqrt{3}} = 181.29 \text{ V}$$

Neglecting magnetising impedance

$$Z_1 = 1 + j2.6 + \frac{0.5}{0.05} + j2.4 = 11 + j5 = 12.083$$

Impedance angle is 24.444°

$$\text{The fundamental current} = \frac{400}{12.083} = 33.104 \text{ A}$$

The instantaneous value of current = 46.80 sin(ωt – 24.44) A

$$\text{5th harmonic current: } s_5 = \frac{1500 \times 5 + 1425}{1500 \times 5} = 1.19$$

$$V_5 = \frac{400}{5} = 80 \text{ V}$$

$$Z_5 = 1 + j13 + j\frac{0.5}{s_5} + j12 = 1 + j13 + 0.42 + j12$$

$$= 1.42 + j25 = 25.04\angle 86.75°$$

Fifth harmonic current

$$= \frac{80}{25.04}\angle -86.75° = 3.195\angle -86.75°$$

$$i_5 = 4.518 \sin(5\omega t - 86.75°)$$

$$\text{Seventh harmonic } s_7 = \frac{7 \times 1500 - 1425}{7 \times 1500}$$

$$= 0.8643$$

$$Z_7 = 1 + j18.2 + 0.5785 + j16.8$$

$$= 1.5785 + j35 = 35.036\angle 87.42°$$

$$V_7 = 57.143 \, V \quad I_7 = 1.631$$

$$I_7 = 2.306 \sin(7\omega t - 87.42)$$

$$s_{11} = 1.0864$$

$$Z_{11} = 1 + 0.4603 + j(28.6 + 26.4)$$

$$= 55.02\angle 88.14$$

$$I_{11} = \frac{400}{11 \times 55.02} = 0.6609\angle -88.14°$$

$$i_{11} = 0.9345 \sin(11\omega t - \delta 88.14°)$$

$$s_{13} = 0.927$$

$$Z_{13} = 1.5394 + j65 = 65.0182$$

$$V_{13} = \frac{400}{13}$$
$$I_{13} = 0.4732\angle -88.64°$$
$$i_{13} = 0.669 \sin \omega t(13\omega t - 88.64)$$

It can be shown that based on approximate equivalent circuit

$$I_{\text{har}} = \sqrt{0.0022}\,\frac{V_1}{(X_1 + X_2)} = 3.752$$

The wave forms may be drawn to determine the net waveform.

The above procedure may be adopted to determine the currents at other frequencies.

4.17 *A 500 V, three phase, 50 Hz star connected induction motor has the following equivalent circuit parameters: Stator impedance (0.13+j0.6)Ω rotor impedance (0.32+ jl.48)Ω magnetising reactance x_m = 20 Ω. The synchronous speed at 50 Hz is 1500 rpm. The full load speed is 1425 rpm. Calculate the approximate value of rms stator current when the motor is supplied by a six pulse VSI at speeds 1425 and 675 rpm. The frequency is varied such that V/f is constant. The stepped voltage applied to the motor has a fundamental equal to rated voltage at 50 Hz.*

Solution At 50 Hz the amplitude of dc voltage

$$V_{\text{dc}} = \frac{500}{\sqrt{3}}\,\frac{\pi}{4} = 216.5 \text{ V}$$

Wave form of the voltage is shown in Fig. p 4.17. Slip of the motor = 5%

$$Z_1 = 0.13 + j0.6 + \frac{0.32}{0.05} + j1.48$$
$$= 6.53 + j2.08 = 6.853\angle 17.67°$$
$$I_1 = 42.13\angle -17.67 \quad i_1 = 59.56 \sin(\omega t - 17.67°)$$

Assuming the approximate equivalent circuit

$$I_{\text{har}}^2 = \frac{0.0022V_1^2}{(X_1 + X_2)^2} = 42.375 \quad I_h = 6.51 \text{ A}$$
$$I = 42.62 \text{ A}$$

At 25 Hz

$$n_s = 750 \text{ rpm}$$
$$\text{slip} = 0.10$$
$$Z_1 = 0.13 + j0.6 + \frac{0.32}{0.1} + j1.48$$
$$= 3.33 + j2.08 = 3.926\angle 31.99°$$

$$I_1 = \frac{250}{\sqrt{3} \times 3.926\angle 31.99} = 36.77\angle -31.99^\circ$$

$$I_{\text{har}}^2 = \frac{0.0022 \times (250)^2}{(X_1 + X_2')^2} = 31.782$$

$$I = 37.2 \text{ A}$$

4.18 *Making suitable assumptions show that the ratio of rms value of no-load current to the rms value of its fundamental is given by*

$$\frac{I}{I_{10}} = \sqrt{1 + \frac{1}{\sigma^2}\left(\frac{5}{486}\pi^4 - 1\right)} = \sqrt{1 + \frac{0.0021153}{\sigma^2}}$$

Also show that

$$\frac{I_{\text{eff}}}{I_{\text{IN}}} = \sqrt{1 + \left(\frac{Z_{\text{IN}}}{\sigma X_{\text{IN}}}\right)^2 \left(\frac{5}{486}\pi^4 - 1\right)}$$

Solution When an induction motor is fed from a voltage source inverter.
At no-load: the slip is assumed to be zero
The rotor is almost open circuit
The impedance is magnetising impedance
Also, from the Fourier analysis

$$\frac{V_\nu}{V_{\text{d}}} = \frac{\sqrt{6}}{\pi \nu}$$

and $$Z_\nu = \nu \sigma X_{\text{m}}$$

The effective value of the harmonic current is

$$I_\nu = \frac{V_\nu}{Z_\nu} = \frac{\sqrt{6} V_{\text{d}}}{\pi \nu^2 \sigma X_{\text{m}}}$$

The fundamental is

$$I_{10} = \frac{\sqrt{6} V_{\text{d}}}{\pi X_{\text{m}}}$$

Therefore $$\frac{I_\nu}{I_{10}} = \frac{1}{\sigma \nu^2}.$$

The total effective value

$$\frac{I_{\text{eff}}}{I_{10}} = \sqrt{1 + \sum_{\nu=5}^{\infty}\left(\frac{I_\nu}{I_{10}}\right)^2} = \sqrt{1 + \frac{1}{\sigma^2}\sum \frac{1}{\nu^4}}$$

The sum

$$\sum_{1,5,7}^{\infty} \frac{1}{\nu^4} = \frac{5}{486}\pi^4$$

Hence

$$\sum_{1,5,7}^{\infty} \frac{1}{\nu^4} = \frac{5}{486}\pi^4 - 1$$

$$\therefore \qquad \frac{I_{eff}}{I_0} = \sqrt{1 + \frac{1}{\sigma^2}\left(\frac{5}{486}\pi^4 - 1\right)}$$

At rated slip the impedance for fundamental $Z_{1r} = \dfrac{V_{1r}}{I_{1r}}$

The rated fundamental current

$$I_{1r} = \frac{V_{1r}}{Z_{1r}} = \frac{\sqrt{6}V_d}{\pi Z_{1r}}$$

Assuming the simplified equivalent circuit for harmonics

$$I_\nu = \frac{\sqrt{6}V_d}{\pi \nu^2 \sigma X_m}$$

giving

$$\frac{I_\nu}{I_{1r}} = \frac{Z_{1\nu}}{\sigma X_m \nu^2}$$

$$\frac{I_{eff}}{I_{1r}} = \sqrt{1 + \left(\frac{Z_{1\nu}}{\sigma X_m}\right)\sum_{\nu=5} \frac{1}{\nu^4}}$$

Using the above relation of $\sum_{5}^{\infty} \frac{1}{\nu^4}$ we have

$$\frac{I_{eff}}{I_{1r}} = \sqrt{1 + \left(\frac{Z_{1r}}{\sigma X_m}\right)\left(\frac{5}{486}\pi^4 - 1\right)}$$

4.19 *A 400 V, three phase, 50 Hz, four pole squirrel cage induction motor has the following data:*

$$r_1 = 1.0\,\Omega \quad X_1 = 2.6\ \Omega \quad r_2' = 0.5\ \Omega$$
$$x_2' = 2.4\,\Omega \quad \text{full load slip } 5\% \quad x_m = 36.4\ \Omega$$

The stator winding is delta connected. The motor is supplied from a six pulse inverter whose dc voltage is such that its fundamental is the rated voltage of the motor at rated frequency. Determine approximately the rms value of current at 50 Hz, 25 Hz and 5 Hz at a rotor frequency of 2.5 Hz

Solution

At 50 Hz slip $= \dfrac{2.5}{50} = 0.05$

$$Z_1 = (1 + j2.6) + \frac{j36.4(10 + j2.4)}{10 + j38.8}$$

$$= (1 + j2.6) + \frac{36.4\angle 90\ 10.284\angle 13.496^\circ}{39.68\angle 75.403}$$

$$= 1 + j2.6 + 9.4339\angle 28.093$$

$$= 1 + j2.6 + 8.3224 + j4.4425$$

$$= 9.3224 + j7.0424623$$

$$= 11.683\angle 44.127^\circ$$

$$Z_1 = 9.3224 + j7.0425\ \Omega$$

$$\frac{I_{\text{eff}}}{I_{1\text{r}}} = \sqrt{1 + \left(\frac{11.683}{\sigma 36.4}\right)^2 \left(\frac{5}{486}\pi^4 - 1\right)} = 1.028$$

$$\sigma = \sqrt{1 - \frac{x_{\text{m}}^2}{x_1 x_2}} = 0.062$$

At 25 Hz $\quad s = \dfrac{2.5}{25} = 0.1$

$$Z_1 = 1 + j2.3 + \frac{j18.2(5 + j1.2)}{5 + j19.4}$$

$$= 1 + j1.3 + \frac{18.2\angle 90^\circ\ 15.142\angle 13.5}{20.034\angle 75.55}$$

$$= 1 + j1.3 + 4.67\angle 27.95$$

$$1 + j1.3 + 4.125 + j2.189$$

$$= 6.199\angle 34.2463$$

$$V_1 = 200 \text{ V}$$

$$\frac{I_{\text{eff}}}{I_{1\text{r}}} = \sqrt{1 + \left(\frac{6.199}{\sigma 18.2}\right)^2 \times 0.0021153} = 1.03143$$

At 5 Hz $\quad s = \dfrac{2.5}{25} = 0.5$

$$Z_1 = 1 + j0.26 + \frac{j3.64(1 + j0.24)}{1 + j3.88}$$

$$= 1 + j0.26 + \frac{3.64\angle 90^\circ\ 1.0284\angle 13.5^\circ}{4.007\ \angle\ 75.55^\circ}$$

$$= 1 + j0.26 + 0.9342\ \angle\ 27.95$$

$$= 1 + j0.26 + 0.438 + j0.825$$

$$= 1.438 + j1.085 - 1.542$$

$$\frac{I_{\text{eff}}}{I_{1\text{r}}} = \sqrt{1 + \left(\frac{1.542}{0.062 \times 3.64}\right)^2 (0.0021153)}$$

$$= 1.0482$$

4.20 *A bhp 230 vdc shunt motor has armature resistance of 0.05Ω and field circuit resistance R_f of 460 Ω. The no load speed is 1000 rpm. Find the speed when line current is (a) 75A (b) 150A, (c) 250A. Assume that the motor has compensating winding.*

Solution Speed at no-load 1,000 rpm
Line current at no load $I_{ao} = 0$
Back emf $E_o = 0V = 230V$

Since the motor has compensating winding, the armature reaction does not cause any demagnetisation and field flux is constant.

Therefore, $E_b \propto N$.

$$I_f = \frac{230}{46} 5A$$

(a) $75 - 5 = 70A$

$$\frac{E_{b1}}{E_{b2}} = \frac{N_1}{N_0} = \frac{1}{1000}$$

$$E_{b1} = V - I_a ra$$

$$230.70 \times 0.05$$

$$= \frac{226}{230} \times 1000$$

$$I_{a2} = 150 - 5 = 145$$

$$E_{b2} = 230 - 145 \times .05 = 222.75V$$

$$\frac{222.75}{230} \frac{N_3}{1000} \text{ or } N_3 = 946.7 \text{ rpm}$$

Problems

4.1 A single phase half wave phase controlled rectifier feeds a separately excited dc motor having a resistance and inductance in the armature circuit of 1.5Ω and 45 mH respectively. At a firing angle of 60° the back emf is 160 V. Determine the expression for the instantaneous armature current. Calculate the average output torque and speed. Neglect the converter losses and supply impedance. The motor is fed from a 220V, 50 Hz source.

4.2 A single phase, half wave controlled rectifier fed from a 220 V, 50 Hz mains feeds a separately excited dc motor. The resistance of the motor can be neglected compared to its inductance. It develops a torque of 35 Nm at a speed of 700 rpm and has an inertia of 0.1 kgm^2. Determine for a firing angle $a = 0$ the speed dip between the conducting periods. Assume a coasting period of π radians.

4.3 A separately excited dc motor is supplied from a 240 V, 50 Hz ac mains by a single phase half wave controlled rectifier. The motor has a resistance and an inductance of 1.5Ω and 45 mH respectively. For a given condition the firing angle is 60° and the motor has a back emf of 160 V. Assuming the devices to be ideal determine the armature current waveform of the motor and the mean value of converted power.

4.4 The motor in Problem 1 develops a rated torque of 15 Nm at a speed of 50 rad/sec. Determine the firing angle of the rectifier when the speed has to be increased to 75 rad/s at the constant torque. Determine also the speed fluctuation both at a = 60° and at the new firing angle. Coasting period is 2π/3.

4.5 A separately excited dc motor is supplied from a 220 V, 50 Hz ac mains by a half wave controlled rectifier. The motor develops a torque of 25 Nm at a speed of 400 rpm. The motor has a constant of 3.5 Vs/rad. It has an armature resistance of 0.5Ω. Determine the firing angle for the rectifier and speed fluctuation of the motor. Assume a coasting period of 150° between the pulses.

4.6 A single phase, half wave controlled rectifier operating on a 220 V, 50 Hz supply, feeds a separately excited dc motor having an armature resistance of 5 Ω. The armature runs at a speed making a back emf of 110 V. The speed fluctuation of the motor can be neglected due to high inertia. Determine the average value of the motor armature current, if the armature inductance is neglected.

4.7 A separately excited dc motor is fed from a single phase half wave controlled rectifier fed from a 220 V, 50 Hz mains. The resistance of the motor is 2.5 Ω. It develops a torque of 35 Nm at a speed of 700 rpm and has an inertia of 0.1 kgm^2. Further it develops a torque of 0.65 times the rated value for which the firing angle is 90°. The coasting period has been observed under these conditions to be 210° and the armature constant is 3.0 V/r/s.

Determine the average speed of the motor and speed fluctuation as a percentage of the average speed.

4.8 A single phase, two pulse, bridge rectifier operating from 220 V, 50 Hz mains feeds a separately excited dc motor which has a rated speed of 1200 rpm at full load torque of 35 Nm. The motor inertia is 0.1 kg m^2. A mechanical load having an inertia of 0.2 kg m^2 is coupled to the motor which makes the motor deliver its rated torque. The thyristors are fired at an angle of 60° and the coasting period has been found to be 80°. Determine the speed drop between the conduction periods.

4.9 A separately excited dc motor having an armature resistance of 3 Ω and an inductance of 25 mH is fed from a two pulse bridge rectifier. When the firing angle is 75° the motor has a back emf of 55 V.

(a) Determine the power output, input voltamperes, and power factor

(b) Assuming perfect smoothing of the load circuit determine the power factor.

4.10 A three phase, three pulse rectifier operating from a 250 V, (line to neutral) 50 Hz supply feeds a dc motor having the following specifications

r_a = 0.05 Ω, L_a = 2 mH, the armature constant = 1.5 V/rad/s. Rated armature current = 400 A.

Rated speed of the motor = 1500 rpm

Assuming continuous conduction determine the firing angle α for rated speed? Determine the range of firing angles to give speeds between 1500 rpm to 750 rpm. If the rated speed has to be obtained at $\alpha = 0$ determine the rms phase voltage of the supply. What is the range of firing angles for this case.

4.11 A 0.75 kW, 500 rpm dc motor has an armature with an inertia of 0.01 Kgm^2 sec^2 and a rated torque of 20 Nm and is fed by a 50 Hz half wave thyristor rectifier. The coasting period is π radians between conduction periods Find the speed drop between conduction intervals. The motor is delivering rated torque at rated speed.

4.12 The motor in Problem 11 has an armature resistance of 8 Ω and inductance of 0.6 H and a motor constant of 4.5 Nm/A or V/rad/s. The converter is fed from a 150 V single phase supply. The motor develops a torque of 40% rated torque. The firing angle is 80° and thyristor conducts for 120°. Determine the average speed and speed fluctuation.

4.13 A 100 hp 1500 rpm dc separately excited motor has an armature resistance of 0.02 Ω and inductance of 1.5 mH. The motor constant is 1.5 Vs/rad. A half wave thyristor controlled rectifier feeds the motor. The input voltage to the rectifier is 440 V line to line. The developed torque is 500 Nm.

(a) Determine the firing angle of the thyristor converter to obtain the rated speed of 1500 rpm. Assume continuous conduction and forward voltage drop 1 V for the thyristors.

(b) The firing angle is set at 90°. The motor is at standstill carrying the rated current. Determine the ripple content of armature current.

4.14 A 230 V dc motor having separate excitation has an armature resistance of 0.3 Ω. Its rated speed is 1150 rpm at an armature current of 36.7 A. This drives a load having a torque speed characteristic given by

$$T = 0.5\,\omega \text{ Nm}$$

When this motor is fed from a 480 V, 50 Hz, three phase three pulse rectifier, determine the firing angle of the converter. If the speed is to be reduced to 500 rpm determine the firing angle and torque developed by the motor.

4.15 A three phase six pulse bridge converter working on a 500 V, 50 Hz, supply feeds a dc motor having a rated voltage of 250 V. The motor is separately excited and draws at 250 V an armature current of 181 A and runs at a speed of 1500 rpm. The motor drives a load whose torque-speed characteristic is given by

$$T_L = 0.64\,\omega^2$$

If the speed control of this motor is required from 1500 to 500 rpm determine the range of firing angles. If the firing angle $a = 0$ for operation at rated speed, determine the line voltage, what is the range of firing angles in this case?

4.16 A separately excited dc motor fed from a three-phase six pulse fully controlled bridge converter develops a full load torque at 1300 rpm when the firing angle is zero. The armature takes a current of 55 A from a 400 V supply having a resistance of 0.70 Ω. Determine the voltage of ac system and range of delay angles for speed control from rated speed to 500 rpm, at constant rated torque being delivered by the motor.

4.17 The motor with specifications given in Problem 16 is fed from a three phase half controlled bridge rectifier. Repeat the problem and determine the voltage of the system required, as well as the range of firing angles for the required speed range at constant torque.

4.18 The armature of a separately excited dc motor is fed from a three phase, fully controlled bridge rectifier which is supplied from a 580 V, 50 Hz three phase system. The motor has an armature resistance of 0.15 Ω and has a motor constant of 1.3 V/rpm. Determine

(a) the firing angle required to give a no-load speed of 600 rpm.

(b) speed at this firing angle when the motor draws a current of 500 A at a given load.

(c) the new firing angle to bring the speed to 600 rpm.

(d) the firing angle at starting the motor against the same load torque.

4.19 A 600 V, dc series motor has a combined armature and series field resistance of 0.16 Ω. The motor draws an armature current of 210 A while running at a speed of 600 rpm. The motor drives a load having a torque speed characteristic

$$T_L = A\ \omega^2$$

Determine the constant A when running at a speed of 600 rpm. This motor is to be controlled by a three phase fully controlled bridge rectifier. Determine the input voltage of the converter if the speed of 600 rpm is required at $a = 0$.

Assuming linear magnetisation characteristic determine the range of firing angles for a speed control range 600 to 0 rpm.

4.20 A three phase half controlled bridge converter feeds a dc, series motor having the following details:

Rated voltage = 240 V $(r_a + r_s) =$ 0.5 Ω

Back emf at 800 rpm = 218 V

The motor drives a load having a torque speed characteristic

$$T = 4.2\,\omega \text{ Nm}$$

Where ω is rad/s. Select a suitable voltage to the converter so that the motor would run at 800 rpm with $a = 0$. If a speed control is required from 800 to 400 rpm determine the range of firing angles. Assume linear magnetisation curve.

4.21 If the rectifier in Problem 20 is fed from a 380 V, 50 Hz supply determine the firing angle for 800 rpm of the motor. What is the range of firing angles for a speed control range of 800 to 400 rpm.

4.22 A 600 V, 150 hp, 600 rpm series motor has a combined series and armature resistance of 0.2 Ω and draws a line current of 200 A. The speed of this motor has to be controlled at constant torque. The motor is fed from a three phase six pulse bridge rectifier which receives its supply from a 580 V, three phase supply. Determine the range of firing angles to achieve speed control from 600 rpm to 200 rpm. Assume linear magnetisation characteristic.

4.23 A 220 V dc series motor has an armature current of 20 A when developing a full load torque at 1500 rpm. The speed is to be controlled by a two pulse single phase bridge rectifier. The load inductance assures continuous conduction. The rectifier is fed from a 220 V, 50 Hz supply. Determine the firing angle of the converter. If the speed of the motor has to be decreased to 1000 rpm requiring only 50% of the rated torque determine the new firing angle. Assume $r_a + r_s = 0.4\ \Omega$ and linear magnetisation characteristic.

4.24 A 250 V three phase supply is used to drive a separately excited dc motor through a three phase, six pulse bridge converter. The armature resistance is 0.2 Ω and the motor draws an armature current of 250 A to run the motor at 1500 rpm at its rated voltage of 220 V.

(a) Determine the firing angle α for the above condition.

(b) Determine the firing angle required to run at 1000 rpm.

(c) If the firing angle is 80° what is the speed?

4.25 A separately excited dc motor is controlled from a dc to dc chopper. The motor has the following data:

$$r_a = 0.7\ \Omega \text{ and } L_a = 10 \text{ mH}$$

The motor has a constant of 1.5 Nm per ampere. The chopper is fed from a 250 V dc supply and has an ON time of 3 ms. The frequency of the chopper is 200 Hz. Determine the average current and torque of the motor.

4.26 The armature voltage of a separately excited dc motor is controlled by means of a chopper having a frequency 250 pulses/s. The input to the chopper is from a 250 V dc source. At a time ratio of 0.9 the motor runs at a speed of 900 rpm. The motor has an armature resistance of 0.1 Ω and an inductance of 20 mH. It has a torque constant of 2.5 Nm/A. Find out the mode of operation of the chopper and the output torque of the motor. Determine the speed and torque for ratio of 0.6 if the motor drives a load having a characteristic

$$T_L \propto N^2$$

4.27 A 250 V dc shunt motor has an armature current of 20 A. It develops its full load torque at a speed of 1500 rpm The speed of the motor must be reduced to 1000 rpm at a torque of 0.8 times full load torque. The motor is fed from a chopper having a frequency of 400 Hz. Determine the ON time of the thyristor chopper, if the field excitation reduces to 0.8 times full excitation. The armature has a resistance of 0.5 Ω.

4.28 A dc series motor is controlled by a dc to dc chopper circuit fed from a 600 V dc supply. The armature and field of the motor have a combined inductance of 0.15 H and negligible resistance. The chopper has a pulse frequency of 80 pulses/s and operates on a time ratio of 0.6. Determine the ripple of the load current. If the ripple has to be limited to half this value determine the pulse frequency keeping the time ratio the same.

4.29 A dc shunt motor rated at 50 hp takes a current of 180.5 A from a 230 V supply. It runs at 1300 rpm. The armature and shunt field resistances are 0.05 Ω and 115 Ω respectively. The above motor is controlled from a dc chopper for speed control in the range of 300 – 1200 rpm at constant load torque. The chopper operates on variable frequency with a ON time of 4 ms. The field is supplied directly from a 240 V supply. Determine the operating frequency range of chopper.

4.30 A 230 V, 25 hp dc shunt motor has armature and field resistances of 0.2 Ω and 215 Ω. It develops full load torque at a speed of 1500 rpm. The speed of the motor is to be controlled using armature voltage control using dc chopper. The field is supplied from a 230 V supply. The chopper is operated at a constant frequency of 1000 Hz. Determine the time ratio of the chopper to run the motor at 1200 rpm developing a torque of 0.6 times full load torque.

4.31 The armature of a separately excited dc motor is supplied from a three phase full wave bridge rectifier. The armature has a resistance of 0.25 Ω. The armature current of the motor is 220 A when running at a speed of 1500 rpm. The armature is supplied from a 240 V dc supply under these conditions. If the input of the rectifier is from a 230 V, three phase supply, determine the firing angle of the rectifier. If the speed of operation is 800 rpm what is the firing angle required? Determine the speeds of operation for firing angles of $\alpha = 30°$ and 65°.

4.32 A small separately excited dc motor is supplied from a 230 V single phase half controlled bridge. The machine is rated for 220 V with a speed of 2000 rpm taking a current of 22 A. The thyristors are fired at 110° and the armature current continues for 50° beyond the voltage zero. Determine the motor speed at half full load torque. The armature resistance is 0.35 Ω.

4.33 The speed of a separately excited dc motor is controlled using a single phase half controlled rectifier operating on a 230 V single phase supply. The armature has an inductance of 0.065 H and a resistance of 0.36 Ω. The rated speed of the motor is 1500 rpm. At this speed the motor requires an armature voltage of 220 V to draw a current of 20 A to develop rated torque. The firing angle of the bridge is 75° when the motor runs at a speed of 1000 rpm. Determine an expression for the armature current. Determine also the torque developed.

4.34 The speed of a separately excited dc motor is controlled by armature voltage control from a half controlled rectifier operating on a 240 V, 50 Hz supply. The motor has a constant of $K_t = 3.65$ Nm/A or Vs/rad. The motor develops its full load torque of 25 Nm at a speed of 45 rad/s. The armature resistance is 0.6 Ω. Determine the variation of firing angle to develop the torque from no-load to full load at a constant speed of 40 rad/s. Determine also the speed fluctuation of the motor at the torque condition of 20 Nm.

4.35 A single phase fully controlled rectifier operating on a 230 V, 50 Hz supply is used to provide a variable voltage to the armature of a dc motor whose inductance and resistance are 0.07 H and 3 Ω. The motor constant is 1.5 Nm/A or Vs/rad. For a firing angle of 75° the motor rotates at a speed of 160 rad/s. Determine the torque developed by the motor.

Sketch the voltage and current waveforms of the motor. Determine the rms value of the armature current and compare the same with the armature loss that would occur on normal dc operation.

Determine the firing angle if the load torque is increased by 30% keeping the speed constant.

4.36 A three phase half controlled bridge rectifier operating from a 250 V, 50 Hz three phase supply provides a variable voltage supply to the armature of a separately excited dc motor whose parameters are given below:

Armature inductance	= 0.025 H
Armature resistance	= 1 Ω
Full load torque	= 60 Nm
Full load speed	= 1500 rpm
Rated voltage of the motor	= 250 V

Determine the torque developed for a firing angle of 30° at speeds of 500, 700, 1000 rpm.

4.37 Repeat Problem 36 if a fully controlled three phase bridge rectifier is used.

4.38 A dc series motor is controlled from a dc chopper fed from a 500 V supply. The series motor has an inductance of 0.15 H and a resistance of 0.15 Ω. The rated voltage of the motor is 250 V. The motor drives a load torque represented by

$$T = 22.5 \times 10^3/\omega$$

Determine the time ratio of the chopper if the motor speed is 50 rad/s.

Determine also the maximum and minimum values of armature current while operating at this load if the chopper frequency is 80 pulses/s.

If the chopper frequency is increased to 100 pulses/s determine the limits of armature current.

4.39 A dc series motor is fed from a single phase fully controlled bridge rectifier. The details of the motor are: rated voltage = 240 V, Full load current = 38 A, Armature and series field resistance = 0.6 Ω, rated speed = 600 rpm. Determine the firing angle of the rectifier to run the motor driving at 300 rpm while load having a torque-speed characteristic given by

$$T = 0.19\,\omega^2$$

Determine also the torque developed. Draw the voltage and current waveforms. The controlled rectifier is fed from 240 V, 50 Hz single phase supply.

4.40 (a) A three phase, mid point (bridge) converter operates on a 250 V, three phase, 50 Hz supply. This provides a variable voltage supply to the armature of a separately excited dc motor whose parameters are as follows: Armature resistance = 0.03 Ω, Armature inductance = 0.002 H. The motor constant = 1.3 Nm/A or Vs/rad. The full load current of the motor = 450 A. Neglecting the thyristor voltage drop and assuming continuous conduction, determine the firing angle required to drive the motor at a speed of 150 rad/s.

(b) If the rectifier is supplied by a transformer having an inductance of 1 mH determine the firing angle for the desired operating point.

4.41 A single phase full wave thyristor bridge operating on a 100 V, 50 Hz supply controls the speed of a separately excited motor whose full load speed is 600 rpm at a full load torque of 30 Nm. The moment of inertia of the load inclusive of the motor is 0.22 kgm^2. The thyristor and when the coasting period of the motor is 80°. (Firing angle is adjusted to 80°). Determine the speed drop of the shaft.

4.42 (a) A two pulse bridge rectifier operating on a 110 V, 50 Hz supply provides a variable dc voltage to the armature of a separately excited dc motor. The resistance of the armature is 6 Ω. The back emf of the motor is 70 V and can be assumed to be constant due to high inertia of the motor and constant excitation. The armature inductance is negligible. Determine the average value of the armature current for a firing angle of 0° and 30°.

(b) What would be the average value of armature current if the armature has an inductance of 2 mH?

(c) If the load inductance is extremely large determine the currents.

4.43 A single phase half wave controlled rectifier is fed from a 220 V, 50 Hz supply and is used to provide a variable voltage to the armature of a dc motor. The thyristor firing angle is 0°. The armature has a resistance of 6 Ω and a back emf of 75 V which can be assumed to be constant due to inertia and constant excitation of the motor. The armature has a negligible inductance. Determine the average value of the current.

4.44 A dc motor having a separately excited field develops a torque of 31 Nm at a speed of 500 rpm. Its inertia is 0.05 kgm^2. This motor has an inductance playing an important role, and negligible resistance. A single phase half controlled rectifier is used to control the speed. If the thyristor is fired at $a = 0$, determine the speed dip if the coasting period is π radians between two conduction periods.

4.45 Repeat Problem 4.44 if the armature resistance is 2.3 Ω and the load torque is 0.7 times the full load value. The thyristor is fired at 80°. The coasting period has been found to be 240° and the motor constant 3.2 V/radian/s.

4.46 A three phase, eight pole, 50 Hz induction motor has the following parameters:

$$r_2 = 0.15\,\Omega \quad x_2 = 0.7\,\Omega$$

Neglect the stator impedance, magnetising current and rotational losses. The motor speed is controlled by varying the applied voltage by an ac voltage controller, which operates from a 380 V, 50 Hz supply. Determine the applied voltages per phase of the motor to have slips of 0.15 and 0.25 respectively. The motor drives a load with a characteristic

$$T_1 = 0.014\,\omega^2 \text{ Nm}$$

Determine also the angles of firing of the converter as well as the conduction angles of the thyristors.

4.47 A three phase, four pole, 50 Hz squirrel cage induction motor has the following equivalent circuit parameters: $r_1 = 0.05\ \Omega$; $r_2' = 0.09\,\Omega$; $x_1 + x_2' = 0.55\,\Omega$. The motor is star-connected and rated voltage is 400 V. It drives a load whose torque is proportional to speed and is given by

$$T_1 = 0.05\,\omega \text{ Nm}$$

Determine the speed and torque of the motor for a firing angle of 45° of the ac voltage controller operating on a 400 V, 50 Hz supply.

4.48 A three phase, 420 V, 50 Hz delta connected induction motor has the following parameters:

$$r_1 = 2.95\,\Omega \quad x_1 = 6.82\,\Omega$$

$$r_2' = 2.08\,\Omega \quad x_2' = 4.11\,\Omega \text{ per phase}$$

Neglect core loss. The motor draws a current 6.7 A at no load. This motor drives a load with torque-speed characteristic given by

$$T_l = 0.11\omega^2 \text{ Nm}$$

Determine the speed and torque for the following cases:

(a) The voltage controller operating on 440 V supply is fired at a firing angle of 30°.

(b) The voltage controller provides an rms voltage of 250 V between the lines of the motor.

4.49 Repeat Problem 4.48 for a rotor resistance of 5 Ω, the other data being unchanged.

4.50 The speed of a three phase slip ring induction motor is controlled by variation of rotor resistance. The full load torque of the motor is 50 Nm at a slip of 0.03. The motor drives a load having a characteristic

$$T \propto N^2$$

The motor has four poles and operates on 50 Hz, 400 V supply. Determine the speed of the motor for 0.8 times the rated torque. This operating condition is obtained with additional resistance in the circuit. The resistance is controlled by chopper in the rotor circuit. Determine the time ratio of the chopper to give an average torque 75% of the rated torque. Determine also the average torque developed for a time ratio of 0.4.

4.51 The motor of Problem 48 has its speed controlled by a chopper controlled resistance. A resistance of r_e Ω has been controlled by the chopper. Determine the value of r_e to get a speed control range of 1500 to 500 rpm, assuming a turns ratio of two between the stator and rotor. The torque and speed of the load are related by

$$T \propto N$$

Determine the characteristic giving the speed vs time ratios of the chopper. Assume a star-connection to the rotor.

4.52 Repeat Problem 4.51 if the motor drives a constant torque load.

4.53 An eight pole, 50 Hz, 380 V, star-connected induction motor has a star-connected slip ring rotor. The stator/rotor turns ratio is 1.25. The speed of the motor is controlled by a converter cascade in the rotor circuit. Determine the firing angles of the inverter to get 600 rpm and 400 rpm at no-load. The inverter is connected to a 380 V, three phase system. Assume no overlap in the rectifier as well as in the inverter. What is the minimum possible speed?

4.54 Repeat Problem 53 for overlap angles of 25° on the diode rectifier and 5° on the inverter. Assume also voltage drops of 1.5 and 2.0 V in the thyristor and diode Determine the minimum possible speed.

4.55 If a transformer is interposed between the inverter and supply with $N_p/N_s = 2$ determine the firing angles for the above two cases. Find out the minimum possible speed in these cases.

4.56 A slip ring induction motor has rotor resistance and standstill reactance of 0.2 Ω and 0.6 Ω per phase. The stator is delta-connected with an operating voltage of 400 V. The rotor is star-connected having an effective stator/rotor turns ratio of 1.3. The inverter is connected to a 400 V, three phase system. The motor has a six pole winding. The firing angle is adjusted such that the motor runs at 600 rpm at no-load.

(a) Determine the firing angle.

(b) The machine is loaded so that a torque of 250 Nm is developed. Determine the speed.

(c) What should be the firing angle of the inverter if the speed under load is to be maintained at 600 rpm?

4.57 Repeat Problem 4.56 taking the stator impedance of $(0.2 + j0.6)\ \Omega$ into consideration.

4.58 A three phase four pole 50 Hz induction motor has the following data:

$r_1 = 0.05\ \Omega \quad r_2' = 0.09\ \Omega$

$x_1 = 0.35\ \Omega \quad x_2' = 0.35\ \Omega$

$V = 440$ V, the stator is star-connected.

The motor operates on a variable frequency variable voltage supply such that V/f is constant. The motor drives a load with a torque-speed characteristic given by

$$T_1 = 0.009\ \omega_r^2 \text{ Nm}$$

where ω_r is the rotor speed in rad/s. Determine the operating speed and torque when the motor has supply frequencies of 5 Hz, 25 Hz and 50 Hz.

4.59 Repeat Problem 4.58 if the motor is controlled to have constant flux, i.e., E/f is constant.

4.60 Consider the motor of Problem 4.58. Find the lowest possible frequency and voltage for proper operation. Determine the starting torque and speed at maximum torque for this frequency as well as 30 Hz and 50 Hz.

4.61 A 50 hp, 440 V, 50 Hz six pole star-connected induction motor has the following equivalent circuit parameters:

$r_1 = 0.1\ \Omega \quad r_2' = 0.12\ \Omega$

$x_1 = 0.3\ \Omega \quad x_2' = 0.3\ \Omega$

$x_m = 15\ \Omega$

For a slip of 0.03 at rated frequency determine the torque developed. If the motor drives a load having a torque-speed characteristic

$T_1 = A\,\omega^2$ Nm ω rad/s Determine A.

The speed of the motor is controlled using variable frequency with V/f constant. Determine the torque and speed of the motor when 220 V is applied to the motor.

4.62 Consider Problem 4.61. What voltage must be applied at an operating frequency of 5 Hz? Determine the performance of the motor at this condition.

4.63 A 15 hp, 220 V, three phase, 50 Hz, six pole star-connected induction motor has the following equivalent circuit parameters

$r_1 = 0.128\ \Omega \quad r_2' = 0.0935\ \Omega$

$x_1 = 0.25\ \Omega \quad x_2' = 0.25\ \Omega$

$x_m = 8\ \Omega$, all at 50 Hz.

(a) Determine the starting torque vs frequency and starting current vs frequency of the motor when the stator frequency is varied from 0 to 50 Hz, the voltage being simultaneously varied using the relation V/f = constant.

(b) The motor is operated by a square wave inverter whose peak amplitude is such that its fundamental has an amplitude equal to rated voltage at 50 Hz.

(c) The motor operates at 40 Hz and has a slip of 0.04. Determine the torque and input current waveform when the voltage is varied such that V/f = constant.

(d) The stator frequency is varied to 5 Hz keeping the rotor frequency the same. The voltage is varied such that V/f is constant. Determine the torque and current waveform.

4.64 Consider Problem 4.63. The motor is controlled such that E/f is constant. Repeat the problem for the cases of the above problem.

4.65 A 50 hp, 440 V, 50 Hz, six pole star-connected induction motor has the following equivalent circuit parameters:

$r_1 = 0.15\ \Omega \quad r_2' = 0.2\ \Omega$

$x_1 = 0.40\ \Omega \quad x_2' = 0.45\ \Omega$

$x_m = 15\ \Omega$, all at 50 Hz.

The motor is fed from a constant current source of 70 A of variable frequency.

Determine torque-speed characteristics at 5 Hz, 25 Hz, and 50 Hz.

4.66 (a) Based on equivalent circuit show that the torque developed and input current of a variable frequency induction motor are independent of stator frequency but depend on rotor frequency for constant airgap flux.

(b) Use the above result to determine the relationship between the stator current and rotor frequency as well as the relationship between torque and rotor frequency.

4.67 A 10 kW, 380 V, three phase, 50 Hz four pole synchronous motor has the following equivalent circuit parameters:

$R_a = 1\ \Omega$ $x_d = 12.5\ \Omega$/phase.

leakage reactance = x_{1s} = 1.5 Ω/phase.

The motor is fed from a VSI and drives fan toad whose torque-speed characteristic can be represented by

$$T_1 = A\ \omega^2 \text{ Nm} \quad \omega \text{ rad/s}$$

Neglecting rotational and excitation losses determine the value of A, assuming that the motor operates at UPF. The speed of the motor is required to be controlled in the range 1500–600 rpm. The E/f of the motor is kept constant in the complete speed range. If the field current is adjusted so that the motor operates at UPF at 1500 rpm, determine the line current and powerfactor at 600 rpm. If the field current is adjusted so that the motor operates at UPF at 600 rpm determine the line current and powerfactor at 1500 rpm. Assume a turns ratio of the field to armature as $N_{re}/N_{sc} = 10$.

4.68 Consider a salient pole synchronous motor operating on a variable frequency, variable voltage supply. Neglecting the effects of stator resistance prove the following based on phasor diagram.

(a) When the applied voltage is varied in proportion, to frequency at constant field current, the motor draws a constant current independent of stator frequency.

(b) The stator current under these operating conditions is a function of torque angle.

4.69 A synchronous motor rated at 100 hp, 500 V, three-phase star-connected has a resistance and synchronous reactance of 0.03 Ω and 3 Ω respectively. The leakage reactance of motor may be taken as 0.7 Ω/phase. The motor has a ratio $N_{re}/N_{se} = 15.00$. The motor drives a load whose speed torque curve is represented by

$$T_L = A + B\,\omega_m \text{ Nm}$$

Above the base speed the terminal voltage is held constant at rated value. The speed control is achieved by frequency variation only and a limit is introduced on the line current as 80% rated line current. In the complete speed range the power factor is maintained at UPF. Determine the maximum speed possible.

4.70 A 11 kV, three phase star-connected synchronous motor has effective resistance and synchronous reactance of 1 Ω and 30 Ω per phase. Leakage reactance is 5 Ω. The field system of the motor is designed for UPF operation. The ratio of $N_{re}/N_{se} = 25$. Neglect rotational and exciter losses. The motor operates from a CSI and is coupled to a load whose

$$T_1 = a + c.N$$

(a) When the motor operates at its rated load and UPF determine E/f, line current I_a field at I_f and the torque angle δ.

(b) The motor operates at field current to keep E/f constant.

The motor operates at the value of δ obtained in (a). Determine the performance of the motor at an armature current of 60 A.

4.71 An induction motor is described as a rotating transformer comment on the statement. With suitable reasons.

4.72 Give a comparison of induction motor and synchronous motor.

4.73 The speed of an induction motor cannot be equal to synchronous speed. Give the reason.

4.74 When the mechanical load on a 3 phase motor is increased, how does input current increase to meet the new conditions?

4.75 The rotating magnetic fields of the stator and rotor are stationary with respect to each other justify the statement.

4.76 A 3 phase induction motor develops a starting torque but a single phase cannot develop a starting torque. Why?

Multiple-Choice Questions

4.1 A three phase ac voltage controller feeding a three phase induction motor has an output of
- (a) constant voltage of variable frequency
- (b) variable voltage of variable frequency
- (c) variable voltage of constant frequency
- (d) constant voltage of constant frequency

4.2 The method of speed control using a three phase voltage controller is suitable for loads
- (a) whose torque is proportional to speed
- (b) having constant torque
- (c) whose torque is proportional to $\sqrt{N}$
- (d) having torque proportional to N^2

4.3 A three phase ac voltage controller feeds an induction motor. The motor has
- (a) very good efficiency and power factor at all speeds
- (b) very good efficiency but poor power factor at all speeds
- (c) poor efficiency but good power factor at all speeds
- (d) poor efficiency and poor power factor at low speeds

4.4 A three phase induction motor having speed control using chopper controlled resistance is characterised by
- (a) poor power factor and good efficiency
- (b) poor efficiency and poor power factor
- (c) good efficiency and good power factor
- (d) good power factor and poor efficiency

4.5 A three phase induction motor with chopper controlled resistance has its torque proportional to
- (a) rotor current
- (b) square of rotor current
- (c) stator resistance
- (d) square root of rotor current

4.6 A three phase induction motor having a combination of diode rectifier and line commutated inverter in the rotor circuit can give
- (a) speeds below synchronous speed only
- (b) speeds above synchronous speed only

(c) both sub and super synchronous speeds
(d) speeds varying 0 to 50% of synchronous speeds

4.7 A three phase induction motor having subsynchronous converter cascade in the rotor circuit is characterised by
(a) good power factor and poor efficiency
(b) good power factor and good efficiency
(c) poor power factor and poor efficiency
(d) poor power factor and good efficiency

4.8 The motor having slip energy recovery scheme can be braked by means of
(a) regenerative braking
(b) plugging
(c) dc dynamic braking
(d) all the methods of (a), (b) and (c)

4.9 The speed of an induction motor can be varied by means of variable frequency supply from a static power converter. A simultaneous voltage variation is also effected in order to
(a) avoid saturation and provide optimum torque capability
(b) the torque pulsations decrease if supplied from variable voltage supply
(c) to limit the peak value of stator current
(d) to minimise the additional losses

4.10 A three phase induction motor is supplied from a variable voltage variable frequency supply such that air gap flux is constant (*E*/*f* is constant). The armature current of the motor now
(a) is independent of stator frequency but depends on rotor frequency
(b) is dependent on stator frequency but does not depend on rotor frequency
(c) is dependent on both stator and rotor frequencies
(d) is independent on both stator and rotor frequencies

4.11 A current source inverter fed induction motor is inherently unstable when it operates on
(a) open loop
(b) closed loop
(c) a variable frequency supply keeping air gap flux constant
(d) a variable frequency supply keeping stator flux constant

4.12 A three phase induction motor operates at constant rotor frequency when the stator frequency is varied from zero to rated value. The torque developed by the motor is
(a) constant from zero to rated speed
(b) proportional to speed
(c) proportional to square of the speed
(d) inversely proportional to speed

4.13 A variable frequency variable voltage induction motor
(a) can be accelerated at constant torque or constant current
(b) suffers from poor starting characteristics as in the case of mains fed motor
(c) has only stepped variation of speed
(d) suffers from stability considerations

4.14 When operated with variable frequency a synchronous motor has an advantage over an induction motor in
(a) that it is free from torque oscillations
(b) that it has very good efficiency
(c) that the line power factor can be improved by varying excitation
(d) that in certain cases the inverter can be of simpler configuration due to the possible load commutation.

4.15 In a three phase synchronous motor operating on variable frequency supply the armature resistance is neglected. The voltage to the motor is varied keeping V/f constant and I_f constant. The stator current would now

(a) increase as the frequency increases
(b) decrease as the frequency increases
(c) increase with frequency up to certain value of frequency and decreases as the frequency is increased beyond this value
(d) is constant as long as the load torque is constant

4.16 By self control of a synchronous motor we mean that
(a) elimination of torque ripple
(b) the speed of the motor is varied in steps
(c) the speed of the motor is a function of input frequency
(d) the input frequency is controlled from the speed of the motor

4.17 The disadvantage of Load commutation is
(a) harmonic torques
(b) loss of efficiency due to losses
(c) the speeds from 0 to 10% of base speed are not possible
(d) the speed control range is limited to 0 – 10% of base speed

4.18 The operation of ac motors fed from current source inverters in characterised by
(a) sinusoidal line voltage
(b) peaky armature voltage
(c) good harmonic torques
(d) sinusoidal armature voltage with spikes

4.19 While operating on a phase controlled converter the commutation capability of the motor deteriorates because of
(a) increase in the RMS current
(b) harmonic content of armature current
(c) peak value of armature current
(d) peak value of armature current, harmonic content of armature current and possible discontinuous conduction

4.20 When only one quadrant operation is required the converter normally preferred is
(a) fully controlled converter
(b) fully controlled converter with FWD
(c) half controlled converter
(d) sequence control of two series connected converters

4.21 A separately excited dc motor designed for operation on phase controlled converter has laminated interpoles
(a) to improve the commutation of the motor
(b) to facilitate easy construction
(c) to decrease the losses in the interpole
(d) to improve the power factor

4.22 For low speed high power reversible drives
(a) cycloconverter-fed ac drives are suitable
(b) current source inverter-fed ac drives are suitable
(c) voltage source inverter (square wave inverter) fed drives are suitable
(d) induction motors fed from ac voltage controller are suitable

4.23 Turn of a thyristor takes place
(a) When anode to cathode voltage is positive
(b) When anode to cathode voltage is negative
(c) When there is positive current pulse at the gate
(d) When the anode to cathode voltage is positive and there is also a positive current pulse at the gate

4.24 When a voltage is applied to a thyristor making anode negative and cathode positive
(a) all the junctions are negatively biased
(b) outer junctions are negatively and inner junctions are positively biased
(c) outer junctions are positively biased and central junction is negatively biased
(d) all the junctions are positively biased

4.25 Turn off time of a thyristor affects its
(a) operating voltage
(b) operating frequency
(c) overload capacity
(d) thermal behaviour

4.26 The sharing of the voltages between thyristors operating in series is influenced by their
(a) *di/dt* capability
(b) *dv/dt* capability
(c) *v.i* characteristics (leakage currents)
(d) junction temperatures

4.27 In the state of saturation a MOSFET acts as
(a) closed switch
(b) an open switch
(c) an amplifier
(d) pure resistor

4.28 The $\frac{di}{dt}$ capability of a thyristor is decided by
(a) The spreading velocity of the current across the junction
(b) The voltage rating of the thyristor
(c) Reverse blocking capability of the device
(d) The steady state forward losses of the device

4.29 When operated with variable frequency, a synchronous motor has an advantage over an induction motor in
(a) that it is free from torque oscillations
(b) that it has very good efficiency
(c) that the line power factor be improved by varying excitation
(d) that in certain cases the inverter can be simpler configuration due to possible load commutation

4.30 By self-control of asynchronous motor, we mean
(a) the elimination of torque ripple
(b) the speed of the motor is varied in steps
(c) the speed of the motor is function of input frequency
(d) the input frequency is controlled from the speed of the motor

4.31 A separately excited dc motor is driving a centrifugal pump whose torque is proportional to the square of the speed. The armature current is 20A at a speed of 500 rpm. The armature current at the speed of 200 rpm is
(a) 6.4 A
(b) 3.2 A
(c) 9.6 A
(d) 8 A

4.32 A variable frequency variable voltage induction motor
(a) can be accelerated at constant torque or constant current
(b) suffers from poor starting characteristics as in the case of mains feds motor
(c) has only stepped variation of speed
(d) suffers from stability consideration

4.33 A three phase induction motor with chopper-controlled resistance has its torque proportional to
(a) rotor current
(b) square of rotor current
(c) stator resistance
(d) square root of the rotor current

4.34 In static Scherbius drive of 3 phase slip ring induction motor converter-1 is a diode rectifier connected to the rotor converter-2 is a thyristor converter connected to 3 phase mains. This drive can't be used for
(a) speeds above synchronous speed only
(b) speeds below synchronous speed only
(c) speeds both above and below synchronous speeds only
(d) speeds varying from 0 to 50% of synchronous speeds only

Solid State DC and AC Drives

4.35 A line commutated converter driving a DC motor can be considered as
(a) linear load
(b) non-linear load
(c) linear load at low speed and non linear at high speed
(d) linear load at high torque & non-linear at low torque

4.36 Starting current of a motor is kept low
(a) to avoid excessive heating
(b) to safeguard the life of the motor
(c) to reduce the fluctuation in the supply voltage
(d) to reduce the acceleration time

4.37 For quick and efficient starting of a load requires
(a) a motor torque to be more that load torque
(b) a motor torque to be equal to load torque
(c) a motor torque to be less than load torque
(d) none of the above

4.38 Ward Leonard method of speed control achieves speed change by
(a) varying field flux
(b) varying armature voltage
(c) varying field voltage
(d) varying armature resistance

4.39 The acceleration speed under steady state condition
(a) 10
(b) 0.999
(c) 1.265
(d) 0

4.40 Determine the steady state condition for $T = -1 - 2W_m$ $T_{load} = -3\sqrt{W_m}$ when $W_m = 1$, $T = -3$
(a) unstable
(b) stable
(c) partially stable
(d) partially unstable

4.41 What is the effect when viscous friction coefficient is negligible?
(a) no oscillations
(b) shaft breaks
(c) smooth operation
(d) no effect

4.42 Load torques which always oppose the and change their sign on the reversal of the motion called
(a) active
(b) passive
(c) reactive
(d) none of the above

4.43 The inverter limit of a line commutated inverter is due to
(a) overlap
(b) turn off time of the thyristors
(c) turn on time of the thyristors
(d) both (a) and (b)

4.44 Regenerative braking is employed in series motor when
(a) armature connections reversed
(b) field winding reversed
(c) field winding is separately excited
(d) (b) and (c)

4.45 What is effect of regenerative braking when input side load is negligible
(a) speed decreases
(b) terminal voltage rises
(c) improves power factor
(d) (b) and (c)

4.46 When friction torque is greater than load torque, then motor
(a) stalls
(b) runs at –50 RPM runs at 50 RPM

4.47 In a discontinuous mode of conduction for converter fed dc drive which is true
(a) peak current increases
(b) average current increases
(c) Rms current decreases
(d) average currents decreases

4.48 What is the condition for continuous mode of conduction for converter fed dc drive
(a) torque is greater than the rated value
(b) torque is lesser than the rated torque
(c) rated torque is greater than torque
(d) rated torque is lesser than torque

4.49 A conducting thyristor can be successfully turned off
(a) by making the current zero
(b) by maintaining the current at zero value for reverse recovery
(c) time (t_{rr}) + gate recovery time ($t_g t$)
(d) by simply applying reverse voltage between anode and cathode

4.50 An amplifying gate thyristor has advantages of
(a) high gate current at low gate drive
(b) a poor $\frac{di}{dt}$ rating even at high gate current
(c) its $\frac{di}{dt}$ improving only at high gate currents
(d) very slow spreading velocity

4.51 Peak inverse rating of a triac
(a) is the same as that of a thyristor
(b) is greater than that of a thyristor
(c) is inferior and very less than that of a thyristor
(d) is not significant due to the nature of its application

4.52 A dc chopper has an ontime T_{ON} = 1ms and its frequency is 500 Hz. The time ratio of chopper is
(a) 100%
(b) 75%
(c) 50%
(d) 25%

4.53 A chopper feeds an RL load comprising a resistance $R = 10\ \Omega$ and an inductance 5 mH chopper frequency of 500 Hz an it fed from 220 VDC. Determine minimum and maximum of load current peak to peak ripple average value of load current

Solution:
Source voltage 220 V, chopper frequency = 500 Hz, chopper period 2 ms, Time ratio of chopper 0.5

$$T_{ON} = 1 \text{ ms } T_{off} = 1$$

$$\text{The load time constant } = \frac{L}{R} = \frac{15}{10} = 1.5 \text{ ms}$$

The maximum values of the ct = I_u

$$= \frac{V_B}{R}\frac{1 - e^{-T_{ON}/\tau}}{1 - e^{-2/1.5}}$$

$$= \frac{220}{10} = \frac{220}{40}\frac{1 - e^{-T_{ON}/\tau}}{1 - e^{-T/\tau}}$$

$$= \frac{22}{10}\frac{(.487)}{073640} = 14.5490$$

The minimum value

$$I_L - 14.5491\, e^{-1/1.5} = 7.4697\text{A}$$

(b) Peak to peak ripple 7.0793 A
(c) Average value of L load ct =

$$\frac{220 \times 0.5}{10} = 11.0\text{A}$$

References

4.1 Bose, B.K., *Adjustable ac drives*, IEEE Press, New York, 1980.

4.2 Leonhard, W., *Control of electrical drives*, Springer Verlag, New York, 1985.

4.3 Bose, B.K., *Power electronics and ac drives*, Prentice-Hall Inc., Englewood Cliffs, 1986.

4.4 Dewan, S.B., G.B. Slemon and A. Stranghen, *Power semiconductor drives*, John Wiley and Sons, Inc. New York 1984.

4.5 Murphy, J.M.D., *Thyristor control ac motors*, Pergamon Press Ltd., Oxford, 1973.

4.6 Kusko, A, *Solid-state dc motor drives*, The M.I.T. Press, 1969 Cambridge Mass and London.

4.7 Ramshaw, R.S., *Power electronics, thyristor controlled power for electric motors*, Chapman and Hall, London, 1973.

4.8 Subrahmanyam, Vedam, *Thyristor control of electric drives*, Tata McGraw-Hill Publishing Company Ltd., New Delhi, 1988.

4.9 Sen, P.C., *Thyristor dc drives*, John Wiley and Sons Inc., New York, 1981.

4.10 Campbell, S.J., *Solid-state ac motor controls, selection and application*, Marcel Dekker, Inc., New York.

4.11 Lipo, T.A, The analysis of induction motors with voltage control by symmetrically triggered thyristors', *IEEE Trans, on PAS*, Vol. PAS, 90, pp. 515–525, 1971.

4.12 Sen, P.C., and KJ. Ma, 'Rotor chopper control of an induction motor drive: TRC strategy', *IEEE Trans. Ind. App*., Vol. 1A–11, pp. 43–19, Jan./Feb. 1975.

4.13 Kleinrath, H. 'Torque pulsation of an induction motor with speed control by a static frequency converter in the slipring circuit', *Proc. ICEM, Brussels*, E3/8-1-E-3/8–9, 1978.

4.14 Smith, A, 'Static Scherbius system for induction motor speed control', *Proceedings IEE*, Vol. 124, 1977, pp. 557–565.

4.15 Meyer, A., 'Network reactions and Oscillating torques caused by subsynchronous converter cascades,' *B.B.C. Review*, 1982, p. 133.

4.16 Sharpenberg, H., and A. Streck, 'Subsynchronous Converter Cascade for Large Centrifugal Pumps', ibid, p. 142.

4.17 Klors, A., J. Heinrich, 'Large Induction Motors of Medium Rating with Current Source Inverters', ibid, p. 163.

4.18 Lipo, T.A., and E.P. Cornell, 'State variable steady state analysis of a current controlled induction motor drive, *Conference Record*, IEEE IAS, 1974.

4.19 Subrahmanyam, V., S. Yuvarajan, B. Ramaswami, 'Analysis of commutation of a current inverter feeding an induction motor load', *IEEE Trans. on IA*, Vol. 1A–16, 1980, pp. 332.

4.20 Yuvarajan, S., B. Ramaswami and V. Subrahmanyam, 'Analysis of a current controlled inverter-fed induction motor drive using digital simulation', *IEEE Trans. on IECI*, Vol. IECI–27, 1980, pp. 67–76.

4.21 Reddy, D.S., M.V.I. Prakash and Vedam Subrahmanyam, 'State space approach for the analysis of multiple inverter-fed three-phase induction motor', *Journal IE (India)*, Vol. 69, 1989, pp. 126–133.

4.22 Subrahmanyam Vedam and K. Surendran, 'stability analysis of variable frequency induction motors using D-decomposition technique,' *Electric Machines and Power Systems*, Vol. 11, 1986.

4.23 Masato, K. M. Yano, I. Kamiyana, and S. Yano, 'Microprocessor-based vector controlled system for induction motors drives with rotor time constant identification', *IEEE Transactions on IA*, Vol. IA–22, No. 3, 1986, pp. 453–459.
4.24 Eder, E., 'Stromrichter zur Drehzahlsteurung von Drehfeldmaschimen,' Teil 1-Teil 4, Siemens, AG, West Germany.
4.25 Meyer, A., H. Schweickardt and P. Strozzi, 'The converter-fed synchronous motor as a variable speed drive system', B.B.C. Review, p. 151, 1982.
4.26 LeHuy, H., R. Perret and D. Roye, 'Microprocessor control of a current fed synchronous motor drive', *IEEE Industry Applications Society Conference Record*, 1980, pp. 562–569.
4.27 Terens, L., J. Bommell and K. Peters, 'The cycloconverter-fed synchronous motor', B.B.C. Review, 1982, p. 122.
4.28 Steigerwald, R.L., and T.A. Lipo, 'Analysis of a novel forced commutation starting scheme for a load commutated synchronous motor drive', *IEEE Trans, on IA*, Vol. IA–15, p. 14–24, 1979.
4.29 Subrahmanyam, Vedam and N. Gopalakrishnan, 'Steady state performance of a current controlled synchronous motor', *ETZ-Archiv*, Vol. 8, 1986.
4.30 Subrahmanyam, Vedam and N. Gopalakrishnan, 'Analysis of a self controlled synchronous motor, *Journal of IE (India)*, Vol. 68, 1987.
4.31 Sen, P.C., and M.L. McDonald, 'Thyristorized dc drives with regenerative braking and speed reversal,' *IEEE Transactions IECI*, Vol. IECI–125, 1978, pp. 347–354.
4.32 Sen, P.C., and S.R. Doradla, 'Evaluation of control schemes for thyristor controlled D.C. motors' *IEEE Transactions IECI*, Vol. IECI–125, 1978, pp. 247–255.
4.33 Subrahmanyam, Vedam 'A state space approach to the performance analysis of a separately excited dc motor using state transition signal flow graph technique', *Journal of IE (India)*, EL, Vol. 67, Pt. ELI, 1986.
4.34 Blaschke, F., 'The principle of field orientation as applied to the new TRANSVECTOR closed loop control system for rotating field machines', *Siemens Review*, Vol. 34, pp. 217–220, May 1972.

5 Rating and Heating of Motors

5.1 REQUIREMENTS OF A DRIVE MOTOR

While selecting a motor for a drive, it must, besides providing suitable torque-speed characteristic and adequate power, satisfy the following requirements:

i. The rating of the motor selected should be in accordance with the mechanical work required by the driven machine. When loaded to do this mechanical work its final steady-state temperature rise must be within the permissible value for the class of insulation used.
ii. The motor selected should be capable of driving the load satisfactorily both under steady-state as well as transient conditions. The choice of the motor rating to suit this requirement is based on the load diagrams which depict the motor torque, power and current as functions of time.
iii. When a motor is selected based on a given load diagram and is fully loaded it must not have excess temperature rise. It must be capable of withstanding short time overloads and must have enough starting torque to accelerate the motor to the desired speed in a given time.

Motors are selected for a given class of duty after giving due consideration to the thermal as well as overload capacity.

Selection of a motor has a bearing on the economic interest of the establishment. A motor chosen should neither be too small nor too big to drive a particular load. In the former case it may not be able to drive the load satisfactorily and may get unduly overloaded with a temperature rise much greater than the permissible value. Some times it may run the risk of damage or even burn out. In the latter case, the drive motor is not fully loaded, operates with poor efficiency and involves capital investment. In the case of ac motor the line pf is poor at light loads.

5.2 POWER LOSSES AND HEATING OF ELECTRIC MOTORS

An electric motor has power losses occurring in its various parts such as copper losses occurring in armature and field, core losses occurring due to hysteresis and eddy currents in the magnetic material used as a core, and mechanical losses due to friction and windage. These unavoidable power losses cause localised heating and are responsible for temperature rise of the motor. The heat flows from the point of origin to the external surface where it is dissipated to the surrounding cooling medium. Hence, heating and temperature rise of an electric machine are a function of the losses occurring in it.

Losses in an electric machine can be broadly classified into constant and variable losses. Constant losses remain independent of load current whereas variable losses vary as the square of the load, expressed as a fraction of the rated load. The losses occurring in a machine are

$$W = W_c + x^2 W_v \tag{5.1}$$

where W_c = constant losses
W_v = variable losses at full load
x = load on the motor expressed as a fraction of rated load.

If there is no cooling and the machine cannot dissipate heat to the external medium its temperature increases to a very high value. Hence, it must be provided with cooling to limit the maximum temperature rise to a permissible value, depending upon the class of insulation employed. The heat generated in the machine is dissipated to the surrounding cooling medium. However, a portion of heat is stored in the material itself causing a temperature rise. In the begining all the heat is stored in the material and no heat is given to the medium. As the temperature rises, the component of heat stored decreases and the component of heat dissipated increases. Finally when the machine attains a steady temperature it can no longer store any more heat and all the heat generated is dissipated. Under these conditions the temperature remains constant, heat generated being equal to heat dissipated. The time taken by the motor to reach these final steady-state values depends upon the effectiveness of cooling. If a machine is well ventilated it reaches the steady-state quickly.

If the machine is switched off or its load is reduced it cools. It cools to the ambient temperature when it is switched off. In the second case of temperature rise drops to a lower value, corresponding to the new load.

An electric machine is normally designed for a given temperature rise as decided by the class of insulation used. The design rating of the motor is called its continuous rating because the final steady-state temperature rise of the motor is within permissible limits if the machine delivers the power continuously over an extended period of time. In case a machine operates at a temperature higher than the specified one this result in thermal breakdown of the insulating materials. This may not occur as an immediate consequence, but deteriorates the quality of insulation used and reduces the life of the motor. The Classes of insulating materials and their permissible temperatures are given in Table 5.1.

An electric machine has sufficient overload capacity. The thermal restrictions as stated above do not allow a continuous overloading of the motor. This is because the losses rise more steeply than the power. The corresponding final steady-state

Table 5.1 *Classes of insulating materials and permissible temperatures*

Class	*Materials*	*Temperatures*
Y	Cotton yarn, fabrics, fibrous materials of cellulose or silk, neither impregnated with nor immersed in a dielectric paper	90°C
A	The above materials but impregnated or immersed in a liquid dielectric	105°C
E	Certain synthetic organic films and other materials having the same thermal stability	120°C
B	Mica, asbestos, or glass fibre base materials with an organic binding agent	130°C
F	The above materials combined with suitable synthetic binding agent as well as impregnating one	155°C
H	Mica, asbestos or glass fibre combined with silicon binding and impregnating agents	180°C
C	Mica, ceramic materials, glass or quartz used without or with inorganic binding agents	exceeds 180° but is limited by the properties of the materials (physical, chemical or electrical)

temperature rise is also more. However, there is a time lag between the losses of the motor and resulting temperature rise. This allows a overloading of the motor for short periods limiting the final temperature rise to the permissible value.

Several classes of duty of an electric motors are possible. Before discussing these classes of duty and selecting a suitable motor rating for a given class of duty let us discuss in detail the heating and cooling of an electric motor under the influence of losses.

5.3 HEATING AND COOLING OF AN ELECTRIC MOTOR

The geometry of an electric motor is too complex to predict accurately the heat flow and temperature distribution. This is a difficult task. A portion of the conductors of armature winding are embedded in slots and some portion is outside the iron material as overhang. The heating calculations are also complicated by the loading of the motor. The direction of heat flow does not remain the same at all loaded conditions. At no-load or lightly loaded conditions heat flows from iron parts to the winding due to the temperature gradient. When once the load increases the gradient changes as the heat (loss) generated in the winding is greater than the heat (loss) generated in iron and consequently heat flows from the winding to iron core. A considerable simplification is therefore necessary to compute the temperature rise of a motor.

The heating and cooling calculation of an electric motor are based on the following simplifications:

i. The machine is considered to be a homogeneous body having a uniform temperature gradient. All the points at which the heat is generated have the same temperature. All the points at which the heat is dissipated to the cooling medium are also at the same temperature.
ii. Heat dissipation taking place is proportional to the difference of temperatures of the body and surrounding medium. No heat is radiated.
iii. The rate of dissipation of heat is constant at all temperatures.

Under these assumptions a machine develops heat internally at a uniform rate and gives it to the surroundings proportional to the temperature rise. It can be shown that the temperature rise of a body follows an exponential law. Assuming the heat developed is proportional to the losses, we have the heat balance equation.

$$W\,dt = A\lambda\theta\,dt + Gs\,d\theta \tag{5.2}$$

where W is the power loss on the motor responsible for heat, Watts.
G weight of active parts of the motor in kg
s specific heat of the material of the body in J/degree/kg
A cooling surface in m^2
λ specific heat dissipation or emissivity in J/s/m^2/degree difference in temperature
θ is temperature rise of the body
$d\theta$ temperature rise in a small interval dt

Equation (5.2) indicates that the total heat generated in the body ($W\,dt$) is equal to the sum of the heat dissipated to the surrounding medium ($A\lambda\theta\,dt$) and the heat stored in the body causing a temperature rise θ above the ambient ($Gs\,d\theta$).

The following results may be obtained by a close examination of Eq. 5.2.

- When the temperature rise reaches a constant value, the body is said to have reached maximum temperature rise θ_m. When this condition is reached $d\theta = 0$. The heat developed in the body is completely dissipated to the surroundings. No more heat is stored in the body and the body attains thermal equilibrium. Therefore,

$$W = A\lambda\theta_m \tag{5.3}$$

and the maximum temperature rise

$$\theta_m = \frac{W}{A\lambda} \tag{5.4}$$

Rearranging Eq. 5.2 we have

$$\frac{d\theta}{dt} = \frac{W}{Gs} + \frac{A\lambda}{Gs}\theta \tag{5.5}$$

If the cooling were not there the machine would heat up enormously and the temperature rise would attain very high values. But θ must be limited to θ_m. The time taken by the machine to reach this temperature rise in the absence of dissipation can be determined using

$$W\,dt = Gs\,d\theta \tag{5.6}$$

which gives a linear relation between θ and t. Therefore

$$\frac{\theta}{t} = \frac{W}{Gs} \tag{5.7}$$

If τ_1 is the time taken to reach θ_m (Fig. 5.1). We have

$$\frac{\theta_m}{\tau_1} = \frac{W}{Gs} \tag{5.8}$$

Substituting for θ_m from Eq. 5.4 we have the value of τ_1, the time taken by the body to reach the maximum temperature rise

$$\tau_1 = \frac{Gs}{A\lambda} \tag{5.9}$$

τ_1 is called the thermal (heating) time constant. In other words this happens to be the time taken by the motor to reach the final steady-state temperature rise if the initial rate of rise of temperature continues.

Heat balance equation (Eq. 5.2) must be solved to obtain a relationship between the temperature rise and time. Rearranging Eq. 5.2 we have

$$\left(W - A\lambda\theta\right) dt = Gs\,d\theta \tag{5.10}$$

from which

$$dt = \frac{Gs\,d\theta}{\left(W - A\lambda\theta\right)} \tag{5.11}$$

Integrating both sides we have

$$t = -\frac{Gs}{A\lambda}\left(\log\left(W - A\lambda\theta\right) - \log C_1\right) \tag{5.12}$$

where $\log C_1$ is a constant of integration. It is evaluated using the initial conditions at the start of heating. This equation can be written as

$$e^{-t\left(\frac{A\lambda}{Gs}\right)} = \frac{W - A\lambda\theta}{C_1} \tag{5.13}$$

For simplicity, it is assumed that the machine is started from cold. Therefore $\theta = 0$, at $t = 0$.

$$C_1 = W \tag{5.14}$$

Substituting for C_1 and rearranging the terms we have

$$\theta = \frac{W}{A\lambda}\left(1 - e^{-t\left(\frac{A\lambda}{Gs}\right)}\right) \tag{5.15}$$

Using relations of Eqs 5.8 and 5.9 we have

$$\theta = \theta_m\left(1 - e^{-t/\tau_1}\right) \tag{5.16}$$

Figure 5.1 depicts the temperature rise curve.

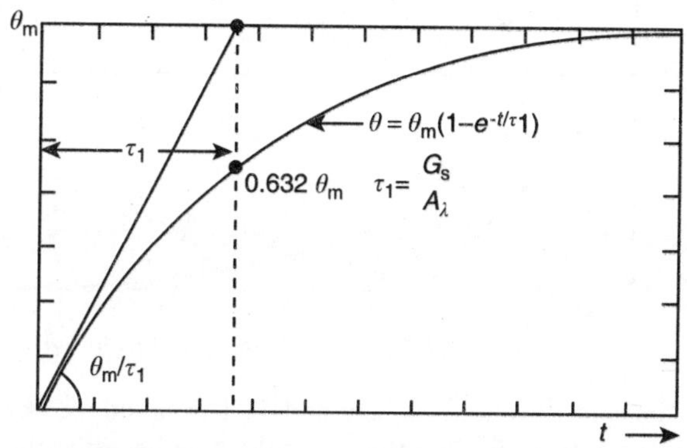

Fig. 5.1 *Temperature rise of an electric motor starting from cold. Time constant = time taken to reach 0.632θ_m*

However while considering the heating and cooling of an electrical machine, in many cases the initial temperature rise is not zero, i.e., at $t = 0$, $\theta = \theta_0$. Substituting the initial conditions in Eq. 15.12 we have

$$C_1 = W - A\lambda\theta_0 \tag{5.17}$$

Using the value of C_1 in Eq. 5.13.

$$e^{-t\left(\frac{Gs}{A\lambda}\right)} = \frac{W - A\lambda\theta}{W - A\lambda\theta_0} \tag{5.18}$$

Rearranging the terms we get

$$\theta = \frac{W}{A\lambda}\left[1 - e^{-t\frac{Gs}{A\lambda}}\right] + \theta_0\, e^{-t\frac{Gs}{A\lambda}} \tag{5.19}$$

Using again $\theta_m = \dfrac{W}{A\lambda}$ and $\tau_1 = \dfrac{A\lambda}{Gs}$ (5.20)

We have

$$\theta = \theta_m\left(1 - e^{-t/\tau_1}\right) + \theta_0 e^{-t/\tau_1} \tag{5.21}$$

The heating curve can be looked upon (Fig. 5.2) as the sum of two curves:

i. heating curve when the machine has a load to give a maximum temperature rise of θ_m.
ii. cooling curve when the machine is disconnected from supply with an initial temperature rise of θ_0.

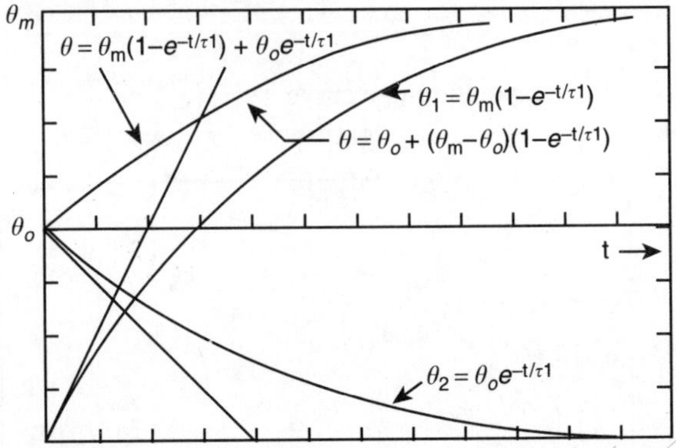

Fig. 5.2 *Temperature rise curve of a motor having some initial temperature*

Heating Time Constant τ_1 τ_1 in Eqs 5.16 and 5.21 is called the heating time constant. It can be seen from Eq. 5.5 that it would have been the time taken by the machine to reach the maximum temperature rise had the cooling been completely ineffective or had the initial rate of rise of temperature been continued, i.e,

$$\left.\frac{d\theta}{dt}\right]_{t=0} = \frac{\theta_m}{\tau_1} \tag{5.22}$$

But the machine has cooling, machine takes longer time to reach the maximum temperature rise and the time is greater than τ_1. Substituting $t = \tau_1$ in equation we see that the temperature rise of the body is 63.2% of θ_m. Therefore it can be defined as the time taken by the machine to reach a temperature rise of 63.2% of maximum temperature rise. These are all depicted in Fig. 5.1.

The value of τ_1 gives an idea of the effectiveness of cooling. Well ventilated machines have smaller time constants. The time constants of open machines are of the order of 25 minutes. The totally enclosed machines have poor ventilation.

Heating time constant can be determined from the experimental curve using the definition that it is the time taken to reach the 63.2% of θ_m. The time constant

$$\tau_1 = \frac{\theta_m}{(\text{initial slope of the curve})}$$

Cooling Curve When a machine is switched off from the mains or when the load on the motor is reduced, the machine cools. In the first case it cools to the ambient temperature and in the second case it cools to a temperature determined by the power losses at reduced load. The cooling curve is also an exponential curve. When the machine is switched off there is no heat generation in the motor and all the heat stored in the machine is now dissipated to the surroundings. In this Eq. 5.2 reduces to

$$0 = A\lambda\theta \, dt + Gs \, d\theta \tag{5.23}$$

Solution of this equation with proper initial values

$$\theta = \theta_{\mathrm{m}} e^{-t/\tau_2} \tag{5.24}$$

Where τ_2 is the time constant during cooling. Fig. 5.3 depicts the curve.

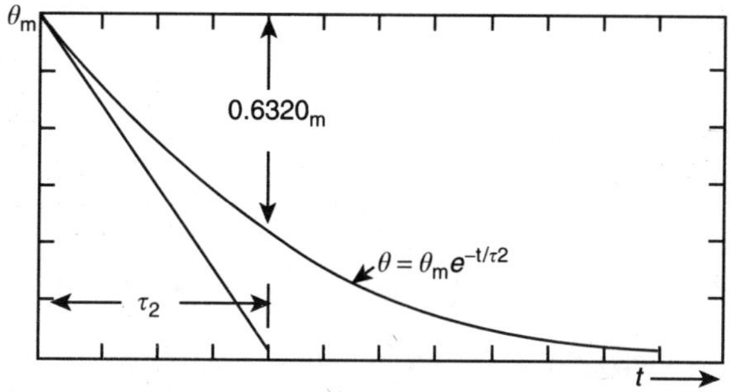

Fig. 5.3 *Cooling curve of an electric motor switched off from the mains*

The machine sometimes is not switched off but load on it is reduced in which case it cools to a temperature

$$\theta_f = \frac{W_1}{A\lambda}$$

where W_1 is the total loss at reduced load. The cooling curve in this case determined as

$$\theta = \theta_f + \left(\theta_{\mathrm{m}} - \theta_f\right) e^{-t/\tau_2} \tag{5.25}$$

which can also be written as

$$\theta = \theta_{\mathrm{m}} e^{-t/\tau_2} + \theta_f \left(1 - e^{-t/\tau_2}\right)$$

The cooling curve is shown in Fig. 5.4. This may also be looked upon as the sum of two curves

i. heating curve as if the machine is loaded to give a maximum temperature rise of θ_{f}.
ii. cooling curve as if the machine is disconnected from the supply when it had a temperature rise of θ_{m}.

τ_2 in the cooling curve is the cooling time constant. The heating time constant τ_1 and cooling time constant τ_2 are not equal in self cooled machines. τ_2 is two to three times greater than τ_1. All other conditions being equal, the time taken by the motor to cool to ambient temperature is longer than the time taken by the motor for heating. If forced cooling is employed $\tau_2 = \tau_1$. In such cases the heating curve and cooling curve are mirror images of each other.

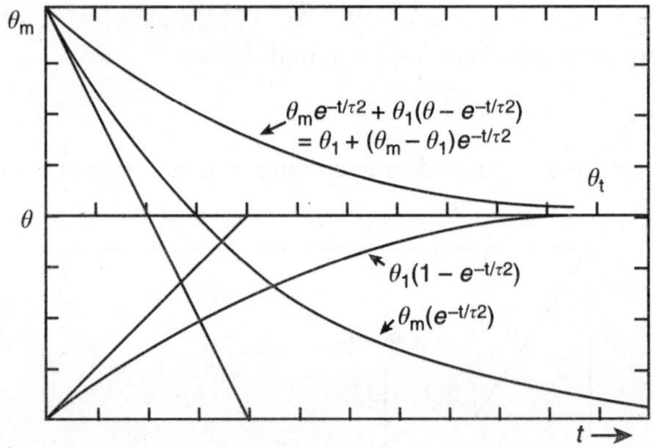

Fig. 5.4 *Cooling curve of an electric motor when the load is reduced*

The power rating of a motor for a particular operating condition is selected based on the thermal rating of the motor, so that it meets the specification regarding the final steady-state temperature rise, i.e., it must be equal to or slightly less than the permissible temperature rise.

As has already been pointed out, an electric motor has a very good overload capacity. The fact that the temperature rise lags the steep rise in power loss due to over loading allows a short time overloading of the motor. The duration of overload is normally till the motor reaches the permissible temperature rise. This overload must be within pull-out torque capability of the motor.

A motor can therefore be selected for a given class of duty based on its thermal rating with due consideration to pull-out torque capability.

5.4 CLASSES OF DUTY AND SELECTION OF MOTOR

Three modes or classes of duty can identified for an electric motor

i. Continuous duty
ii. Short time duty
iii. Intermittent load cycle

5.4.1 Continuous Duty

There are two types of continuous duty—continuous duty at constant load and continuous duty with variable load cycle. In the former the load torque remains constant for a sufficiently longer period corresponding normally to a multiple of time constant of the drive motor. The drive motor is therefore loaded for a sufficient amount of time continuously, till it attains thermal equilibrium.

While driving such a load a motor should have a rating sufficient to drive it without exceeding the specified temperature. The rating of the motor selected for this duty is called its continuous rating or design rating. By continuous rating one means that it is the maximum load that the motor can give continuously over a period of time, without exceeding the temperature rise. Also, the motor selected should be able to withstand momentary overloads. Therefore, the selected motor may sometimes have a rating slightly greater than the power required by the load.

The load diagram and the temperature rise curve of the motor selected for the purpose are shown in Fig. 5.5. Centrifugal pumps, fans, conveyors and compressors are some types of loads where this type of continuous duty at constant load is required.

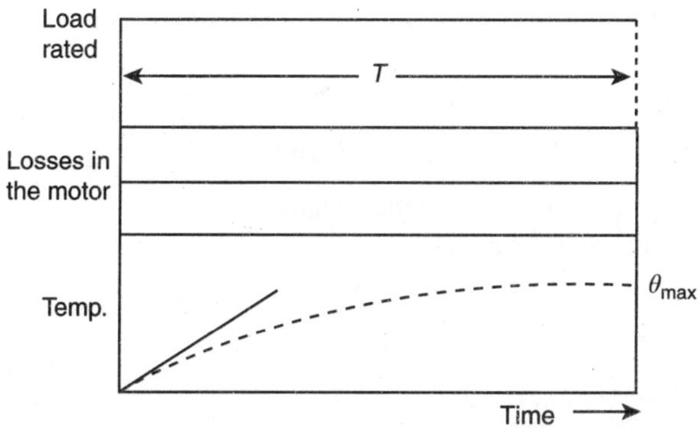

Fig. 5.5 *Pertaining to continuous duty of a motor delivering its rated load (constant)*

Selection of a motor for this class of duty is rather simple and straightforward. From the load characteristics or requirements one can determine the continuous input required to the mechanical load. A suitable motor may be selected from the catalogue of series manufactured motors. These need not be checked again for thermal or overload capacities. The design rating normally takes care of heating and temperature rise and the motor normally has a short time overload capacity.

While selecting a motor for this type of duty it is not necessary to give importance to the heating caused by losses at starting even though they are more than the losses at rated load. This is because the motor does not require frequent starting. It is started only once in its duty cycle and the losses during starting do not have much influence on heating. However, sometimes it may be necessary to check whether the motor has sufficient starting torque, if the load has considerable amount of inertia.

For most types of loads where the torque and speed are known, the power output of the load

$$P_{out} = \frac{2\pi}{60} TN \text{ W} \tag{5.26}$$

If the efficiency of load and transmission is η the power input to the load

$$P = \frac{2\pi}{60} \cdot \frac{TN}{\eta} \text{ W} \tag{5.27}$$

In case of linear motion the rating of the motor corresponds to

$$P = \frac{F \times V}{0.102\eta} \text{ W} \tag{5.28}$$

where F is the force exerted by load in kg
V is velocity of motion in m/s
η efficiency

Using these expressions the rating of the motor is decided. For example the rating of a motor for an elevator is

$$P = \frac{F \times V}{2 \times 0.102\eta} \tag{5.29}$$

2 in the denominator is by virtue of the counter weight.

The rating of the motor for a pump is

$$P = \frac{HQ\rho}{0.102\eta} \text{ W} \tag{5.30}$$

where ρ density of liquid being pumped.
H head comprising suction, delivery, friction and velocity
Q delivery of pump
η combined efficiency of pump and transmission

The rating of a fan motor

$$P = \frac{QH}{0.102\eta} \text{ W} \tag{5.31}$$

where Q volume of air in m^3/s.
H pressure of air in mm water or kg/m^2

Depending upon the work of a driven machine the rating of the drive motor can be determined.

Continuous Duty—Variable Load In this type of duty the load is not constant, but has several steps in one cycle. This cycle of loading repeats for a longer time. If the load variations are slight the motor of continuous rating of the highest load may be chosen from the available catalogue.

However, if the variations in the load cycle are large the machine undergoes a continuous change of temperature. However, after several cycles of operation the motor selected may attain a steady-state value. The thermal calculations of the motor are involved. The selection of a motor based on heating is rather involved and a difficult task. Therefore some simplified criteria may be evolved for selecting a motor for this duty.

If for such a load cycle a motor is selected according to the lowest load, it may not be able to drive the load satisfactorily; the temperature rise of the motor will be exceedingly high and it may not have sufficient capacity to drive the highest load. If the motor is selected according to the highest load, it becomes overrated and may have poor efficiency. If it is an ac drive motor the p.f. is also poor. The motor is underutilised.

The choice of the motor may be based on the average power or average current. At the outset, this method seems to be applicable. It has, however, a disadvantage in that it does not consider the variation of losses. The motor chosen will be smaller for the load cycle, and of insufficient capacity. This may have increased temperature due to overloads, where the losses increase. This happens if the load fluctuations are considerable. The method may give a satisfactory motor if the load fluctuations are relatively small.

However, a method based on average losses of the motor for the load cycle seems to be more appropriate for selecting a motor for a continuous duty, variable load. A motor having its rated losses equal to the average of the losses of the motor for the variable load cycle is chosen to drive the load. In this case the final steady-state temperature rise of the motor under variable load is the same as the temperature rise of the motor with constant load. The motor therefore operates with permissible temperature rise.

The selection of a motor based on average losses requires an iterative procedure. A motor whose losses at its rated load are equal to or somewhat greater than the average losses is suitable for the job. However, it may be expected that the motor will have short time peaks of temperature and these may not be very detrimental to the life of the motor. The method does consider maximum temperature rise of the motor under variable load.

A typical load diagram for continuous variable load is shown in Fig. 5.6. The following steps are involved in the choice of the motor.

i. The average power is determined. The foregoing discussion shows that a motor of this rating is of insufficient capacity. Therefore, a motor is selected from the catalogue, which has a rating 15 to 30% greater than the average power.
ii. For the loads of the load cycle the loss diagram is determined using the efficiency curve of the motor.

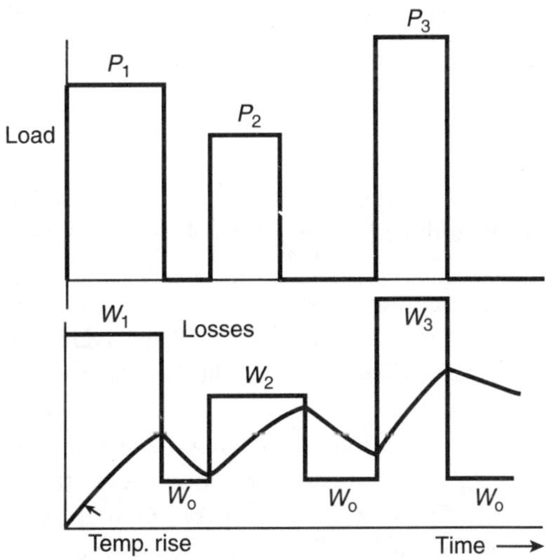

Fig. 5.6 *Heating of a motor having a typical variable load (one cycle of operations)*

iii. The average losses are determined using the formula

$$W_{av} = \frac{W_k t_k}{t_k} \tag{5.32}$$

iv. These losses are compared to the rated losses of the motor (W_r). If W_r is equal to or somewhat greater than W_{av}, the motor selected is sufficiently good. If W_r is very much greater or less than $W_{av,}$ the calculations are repeated for a new motor until a right motor is obtained.

v. A check on the overload capacity of the motor must be made. If the motor chosen does not satisfy the overload requirement as per the load cycle, a motor of higher capacity having the overload requirement may be chosen. The basis of heating is disregarded. The motor will however have the thermal rating.

Sometimes it is more convenient to base the selection of motor on equivalent current, torque or power. These equivalent values are the rms values.

In the equivalent current method, a motor having a rated current equal to the rms value of the variable current of the load cycle is chosen. Let us assume that I_{eq} is the current rating of such motor. The losses of the motor at rated load are

$$W_r = W_c + I_{eq}^2 .R \tag{5.33}$$

When the motor operates on the given load cycle having corresponding current load diagram as shown in Fig. 5.7, the losses occurring in the motor with variable load are

$$W_{av} = \sum \frac{\left(W_c + I_k^2 R\right) t_k}{\sum t_k} \tag{5.34}$$

which after simplification is given by

$$W_{av} = W_c + \frac{\sum I_k^2 R t_k}{\sum t_k} \tag{5.35}$$

Equating these average losses of the motor to its rated losses and simplifying we get

$$I_{eq} = \sqrt{\frac{\sum I_k^2 t_k R}{\sum t_k}} \tag{5.36}$$

Sometimes the load cycle may have a rest period. If the rest period is very large, I_{eq} determined in the above is on the higher side. Therefore this must be properly taken care of in the calculation. A load diagram with rest period is as shown in Fig. 5.8. The equivalent current for this case is obtained using the relation

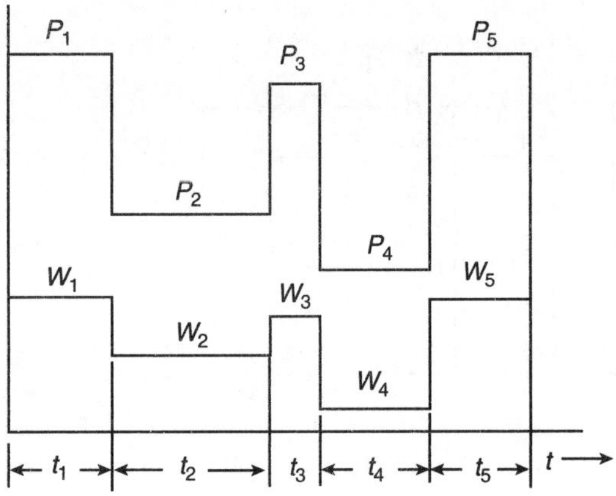

A typical simplified diagram for variable load.

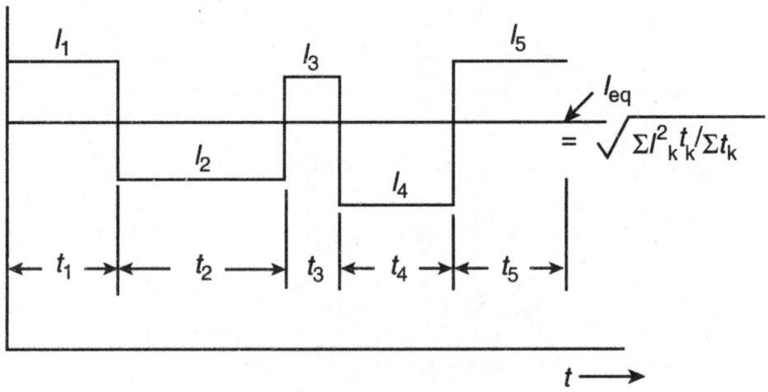

Current load diagram

Fig. 5.7(a) *Equivalent current criterion to select a motor for variable load*

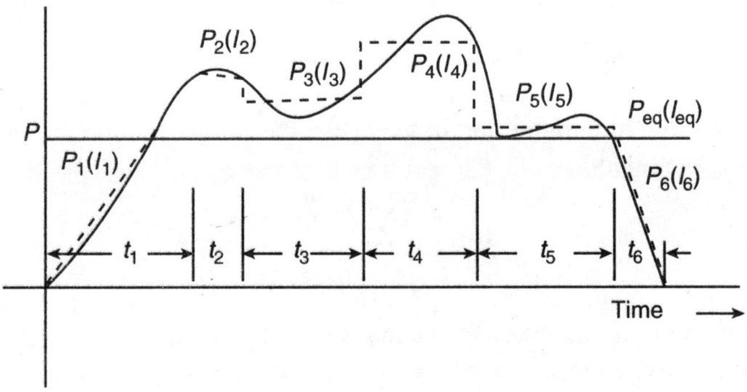

Fig. 5.7(b) *Simplification of a typical variable load. The Currilinear portion of load diagram are presented by segments of straight lines*

$I_{eq} = \frac{1}{t_k} \int \sum \int_0^{t_k} i_k dt$ *I_1, I_2 are the currents corresponding to P_1, P_2*

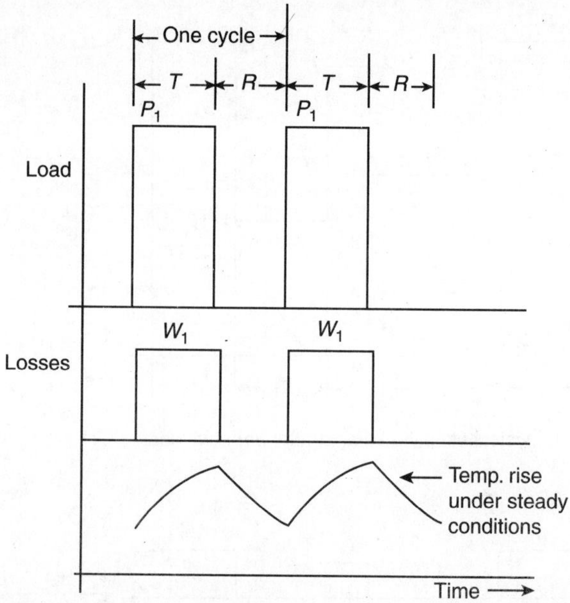

Fig. 5.8 *Load diagram of a load with intermittent load having rest period*

$$I'_{eq} = \sqrt{\frac{\sum I_k^2 t_k R}{\sum t_k}} \cdot \left[\frac{1}{\sqrt{1 + r(1-\beta)}}\right] \tag{5.37}$$

where r is the loss ratio W_c/W_r and

$$\beta = \frac{\sum t_k}{t_0 + \sum t_k} < 1$$

This relation may be easily obtained equating the losses of the load cycle taking the rest period into consideration. The derivation is left to the reader.

If there is a free running period in the cycle, the equivalent current can be calculated using the relation

$$I''_{eq} = \sqrt{\frac{\sum I_k^2 R t_k}{t_0 + \sum t_k}} \tag{5.38}$$

where t_0 is the free running period. It can be seen that the motor chosen will have a rating which lies between the two cases discussed above.

Sometimes the loads during intervals of the cycle may not be constant and may vary with time. The equivalent current for such cases can be determined using integration. $\sum I_k^2 t_k$ in the above equations is replaced by $\int_0^{t_k} i_k^2 dt$ (Fig. 5.10).

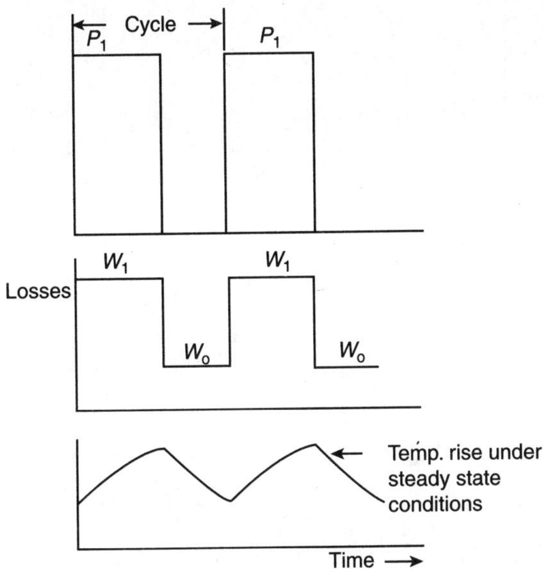

Fig. 5.9 *Load diagram of a motor having intermittent load with free running*

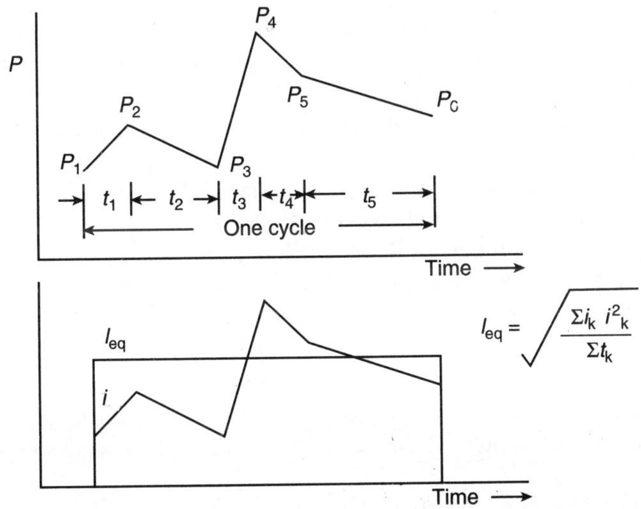

Fig. 5.10 *Load varying with time in a cycle*

If the load diagram shows some arbitrary variation of load during intervals, as shown in Fig. 5.11, some kind of approximation may be made to obtain equivalent current. The diagram may be approximated by rectangular figures. The calculation will be more accurate if i^2 diagram can be drawn before applying the approximation.

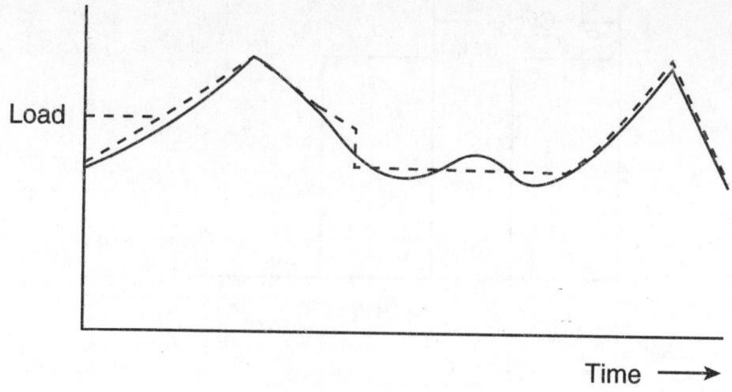

Fig. 5.11 *Pertaining to simplification of an irregular load*

After determining the equivalent current, a motor is selected whose rated current is equal to, or slightly greater than, this value.

The motor must be checked for its overload capacity, i.e., one should check that the maximum load in the cycle is less than or equal to permissible overload of the motor. Therefore a chosen motor should have

$$I_r \text{ equal to or slightly greater than } I_{eq}$$
$$T_{br} \geq T_{max} \text{ of the load cycle} \tag{5.39}$$

If the second condition is not satisfied the motor would be insufficient to drive the load. In this case a motor of higher capacity may be selected according to overload capacity required as per the second condition. The basis of heating need not be considered and the machine will be underutilised as far as the thermal capacity is concerned.

The method of equivalent current gives a right selection if the speed of the motor is constant over the operating cycle. If the speed changes, at lower speeds the cooling will be insufficient in self cooled machines and the motor may prove thermally ineffective to do the job. The speed variations may have to be taken into consideration in the selection of the motor. If, however, forced ventilation is used, the correction for speed variation may not be necessary. This method is also not applicable to motors having a variation in losses due to skin effect or any other reason. The losses of such motors vary sharply during starting and braking. The method cannot be applied if the 'constancy' of so-called constant losses is in question.

When the selection of the motor is based on the equivalent torque criterion, a motor is chosen whose rated torque is equal to (referring to Fig. 5.12).

$$T_{eq} = \sqrt{\frac{\Sigma T_k^2 t_k}{\Sigma t_k}} \tag{5.40}$$

If we can assume that the shaft torque is equal to the electromagnetic torque developed, the motor selected as above will be sufficient for the job. If this assumption is not true due to friction torque or any other loss torque, the motor

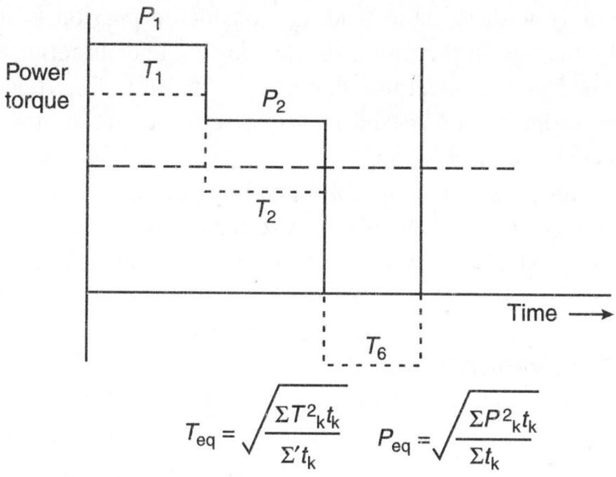

Fig. 5.12 *Equivalent torque and power criteria*

may be chosen to have its rated torque greater than T_{eq} accordingly. The selection of the motor under this criterion also assumes constant flux conditions in the motor. Under this assumption the torque will be directly proportional to the armature current. Also, this method assumes constancy of power factor at all loads if ac motors are employed. This method cannot be applied to motors having variable flux conditions, e.g., series motors or squirrel cage motors under starting and braking. Proper corrections may be applied to take into consideration the variable flux condition. The method of equivalent torque is not applicable if the equivalent current method does not give the correct result.

The selection of motor may sometimes be based on equivalent power rating (Fig. 5.13). When this is done, the motor rating is equal to or slightly greater than

$$P_{eq} = \sqrt{\frac{\Sigma P_k^2 t_k}{\Sigma t_k}} \tag{5.41}$$

The method assumes constant speed at all loads. This method cannot be applied if there are starting and braking periods in the load cycle.

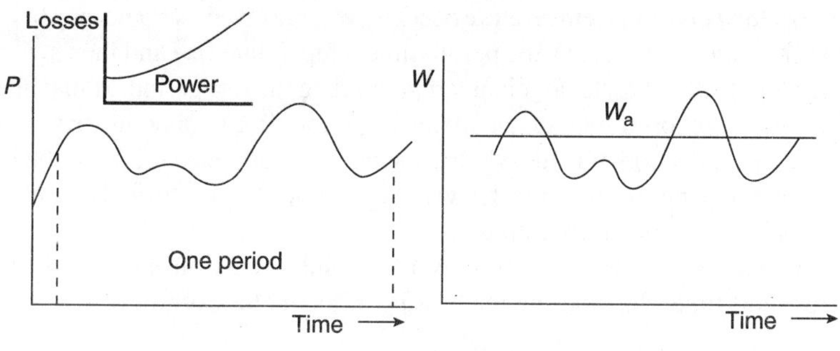

Fig. 5.13 *Equivalent (average) loss criterion*

Continuous duty with variable loading, considered previously, has not taken into account the heating of the motor during starting and stopping. If the starting and braking losses are considerable enough to effect the final temperature rise, they have to be properly considered in the selection of the motor. The relevant load diagram is depicted in Fig. 5.6.

This class of duty may have variable speed operation also. The variable loads may be followed by variation in speeds. The load diagram is shown in Fig. 5.6. The variation of speeds if any must be taken into consideration in the selection of motor.

5.4.2 Short Time Intermittent Duty

Another class of duty for normally occurring loads is short time intermittent duty in which the load requires a constant power for a short interval of time and rests for sufficiently longer time. When a motor is used for this purpose, duration of the load on the motor is less than the heating time constant of the motor or the time required for obtaining thermal equilibrium. The period of rest is sufficient enough to cool the motor to the ambient temperature. The next cycle begins therefore from a cold condition. Such loads occur in some crane drives, household appliances, opening and closing of weirs, lockgates, bridges, etc. The load diagram for this class of duty is shown in Fig. 5.14.

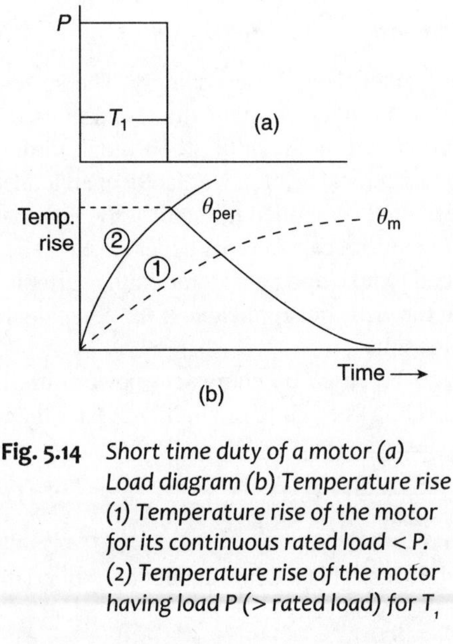

Fig. 5.14 *Short time duty of a motor (a) Load diagram (b) Temperature rise (1) Temperature rise of the motor for its continuous rated load < P. (2) Temperature rise of the motor having load P (> rated load) for T_1*

An electric motor, as has already been pointed out, will have steeply rising losses when it is overloaded. However, there is a time lag between the losses taking place and temperature rise. Therefore, a given machine may be overloaded for a time till it reaches the permissible temperature rise and then switched off, allowing it to cool to the ambient temperature in the rest period. A machine of suitable rating is chosen such that it attains its permissible temperature rise during the period of application of load. When a machine is overloaded this way, the internal hot spot temperatures may reach a very high value. The simplified model gives a rough estimate of the motor rating.

A machine of smaller capacity may be advantageously used to drive these loads for short time. Considering the heating of the motor over a cycle.

$$\theta = \theta_{max}(1 - e^{-t_1/\tau_1}) \tag{5.42}$$

If a motor is selected having a rating equal to the amplitude of the power pulse represented by load the diagram, the temperature rise follows curve 1 of Fig. 5.14. From this it is very clear that at the end of duration of load t, when the motor is switched off its temperature rise is well below the permissible value. The motor is underutilised with regard to its thermal capacity. If, on the other hand, a motor of smaller capacity is chosen, when loaded to give an output corresponding to power pulse of load diagram, it is overloaded and has a temperature rise curve shown by curve 2 of Fig. 5.14. From the figure it is clear that the temperature of the motor rises rapidly towards another maximum value, as decided by the increased losses of the motor. In so doing it attains the permissible temperature rise at the end of t_1. The motor is switched off at this instant. The rating of the motor is called short time rating. The motors may have 10 minute, 30 minute or 60 minute rating based on this criterion. The motor is thermally well utilised. A short time rating of an electric motor can be defined as the extrapolated overload rating of the motor which it can supply for the specified short time without getting overheated. Now a days the machines are being designed and manufactured for short duration having sufficient overload torque capability.

Using the temperature rise curve the short time rating of the given motor can be determined. Referring to Fig. 5.14 the machine having a continuous rating of P_r is used to drive the load P so that it reaches θ_{max} (θ_{per}) at the end of t_1 according to curve 2 of the figure. The temperature curve of the motor is curve 1, shown in Fig. 5.14. The curve 2 extends exponentially toward θ'_{max} which would be the maximum temperature rise of the motor if the drive motor were loaded continuously with P. Therefore

$$\theta = \theta'_{max}(1 - e^{-t/\tau_1}) \tag{5.43}$$

The ratio (P/P_r) is so chosen that at the end of t_1 the temperature rise of the motor is θ_{max} (θ_{per}). Using these values in Eq. 5.43 we have

$$\frac{\theta'_{max}}{\theta_{max}} = \frac{\theta'_{max}}{\theta_{per}} = \frac{1}{1 - e^{-t_1/\tau_1}} \tag{5.44}$$

The heating and temperature rise of the motor are proportional to losses and therefore

$$\frac{\theta'_{max}}{\theta_{max}} = \frac{W'_L}{W_L} \tag{5.45}$$

where W'_L is the total losses at the overload (P)

W_L is the losses at rated P_r

Again using $r = \dfrac{\text{constant losses}}{\text{variable loss at } P_r} = \dfrac{W_c}{W_v}$

we have $W_L = (1 + r)W_v$ (5.46)

$$W'_L = [r + (P/P_r)^2]W_v$$

Using these relations and simplifying, the rating P_r of the motor to drive load of P can be obtained as

$$P_r = \frac{P}{\sqrt{\dfrac{(1+r)}{1-e^{-t_1/\tau_1}} - r}} \tag{5.47}$$

Sometimes it may be required to estimate the time of operation of a given motor of continuous rating P_r to drive the load P. For this we know (P/P_r). The value of t_1 can be obtained from the equation

$$t_1 = \tau_1 \log e\left(\frac{(P/P_r)^2 + r}{1+r} + 1\right) \tag{5.48}$$

The ratio of losses given by Eq. 5.44 represents the thermal overload and the ratio (P/P_r) from Eq. 5.47 represents the mechanical overload. These are shown in Fig. 5.15. From the figure as well as equations, it can be seen that increase in power output is possible by decreasing t_1.

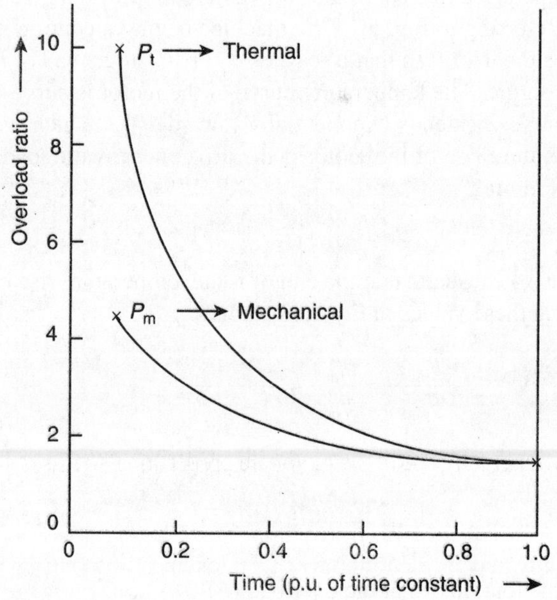

Fig. 5.15 *Typical thermal and mechanical overloads of an electric motor for short time duty*

An overload ratio must be chosen within the available torque range. The machine must be capable of providing the overload without stalling, because there is an absolute limit for motors exhibiting pull-out effect. In commutator machines the commutator presents another limitation.

However, to drive high torque (power) pulses in this class of duty, the motor may be equipped with a flywheel having sufficient inertia. This is brought to the required speed before the load is applied. The energy stored in the flywheel during acceleration is given to the load.

5.4.3 Periodic Intermittent Duty

The class of duty making use of the ability of electric motors to operate with overload is called periodic intermittent duty. The load on the motor is a sequence of identical duty cycles as shown in Fig. 5.17. The drive motor is loaded at constant load for a period t_{on}. At the end of t_{on} the machine is switched off for a period of t_{off}. The machine heats during t_{on} and cools during t_{off}. The time t_{off} is insufficient for the machine to cool to ambient temperature. Also one cycle comprising t_{on} and t_{off} is insufficient for the motor to attain thermal equilibrium. This load cycle repeats and there is temperature fluctuation. The machine attains thermal equilibrium after a number of load cycles. The mean temperature attains a steady value. The load diagram for this class of duty is shown in Fig. 5.16 and the ratio $t_{on}/(t_{on}+t_{off})$ is called the duty factor. The effects of losses during starting on heating, are neglected.

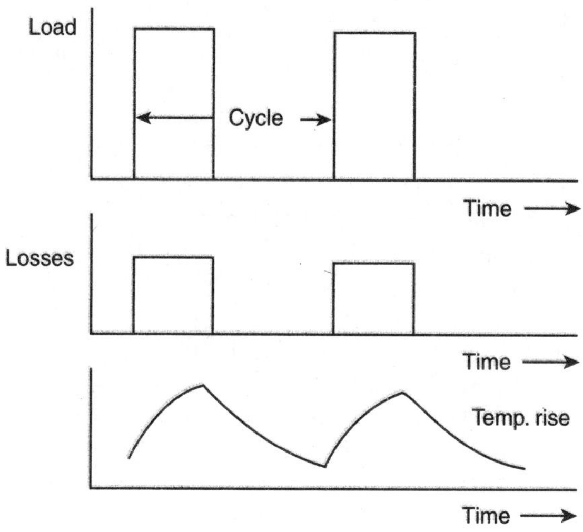

Fig. 5.16 *Periodic intermittent duty of a motor*

Such a class of duty may be identified when the losses during starting affect the motor heating and have to be considered. The load diagram for this duty is shown in Fig. 5.17. The duty cycle includes starting period, period of constant load and rest period. The duty factor is given by $(t_{st}+t_{on})/(t_{st}+t_{on}+t_{off})$. In this duty the motor stops during t_{off} by natural means and stopping does not include any additional losses. This takes place if t_{off} is sufficiently greater than the time required for stopping.

When the machine has special methods of braking, the losses during braking must also be considered in the choice of motor rating. This is another sub class of duty. The duty cycle contains a braking period besides a starting period, period of constant load and rest period as shown in Fig. 5.18. The duty factor $(t_{st}+t_{on}+t_{br}+t_{off})$.

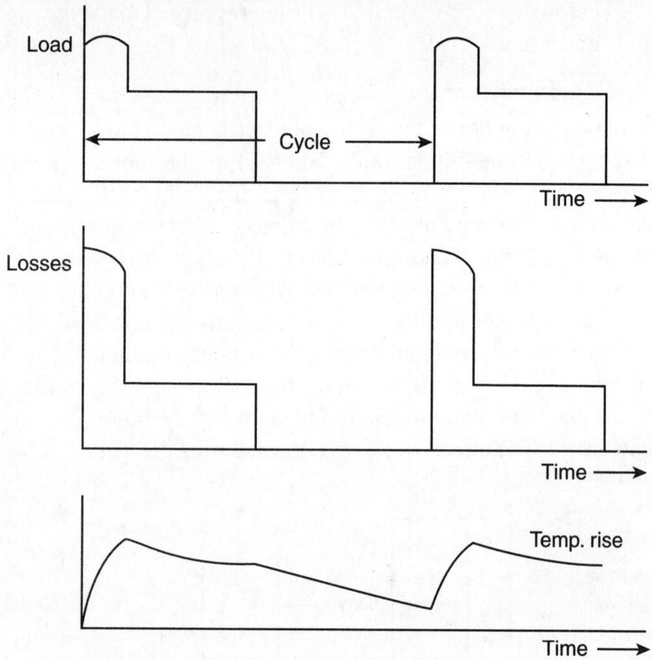

Fig. 5.17 *Periodic intermittent duty with starting*

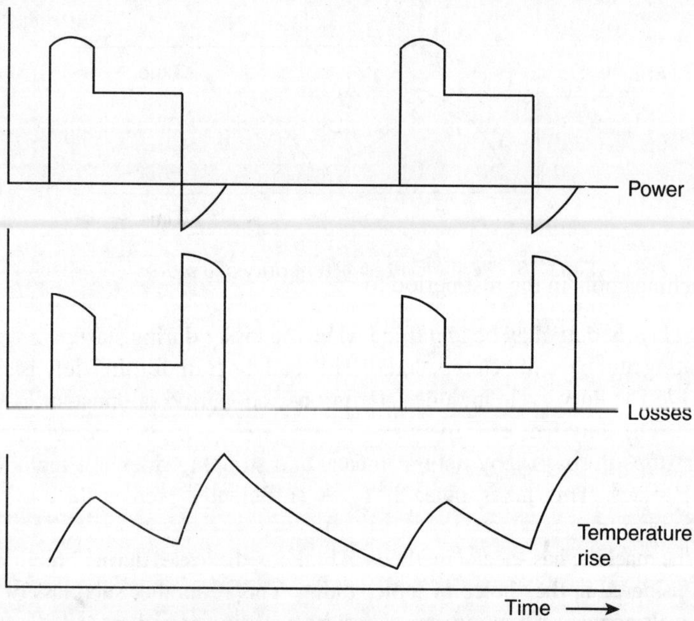

Fig. 5.18 *Load diagram for intermittent duty with starting and braking*

Let us consider the selection of a motor for duty factor having t_{on} and t_{off} (no starting and braking). The starting losses do not influence the heating. The load cycle and heating curve are as shown in Fig. 5.18. Here again a motor of rating equal to the constant load in the load cycle would prove uneconomical because it would have excess heating capacity. A motor of smaller capacity would be sufficient due to rest periods in between successive loadings, during which the machine has an opportunity to cool. A ratio (P/P_r) can be estimated for this class of duty also.

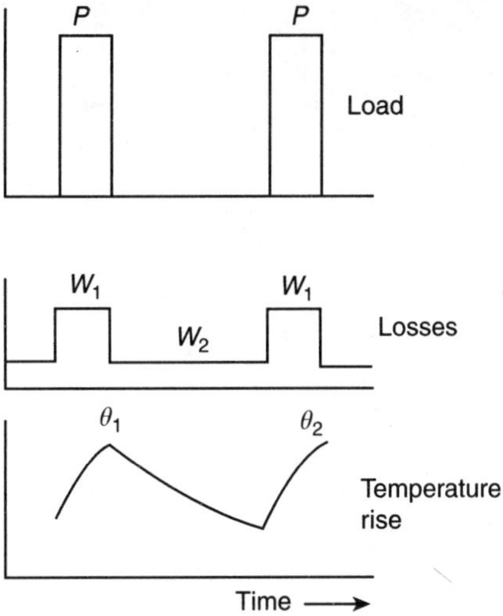

Fig. 5.19(a) *Load diagram for intermittent periodic loading*

Referring to Fig. 5.19 let θ_{n1}, θ_{n2}, θ_{n3} be the temperature rises of the motor at the end of successive load pulses. Similarly, $\theta_{c1}, \theta_{c2}, \theta_{c3}$ are the temperature rises at the end of successive rest periods. If θ_{max} is the maximum temperature rise when the machine has a continuous load of P, the machine has a temperature of rise of θ_{n1} at the end of first pulse given by

$$\theta_{n1} = \theta_{max}(1 - e^{-t_1/\tau_1}) \tag{5.49}$$

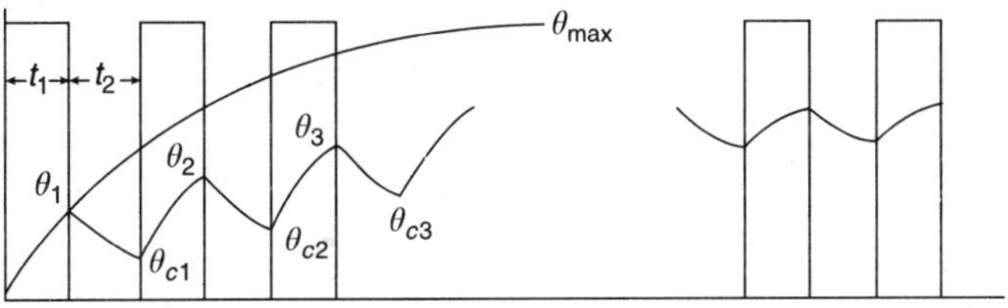

Fig. 5.19(b) *Temperature rise for intermittent loading transient condition to steady-state*

The machine cools in the rest period to

$$\theta_{c1} = \theta_{n1} e^{-t_2/\tau_2} \tag{5.50}$$

Similarly we have

$$\theta_{n2} = \theta_{c1} e^{-t_1/\tau_1} + \theta'_{max}(1 - e^{-t_1/\tau_1}) \tag{5.51}$$

$$\theta_{c2} = \theta_{n2} e^{-t_2/\tau_2} \tag{5.52}$$

$$\theta_{n3} = \theta_{c2} e^{-t_1/\tau_1} + \theta'_{max}(1 - e^{-t_1/\tau_1}) \tag{5.53}$$

$$\theta_{c3} = \theta_{n3} e^{-t_2/\tau_2} \tag{5.54}$$

and so on.

Using these relations we have

$$\theta_{n3} = \theta'_{max}(1 - e^{-t_1/\tau_1})\ (1 + e^{-t_1/\tau_1}\, e^{-t_2/\tau_2} + e^{-2t_1/\tau_1} e^{-2t_2/\tau_2}) \tag{5.55}$$

Temperature at the end of n such cycles

$$\theta_{nn} = \theta'_{max}(1 - e^{-t_1/\tau_1})(1 + e^{-t_1/\tau_1}\ e^{-t_2/\tau_2} + e^{-2t_1/\tau_1}\ e^{-2t_2/\tau_2} + \cdots + e^{-(n-1)t_1/\tau_1}\ e^{-(n-1)t_2/\tau_2}) \tag{5.56}$$

The third expression on the r.h.s. is a geometric series and simplifying this summation we have

$$\theta_{nn} = \theta'_{max}(1 - e^{-t_1/\tau_1})\frac{1 - e^{-nt_1/\tau_1} e^{-nt_2/\tau_2}}{1 - e^{-t_1/\tau_1} e^{-t_2/\tau_2}} \tag{5.57}$$

For sufficiently large number of duty cycles; $n \to \infty$ and $e^{-nt_1/\tau_1} e^{-nt_2/\tau_2} \to 0$. Also the value of θ_{nn} should be limited θ_{per} of the motor having continuous load P_r. Therefore

$$\theta_{max} = \theta_{per} = \theta'_{max}\left[\frac{1 - e^{-t_1/\tau_1}}{1 - e^{-t_1/\tau_1} e^{-t_2/\tau_2}}\right] \tag{5.58}$$

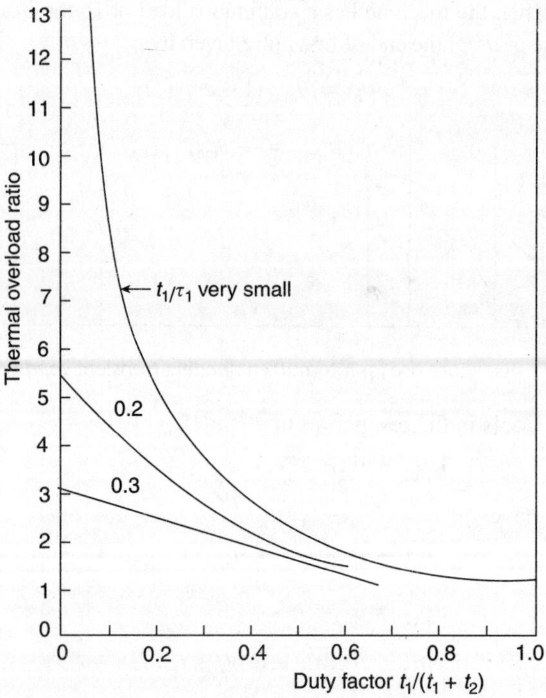

Fig. 5.20 *Thermal overload ratio for intermittent loading (P_h) mechanical overload ratio $P_m \simeq \sqrt{P_h}$*

or thermal overload ratio is

$$\frac{\theta'_{\text{max}}}{\theta_{\text{per}}} = \frac{1 - e^{-t_1/\tau_1} . e^{-t_2/\tau_2}}{1 - e^{-t_1/\tau_1}} \tag{5.59}$$

The mechanical overload ratio can be determined assuming the loss ratio $r = W_c/W_v$

$$\text{Losses at } P_r = W_r(1 + r) \tag{5.60}$$

$$\text{Losses at } P = W_v[r + (P/P)_r{}^2]$$

$$\frac{\theta'_{\text{max}}}{\theta_{\text{per}}} = \frac{r + (P/P_r)^2}{(1 + r)} \tag{5.61}$$

Using this relation in Eq. 5.59 and simplifying we have

$$\frac{P}{P_r} = \sqrt{\frac{(r + 1)(1 - e^{-t_1/\tau_1}\ e^{-t_2/\tau_2})}{1 - e^{-t_1/\tau_1}} - r} \tag{5.62}$$

From this relation the rating P_r of the motor to drive the load can be determined. For duty ratio > 0.6 the expression shows that a motor having a continuous rating of P may be selected. (P/P_r) tends to unity for load ratios > 0.6 assuming the same heating and cooling time constants. The motor size becomes smaller only for small duty ratios. Also the 'on' period should be very small compared to the time constant to get the benefit of a smaller motor.

For classes of duty where starting and braking periods are there, the motor selection should take these into consideration.

The motor selected must be checked for its overload capacity. If the duty factor is very small, this may pose a problem and a motor of higher capacity may be chosen. For larger duty ratios the value P/P_r will be such that the machine does have overload capacity to drive the given load cycle.

Worked Examples

5.1 *The temperature rise of an electric motor is 40°C after 1 hour and 60°C after 2 hours. The motor current is 100 A. Determine approximately its final temperature rise when it works on load cycle of 4 minutes working, 8 minutes rest with a current of 125 A. Neglect the effects of iron losses.*

Solution

$$\theta_1 = \theta_m(1 - e^{-1/\tau})$$
$$\theta_2 = \theta_m(1 - e^{-2/\tau})$$

$$\frac{\theta_1}{\theta_2} = \frac{1 - e^{-2/\tau}}{1 - e^{-1/\tau}} + 1 + e^{-1/\tau} = \frac{60}{40} = 1.5$$

$$e^{-1/\tau} = 0.5$$

$$\frac{1}{\tau} = 0.693$$

$$= 1.442 \text{ hrs}$$

$$\theta_m' = \frac{40}{0.5} = 80°$$

With 125 A $\theta_m' = 80\left(\frac{125}{100}\right)^2 = 125°$

$$\frac{N}{\tau} = x = \frac{4}{60 \times 1.443}$$

$$e^{-x} = 0.955$$

$$Y = \frac{R}{\tau} = \frac{8}{60 \times 1.443} e^{-Y}$$

$$= 0.912$$

$$\theta_m = 125 \frac{1 - 0.955}{1 - 0.955 \times 0.912}$$

$$= 125 \frac{0.045}{0.1293}$$

$$= 43.51°$$

5.2 *A motor has a heating time constant of 90 minutes. If the temperature rise of the motor is 100° C when it is continuously loaded with its rated load determine the temperature rise of the motor after 2 hours of its rated load. If the temperature after 2 hours reaches the maximum permissible temperature (final steady-state temperature with rated load applied continuously) after it is over-loaded, determine the permissible overloading. Assume constant losses = 0.5 of full load copper losses.*

Solution Temperature rise after 2 hours of continuous loading

$$= 100\left(1 - e^{-\frac{2}{1.5}}\right) = 73.64°\text{C}$$

$$100 = \theta_m \left(1 - e^{-\frac{2}{1.5}}\right) = \theta_m \times 0.7364$$

$$\theta_m = 135.8° \quad \frac{W_c}{W_{cu}} = a$$

$$\frac{135.8}{100} = \frac{W_c + x^2 W_{cu}}{W_c + W_{cu}}$$

$$= \frac{a + x^2}{a + 1}$$

$$1.358(a + 1) = a + x^2$$

$$0.358a + 1.358 = x^2$$

$$a = 0.5$$

$$x = 1.24$$

An overloading of 24% can be allowed

5.3 *The heating and cooling time constants of an electric motor are 100 and 150 minutes respectively. The rating of the motor is 125 kW. If it is working on duty cycle of 15 minutes on load and 30 minutes on no-load determine the permissible overloading of the motor. Assume the losses are* $P_c + x^2 P_{cu}$ *and* $P_c / P_{cu} = a = 0.4$.

Solution The ratio

$$\frac{P_x}{P_r} = \sqrt{\frac{(a+1)(1 - e^x \cdot e^y)}{1 - e^x}} - a$$

$$e^x = e^{-\frac{15}{100}} = 0.861$$

$$e^y = e^{-\frac{30}{150}} = 0.819$$

$$\frac{P_x}{P_r} = \sqrt{\frac{(1.4)(0.295)}{1 - 0.86}} - 0.4$$

$$= 1.596$$

It can be overloaded by 0.596
The intermittent rating = 199.55 kW

Problems

5.1 A motor with a thermal time constant of 45 min. has a final temperature rise of 80°C on continuous rating.

(a) What is the temperature rise after 1 hr at this load?

(b) If the temperature rise on 1 hr rating is 80°C determine the final steady temperature at this rating.

(c) What is the 1 hr rating of the motor as a percentage of its normal rating?

(d) When working at this 1 hr rating how long does it take for the temperature to increase from 60° to 80°?

5.2 (a) A motor has a heating time constant of $\tau_h = 2.2$ hrs and a cooling time constant of $\tau_c = 3.5$ hrs. The motor has a final steady temperature rise of 65°C and losses are proportional to (load)^2. The motor is started from cold runs on duty cycle of rated load for 2 hrs and the motor is switched off for 1 hr and then loaded to 1.5 times rated load for 1 hr. Determine the temperature rise at the end of the cycle.

(b) The above motor operates on a repeated duty cycle of 1.5 rated load for 0.6 hr followed by a 1 hr shutdown. Determine the final steady-state temperature rise.

5.3 A motor has a heating time constant of 90 minutes and cooling time constant of 120 minutes and final steady-state temperature rise on full load of 60°C. The motor has repeated load cycle of full load for 30 minutes followed by stationary period of 30 minutes. Determine the maximum and minimum temperatures. Determine the overload on the motor that can be allowed on this cycle such that the maximum temperature rise does not exceed the permissible value of 60°C.

5.4 A motor is continuously rated at 50 kW. It has a heating time constant of 100 minutes. Determine the 1 hr rating of the motor. The motor losses can be expressed as

0.6 full load copper losses + x^2 full load copper loss

where x is load as a fraction of full load.

5.5 A motor has a heating time constant of 45 minutes and cooling time constant of 75 minutes. The motor has final steady temperature rise of 50°C while delivering its continuous rating of 25 kW.

(a) Determine the load the motor can deliver for 15 minutes so that the temperature rise does not exceed 50°C.

(b) The motor delivers 35 kW for a period of 15 minutes followed by a shutdown for 15 minutes. Determine the maximum temperature rise.

(c) If the maximum temperature rise is limited to 50°C determine the maximum load the motor can deliver during ON time of 15 minutes in (b).

5.6 A motor has a cyclic loading as given below:

250 Nm for 15 minutes
350 Nm for 20 minutes
100 Nm for 15 minutes
No load for. 10 minutes

The motor runs at a constant speed of 500 rpm. Determine the rating of a suitable motor.

5.7 The following load cycle repeats for a particular drive having TN = constant.

250 Nm 15 minutes at a speed of 600 rpm
350 Nm 15 minutes at a speed of 500 rpm
100 Nm 20 minutes
200 Nm 10 minutes

Determine a suitable rating of the motor for the purpose.

5.8 The heating and cooling time constants of a 150 kW electric motor are 100 and 150 minutes respectively. The motor is subjected to a duty cycle of 20 minutes of loading followed by 40 minutes of no-load. The losses of the motor can be expressed as

$(x^2 + 0.5)$ full load copper loss

Determine the rating of the motor.

Multiple-Choice Questions

5.1 The heating time constant of an electrical machine gives an indication of its
(a) cooling
(b) rating
(c) overload capacity
(d) short time rating

5.2 Short time rating of an electrical machine
(a) is equal to name plate rating
(b) is less than the name plate rating
(c) is greater than the name plate rating
(d) has no bearing to its name plate rating

5.3 All the physical dimensions of two electric machines are in the ratio K. The iron losses of the machines are in the ratio (assuming constant flux density in both the cases)
(a) K
(b) K^2
(c) K^3
(d) K^4

5.4 Class B insulation can withstand a maximum temperature of
(a) 145°C
(b) 105°C
(c) 135°C
(d) 120°C

5.5 The rating of a motor for a given industrial load cycle should have
(a) sufficient thermal capacity
(b) sufficient over load capacity
(c) both of the above
(d) sufficient starting torque

5.6 A machine driving pulsed torque load is equipped with a flywheel in order to
(a) equalise the current demand during the operation
(b) equalise the torque requirement
(c) reduce the mechanical overload
(d) make the motor thermally suitable to drive the load

5.7 Two motors of the same name plate details have different thermal time constants.
(a) The short time ratings of the two motors are the same
(b) The short time rating of the motor with large time constant is large
(c) The short time rating of the motor with large time constant is small
(d) Overload capacity of the motor with large time constant is large.

References

5.1 Leonhard, W., Control of electrical drives, Springer Verlag, Berlin, 1985.
5.2 Pillai, S.K., A first course on electrical drives, Wiley Eastern Limited, 1982.
5.3 Chillikin, M., Electric drive, Mir Publishers, Moscow, 1965.
5.4 Deshpande, M.V., Electric motors: Applications and control, A.H. Wheeler. 1985.

6 Control Techniques for Electrical Drives

6.1 INTRODUCTION

An electric drive is a well established industrial drive as it has several advantages and special features. Its control consists in starting, speed control, braking and speed reversal, and also maintaining the drive conditions required by the process or work being performed by the drive.

Modern electric drives employ thyristors and thyristor power converters for feeding the electric motor for the purpose of speed control, e.g., they provide a variable voltage to the armature of a dc motor; dc link converters or cycloconverters are used to provide variable voltage variable frequency supply to ac motors. These converters are static devices and their use makes the drive system compact, small in size, light weight, and less bulky. They have a high amplification factor. The overall efficiency of the drive improves because of insignificant losses in the static equipment. These drives employ automatic closed loop control. The automatic control of the drive has the following advantages:

i. It permits increased productivity and improves the quality of production.
ii. It reduces running costs and hence production is economical.
iii. It reduces the expenditure on electrical energy
iv. It improves the reliability of the system
v. It provides better working conditions
vi. It simplifies the operation of the equipment
vii. It makes remote control possible, particularly when the drive is inaccessible and the local control is difficult.

Thyristor controlled electrical drives having automatic control of current and speed are very popular. They can be controlled during starting, speed control, regenerative braking and speed reversal. It is well known that soft starting of an electric motor is possible with these drives. Using proper control the drive can be started and accelerated at constant torque and current. The motor does not see its blocked rotor behaviour. In an ac motor this is possible by simultaneously controlling the frequency and voltage of the motor using what is called slip control, in which slip frequency is kept constant. To maintain constant current during acceleration closed loop control is necessary. Speed control of the drive motor requires the simultaneous control of voltage and frequency to maintain constant flux conditions in the motor. This also requires closed loop control. During braking up to zero speed and speed reversal from thereon, the thyristor converters are so controlled that the kinetic energy of the motor is returned to the mains during braking, and soft starting is made available for acceleration in the opposite direction. Control during braking and speed reversal also require closed loop automatic control.

Therefore modern electric drive systems employ closed loop controls and the principles of feedback control theory. They are found to be versatile and are becoming very popular. They are also becoming price competitive, as the price of thyristors is coming down. Very sophisticated drive systems are being developed with excellent dynamic and steady state response.

The performance of the closed loop drive is of primary interest. A suitable drive system using closed loop control using speed feedback must be stable. It should provide acceptable transient and steady-state response to input commands. The system must be less sensitive to parameter variations. The steady-state error, which is a measure of steady-state response and ability of the control system to follow the input, should be minimum for the inputs. The system must be able to eliminate the effects of undesirable disturbances.

Even though a feedback control system is complex and costly, one of the foremost and fundamental reasons to employ the feedback in the drive systems is its improved performance with regard to reduction of steady-state error of the system. A closed loop system has a steady-state error which is several times less than that of open loop systems. However it is impossible to realise an optimum control system having all these requirements. Some adjustments may have to be made to improve the performance. Sometimes a compromise may have to be arrived at between conflicting and demanding specifications in choosing the system parameters to provide acceptable performance.

Therefore, in order to assess the behaviour of these drives, the techniques of conventional feedback control theory have to be applied. The analysis and synthesis of drive systems form a special case of conventional feedback theory. Conventional transfer function methods can be applied to determine the time domain and frequency domain behaviour of the system. The stability of the drive which is a necessary but not sufficient condition may be analysed using the conventional Routh-Hurwitz and Nyquist stability criteria. Based on these methods, the design of the controllers for stabilization of the system is possible both in the time domain using root locus techniques and frequency domain using Bode plots. The ac drive systems utilising induction and synchronous motors may be considered to be multivariable systems. These can be analysed using the methods of modern control using state space techniques to determine the drive behaviour. The controllers may be designed based on these methods.

In the following, the aspects of feedback control as applied to drive problems are discussed. The area of the control of electric drives is a practical application of feedback control theory. The following discussion gives an insight into the drive problems, so that simplified but sufficiently accurate procedures of control theory may be applied to solve them.

6.2 BASIC FEATURES OF AN ELECTRIC DRIVE

Before applying advantageously the control principles to electric drive systems to adjust or improve their behaviour as a special case of control problem, it is necessary to have a knowledge of the specific features of the drive, which can be summarized as follows with reference to its electrical characteristics.

i. Electrical drive offers energy transformations. These systems have reasonably high efficiency and are of special interest.
ii. The control components are used to limit the amplitude or rate of change of variation of individual quantities, e.g., the armature current of a dc motor.
iii. The finite inertia (element of energy storage) of the system does not allow the instantaneous speed changes there by resulting in a finite acceleration, and the drive takes a definite time to follow the speed changes.
iv. The characteristics of almost all the control components are more or less non-linear. Normally this non-linearity is introduced by saturation. These are approximated as linear elements or linearized about an operating point.
v. The static and dynamic parameters of control components can be obtained only approximately or sometimes be estimated. In many cases they change during operation.
vi. When fed from thyristor power converters, the input voltage and current differ from the conventional input to the motors. In the case of dc motors the input is superimposed by ac components. In the case of ac motors the input is non-sinusoidal having harmonic components. If the measuring equipment of the control system is very much affected by these disturbances the control properties of the drive may not be satisfactory. Therefore care must be exercised to reduce or even eliminate, if possible, the effects of these disturbances on the measuring equipment.
vii. The power circuit of thyristor converter receives the thyristor control pulses from a control unit. The operation of this may be disturbed by the power pulses created by the protective equipment of the drive. Necessary care must be exercised here also in designing the control unit to reduce these effects.

viii. The control of the drive system should be designed to protect the same from dangerous operating conditions or overloads. In the first case the system must be brought to standstill and in the second case the drive should be able to operate within its limits.

ix. In many cases control of one quantity may depend on the control of the other. Speed control may depend on voltage control. The control must be simple. The system must be insensitive to parameter variations, reswitching operations, etc.

The above discussion makes it very clear that many of the control circuits offered by the conventional theory of control may be used but with a caution in the area of electrical drives, so that they give satisfactory dynamic and steady-state performance with simpler design, very little complexity, maximum reliability and safety, as demanded by the industrial user.

6.3 BLOCK DIAGRAM REPRESENTATION OF DRIVE SYSTEMS

It is normal practice in control engineering to represent a control system by means of a block diagram, with a systematic connection of blocks in the direction of signal flow, showing the functions performed by each component of the system. A control system is dynamic and its performance is represented by a set of differential equations. Each component is represented by a block having a definite input-output relation. In effect a block diagram is a graphical representation of the basic equations of a physical system. The direction of signal flow is specified and each block is unidirectional, arrow heads in the diagram representing signal flow.

The drive system employing the principles of feedback control theory also have several components performing individual functions (Fig. 6.1). An electric motor drives the mechanical load. The motor is supplied from a thyristor power converter. The converter has necessary controls to provide the required supply to the motor, e.g., variable voltage variable frequency supply, if the drive motor is ac motor. In addition to these there are speed and current controllers and limiters in the systems. Furthermore, there are devices for speed and current measurement. Therefore, the system can be advantageously divided into several parts, each part being represented by a block. All the blocks are interconnected. The drive system with all its components may be represented by a set of differential equations which describe the dynamic and steady-state behaviour. When once it is represented by a block diagram, each block having its own input-output relationship, it gives a valuable insight into the nature of the system. The system simplification using block diagram reduction techniques is very easy. The process is mechanical and yields a single overall block to represent the performance of the system. All the blocks being unidirectional are connected in the direction of signal flow. The block

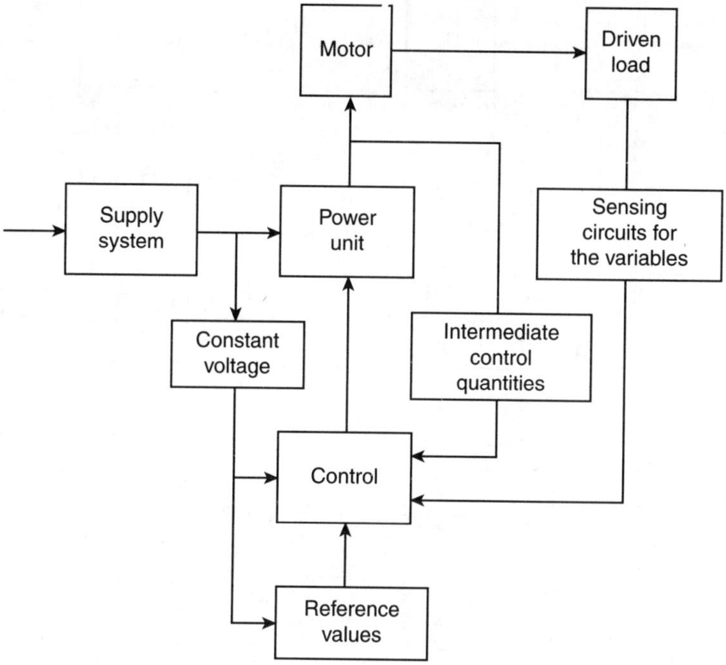

Fig. 6.1 *Representation of a controlled drive by blocks performing individual functions*

diagram, therefore indicates realistically the flow of signals, apart from representing the system by a mathematical model. It is an easy representation to obtain the information regarding the dynamic behaviour. However it does not give any information about the physical nature of the system. A single block diagram may represent different dissimilar unrelated systems.

The system analysis utilising the method of block diagram reduction gives a better understanding of the system and the contribution of each component to the performance of the system. It is therefore logical to expect an improvement or alteration in the system performance by changing the input-output relation of one or several blocks. Sometimes new blocks may be added to improve the performance. The process is rather easier than manipulating machine equations.

However, only linear systems may be represented by a block diagram. The drive systems invariably have non-linear elements. These have to be linearised about an operating point to be able to apply the block diagram techniques.

Figure 6.2 summarises various reduction techniques of block diagram for simplification of the system. The block diagram of a typical dc motor speed control system is given in Fig. 6.3.

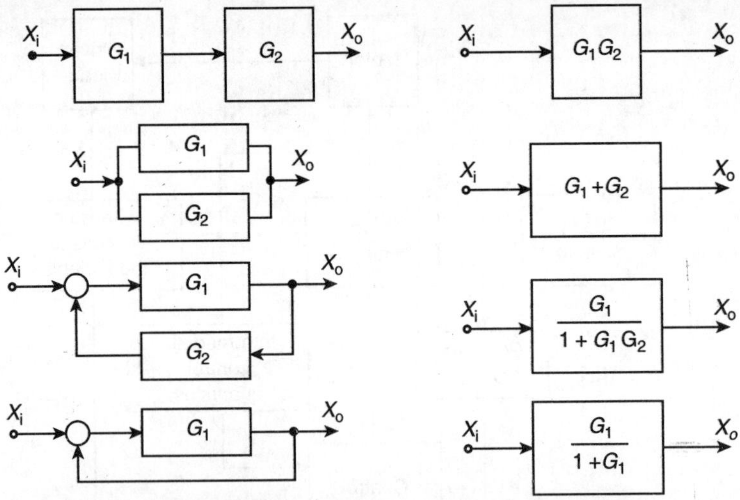

(a) Multiplication and addtion in block diagram technique

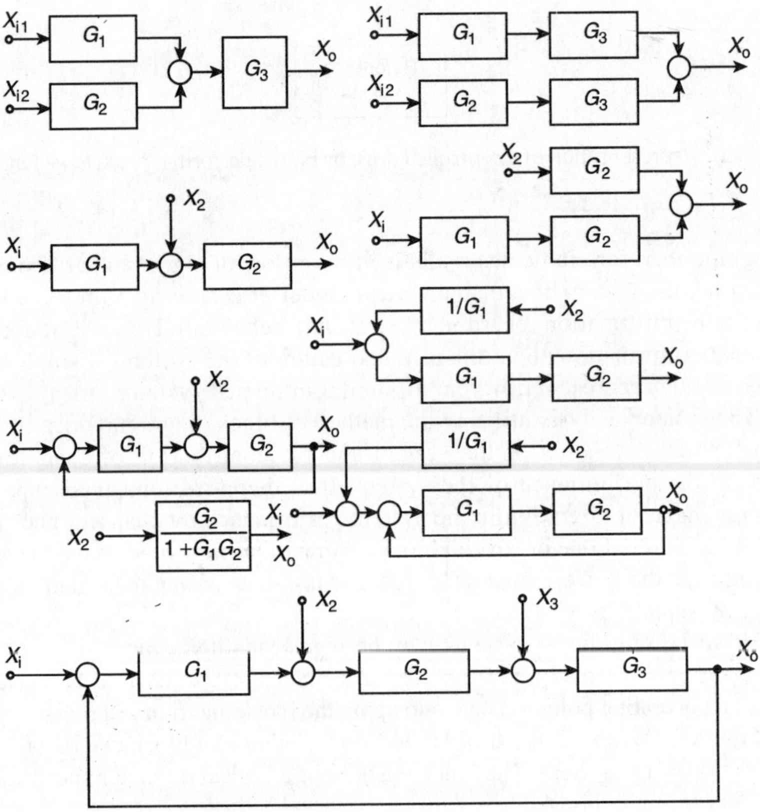

Fig. 6.2 *Block diagram simplifications*

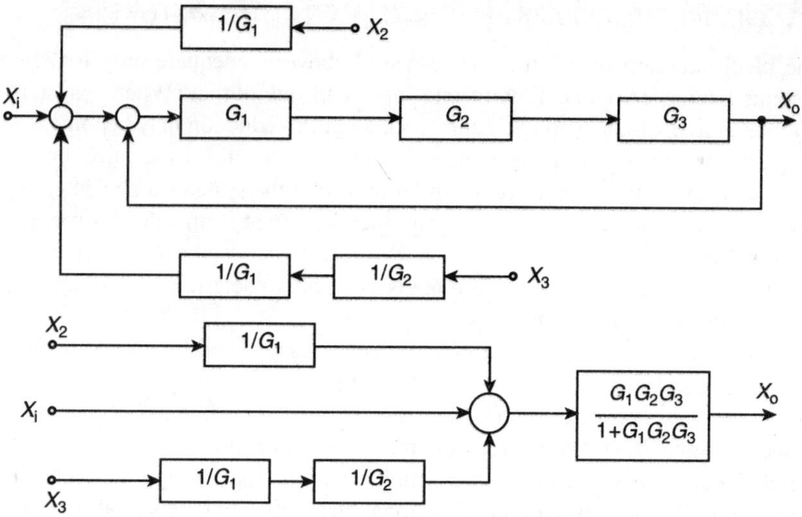

Fig. 6.2 *(Continued)*

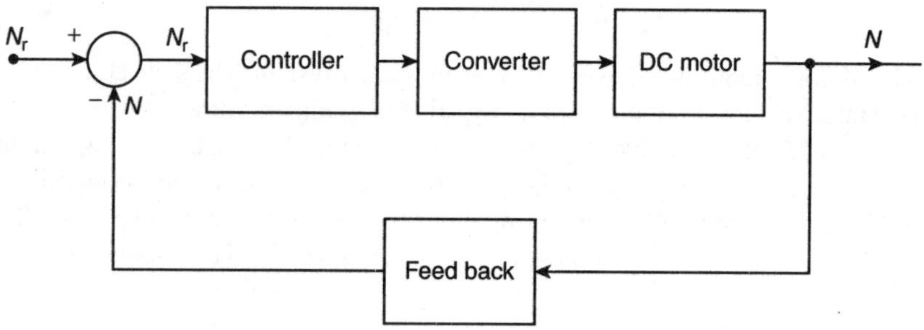

Fig. 6.3(a) *Speed control of a dc motor with speed loop alone*

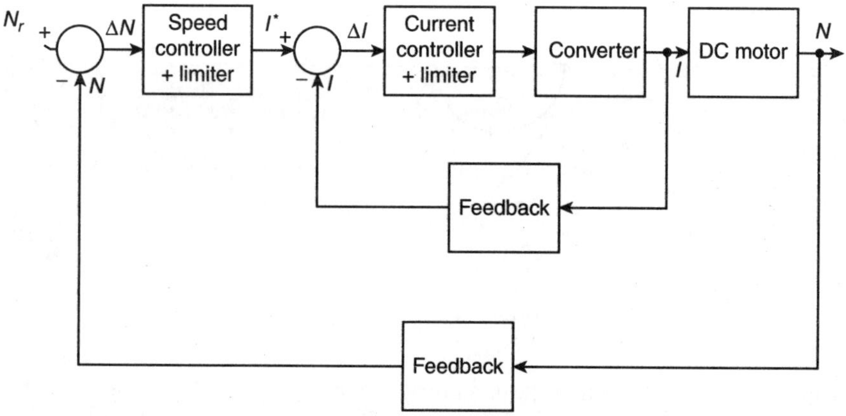

Fig. 6.3(b) *Speed control of a dc motor having both speed and current loops*

6.4 SIGNAL FLOW GRAPH REPRESENTATION OF THE SYSTEMS

The block diagram representation discussed above is adequate only for simple systems having components with one input and one output. When a system is complex, having many variables, the block diagram reduction may become a very difficult task to perform. In such cases it may be advantageous to have a method using which the overall input-output relationship of the system can be determined without needing to perform a reduction. Such a method is afforded by representing a system by its signal flow graph which has line segments directed in the direction of signal flow. A gain formula is available to determine the input-output relationship between the variables.

As has already been stated, a system can be represented by a set of differential equations. The pictorial representation of these equations by means of nodes which are interconnected by means of directed line segments called branches, is called the signal flow graph of the system. A node of the graph represents a system variable and a branch acts as a single multiplier with the multiplying factor indicated on it. The branches are unidirectional elements allowing signal flow in the direction of arrow head and are equivalent to blocks in a block diagram.

A multivariable system is represented by state equations of the form

$$\dot{X} = AX + Bu \tag{6.1}$$

A signal flow graph can be used to represent such a system. The state equations of a separately excited dc motor and its signal flow graph are shown in.

A signal flow graph gives the same information given by a block diagram, but has several advantages. Multivariable systems can be represented very easily. A gain formula is available to obtain relationships between variables. Simple flow graphs allow writing down of these relations by inspection. Time domain solution

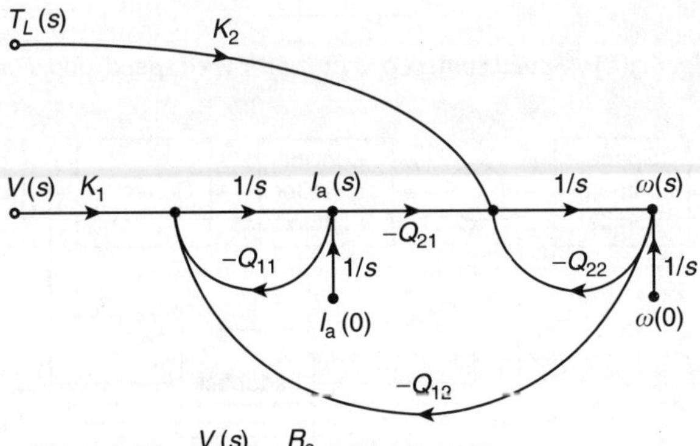

$$sI_a(s) - I_a(0) = \frac{V(s)}{L_a} - \frac{R_a}{L_a} I_a(s) - K\omega(s)$$

$$s\omega(s) - \omega(0) = \frac{T_L(s)}{J} + \frac{K}{J} I_a(s) - B\omega(s)$$

Fig. 6.4 *Signal flow diagram of a dc motor. Constants Q_{11}, Q_{22}, Q_{12}, Q_{21} can be read out from equations* of the equations of multivariable systems using signal flow is straightforward. There is no need for evaluating the state transition matrix and convolution integrals. The effect of input is normally taken care of in the formulation of input-output relations of the graph.

Mason's Gain Formula relating an output variable to an input variable is given by

$$\frac{C(s)}{R(s)} = \frac{\Sigma P_k \Delta_k}{\Delta} \tag{6.2}$$

where P_k is the loop gain defined as the continuous succession of branches in the direction of signal flow, encountering a node only once.

Δ_k cofactor of the path P_k
Δ determinant of the graph

6.5 TRANSFER FUNCTIONS

The dynamic behaviour of a system is described by a set of differential equations. On many occasions the solution of these equations requires the evaluation of convolution integrals which may be a difficult task. In such cases a simpler analysis is made feasible by the use of transfer functions making use of Laplace transforms. A transfer function is defined for linear time invariant systems as the ratio of the Laplace transform of the output variable to the Laplace transform of the input variable, assuming all initial conditions to be zero. Non-linear systems with one or more time varying parameters cannot have transfer functions as the Laplace transform does not exist for these. However, drive systems having non-linearities can be linearised and linear feedback theory can be applied. These systems may be represented by block diagrams or signal flow graphs. In the block diagram approach a block represents the function of a component with its input and output. The transfer function of the component is written in the block. Thus the transfer function is of the form

$$G(s) = \frac{Y(s)}{X(s)} = \frac{a_0 s^{\mathrm{m}} + a_1 s^{\mathrm{m}-1} + \cdots + a_{\mathrm{m}}}{b_0 s^{\mathrm{n}} + b_1 s^{\mathrm{n}-1} + \cdots + b_{\mathrm{n}}} \tag{6.3}$$

while defining the transfer function, Laplace transform has been used because it transforms all the differential equations to simple algebraic equations. Fig. 6.5 depicts a block diagram of a dc motor in a drive system. The motor has armature voltage as input variable and speed as output variable. The transfer function is written in the block.

The transfer function is derived from the set of differential equations which describe the behaviour of the system. It is used to determine the dynamic behaviour of the system as it gives locations of poles and zeros on the s-plane.

It, however is not concerned with the internal physical structure of the system. Dissimilar physical systems may have similar transfer functions and also similar dynamic behaviour. The transfer function is in terms of parameters of the system

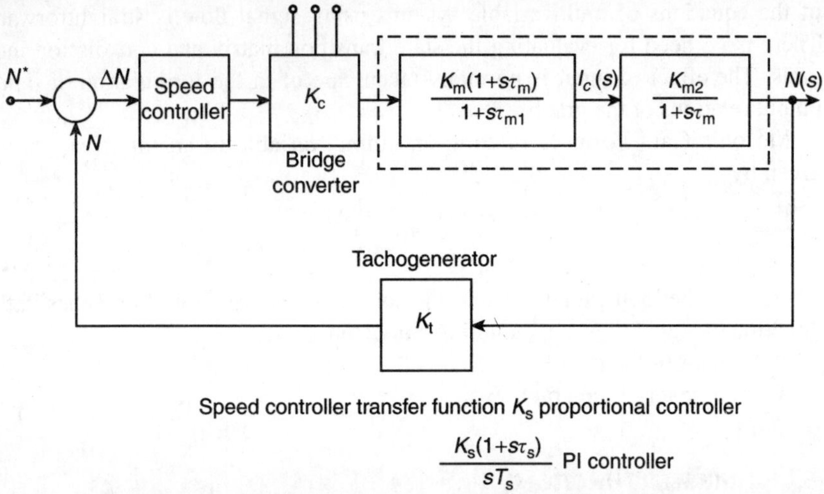

Fig. 6.5(a) *Block diagram of speed control with only speed loop*

and is a property of the system. It does not depend on the magnitude or nature of the input. The highest power of s in the denominator represents the order of the system.

A simple example is given in the following to show the simplicity afforded by the transfer functions.

Example Two RC circuits are connected in cascade, as shown in Fig. 6.6. The input is a step voltage

$$u_0 = 0 \quad t < 0$$
$$u_0 = 1 \quad t \geq 0$$

The amplifier is used to function as dc coupling between them. Determine the time variation of u_2. We know

$$u_1 = \frac{1}{c_1}\int_0^t i_1\,dt \qquad u_2 = \frac{1}{c_2}\int i_2\,dt$$

$$i_1 = \frac{u_0}{r_1}\exp(-t/T_1) \quad i_2 = \frac{u_1(t)}{r_1}\exp(-t/T_2)$$

Substituting

$$u_1 = u_0(1 - e^{-t/T_1}) \qquad u_2 = u_1(t)(1 - e^{-t/T_2})$$

The ratio u_2/u_0 is obtained by evaluating the convolution integral

$$\frac{d}{dt}\int_0^t (1 - e^{-\tau/T_1})\,(1 - e^{-(t-\tau)/T_2})\mathrm{d}\tau$$

as

$$1 - \frac{T_1}{T_1 - T_2}e^{-t/T_1} + \frac{T_2}{T_1 - T_2}e^{-t/T_2}$$

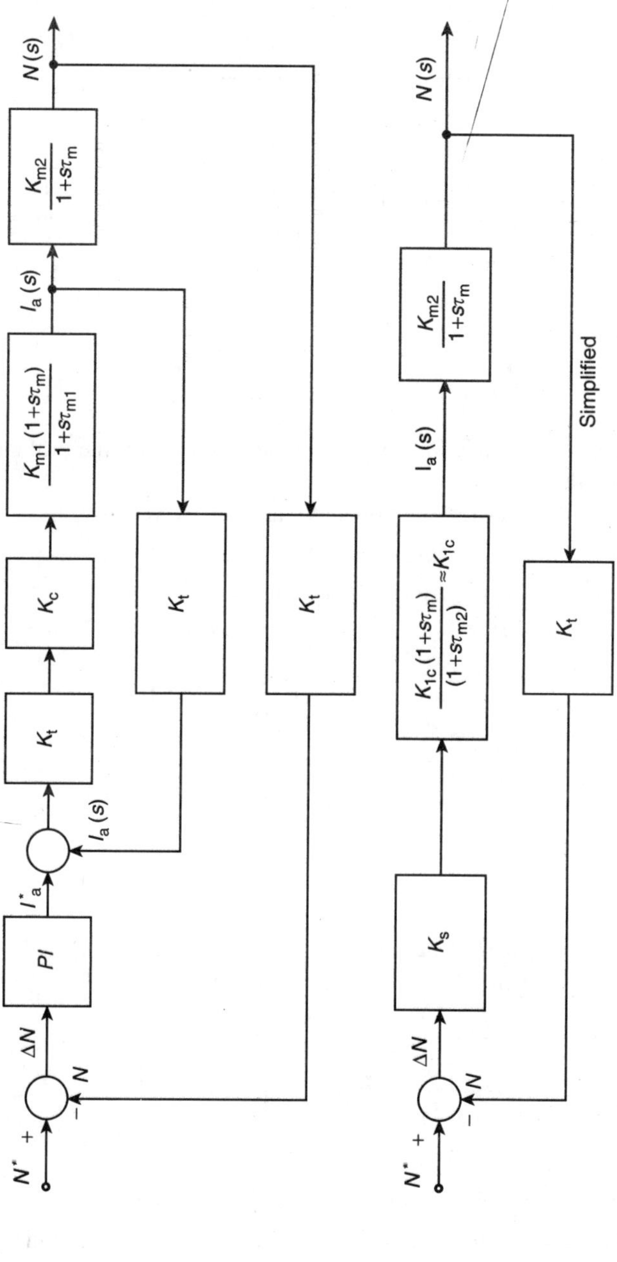

Fig. 6.5 (b) *Block diagram of a.d.c. motor speed control having both speed and current control.*

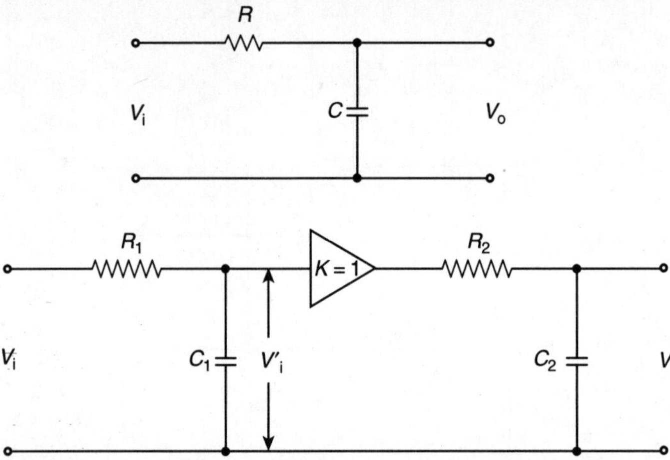

Fig. 6.6(a) *Two RC circuits in cascade*

If one more RC circuit is added in cascade to the given circuit the evaluation of u_2/u_0 is very difficult.

On the other hand if transfer function approach is used we have

$$\frac{u_1(s)}{u_0(s)} = \frac{1}{(sT_1 + 1)}$$

and

$$\frac{u_2(s)}{u_1(s)} = \frac{1}{(sT_2 + 1)}$$

Using these equations, we have

$$\frac{u_2(s)}{u_0(s)} = \frac{1}{(sT_1 + 1)} \; \frac{1}{(sT_2 + 1)}$$

But $\quad u_0(s) = u_0/s$

Therefore

$$\frac{u_2(t)}{u_0} = L^{-1} \frac{1}{s} \frac{1}{sT_1 + 1} \; \frac{1}{sT_2 + 1}$$

can be very easily evaluated as

$$1 - \frac{T_1}{T_1 - T_2} e^{-t/T_1} + \frac{T_2}{T_1 - T_2} \; e^{-t/T_2}$$

using partial fraction expansion.

6.5.1 Transfer Function of an Armature Controlled dc Motor

The speed of a dc motor can be controlled by varying the voltage applied to the armature of a dc motor. A separately excited dc motor with variable armature voltage finds application as a drive motor in a variable speed drive. The variable

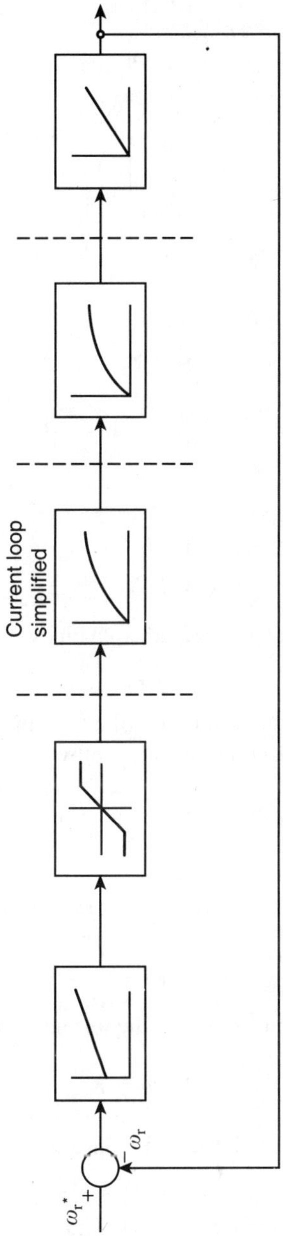

Fig. 6.6(b) *Simplified block diagram of closed loop control of a dc motor*

armature voltage is provided by a phase controlled rectifier. The schematic of an armature controlled dc motor is shown in Fig. 6.7.

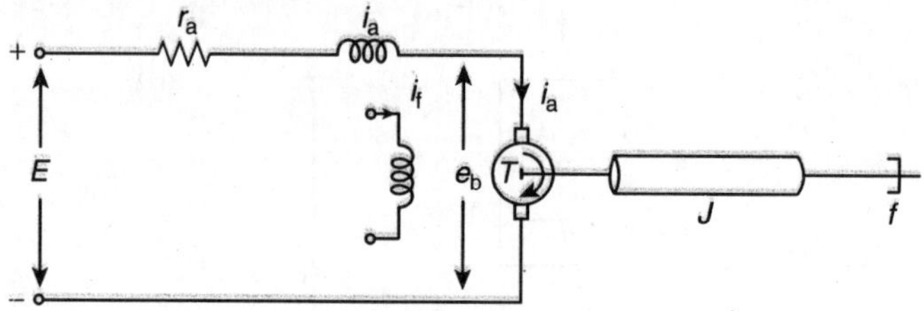

Fig. 6.7 *Armature controlled dc motor*

The torque developed by the dc motor

$$T_d = K\phi i_a \tag{6.4}$$

where ϕ is air gap flux
i_a is armature current
and K is a constant

Neglecting the affects of saturation and armature reaction we have the air gap flux proportional to the field current. That is

$$\phi = K_f i_f \tag{6.5}$$

Because i_f is constant the torque developed is given by

$$T_d = K_t i_a \tag{6.6}$$

where K_t is motor constant. The armature voltage e_a is supplied by the thyristor converter. The armature circuit equation is given by

$$e_a = i_a r_a + L_a \frac{di_a}{dt} + e_b \tag{6.7}$$

e_b in Eq. 6.7 is the rotational (back) emf induced in the armature and is proportional to the product of speed and flux. But, the flux of the motor is constant. Therefore,

$$e_b = K_e \omega \tag{6.8}$$

The dynamic equation of the motor giving the torque balance can be written as

$$J\frac{d\omega}{dt} + f \cdot \omega = K_t i_a \tag{6.9}$$

Assuming the initial conditions to be zero, Laplace transforms of Eqs 6.7, 6.8 and 6.9 can be written as

$$E_a(s) = r_a I_a(s) + sL_a I_a(s) + E_b(s) \tag{6.10}$$

$$E_b(s) = K_e \omega(s) \tag{6.11}$$

$$sJ\omega(s) + f\omega(s) = K_t I_a(s) \tag{6.12}$$

Taking $E_a(s)$ as the input and $\omega(s)$ as the output, the transfer function $\omega(s)/E_a(s)$ can be obtained by eliminating $I_a(s)$ from the equations and is given by

$$\frac{\omega(s)}{E_a(s)} = \frac{K_t}{L_a J s^2 + (L_a f + r_a J)s + (r_a f + K_t K_e)} \tag{6.13}$$

The block diagram given in Fig. 6.8(a) represents Eq. 6.13. This can be finally reduced to a single block given in Fig. 6.8(b).

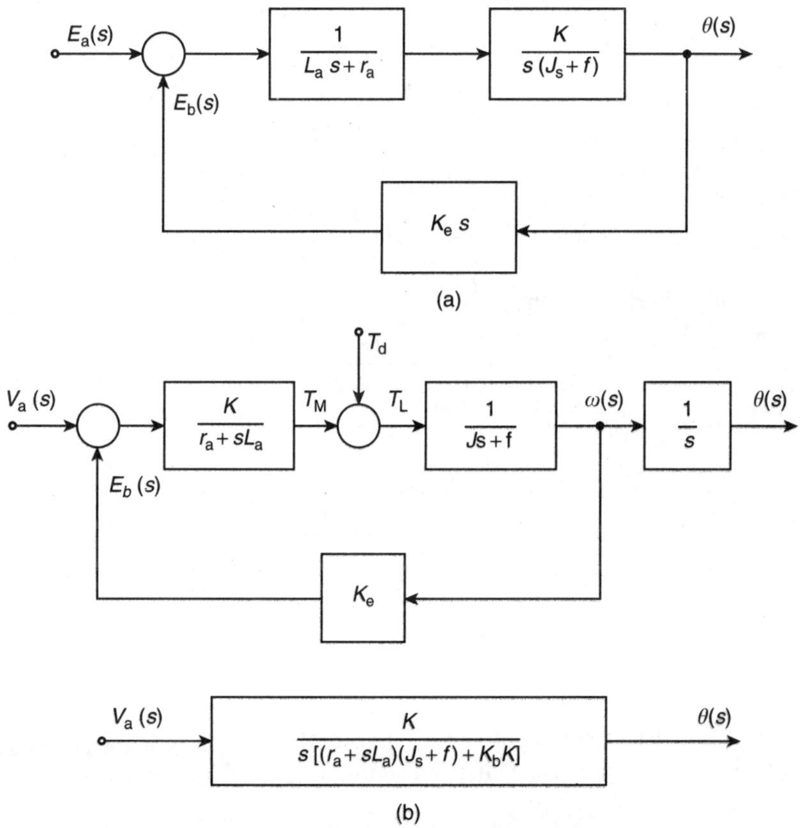

Fig. 6.8(a & b) *Block diagram of an armature controlled dc motor and its simplification*

Normally the armature inductance L_a is very small and may be neglected. The transfer function in this case is given by

$$\frac{\omega(s)}{E_a(s)} = \frac{K_m}{(T_m s + 1)} \tag{6.14}$$

where

$$K_m = \frac{K_t}{(r_a f + K_t K_e)}$$

$$T_m = \frac{r_a J}{(r_a f + K_t K_e)}$$

It can be seen that the back emf affects the damping of the system. A transfer function between the speed and load torque can be derived by assuming the other input e_a to be zero. In this case the dynamic equation would be

$$T_d(s) = sJ\omega(s) + f\omega(s) + T_\omega(s) \tag{6.15}$$

from which

$$\omega(s) = \frac{T_d(s) - T_\omega(s)}{sJ + f} \tag{6.16}$$

But from Eqs 6.10 and 6.11 we have

$$T_d(s) = \frac{K_t K_e \omega(s)}{(r_a + sL_a)(sJ + f)} \tag{6.17}$$

Substituting in Eq. 6.15 and simplifying we get

$$\frac{\omega(s)}{-T_\omega(s)} = \frac{(r_a + sL_a)}{(r_a + sL_a)(sJ + f) - K_t K_e} \tag{6.18}$$

Substituting $T_a = L_a / r_a$

$$T_m = J/f$$

we have

$$\frac{\omega(s)}{-T_\omega(s)} = \frac{(sT_a + 1)(1/f)}{(sT_a + 1)(sT_m + 1) - K_t K_e /(r_a f)} \tag{6.19}$$

$$\frac{\omega(s)}{-T_\omega(s)} = \frac{K(sT_a + 1)}{(sT_1 + 1)(sT_2 + f)} \tag{6.20}$$

where K is constant. If the poles of this transfer function are complex conjugates the speed change for a change in the load torque is oscillatory.

6.5.2 Transfer Function of a Field Controlled dc Motor

The speed of a dc motor can be varied by varying the field current. The speed can be increased beyond base speed by decreasing the field current (Fig. 6.8(c)). In this type of control, constant torque operation is not possible, as the armature current would increase to dangerous values at low fluxes. It is therefore necessary to maintain the armature current at a constant value at all flux levels. The field current is varied. The armature is also supplied by means of a phase controlled rectifier to maintain constant armature current. While deriving the transfer function the effects of saturation and armature reaction are neglected.

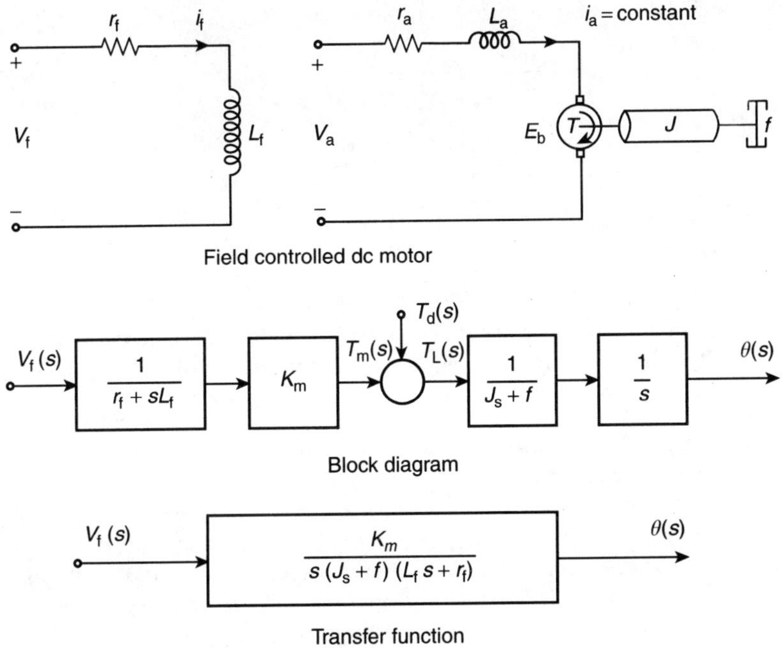

Fig. 6.8(c & d) *Block diagram of field controlled dc motor [Armature current constant]*

The torque developed by the motor

$$T_d = K\phi i_a \tag{6.21}$$

In a field controlled dc motor as discussed above, the armature current is constant and field current is variable. Therefore we have

$$T_d = K_2 i_f \tag{6.22}$$

The equation of the field circuit is given by

$$r_f i_f + L_f (di_f dt) = e_f \tag{6.23}$$

The dynamic equation of the motor is

$$J\frac{d\omega}{dt} + f \cdot w = K_2 i_f \tag{6.24}$$

The Laplace transforms of Eqs 6.22 – 6.24 with zero initial conditions are given by

$$(L_f s + r_f) I_f(s) = E_f(s) \tag{6.25}$$

$$(Js + f) w(s) = K_2 I_f(s) \tag{6.26}$$

Eliminating $I_f(s)$ from Eq. 6.26 and simplifying we get

$$\frac{\omega(s)}{E_f(s)} = \frac{K_2}{(sL_f + r_f)(Js + f)} \tag{6.27}$$

$$= \frac{K_m}{(T_f s + 1)(T_m s + 1)} \tag{6.28}$$

where $K_m = K_2/(r_f \cdot f)$ motor gain constant
$T_f = L_f/r_f$ field time constant
$T_m = J/f$ mechanical time constant.

Since the field inductance cannot be neglected, field controlled dc motor having speed as variable quantity is a second order system. The block diagram of the system and its transfer function are shown in Fig 6.8(d).

6.5.3 A Comparison of Armature Controlled and Field Controlled dc Motors

Armature controlled	*Field controlled*
1. Second order system can be reduced to first order due to negligible armature inductance L_a	Second order system Approximation to first order is not possible due to large value of L_f
2. Speed control is at constant torque up to based speed	Speed control at variable torque, i.e., at constant power beyond base speed.
3. Maintenance of i_f at constant values is easy. A converter on the field side is not necessary.	Armature current has to be maintained at constant value. This can be done either by means of a controller on the armature circuit or by extra resistances. In the former case the converters are required on the armature as well as field sides. In the latter case the efficiency is poor.
4. Back emf provides damping	This is not the case and necessary damping may be provided.
5. Time constants are small. The constants of the power amplifier must also be considered.	Time constants are large.

6.5.4 Transfer Function of a Ward Leonard Drive

In the classical Ward Leonard, system, the dc drive motor having constant excitation is supplied from a variable voltage generator, as shown in Fig. 6.9. The generator is driven at constant speed. The field current of the dc generator is varied by varying the field voltage to have variable voltage at the terminals, which is ultimately fed to the motor. A transfer function relating the speed of the motor and generator field voltage is required.

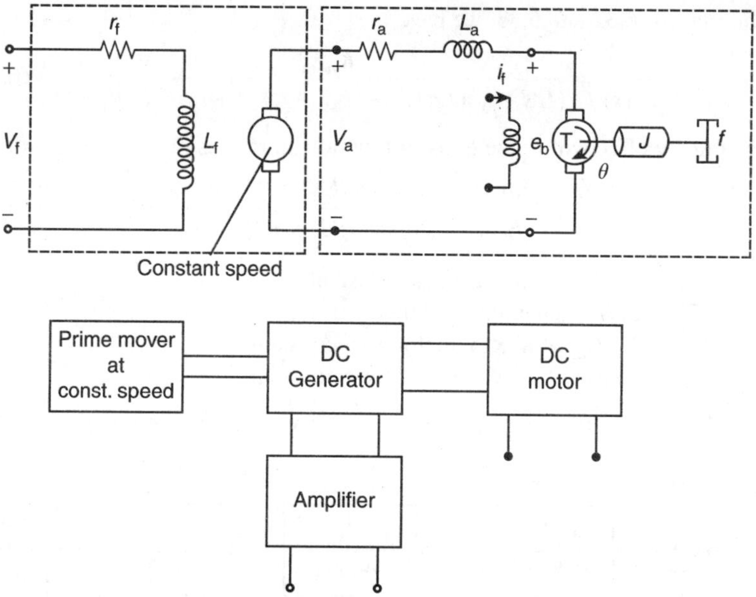

Fig. 6.9 *Ward Leonard control of a dc motor*

The system equations are

$$L_f \frac{di_{fg}}{dt} + r_f i_{fg} = e_{fg} \tag{6.29}$$

The voltage induced in the armature of the generator is given by

$$e_g = K\omega_g \varphi_g = K_a i_{fg} \tag{6.30}$$

where K_a is a constant which includes the constancy of generator speed and the proportionality constant between field current and air gap flux. As usual the effects of saturation are neglected. Eliminating i_{fg} from Eq. 6.30 we have

$$L_f \frac{de_a}{dt} + r_f e_a = e_{fg} \tag{6.31}$$

The transfer function between the armature voltage and field voltage of the generator is given by

$$\frac{E_a(s)}{E_f(s)} = \frac{K_a}{(sL_f + r_f)} \tag{6.32}$$

The transfer function of the armature controlled motor is

$$\frac{\omega(s)}{E_a(s)} = \frac{K_t}{(L_a Js^2) + (L_{af} + r_a J)s + r_{af} + K_t K_e} \tag{6.33}$$

Combining Eqs 6.32 and 6.33, we have

$$\frac{\omega(s)}{E_f(s)} = \frac{K_a K_t}{(L_f s + r_f)(LaJs^2 + (L_{af} + r_a J)s + r_{af} + K_t K_e)} \tag{6.34}$$

Neglecting the effects of L_a the transfer function simplifies in the second order to

$$\frac{\omega(s)}{E_f(s)} = \frac{K_g K_m}{(T_f s + 1)(T_m s + 1)} \tag{6.35}$$

where $K_g = (K_a / r_f)$ generator gain constant
$T_f = L_f / r_f$ generator field time constant
$K_m = K_t / (r_{af} + K_t K_e)$ motor gain constant
$T_m = r_a J / (r_{af} + K_t K_e)$ motor time constant

The block diagrams of a generator and a motor are shown in Fig. 6.10.

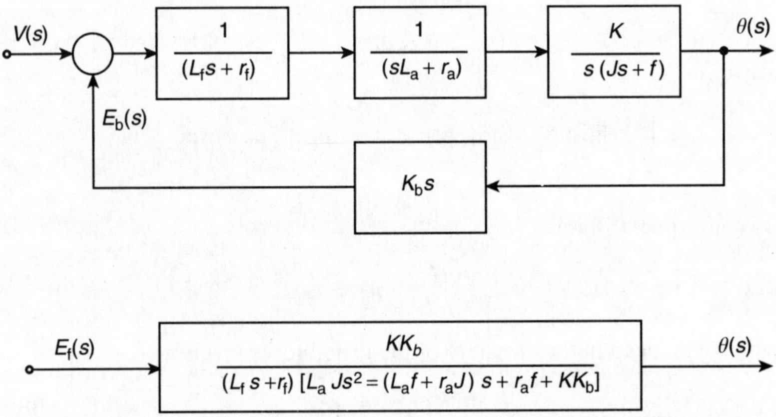

Fig. 6.10 *Block diagram and transfer function of Ward Leonard control*

Alternately The equations of the system may be derived in the state variable form taking i_f, i_a and ω as the state variables. A signal flow graph may be drawn to represent these equations. Using Mason's gain formula the transfer function may be derived.

Using the schematic of Fig. 6.9 representing a Ward Leonard speed control system we have for the generator field circuit

$$e_{fg} = r_f i_{fg} + L_f \frac{di_{fg}}{dt} \tag{6.36}$$

which can be rearranged as

$$\frac{di_{fg}}{dt} = \frac{e_{fg}}{L_f} - \frac{r_f}{L_f} i_{fg} \tag{6.37}$$

For the loop of armatures the circuit equation is given by

$$e_a = r_a' i_a + L_a' \frac{di_a}{dt} + e_m \tag{6.37}$$

e_a is the voltage induced in the armature of the generator and is given by $K_g i_{fg}$. r_a' and L_a' include resistance and inductance of both generator and motor armatures respectively.

e_m is the back emf of the motor given by $K_m\omega$. Substituting in Eq. (6.37) and rearranging the terms we have

$$\frac{di_a}{dt} = -(r_a'/L_a')i_a + (K_g/L_a')i_{fg} - (K_m/L_a')\omega \tag{6.38}$$

The torque balance equation gives the dynamics of the motor. It is

$$T_d = -J\frac{d\omega}{dt} + f\cdot\omega + T_L = K_m i_a \tag{6.39}$$

Rearranging the terms we have

$$\frac{d\omega}{dt} = (K_m/J)i_a - (f/J)\omega - (T_L/J) \tag{6.40}$$

Equations 6.40 are the state equations of the system and given in matrix rotation as

$$\frac{d}{dt}\begin{bmatrix} i_a \\ i_{fg} \\ \omega \end{bmatrix} = \begin{bmatrix} -r_a'/L_a' & K_g/L_a' & -K_m/L_a' \\ 0 & -r_{fg}/L_{fg} & 0 \\ K_m/J & 0 & -f/J \end{bmatrix}\begin{bmatrix} i_a \\ i_{fg} \\ \omega \end{bmatrix} + \begin{bmatrix} 0 & 0 \\ 1/L_{fg} & 0 \\ 0 & -1/J \end{bmatrix}\begin{bmatrix} e_{fg} \\ T_L \end{bmatrix} \tag{6.41}$$

6.6 TRANSIENT RESPONSE OF CLOSED LOOP DRIVE SYSTEMS

The time response of closed loop drive systems is of fundamental importance and must be investigated in detail. The time response of a system to specified inputs gives information with respect to the following characteristics:

i. Whether the system is stable or unstable. The system is said to be stable if it reaches a steady-state condition which has the same form as the input disturbance. It is said to be unstable if it does not attain an appropriate steady-state condition.

ii. The possible oscillatory nature of the system and the peak amplitudes of the oscillations.

iii. The damping of the system oscillations which is a measure of the speed or sluggishness of the system. An underdamped system even though it gives an oscillatory output has a fast response. On the otherhand an overdamped system is sluggish and does not have oscillatory behaviour.

iv. Steady-state error of the system which is actually a measure of the 'goodness', indicating how exactly the output can follow the input. Closed loop systems have very small steady-state error and thus have an output following the input very closely, compared to open loop systems.

The parameters of the system actually influence the above characteristics. It is possible to adjust the parameters to give the desired performance with respect to the above characteristics. If the system has parameters already adjusted to the extent possible and the system is not giving the desired performance, additional controllers are added to the given system to improve the performance.

The time response of the system has two parts; the transient and steady-state response. The transient response is the time variation of the output variable while going from one state to the other. The steady-state response of the system is the time variation of the output variable as the time approaches infinity.

A stable linear time invariant system comes back to its original state of equilibrium when the system is subjected to a disturbance, whereas an unstable system has its output variable in a continuous state of oscillation or it diverges very much without bounds from the state of equilibrium. The characteristic nature of stability (either absolute or relative) of the system can be very easily assessed without need for determining the time response. Stability tests are available and can be applied to decide whether the system is stable or not.

The time domain response can be obtained by directly solving the equations of the system. It is normally determined for step input, because it enables us to determine the time response of the system for any other type of input. Thus, the step response of a system is very important. However, the direct solution of the equation is rather tedious when the order is more. The method of Laplace transform can be applied to solve the equations. It has been very well utilised to obtain the transfer function, which is the ratio of the Laplace transform of output variable to the Laplace transform of input variable. The inverse transform of the output variable after substituting the Laplace transform of input variable gives the time response of the system. The procedure is illustrated with respect to a typical transfer function.

Let the transfer function be

$$Z(s) = \frac{C(s)}{R(s)} = \frac{A(s)}{B(s)} \tag{6.42}$$

The denominator polynomial is solved and its roots are found out. These are the poles of the transfer function. Therefore

$$C(s) = R(s)\frac{A(s)}{(s - s_1)(s - s_2)(s - s_3)} \tag{6.43}$$

The system has step input. Therefore

$$R(s) = \frac{1}{s} \tag{6.44}$$

Using this relation

$$C(t) = L^{-1}\left[\frac{1}{s}\frac{A(s)}{(s-s_1)(s-s_2)(s-s_3)}\right] \quad (6.45)$$

which is the sum of the residues of the function at poles s_1, s_2, s_3. Assuming that these do not repeat, which is normally the case

$$C(t) = \frac{A(O)}{(-s_1)(-s_2)(-s_3)} + \frac{A(s_1)}{(s_1-s_2)(s_1-s_3)}e^{s_1 t} + \frac{A(s_2)}{(s_2-s_1)(s_2-s_3)}e^{s_2 t} + \frac{A(s_3)}{(s_3-s_1)(s_3-s_2)}e^{s_3 t} + \ldots \quad (6.46)$$

If the poles are real, the time response $C(t)$ is made up of constants and a sum of exponential functions. The output is an aperiodic function having no oscillations. When the poles are complex, they appear in complex conjugate pairs. The time response in this case shows harmonic oscillations. The time response may result in a decreasing, neutral, or increasing response depending upon the nature of the roots. The transient response of $C(t)$ decreases and oscillations if any are damped out, if the poles of the transfer functions have negative real parts. It has sinusoidal oscillations of constant magnitude if the real parts are zero. The transient response of the system, on the other hand, is unbounded in its magnitude if one or more roots have positive real parts. This nature of the transient response clearly defines the stability of the system. In the first case the system is stable, in the second case it is neutral and in the third case it is unstable. These are depicted in Fig. 6.11. We can also see that in the first case the roots lie on the left hand side of the s-plane, in the second case they lie on the imaginary axis and in the third the roots with positive real parts lie on the right hand side of the s-plane. Therefore the stability of the system may be related to the location of poles on the s-plane. For absolute stability of a control system it is a necessary and sufficient condition that all the

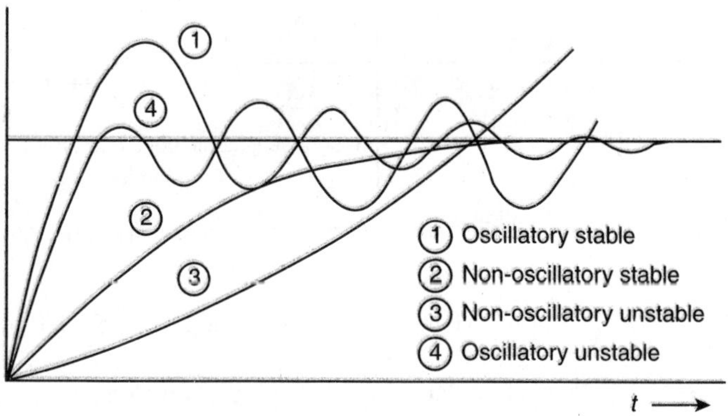

Fig. 6.11(a) *Typical transient responses of systems*

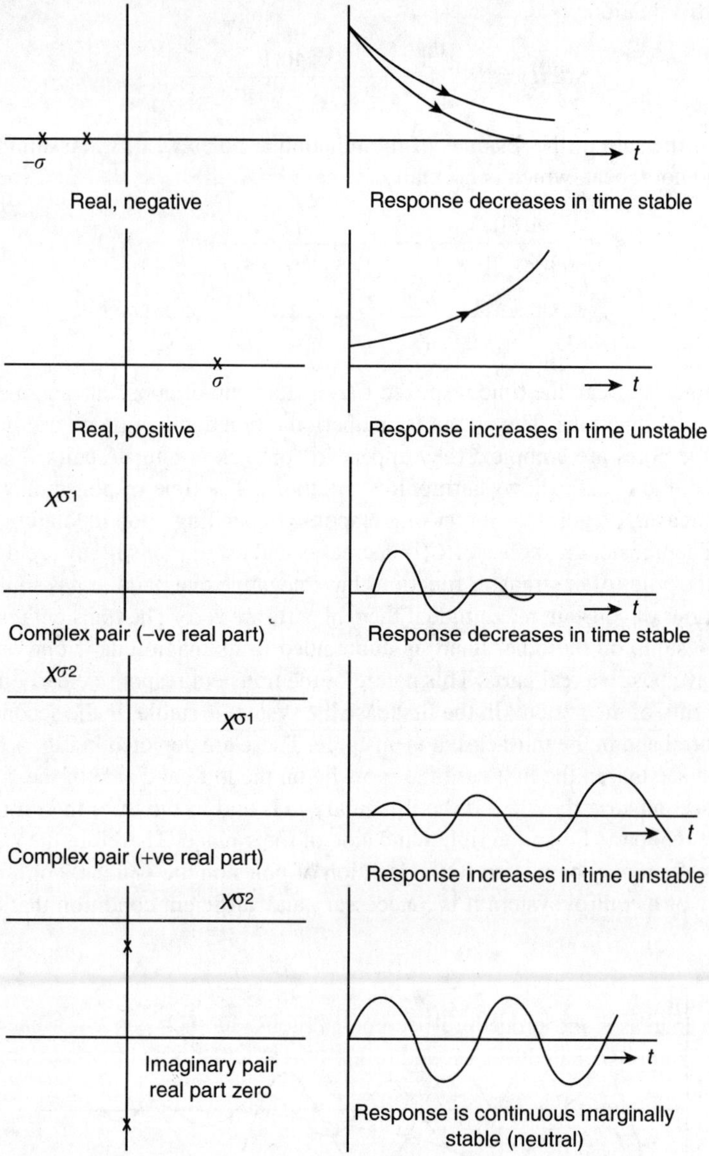

Fig. 6.11(b) *Basic root locations transient response and stability of the systems*

poles of the transfer function have negative real roots and lie on the left hand side of the s-plane.

When once the system is found to be stable having a bounded output, the quality of the time response is judged by certain time domain specifications, such as overshoot, rise time, peak time, swiftness of response, settling time, etc. The system must have fast response so that it settles down to its new equilibrium position quickly. As a fast response is always accompanied by oscillations this cannot be achieved in systems where oscillations cannot be tolerated. The system must also have a smaller overshoot. To achieve this it must have sufficient damping. Two typical time responses are depicted in Fig. 6.12. From the figure it is clear that a fast response is possible with smaller damping resulting in smaller rise time and larger overshoot. An increase in damping results in an increase in rise time and decrease in overshoot. Therefore any attempt to make the system faster it is followed by increased overshoot. The system becomes sluggish if an attempt is made to decrease the overshoot. It is therefore difficult to have a system with smaller rise time as well as overshoot. A compromise must be made between these two specifications.

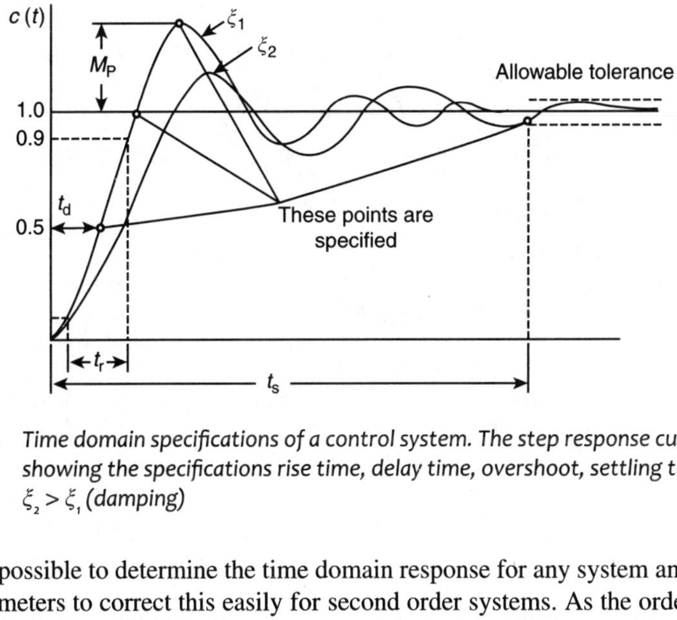

Fig. 6.12 *Time domain specifications of a control system. The step response curve showing the specifications rise time, delay time, overshoot, settling time* $\xi_2 > \xi_1$ *(damping)*

It is possible to determine the time domain response for any system and adjust its parameters to correct this easily for second order systems. As the order of the system increases the procedure becomes tedious. In such cases a higher order system is approximated to a second order system having the effects of dominant pole pair.

Let us consider a third order system of normalised natural frequency, in which case it has a closed loop transfer function given by

$$Z(s) = \frac{1}{s^2 + 2\xi s + 1} \frac{1}{(\beta s + 1)} \tag{6.47}$$

The effect of the pole $s = -1/\beta$ on the transient response is shown in Fig. 6.13. The effect of this pole is insignificant if $|1/\beta| \geq 10\,|\omega_n \xi\,|$ and the system has a response similar to that of a second order system. The location of the pole is significant, i.e., as it moves nearer to the imaginary axis the time response is altered.

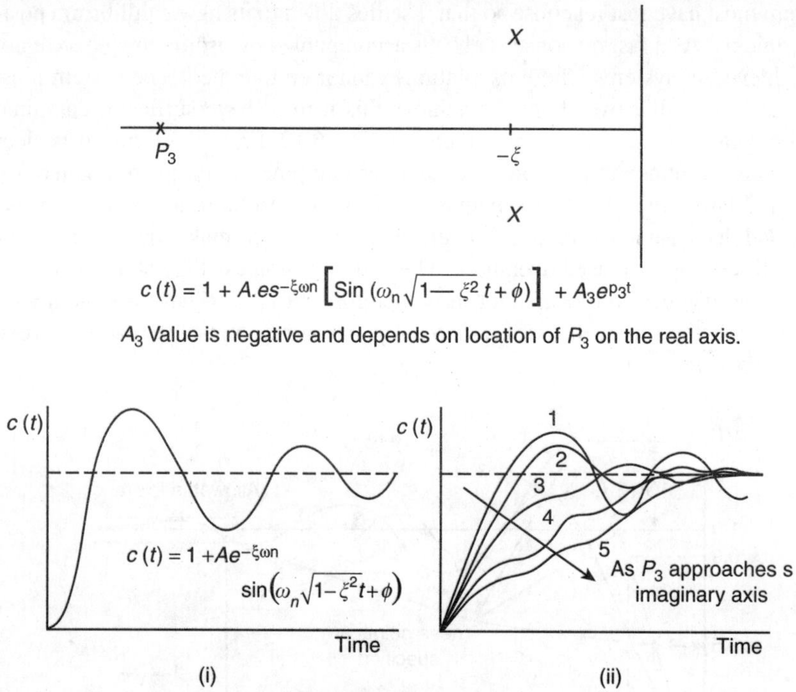

Fig. 6.13(a) *Effect of an additional real pole on the transient response*
(i) Second order system or a third order system with its real pole having insignificant effect by being far off the imaginary axis
(ii) Addition of real pole. Transient response alters from 1 to 5 as the additional pole approaches imaginary axis

Further, the presence of a zero in the transfer function also alters the transient performance. Its relative position with respect to the additional pole is also important. If this zero is to the left of the pole the system behaves as if it has only complex poles but with smaller peak overshoot. If the zero is to the right of the pole the overshoot is greater than that of the system with complex poles.

Therefore the approximation of a higher order system must be made with caution. However they may be approximated to a system having (a) two complex poles (b) two complex poles and one real pole and (c) two complex poles, one real pole and one real zero. In the first case it is a second order system for which the time domain specifications can be determined. In the second case depending upon the location of the third pole the performance can be corrected to achieve the desired performance. In the third case their effect will be small if they are very near to each other.

The response of a second order system is therefore very important. A detailed inspection of this response with reference to time domain specification is given in the following.

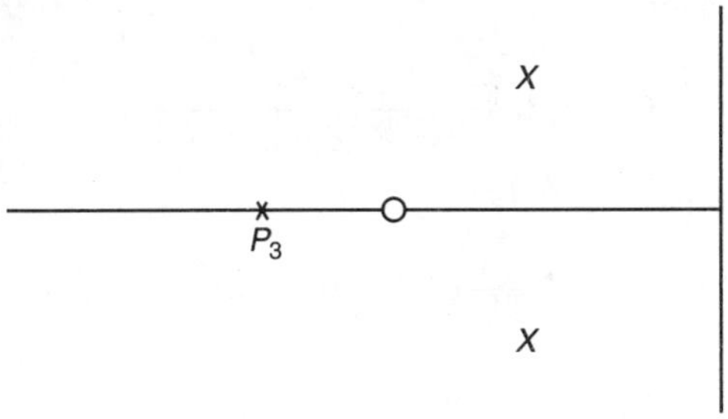

$$Z(s) = \frac{K(s-Z)}{(s^2+2\xi\omega^2{}_n s+\omega^2{}_n)(s-P_3)}$$

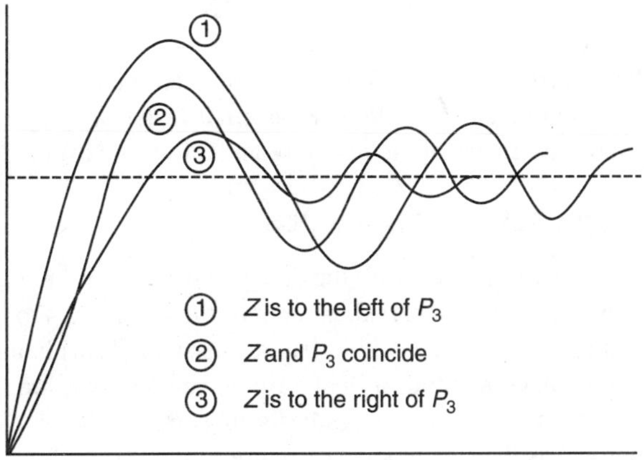

$$C(t) = 1 + A_1 e^{-\xi\omega}[\sin(\omega_n\sqrt{1-\xi^2}\,t+\phi)] + A_3 e^{p_3 t}$$

A_3 is negative for case ①

A_3 is zero for case ②

A_3 is positive for case ③

Fig. 6.13(b) *Alteration of transient response by the location of zero*

Using generalised notation the closed loop transfer function of a second order system is given by

$$\frac{C(s)}{R(s)} = \frac{\omega_n^2}{s^2 + 2\xi\omega_n s + \omega_n^2} \tag{6.48}$$

where ω_n is natural frequency of the system
ξ is damping ratio.

Time domain response is normally defined for step input in which case

$$R(s) = \frac{1}{s} \tag{6.49}$$

The time response

$$C(t) = L^{-1}\frac{\omega_n^2}{s(s^2 + 2\xi\omega_n s + \omega_n^2)} \tag{6.50}$$

which can be obtained as

$$C(t) = 1 - \frac{1}{\beta}e^{-\xi\omega_n t}\sin(\omega_n \beta t + \theta) \tag{6.51}$$

where $\beta = \sqrt{1-\xi^2}$ and $\theta = \tan^{-1}(\beta/\xi)$.

The transient response of a second order system is shown in Fig. 6.12. As the value ξ decreases the closed loop roots approach the imaginary axis and the response is oscillatory and the % overshoot also increases.

$$\%\ \text{Overshoot} = \exp(-\xi\pi/\sqrt{1-\xi^2} \times 100) \tag{6.52}$$

Normalised peak time $\omega_n t_p = \pi/\sqrt{1-\xi^2}$

From these two equations it is clear that the smaller the value of ξ the larger the overshoot and smaller the peak time. The system is faster. A compromise has to be made between the swiftness of the response and allowable overshoot.

6.7 FREQUENCY RESPONSE APPROACH

The previous sections show that with the time response of a system, even though it is a direct method of analysis, the adjustment of the parameters to give a satisfactory time domain performance is rather tedious particularly with higher order systems. On the other hand methods utilising frequency response are extremely easy and practical for the analysis and synthesis of control systems. The frequency response is defined as the steady-state response of the system for sinusoidal excitation over a range of frequencies. It is well known, that sinusoidal excitation offers advantages in the analysis that, in a linear system, the output or signal at any other point of the system is sinusoidal; however it differs from the input signal in phase and magnitude.

The frequency response method has the following advantages:

i. The experimental determination of frequency response of a system is very easy, because sinusoidal signal of varying frequencies and amplitudes are readily available. The method is reliable and uncomplicated for the experimental analysis of a system. This data can be used to formulate the transfer function of the system. The frequency response provides information regarding the response of the system to noise and disturbance. Bandwidth can be controlled. The system can be designed so that the effects of these are minimum.
ii. The transfer function describing the frequency response is very easily obtained by substituting $j\omega$ for s in its transfer function in s domain. It is a complex function of ω having real and imaginary parts. It has a magnitude and phase, the plot of which gives the frequency response of the system.

iii. One such plot is the Nyquist plot which gives the absolute and relative stabilities of the closed loop system response from the knowledge of the open loop frequency response. The actual roots of the system need not be determined.

iv. Design methods in the frequency domain are simple.

However to design a control system in the frequency domain satisfying the given time domain specifications, a correlation between the time response and frequency response is necessary. There is no such direct correlation except for second order systems. While designing a closed loop system we may adjust the frequency response characteristics by using several design criteria in order to obtain the desired transient or time response.

In order to utilise the frequency response approach effectively, it is necessary to understand the indirect correlation between the transient and frequency response. The indirect correlation actually interprets the desired dynamic response in terms of the frequency response characteristic. Changes are made in the frequency response, so that the modified one gives the desired time response.

6.7.1 Representation of Frequency Response

There exist many possibilities of representing a frequency response:

One of them is the polar plot representation. The transfer function in the frequency domain is obtained by substituting $j\omega$ for s as complex function of ω. It is separated into real and imaginary parts which are plotted on the real and imaginary axes to get the locus on the complex plane, as ω is varied. It is also possible to transform the complex function into its polar form having magnitude and phase. For example

$$\begin{aligned} Z(j\omega) &= Z(s)_{s=j\omega} = R(\omega) + jX(\omega) \\ &= Z(\omega)\exp(j\phi(\omega)) \end{aligned} \tag{6.53}$$

where

$$Z(\omega) = (R(\omega)^2 + X(\omega)^2)^{1/2}$$

$$\phi(\omega) = \tan^{-1}\frac{X(\omega)}{R(\omega)} \tag{6.54}$$

A typical polar plot is shown in Fig. 6.14. It can be very easily seen that this kind of representation has a limitation. The investigation of the effects of changing parameters or adding poles or zeros to the existing transfer function requires recalculation of frequency response. The adjustment of parameters for the desired performance is rather tedious. The effects of individual poles or zeros are not separately indicated by the method.

Another method of portraying the frequency response, which is relatively simple and is widely used, is to represent separately the magnitude and phase of the transfer functions as functions of logarithmic frequency. These frequency response plots are called Bode plots. We know that for a given transfer function

$$Z(j\omega) = Z(\omega)e^{j\phi(\omega)} \tag{6.55}$$

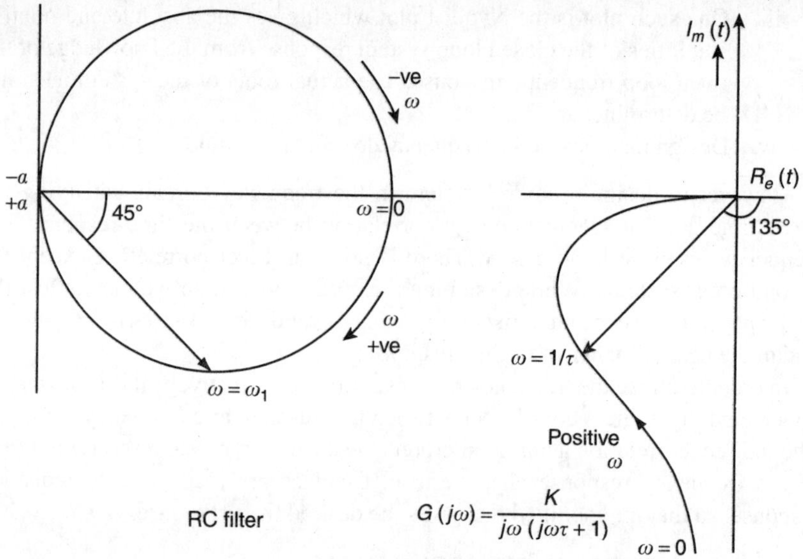

Fig. 6.14 *Typical polar plots for frequency response*

where $Z(\omega)$ is the magnitude and
$\phi(\omega)$ is the phase.

The natural logarithm is

$$\ln(Z(j\omega)) = \ln(Z(\omega)) + j\phi(\omega) \tag{6.56}$$

The log magnitude transformed to the base 10 becomes

$$= 20 \log_{10} Z(\omega) \tag{6.57}$$

The logarithmic magnitude (gain) and phase angle are drawn as functions of frequency on a separate set of axes. The frequency is represented on a logarithmic scale as it is the most convenient one for this case. By this magnitude the plot becomes a plot of asymptotic lines, each of which has a slope –20 db/decade of frequencies.

An immediate advantage that follows from these plots is that all the multiplicative factors become additive ones. Therefore, by drawing the plots of individual factors and adding them we get the frequency response plot of the overall transfer function. Figure 6.15 gives the summary of the Bode plots of the important basic functions.

However the frequency response of a second order system cannot always be represented by a pair of asymptotic lines. The transfer function of a typical second order system in the complex frequency plane is given by

$$Z(j\omega) = \frac{\omega_n^2}{(j\omega)^2 + 2\xi\omega_n(j\omega) + \omega_n^2} \tag{6.58}$$

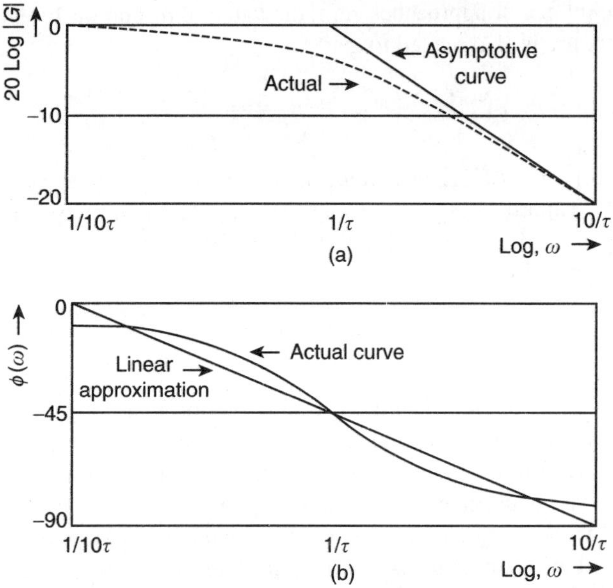

Fig. 6.15 *Bode diagrams for frequency response. (a) Magnitude plot (b) Phase plot of 1/(1 + jωr)*

The magnitude and phase plots are shown in Fig. 6.16. From these plots it is very clear that the log-magnitude and phase plots of a second order system depend very much on the damping ratio. The system may be represented by asymptotic lines if $\xi > 0.707$. The deviation of actual response from the asymptotic one is a function of ξ and must be taken into consideration if $\xi < 0.707$, The frequency at which the peak value of magnitude (M_{pu}) occurs is called the resonant frequency (ω_r). A relationship between ω_r and ω_n can be derived as

$$\omega_r = \omega_n\sqrt{1-\xi^2} \tag{6.59}$$

$$\frac{1}{1+2\xi(j\omega/\omega_n)+(j\omega/\omega_n)^2}$$

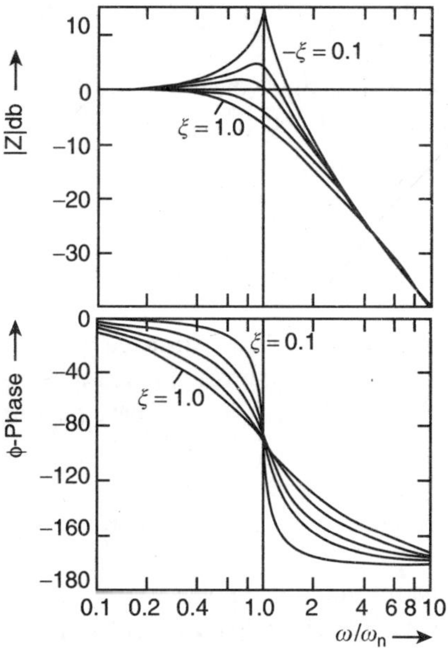

Fig. 6.16 *Magnitude and phase plots of a second order system*

As the damping factor approaches zero, the value of ω_r approaches ω_n. The peak value of response is also related to ξ as

$$M_{pu} = \frac{\xi}{\sqrt{1-\xi^2}} \tag{6.60}$$

M_{pu} is unity for $\xi = 0.707$. A clear value of ω_r at which $M_{pu} > 1$ can be identified for $\xi > 0.707$. As ξ approaches zero the value M_{pu} approaches infinity. These relations are shown graphically in Fig. 6.16. The phase of the transfer function at resonant frequency is given by

$$\phi(\omega_r) = -90 + \sin^{-1} \frac{\xi}{\sqrt{1-\xi^2}} \tag{6.61}$$

We discussed and concluded in the previous section that a higher order system having a dominant pair of complex conjugate poles can be represented by an equivalent second order system. These relations are useful for determining the damping ratio for experimental frequency response.

From the above discussion it is clear that a second order system cannot always be represented by asymptotic approximation. In terms of ξ we decided when this would be feasible. It is also helpful to decide the ratios of time constants of the system at which this approximation would be feasible. To solve this problem let us consider the normalised transfer function of a second order system in the frequency domain as

$$Z(j\omega) = \frac{1}{1 + j\omega T_m + (j\omega)^2 T_A T_m} \tag{6.62}$$

Let us investigate the errors involved in representing this by two first order delays $(1 + j\omega T_A)$ and $(1 + j\omega T_m)$.

$$\begin{aligned} Z(j\omega) &= \frac{1}{(1 + j\omega T_A)(1 + j\omega T_m)} \; \frac{(1 + j\omega T_A)(1 + j\omega T_m)}{1 + j\omega T_m + (j\omega)^2 T_A T_m} \\ &= Z_1(j\omega) \cdot Z_2(j\omega) \end{aligned} \tag{6.63}$$

From this $Z_2(j\omega)$ is the error involved and it provides the correction required when $Z(j\omega)$ is approximated by $Z_1(j\omega)$. Let us investigate for what ratios of T_m/T_A the approximation is feasible. Assuming $T_m/T_A = t$ and introducing a variable $K = T_A\sqrt{t} \cdot (j\omega)$ we have the transfer function relating to the error

$$Z_2(K) = \frac{1 + \dfrac{1+t}{\sqrt{t}} K + K^2}{1 + \sqrt{t} \cdot K + K} \tag{6.64}$$

The magnitude and phase Z_2 as function of K are shown in Fig. 6.17. From the figure it is clear that the error in magnitude as well as phase plots increases as the value of t decreases. For larger values of t the deviation of the asymptotic approximation from the actual response becomes negligible and $Z_2(K) = 1$. Thus the ratio of T_m/T_A decides whether the approximation would be valid or not.

The following procedure may be adopted for drawing the log-magnitude and phase plots to represent the frequency response:

i. Identify the corner frequencies of the factors of the sinusoidal transfer function.
ii. Draw the asymptotic log-magnitude curves with proper slopes at corner frequencies.
iii. The proper corrections are made to these asymptotic curves to arrive at the exact curve.
iv. The phase angle plot is obtained by adding the individual phase angle curves of the factors.

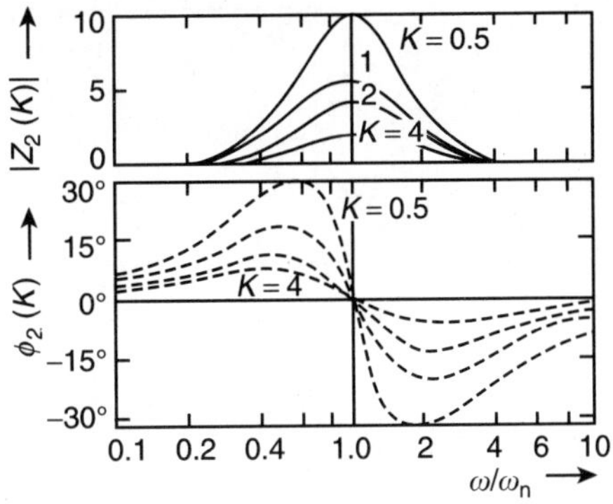

Fig. 6.17 *Error involved in representing a second order system by two first order cascaded systems*

The advantages of this method are obvious. It takes much less time to draw these curves compared to the other methods used for determining the frequency response. The curves of a given transfer function can be easily modified to take into account addition of poles and zeros for improving the performance. Once one becomes used to these plots, the composite asymptotic plot can be directly drawn doing the summing process mentally.

6.7.2 Correlation Between Frequency and Transient Response

As has been stated, the use of frequency response for the design of control systems requires a correlation between the time and frequency response. Time response specifications are available for the performance of a system. These must be translated to frequency response. There must be frequency domain specifications also, corresponding to the time domain specifications, such as overshoot, settling time, etc. However, it is easy to have a direct correlation between the transient and frequency responses of a second order system. A typical magnitude plot of a second order system is shown in Fig. 6.17. The resonant peak (maximum amplitude) $M_{p\omega}$ of the magnitude plot depends upon the damping ratio of the system. The resonant frequency also depends upon the damping ratio. These relations are given by Eqs 6.59 and 6.60 and represented graphically in Fig. 6.18. The resonant frequency ω_r and band width of the frequency response relate to the fastness of response. Smaller the values of bandwidth and resonant frequency smaller is the rise time of the transient response and faster is the response. The overshoot of the time response can be related to the resonant peak of the frequency response. This resonant peak also indicates the relative stability of the systems. The bandwidth is

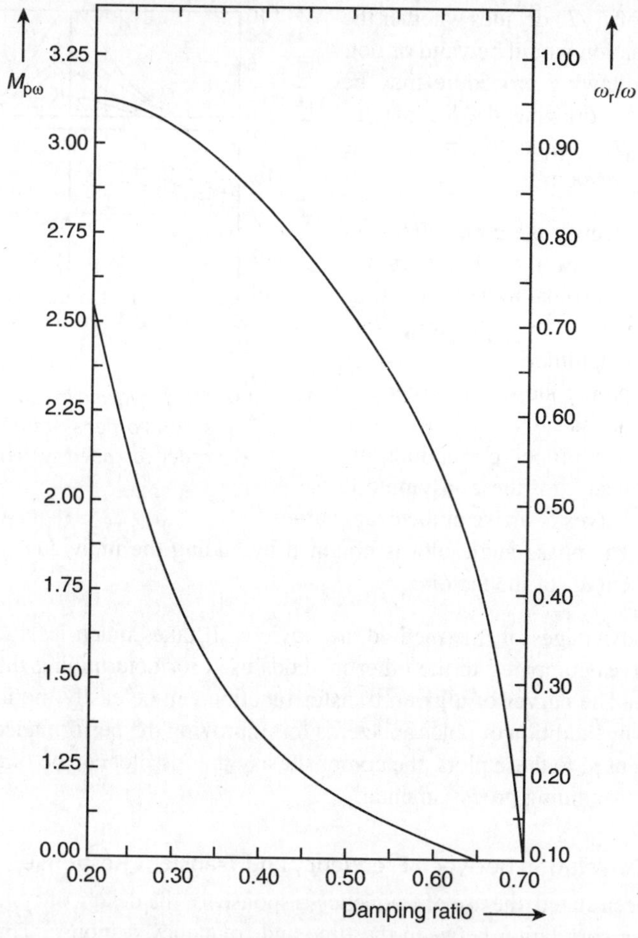

Fig. 6.18 *Peak of frequency response and resonating frequency of second order system as a function of damping ratio*

related to the natural frequency ω_n of the system. For a given ξ greater the value ω_n faster is the response. The value of ξ must be chosen to compromise between $M_{p\omega}$ and ω_r.

The frequency domain specifications are therefore

i. The peak amplitude and the frequency at which this occurs. The peak amplitude must be normally less than 1.5. The acceptable range of peak amplitude corresponds to damping rates of 0.4 to 0.7.

ii. Relatively large resonant frequency and hence large bandwidth of the frequency response. The system will have relatively small time constants. The system becomes faster.

iii. The closeness of the polar plot of frequency response to (–1, 0) point indicates the peak overshoot of the time response. This also gives the relative stability.

iv. The steady-state error can be related to frequency response also. The gain and number of integrations involved in the open loop system indicate the influence of steady-state error.

The correlation between the transient and frequency response of higher order systems is not so simple and straightforward as it is for second order systems. The mathematical treatment of higher order systems for such a correlation is rather involved and laborious. However, a higher order system can be represented by a second order system if it has a pair of dominant complex conjugate poles. The frequency and time responses of this system is influenced by this pair of dominant poles. In such a case the correlation existing between the transient and frequency response for a second order system can be very easily extended for higher order systems.

For higher order systems having a dominant pair of complex conjugate poles, the following correlation exists between the transient and frequency responses:

i. The peak magnitude of frequency response indicates the relative stability. A system having a peak amplitude in the range of 1 to 1.4 would have a time response with an effective damping ratio in the range of 0.4 to 0.7.

ii. If the peak amplitude of the frequency response is greater than 1.5 the time response is oscillatory, having a large overshoot.

iii. The resonant frequency, i.e. the frequency corresponding to peak amplitude, is a measure of the fastness of response. Larger the value of resonant frequency, faster is the response, i.e. the smaller is the rise time.

iv. The system is highly damped if the resonant frequency and damped natural frequency are close to each other.

v. Larger values of ω_r characterise larger bandwidth. However, in view of the noise the system should not have large bandwidth. Larger the bandwidth costlier is the system. A compromise is required.

vi. Cut off frequency (frequency at which the amplitude is 3 db below the zero frequency value) characterises the filtering characteristics of high frequency components.

vii. The slope of the log-magnitude curves known as cut off rate gives the ability of the control system to distinguish between noise and signal.

Using the above correlation, the time domain specifications can be translated to frequency domain specifications. The design of the control system is carried out in the frequency domain to meet the required specifications.

6.8 STABILITY OF CONTROLLED DRIVES

The concept of stability is very well known. By the term stability one implies the ability of a system to return to its original position or attain a new steady-state condition when there is some disturbance or change in the input condition of the system. Following changes in input, or any disturbance, a controlled drive system has a time response made up of steady-state and transient responses. The former is of the form of input or noise and does not reveal any information about the stability of the system. The nature of the system with regard to stability is revealed by the transient response. If the transients are damped out and the system successfully takes up a steady-state operating point the system is said to be stable. It is always necessary that a drive system must be stable.

The stability of the system is associated with its characteristic equation. For the transient to die down the roots of the characteristic equation must lie on the left hand side of the s-plane, i.e., the roots must be negative if they are real or they must have a negative real part if they are complex. The presence of one or more positive roots or complex roots with positive real parts indicates instability, as the transient response associated with these roots increases without bounds with time. Obviously such roots lie on the right hand side of the s-plane.

One of the requirements of a controlled drive system is stability. From the above discussion, a linear control system is said to be stable if (a) it attains a steady-state condition which is unique and repeatable for a special input and is of the form of the input, (b) the response dies away when the input is removed. On the other hand it is unstable if its response increases continually with time, the system is self sustained and the response does not die down when the input is removed. Sometimes the system also shows a stable oscillatory behaviour, which is just the borderline case between the stable and unstable responses. This is also not desirable. Some systems are conditionally stable, i.e., the system is stable for a range of values of a parameter and for other values it is unstable.

A direct and straightforward method for ascertaining the stability of a system is to determine the roots of the characteristic equation and to examine them for the negative real parts. The nature of a system's stability is also revealed by the determination of the system's response to specified inputs.

However, these two methods are very tedious and are difficult to apply when the order of the system is large. It is, therefore, desirable to have some indirect methods leading to an investigation of the stability of a system using some criteria without needing to evaluate the roots. These save both time and labour. Two such criteria are (a) Routh-Hurwitz criterion and (b) Nyquist stability criterion. The Bode frequency response plots may also be used to ascertain the nature of system stability. Sometimes a study of the effect of variation of parameters of a drive on its stability may be required for a judicious choice of the parameters or for correcting the parameters already existing. In such cases we may use root-locus techniques, which are developed to ascertain the variation of roots of a characteristic equation when the drive parameters are varied. The parameter plane method due to Siljak, and the domain decomposition method are very powerful for this purpose. These methods give the boundary between the stable and unstable operating regions when a pair of parameters are varied at a time.

The study of absolute stability is required in the drive technology where a drive motor is controlled and the control system has several components. It is also necessary to study the effect of variation of parameters on the stability for a suitable design of controllers, to improve the performance of the system.

Before going into the details of the controller designs to improve the system stability, a brief outline of the stability criteria is given below.

6.8.1 Routh-Hurwitz Criterion

The Routh-Hurwitz stability criterion states that the system is stable if the Rouths table has no negative elements in the first column. The roots of the characteristic polynomial are negative if they are real or contain negative real parts if the elements of the first column of the Routh's Table are positive. The negative elements indicate positive real roots or roots with positive real parts. The number of changes in the sign of the first column actually indicate the number of roots on the right hand side of the s-plane.

Using this method the stability of a linear system can be investigated very easily and quickly. It also indicates the number of roots, if any, on the right hand side of the s-plane. However, the method does not provide any information regarding the relative stability of the system. The method can be used to select the parameters which would make the system stable, mostly the gain constant of the system. In the case of controlled variable frequency drives this can be used to study the effect of variation of parameters on the stability of the drive system.

Before applying the stability criterion to the characteristic equation, the first and foremost condition to be satisfied is that all the coefficients of the characteristic equation are present and positive. If any coefficient is missing or is negative, positive roots may occur.

The coefficients of the characteristic equation are used to write down the first two rows of the Routhian array using alternate powers of s, in the order of descending powers. The missing terms are represented by zeros. These zeros cause sign changes in the first column indicating positive roots or complex roots with a positive real part. These two rows are used to build the complete array. The arithmetic manipulations in the development of the array may be made easier by dividing any row by a constant. The array is developed until a row of zeros appears, after a systematic formation of zeros in a step pattern in pairs. If the row of zeros appears before the array is complete, the subsidiary equation before the row of zeros is differentiated and substituted for the row of zeros. If the intermediate rows of the array has a zero in the first column, it presents a problem in developing the subsequent rows. This difficulty is overcome by multiplying the characteristic equation by $(s + K)$. This does not alter the situation because $s = -K$ is a negative root.

The application of the Routh's criterion is discussed in the following by means of an example.

Example The open-loop transfer function of a unity feedback control system is given by

$$G(s) = \frac{K}{s(sT_1 + 1)(sT_2 + 1)}$$

Use R-H criterion to establish a relation between K, T_1, T_2 so that the system is stable.

Solution The characteristic equation of the system is

$$s(sT_1 + 1)(sT_2 + 1) - K = 0$$

$$s(T_1T_2s^2 + (T_1 + T_2)s + 1) + K = 0$$

$$T_1T_2s^3 + (T_1 + T_2)s^2 + s + K = 0$$

i.e.,

The Routh array is

$$\begin{array}{lll} s^3 & T_1T_2 & 1 \\ s^2 & T_1 + T_2 & K \\ s^1 & \dfrac{(T_1 + T_2) - K(T_1T_2)}{T_1 + T_2} & \\ s^0 & K & \end{array}$$

For the system to be stable

i. $K > 0$

ii. $\dfrac{(T_1 + T_2) - KT_1T_2}{T_1 + T_2} > 0$

$T_1 + T_2 > KT_1T_2$

$K < \left(\dfrac{1}{T_1} + \dfrac{1}{T_2}\right)$

Therefore $K > 0$ and $< \dfrac{1}{T_1} + \dfrac{1}{T_2}$.

6.8.2 Nyquist Criterion

This stability criterion enables one to establish the stability of a system using a graphical procedure in the frequency domain. Here also there is no necessity for the evaluation of the roots of the characteristic equation. The stability of a closed loop system is revealed by subjecting the open loop transfer function to a frequency response analysis. The stability criterion can be stated as

> A closed loop control system is stable if the polar frequency locus of the systems open loop transfer function does not encircle the point (–1, 0) in a clockwise sense for all real values of frequency $\omega = \pm\infty$. If it encircles the point (–1, 0) the system is unstable. If the locus passes through the point (–1, 0) the system is marginally stable, i.e., it is on the border line between stability and instability.

While drawing the Nyquist plot (polar frequency locus) it is sufficient to calculate the response for positive values of ω only, as ω changes from 0 to ∞. The plot for negative values of ω, i.e., as ω changes from 0 to $-\infty$, is just a mirror image or conjugate of the above. However while establishing the system it is necessary to draw the complete locus for all frequency values varying from $-\infty$ to $+\infty$ (Fig. 6.19).

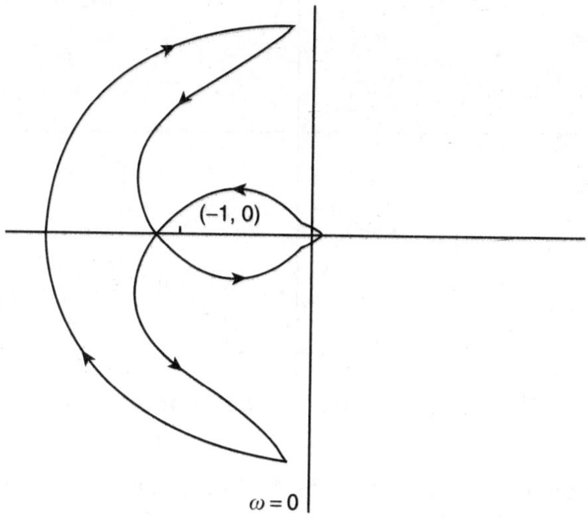

Fig. 6.19 *Nyquist plot of* $\frac{K_1(1 + K_2 s)}{s(s-1)}$ *to ascertain stability*

Application of the Criterion In short, the following steps may be followed to apply the criterion to investigate the nature of stability of a closed loop system:

i. Obtain the open loop transfer function of the system.
ii. Determine the frequency response plot of the function for positive values of ω starting from $\omega = 0$ to ∞. The locus starts on the real x-axis and ends at the origin for minimum phase functions. If the transfer function contain factors of $1/s$, $1/s^2$, the locus starts ∞ on the real axis for $\omega = 0$ and swings through $\pm 90°$ to reach the appropriate axis for $\omega = 0^+$.
iii. Draw the mirror image of the locus which gives the locus for the frequency in the range $\omega = -\infty, 0$.
iv. If the locus encircles the point (–1, 0) the system is unstable. The system is on the boundary stability if the locus passes through (–1, 0).

Nyquist stability criterion has the following features:

i. The stability as well as performance of a closed loop system can be investigated using open loop data and a simple graphical procedure. The use of an open loop transfer function is simpler than the use of a corresponding closed loop transfer function.
ii. The method can be used to ascertain the relative stability of the system (Fig. 6.20). This enables one to improve both the transient and steady-state responses by properly adjusting the parameters of the system.
iii. Only a few points on the locus need to be calculated. The full plot of the locus can be drawn using these values. It is often necessary and sufficient if the locus near the point (–1, 0) is drawn.

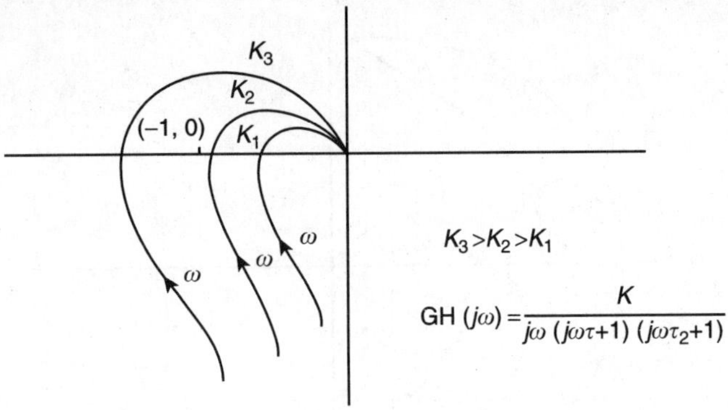

Fig. 6.20 *Nyquist plot to ascertain relative stability*

iv. For minimum phase functions the locus starts on the real axis for $\omega = 0$ and ends at the origin for $\omega = -\infty$.

v. The Nyquist plot can be very simply sketched if the transfer function has no terms containing s in the numerator. A pole or a factor in the denominator of nature s or $(1 + s\tau)$ causes a rotation of the locus by –90°. One or more such factors in the denominator tend to make the system unstable. Based on these considerations the first and second order system are never unstable.

vi. The Nyquist plot of a transfer function having s factors in the numerator cannot be very easily sketched. This is because the phase of the function need not continually change with continuous change in ω. However a factor of the nature s or $(1 + s\tau)$ in the numerator has the effect of nullifying the effect of a similar factor in the denominator by contributing or rotating the frequency by +90°.

vii. The application of Nyquist criterion to a system also reveals the number of roots on the right hand side of the s-plane. This is given by the number of times the Nyquist locus encircles the (–1, 0) point.

viii. However, for complicated systems, sometimes the encirclement of the point by the locus is not directly obvious. The following procedure may be used to check whether the locus has encircled the point (–1, 0).

(a) Draw a line in any direction from (–1, 0) to intersect the locus.

(b) Put an arrow on the locus touching the line in the direction of locus ascertained in the sense of 0^+ to $\infty +$ at each intersecting point.

(c) If the number of anticlockwise arrows and clockwise arrows are equal, it may be concluded that the locus does not encircle the point (–1, 0), otherwise the point (– 1, 0) is encircled and the system is unstable.

6.8.3 Relative Stability from the Nyquist Plots

The considerations discussed above provide information about the absolute stability of the system, i.e., whether the system is stable or not. An equally important system behaviour to be considered is the relative stability. The relative stability is indicative of how various poles of the system affect the transient behaviour of a system, in other words, how oscillatory the transient behaviour of the system is. Because of the energy storage elements of the system, the system has so-called time constants and it cannot follow the input instantaneously. It exhibits a transient behaviour which dies down in a stable system. A practical system has damped oscillations before reaching the new steady-state behaviour. This behaviour of the system can be identified by the term relative stability.

The Nyquist plot (polar frequency response plot) of the open loop transfer function of a closed loop system can be used to ascertain the relative stability (or degree of stability) of the system, besides its absolute stability. This feature of the system is revealed by the nearness of the polar plot to (–1, 0) point. If the locus is very close to the point, the system is very close to instability and the transient response is oscillatory. The damping ratio of the system decreases. On the other hand if the locus is away from (–1, 0) the system has more damping and less oscillatory behaviour and has better stability conditions.

The nearness of the polar plot, which reveals the relative stability of the system, is characterised by phase margin and gain margin. These definitions are used with simple shapes of locus which cross the x-axis only once. For complex shapes these are ambigious.

The gain margin of a system is used to describe the nearness of the point of intersection of the locus and the x-axis to (–1, 0) point. Referring to Fig. 6.21 a is referred to as gain margin. For minimum phase stable system $a < 1$ and unstable systems $a > 1$. The gain margin a is the reciprocal of the gain of the system $(|G(j\omega)|)$ at the frequency at which the locus cuts the x-axis or has a phase angle of –180°. This frequency is the phase cross-over frequency. Smaller the gain of the system at the phase cross-over frequency, more stable is the system, or better

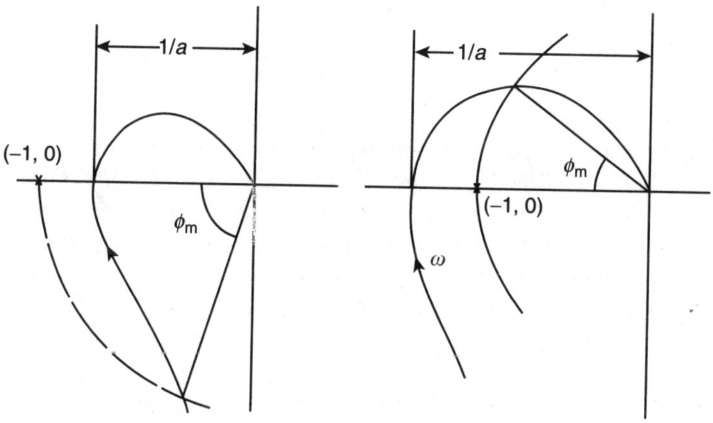

(a) Stable system $a > 1, \phi_m$ positive (b) Unstable system $a < 1, \phi_m$ negative

Fig. 6.21 *Gain and phase margins from Nyquist plot*

is the system with respect to relative stability. As the gain (at phase cross-over) increases, the locus comes nearer to (–1, 0) and the system tends to be less stable. For a stable system the gain margin must be positive, and negative for unstable ones. In other words the gain margin of a stable system indicates by how much the gain must be increased before the system is unstable. For unstable systems it indicates how much it must be reduced to make the system stable. The gain margin of first and second order systems is infinite as their polar plots do not cross the x-axis. Therefore, theoretically speaking they cannot become unstable. In practice these systems may have small time lags which are neglected in the derivation of transfer functions. When these are taken into consideration the systems may become unstable. However, increase of gain constant brings the locus nearer to the point (–1, 0) making the transient response more oscillatory. This is depicted for a second order system in Fig. 6.20 which gives the Nyquist locus and transient response of the system for different values of gain constant.

The phase margin of the system ϕ_m is also used to describe the closeness of the Nyquist locus to (–1, 0) or in otherwords to describe the relative stability. The phase margin can be defined as the angle by which the locus may be rotated in the clockwise direction so that the locus passes through the point (–1, 0). The phase margin is, therefore, the angle enclosed by the negative real axis and the line joining the origin and the point of intersection of the locus and the unit circle drawn with the centre as the origin. Referring to Fig. 6.21 the phase margin

$$\phi_m = 180° + \phi_0 \tag{6.65}$$

where ϕ_0 is the phase angle of the open loop transfer function when its magnitude is unity. The frequency at which the magnitude is unity is called the gain cross-over frequency. The phase margin is the additional phase lag at the gain cross-over frequency to make the locus pass through the point (–1, 0), so that the system is on the verge of instability. From the definition, it is clear that the phase margin (ϕ_m) is positive for stable systems. As it approaches zero, the system becomes less stable and its transient response becomes oscillatory. When ϕ_m is zero the system is on the verge of instability. The value of ϕ_m is negative (< 0) for unstable systems.

The specification of gain or phase margin above does not indicate the degree of stability or relative stability of the system. The relative stability of the system is decided by the specification of both the margins. For a system to have adequate stability some minimum values of these margins have to be specified, i.e. $a > 1.5$ and $\phi_m > 45°$. These minimum values may be set depending upon the degree of stability required. They ensure the stability of a system against the variations in the parameters of the system or components of the system, e.g., open loop gain constant or time constants. These margins being a measure of the closeness of the plot to (–1, 0) indicate the effective damping ratio of the system. Closer the response to (–1, 0), lesser is the effective damping ratio. The phase margin and damping ratio are very closely related, as shown in Fig. 6.22. A phase margin of 60° corresponds to a damping ratio of 0.6. Even though the gain margin can also

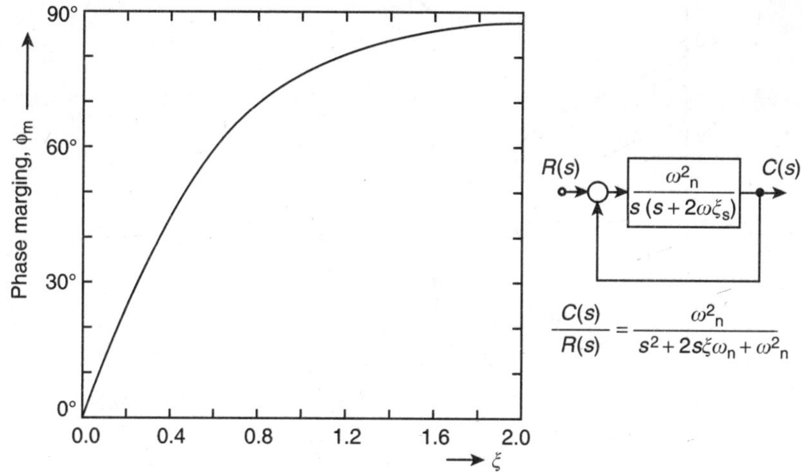

Fig. 6.22 *Phase margin vs damping factor for a second order system*

be related to the damping ratio, phase margin is the better estimate of damping ratio, since it influences the transient overshoot more than gain margin.

These margins can therefore be used as design criteria. They offer a means of adjusting the gain and time constants of the system to obtain the required behaviour. The design of various controllers to improve the stability performance is based on providing specified values of phase and gain margins. Their use in the design of controllers will be explained later.

The above discussion is applicable to systems which are absolutely stable. It is not applicable to conditionally stable systems.

It is necessary to note that the stability condition for non-minimum phase functions is satisfied only if the Nyquist plot encircles (–1, 0). For these systems therefore, the phase and gain margins are negative.

Stability from Bode Plots of Open Loop Transfer Function The Nyquist criterion details how the open loop polar plot can be used for establishing the stability of a closed loop system. In terms of phase and gain margins it has been concluded that a minimum phase function should have positive values of phase and gain margins for stable operation.

Bode plots of open loop transfer function can also be used to ascertain the stability (both absolute and relative) of a closed loop system. The phase and gain margins can be easily determined from these plots. Referring to Fig. 6.23 the gain margin is the reciprocal of the open loop magnitude ratio of the transfer function at a phase of –180°. When measured in decibels it is the negative of the actual magnitude ratio at the phase of –180° or phase cross-over. A stable system has positive gain margin. From the same figure it can be seen that the phase margin is 180°+ phase of the open loop transfer function at the gain cross-over frequency (when the gain of the function is unity or 0 db).

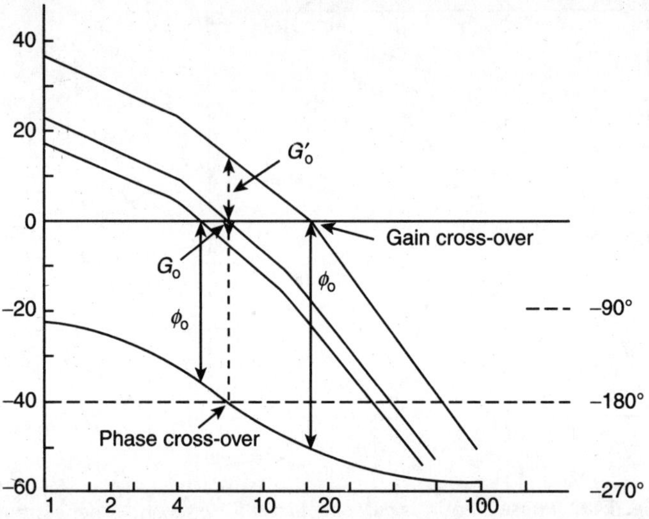

Fig. 6.23 *Gain and phase margins from bode plots. Gain margin a = G_o'. Phase margin $\phi_m = 180° + \phi_o$*

With the knowledge of these margins and phase and gain cross-over frequencies, Bode plots can be used to ascertain the stability of the system. Using the fact that the phase margin is positive for a stable system, we may conclude that such a situation occurs if the gain cross-over occurs earlier than the phase cross-over or in other words the gain cross-over frequency is less than phase cross-over frequency. The system is marginally stable if both occur simultaneously and the system is unstable if the phase cross-over occurs before gain cross-over. The increase in the gain constant of a system has a tendency to raise the log-magnitude plot vertically upwards without altering the phase plot.

This shifts the gain cross-over to the left, or effectively the gain cross-over frequency decreases. The system therefore tends to become unstable.

The Bode plots are therefore used to adjust the parameters, such as gain and time constants of the system, to obtain the required phase and gain margins. They can also be used to design the controllers to improve the system performance by bringing the margins to the specified minima. It has been stated that for adequate stability the phase margin must be greater than 45° and gain margin > 1.5.

The shape of the Bode plots and its slope at the gain cross-over also indicate the stability of the system. A factor of the nature $1/(1 + j\omega T)$ has a slope of –20 db/decade and an angle changing from 0° at lower frequencies to –90° at higher frequencies. Several such factors affect the slope of the log-magnitude curve and the phase angle. A slope of –20 db/decade may be related to –90° whereas a slope of –40 db/decade to an angle of –180°. Hence for a stable system the log-magnitude curve should be more gradual than –40 db/decade. The preferable slope is –20 db/decade at gain cross-over frequency. The system is unstable if the slope is –60 db/decade.

The shape of the curve at low frequencies determines the type of system and steady-state accuracy. The speed of response of a system is indicated by the frequency at phase margin.

Stability from Log-magnitude—Angle Diagram Yet another way of portraying the frequency response is to draw a curve showing the log-magnitude in decibels on the *Y*-axis and phase angle on the *x*-axis. The frequency ω is varied from $-\infty$ to ∞ by combining the log-magnitude and phase plots (Fig. 6.24). Thus the frequency is a parameter. This will be a closed contour for minimum phase functions. The stability can also be ascertained from this diagram, when one is interested in stability, it is sufficient to draw the curve in the range of frequencies from 0 to ∞ for minimum phase functions. Translating the Nyquist criterion (Bode plots) which states that the phase margin and gain margin must be positive for minimum phase functions, a thumb rule can be given for determining the stability of the system decided by the disposition of this curve with respect to the (0 db, –180°) point. As the curve is traced in the direction of increasing frequency, if this point lies to the right of the curve the system is stable. The relative stability of the system can also be ascertained from the nearness of this curve to the point. Nearer the curve to this point, the gain and phase margins also decrease. The system approaches cross-over; phase cross-over frequencies are marked on the figure.

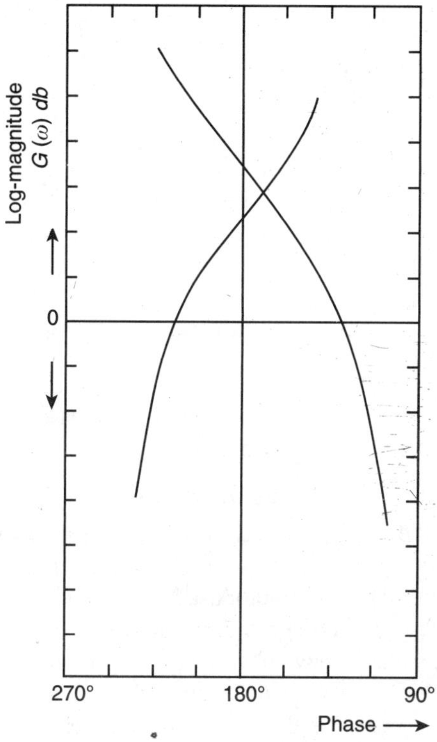

Fig. 6.24 *Log-magnitude vs. phase plots*

The advantages of the curve are obvious: The curve affords a quick estimation of relative stability and the design of compensation can be worked out easily. The stability of conditionally stable systems can be investigated very easily using these plots. Log-magnitude versus angle plots of a conditionally stable system cross the –180° axis more than once. Stability analysis and range of parameters (e.g., gain) for stable operation may be obtained using this curve. The effect of changing the parameters on the stability can be investigated using this curve. For example, increasing the gain shifts the curve upwards without changing the angle

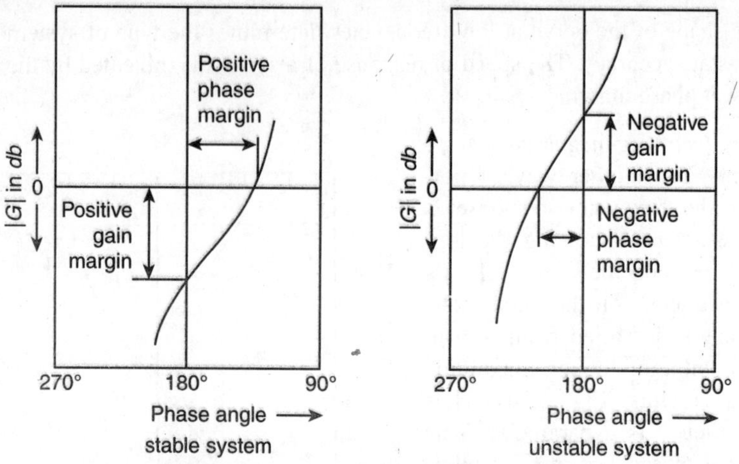

Fig. 6.25 *Phase and gain margins from log-magnitude vs. phase plots*

characteristic. This results in a decrease of both phase and gain margins and the system becomes less stable.

6.8.4 Root Locus Techniques

In the electrical drives employing closed loop control techniques, it is often necessary to investigate the effects of changing the parameters of the system on its stability. The drive motor used itself has several parameters and time constants which affect the stability of the motor itself. The controls used also introduce some time lags which may affect the system stability. It is also required sometimes to investigate the sensitivity of the drive system to parameter variations. The feedback control used makes the system less sensitive to parameter changes. The root locus technique is very useful in investigating

i. the stability of the system,
ii. the effect of variation of parameters so that the limits of these parameters can be determined for stable operation.

The method is particularly suitable in the design stage of the system when there is some latitude in the choice of system parameters.

The root locus technique can be applied to determine the dynamic response of the system. This method associates itself with the transient response of the system and is particularly useful in the investigation of stability characteristics of the system. It can also be used to determine the stability boundaries of the system. Selection of suitable parameters may be made using the root locus analysis.

It is a well established fact that the condition of roots of the characteristic equation governs the transient behaviour and hence the stability of the system. In the root locus technique which is essentially a graphical procedure, the locus of the roots on the s-plane is afforded by the variation of the suitable parameters.

The root locus plots are drawn in the complex plane, as the roots may be real or complex.

In this technique also the open loop transfer function may be used to simplify the procedure. The plots drawn and the information obtained from these plots pertain to a closed loop system. The plots clearly illustrate the effects of variation of parameters on system stability and the nature of response. The branch of a root locus shows all the possible values of one root when a parameter is varied. All the separate branches of the root locus plot give the effect of variation of parameters.

The root locus technique is essentially a time domain technique. Frequency response data can be obtained from the root locus plot. Using this technique the open loop poles and zeros can be modified to satisfy the requirements to be met by closed loop poles and zeros. It may be noted here that a slight change in the pole zero configuration may cause a significant changes in the root locus configuration. The method of root locus is illustrated by an example.

6.9 COMPENSATION AND THE USE OF CONTROLLERS TO IMPROVE THE PERFORMANCE

6.9.1 Performance Indices of a Control System

From the foregoing discussion on control systems it can be seen that the behaviour of the control systems can be specified, based on several time domain specifications. To provide a basis for comparison of several types of control and solution, the performance indices are defined. These help as a quantitative measure of the system performance in which the system specifications are emphasised. The practical requirements of a system can be dictated by these indices. One must be able to derive these indices from the requirements of the system and to determine them from experiments. In a linear control system the performance indices are normally independent of the operating point. Sometimes a compromise may be required between the parameters. The design of the power circuit and control depends mainly on satisfying these performance indices.

The time domain specifications of a control system are normally defined from its transient response to different inputs. These are:

i. *Steady-state error:* This is the deviation of the actual output from the desired one in the final steady-state condition. For a step input, the output follows the input and hence the steady-state error is the deviation of the output from the input. This is also called offset. Steady-state error depends on gain in the case of proportional control. It is theoretically zero for integral control. For ramps and parabolic inputs velocity and acceleration errors defined.

ii. *Rise time:* The time period between the instants of the change in the input or disturbance and the output attaining the value of the input for the first time; $x = 0$ for the first instance for under damped systems. For overdamped systems this is the time period between the output values 10–90% of the input. This characterises the speed of response.

iii. *Peak time:* The peak time is the time period between the instants of the change in the input or disturbance and the first peak overshoot of the response.

iv. *Settling time:* The period of time which starts when the change in the input occurs and ends when the output differs from the input with an acceptable tolerance. This is related to the time constant of the system.

v. Overshoot M_p of the controlled quantity is the maximum deviation of the output from the desired one (input) during the control when the change in the input occurs. This is defined by

$$\text{Maximum \% overshoot} = \frac{C(t_p) - C(\infty)}{C(\infty)} \times 100$$

This is directly a measure of the relative stability of the system. It must be within 3 to 5%.

The quantity of the dynamic response which is specified by the above can be represented by the following performance indices (Fig. 6.26).

$$A_1 = \int_0^\infty (x(t) - x(\infty))dt \tag{6.66}$$

$$A_2 = \int_0^\infty (x(t) - x(\infty))^2\, dt \tag{6.67}$$

$$A_3 = \int_0^\infty |(x(t) - x(\infty))|\, dt \tag{6.68}$$

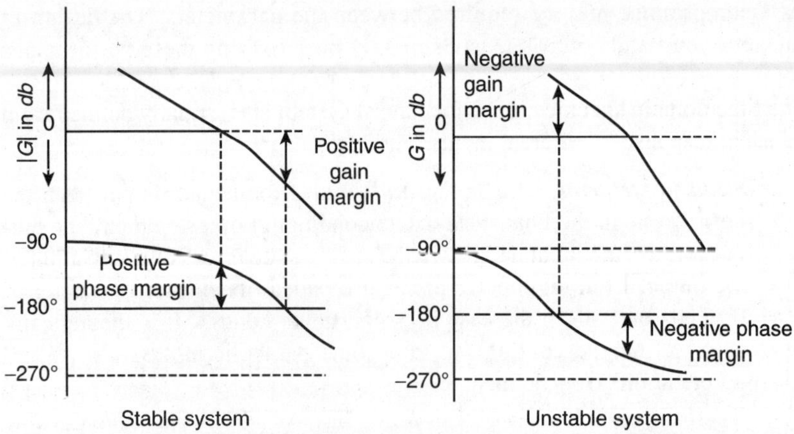

Fig. 6.26 *Gain and phase margins from bode plots*

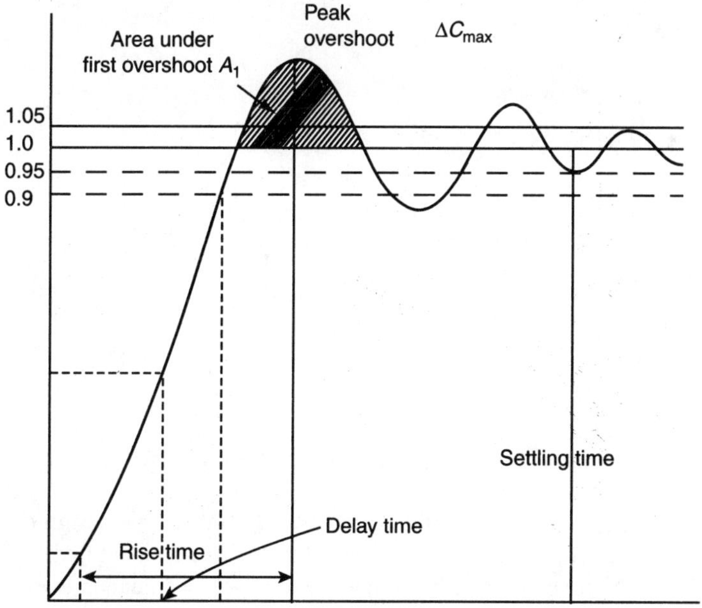

Fig. 6.26(a) *Performance indices pertaining to dynamic response*

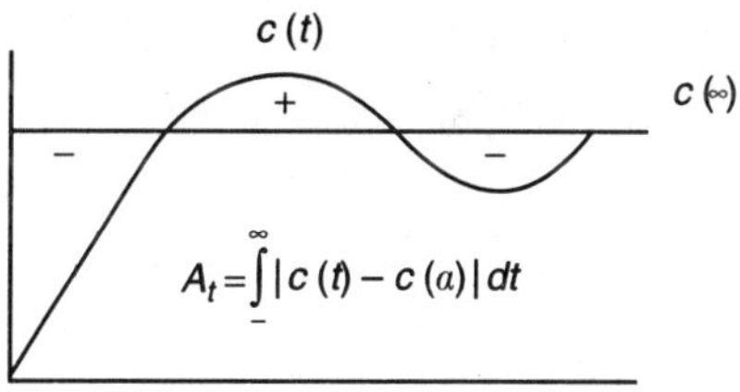

Fig. 6.26(b) *Pertaining to determination of $A_1A_2A_3$*

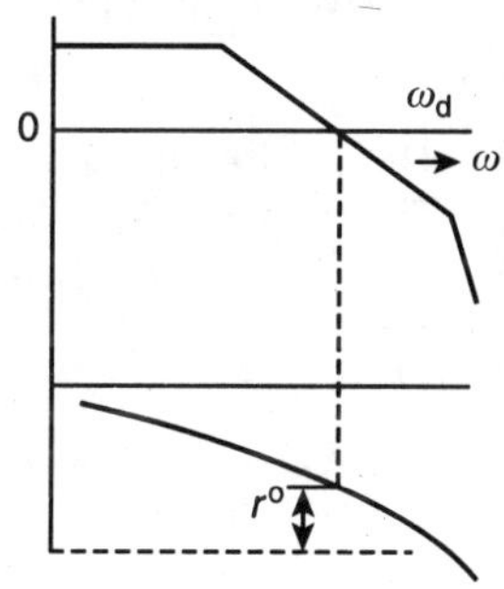

Fig. 6.26(c) *Performance indices in frequency response in open loop operation*

The first one is useful for representing a periodic behaviour, whereas the others are suitable for evaluating analog computer results. In the design of the required control the area under the first overshoot is considered and it is

$$A = \int_{t_1}^{t_2} [x(t) - 1]\, dt \tag{6.69}$$

The comparison of several performance indices is made in Table 6.1. These are normally useful in the design of control systems, to meet the required performance.

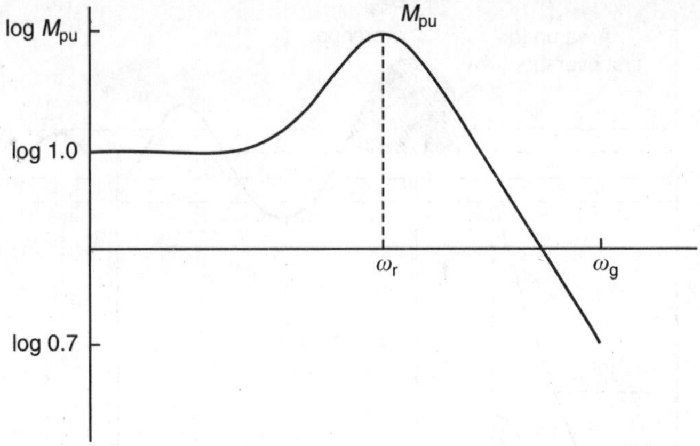

Fig. 6.26(d) *Performance specifications of a closed loop system in frequency domain*

The frequency response is advantageously employed in the design of control systems. A set of frequency domain specifications are also identified. A correlation between the time domain and frequency domain specification is established for the control systems. The speed of response of a control system is characterised by the gain cross-over frequency of the frequency response. The phase margin qualifies the overshoot of the system. Therefore the frequency response specifications are

i. The limiting frequency ω_g at which the closed loop gain > 0.7.
ii. The frequency ω_0 where the amplitude attains peak value.
iii. The peak amplitude of the frequency response $(M_{p\omega})$.

The typical specifications of the drives are given in Table 6.2.

Table 6.2 *Typical specification of drives*

	Damped natural frequency ω_a	*Rise time*	*% overshoot*	a/x_ω
Current controller	250 to 50 rad/s	0.010 to 0.050 s	1%	
Speed controller	125 to 25 rad/s	0.020 to 0.100 s	0.01 to 0.5%	
Position controller	100 to 10 rad/s	0.025 to 0.20 s	× 0.001 mm … 0.01 mm	10 … 100 rad/s

6.9.2 Compensation

A drive system having closed loop control may not be satisfactory with regard to its stability characteristics, speed of response and steady-state accuracy. The system may be oscillatory or even unstable. It may have either extremely fast or very sluggish response. The errors under steady-state between the actual and desired values may be excessive and not acceptable. Therefore, a necessity arises to modify the system or system parameters to provide the de sired performance with respect to the above characteristics. This has certain practical limitations, such as size, range, and cost of available components. Sometimes there may not be any room to effect the change in the parameters.

In such cases the performance of the drive is improved by adding additional components to it. The method of improving the performance in this way is called compensation. The additional component changes the transfer function of the overall system and gives the desired performance. In feedback systems the compensation added is simple and less expensive and provides substantial improvement in the performance.

The design of a compensation network to achieve the required improvement in the performance can be based on the frequency response plots or the root locus techniques discussed.

The compensations introduced in the closed loop system to improve the performance reshapes the root locus or frequency response of the system. With this the system is stable, has satisfactory transient response, and has a steady-state error within the acceptable tolerance level, or even sometimes zero. The purpose of providing a compensation, either by means of reshaping the frequency response or root locus may generally fall into one of the following categories:

i. The system may be stable having the desired transient response but its steady-state error is large.
ii. The given system may be stable, having a time response which is not acceptable.
iii. The given system may be stable, but both the transient response and the steady-state error are not acceptable.
iv. The system may be unstable for all values of gain.

Ideal integral compensation increase the type of the system. An increase in the type of system completely eliminates the steady-state error. Therefore for a system having steady-state error with satisfactory transient response, a PI compensation is used (proportional and integral). Proportional control improves the stability. The improved system will have acceptable transient response.

PD compensation is used if the transient response has to be improved. It is equivalent to introducing an anticipation into the system. The derivative action increases the speed of response when the input is rapidly changing.

It has already been shown and brought out how

i. the response of a second order system is altered by variation of damping ratio.
ii. the stability criteria such as Nyquist and Routh Hurwitz criteria can be used to study the effect of variation of parameters on the system stability.
iii. how the root locus based on open loop transfer function can be used to study the variation of parameters to improve the performance of the system by investigating the sensitivity of the system for parameter variations.

The design of controllers and compensators to improve the system performance will be discussed in the following. The design may be based on Bode plots or root locus techniques. The dynamics of the control system is altered or improved by the adjustments of the controller parameters and not the plant parameters. The adjustments change the controller transfer function.

The controller providing compensation can be placed either in series with the plant in the forward path or in the feedback path. In the former it is called series (or cascade) compensation and in the latter parallel compensation. These are depicted in Fig. 6.27. The location of the compensation depends on the control system and the modifications required on the response.

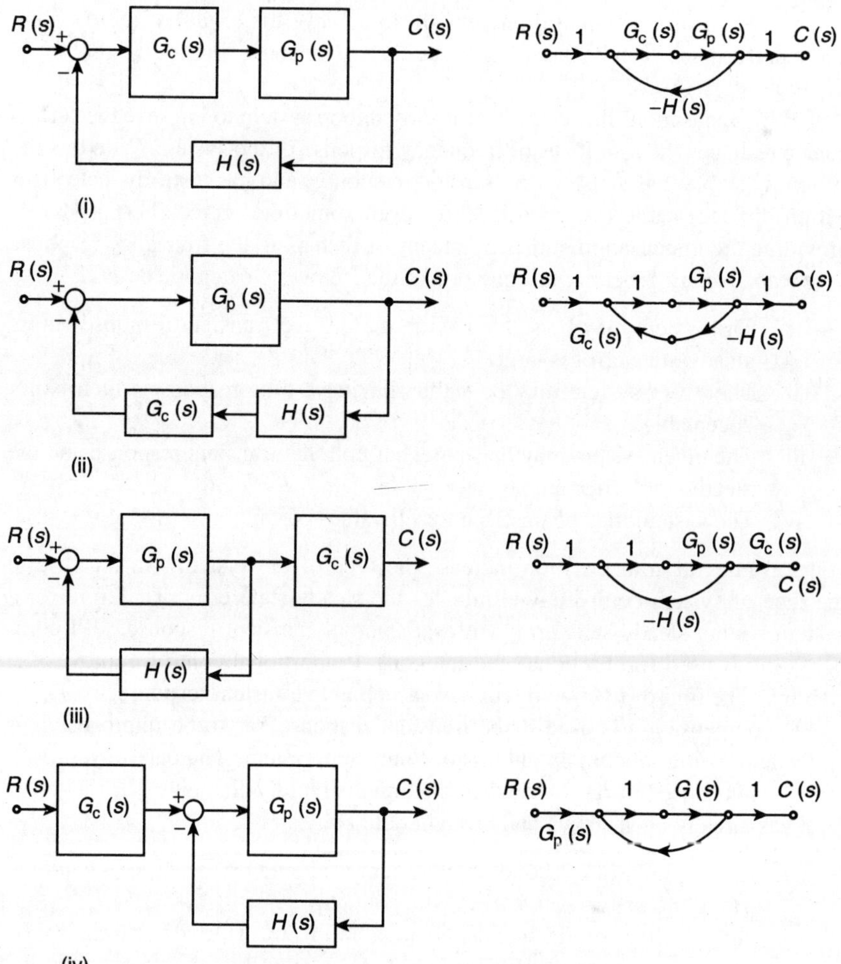

Fig. 6.27 *Methods of compensation. (i) Cascade compensation (ii) Feedback compensation (iii) Load circuit compensation (iv) Input circuit compensation*

A controller used in a feedback control system provides a corrective action depending upon the error between the actual and desired value of the quantity. The output of the control depends upon the state of the error. The controller component is identified in Fig. 6.28. The relationship between the output and the input of the controller is known as controller action $\theta_c(s)/\theta_i(s) = f_1(s)$ where $\theta_c(s)$ and $\theta_i(s)$ are the output and input of controller respectively. The basic types of controllers are compared in Table 6.3.

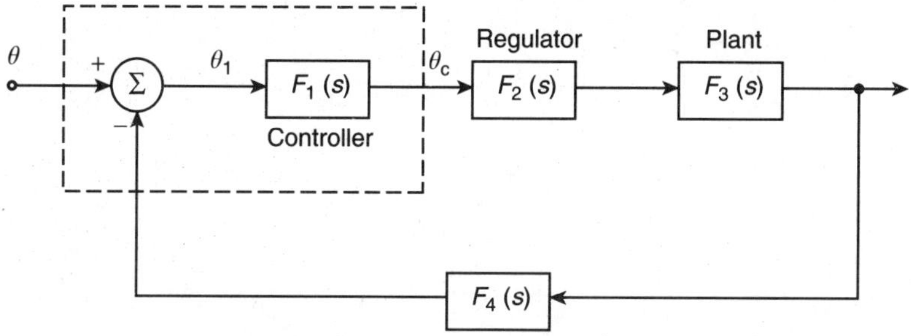

Fig. 6.28 *Pertaining to action of a controller in a closed loop system*

6.9.3 Design of Controllers

The foregoing discussion makes it clear that there exists a possibility of improving the performance of a given control system using a controller or compensating network. The controller frequency response (or root locus) alter; or reshapes the frequency response (or root locus) of the original system to the desired one. Thus the ideal goal of compensation is to achieve a control system which has a desired performance, viz. having zero steady-state error, optimised dynamic performance, (which is neither too fast nor too slow) assuring the stability of the system, which is a necessary condition of a control system. The transient response must be damped out satisfactorily.

Basically employing a controller improves the behaviour by

i. compensating or modifying one or more large time constants
ii. limiting the upper frequency so that a number of small time constants may lose their significance as well as their effect on the control system.

A system is in general considered to be optimum as far as its dynamic behaviour is concerned if it transmits all the operating frequencies in a similar way. This can be satisfied if the system has a transfer function

$$G(s) = \frac{a_0 + \Sigma\, a_\nu s^\nu}{b_0 + \Sigma\, b_\mu s^\mu} \tag{6.70}$$

whose magnitude is constant for all frequencies, or at least for a large range of frequencies. Further, for the transfer function to satisfy this condition it must have all possible derivatives with respect to ω zero.

Table 6.3 *Comparison of controllers*

Type of controller	*Action*	*Characteristics*	*Features*	*Application*
1. Proportional	$\frac{\theta_c(s)}{\theta_i(s)} = K$	The output of the controller is proportional to the input	This provides a simple connection between the error and output. It is essentially an amplifier with adjustable gain, whatever may be the actual mechanism or operating point. The controller has a response characterised by medium rise time, stable response and undesirable offset. This undesirable offset, however, can be decreased by increasing the gain. There is an upper limitation of gain. Improvement of the steady-state accuracy by increasing gain results in oscillatory behaviour. The effect can be minimised by good design. The high inertia loads have a tendency to oscillate. The response of the controller is shown in Fig. 6.29(a).	Due to its simplicity and good stability characteristics it is employed when the offset can be accepted.
2. Integral controller	$\frac{\theta_c(s)}{\theta_i(s)} = \frac{K}{s}$	The output of the controller is proportional to time integral of the input or the rate of change of control output is proportional to the input. The output remaining constant for zero actuating error.	This does not provide rigid connection between the input (error) and the output (control action). The controller is characterised by a response free from offset, having a very large time constant making the response very slow or sluggish due to soft connection. The response is oscillatory with hunting. The addition of an integral controller to a second order system converts it to a third order one which may become unstable. The response of the controller is shown in Fig. 6.29(b).	Because of the un-favourable characteristics this is not used alone.

3. Derivative controller	$\dfrac{\theta_o(s)}{\theta_i(s)} = Ks$	The output is proportional to the rate of change of input	A derivative controller has a response characterised by a short rise time indicating for rapidly varying input. If the input is varying slowly, there is a drift due to surge of the action. The response of the characteristic is shown in Fig. 6.29(c).	Because of the un-favourable drift for slowly varying inputs the derivative controller is not used alone.
4. PI controller	$\dfrac{\theta_c(s)}{\theta_i(s)} = \left(K_1 + \dfrac{K_2}{s}\right)$	The output is proportional to the input as well as its time integral	This combines the advantages of proportional and integral controllers. The simple and direct connection between the input and output is present. The controller is characterised by good stability characteristics and offset free response. The proportional constant can be smaller than that in the case of pure proportional controller. The response is oscillatory. It is characterised by medium rise time, zero-steady state error, and oscillatory response without hunting. Proportional action stabilises the system and integral action removes the offset. The response of the controller is shown in Fig. 6.29(d).	Because of its favourable characteristics he controller is very widely used in the compensation.
5. PD controller	$\dfrac{\theta_c(s)}{\theta_i(s)} = K_1 + K_3 s$	The output is proportional to both the input and its rate of change	In this also proportional action imparts good stability characteristics to the controller. The drift of the pure differential action is eliminated. The undesirable offset of proportional action is still present and is not eliminated. The controller	PD controller is employed with high inertia loads.

(*Continued*)

Type of controller	Action	Characteristics	Features	Application
			response is characterised by short rise time indicating rapid response, stable output, with un desirable offset. The nature of derivative action to rapidly changing input is however is present and it extends to slowly changing input. The derivative control stabilises the oscillating tendency of high inertia loads with proportional control. Fig. 6.29(e) depicts the response of the controller.	
5. Proportional plus integral plus derivative	$\frac{\theta_c(s)}{\theta_i(s)} = K_1 + \frac{K_2}{s} + K_3 s$	A general three term controller having output proportional to the input, its time integral and its rate of change. The degree of each action depends upon the constants K_1, K_2, K_3	The proportional action which has direct physical and simple connection between the output and input imparts good stability characteristics. The offset is eliminated by integral action. The derivative action imparts swiftness of response by providing the necessary damping for hunting. This controller is characterised by short rise time imparting a very rapid response, non-oscillatory stable, offset free output. The response of the controller is depicted in Fig. 6.29(f).	A general purpose controller and finds application.

Compensation using passive networks

Type of network	*Transfer function*	*Features*	*Remarks*
Lag network	$G_c(s) = \frac{1+Ts}{1+aTs}\ a > 1$ Pole and zero are close together α is equal to the desired increase in gain	The compensation provided is similar to PI. The necessary gain is provided by the amplifier which also prevents any loading of the original system by providing isolation. A large increase in gain constant K is provided by the compensator. This reduces the steady-state error. However, there is small increase in rise time due to decrease in ω_n. This is basically a low pass filter. Lower frequencies are passed and higher ones are attenuated. This property makes it useful and permits increase in gain.	Schematic of the network and its frequency response are shown in Fig. 6.30(a).
Lead network	$G_c(s) = Aa\frac{1+Ts}{1+aTs}$	Sensitivity increases as $a < 1$. Very small values of a means a very large A. Practical value is 0.1. This result in a moderate increase in gain constant K which improves steady-state accuracy, ω increases resulting in reduction in settling time. As a is 0.1 the additional gain K is larger than increase in K of the system. It is basically high pass filter, high frequencies are passed and low frequencies are attenuated	Used to improve the system falling category 2. Schematic of the network and its frequency response are shown in Fig. 6.30(b).
Lag lead	$G_c(s) = \frac{A(1+T_1s)(1+T_2s)}{(1+aT_1s)\left(1+\frac{T_2}{a}s\right)}$ $a1, T_1T_2$	Results in a large increase in K improving the steady-state accuracy. Improves the transient response by effectively decreasing the settling time due to increase in ω_n. This combines the desirable characteristics of lag and lead networks.	The response and the schematic are shown in Fig. 6.30(c).

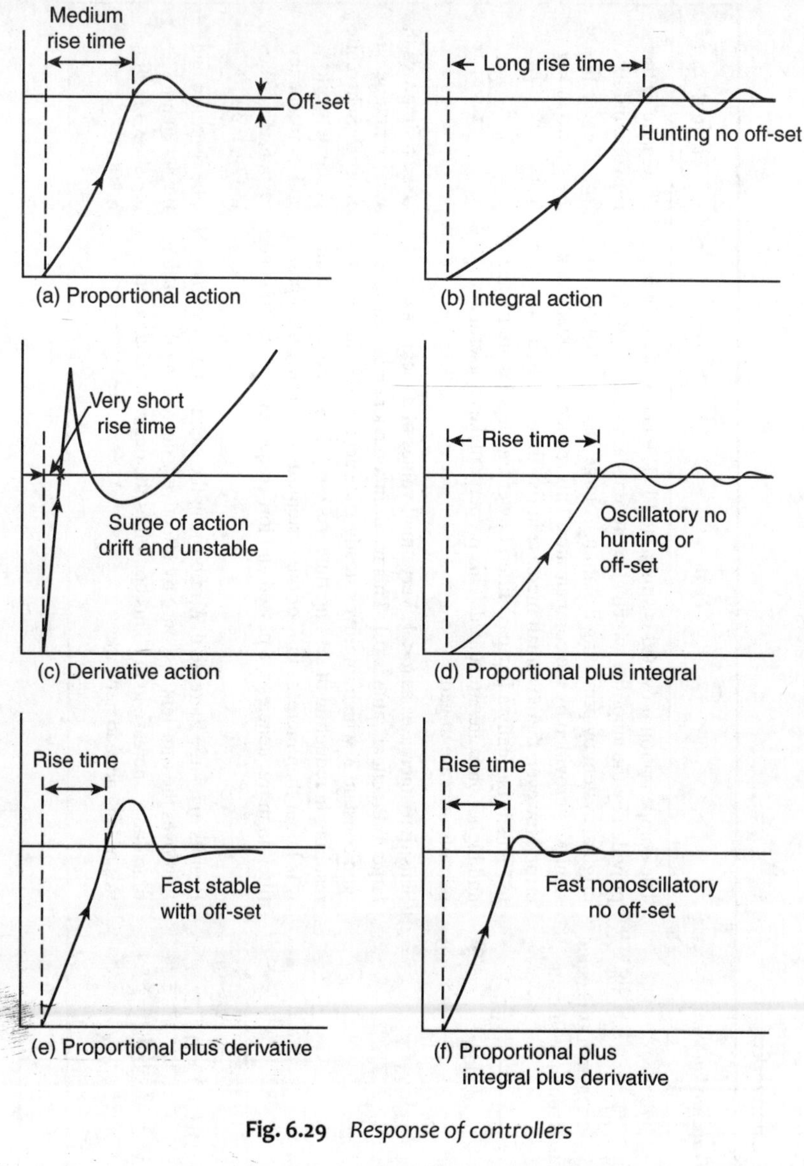

Fig. 6.29 *Response of controllers*

$$\underset{\omega \to 0}{\mathrm{Lt}} \frac{d}{d\omega} |F(j\omega)| = 0 \tag{6.71}$$

Using this criterion on the transfer function indicated in Eq. 6.70, the constants $a_0, a_1, a_2, \ldots, b_0, b_1, b_2, \ldots$ etc. can be evaluated. The controller transfer function obtained this way gives optimum behaviour and the principle is called magnitude optimum. This procedure optimises a system so that it can pass a range of frequencies.

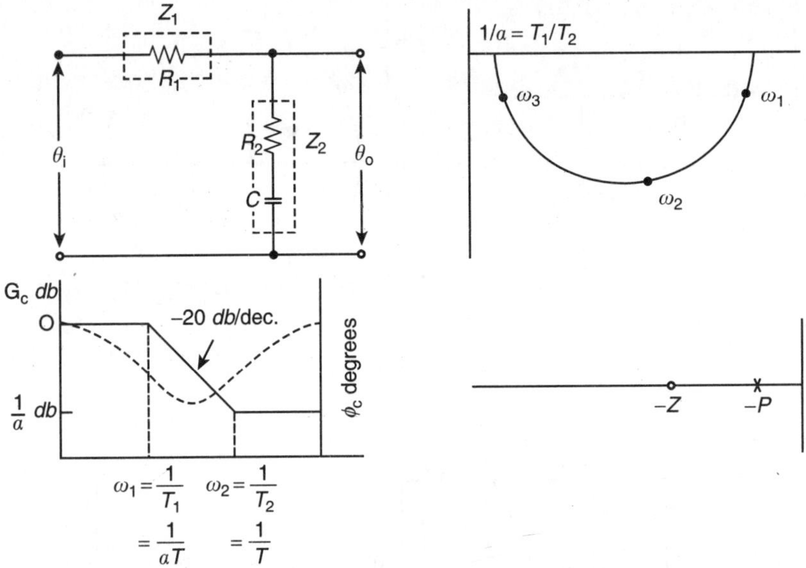

Fig. 6.30(a) *Lag network and its frequency response*

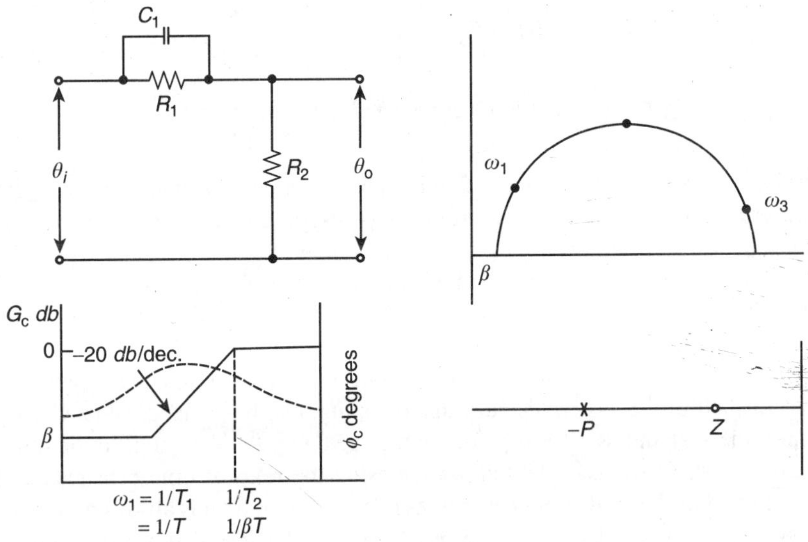

Fig. 6.30(b) *Lead network and its frequency response*

The resulting transfer function can be represented by Bode plots, or the time response can be determined if possible, to check whether the modified system gives the required performance.

Bode plots can also be employed to design a controller to impart the desired performance to the system. The design criteria in this case can be taken as phase

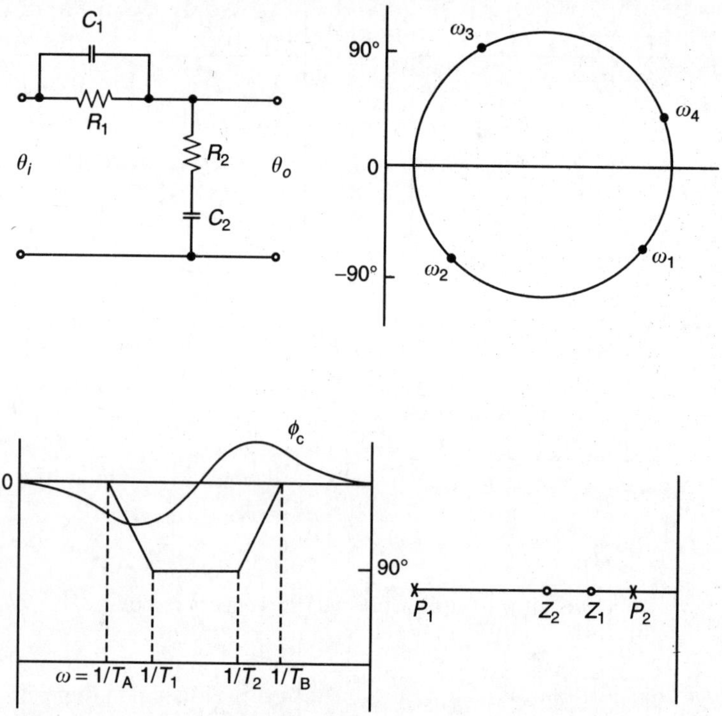

Fig. 6.30(c) *Lag-lead network and its frequency response*

margin, $\varphi = 180 - \varphi_c$. This assures that the phase angle at the gain cross-over has a prescribed minimum value. The frequency, response of the original system is reshaped by that of the controller to the desired one.

A close examination of the methods confirms that the designs are not very much different. They are effective in compensating large time constants. The method based on magnitude optimum gives a system having a definite damping. The ratio of the integrating time constant of the controller and uncompensated time constant of the system decides this damping, and has a fixed value. Consequently the system has a fixed phase margin given by the damping. In the other method the designer has a latitude in choosing the ratio of time constants and hence has control over damping as well as phase margin. As it is always necessary for both the phase and gain margin of a system to be specified simultaneously, the controller designed using phase margin criterion will also have a satisfactory gain margin. Both the methods are also suitable if frequency response results are available in a graphical form from an experiment and cannot be represented by analytical expressions. The methods are discussed in detail in the following.

A controller used can compensate maximum of two large time constants. The rise time is decided by the sum of all time constants which are not compensated and are significant in the working range of frequencies. However if the original control system has more than two large time constants the uncompensated time constants make the dynamic response slow and it may not be to the desired level. In such cases an inner subordinate loop can be employed to compensate for this time constant.

Generally in the control of drives, besides the outer speed loop there is also an inner current loop, to have improved performance. Without introducing any appreciable error the controller transfer function of the inner loop can be approximated to a first order system.

The compensation described above, viz. compensating two large time constants, does not give satisfactory results if the system also has an integration. The design of controller based on the above criteria for use with such a system imparts an oscillatory behaviour. Hence the basis of the design of the controller must be changed in such a case. The controller designed in this case also must be such that the magnitude of frequency response is unity for a wide range of frequencies, i.e., the system allows a wide range of frequencies to pass through. This gives an optimised controller and the method is called symmetrical optimum. The method is given the name because the frequency response is symmetrical with respect to gain cross-over frequency.

The above methods are illustrated in the following discussion.

6.9.4 Magnitude Optimum

The design of a controller based on the principle that it allows all the frequencies to pass through in a similar way for a simple system is shown in Fig. 6.31.

The system has an input $R(s)$ and output $C(s)$. The closed loop transfer function

$$G = \frac{C(s)}{R(s)} = \frac{G_c(s)G_p(s)}{1 + G_c(s)G_p(s)} \tag{6.72}$$

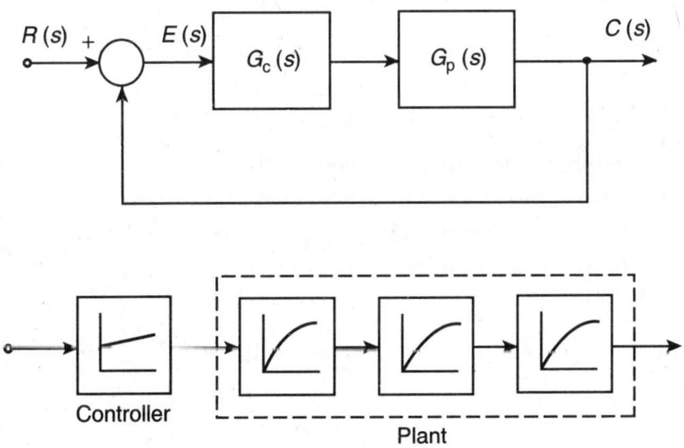

Fig. 6.31 *Typical closed loop system to illustrate the design of controller*

The system has to be optimised so that the overall system has a similar behaviour at all frequencies.

The plant transfer function $G_p(s)$ is assumed to be having linear factors and is given by

$$G_p(s) = \frac{V_s}{\prod_{k=1}^{m}(1 + sT_{pk})(1 + sT_e)}$$

where V_s is the gain of the plant
T_{pk} large time constants of the system
T_e sum of all parasitic time constants.

These parasitic time constants of the control system are introduced by the non-ideal nature of the devices and the elements used for harmonic elimination and smoothing. The value of T_e is of the order of 2 to 5 ms and it is very much smaller than the least of the large time constants.

The controller designed is a PID controller which has a general transfer function given by

$$G_c(s) = \prod_{i=1}^{n}\left(\frac{1 + sT_{ci}}{V_r sT_{co}}\right) \tag{6.74}$$

Using Eqs 6.73 and 6.74 in Eq. 6.72 we have the overall closed loop function given by

$$G(s) = \frac{V_s \prod_{i=1}^{n}(1 + sT_{ci})}{|V_r|\, sT_{co} \prod_{k=1}^{\infty} i(1 + sT_{pk})(1 + sT_c) + V_s \prod_{i=1}^{n}(1 + sT_e)} \tag{6.75}$$

An optimised controller can be obtained on the principle described only if the large time constants are compensated and T_0 and T_e only effect the transient behaviour. Hence the rules of optimization can be stated as

i. the number of large time constants in the numerator and the denominator must be the same.
ii. the values of time constants must be the same.
iii. The time constant of integration of controller must be fixed as $T_0 = (V_s/V_r)2T_e$.

For a specific case of a second order system having a transfer function

$$G_p(s) = \frac{V_s}{(1 + sT_{P1} + s^2 T_{P1}T_{P2})(1 + sT_e)} \tag{6.76}$$

the controller must have a transfer function

$$G_c(s) = \frac{1 + s(2d)T + s^2T^2}{sT_0} \tag{6.77}$$

Using the above rules of optimization

$$\begin{aligned} T^2 &= T_{P1} \cdot T_{P2};\ 2d = \frac{T_{P1}}{\sqrt{T_{P1}T_{P2}}} \quad d < 1 \\ T_0 &= 2T_e; \qquad V_s = V_r \end{aligned} \tag{6.78}$$

If the controller suits the plant with respect to its structure as well as parameters, typical transfer functions are given both for open and closed loops.

$$G_0(s) = \frac{1}{s2T_e + s^2 2T_e^2} \tag{6.79}$$

$$\begin{aligned} G(s) &= \frac{G_0}{1 + G_0} = \frac{1}{1 + 1/G_0} \\ &= \frac{1}{1 + s2T_e + s^2 2T_e^2} \end{aligned} \tag{6.80}$$

This can be obtained directly from the transfer function by applying the criterion that the overall transfer function should have a magnitude which is the same for all ω.

The plant transfer function is

$$G_p(s) = \frac{V_s}{(1 + sT_{K1})(1 + sT_{K2})(1 + sT_e)} \tag{6.81}$$

The controller transfer function is

$$G_c(s) = \frac{V_R(1 + sT_{i1})(1 + sT_{i2})}{sT_0} \tag{6.82}$$

The open loop transfer function is

$$G_0(s) = \frac{V_R(1 + sT_{i1})(1 + sT_{i2})V_s}{sT_0(1 + sT_1)(1 + sT_2)(1 + sT_e)} \tag{6.83}$$

Making $T_{i1} = T_1$, $T_{i2} = T_2$ so that controller completely compensates both the large time constants of the plant we have

$$G_0(s) = \frac{V_R V_S}{sT_0(1 + sT_e)} \tag{6.84}$$

The closed loop transfer function

$$G_c(s) = \frac{G_0(s)}{1 + G_0(s)} = \frac{V_R V_S}{V_R V_S + sT_0 + s^2 T_0 T_e} \tag{6.85}$$

Its frequency response

$$G_c(j\omega) = \frac{V_R V_S}{V_R V_S + (j\omega)T_0 - \omega^2 T_0 T_e} \tag{6.86}$$

$$|G_c(j\omega)|^2 = \frac{(V_R V_S)^2}{(V_R V_S)^2 + \omega^4 T_0^2 T_e^2 - 2V_R V_S \omega^2 T_0 T_e + \omega^2 T_0^2} \qquad (6.87)$$

must be unity to satisfy the magnitude optimum. This happens only if

$$\omega^4 T_0^2 T_e^2 - 2V_R V_S \omega^2 T_0 T_e + \omega^2 T_0^2 = 0$$

$$\omega^2 (\omega^2 T_0^2 T_e^2 - 2V_R\, V_S T_0 T_e + T_0^2) = 0 \qquad (6.88)$$

At $\omega = 0$

$$T_0^2 = 2V_R V_S T_0 T_e$$

$$G_c(s) = \frac{1}{1 + s2T_e + s^2 2T_e^2} \qquad (6.89)$$

The compensation of the two large time constants of the plant by the controller is depicted in Fig. 6.32. The Bode plot of the open loop transfer function corresponds to a first order system with integration. The gain cross-over frequency

$$\omega_c = \frac{1}{2T_e} \qquad (6.90)$$

The phase margin at gain cross-over is 65°.

The time response of the modified system with the controller can now be determined. The closed loop transfer function

$$\frac{C(s)}{R(s)} = \frac{1}{1 + s2T_e + s^2 2T_e^2} \qquad (6.91)$$

$$R(s) = C(s)(1 + s2T_e + s^2 2T_e^2) \qquad (6.92)$$

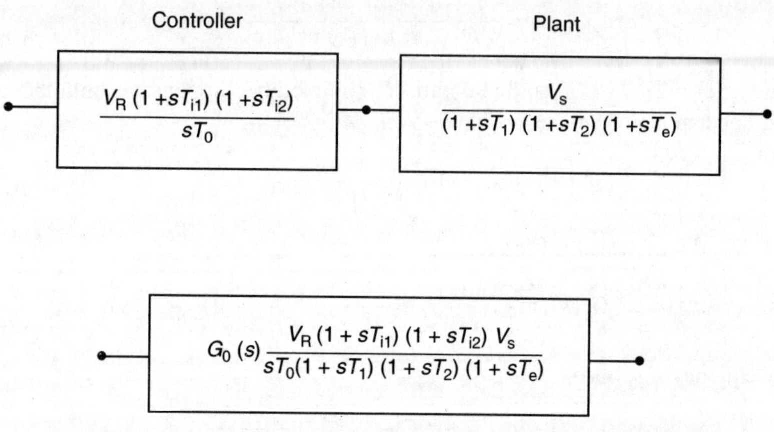

Fig. 6.32 *Compensation of two large time constants by a converter*

$$R(t) = C(t) + 2T_e \frac{dC(t)}{dt} + 2T_e^2 \frac{d^2C(t)}{dt} \tag{6.92}$$

Assuming zero initial conditions, i.e.,

$$t = 0, \quad C(t) = 0, \frac{dC(t)}{dt} = 0 \tag{6.93}$$

We have the time response (Fig. 6.33)

$$\frac{C(t)}{R(t)} = 1 - e^{-t/2T_e}\left(\cos\frac{t}{2T_e} + \sin\frac{t}{2T_e}\right) \tag{6.94}$$

i. The rise time t_r is the time taken by the output to attain the input value for the first time

$$\cos\frac{t}{2T_e} + \sin\frac{t}{2T_e} = 0$$
$$t_r = 4.7T_e \tag{6.95}$$

ii. The settling time is determined using the fact that it takes several cycles for the output quantity to reach the steady-state value

$$\frac{t_s}{2T_e} = \frac{7\pi}{4}$$
$$t_s = 11T_e \tag{6.96}$$

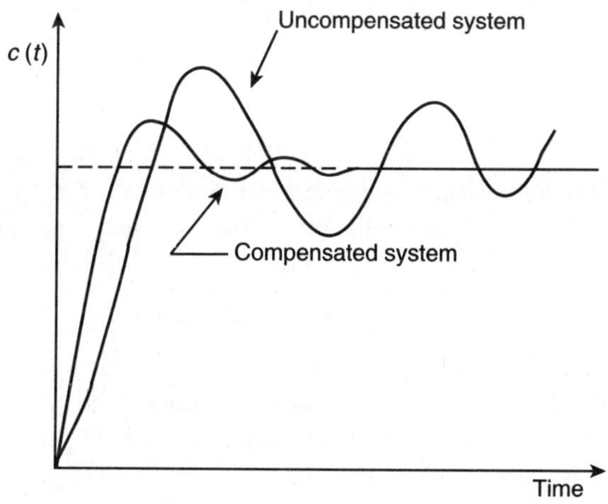

Fig. 6.33 *Time response of compensated system*

iii. The overshoot is determined using the condition

$$\frac{dC(t)}{dt} = 0$$

$$\frac{t'}{2T_e} = \pi$$

$$\frac{C_p}{R} = 1 - e^{-\pi}(\cos \pi + \sin \pi) = 1 + e^{-\pi} \tag{6.97}$$

$$M_p = \frac{C_p - R}{R} = 0.043$$

$$M_p = 4.3\%$$

The resulting system will have a damping factor 0.707 as decided by the ratio of time constants. This damping is rather high and may not be suitable for all applications.

6.9.5 Phase Margin Optimum

The method can be based on the phase margin in which the controller is designed to provide a minimum phase margin. The design makes use of Bode plots. The ratio of time constants can be decided depending upon the phase margin required, which directly influences the damping. Therefore this method has an advantage over the previous one. As the damping factor increases the phase margin also increases. Therefore a clear dependence of these quantities and the effect of the ratio of time constants on the time response of the system is clearly brought out in the following, so that a proper design of the controller to give the desired performance can be made.

The same example as above is taken here also. The plant has a transfer function

$$G_p(s) = \frac{V_s}{(1 + sT_{p1} + s^2T_{p1}T_{p2})(1 + sT_e)} \tag{6.98}$$

A PID controller is designed to improve the performance. The transfer function of the controller is given by

$$G_c(s) = \frac{(1 + sT_{p1} + s^2T_{p1}T_{p2})}{V_0 sT_0} \tag{6.99}$$

The zeros of $G_c(s)$ are the same as the dominant poles of $G_p(s)$, so that they compensate the large time constants of the plant. The open loop transfer function is

$$G_0(s) = \frac{1}{(1 + sT_e)sT_0} = \frac{1}{sT_0 + s^2T_eT_0} \tag{6.100}$$

The design based on magnitude optimum gave a controller having its integrative time constant equal to twice the uncompensated time constant of the plant, i.e., $T_0 = 2T_e$.

To design a controller based on phase margin the ratio of time constants T_0/T_e is assumed to be K_0. Using this ratio we have

$$G_0(s) = \frac{1}{sK_0T_e + s^2K_0T_e^2} \tag{6.101}$$

The closed loop transfer function

$$G(s) = \frac{1}{1 + sK_0T_e + s^2K_0T_e^2} \tag{6.102}$$

In the frequency domain

$$G(j\omega) = \frac{1}{1 + (+j\omega)K_0T_e - \omega^2K_0T_e^2} \tag{6.103}$$

The phase margin of the system can be obtained as

$$\varphi = \arctan \frac{1}{\sqrt{-\frac{1}{2} + \sqrt{\frac{1}{4} + \frac{1}{K_0^2}}}} \tag{6.104}$$

If the phase angle of the transfer function is φ_c, the phase margin is

$$\varphi = 180 - \varphi_c \tag{6.105}$$

Depending on the value of required φ the value of K_0 can be chosen. For a system having a known value of K_0 the phase margin is fixed. The method based on magnitude optimum assigns a value of 2 for K_0 and the corresponding phase margin is 65°.

In the present method there is a flexibility in the choice of K_0. The relation between K_0 and φ is depicted in Fig. 6.34. From this it can be concluded that the integration time constant actually decides the system improvement or its damping

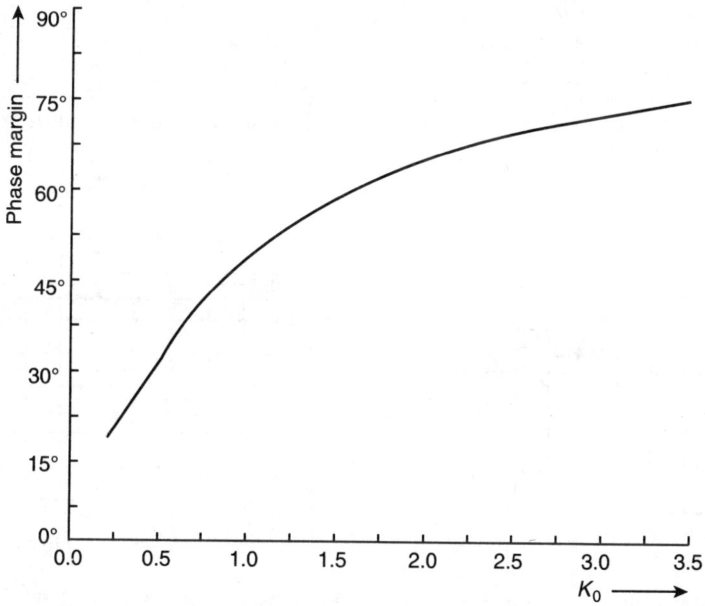

Fig. 6.34 *Dependence of phase margin on the ratio of time constants K_0*

and this is decided from the value of K_0 (ratio of time constants). Smaller the value of K_0 smaller the value of damping. At smaller phase margins the system becomes oscillatory.

The time domain specifications can now be related to the value of K_0. These are the initial rise time, the maximum overshoot and the area under the first overshoot. The time response of the second order system for different values of K_0 is shown in Fig. 6.35. The time response can be given by equation

$$\frac{C(t)}{R(t)} = 1 - e^{-t/K_0T_e}\left(\cos\frac{t}{K_0T_e} + \sin\frac{t}{K_0T_e}\right) \tag{6.106}$$

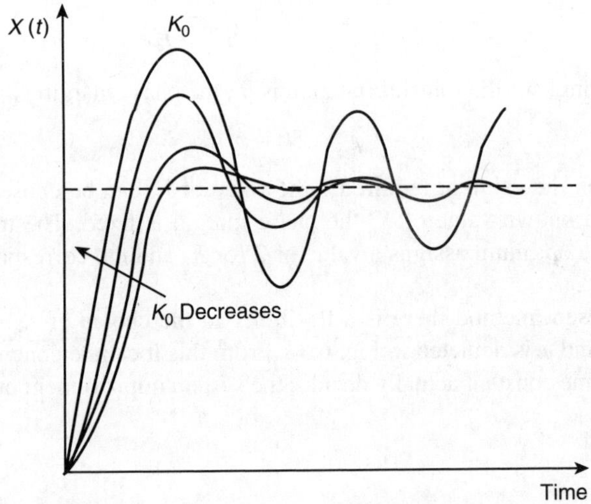

Fig. 6.35(a) *Time response of a second order system for different values of K_0*

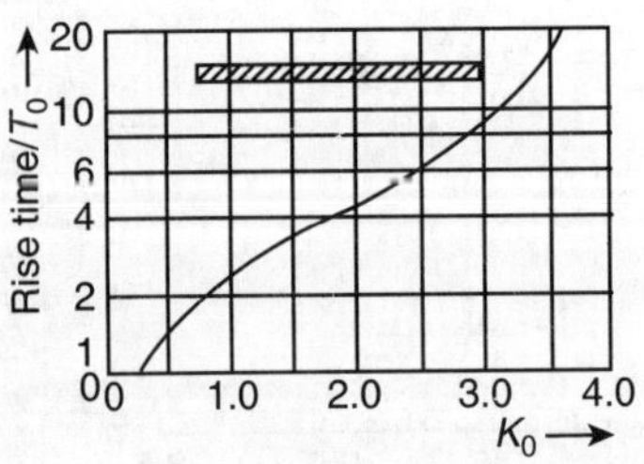

Fig. 6.35(b) *Rise time as a function of K_0*

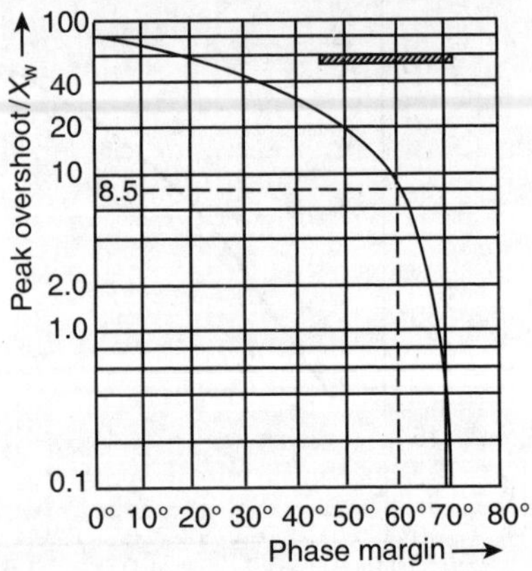

Fig. 6.35(c) *Peak overshoot as a function of phase margin*

The damping of the response increases with the value of K_0. From the time response it can be seen that the proper value of K_0 lies between one and four. For a value of $K_0 < 1.5$ the $t_r < 4T_e$. The settling time can also be determined.

The maximum overshoot as a function of K_0 can be determined and is depicted in Fig. 6.35(b). From the figure it is clear that overshoot decreases as the value of K_0 increases. For $K_0 = 1.5$, % overshoot = 8.5.

The area under the first overshoot as a function of K_0 is also shown in Fig. 6.35(c). This is obtained by evaluating the integral.

$$A_1 = T_e \int_{T_2}^{T_1} (x(T) - 1)dT \tag{6.107}$$

The integration results in

$$A_1 = T_e\sqrt{1-(K_0/4)}(\exp-(\pi + \varphi/t_g\varphi_0^*) + \exp-(2\pi + \varphi_0^*/t_g \cdot \varphi_0^*)) \tag{6.108}$$

where $t_g\varphi_0^* = \sqrt{(4/K_0)-1}$

A_1/T_e drawn as a function of phase margin is depicted in Fig. 6.35(d). The first overshoot dominated for the thick portion of the characteristic.

To determine a controller of suitable frequency response a detailed discussion of the effects of controller amplitude characteristic on the modified system is required. The ratio $K_0 = T_0/T_e$ can be freely chosen while designing a controller using phase margin criterion. It has been established that the phase margin increases as the value of K_0 increases. In other words the phase of the system transfer function at the gain cross-over frequency decreases. As the phase margin is related to the damping of the system the value of K_0 and hence the value of T_0 can also be related to damping, The damping of the system is more as the value of K_0 or T_0 increases.

To have an insight into the amplitude characteristics, compensation of the plants by PI and PID controllers is examined in the following. In the former the

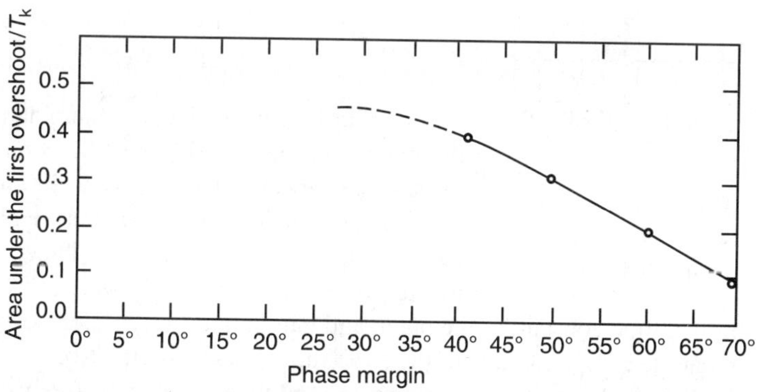

Fig. 6.35(d) *Area under the first overshoot as a function of phase margin*

plant transfer function has only one large time constant to be compensated and one small time constant to be uncompensated. A PI controller to modify the performance will have a time constant the same as a large time constant of the plant. The integration time of the controller T_0 however is at the designer's choice. The transfer function of the controller is

$$G_c(j\omega) = \frac{(1 + j\omega T_e)}{j\omega T_0} T_0 < T_c \tag{6.109}$$

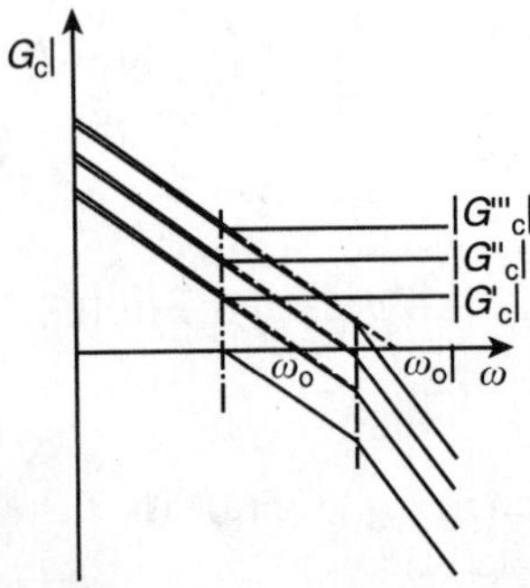

Fig. 6.36 *PI controller for compensation*

The value of T_0 can be chosen freely. Its effect on the magnitude plot of frequency response is similar to the effect gain. Thus change in value of T_0 can be regarded as change in gain and the magnitude plot moves along ordinate. For smaller values of T_0 it moves upward, increasing the value of gain cross-over frequency. The typical amplitude plots of a PI controller used to improve the performance of the plant having one large time constant, are shown in Fig. 6.36. It clearly shows that the cross-over frequency depends upon T_0 and hence K_0. The relationship between ω_c/ω_0 and K_0 is shown in Fig. 6.37.

If the plant has two large time constants and one small one, the two large time constants are compensated by a PID controller. In this case also the integration time of the controller is freely chosen. As far as the effect of T_0 on the magnitude plots is concerned it is same as in the case of PI controller. The magnitude plot of a PID controller having a transfer function given by

$$G_c = \frac{(1 + j\omega T_{c1})(1 + j\omega T_{c2})}{j\omega T_0} \tag{6.110}$$

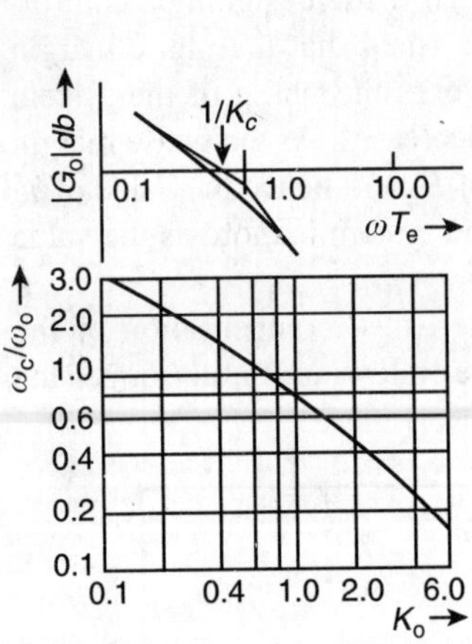

Fig. 6.37 *Dependence of cross-over and corner frequencies on* K_0

is shown in Fig. 6.38. A close examination of the magnitude plots of PI anc PID controllers reveals that they are almost similar. The controller transfer function should have a magnitude plot such that it has a slope of 20 db/decade up to the corner frequency $1/T_e$, where T_e is the uncompensated time constant of the plant, and later it must have a slope of 40 db/decade so that the compensations of the two time constants is effective.

Therefore, while designing a controller its amplitude plot may be freely chosen to have the necessary gain cross-over frequency. This can be achieved by shifting the amplitude plots along the y-axis until the gain cross-over occurs above the required phase angle (decided by phase margin). The phase angle characteristic is the resultant of the phase angle characteristics of the plant and the controller. The proper value of T_0 is arrived at. The resultant open loop transfer function of the modified system can be determined. There will be a small difference between the value of K_0 determined using the asymptotic plots and the actual calculation. If the transfer function of the plant has a gain of K_0 the controller will have a integration time KT_0.

To summarize:

i. The transfer function of the plant to be modified is given by

$$G_p(s) = \frac{V_s}{(1+sT_e)\underset{n}{\pi}(1+sT_{pn})} \tag{6.111}$$

where $T_e \ll T_{pn}$, $n = 1, 2, \ldots$

T_e is the sum of several parasitic time constants of the system.

ii. The controller transfer function is chosen to compensate all T_{pn}. Therefore with respect to these time constants it has a form

$$G_c(s) = \frac{\overset{n}{\pi}(1+sT_{pn})}{V_s s T_e K_0} \tag{6.112}$$

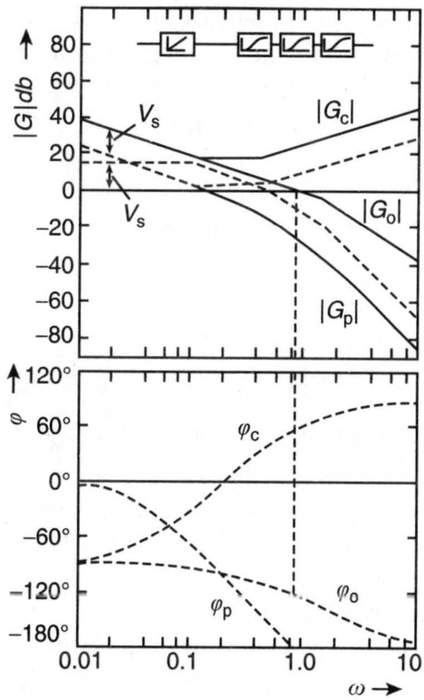

$$G_c = \frac{(1+j\omega 10)\,(1+j2.5\omega)}{|V_s|\,(j\omega/0.87)}$$

$$G_p = \frac{|V_s|}{(1+j10\omega)\,(1+j2.5\omega)\,(1+j\omega/1.5)}$$

$$G_o = G_s\,G_p = \frac{1}{(j\omega/0.87)\,(1+j\omega/1.5)}$$

Fig. 6.38 *Use of PID controller for compensation. Magnitude and phase plots*

The constant K_0 is decided so that the resultant open loop transfer function

$$G_0(s) = \frac{1}{sT_e K_0(1 + sT_e)} \tag{6.113}$$

has the specified phase margin. The closed loop transfer function has a time response of the second order system, having a damping ratio of $\sqrt{K_0}/2$.

6.9.6 System Performance When the Controller Transfer Function Does not Match the Plant

While a controller is designed to improve the behaviour of a given control system, in practice it is very difficult to realise such a controller because of the availability of components. The actual time constant of the controller may deviate from the theoretically designed value. Also the operating conditions may cause a change in plant time constants or even controller time constants. Thus the performance of the system may deviate from the desired or expected one. This incorrect matching may cause differences in the dynamic performance also.

The large time constants of the original system are normally compensated by the time constants of the controller. In other words, the effect of the dominant poles of the plant transfer function is nullified by the zeros of the controller transfer function. This compensation may not be to the desired extent when the actual time constants of the plant vary during operation or the time constants of the controller deviate from the desired ones, for one reason or other.

We shall discuss in the following how the performance of the resulting system gets altered in case of incorrect matching, following a variation in the time constants. We assume that one of the large time constants of the plant varies by a factor K_e leading to the incorrect matching. The transfer function is

$$G_p'(j\omega) = \frac{V_s}{(1 + j\omega T_{p1} K_e)(1 + j\omega T_{p2})(1 + j\omega T_e)} \tag{6.114}$$

which can be written as

$$G_p'(j\omega) = G_p(j\omega)\frac{1 + j\omega T_{p1}}{1 + j\omega T_{p1} K_e} \tag{6.115}$$

This means that the actual transfer function is the desired transfer function multiplied by a correction factor which is the ratio of $(1 + j\omega T_{p1})/(1 + j\omega T_{pl} K_e)$. The magnitude plot of this ratio depicted in Fig. 6.39 clearly shows that the gain cross-over frequency is above the corner frequency $1/T_{p1}$, $1/T_{p2}$. The time response is very much decided by the cross-over frequency. Therefore, the choice of the cross-over frequency plays a significant role in the design of the controller. The frequencies $\omega T_{pl} > 1$ need be considered. In this frequency range the factor $1 + j\omega T_{pl}/1 + j\omega T_{pl} K_e$ can be simplified as $1/K_e$. The transfer function $G_p'(j\omega)$ simplifies to

$$G_p'(j\omega) = G_p(j\omega)/K_e$$

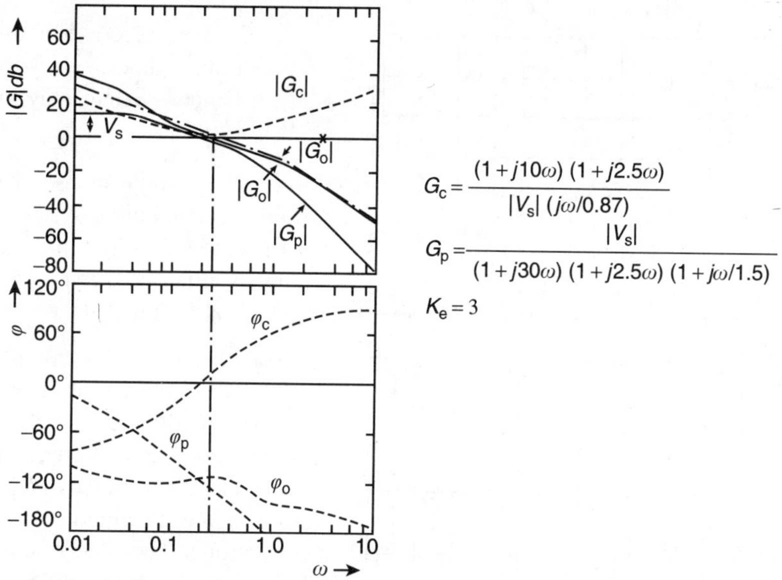

Fig. 6.39 *Mismatching $K_e > 1$*

The open loop transfer function of the system is

$$G_0'(j\omega) = \frac{1}{j\omega K_e T_0 (1 + j\omega T_e)} \tag{6.116}$$

Using the ratio $K_0 = T_0/T_e$ this can be written as

$$G_0'(j\omega) = \frac{1}{j\omega K_e K_0 T_e (1 + j\omega T_e)} \tag{6.117}$$

From Eq. 6.117 it can be seen that the ratio K_0 (ratio of integration time of the controller to the uncompensated time constant of the plant) is effectively multiplied by the factor K_e. Thus the change in the time constant is equivalent to a change in this ratio.

***Case (i)* $K_e > 1$** For this case of $K_e > 1$, the time constant of the plant increases by this factor. This is equivalent to increase in the ratio K_0 to value $K_0' = K_0 K_e$. This is equivalent to increase in the integration of the controller. The Bode plots of the functions of the actual and desired transfer functions are shown in Fig. 6.40(a). From the Bode Plots the effects of increase in the time constants can be observed to be the following:

i. Following an increase in the K_0 the Bode plot moves to the left. This results in smaller values of gain cross-over frequency.
ii. The phase of the system at the gain cross-over decreases and hence phase margin increases.
iii. The value of K_0 as well as the phase margin are related to damping of the system. Increased phase margin results in an increased damping or directly damping $= \sqrt{K_0'/2}$.

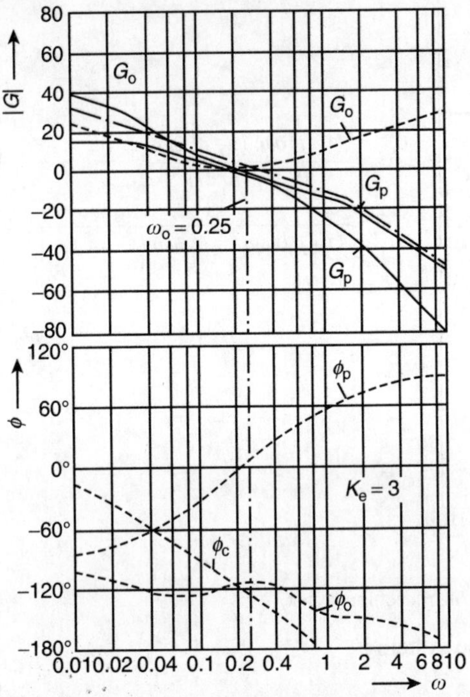

Fig. 6.40(a) *Mismatching $K_e > 1$*

iv. Peak overshoot decreases. Some times an over-damped system may also result.

A typical time response of the resulting second order system for $K_e > 1$ is shown in Fig. 6.40(b). For comparison the time response for $K_e = 1$, is also shown in the figure.

Case (ii) $K_e < 1$ For this case of $K_e < 1$ the time constant of the plant decreases by the factor K_e. In this case the situation is different and it is equivalent to decrease in K_0 and hence in the integration time of the controller. The effect of this on Bode plots shown in Fig. 6.41(a) can be seen to be the following:

i. The amplitude plots of frequency response move towards the right following

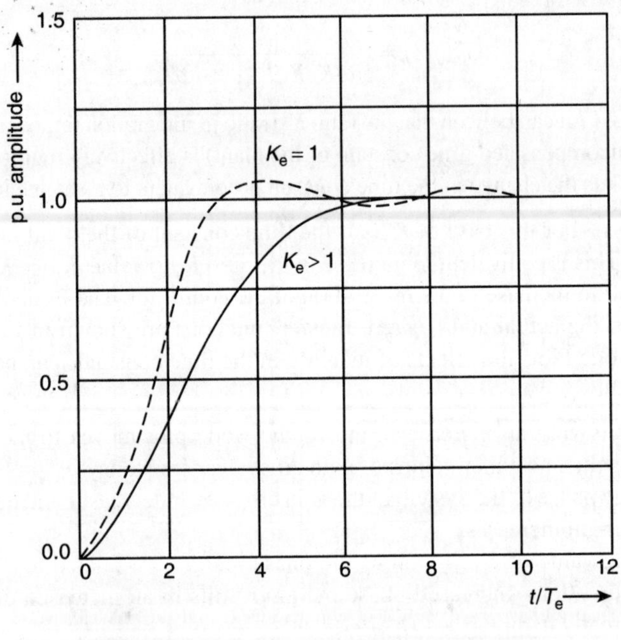

Fig. 6.40(b) *Typical transient response $K_e > 1$*

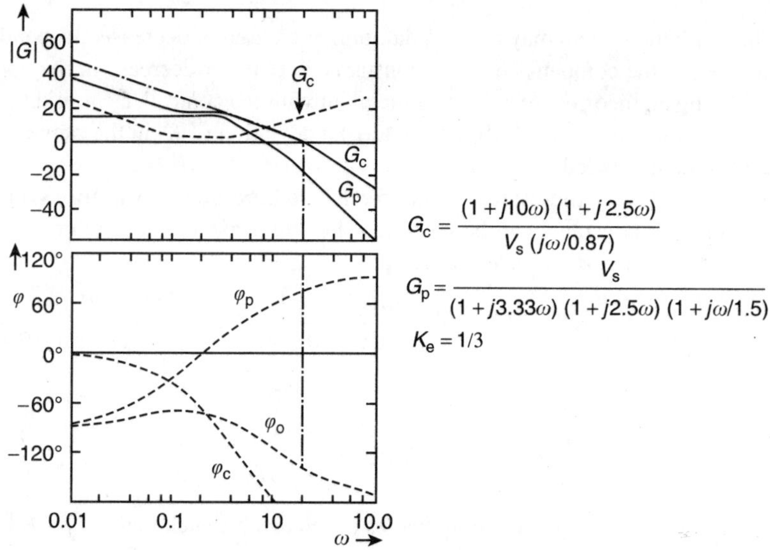

Fig. 6.41(a) *Mismatching* $K_e < 1$

a decrease in K0. The gain cross-over for this case occurs at higher frequencies. The gain cross-over frequency increases.

ii. The phase of the transfer function at gain cross-over increases. The phase margin, therefore, decreases.

iii. The damping of the system decreases, resulting in an oscillatory response.

iv. The peak overshoot is greater than that of the desired one.

A typical time response for this case is depicted in Fig. 6.41(b).

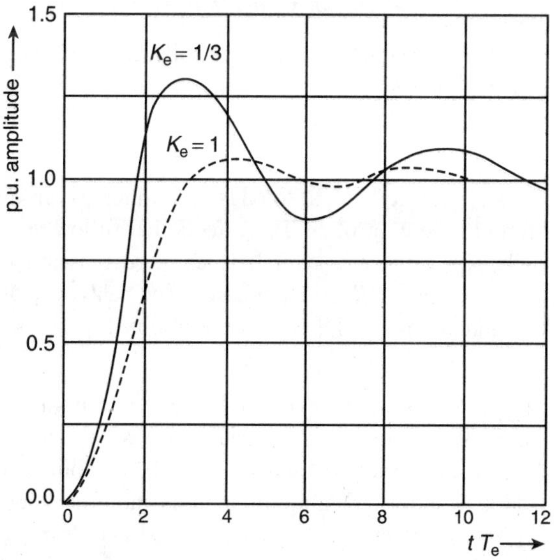

Fig. 6.41(b) *Typical transient response of second order system* $K_e < 1$

Therefore the system may have its damping increased or decreased, depending upon whether the compensated time constants increase or decrease in the operation, causing an incorrect matching of the plant with the controller. From the time responses shown in Figs 6.40(b) and 6.41(b) a deviation of K_e in the range $0.5 < K_e < 2.0$ can be allowed.

In a similar way the effect of incorrect matching due to variations in the controller time constants can be modified. For this case the controller transfer function due to incorrect matching is

$$G_c'(j\omega) = \frac{(1 + j\omega K_c T_{p1})(1 + j\omega T_{p2})}{V_s j\omega T_0} \tag{6.118}$$

which can be written as

$$G_c'(j\omega) = G_c(j\omega)\frac{1 + j\omega K_{c1} T_{p1}}{1 + j\omega T_{p1}} \tag{6.119}$$

By the same argument that the frequencies of interest are such that $\omega T_{p1} \gg 1$ the correction factor $(1 + j\omega K_c T_{pl})/(1 + j\omega T_{pl})$ can be replaced by K_e. Therefore

$$G_c'(j\omega) = G_c(j\omega) \quad K_e$$

The open loop transfer function of the system is

$$G_0'(j\omega) = \frac{K_f}{(1 + j\omega T_e) j\omega K_0 T_e} \tag{6.120}$$

which can be written as

$$G_0'(j\omega) = \frac{1}{j\omega (K_0/K_f) T_e (1 + j\omega T_e)} \tag{6.121}$$

Thus the change in the time constant of the controller may be studied by an equivalent change in the value of K_0.

A study shows that the effect of reduction in the controller time constant has the same effect as an increase in plant time constant.

Case (i) $K_f < 1$ This case is equivalent to causing an increase in K_0 or increase in the integration time of the controller. The effects that follow are the same those caused for $K_e > 1$. The gain cross-over frequency decreases and phase margin increases. The damping $(d = \sqrt{K_0/2})$ increases. The peak overshoot decreases if the system is still underdamped. There is a possibility for the system to become overdamped (Fig. 6.42).

Case (ii) $K_f > 1$ The effect is equivalent to a decrease in the integration time. The effects for the case are the same as those caused for $K_e < 1$, viz., the gain cross-over frequency increases, phase margin decreases, the damping decreases causing an increase in peak overshoot. The magnitude plots and time responses are shown in Fig. 6.43 for different values of K_0. These responses confirm the above conclusions. A very low value of K_0 results in a very lightly damped system. Allowable variation K_0 is in the range $1 - 1.5 (d = 0.5)$.

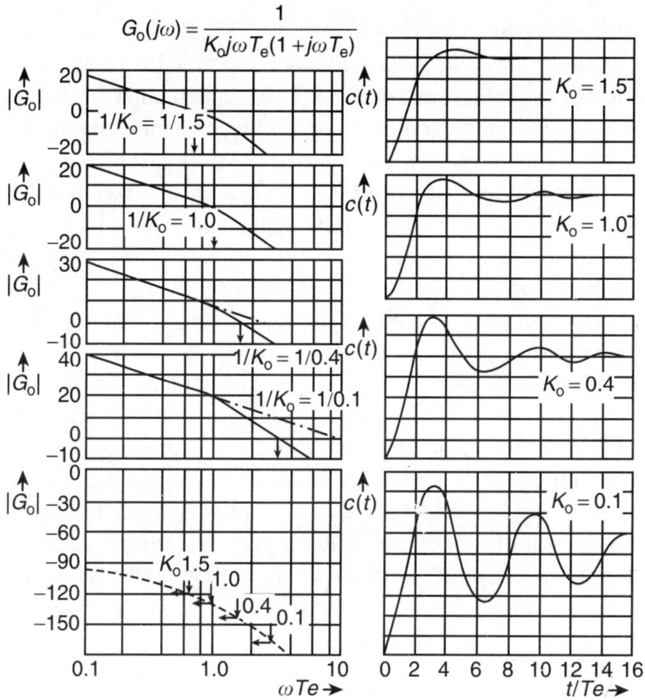

Fig. 6.42 *Magnitude plots and time response $K_f > 1$, $K_f < 1$*

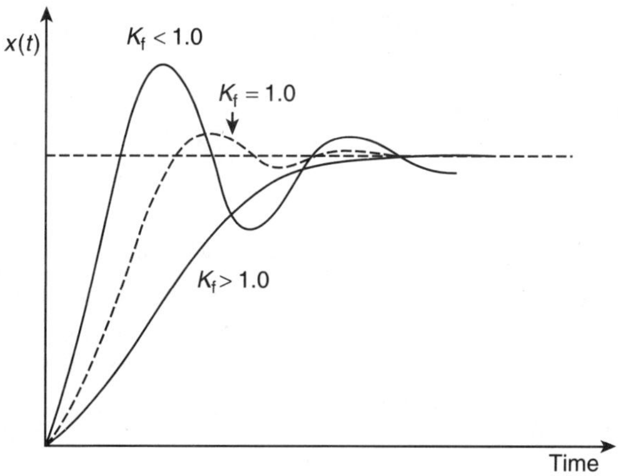

Fig. 6.43 *Time responses for K_f representing the mismatch of the controller*

From the above discussion it is clear that the effect of incorrect matching due to increase in the time constants compensated is the same as those due to decrease in the compensating constant of the controller.

6.9.7 Design of Controllers for Linearly Varying Inputs

The design of controllers discussed in the foregoing sections are for the step input. The performance has been found to depend very much upon the value of integration time in relation to the uncompensated time constant of the plant. A control system is normally employed in situations where the controlled quantity (output) has to follow its reference value (input). If the changes in the input are very fast a controller having a small value of K_e (and hence small value of T_0) is prohibited, as this results in a lightly damped system with a large overshoot. The oscillatory behaviour does not provide the necessary requirement if the input changes very fast. Also in some drive applications, the input is not a fast changing quantity. The effect of varying inputs must be analysed in detail, so that their variation can be taken into consideration in the design. The dynamic behaviour of the load influences the operation of the controller. The dynamic performance of load is, however, taken to be very fast. From the above considerations, the rate of change of the input is limited for the system having a poor damping, i.e., very steep variation of the input is prohibited. A constant variation of the input to the controllers is achieved by means of limiters. A limiter is provided before the controller to see that the input to the controller does not change very fast but provides a suitable rate of change. The block diagram of such an arrangement is shown in Fig. 6.44. The transfer relation of the limiter is

$$\begin{aligned}\frac{\Delta x}{\Delta R} &= t/T_w \quad 0 < t < T_w \\ &= 0 \quad \text{for } t > T_w\end{aligned} \tag{6.122}$$

The transfer function is

$$\frac{\Delta x(s)}{\Delta R(s)} = 1/s^2 T_w \tag{6.123}$$

The transfer function of the total system is

$$\begin{aligned}\frac{C(s)}{R(s)} &= \frac{1}{s^2 T_w (1 + sK_0T_e + s^2K_0T_e^2)} \qquad \text{for } 0 < t < T_w \\ &= \frac{1}{(1 + sK_0T_e + s^2K_0T_e^2)} \qquad \text{for } t > T_w\end{aligned} \tag{6.124}$$

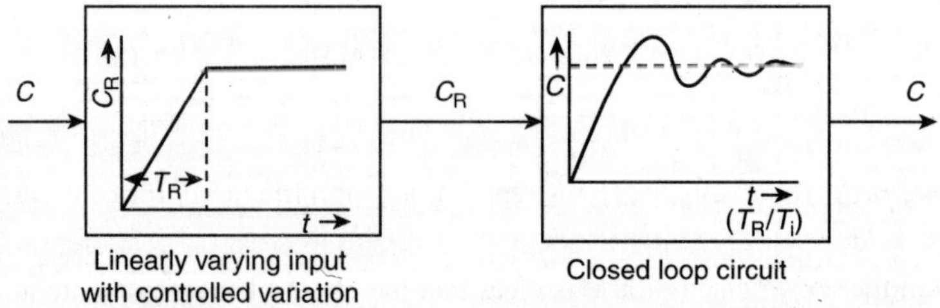

Fig. 6.44 *Limiting the variation on the reference value (Block diagram)*

Using the convolution integral the output of the system is

$$\frac{C(t)}{R(t)} = \frac{T_e}{T_w}\left(\frac{T}{T_e} + \frac{1}{\sqrt{1-(K_0/4)}}\sin\sqrt{1-(K_0/4)}\,\frac{t}{t_e}\,e^{-\frac{\sqrt{K_0}}{2}\frac{t}{t_w}}\right)$$

$$\text{for}\quad t < T_w \tag{6.125}$$

and

$$\frac{C(t)}{R(t)} = 1 + \frac{T_e}{T_w}\left(\frac{1}{\sqrt{1-\frac{K_0}{4}}}\sin\sqrt{1-\frac{K_0}{4}}\,\frac{t}{t_e}\,e^{-\frac{\sqrt{K_0}}{2}\frac{t}{T_w}}\right) \tag{6.126}$$

The damped periodic transient term of the response depends upon T_e/T_w. From the expression it can be seen that the transient term is absent if the input value changes very slowly. In this case the controlled quantity follows the input very closely. The oscillatory behaviour of the system is completely eliminated. The system can be designed for low damping if the input changes slowly. Thus, a lightly damped system can be made to have a linearly changing input to the controller. The controller thus designed has a lower phase margin, is generally sensitive to errors and is always in stability limits. The time responses of a typical system for step as well as linearly varying inputs are shown in Fig. 6.45. The effect of variation of T_w (in relation to T_e) is very clear from the figure. A proper choice of T_w controls or even eliminates the overshoot and the output is made to follow the input. If the variation of the input is slow the transient conditions in the system are the least.

6.9.8 Exponential Variation of the Input to the Controller

The linear variation of the input to the Controller, discussed in the foregoing section, is too involved to achieve by means of circuits using discrete elements. On the other hand a simple RC circuit connected at the input of the controller provides an input which varies exponentially. This RC circuit functions as a limiter having a time constant T_c. The cascading of this with the modified system, designed using the above criteria, is shown in Fig. 6.46 by a block diagram.

The closed loop transfer function of the total system is

$$G(j\omega) = \frac{1}{(1 + j\omega T_c)(1 + j\omega K_0 T_e + K_0 j\omega T_e^2)} \tag{6.127}$$

The magnitude plots of this function drawn for a given value of K_0 and different value of T_c are shown in Fig. 6.47(a). For comparison, the plot for RC = 0 is also shown which corresponds to a step input. The value of K_0 is chosen to provide a lightly damped system. So, the magnitude plot shows a resonant peak which

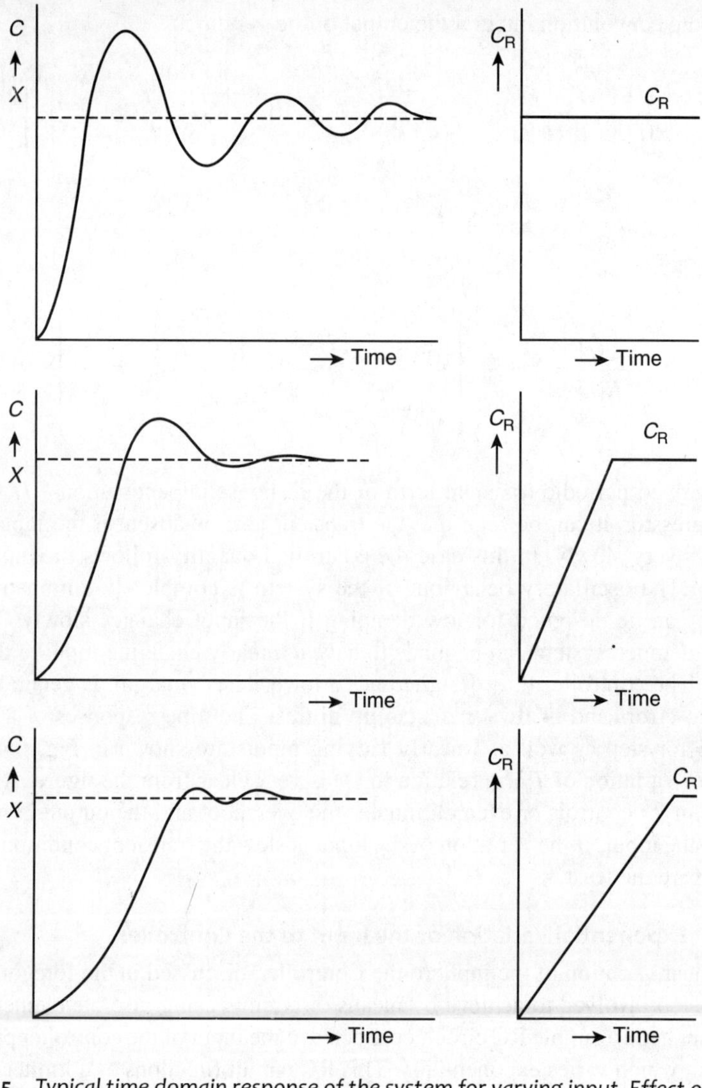

Fig. 6.45 *Typical time domain response of the system for varying input. Effect of the shape of the input on the response*

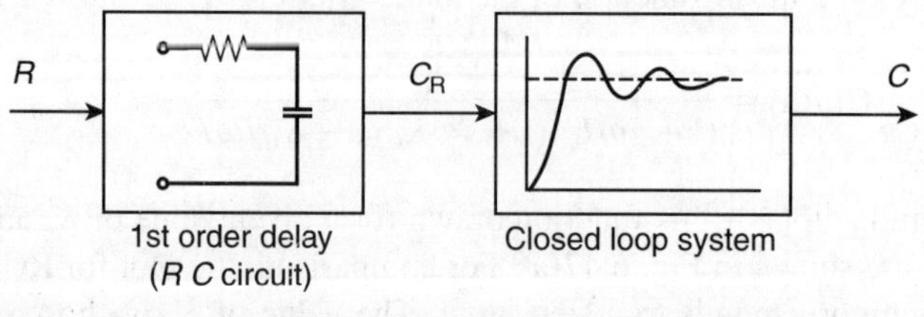

Fig. 6.46(a) *Cascading of RC network to the system to delay the input variation*

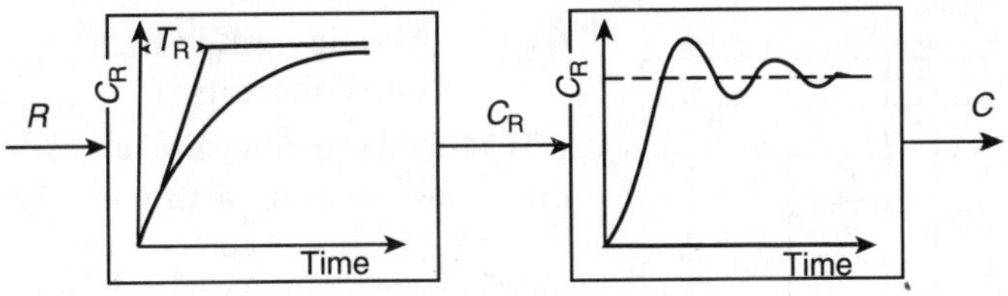

Fig. 6.46(b) *Experimental variation of the input to the system by cascaded RC network*

is representative of overshoot in the transient response. As $T_c > 0$ the resonant peak disappears and the slope beyond the corner frequency increases. The time response is similar to that having increased damping. The time responses of the system for different T_c, shown in Fig. 6.47(b), support the conclusion. The overshoot as a function of K_0, T_c being the parameter is depicted in Fig. 6.47(c). The offer helping tools in the suitable choice of T_c for given K_0 or vice versa. For a K_0 of 0.8 the value of T_c can be in the range of 1.6 to 1.3.

Hence from the above discussion it may be concluded that the transient behaviour of a lightly damped system can be improved by having a variation in the input to the system. The variation can be linear or exponential. The latter is easier to realise with simple RC circuits. The time of variation of the input quantity, as decided by the limiter in the former and by the time constant T_c in the latter, influence the transient response very much.

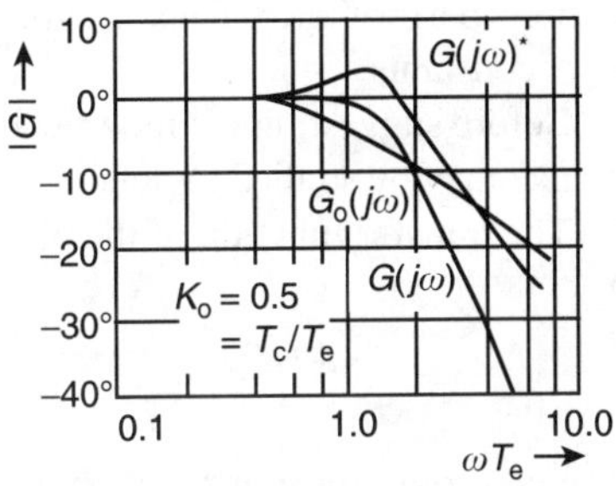

Fig. 6.47(a) *Magnitude plot of the control circuit*

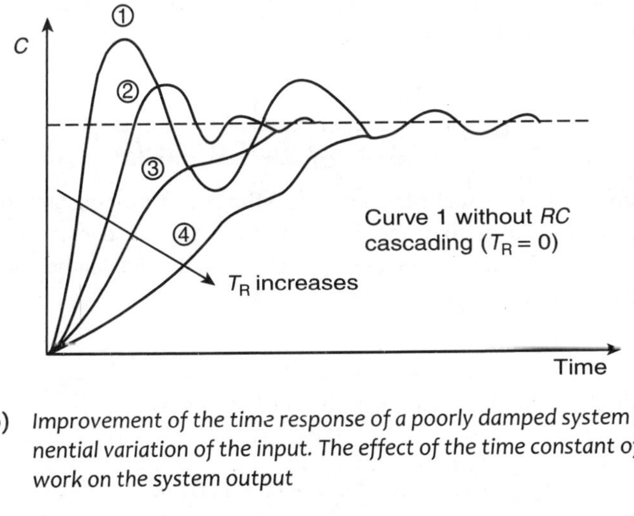

Fig. 6.47(b) *Improvement of the time response of a poorly damped system by exponential variation of the input. The effect of the time constant of RC network on the system output*

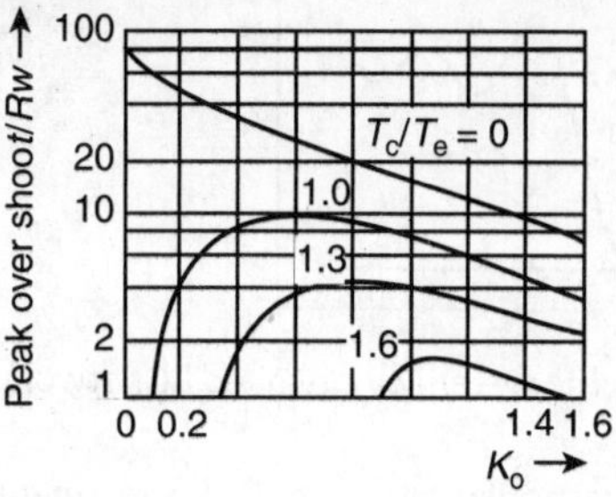

Fig. 6.47(c) *Peak overshoot with delayed input*

6.9.9 Uncompensated Large Time Constants

It is possible to compensate only one large time constant using a PI controller. A PID controller is used to compensate for two large time constants. On technical grounds, compensation of more than two constants using a PID controller is not possible, as only simple differentiation can be performed.

Therefore if a transfer function has more than two dominant poles, only the effect of two poles can be compensated and the rest of the dominating poles do not assure desired performance.

It is therefore necessary to suggest modifications to the controller design, which takes care of this uncompensated large time constant and still provides a desired performance. This section deals with considering such uncompensated time constants.

Before suggesting the modifications the effect of these constants on the performance may be studied in detail.

The transfer function of the plant requiring the compensation is

$$G_p(j\omega) = \frac{V_s}{(1 + j\omega T_e)(1 + j\omega T_1)(1 + j\omega T_2)(1 + j\omega T_3)} \tag{6.128}$$

The controller transfer function is

$$G_c(j\omega) = \frac{(1 + j\omega T_1)(1 + j\omega T_2)}{V_s j\omega T_0} \tag{6.129}$$

Hence in the simplest case the time constant T_3 cannot be compensated by the controller. The open loop transfer function of the cascaded plant and controller is

$$G_0(j\omega) = \frac{1}{(1 + j\omega T_e) j\omega T_0 (1 + j\omega T_3)} \tag{6.130}$$

The time constant T_3 would be the least of all the large time constants of the plant. T_e has already been taken to be the sum of all parasitic time constants of the plant. The time constant T_3 can also be considered one such, and can be added on to T_e so that the new value of T_e is

$$T_e' = (T_e + T_3) \tag{6.131}$$

with this modification the open loop transfer function is

$$G_0(j\omega) = \frac{1}{(1 + j\omega(T_e + T_3)) j\omega T_0} \tag{6.132}$$

The ratio K_e would now change to K'_e where

$$K'_e = \frac{T_0}{T_e + T_3} \tag{6.133}$$

Substituting these relations in Eq. 6.132 we have

$$G_0(j\omega) = \frac{1}{j\omega(T_e + T_3)K'_e(1 + j\omega(T_e + T_3))} \tag{6.134}$$

The value of K'_e is less than that of K_e. The effect of T_3 as a parasitic time constant adding on to T_e is to effectively decrease the ratio K_e to K'_e. The effects of decreasing the value of K_e have already been discussed. The effects due to T_3 are as follows:

i. The gain cross-over frequency increases.
ii. The phase margin decreases. The damping of the system is less than the desired one.
iii. The peak overshoot will be more than the desired one.

The investigation of these effects as caused by the uncompensated time constant may be done assuming that $T_3/T_e = g$ and $j\omega T_e = q$. The open loop transfer function given by Eq. 6.130 may be written as

$$G_0(j\omega) = \frac{1}{(1 + gq)K_e(g)(1 + q)} \tag{6.135}$$

The block diagram of the system with uncompensated time constant T_3 is shown in Fig. 6.48. The Bode plots (magnitude and phase plots) of this function are shown in Fig. 6.49 for different values of g. Assuming a given value for T_0, the integration time of the controller, and the controller constant, the effect of T_3 can be seen to be to decrease the phase margin. For this case the gain cross-over frequency remains the same and phase margin decreases as shown by the vertical line for different values of g $(= T_3/T_e)$. This falls in line with the effects of change in the value of K_e to K'_e, listed above. Therefore the effect of T_3 can be considered as a change in T_e to T'_e. The value of $T'_e = T_e(1 + g)$. Using this the open loop transfer function can be written as

$$G_0(q) = \frac{1}{K_{e0}(1 + g)q(1 + (1 + g)q)} \tag{6.136}$$

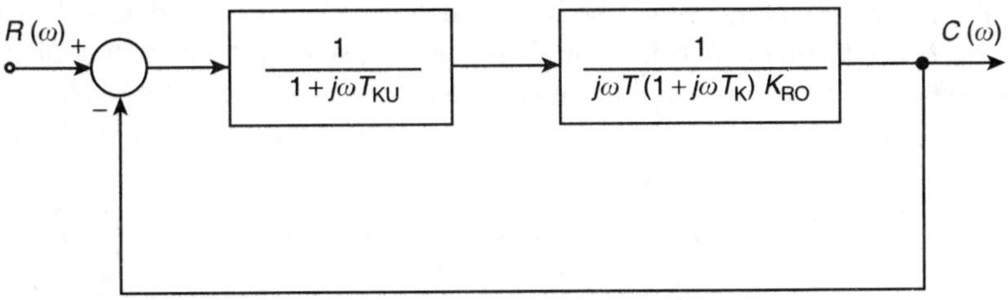

Fig. 6.48 *Block diagram of a system having an uncompensated time constant*

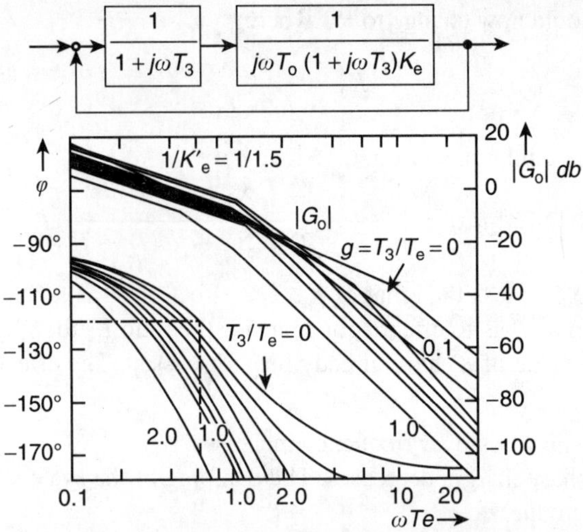

Fig. 6.49 *Magnitude plots q as a parameter*

The actual transfer function given by Eq. 6.135 can be approximated by the transfer function of Eq. 6.136.

For the damping to be the same as the desired one the phase margin must be maintained at the value corresponding to $q = 0$. The Bode plots are drawn so that the phase and magnitude plots for $q \neq 0$ are such that the phase margin is maintained at the value corresponding to $g = 0$. The effect of this is to shift the magnitude plots towards the left (Fig. 6.49), i.e., the gain cross-over frequency decreases. This amounts, effectively, to increasing the controller constant to maintain the value of phase margin at a reduced value of gain cross-over frequency. This can be achieved by increasing the integration time of the controller. If on the other hand the integration time does not change, the phase margin decreases, causing reduced damping and increased peak overshoot. Hence the performance is poorer than the desired one.

Therefore the controller designed to compensate only two large time constants does not give satisfactory performance. To get the desired performance its design must be modified. The above discussion gives the direction in which the modification may be oriented. As the damping is decided by the phase margin, the various constants of the controller must be chosen such that phase margin does not change when T_3 is taken into consideration. This can be easily achieved by introducing a correction by which the phase of the function changes, bringing the phase margin to the desired value. Using this factor the open loop transfer function of Eq. 6.136 is given by

$$G_0(q) = \frac{1}{K_g K_{e0}(1+g)q(1+(1+g)q)} \tag{6.137}$$

It is equivalent to changing to the value of gain cross-over frequency to $1/(K_g K_{e0}(1 + g))$. The value of K_g can be chosen such that the magnitude and phase plots for the value of g are such that the phase margin of the transfer function is same as that of the transfer function for $g = 0$. A relation can be derived between g and K_g to achieve the desired phase margin. Typical curves are shown in Fig. 6.50 with the phase margin as a parameter. This is equivalent to changing the integration time of the controller. The gain cross-over changes to $1/(K_g(1 + g)K_{e0})$.

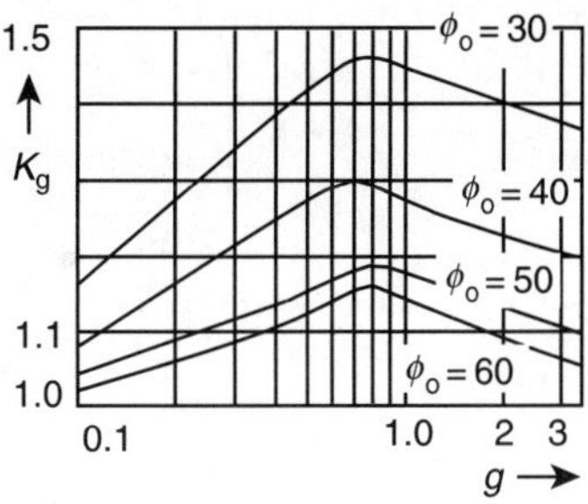

Fig. 6.50 *Relation between g and the correction factor K_g*

Typical time responses of the system are shown in Fig. 6.51. Curve 1 indicates the time response of the system with its uncompensated time constant. This shows a characteristic increase in peak overshoot due to T_3. Curve 2 indicates the time response of the system with the correction factor. Curve 3 indicates the desired time response, or time response with $T_3 = 0$. A close examination shows that the correction introduced improves the performance so that the peak overshoot is within the desired value. This is because the phase margin decides the damping and the overshoot very much and it is maintained constant by a proper correction. However the rise time changes. Therefore, while designing a controller for a system having three time constants of which two can be directly compensated, the integration time should be properly chosen so that the effect of the third time constant is minimal. The integration time of the controller is

$$T_0 = K_g(1 + g)K_{e0}T_k \tag{6.138}$$

and $$K_e = K_g(1 + g)K_{e0}$$

The value of K_g can be read from Fig. 6.50.

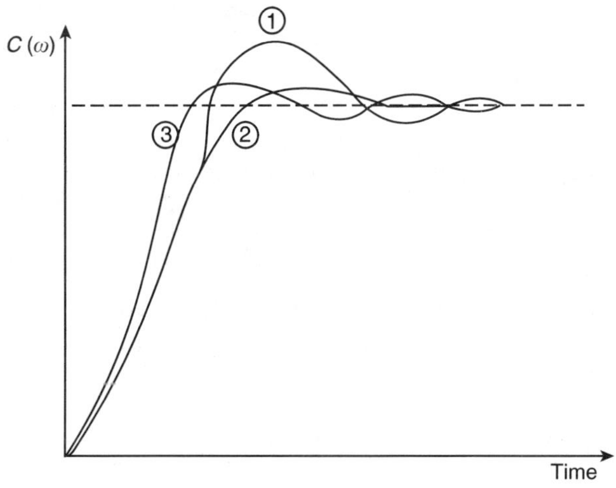

Fig. 6.51(a) *Typical time response of the system with uncompensated time constant (constant phase margin)*

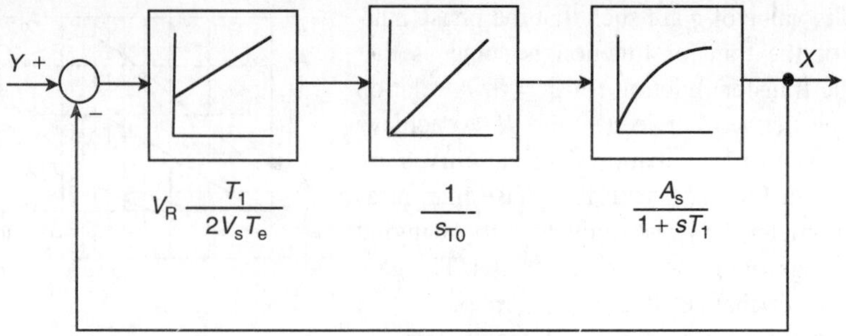

Fig. 6.51(b) *A typical control system with integrating element and compensation by PI controller*

6.9.10 Symmetrical Optimum

Sometimes automatic control systems contain integrating also besides first order delay elements, proportional elements and deadzones. Such a system when compensated on the basis of magnitude optimum discussed in the previous sections will become oscillatory with zero damping. As has already been explained the magnitude optimum utilises a PI controller to compensate the effects of largest time constant of the system. The other first order delays present are replaced by a single delay which is the sum of all the other uncompensated delays. The rise time of the system decreases. The amplification of the proportional element is

$$V_R = \frac{T_1}{2V_sT_e} \tag{6.139}$$

If this principle is employed to compensate the system with integrating element, the integrating element of the compensator combines with that of the system resulting in a double integration. The characteristic equation of the closed loop system contains only s^2 and constant terms showing that the system has no damping. The system has sustained oscillations.

This can be illustrated by considering a system having an integrator and a first order delay in cascade. A closed loop with PI controller is shown in Fig. 6.51.

The open loop transfer function of the system is

$$G_0(s) = V_R \frac{1+sT_n}{sT_n} \frac{1}{sT_0} \cdot \frac{V_s}{1+sT_1} \tag{6.140}$$

Using the principle of magnitude optimum $T_n = T_1$. Therefore,

$$G_0(s) = V_R V_s \frac{1}{s^2 T_0 T_1} \tag{6.141}$$

The closed loop transfer function is

$$G_c(s) = \frac{1}{s^2 T_0 T_1 + 1} \tag{6.142}$$

assuming $V_R V_s = 1$ (as a special case). The term containing s is not present in the characteristic equation which clearly shows that the damping is zero.

Also we know that

$$\frac{x(s)}{w(s)} = \frac{1}{s^2 T_0 T_1 + 1} \tag{6.143}$$

The differential equation representing the system is

$$T_0 T_1 \frac{d^2 x(t)}{dt^2} + x(t) = w(t) \tag{6.145}$$

For a step input the response of the system is given by

$$x(t) = e^{-j\frac{t}{\sqrt{T_0 T_1}}} + e^{j\frac{t}{\sqrt{T_0 T_1}}} \tag{6.146}$$

is representative of sustained oscillations.

Therefore the controller design to compensate the effect of the dominant poles of the system is based on the principle of symmetrical optimum. The frequency response of the compensated system is symmetrical about gain cross over frequency.

The criteria for this optimum are derived in the following. The system contains an integration time of T_0 and a first order delay of T_1 (Fig. 6.52). A PI controller is used to compensate the system. In magnitude optimum the criteria have been developed assuming that the amplitude of transfer function is unity in the complete range of frequencies. In the present case, the criteria for symmetrical optimum are developed so that the transfer function has an amplitude of unity over a restricted range of frequencies.

The open-loop transfer function of the compensated system

$$G_0(s) = V_R \frac{1 + sT_n}{sT_n} \frac{1}{sT_0} \frac{V_s}{1 + sT_1} \tag{6.147}$$

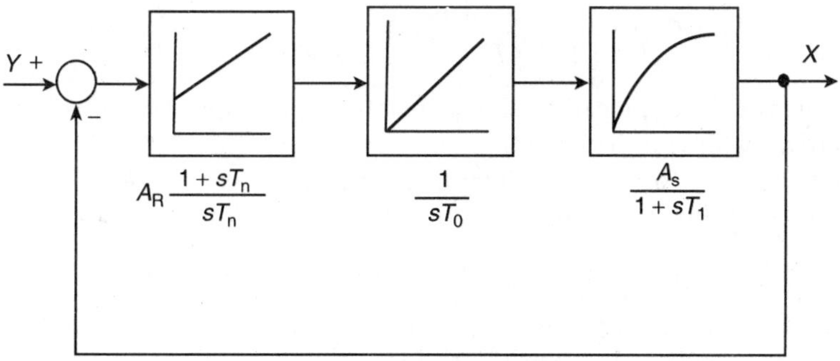

Fig. 6.52 *Pertaining to symmetrical optimum*

The closed loop transfer function is

$$G_c(s) = \frac{V_R V_s(1 + sT_n)}{V_R V_s + sV_R V_s T_n + s^2 T_n T_0 + s^3 T_n T_0 T_1} \tag{6.148}$$

The amplitude of frequency response

$$G_c(j\omega) = \Big[(V_R V_s)^2 (1 + T_n^2 \omega^2) / ((V_R V_s)^2 - \omega^2 (2V_R V_s T_n T_0 - (V_R V_s T_0)^2) + (\omega^4)(T_n^2 T_0^2 - (V_R V_s T_n^2 T_0 T_1)^2) + \omega^6 (T_n T_0 T_1)^2 \Big]^{\frac{1}{2}} \tag{6.149}$$

The system has apparently a damped response as the characteristic equation has all powers of s present and the coefficients are positive. The amplitude is unity at zero frequency. However, it approaches unity in a certain frequency range if the following conditions are satisfied.

$$\begin{aligned} (V_R V_s T_0)^2 &= 2V_R V_s T_n T_0 \\ T_n^2 T_0^2 &= 2V_R V_s T_n^2 T_0 T_1 \end{aligned} \tag{6.150}$$

These lead to a result that the reset time of the integrator.

$$T_n = 4T_1 \tag{6.151}$$

and the proportional amplification is

$$V_R = \frac{T_0}{2V_s T_1} \tag{6.152}$$

The integration time constant will always include the amplification of the integrator also. Therefore $\frac{T_0}{V_s}$ can be considered as T_i, the integration time.

Thus,

$$V_R = \frac{T_i}{2T_1} \tag{6.153}$$

The closed loop transfer function of the system is

$$G_c(s) = \frac{1 + s4T_1}{1 + s4T_1 + s^2 8T_1^2 + s^3 8T_1^3} \tag{6.154}$$

This is the general form of the transfer function of the compensated system based on symmetrical optimum.

The amplitude plot of the open loop transfer function of the compensated system is shown in Fig. 6.53. The plot shows a symmetry about the corner frequencies $\frac{1}{4T_1}$ and $\frac{1}{T_1}$ with respect to gain cross over frequency $\frac{1}{2T_1}$ on the $0-dB$ axis. This property justifies the name symmetrical optimum.

The system compensated by this principle has its behaviour decided by the time constant T_1 (the sum of all parasitic time constants).

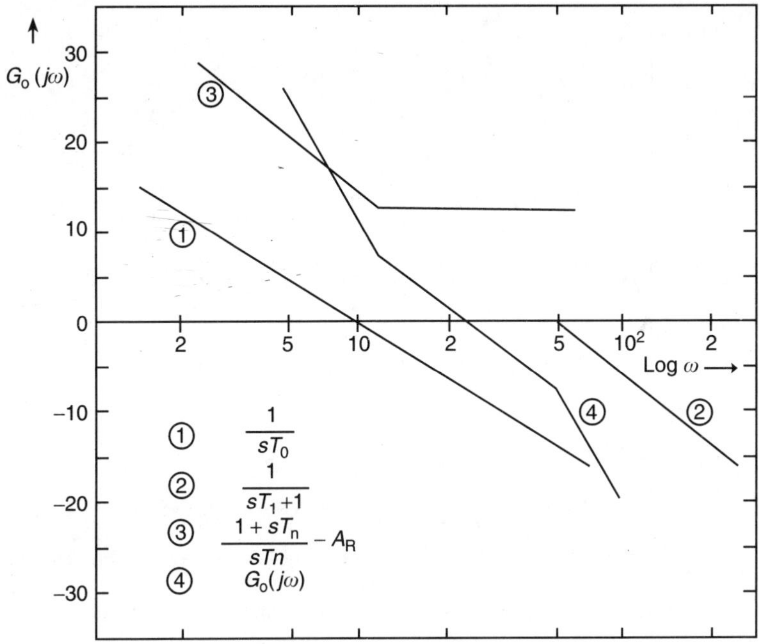

Fig. 6.53 *Amplitude plot of a system designed on the basis of symmetrical optimum*

The time response of the system is given by solving the differential equation

$$4T_1 \frac{d\omega_1(t)}{dt} + \omega(t) = x(t) + 4T_1 \frac{dx(t)}{dt} + 8T_1^2 \frac{d^2x(t)}{dt^2} + 8T_1^3 \frac{d^3x(t)}{dt^3} \tag{6.155}$$

For a step input the time response of the system is given by

$$\frac{x(t)}{\omega(t)} = 1 + e^{-t/2T_1} - 2e^{-t/4T_1} \cos \frac{\sqrt{3}t}{4T_1} \tag{6.156}$$

and it is shown in Fig. 6.54. It has a rise time of 3.1 T_1, a peak overshoot of 43.4%, and a settling time of $16.5T_1$.

This large overshoot of the system is due to the factor $1 + s4T_1$ in the numerator. If the reference input to the controller is delayed by a first order delay of transfer function $\frac{1}{(s4T_1 + 1)}$, the overshoot can be reduced to 8%. This is shown in Fig. 6.55. The rise time in this case increases to $7.6T_1$. The PI controller with delay in the reference input is shown in Fig. 6.56.

If the system contains delay elements such that the largest time constant is greater than the four times the sum of all other small time constants this method can be employed for compensation. If T_1, T_2, T_3, T_4 … are time constants of the system such that

$$T_1 \geq 4(T_2 + T_3 + T_4 \ldots) \tag{6.157}$$

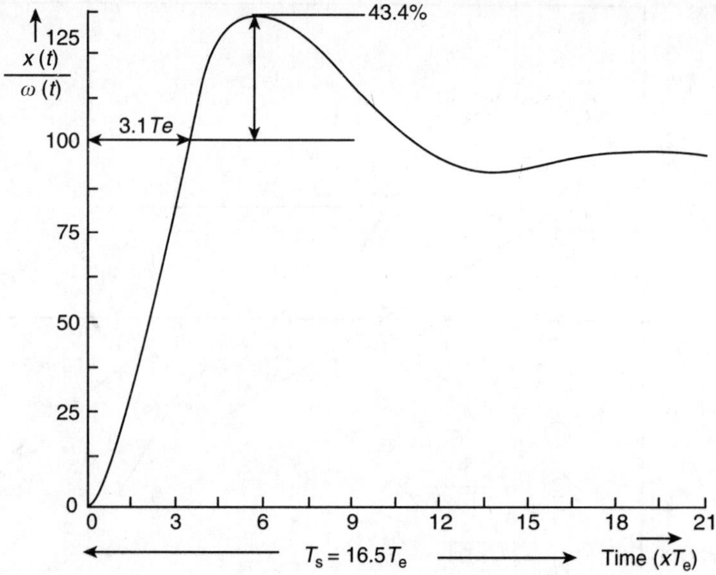

Fig. 6.54 *Time response of a symmetrically optimised system*

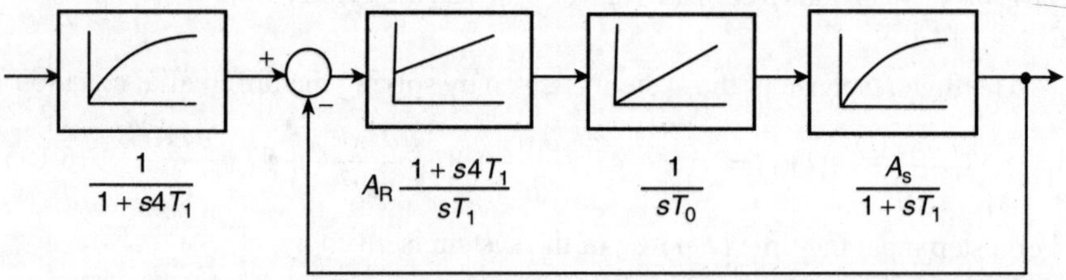

Fig. 6.55 *Use of fiirst order filter in the reference circuit to reduce the peak overshoot*

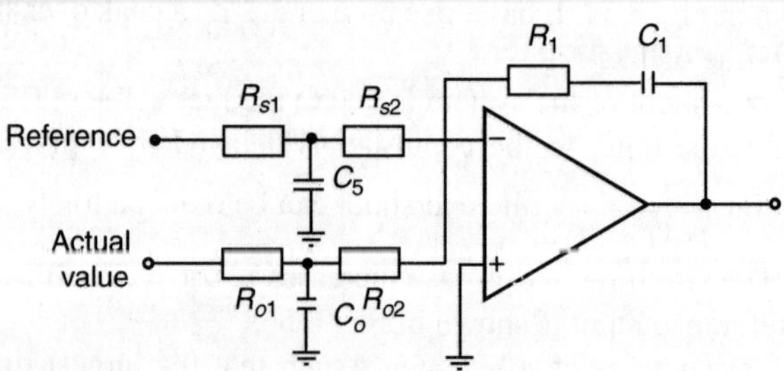

Fig. 6.56 *PI controller with a filter in the reference value*

the first order delay with time constant T_1 can be considered an integration in the system compared to the other time constants.

Worked Examples

6.1 *A Ward Leonard System has the following parameters:*

Generator: Armature resistance = 0.4 Ω
Armature inductance = negligible
Field resistance = 80 Ω
Field inductance = 50 H
Rotational inductance = 1.0 H
Speed is 120 rad/s

Motor: Field current 2.2 A, Armature resistance r_a = 0.7 Ω, armature inductance is negligible.
Rotational inductance = 0.5 H, Moment of inertia 2 kg m²

Determine expressions for the motor speed ω_γ and armature current i_a following an application of 50 V to the generator field winding with the motor initially at rest and unloaded. Neglect rotational losses.

Solution The dynamic behaviour is mainly decided by the speed voltage of the generator and motor, The armature current is given by $(e_g - e_m)/(r_g + r_m)$.
The rotational voltage of the generator

$$e_g = M_g i_f \omega_g = 120 i_f$$

The rotation voltage of the motor

$$e_m = M_m i_f \omega_m = 1.1\,\omega_m$$

$$i_a = (120 i_f - 1.1\omega_m)/(1.1) = \frac{120}{1.1} i_f - \omega_m.$$

The torque is $J\dfrac{d\omega_m}{dt} = M_m i_{fm} i_a$

$$i_a = \frac{J}{M_m i_{fm}} \frac{d\omega_m}{dt} = \frac{2}{0.5 \times 2.2} P\omega_m$$

$$= \frac{2}{1.1} P\omega_m$$

$$= 1.82 P\omega_m$$

Substituting for i_a in equation for i_a

$$1.82 P\omega_m = 109.1 i_f - \omega_m$$

or

$$109.1 i_f = (1 + 1.82P)\omega_m$$

But $$i_f = \frac{V_f}{R_f + L_f p} = \frac{V_f / L_f}{p + R_f / L_f}$$

$$\omega_m(s) = \frac{109.1 \times 50/50 \times 1.82}{s(s+1.6)(s+0.549)} = \frac{60}{s(s+1.6)(s+0.55)}$$

$$\omega_m = A + Be^{-1.6t} + Ce^{-0.55t}$$

$$A = \frac{60}{1.6 \times 0.55} = 68.2 \quad C = \frac{60}{-0.55 \times 1.05}$$

$$B = \frac{60}{+1.6 \times 1.05} = 35.71 = -103.9$$

$$\omega_m = 68.2 + 35.71e^{-1.6t} - 103.9e^{-0.55t}$$

$$i_a = 1.5(-1.6 \times 35.7e^{-1.6t} + 103.4 \times 0.55e^{-0.55t})$$

$$= 85.305e^{-0.55t} - 85.68e^{-1.6t}$$

The current is zero at $t = 0$.

6.2 *The forward loop transfer function of an induction motor whose speed is controlled by a voltage regulator is*

$$G(s) = \frac{K}{s(1 + s\tau)}$$

Typical values of τ and K are 0.20 and 15. Draw the Bode diagrams showing Amplitude vs ω and phase angle vs ω. Determine the cross-over frequency for $|G(j\omega|$. How can this frequency be increased?

Solution The block diagram of the system is shown in Fig. E6.2(a) The transfer function is

$$G(s) = \frac{K}{s(1 + s\tau)}$$

The frequency response is represented by

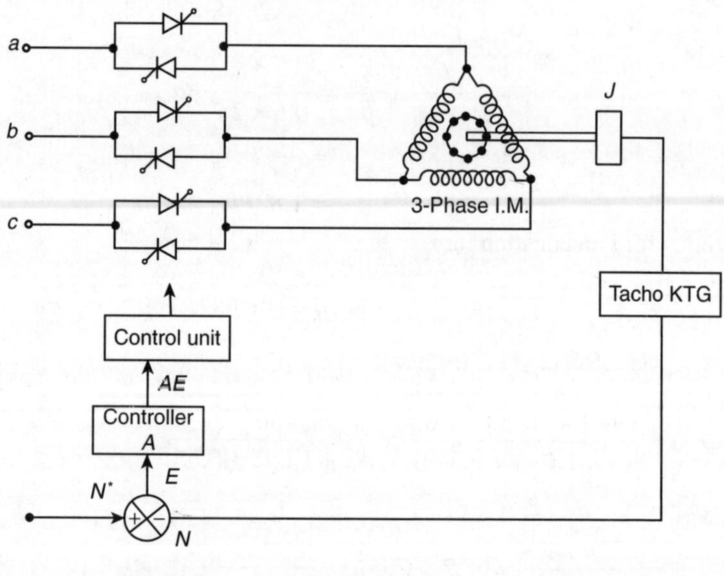

Fig. E6.2(a) Block diagram pertaining to example

$$G(j\omega) = \frac{K}{j\omega(1 + j\omega\tau)}$$

Substituting the values of K and τ

$$G(j\omega) = \frac{15}{j\omega - \omega^2\tau} = \frac{15}{j\omega - 0.2\omega^2}$$

$$= \frac{15}{j\omega(1 + j0.2\omega)}$$

$G(j\omega)$ has a break over frequency

$$\frac{1}{T} = \frac{1}{0.2} = 5 \text{ rad/s}$$

$$\text{At } \omega = 1 \qquad |G(j\omega)| = 15$$

The gain in db at $\omega = 1$ is $20 \log |G(j\omega)| = 23.522$

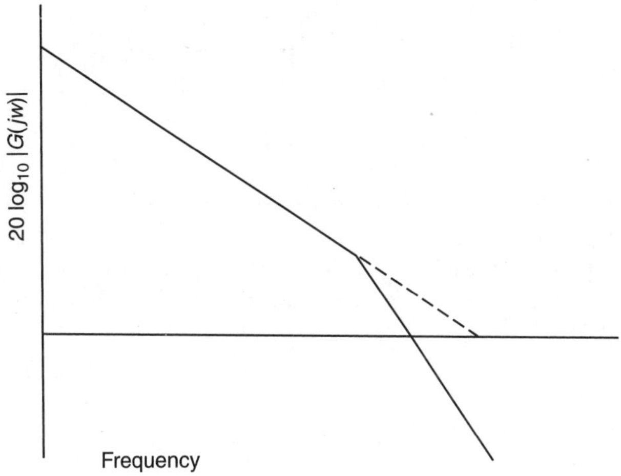

Fig. E6.2(b) *Straight line approximation of magnitude plate*

The straight line approximation to Bode diagram shown in Fig. E6.2(b) has a slope of 20 db/decade till it reaches the frequency of 5 rad/s. Thereafter the slope is 40 db/decade. Drawing these approximations, the gain cross-over frequency is 8.5 rad/s.

6.3 *Sketch the Nyquist diagram to represent the frequency response of the motor of Problem 6.3.*

Solution

$$G(s) = \frac{K}{s(1 + sT)}$$

$$G(j\omega) = \frac{K}{j\omega(1 + j\omega T)}$$

Rationalising we get

$$G(j\omega) = \frac{-KT}{1+\omega^2T^2} - \frac{jK}{\omega\left(1+\omega^2T^2\right)}$$

Substituting the numerical values for K and T

$$G(j\omega) = \frac{-3}{1+0.04\omega^2} - \frac{j15}{\omega\left(1+0.04\omega^2\right)}$$

$\omega = 0$	*Real part* – 3	*Im. part* ∞	*Magnitude* ∞	*Phase* – 90°
1	– 2.885	– 14.42	14.7	– 101.32
5	– 1.5	– 1.5	2.121	– 135
7	– 1.014	– 0.724	1.246	– 119.915
8	– 0.843	– 0.527	0.994	– 147.90
10	– 0.6	– 0.3	0.671	– 153.43
20	– 0.176	– 0.044	0.1814	– 165.96
00	0	0	0	– 180

The Nyquist plot is shown in Fig. E6.3. The curve moves towards the left and becomes more oscillatory if the constant K increases. However it does not embrace (– 1, 0).

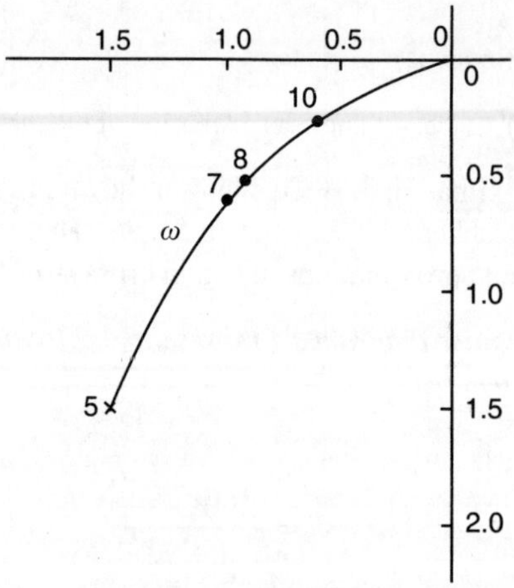

Fig. E6.3 *Nyquist plot*

6.4 *A closed loop control system of an induction motor fed from a thyristor power controller is represented in Fig. E6.4. The open loop transfer function of the motor is given by*

$$G(s) = \frac{\omega(s)}{e(s)} = \frac{K_c K_e K_m}{(1+sT_c)(1+sT_e)(1+sT_m)}$$

K_c represents the overall gain of the amplifier, control unit and thyristor controller configuration, and is adjustable to give the desired performance. $K_e = 3.0$ and $K_m = 10$. The time constants are $T_c = 0.002$ s, $T_e = 0.05$ s, $T_m = 0.15$ s. The tachometric feed back has a gain constant of 1 V/rad/s. Determine the maximum permissible K_c.

Solution The value of K_c can be found out by drawing the Nyquist diagram. A Nyquist plot drawn represents the stability frequency response of a closed loop system. If the frequency response of open loop transfer function encloses (– 1,0) the closed loop system becomes unstable.

$$G(j\omega) = \frac{K_c K_e K_m}{(1+j\omega T_c)(1+j\omega T_e)(1+j\omega T_m)}$$

Rationalising we get

$$G(j\omega) = \frac{K_c K_e K_m}{\left[1-\omega^2\left(T_c T_m + T_c T_e + T_e T_m\right)\right] - j\omega\left(\omega^2 T_c T_e T_m - T_c - T_e - T_m\right)}$$

$$= K_c K_e K_m \frac{\left[1-\omega^2\left(T_c T_m + T_c T_e + T_e T_m\right)\right] + j\omega(\omega^2 T_c T_e T_m - T_c - T_e - T_m)}{\left[1-\omega^2\left(T_c T_m + T_c T_e + T_e T_m\right)\right]^2 + \omega^2\left(\omega^2 T_c T_e T_m - T_c - T_e - T_m\right)^2}$$

The locus of $G(j\omega)$ as ω changes from 0 to ∞ is drawn. The value of K_c is found out when it crosses (– 1, 0). Therefore

$$K_c K_e K_m \omega\left(\omega^2 T_c T_e T_m - T_c - T_e - T_m\right) = 0$$

from which

$$\omega^2 = \frac{T_c + T_e + T_m}{T_c T_e T_m}$$

The real part is – 1

$$\frac{K_c K_e K_m}{1-\omega^2\left(T_c T_m + T_c T_e + T_e T_m\right)} = -1$$

substituting for ω^2 and the numerical values we get

$$|0.429 K_c| = 1$$

$$K_c = \frac{1}{0.429} = 2.33$$

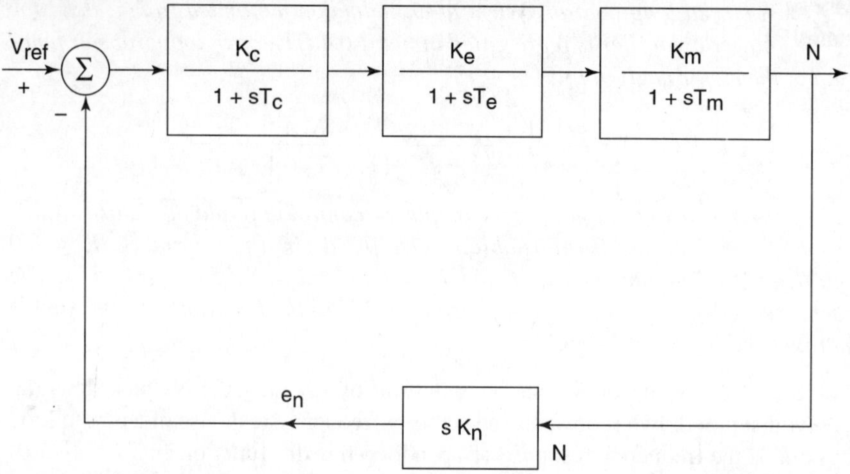

Fig. E6.4 *Closed loop control system of example 6.4*

For stable operation of the system K_c is chosen to be 0.5 to 0.8 times this critical value. Typical Nyquist locus is shown in Fig. E.6.4.

6.5 *A control system has four cascaded single delay elements having time constants of* $T_1 = 400$ *ms,* $T_2 = 80$ *ms,* $T_3 = 15$ *ms,* $T_4 = 5$ *ms. This system has to compensated by a controller. Compare the behaviour of the compensated system with (a) integral (b) proportional and integral control and (c) proportional, integral and differential (PID) control.*

Solution The four delay elements in cascade are shown in Fig. E6.5. Consider the effect of integral control. The integration time constant of the integral controller

$$T_I = 2A_s T_a$$

where T_a is the equivalent time constant of the system .

$$T_a = T_1 + T_2 + T_3 + T_4 = 500 \text{ ms}$$

The rise time is

$$t_r = 4.7T_a = 4.7 \times 500 = 2350 \text{ ms}$$

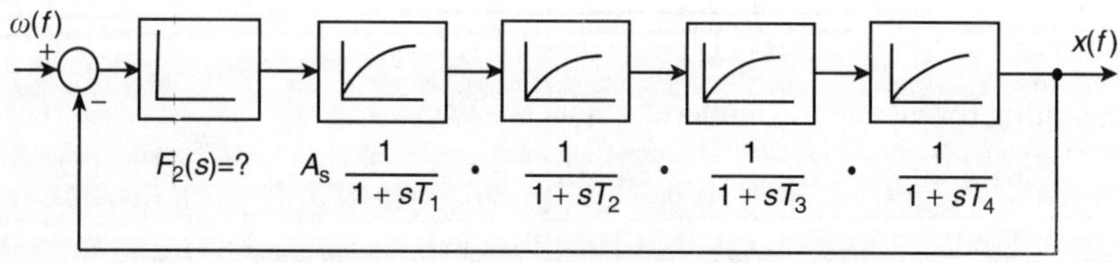

Fig. E6.5 *Control system of example 6.5*

A PI controller on the other hand can be used to compensate the largest time constant. The other three time constants which are not compensated only contribute to the behaviour. The reset time constant of the controller is the largest time constant being compensated. The sum of the small time constants

$$T_b = T_2 + T_3 + T_4 = 100 \text{ ms}$$

The proportional amplification of the controller

$$A_R = \frac{T_1}{2A_sT_b} = \frac{2}{A_s}.$$

The rise time t_r in this case

$$t_r = 4.7T_b = 470 \text{ ms}$$

which is 1/5 of the rise time of integral use.

A PID controller improves the performance further by introducing a differentiating phase advance to compensate for one of the time constants. Thus two large time constants are compensated. The reset time of the controller is T_1 where as the phase advance provided by the controller is T_2, i.e., second largest among the time constants.

The equivalent time constants of the compensated system is

$$T_e = T_3 + T_4 = 20 \text{ ms}$$

The proportional amplification is

$$A_R = \frac{T_1}{2A_sT_e} = \frac{10}{A_s}$$

The amplification is increased and the rise time is reduced.

The rise time = 4.7 T_e = 94 ms

This seems to suggest that introduction of another differentiating element would make the response faster by compensating third time constant. This is not advantageous as the double differentiation would not allow a stable control.

A proportional controller has a characteristic feature of steady-state error. The steady-state error depends upon the proportionality constant (proportional amplification) of the controller. Larger the value of this amplification smaller is the steady-state error.

$$A_R \text{ for this case } = \frac{T_1}{2A_sT_e}$$

where T_1 is the largest time constant and T_e is the sum of all the other small time constants. A_R would be sufficiently large if A_s is small and T_1/T_e is large.

A proportional controller would be effective if T_1/T_e is very large. For the case illustrated

$$A_R = \frac{4}{2A_s} = \frac{2}{A_s}$$

As A_s increases A_R decreases and the steady-state error increases. Further second large delay may be compensated by phase advance. A P controller is justified if T_1/T_e is very large. Therefore a PD controller is used in such cases. Other wise controller with integrating action may be used.

6.6 *A 400 V, 2800 rpm, 80 A dc motor is fed from a phase controlled converter for speed control. Its electrical and mechanical time constants are 50 ms and 580 ms respectively. Design current and speed controllers for the closed loop speed control system.*

Solution The closed-loop speed control system is depicted in Fig. E6.6(a). It has an outer speed loop and an inner current loop. Besides armature time constant there are other parasitic time constants in the current control loop such as converter dead zone, the delay in the current etc. In the speed loop besides mechanical time constant the delay in the speed measurements etc will be present. With these parasitic time delays the block diagram of speed control is shown in Fig. E6.6(b). Once the current controller is designed, the current loop can be simplified as one first order delay and this is used in the design of speed controller.

Design of current controller:

The dead zone of the converter depends on the number of pulses of the converter. It is determined from the relation

$$T_c = \frac{1}{2}\frac{T}{P}$$

where T is the time period of the input voltage and P is the pulse number. The value of $T_c = 1.7$ ms assuming a 6 pulse converter being fed from a 50 Hz supply. A delay of 2.5 ms due to filtering of the measured value of current is considered. The current control loop has three time constants 50 ms, 2.5 ms and 1.7 ms. The controller compensates the largest time constant of 50 ms. Sum of other time constants is 1.7 + 2.5 = 4.2 ms.

If an integral controller is used, it suffers from a large rise time of 54.2 ms whereas a proportional controllers has a disadvantage of large offset. A PI con-

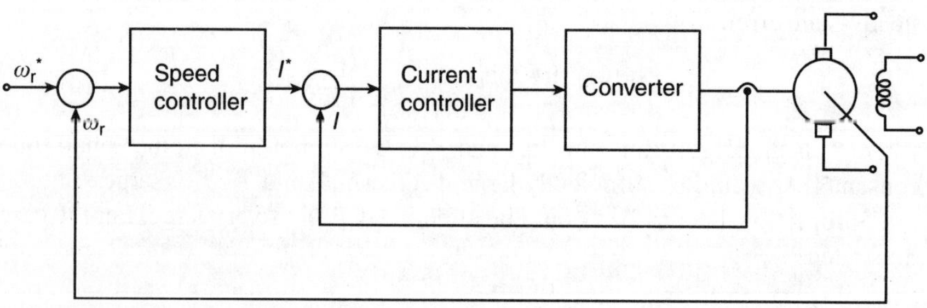

Fig. E6.6(a) *Closed loop control of dc motor with outer speed loop and inner current loop*

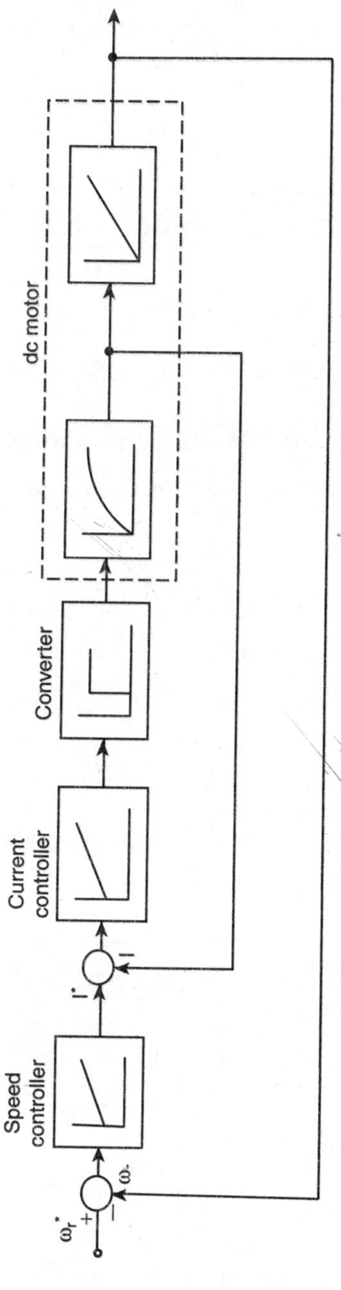

Fig. E6.6(b) *Block diagram representation of dc motor with closed loop control*

troller is used. Using the criteria of the design based on magnitude optimum, the proportional amplification of the controller is

$$V_R = \frac{T_A}{2V_sT_e} = \frac{50}{2V_s \times 4.2}$$

Assuming a gain of $V_s = 10$

$$V_R = 0.6$$

The transfer function of the controller

$$= \frac{V_R\left(1 + sT_n\right)}{sT_n} = \frac{0.6\left(1 + 0.05s\right)}{0.05s}$$

The PI controller parameters are such that (Fig. E6.6(c)).

$$V_R = \frac{R_1}{R_0}, \quad T_1 = R_1C_1$$

The maximum value of $V_1 = 10$ V Assuming a controller current of 0.5 mA. We have $R_0 = \frac{10}{0.5 \times 10^{-3}} = 20\text{K}\Omega$. This divides equally as $R_{01} = R_{02} = 10$ KΩ. The time constant of RC filter

$$2.5 = \frac{R_{01} \cdot R_{02}}{R_{01} + R_{02}} C_0 = \frac{10 \times 10}{2 \times 10} C_0$$

$$C_0 = 0.5\ \mu F$$

The value of $R_1 = V_R \quad R_0 = 12$ K!

and $$C_1 = \frac{50 \times 10^{-3}}{R_1} = 4.2\mu F.$$

Speed controller

The speed control loop comprises an integration time between current and speed and also the time constants such as the delay due to filtering of the measured

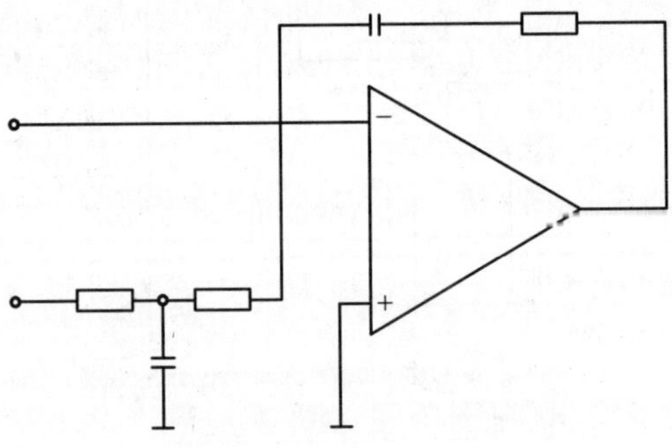

Fig. E6.6(c) *P.I. controller*

speed, the equivalent time constant of the current loop etc. These are 2 ms and 8.4 ms respectively. Symmetrical optimum is used. The criteria are

$$V_R = \frac{T_0}{2V_s T_1}; \quad T_n = 4T_1$$

As the integration time constant includes the amplification is

$$T_i = \frac{T_0}{V_s}$$

The value of T_i is fixed at 500 ms.

$$\text{The amplification } V_R = \frac{500 \times 10^{-3}}{2 \times 10.4} = 24.$$

The reset time of the integrator

$$T_n = 4T_1 = 4 \cdot 1.6\text{ms}.$$

The PI controller is shown in Fig. E6.6(d) with a delay in the reference input. The constants are

$$R_0 = 20 \text{ K}\Omega, \quad R_1 = V_R R_0 = 480 \text{ K}\Omega$$

$$R_s = 20 \text{ K}\Omega, \quad L_1 = \frac{4T_1}{R_1} = 86.7\text{nF}$$

$$C_0 = \frac{R_{01} + R_{02}}{R_{01} \cdot R_{02}} \tau = 0.4 \text{ ms with } \tau = 2 \text{ ms}$$

$$C_s = \frac{R_{s1} + R_{s2}}{R_{s1} \cdot R_{s2}} \tau_\Delta = 8.3\mu \text{ F where,}$$

$$T_s = 4T_1.$$

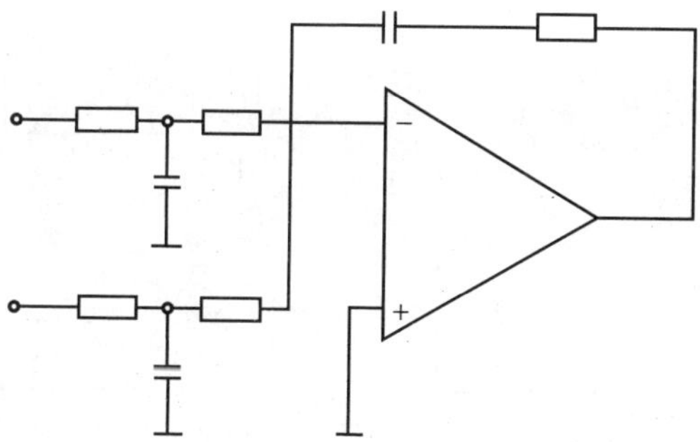

Fig. E6.6(d) *P.I. controller with filter in the reference circuit*

Problems

6.1 A separately excited dc motor has an emf constant 1.2 Vs/rad. The motor is controlled by variation in armature voltage. The resistance and inductance of the armature are 0.5 Ω and 0.2 H. The inertia of the motor and coupled load is 0.5 kg m². The coefficient of viscous damping is 0.050 Nm/rad/s.

(a) Determine the transfer function relating the armature angular velocity to armature applied voltage.

(b) Determine the step response when a voltage of 20 V is applied when the machine is at rest.

(c) Determine the response to a sinusoidal input of 20 V (rms) over a range of 0–5rad/s of ω_r.

6.2 A separately excited motor has its armature fed from a constant current. The resistance and inductance of the field circuit are $r_f = 40\ \Omega$ and $L_f = 20$ H. The torque of the motor is $K_t I_f$ where the constant K_t is 65 Nm/A of field current. The damping coefficient is 0.04 Nm per rad/s and the inertia of the motor is 0.3 kg m² and that of load is 50 kg m². The load has a torque speed characteristic represented by

$$T = 50\,\omega \text{ Nm}$$

The control of the motor is effected by field voltage. Determine the transfer function ω_r/V_f when direct coupled to the load through a matched gearing to give maximum acceleration.

6.3 A conventional Ward Leonard system has the following parameters:
Generator: provides variable voltage to the armature of the drive motor. Field resistance 120 Ω, field inductance 65.0 H. The armature resistance = 0.5 Ω rotational inductance 0.9 H Speed ω_g = 120 rad/s.
Motor: variable speed motor. Field current = 2.5 A; armature resistance $r_a = 0.75\ \Omega$; $L_a = 0.0035$ H, rotational inductance 0.6 H, inertia 2.0 kg m². Determine how the motor speed varies when a step voltage of 50 V is applied to the generator field assuming the motor is initially at rest and on no-load. Determine also the motor current variation.

6.4 In a Ward Leonard system shown in Fig. P6.4 a field controlled dc generator driven at constant speed feeds the

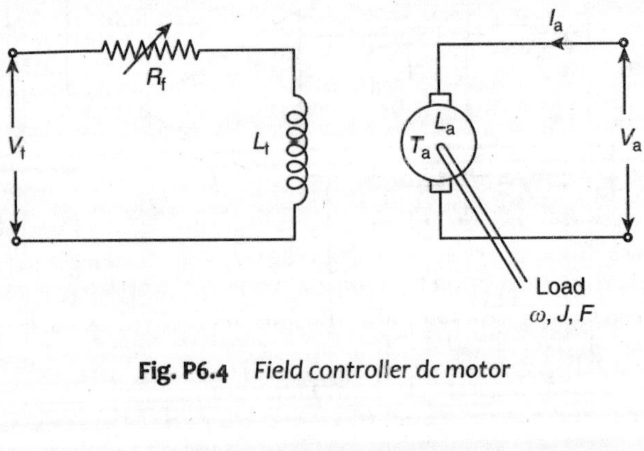

Fig. P6.4 *Field controller dc motor*

armature of an armature controlled dc motor for speed control. The motor speed is ω. T_L, J, B are the load torque, inertia and friction coefficient of load. Develop the state equations of the system. Determine the transfer function relating the motor speed and field voltage V_f.

The values of parameters are: $R_f = 15\ \Omega$, $L_f = 120$ mH; $K_g = 110$ V/A, $R_a = 2.5\ \Omega$; $L_a = 12$ H; $K_m = 3.5$ V/rad/s, $J = 8.0$ kg m^2, $B = 7.5$ Nm/rad/s.

6.5 Two systems A and B have unit step responses $1 - e^{-3t}$ and $2(1 - e^{-t})$ respectively. These two systems are connected in cascade using a unit gain buffer amplifier such that there is no loading effect. Determine the step response of the combination.

6.6 A Ward Leonard system which is a combination of a generator and motor is used to control large powers of the motor using relatively low power at the field of the generator.

(a) $r_{fg} = 5$ and $L_{fg} = 2.5$ determine the generator voltage e_{fg} for $V_{fg} = 100\, u(t)$.

(b) The motor in the combination is armature controlled, having constant excitation. Determine its transfer function relating the angular position $\theta_m(s)/V_m(s)$ in terms of the parameters of the motor and its torque constant.

(c) Determine the overall transfer function $\theta_m(s)/V_f(s)$ for representative values of r_{ag}, r_{am}, K_{e2} and K_t.

(d) What is the difference between this example and example 5?

(e) If $V_f(s) = \dfrac{20}{s}$ determine $\omega(t)$ as a function of time.

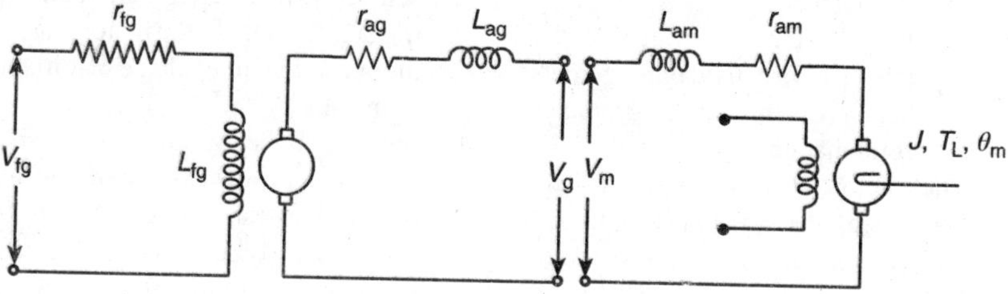

Fig. P6.6 *Circuit diagram of Problem 6.6*

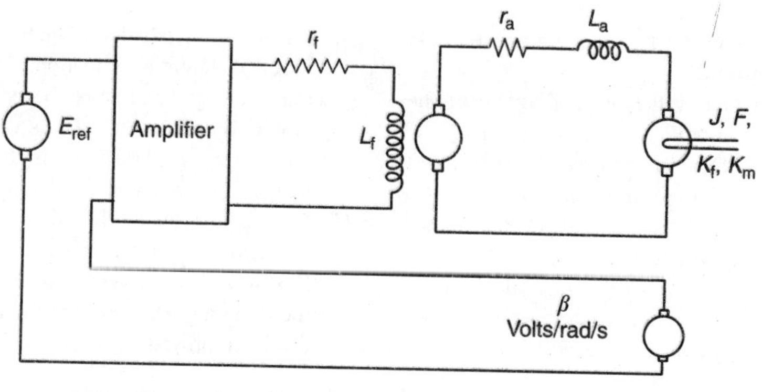

Fig. P6.7 *Closed loop controlled Ward Leonard System*

6.7 A closed loop Ward Leonard control system is shown in Fig. P6.8. Draw the complete block diagram of the system indicating the transfer function of each block. Identify the state variables and write down the system equations in state variable form in matrix notation. Determine the characteristic equation of the system.

6.8 A 10 hp armature controlled dc motor drives a load of viscous friction coefficient 2 lb ft/rad/s and angular inertia is 17.5 lb ft s². The field is separately excited and is constant. The motor parameters are $r_a = 0.762\ \Omega$, $L_a = 12.5$ mH. The motor constant = 3.0 V/rad/s.
(a) Compute the steady-state speed for a step input to the armature of 220 V.
(b) Determine the time taken by the motor to reach within 95% steady-state speed.
(c) What is the total value of the effective viscous damping coefficient of this motor-generator combination?

6.9 The motor of Problem 8 is field controlled, the armature being fed with a constant current. The motor field parameters are $R_f = 60\ \Omega$ and $L_f = 20$ H. The motor operates at constant torque. The torque constant = 60 lb.ft/ampere of field current or 80 Nm per ampere of field current.
(a) Draw the block diagram of the motor.
(b) Determine the steady-state speed of the motor for a step input to the field circuit of the motor equal to 120 V.
(c) Determine the time taken for the motor to reach a speed of 95% of the final speed.
(d) Write down the system equations in matrix from

6.10 Determine the transfer function $\theta(s)/V(s)$ of an armature controlled dc motor. Assume $e_a = 25$ V, $r_a = 0.09\ \Omega$, $L_a = 2.5$ mH, $J = 1.82$ kg m², $F = 0.275$ Nms/rad
Determine the variation of speed as a function of t.

6.11 Determine the transfer function $\theta(s)/E_f(s)$ of a field controlled dc motor. Assume that $e_f = 120$ V, $i_a = 20$ A, $\omega_{ss} = \theta_{ss} = 1300$ rpm, $R_f = 100\ \Omega$, $L_f = 20$ H, $J = 2$ kg m², $F = 0.275$ Nms/rad.
Determine the variation of speed as a function of time.

6.12 Write down the basic equations of a three-phase induction motor in space phasor notation. Convert these equations into state variable form. Deduce the signal flow graph of the motor. Assuming the motor is fed from a square wave inverter determine the transfer function of the motor relating the speed to input voltage of variable frequency.

6.13 Show that the dynamics of a voltage source inverter fed induction motor can be represented by a second order system. Determine the transient response of the motor.

6.14 Consider a CSI fed induction motor. Show that the dynamics of this motor can be completely described by a first order differential equation. Determine the speed response of such a motor. Based on the results discuss how the transient response is affected by the rotor frequency of the motor.

6.15 Derive a transfer function of an induction motor fed from a CSI in flux weakening mode. Discuss how the dynamic behaviour of the motor can be improved.

Multiple-Choice Questions

6.1 A system has a transfer function given by $G(s) = 1/(s^2 + 3s + 6)$. When it is excited by a step input of 4 $u(t)$, the steady-state output of the system is
(a) 0.5
(b) 2/3
(c) 1
(d) 3/2

6.2 The response of a system in s-domain to a unit impulse input gives
(a) gain of the system
(b) transfer function of the system
(c) time response of the system
(d) error under steady-state conditions

6.3 A system has a transfer function given by $G(s) = 25/(s^2 + 6s + 9)$. When it is excited by a step input, it reaches its final steady-state response
(a) immediately after the input is applied
(b) after going through a set of oscillations
(c) after a very long time without going through any oscillations
(d) after a very short interval of time however without oscillations

6.4 The imaginary part of the conjugate poles of a second order system represents
(a) natural frequency of the system
(b) overshoot of the system
(c) damped frequency of oscillation
(d) damping factor of the system

6.5 The transfer function having least steady-state error is
(a) $5/(s^2 + 3s + 5)$
(b) $10/(s^2 + 3s + 10)$
(c) $25/(s^2 + 3s + 25)$
(d) $15/(s^2 + 3s + 15)$

6.6 Integral control of a system
(a) improves the transient response of the system
(b) improves the steady-state stability of the system
(c) increases the order of the system
(d) decreases the order of the system

6.7 The derivative action of a PD controller
(a) increases the rise time
(b) decreases the rise time
(c) decreases the steady-state error
(d) has no effect on the rise time

6.8 Routh's stability criterion gives
(a) the degree of stability of a given system
(b) the indication of stability condition of a linear system without actually finding the roots of the characteristic equation
(c) the indication to improve the stability of a system
(d) the indication of stability condition only if the roots of the characteristic equation are known.

6.9 Bode plots of a certain transfer function show that gain cross-over frequency is greater than phase cross-over frequency. This system is
(a) stable
(b) unstable
(c) marginally stable
(d) the information is not sufficient to indicate stability condition.

6.10 A third order system is compensated by a PI controller. The ratio of the controller time constant to the uncompensated time constant of the system is K_0. When the value of K_0 is increased
(a) phase margin of the overall system increases
(b) phase margin of the overall system decreases
(c) phase margin remains the same
(d) the damping of the system is unaffected.

References

6.1 Dorf, R.C., *Time domain analysis and design of control systems*, Addison-Wesley, 1965.

6.2 Ogata, K., *Modern control engineering* (International Edition) Prentic-Hall of India, New Delhi, 1974.

6.3 D'Azzo, J.J., and C.H. Houpis, *Feedback control system analysis and synthesis*, McGraw-Hill Book Company, Inc., New York, 1966.

6.4 Dransfield, P., *Engineering systems and automatic control*, (International Edition), Prentice-Hall of India, New Delhi, 1974.

6.5 Kuo, B.C., *Automatic control systems*, Prentice-Hall Inc., Englewood Cliffs N.J., 1962.

6.6 Kümmel, F., *Elektrische antriebstechnik*, Springer Verlag, Berlin, 1971.

6.7 Das Gupta, S., *Control system theory*, Khanna Publishers, New Delhi, 1983.

7 Drives for Specific Applications

The discussions in the previous chapters were oriented towards determining the torque-speed characteristics of the electric motors during motoring, braking and starting; the choice of a motor for a given loading condition; the characteristic features of the solid state converters employed in the drive practice and also the various dc and ac drives with their characteristics while operating on these power converters for speed control, reversing and braking.

The industrial applications of electrical drives are numerous. To arrive at a particular motor of suitable size for a given application, a knowledge of the following is very much necessary:

i. The torque-speed curve of the industrial load is required. This fixes up the type of motor required to drive the load.
ii. The environmental conditions in the industry where the motor is located. This also decides the ambient temperature at which the motor operates.
iii. Duty cycle of the load, and the frequency of starting and braking. The kW rating of the motor is decided by the load cycle, taking into consideration both the thermal as well as mechanical overload capacity of the motor. The starting and braking of the motor, if employed very frequently alters the kW rating of the motor because of heating of the motor due to extra losses ocurring during these conditions. The methods of starting and braking must also be known. The motor rating must be so chosen that it is capable of driving the highest loading in the duty cycle without getting thermally overloaded, providing the necessary torque at all conditions.
iv. The speed control needed also decides the type of motor. Smooth speed control over a wide range is not a problem with dc motors but had been a problem with ac motors till the advent of thyristor power converters. With the recent developments in the area of these power converters, the ac motors also are becoming popular as variable speed drive motors. The economics involved in the methods of speed control employed also need due consideration.

v. The necessary information is required to decide between the individual and group drive. In a group drive, as has already been mentioned, one single machine is used to do several jobs. It is therefore economical to employ a group drive. In a group drive there are serious disadvantages, viz., servicing or repair of the motor requires complete stoppage. Transmission losses occur and efficiency is poor. Line shaft makes the place untidy. In an individual drive on the other hand, separate motors are employed to perform separate jobs. Basically it is very costly. But it has facility for overtime working, complete shut down or stoppage is not required, in case of repair and absence of line shaft makes the place clear. Hence taking into consideration all these factors, a proper choice between these two types may be made.

vi. To meet certain specific duties in a particular application special designs of the motors may be required. Special controls and protection circuits may be needed. The motors may require protection against single phasing, overloading, overheating, etc. The rating of the motor must be suitable to do the job under the worst conditions available. Based on the above information it is possible to decide the requirements of a drive motor to suit a given application. Of the numerous industrial applications in which electric motors are employed as drive motors, only a few are discussed here:

(a) Textile mills
(b) Steel rolling mills
(c) Cranes and hoists
(d) Cement mills
(e) Paper mills
(f) Sugar mills
(g) Machine tool applications
(h) Coal mining
(i) Centrifugal pumps
(j) Turbo compressors

Some industrial applications may have several stages by the time the end product is ready from the basic raw material. While selecting motors for operations in these stages, one has to give due consideration to the above factors, such as environmental conditions of the stage of operation, electrical features of the motor, i.e., starting torque, braking, speed control, etc., and the ambient temperature. The requirements of the drive must be very well defined to properly select the motor.

7.1 DRIVE CONSIDERATIONS FOR TEXTILE MILLS

There are several processes involved by the time the finished cloth comes out of a mill from its basic raw material, cotton picked up from the fields. The requirements of the motors are different for different processes. These mainly depend on the nature of the process. The several stages in the textile mills and the requirements of a drive motor for each stage are discussed in the following:

Ginning The process of separating seeds from the raw cotton picked from the field is called ginning. This may be done in the mills located near the fields or in the industrial location itself. In the former the ginned cotton is transported to the industrial area in the form of bales. The ginning motors must have speed ranges of 250 to 1450 rpm. The load speeds are fairly constant. No speed control is required.

Commercially available squirrel cage induction motors may be employed.

Blowing The ginned cotton in the form of bales is opened up and is cleaned up very well in a blowing room. Normal three-phase induction motors may be used for the purpose. No speed control is required. The motors having synchronous speed of 1000 or 1500 rpm may be employed.

Cording The process of converting cleaned cotton into laps is done by lap machines which are normal three-phase standard squirrel cage motors. These laps are converted to slivers by a process called cording. A motor used for cording is required to accelerate a drum having a large moment of inertia. It is required to withstand prolonged accelerating periods. To meet these requirements the motor selection must be made. The motor selected must have a very high starting torque and low starting current so that starting losses are kept to a minimum. The motor must have sufficient thermal capacity to withstand the heat produced by the losses occurring during prolonged acceleration. These cord motors are standardised in IS:2972 (part II) 1964 which gives the specifications for cord motors.

Normally, three-phase totally enclosed or totally enclosed fan cooled squirrel cage induction motors with high starting torque may be employed. The rating of the motor depends upon the type of fabric. Smaller rating motors in the range 1.1 to 1.5 kW may be used for light fabric. For heavy fabric the rating increases to 2.2 to 5.5 kW. The operating speeds of these motors are in the range 750–1000 rpm. Squirrel cage motors (8/6 poles) having synchronous speeds in the range of 750–1000 rpm are normally employed. The motors may be started directly from the line to achieve good starting torque. Slip ring motors may be advantageously used with rotor resistance starters as they give high starting torque at low starting currents. Once started the operation is continuous and uninterrupted.

These slivers are converted to uniform straight fibre by means of drawing machines. These are also normal standard motors. However, the motor selected must be capable of stopping instantaneously, in case of sliver breaking. The drawing machines are sometimes self brake motors which satisfy this requirement. The motor is subjected to inching to place up the broken sliver again. The inching operations may amount to 20. When the brake forms an integral part of the motor there is no necessity for a clutch and the motor becomes compact.

Combing and lap operations take place after slivers are straightened in a drawing machine. The motors used for these operations are also normal squirrel cage motors. The combing process upgrades the fibre. The slivers are converted into laps before combing.

The next process is spinning. Before the thread is ready for final spinning it is thinned down in two or three stages by processing it on a fly or speed frame. A motor with smooth acceleration is necessary to drive this frame. The drive motor should be capable of working in high ambient temperatures. The motor must be totally enclosed, with a clean floor construction. This is to prevent the cotton fluff from getting deposited on the motor surface, which may lead to poor cooling of the motor as well as burning of fluff due to motor heating. The motor must have uniform acceleration having thermal reserve.

During spinning process the yarn is twisted and made to have sufficient strength. These spinning motors are standardised in IS:2972 (part III) 1965, which gives the specifications for spinning motors. The strengthened yarn is wound on bobbins. The spinning motor must be capable of drawing, twisting and winding operations. The breakage must be minimum and the yarn produced must have uniform tension. These motors also must have uniform acceleration to avoid yarn breakage. For a motor to have uniform acceleration, its speed-torque characteristic must be as shown in Fig. 7.1. Its starting torque must be 150–200% and the peak torque 200–250%. The difference between these torques must be constant as the motor speeds up and must be small to ensure uniform acceleration. The acceleration is also slow and smooth. This is to avoid yarn breakage. A normal motor is, therefore, not suitable for this process. The motor must have an acceleration time of 5 to 10 s. The operating speed is 500 rpm. The kW rating of the motor is decided by ring frame; number of spindles, ring diameter and spindle speed. When once the initial build is over the motor may run at a higher speed. A two speed pole change motor may be used. It must have constant torque operation at both the speeds so that uniform tension is assured both at starting and running. One can employ two different motors, one for starting and low speeds and the other for high speeds. In either case the motors are costly. Two speed pole change motors are bulky and costly. However, the increased uninterrupted production may compensate for the

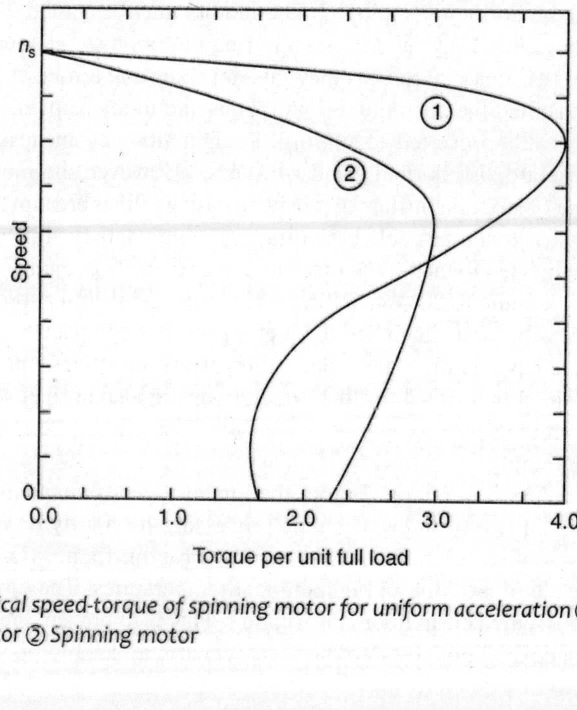

Fig. 7.1 *Typical speed-torque of spinning motor for uniform acceleration ① Normal motor ② Spinning motor*

cost. A two motor drive is also costly but has several advantages. It allows setting of any speed difference by adjusting the pulley diameters and speed ratios. The yarn tension can be adjusted independently. There is no interruption in production even when one motor fails.

For mule spinning, a group drive may be employed. The motor employed should have high starting torque as well as high operating slip. A slip ring motor with rotor resistance control or high torque cage motors may be employed for the purpose. When an individual drive is used, the motor chosen should be able to take care of peak power demands and must have higher slip.

For operations like winding, warping and sizing, normal motors may be employed. Low speed motors are required. Reduction in speed using a gearing unit may be accomplished. When the yarn is transferred from the bobbin, a speed drop of nearly 100 rpm is required. So for these operations high slip motors are preferred.

Looms The weaving of yarn into cloth is done in looms. The drives may be either semi group drives or individual drives depending upon the quality of the cloth required. The speed of operation is normally 600 to 750 rpm. Requirements of a loom motor are:

i. Starting torque must be high to complete the pick up job in a very short time.
ii. The duty cycle consists of frequent starting and stopping. The load on the loom motor is variable and intermittant. To avoid frequent starting and stopping the motor may be decoupled (or coupled) from the load by means of a clutch.
iii. The operation requires a reciprocating mechanism. Actually rotary motion must be converted to linear reciprocating motion. The current and torque pulsations are present. A flywheel is required for smoothing.
iv. These are also located in places where dust accumulates on the motor. The cotton fluff should not get collected on the motor surface to avoid burning of the same due to motor heating.
v. Loom motors must withstand the effects of humidity.

To suit to the above requirements the loom motors are normally totally enclosed three-phase induction motors with high starting torque. Fan cooling of the motors is also employed. The fan cooling helps to avoid the collection of cotton fluff on the motor surface. The motor must be designed, taking into consideration the possible torque and current pulsations due to reciprocating motion. The surface of the motor must be such that it does not collect any cotton fluff. The kW rating of the motor selected must be decided taking into account the frequent starting and stopping in the duty cycle. The size of the loom motor depends upon the fabric. For light fabric, motors of rating up to 1.5 kW and for heavy fabrics motors of rating 2.2 to 3.7 kW are employed. Speeds of motors are in the range of 100 to 750 rpm. Brake motors may be used here so that motor stops automatically in case the thread breaks.

The loom motors are standardised under IS: 2972 part I (1964).

7.1.1 Control of ac Motors to Have Torque Control

From the foregoing discussion it is clear that the motors used for textile applications must have

i. high starting torque
ii. torque control providing uniform acceleration so that the breakage of the yarn is minimum and the quality of the product is improved.

The production increases if the down time of the process and drive is reduced.

In conventional drive systems the torque control is achieved by variation of applied voltage to the squirrel cage motors. In slip ring induction motors torque control is possible by means of rotor resistance control. The variation of applied voltage is realised by series connected reactors on the stator side. The torque control in stepless. The accelerating energy keeps the constant tension in the yarn. The resistances, even though they have advantages, are not employed as their heating may cause fire hazards.

With the development of thyristor power converters the applied voltage variation to an induction motor for torque control is accomplished by means of ac voltage controller in the supply circuit. Closed loop control can be employed (Fig. 7.2).

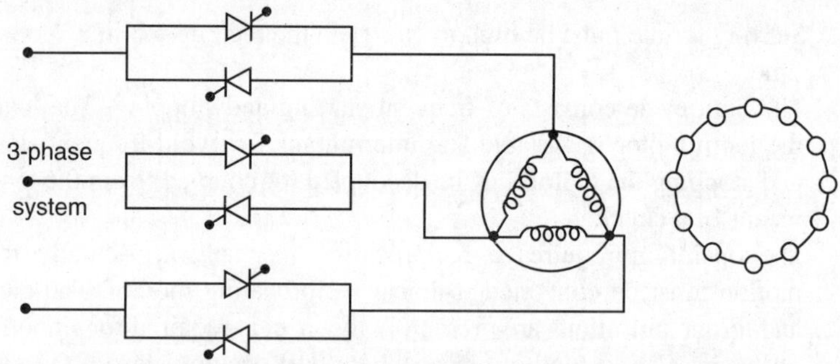

Fig. 7.2(a) *Ac voltage controller feeding a high resistance rotor*

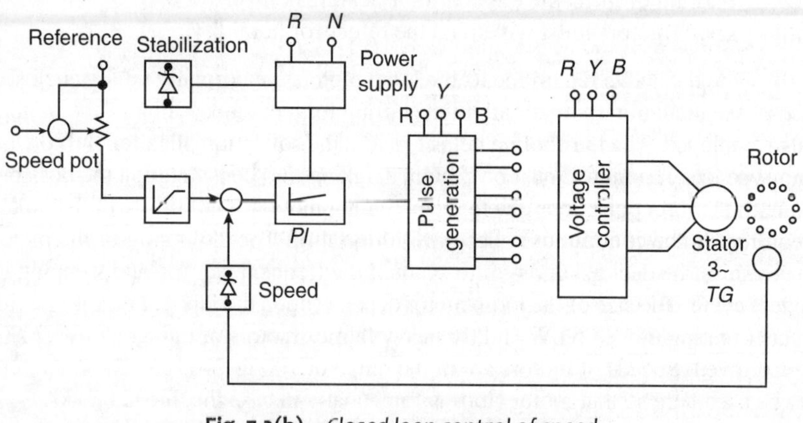

Fig. 7.2(b) *Closed loop control of speed*

This method of using reduced voltage for torque control, however, has disadvantages of poor power factor at reduced voltages.

The torque control as well as speed control is obtained using induction motors fed from variable frequency inverters. The inverter is controlled such that the motor operates with constant flux, to achieve the best possible torque capability of the motor. Both voltage source and current source inverters can be employed. Slip controlled drives assure uniform starting torque. For smooth low speed operation cycloconverters are preferred (Fig. 7.3).

With slip ring rotors, chopper controlled resistance may be used for torque control.

Converter fed synchronous motors are also becoming very popular. They have the advantages that the converter can employ machine commutation, in which case the inverter configuration is simple. A self control of the motor converter imparts dc motor characteristics, where the problems of hunting and stability are not there. For low speeds cycloconverters are preferred. Voltage source and current source link inverters can also be used for control of synchronous motors.

For other processes, such as bleaching, drying, printing, finishing, etc., in textile mills, the normal motors are used. When very low speeds are required, gearing may be employed. Ward Leonard dc drive, ac commutator motors are employed. The thyristor control may be used where efficient, smooth speed control over a wide range is required.

Special design of textile mill motors are required owing to the location of the motor, running conditions and also the torque requirements while starting. In textile mills the motors are located in places where there is a lot of dust. The cotton fluff may be deposited on the motor, which affects natural cooling of the motor. Consequently the temperature rise of the motor is more, which may heat the cotton fluff leading to fire. To prevent this situation and also prevent the dust and fluff entering the interior of the motor the motors must be totally enclosed and fan cooled (to keep the surface clean). Also the motors employed for the job of bleaching, etc., must have proper enclosure to prevent any possible damage. If a group drive is employed the motor may be located outside the room. The motors must be designed to withstand thermal variations under the worst conditions.

7.2 STEEL ROLLING MILLS

Steel rolling mills are either hot rolled or cold rolled. These may be either reversing type or continuous type. The motors used for reversing mills need operation in both the directions of rotation. A four quadrant operation of the motor may be required. In the continuous type the motor rotates only in one direction.

Steel rolling mills, where the cross section of steel is transformed to required sizes, are classified depending upon the end product required. The choice of the motor to meet the requirements and the choice of the mill-stand depends upon the products required. In blooming mills the end product is steel blooms—steels bars of any length with a definite cross section. The slabbing mills produce slabs

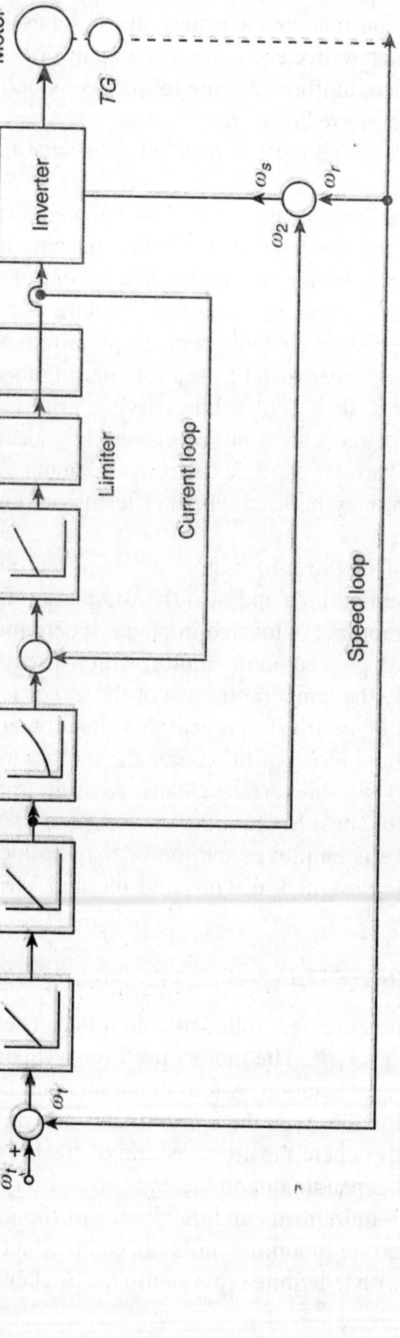

Fig. 7.3(a) Closed loop slip controlled drive with current control

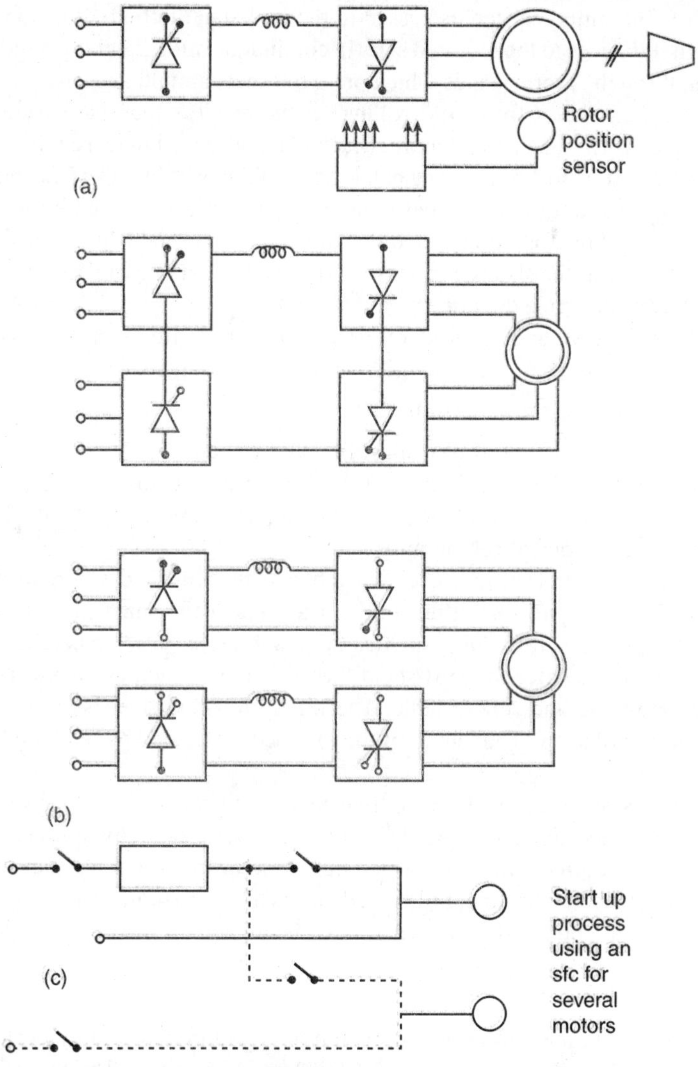

Fig. 7.3(b) *S.F.C. for synchronous motor (a) 6-pulse operation (b) 12-pulse converters (c) start up converter*

which are rolled metals of rectangular section. The blooms of reduced cross section are called billets which can further be rolled into bars, squares and angles. The mills that produce these materials of different sizes are called merchant mills. Slabs are converted to plates in plate rolling mills. These plates have less thickness than width. Strip mills convert these plates into strips which are transformed to sheets in a sheet mill. Cold rolling is normally used for producing sheets of good quality of uniform gauge. Hot rolling is used to make blooms, slabs and billets.

In reversing mills the steel is passed in the mill-stand in both directions alternately, till it reduces to the required size. In continuous mills, steel is passed in one direction through several stands which press the sheet simultaneously.

The finished sheets from cold rolling reversing mills may have a thickness ranging from 0.15 to 2 mm and more. Black plate produced in these mills may be of the order of 0.07 to 1.3 mm. The bands have a thickness of 0.0015 m and width 1 m. The production of sheet steel in a cold rolled mill is limited by low sheet speed because of the forward acceleration and retardation and adjustments of gaps in the mill-stand as the sheets are pressed. The production rate can be increased by passing the sheet in one direction.

The billets, strips, and the products of merchant mills are produced in continuous mills. Blooming and slabbing mills are of reversing type.

7.2.1 Reversing Hot Rolling Mills

Hot ingots from the soak pits or from steel making shops are rolled in these mills. These are transported to mill body by means of a car. A crane is used to load the car with these ingots. The mill receives these ingots and processes them. The mill bed comprises a series of rolling mills.

The ingots are fed on the receiving table of the mill bed wherein they are weighed. Ingots travel on rolling mills. They reach the main mill-stand after passing through several tables, such as approach table, intermediate table, and front work tables. After the mill-stand there are tables such as backwork table, intermediate table and runout table. The finally finished steel is cut in a spear table to standard sizes. The length of the mill bed is decided by the length of the product.

The ingots are passed in the mill-stand in both the directions till they are pressed to desired thickness. As the thickness decreases, automatic adjustment of the gap is required, which is carried out by screwing down machanism. This adjustment is made when the mill is made ready for reverse motion.

The metal at entry positions of the mills is aligned by means of manipulator slide guides.

Based on the process discussed above the nature of the drive is as follows:

i. A wide range of speeds of operation is required. The duty cycle of the load has frequent starts and (stops) speed reversal. The motor and its control must be selected taking this into consideration. To increase the production rate the dynamic behaviour during speed reversal must be fast.
ii. The direction of rotation must be reversible without causing serious disturbances to power handling circuits. The method employed should be such that the starting and speed reversal take place without any large dip in the terminal voltage.
iii. Reliability and accuracy are imperative.

The transport of ingots from the hot chamber to the car, conveying of the finished blooms or slabs, and the mechanical adjustments of the mill-stand are also carried out by several drive motors. These may be integrally controlled with the above mill-stand.

A motor selected should meet the above requirements. Its speed must be controlled over a wide range. The kW rating must be sufficient to drive the intermittant continuous load having the definite duty cycle with frequent starts and reversal. Braking may be required to stop the mill bed if required. Regenerative braking may not be advantageously employed. Plugging may not be suitable here due to peaks of current during speed reversal. These peaks cause voltage dips and hence must be avoided. Accurate speed control using principles of automatic control is also possible and is a reliable method.

Ward Leonard control of dc motors is very much suitable here. The regenerative speed reversal is possible. Armature current control can be employed for fast retardation. Armature voltage variation in a smooth manner enables a wide range of speed control. Flux weakening of the motor increases this range on the upper side. Load equalisation is possible by means of a flywheel. The speeds can be very accurately set and the system has a very high reliability. This allows closed loop automatic speed control.

Ac motors with conventional methods of speed control are not suitable. An ac commutator motor may be used. But braking may have to be employed using the methods of plugging; dc dynamic braking has been done for normal three-phase motors. This may result in dips of supply voltage.

The advent of thyristor power converters has made the speed control of both induction and synchronous motors very simple. Ac motors employing a variable frequency supply for speed control may be employed. These are becoming competitors to dc motors. The cycloconverters have advantages at very low speeds over the dc link converters. So, cycloconverter fed synchronous motors are used very commonly for driving steel mills. The converters facilitate four quadrant operation. These drives meet all the requirements mentioned above.

7.2.2 Continuous Hot Rolling Mills

Billets or strips are produced in these mills. They operate in the forward direction only. The mill-stands are of two kinds here, roughing mills-stands and finishing mill-stands. These stands are also two or four high depending upon the number of rolls the stand has. In a four high stand inner rolls are smaller than the outer ones. The gap of rolling is maintained by the outer ones. The metal is worked simultaneously in the finishing stands. The roughing operation is not simultaneous.

The basics of the process described define the requirements of the drive required for continuous hot rolling:

i. When a mill has to produce billets of different sizes the gap between working rolls of the mill-stand must be adjustable.

ii. To be able to reduce the thickness of the metal gradually the motors of consecutive mill-stands must have differing speeds. This requires that the motor must be capable of speed control in the range 1.5:2. Speed control should be accurate.

iii. The sag of the metal between two stands must be avoided. This sag may occur when there is a slight difference in the operating speed. The speed drop may occur due to sudden application of load, which normally happens when the metal comes into contact with the rolls. A closed loop control must assure quick restoration of the speed of the motor. The motor must have a very fast dynamic response to avoid sag.

Based on the above, a motor to suit the job may be selected. The motor must have a constant speed at a given setting. It must have its speed controlled over a given range. The dc motors controlled by Ward Leonard control, ac commutator motors and ac motors fed from thyristor converters may be advantageously employed here.

7.2.3 Reversing Cold Rolled Mills

In these mills the metal in the form of a reel is used to feed the mill-stand. On one side of the mill-stand there is a delivering reel and on the other side there is a receiving one. The mill-stand may be two or four high. When the receiving mandrel is empty, the threading of the metal on to this empty one is done manually. The speed of the motor should increase with uniform acceleration, ensuring the required tension and pressure. Otherwise the sheet would break. The sheet is allowed through the mill-stand in both forward and backward directions till the metal of desired thickness is obtained.

The drive, requirements immediately following the above process are the following:

i. The drive must be capable of reverse rotation. A four quadrant operation must be possible.
ii. One or two individually driven motors may be used. The work rolls may be driven directly. The back up rolls are provided with motion whereas the working rolls move by friction.
iii. The coiling motors besides the driving coilers ensure the desired tension of the strip between the coilers and mill-stand. This is necessary to prevent looping of the strip and/or breaking.
iv. The gap adjustment must he made simultaneously with the reversing. The latter is accomplished by screwing down the upper rolls.
v. The inertia of the motor must be kept low and lower than that of the rollers.
vi. Torque control as well as speed control must be possible to maintain constant tension of the strip. In a dc motor the torque control is possible both by field control as well as armature current control. As the diameter of the roller decreases the torque must also decrease. This is achieved by field weakening. However, field weakening in dc motors is limited by commutation and armature reaction effects. It is also limited by stability conditions of the motor. The armature current control may be employed beyond this limit.
vii. The acceleration of the drive must be uniform to avoid breaking.

The motor selected for this purpose must have its torque developed, causing a smooth acceleration. It should be capable of four quadrant operation with smooth speed reversal. Torque control at different speeds must be possible. To suit these requirement, a versatile motor is a dc motor controlled by Ward Leonard control with flywheel effect. Static Ward Leonard control may become economical with the availability of thyristors at reasonable rates. Three-phase ac commutator motor or cycloconverter fed synchronous motors are suitable for the job.

7.2.4 Continuous Cold Rolling Mills

These work only in the forward direction and no reversing is required. The metal passes in one direction only in different stands till the product has the required thickness. The mill may be two high or four high. The coiler roller requires accurate torque and speed control. The strip tension must be constant and large. Low speed operation is required while threading the steel into the rolls. Immediately after the threading the speed of the motor must be increased. The speed must be brought down while the metal leaves the mill-stand. A large variation in the speed of the mill drive is required.

7.2.5 Motors for Mill Drive

Dc motors are very versatile as motors for mill drives due to their characteristics of high starting torque, capability for wide range of speed control, precise speed setting, large overload capacity and pull-out torque. Care must be taken to have satisfactory commutation in the complete working range. They can be accelerated, braked and reversed very fast. Further, the inertia of the motor must be very small. The motors for mill operations are normally TEFC motors with a high class of insulation.

The speed control of dc motors is accomplished by Ward Leonard control with flywheel effect. Dc dynamic braking may be employed for quick stopping and braking at a controlled rate. Sometimes mechanical brakes are also employed. Conventional Ward Leonard systems may be replaced by thyristorized units.

When smooth speed control is required, ac motors with conventional methods of speed control are not suitable. Ac commutator motor, such as Schrage motor may be employed. The thyristor power converters provide a variable frequency supply which can be used for speed control of ac motors. Both torque control and speed control are possible. For low speeds, cycloconverters can be used to give a smooth speed control. Thyristorized dc drives can be used in the place of Ward Leonard dc drives.

7.3 CRANES AND HOIST DRIVES

7.3.1 The Requiremtents of the Drive

i. The motion of the crane hook is in all three dimensions.
ii. In crane drives, the acceleration and retardation must be uniform. This is more important than the speed control.

iii. For exact positioning of the load creep speeds must be possible.
iv. When the motion is in the horizontal direction braking is not a problem. This is a problem if the load overhauls the motor in vertical motion. In the case of vertical motion the movement of the empty cage has to be carefully considered. The speed must be constant while lowering the loads. The steady braking of the motor against overhauling must be possible.
v. The drive must have high speeds in both the directions. The motor must have high speeds at light loads.
vi. Mechanical braking must be available under emergency conditions.
vii. Power lowering may be used when an empty cage or light hooks are lowered.

The duty cycle of cranes depends upon some requirements. These are:

i. it must be able to perform strenuous duty
ii. it must withstand high ambient temperature
iii. it must be able to work in a dusty atmosphere
iv. it must provide trouble free operation
v. it should have rigid safety measures.

Cranes are classified depending upon the nature of duty they have to perform and also the duty cycle.

These are light duty, medium duty, heavy duty and continuous duty and are tabulated in Table 7.1, along with their applications.

7.3.2 Drive Motors for Cranes

The crane motors can be either dc or ac motors.

They are compared in Table 7.2.

Even though the electrical characteristics make dc motors suitable as crane motors, the lower inertia and simple and economical construction of cage motors favour them as crane motors. With the advent of thyristors and associated power converters it is possible to have torque control during starting, running and braking.

7.3.3 dc Systems for Cranes

Speed control is achieved by means of Ward Leonard system with all facilities of speed control in forward and backward directions with regeneration.

Among the dc motors, series motors are extremely suitable for crane operation. They have the following features:

i. They have very good starting torque, high torque capability at low speeds and light torques at high speeds.
ii. Simple arrangements for braking of the motor.

Table 7.1 *Summary of requirements of motors for different industrial drives*

	Starting torque	*Overload capacity*	*Acceleration*	*Speed range*	*Speed control method*	*Type of motor*
1. Textile mills ginning	Standard	Standard		Operating speed range 200 to 1500 rpm. The operation is however at constant speed.	No speed control is required as the operation is at constant speed and load.	Standard TEFC squirrel cage motor.
Cording	High starting torque. 2.75 to 3.5 times rated value.	High overload capacity 3.75 to 3.00 times rated value.	Should be capable of prolonged acceleration as the starting period is large proper rating must be there for the motor due to losses at starting.	Operating speed is 600 rpm or 700 rpm. Constant speed operation.	No speed control.	Slip ring rotor with rotor resistance starting. High torque motors with DOL starting.
Spinning	150–200% of rated value	200– 275% of rated value.	Acceleration must be constant or uniform so that there is no breakage of thread.	Constant speed operation.	Two speed motors are preferred.	Ring frame motors.
Looms	2 to 2.3 times the rated torque.	2.3 to 2.7 times rated torque.	Frequent starts and stops.	Constant speed of operation.	No speed control.	Totally enclosed high torque squirrel cage motors 0.3 to 2.2 kW.

(*Continued*)

	Starting torque	*Overload capacity*	*Acceleration*	*Speed range*	*Speed control method*	*Type of motor*
2. Cranes			Acceleration and retardation must be suitable for accurate positioning. Creep speeds should be possible.	Speed control of 1:20.		Ward Leonard dc systems.
3. Steel mills reversing hot rolling mills	High starting torque.	3.00 times the rated torque.	Frequent starts and stops. Speed reversal.	Wide range of precise speed control.	Ward Leonard, static frequency control of ac motors.	Ward Leonard systems with flywheel. Fast response drives with precision speed control. Thyristor converter controlled preferably cyclo converter fed ac motor.
Continuous hot rolling mills	High starting torque					Fast response systems. Thyristorized power converter controlled induction motor, cyclo converter fed ac motor.
Reversing cold rolling mills	High starting torque					Fast response systems. Thyristorized power converter controlled ac motors, preferably cyclo converter fed ac drive.

4. Sugar mills	Depends upon centrifugal action. High starting torque.	Standard		500–1000 rpm	Stator voltage variation, rotor resistance control, pole changing methods. As the speed control is frequent there may be possibilities for regeneration. Static frequency control of induction motors. Normal *V/f* control giving due consideration to high starting torque if necessary.	Two speed pole change motors. Solid rotor induction motors controlled from ac voltage controller, dc link inverter, rotor chopper control.
5. Compressors	High starting torque. Onload starting soft starting may be employed as the conventional methods cause peaks of starting current and voltage dips.	Should be sufficiently good.	Controlled acceleration.	Speed range 700–1000 rpm	Static frequency variation for speed control. Slip energy recovery schemes if slip ring induction motor is used.	SFC controlled induction motors or synchronous motros. These provide soft starting facilities.

(*Continued*)

	Starting torque	*Overload capacity*	*Acceleration*	*Speed range*	*Speed control method*	*Type of motor*
6. Pumps and blowers	The torque is proportional to square of speed. Standard requirements of starting torque are 200 to 300% of rated torque for small motors and 150–200% for large motors	Standard		Speed range 1:5	Slip energy recovery schemes, stator voltage control SFC feeding.	Three phase slip ring induction motor, solid rotor induction motor, SFC fed induction or synchronous motor.
7. Cement mills Raw mills and cement mill drives	125% of rated torque with relatively small starting current to avoid voltage dips.	250% of rated torque. The motor should be able to withstand 50% frequently oc-curing overload in an hour.	Frequent starts and stops in a duty cycle.	Mill speed is about 15 rpm.	Speed reduction by gearing. Gear-less drives use SFC with dc link converter or cyclo converter. Use of Twin motor may eliminate huge gear boxes.	Slip ring rotor with rotor resistance starting, synchronous motor with SFC. Squirrel cage motors fed from SFC having soft start facility.
Kiln drives	200–250% of rated torque	200–250% of rated torque	Full load acceleration in about 15 seconds.	1:10	Ward Leonard control. Static Ward Leonard using thyristors has advantages.	Thyristorized dc drives, Twin motor dc drives.
Crusher motors	160% of rated torque. The motor must withstand locked rotor current.	200–250%; 15% overload for 15 seconds.				Slip ring motor with rotor resistance control.
Fan drives	120% of rated torque	200–250%		1000 to 750 rpm	Rotor resistance control, thyristorized control.	TEFC slip ring motor, thyristorized dc drives.

Table 7.2 *A comparison of dc and ac drive for crane applications*

DC drive	*AC drive*
1. Fast acceleration and smooth speed control	All ac motors are not self starting
2. Speed-torque curve can be modified to suit the purpose in a simple manner with resistance in the armature or field circuit.	Induction motors are self starting. Squirrel cage motors require special designs for improved starting performance. Slip ring induction motors can be made to have improved starting performance by means of rotor resistance control.
They reduce the time cycle.	They have large time cycle.
3. Dc series motors have very high torque at low speeds particularly at zero speed, which is more attractive for crane operation.	
4. Frequent maintenance of the motor is required due to commutator	Less maintenance.
5. The drive motor and its control are costly	Squirrel cage motors are cheaper, simple in construction and robust.
6. Dc source is required. A rectifier may be used to convert ac to dc	
7. Inertia of the motor is high.	Inertia of cage motors is less.

iii. Power demand under highly loaded conditions decreases due to fall in speed.
iv. Electrical braking is possible even at low speeds due to low critical speeds.
v. Very light conditions are not possible. Suitable changes may be made in the circuitry to make it to run at low speeds and light load conditions.
vi. The lowering speeds increase with the load. Regenerative braking is not possible to limit this speed. The speed of the empty cage can be limited by limiting the current to full load value.

However, the speed-torque characteristics of dc series motors may be modified to suit all the phases of crane control, and are depicted in Fig. 7.6.

7.3.4 ac Systems for Cranes

Among ac motors, squirrel cage motors are normally used. They have the following features:

i. No speed control is required. However with the development of variable frequency converters which solve the problems of speed control of induction motor, inverter fed motors are used in situations requiring wide range of speed control.

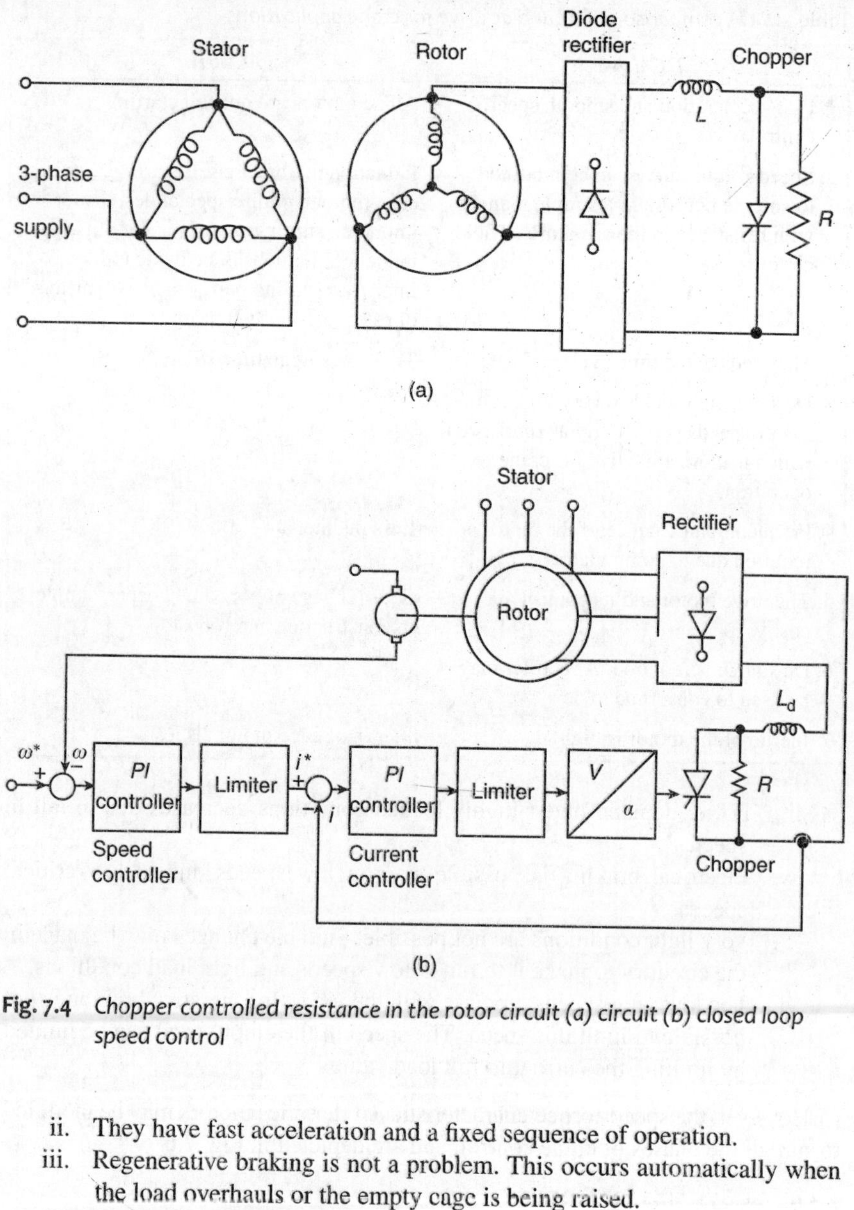

Fig. 7.4 *Chopper controlled resistance in the rotor circuit (a) circuit (b) closed loop speed control*

ii. They have fast acceleration and a fixed sequence of operation.
iii. Regenerative braking is not a problem. This occurs automatically when the load overhauls or the empty cage is being raised.
iv. The motor is simple and robust.

These are not suitable when a large number of startings and brakings are required. A starting torque up to 250% of full load torque may be obtained. In cases of very high starting torque these are not suitable. Non-uniform sequence of operation cannot be handled and conventional methods of speed control are not suitable if precise speed control is necessary.

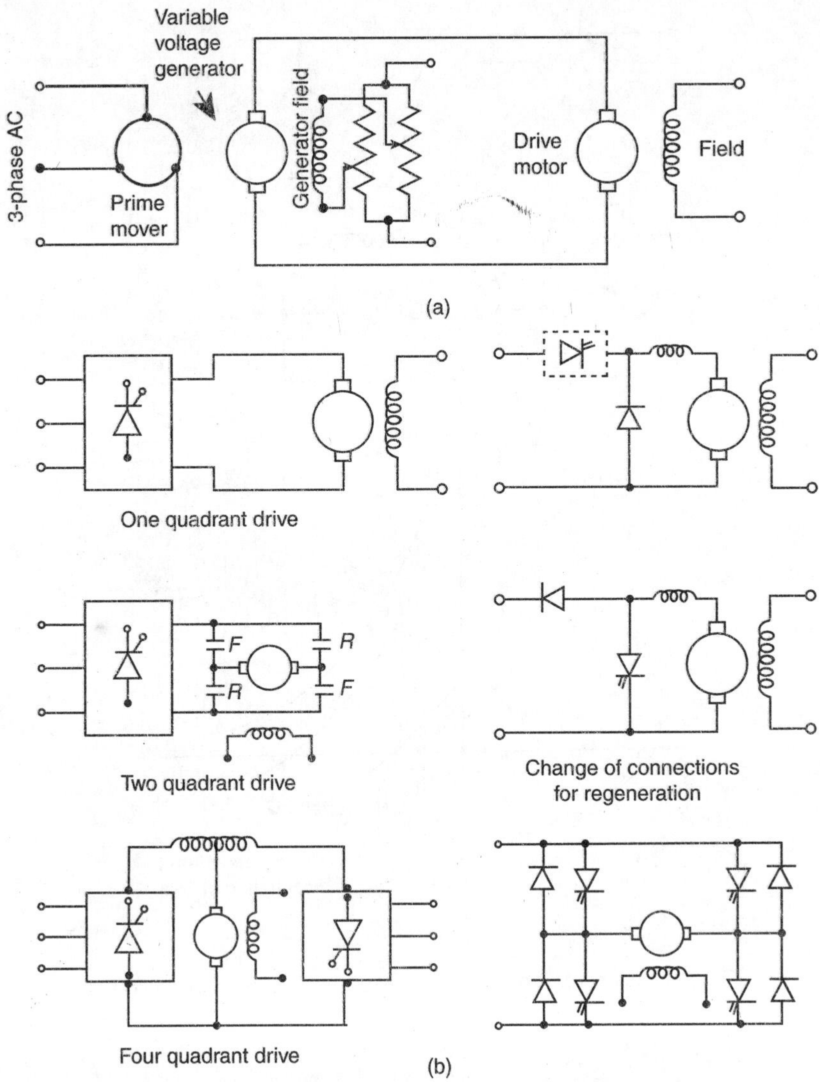

Fig. 7.5 *Ward Leonard control of a dc motor (a) conventional circuit (b) static power converters*

Slip ring induction motors for cranes and hoists have the following features:

i. The speed-torque curve can be modified by suitably altering the rotor resistance. The starting torque can also be varied to the required value.
ii. Regeneration is possible. Reverse current braking can be employed limiting the current to the desired value by rotor resistance. DC dynamic braking may also be employed.

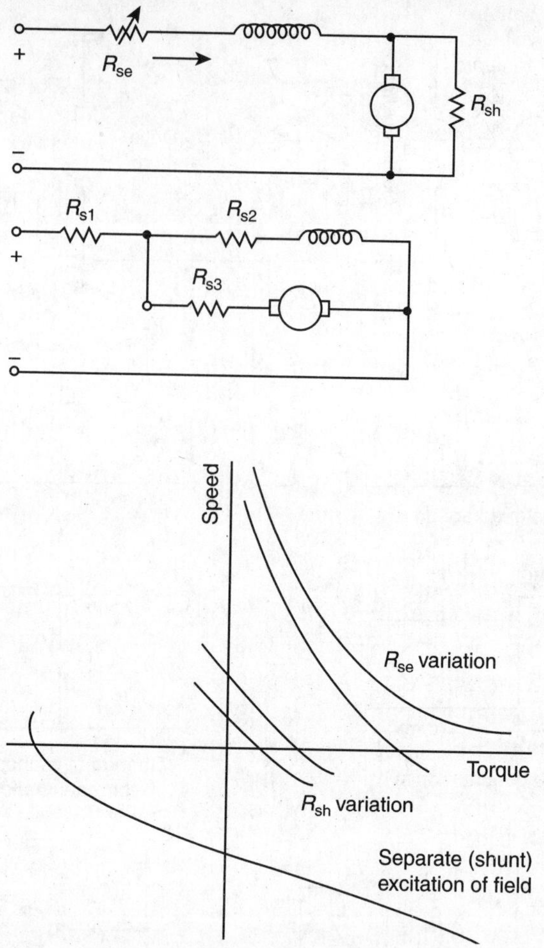

Fig. 7.6 *Connection of series motor to have all phases of control*

While selecting a motor for crane duty the following points require consideration:

i. Breakdown torque must be greater than 250% (in the range of 275–300)
ii. The inertia must be small.
iii. It must withstand large frequency of starts.
iv. The motor must have sufficient running torque, starting torque and thermal capability for the given duty cycle.
v. The duty cycle must be specified.

7.4 CEMENT MILLS

7.4.1 Stages in Cement Production

The raw materials of cement are lime and silica. Alumina and ferric oxide are used as fluxing agents.

i. Collection of raw materials such as lime stone. This is transported to the mill site and crushed there if the quarry is far off. If the quarry is nearer it is crushed at the quarry itself and transported to the mill site.
ii. Grinding of this crushed lime stone after the addition with bauxite, iron ore, etc. By passing air through bottom the lime powder is homogenised
iii. This is fed to the cement kilns where the cement clinker is made at high temperature. The process where dry powder is used is called dry process.
iv. Wet processes of cement making are also popular. Here the slurry is made by crushing or grinding the lime stone, bauxite with water. This is then fed to the kiln through the kiln feed tank.
v. Dry process is preferred to wet process due to the reduced quantity of fuel required. However the latter becomes economical if the materials are already wet and drying them may not be economical.
vi. The hot clinker is then air cooled in special types of coolers and made ready for storage.
vii. After storing for a few days gypsum is added in required quantities and ground to the required fineness.

Every stage has its own drive. Several drives in cement making are raw mill drives, cement mill drives, kiln drives, crusher drives, waste gas fan drives and compressor drives.

7.4.2 Requirement of Mill Motors

i. They should have high starting torque. The starting current must be limited to a maximum of two times full load value to minimise voltage dips. The breakdown torque should also be high so that sufficient overload capacity is available.
ii. An overload capacity of 50% for one minute may be necessary, occurring for four times in an hour.
iii. Three starts from cold conditions and two consecutive starts from hot conditions per hour against full load.

These are very well met by a three-phase slip ring induction motor. Suitable starting torques may be accomplished at reduced starting currents by means of rotor resistance. The motor must have sufficient thermal rating to have frequent starts both under cold as well as hot conditions. The power factor may be improved by capacitor bank.

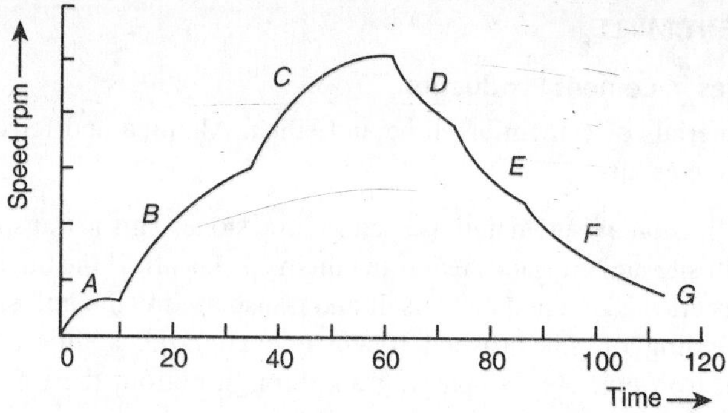

Fig. 7.7 *A typical load cycle of a centrifuge A—Charging; B—Intermediate spinning; C—Spinning; D—E—F—Regenerative braking; G—Plugging for ploughing*

The power rating of the motor is rather high. As the motors have large power ratings and power transmission using gear boxes at this power level is not practical, two motors of identical capacity are used. Both of them may be equipped with identical rotor starters.

Gearless drives are preferable here. Gearless drives using the principle of levitation may be employed. Variable frequency drives using either cycloconverters (Fig. 7.7) or dc link converters may also be used. As the price of thyristors is becoming less, these thyristorized drives are becoming very popular.

7.4.3 Kiln Drives

The rotary kiln drives depend upon the type of cement making process (wet or dry). These are, in general, tubular tilted from the horizontal position with a ring fitted around them. This ring gear engages with one or two pinions. A variable speed motor drives the pinion.

The requirements of a kiln motor are the following:

i. Power requirement is very high.
ii. Speed control ratio is 1:10. Very low creeping speeds of 1 rpm may be required.
iii. Starting torque should be in the range 200 to 250% of full load torque.
iv. The acceleration of the drive should be completed in about 15 s.
v. For small periods an overload capacity of 200–250% may be required.
vi. The motor must have suitable control for inching and spotting during maintenance.

The motors that meet the above requirements are ac commutator motors and Ward Leonard controlled dc motors. These have the disadvantage that the highest rating is limited by the commutator. Speed range is 1:2 in ac commutator motors whereas a speed range of 1:10 with crawling speeds is possible with dc motors. Capital outlay, lower efficiencies and limitations due to commutator, either in the Ward Leonard or ac commutator motors, may be overcome by the use of thyristorised drives. These have a wide range of speed control as against ac commutator motors.

When two motors are used to deliver the power they must be designed to have equal load sharing without overloading any one of them. They must maintain the same speed. They may be series connected or parallel connected with closed loop speed control.

Converter fed dc motors in static Ward Leonard control have limitations on maximum operating speed and power rating due to the presence of a mechanical commutator. The ripple content in the armature current and possible discontinuous conduction further affect the commutating capability of the motor. Therefore for power ratings beyond this value, ac drives are suitable. Tubular mills for cement are very slow speed high power loads. A cycloconverter fed synchronous motor meets the requirements of a drive motor. The salient features of this system are:

i. The drive can be controlled to have an excellent dynamic behaviour with fast regenerative reversal.
ii. The converter can be used for synchronous starting by varying the frequency with smooth acceleration up to desired speed. The disadvantages, such as peak starting current and consequent voltage dips in the supply, can be completely eliminated.
iii. A continuous gearless drive is possible at crawling speeds.
iv. A motor with self control has the characteristics of a dc motor with respect to both steady-state and dynamic behaviour. It is free from hunting. It is mechanically strong, requiring little maintenance and can be built to have a rating several times that of converter fed dc motors. The low inertia of the motor is also responsible for the fast response.
v. A four quadrant operation is simple and straightforward.
vi. Poor line power factor similar to that of converter fed dc motor.
vii. Field weakening is possible above base speed.
viii. The smooth speed control with minimal torque pulsations particularly at low speed is an added attraction, mainly when compared to synchronous motors or dc link converters.

7.4.4 Crusher Drives

The requirements of a crusher drive are as follows:

i. The starting torque is of the order of 160% of full load torque.
ii. The breakdown torque is of the order of 200–250% of full load torque.
iii. The rotor must be capable of withstanding a locked rotor current without any limiting equipment, for one minute. This kind of situation may occur in case of jamming of the crusher due to very big boulders.
iv. Adverse conditions of loading may be encountered.
v. Overload capacity of 15% for 15 s and 20% for 10 s may be required.

Slip ring induction motor with rotor resistance starters and speed control may be suitable for crusher drives. A dc chopper can be used to control the resistance.

7.4.5 Fan or Blower Drives

The drive motors are located in outdoor or semioutdoor locations. Totally enclosed fan cooled motors are suitable, depending upon the location of the motor. The torque requirements are: Starting torque is 120% full load torque; breakdown torque 200–250% full load torque. The speeds are in the range of 1000–1500 rpm.

Slip ring induction motors with rotor resistance control are suitable as drive motors. A subsynchronous converter cascade in the rotor circuit may also be employed for speed control. The latter has improved efficiency of operation. The following are the features of this drive:

i. The design rating of the converter depends on the speed range required. The converter must be capable of handling high currents at high speed and high voltages at low speed. Using a switchable converter cascade the rating may be decreased.
ii. When the lowest speed is other than zero, starting equipment is required. Under emergency conditions the motor may be required to operate with this resistance as a conventional motor. The resistance must be able to withstand the operation under running conditions.
iii. The torque developed has pulsations and harmonics causing heating. A 12 pulse converter may be used to reduce these effects. The line side inverter presents several harmonics to the line which may cause the distortion of the input voltage to the stator.
iv. It has a very poor power factor. Methods are available to improve the line p.f.

7.4.6 Compressor Drives

The drive motors for compressors have a rating in the range of 300–450 kW. Compressors have to be started on load or sometimes there may be means of unloading for starting. For starting on load, high starling torque is required. Normally care must be taken while choosing a starting equipment to limit the starting current. Starting current peaks are not permitted as they cause disturbances such as voltage dips. Totally enclosed fan cooled motors capable of operating in the speed range of 750–1000 rpm are employed.

Conventional squirrel cage motors may be used if starting on no-load can be accomplished. The starting method may be reduced voltage starting, to limit the starting current. If the compressor has to start on the load, high starting torque at reduced starting current may be required. A slip ring induction motor with rotor resistance starter may be used.

The ac drives making use of an induction motor or synchronous motor whose speed is controlled by means of a static frequency converter may also be used. These converters can be used for starting purposes also. They provide better starting conditions. A sufficient amount of accelerating torque with a current of 1.5

times full load current may be achieved. Locked rotor current and torque loose their significance and it is just sufficient to have reserve torque for acceleration. Consequently, voltage dips and the severe burden on the mains may be avoided. When synchronous motor are used, the line side converter uses line commutation and the machine side converter uses machine voltages for commutation. Therefore a simple converter is sufficient. Only at low speeds is the commutation assistance required.

Four quadrant operation is straightforward. Very high speeds may also be obtained. Direct coupling to load is possible. The operation of compressor is variable torque operation. At low speeds torque required is very low and simple control of V/f may be sufficient. No voltage offset for stator resistance compensation is required.

7.5 SUGAR MILLS

The sugar crystals are separated from the syrup by means of a centrifuge. The separation is accomplished by the centrifugal forces set up. The centrifuge is started to a speed of around 200 rpm at which the charging of syrup takes place. During charging the motor is disconnected from the supply. The centrifuge is spun at speeds of 500 and 1000 rpm. The speed is then reduced in steps to about 50 rpm, at which ploughing takes place. To reduce the energy lost during starting and braking of the motor the centrifugal action is performed at different speeds. A typical load cycle of a centrifuge is shown in Fig. 7.7.

The motor used to drive the centrifuge must be a variable speed motor. The centrifugal action is performed to reduce the energy loss due to acceleration and braking. The regenerative braking may be advantageously employed. Normally two stage acceleration and braking are employed. While reducing the speed the regenerative braking is employed till the lowest speed is obtained, where ploughing takes place.

Pole changing induction motors are suitable for the purpose. A large pole winding is switched on to get a speed of 200 rpm, at which charging takes place. After the charging the next pole winding is switched on to get the next higher speed. In two stages the highest spinning speed is achieved. The motor can then be switched on to lower speeds and the regenerative brakings may be obtained in stages. Thus pole change motors give definite speeds of operation. Also, when switched from higher speed to lower speed regenerative braking may be accomplished.

A high resistance rotor induction motor with stator voltage control is suitable to drive the centrifuge. Due to high rotor resistance, stable operation of motor is possible in the complete speed range from zero to rated (synchronous) speed. The voltage control is used for speed control in the first quadrant. Solid rotor induction motors have high inherent rotor resistance and are economical when employed as drives for centrifuges. The stator voltage control can be accomplished by means of an ac voltage controller using antiparallel thyristors. The motor can be controlled in speed and run at desired speed; it can be accelerated and braked (Fig. 7.2).

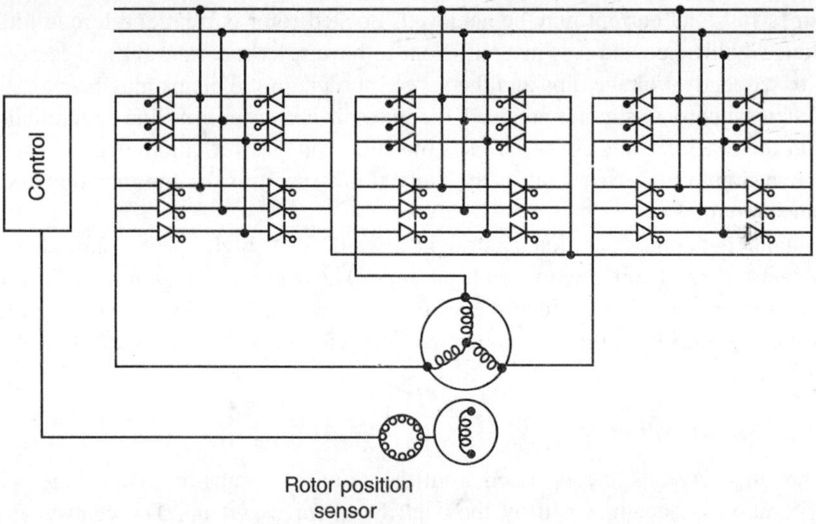

Fig. 7.8 *Cycloconverter feeding a synchronous motor*

The high rotor resistance of a squirrel cage motor, even though it provides a stable characteristic, has poor efficiency. Hence slip ring induction motors with variable resistance in the rotor circuit and variable voltage on the stator are a satisfactory drive system giving a reasonable efficiency under running conditions. The schematic and features of this drive have been discussed in Chapter 4. The braking when these drives are employed can be dc dynamic braking. No regenerative braking is possible since the energy is lost in the external resistance and cannot be regenerated to the mains.

With all the types discussed above an automatic speed control system to suit the duty cycle can be employed.

Alternately, converter fed induction or synchronous motors may also be used. A static frequency converter (dc link or cycloconverter) may be employed to control the drive motor. By this, soft starting and operation at definite speeds are possible. The speed control range is wide and regenerative braking is possible.

Motor Design When pole change motors are used, they require special design to withstand the fluctuations of current when switching over takes place. The line also is subjected to disturbances. Special starting equipment is required. When pole change motors or high resistance cage motors are used the current fluctuations may cause heating of the motor. Protective devices may be required against overheating. Special starting current peaks may be there when an induction motor controlled from a voltage controller is employed. Because of low flux levels at lower speeds, considerable derating of the motor is required. The derating is also due to non-sinusoidal currents. A considerably large size motor is required. Special starting may not be required as the voltage controller employed can be used for starting also. A slip ring induction motor with rotor resistance does not have high peak currents at starting. No special design may be required.

When fed from a variable frequency supply using static frequency controller, the motor is versatile and free from all the difficulties stated above. The motor does not require any special design. No protective devices are required if there is no overheating.

The mounting of the motor is vertical. As oscillations are present on the rotor, the air gap must be large. The oscillations may be present in pole change motors due to switching. In motors supplied from thyristor converters these are presented by the non-sinusoidal current waveforms.

The motors must have humid proof insulation.

7.6 MACHINE TOOLS

The requirements of motors used for machine tools are:

i. The motors must be reliable and low cost, requiring less maintenance.
ii. They must be capable of speed control. Some applications may require operation at fixed speeds. Sometimes stepless and smooth speed control is required for better machining timings and surface finish.
iii. Starting torque may vary from about 10% to 250% of full load torque depending upon the type of machine tool.
iv. The acceleration of the motor should be sufficiently fast to avoid motor heating during starting. The braking must also be effective and fast. This is because frequent starting and stopping are required. In view of safety reasons, braking must be such that exact stopping of the tool is assured.
v. The duty cycles are specified for the machine tool operations. The design of the machine tool depends upon the duty cycle.
vi. Peak loads of short duration may alternate with light load in certain operations. A smaller size motor may be selected if a flywheel is used. The motor should have sufficient speed regulation to make use of flywheel. High slip induction motors or cumulatively compound motors are used.
vii. Variable speed operation with constant torque at all speeds may be required in machine tools, such as grinder, planer, polishing, rapid reversing, etc. Variable speed operation with constant power also finds application.
viii. Some machine tools require very high speeds of operation. These are high speed grinders.
ix. Numerically controlled machine tools are being preferred to conventional machine tools. These have increased utilisation of the machine. The production rate can be increased. These are costly at present, but are expected to become economical.
x. The requirements of the drive motor are fast response, wide range of speed control, low vibrations, better thermal capacity, low maintenance, etc. To have fast response the inertia must be low. They must give precise positioning.

A choice of the motor must be made to meet the above requirements.

Due to the simple, economical and robust construction, realiability and less maintenance, squirrel cage motors are suitable for driving machine tools. Till the advent of thyristor power converters which are capable of providing variable voltage, variable frequency supply, the speed control of these motors was a problem. Using power transmission equipment, such as gears, speed variation in steps is achieved. Fixed speeds of drive are possible. The vibrations caused affect the accuracy of the output. A four pole or six pole motor may be used. A motor with larger number of poles has poor power factor, smaller starting torque and lesser efficiency. Two pole motors are noisy and have vibrations.

Pole change multi speed motors are available when definite stepped speed operation is allowed. These connections are available for constant torque as well as constant power operation. They provide high and low speeds.

When smooth speed control is required for better finishing and machine timing, suitable controls are necessary. The advent of thyristors and associated power converters has paved the way for the smooth speed control of ac motors. Both constant torque as well as constant power operations are possible. The variable frequency converter used for speed control may also be used for starting purposes. Variable frequency starting has advantages and imparts a very good starting behaviour. The starting or accelerating torque can be controlled. These are also costly at present. The availability of thyristors at reduced prices and in higher ratings may make these drives popular in future.

The kW rating of the motor is decided by the duty cycle. A suitable rating of the motor must be selected, having sufficient mechanical overload as well as thermal overload. The motors for duty cycles S1–S6 are listed.

S1 Hydraulic pump motors, lubrication pump motor, coolant pump motor.

S2 Rapid transverse motor.

S3 Main motor for gear shaper, and for drilling machine.

S4 Main drive motor in lathes without clutch in the drive, work head motor in grinding machine, main drive motor in gear nobbing machine, coolant pump motor with frequent starting and stopping.

S5 Work head motor in grinding machine with electric braking.

S6 Main and feed drive motors with clutch in the drive.

The frequency of starting and stopping decides the type of starter as well as the braking method. Slip ring induction motors with rotor resistance starting are desirable. If plugging is employed for frequent stopping the motor gets thermally overloaded. The motor must be disconnected at zero speed. DC dynamic braking can also be employed. This takes more time for braking compared to plugging, but the heating is decreased. For safety reasons, along with electrical braking, mechanical brakes must also be provided as a standby.

High slip motors are required for use with a flywheel, to drive a load which alternates with no-load conditions. When the load is applied the speed should fall by 8–10% so that the flywheel can provide a part of power required by the load from its stored energy. During no-load condition the speeds goes up and flywheel stores this energy. Slip ring motors with additional rotor resistance are advantageous. The resistance causes necessary slip. The motor does not get heated. The inertia must be specified. When dc motors are employed compound motors with cumulative compounding are used.

Motors for numerically controlled machine tools may be specially designed dc motors, inverter fed motors or permanent magnet motors.

7.7 PAPER MILLS

The raw materials for paper making undergo two processes before the paper is available: The pulp is made from the raw materials. This pulp is converted to paper in paper making machines. The drives required for making pulp are different from those required for paper making.

The conversion of raw materials to pulp is accomplished either completely by mechanical process or by a combination of chemical and mechanical processes.

When the pulp is made by purely mechanical means, the logs of wood cut to 1 m length are ground in big grinding machines. The grinders operate at a constant speed, the speed of operation being in the range of 200–300 rpm. No speed control is required. No load starting of the grinders is possible. The power ratings of the motor required are relatively large. The grinders require a very large power and low speed. For such operation, synchronous motors are useful. Geared drives may be used for reduction in speed. A cycloconverter fed synchronous motor can be operated at low speeds and the gears can be completely dispensed with. The problems of starting can also be avoided. A converter fed synchronous motor is also suitable. The motors can drive the load from a separate chamber and are protected from the humid atmosphere.

In the second process, which combines mechanical and chemical processes, the logs of wood are first chopped into smaller pieces. These are treated with suitable chemicals, simultaneously beating the pieces to pulp by means of beaters. The beaters require starting on load and against a large inertia, due to a large disc on which the knives are mounted. The load characteristics of the beaters are also random. The speed of operation is less than 200 rpm.

The rating of the motor ranges to thousands of watts. For driving beaters, slip ring induction motors are suitable. The desired starting torque of the motor can be achieved by a proper rotor resistance. For processes, such as chipping and refining, synchronous motors are employed as they are available in large power ratings. An s.f.c. fed synchronous motor may be employed for beaters.

The conversion of pulp is effected in several stages or sections. In these sections the water is removed from the pulp and it is pressed to sheets of paper which are finally wound up on a mandrel. These sections are wire (couch) section, pressing section, dryer, calender and reel section.

The paper making machine should satisfy the following requirements.

i. The speed of the paper machine must be constant in view of economy while forming the sheets of paper.
ii. A speed control range of 10:1 is required so that it is suitable for performing several jobs.
iii. The speeds of individual sections should be varied independently to allow an elongation of 5% of the web on the wet end of the paper. The quality of the paper is decided by this elongation. To allow free hanging of the web between sections, at the dry end of the paper a definite amount of tension is required and it must be regulated. The successive sections must be run at speeds with a definite difference. This relative speed between the sections also affects the pull on the paper. It must be regulated so that there is no tearing of the paper.
iv. The arrangement should be capable of taking up sag.
v. Even with correct speeds in the last two sections, uneven drying of the paper may cause variations in tension, which must be taken care of by suitable tension control.
vi. The motor must be capable of inching in order that the wire be cleaned up.
vii. Every section must be able to run at crawling speeds.
viii. The starting and acceleration of the sections must be smooth as well as quick. The starting system should be such that peaks of starting current may be avoided, besides obtaining sufficiently high accelerating torques for fast acceleration.

The paper making may employ either group drive or individual drive for several sections. In the group drive a line shaft is driven by the motor with different gear arrangements or belts as power transmitting equipment to drive different stages of paper making.

The drive motor may be either a dc motor or ac motor. DC motors having Ward Leonard speed control provide a lossless smoothly variable speed for the sections. Ac motors controlled from variable frequency sources are available now-a-days and they can provide the required lossless smooth speeds. In conventional ac systems an ac commutator motor may be used for stepless speed control. This is more compact when compared to a Ward Leonard controlled dc motor. The constancy of speed required for a paper mill cannot be maintained with an ac commutator motor, as the speed of the motor falls with load on the motor. The speed range is limited. It has sluggish transient behaviour compared to a dc drive.

With dc motors, static Ward Leonard control can also be employed.

In the case of individual drives each section has its own drive motor. The speeds of these motors are varied by means of voltage variation. The changes in the speeds of a motor with respect to the others can be achieved by field control.

A comparison of group drive and individual drive for paper making is given below:

Group drive	*Individual drive*
1. Only one motor to do all the jobs. Cost is low as far as the motor is concerned. But the additional transmitting devices affect this saving.	Every section has its own drive motor. Cost is more as far as the motors are concerned.
2. Automatic closed loop control for correcting the speed variations is not possible. The speed variations may be caused by the slipping of belts.	Automatic control is possible for correcting any speed variations.
3. The speed difference between sections (draw) must be constant in paper mills. If the speed difference changes due to any of the reasons the correction is difficult.	This can be accomplished very easily.
4. Draw actually decides the quality of the paper. Or depending upon the quality of the paper draw is set. It is difficult to set the draw repeatedly.	It is possible to set the draw setting required for a desired quality of paper repeatedly.
5. In case of formation of loop it can be removed only by means of repeated trial and error correction by affecting draw setting. The rate at which the loop is removed depends on speed.	It can be removed by changing the speed of one section without affecting draw setting. It is constant at all speeds.
6. Crawling speeds for checking up repaired sections are difficult.	Are easily accomplished.
7. Inching reverse facility to clear fanning of paper in dryer is not there.	It is possible to have this facility.
8. Dangerous for the operator.	The protection of each section can be very easily accomplished. The section cannot be put into work unless it has sound safety conditions. Only that section which is faulty needs shut down.
9. Overloading of a particular section overloads the drive motor continuously till the motor breaks down.	Overloading of a section affects that section only and can be switched off once it goes beyond safety conditions.

7.8 COAL MINES

The motors used for coal mining must be flame proof. They operate at high ambient temperatures. Sometimes the environment may be humid and the motors should have humid proof insulation. The motor must satisfy very stringent specifications.

The motors used in coal mines can be classified into two groups. The motors in the first group are drives for mine auxiliaries such as compressors, pumps, etc. The motors of the second group are used to drive the cutters, drillers, etc., which are directly used in the process of mining.

To determine the rating of the motor the load diagram must be known. One of the criteria discussed in Chapter 5 may be used to determine the rating.

The motors used for direct mining process must satisfy the following requirements:

i. Coal cutting or drilling machines do not require any speed control.
ii. High starting torque may be required. High torque squirrel cage motors (double cage) find application as drives for coal cutting equipment.

For haulage purpose the motors must be able to start a large drum. They also require a high starting torque. The load is hauled up a gradient. The motor must be capable of frequent starts and stops. The hauling takes place in stages. If a clutch is used, a heavy starting torque may not be required. Constant torque and constant power operations of the motor may be required. To meet these requirements the best suited motor is a slip ring motor started with rotor resistance starter, which limits the starting current, besides giving a high starting torque.

As mine winder (motor) the following motors find application:

Ward Leonard controlled dc motor. Slip ring motor with rotor resistance control.

The drive speed of the mine winders must be precisely controlled. The choice of the motor rating is based on the load diagram or duty cycle of the mine winder. A typical load diagram of a mine winder is shown in Fig. 7.9. The duty cycle is such that the mine winder starts from zero speed. It is accelerated to rated speed, which is maintained constant for a period of time. The motor is braked to decelerate the load to zero speed. The next cycle starts after a rest period. The drive motor should have sufficient starting torque to start the winder on load. This torque has a constant component and accelerating torque. Some regeneration is possible during braking. The torque due to the rope weight must be balanced. The torque/time diagram in Fig. 7.9 is simplified and does not consider this. The power diagram can be determined, using which the rating of the motor can be determined, based on the methods described in Chapter 5. While deriving this power diagram, speed changes must be taken into consideration. The typical power curve shown in Fig. 7.9 assumes constant speed and hence it is made up of striaght lines. Based on this also, a slip ring motor with rotor resistance starter is best suited.

The centrifugal pumps are used in mines for several jobs. The drive motors used for these pumps have been discussed in detail. For fan drives normal squirrel cage motors may be used.

7.9 CENTRIFUGAL PUMPS

Centrifugal pumps are used as boiler feed pumps and for pumping water in water pipe lines. The former must be adapted to the variable output of the steam generator. In the latter, varying water requirements of an area decide the delivery rate.

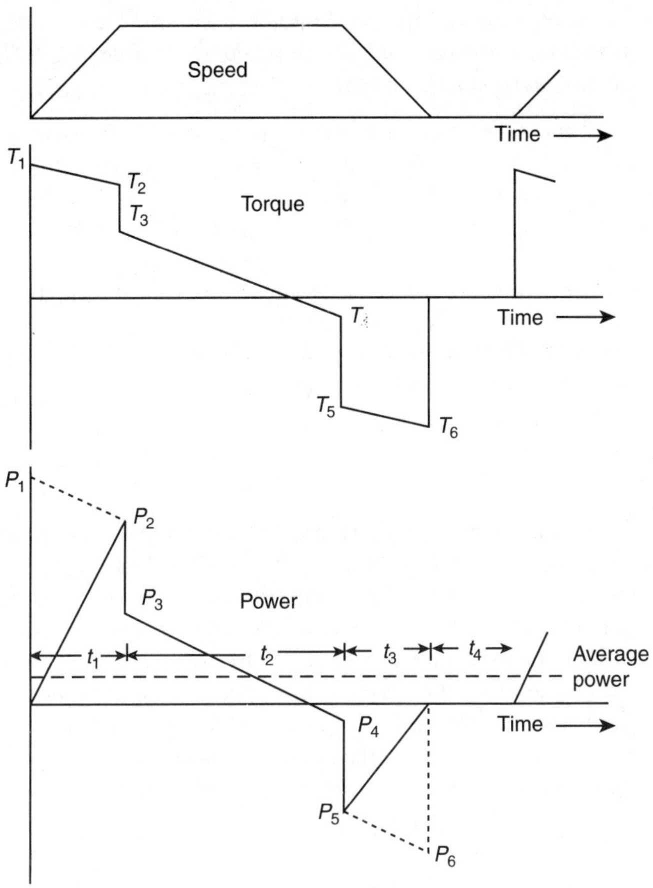

Fig. 7.9 *Typical duty cycle of mine hoist (winder)*

The delivery rate of a pump is regulated by its speed. A variable speed drive system is required to drive the centrifugal pumps.

The requirements of a centrifugal pump load are:

i. ***Starting:*** A smooth starting of the load may be imperative so that no pressure surges exist in the hydraulic system. Also, the starting of the motor must avoid current peaks in the supply system while providing starting torque required by the load.

ii. ***Operating safety—Flexibility:*** The system used should have operating safety and flexibility. In case of any failure in the system the system must be able to drive the load so that continuity is maintained.

iii. ***Available power and speeds:*** The centrifugal pumps are developed for all powers and speeds. Speeds up to 1800 rpm are there. In some exceptional cases speeds of 3600 rpm, 6000 rpm are required.

iv. **Power Consumption:** The drive motor used is based on the power requirement. The speed control is required over a wide range. The power consumption must be optimum.

Based on the above requirements, wound rotor induction motors with sub-synchronous converter cascade in the rotor circuit and converter fed synchronous motors are used very widely as the drives for centrifugal pumps. Even though dc motors fed from converters are versatile in smooth starting and smooth speed control they cannot be developed for high power and high speeds due to the commutator. Hence they are very seldom used as drives for centrifugal pumps.

7.9.1 Suitability of a Wound Rotor Induction Motor with Subsynchronous Converter Cascade in the Rotor Circuit

The following points favour the application of a slip ring motor having static slip energy recovery scheme in the rotor circuit, to drive a centrifugal pump (Fig. 7.10).

i. The delivery rate control of the pump is accomplished by speed control of the pump. The speed control of the motor is done by recovering the slip energy to the mains. The system has no slip losses, i.e., losses occurring due to drop in speed. Therefore the power requirement of the drive is to drive the load, plus converter and motor losses only. Based on the power consumption the system is preferred as it is very efficient.
ii. A very smooth stepless control is possible, to avoid surges in the hydraulic system due to switching the pump sets on and off.
iii. Design capacity of static converter; the advantage of subsynchronous converter cascade in the rotor of an induction motor over the stator fed

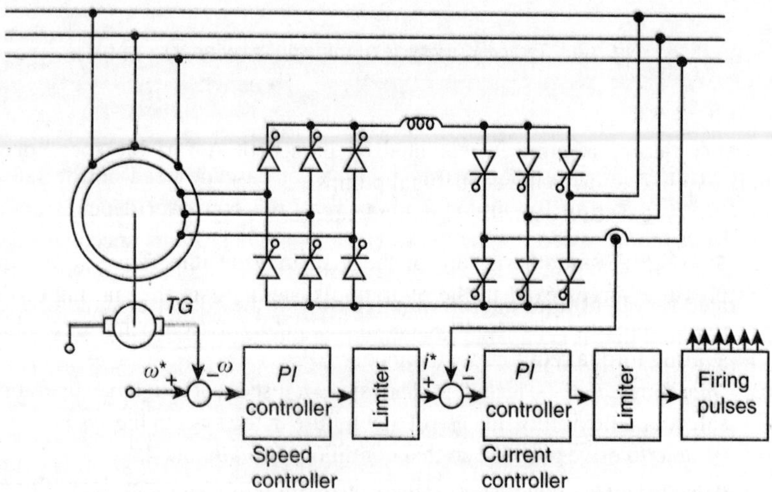

Fig. 7.10(a) *Closed loop control of slip energy recovery schemes*

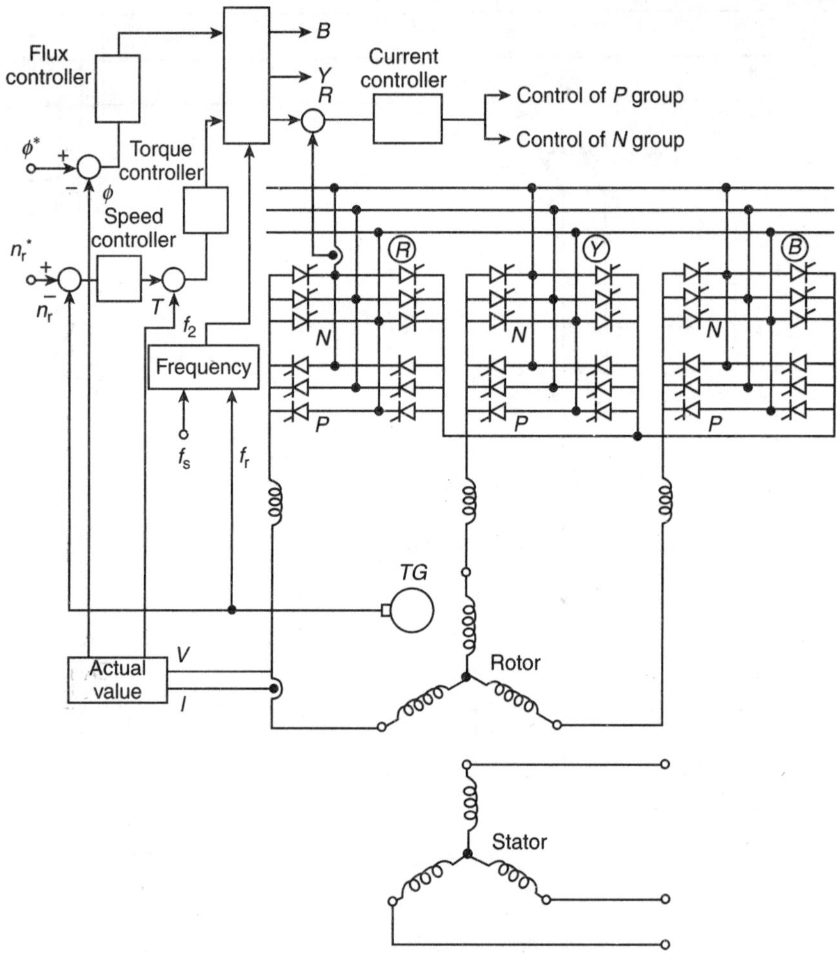

Fig. 7.10(b) *Static Scherbius scheme using cycloconverters*

converter drives is that the static converter cascade need not be rated for full rating of the motor. The rating of the converter depends upon the range of speed control required below synchronous speed. As the speed range is limited to 30–50% the converter cascade also needs to be rated for 30–50% of the full rating. However the converter should be so designed that it is capable of conducting the highest possible current at high speed and withstanding the highest possible voltage at the lowest speed. To accomplish this the actual design rating of the converter cascade is slightly greater than the slip power.

As the highest current and highest voltage do not occur simultaneously, the design rating of the converter cascade can be reduced by switchable cascades.

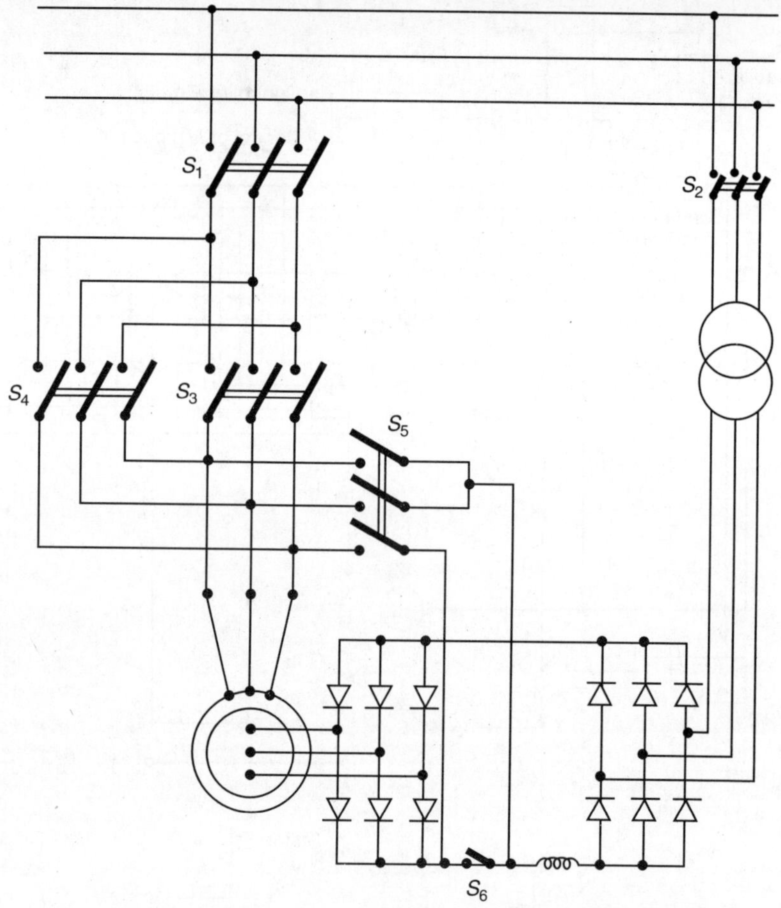

Fig. 7.10(c) *Static Kramer control with a possibility for reversal and regeneration*

iv. ***Starting:*** Smooth starting of the motor is possible. The torque can be regulated by means of link current of cascade. The starting current peaks are not there.

v. ***Operating safety and flexibility:*** The system has increased operating safety and provides flexible availability of the installation. In case of cascade converter failure and in any other emergency condition, it is possible to run the motor having speed control by means of varying the rotor resistance employed for starting. However the rotor resistance must have been designed to serve both the jobs. This possibility is not there with other systems.

There is another flexibility of operation in using a converter cascade. A given unit can be used with several induction motors and hence fewer converter cascades than pumps may be used. This is feasible if all pumps do not require speed control at the same time. Only that pump which requires speed control is used with the cascade and the pump which does not require speed control is used with normal induction motor. The pump to be regulated can be chosen at will and consequently the installation has a very highly flexible control.

vi. The system can be developed for all power ranges and speeds of centrifugal pumps. The highest speed that is available is 1800 rpm and in exceptional cases this can be extended to 3600 rpm. Subsynchronous converter cascade is superior to converter fed dc motors with respect to this criterion. It is however inferior to converter fed synchronous motors which are available for speeds up to 6000 rpm.

vii. To improve the performance, the following modifications are made in the development of the converter. A 12 pulse converter (Fig. 7.11) may be used to decrease the amplitude of pulsating torques and harmonic losses in the motor. The 12 pulse inverter presents only higher order harmonics and the line voltage distortion will be small. The pulsating torques are affected by the number of pulses of the rectifier. The 12 pulse arrangement of the inverter decreases the torque pulsation by decreasing the stator current and voltage distortion. 12 pulse inverters are required if the short circuit rating of the mains is lower than that of the motor. It is also required if the line is such that the harmonics of the inverter are capable of causing distortion.

viii. The main disadvantage of converter cascades is poor power factor due to phase control of the line side inverter. Methods have been developed to

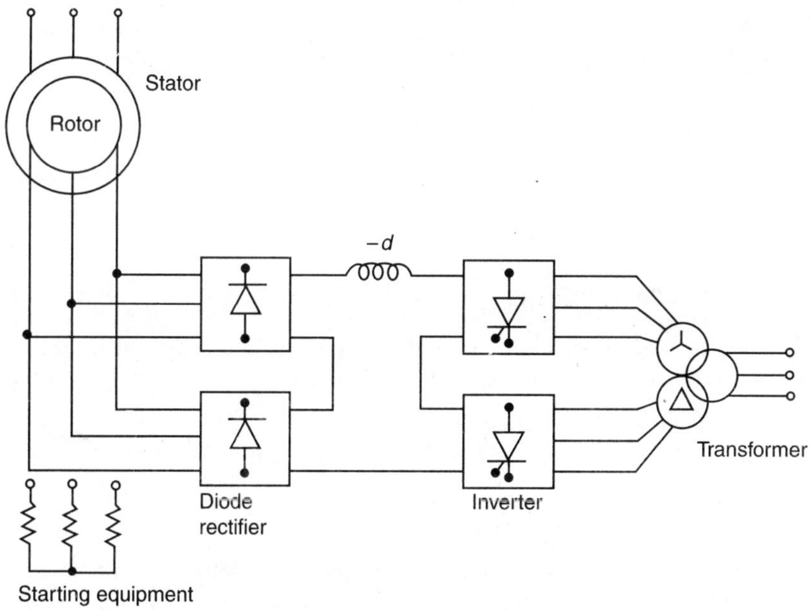

Fig. 7.11(a) *Twelve pulse inverter in the rotor circuit for slip power recovery*

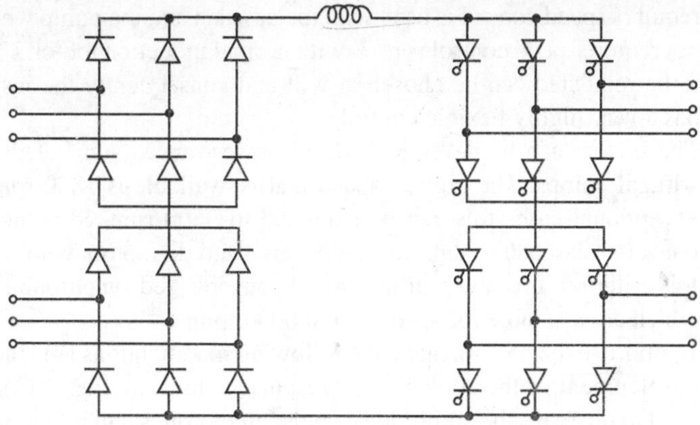

Fig. 7.11(b) *Schematic of a 12-pulse converter to be used with ac drives*

improve the power factor. One of these methods is the sequential control of the inverter on the line side. Sequential operation improves the power factor by requiring reduced reactive power.

7.9.2 Converter Fed Synchronous Motor Drive for Centrifugal Pumps

Converter fed synchronous motor satisfies all the above discussed criteria and has application to driving centrifugal pumps. Compared to the slip ring rotor with converter cascade this has advantages of:

i. higher speed of operation up to 6000 rpm
ii. the simple converter due to machine commutation (with commutation assistance only at low speeds)
iii. the line power factor is better than converter cascade
iv. The features of soft starting and possibility of achieving a gearless drive both at very low and high speeds.

7.10 TURBOCOMPRESSORS

The turbocompressors and blowers in the industry require drives rated up to 40 MW. Drives of large rating are also required

—in steel industry blast furnace blowers
—in natural gas pipelines and liquefaction processes
—in wind tunnels
—chemical process industry.

In the upper power range a converter fed synchronous motor is a suitable drive for compressors. The application fields of synchronous motors as variable speed drives are increasing due to variable frequency converters.

When used with turbocompressors the frequency converter may be used for performing one or more of the following:

i. only starting the compressor
ii. for operating the compressor at a speed higher than the one corresponding to rated frequency.
iii. for speed control of the motor.

Dc link converter fed synchronous motor can be used for the above applications. A dc link converter performs the function of a variable frequency converter. Two converters one on the line side and the other on the motor side are connected by a dc link inductance which keeps the current constant and acts as a decoupling agent between the converters. The converter on the mains is a line commutated phase controlled converter. The machine side converter is simple in its structure, as it makes use of machine voltages for its commutation. Thus the forced commutation circuit is dispensed with. The converter facilitates power flow in the reverse direction also, i.e., from machine to line. In this case the functions of the converters interchange.

The features of this drive are:

i. Four quadrant operation is straightforward. Speed and torque available in both directions.
ii. At low speeds the machine commutation is not possible. Special methods of commutation may be required up to 5 Hz. Beyond 5 Hz machine commutation takes over.
iii. Smooth variable frequency starting of the motor is possible. Frequency variation in the range 0–100 Hz is possible.

Large synchronous motors are required to drive the compressors. Problems arise with synchronous motors as their power ratings increase, especially when the motor rating is greater than the short circuit rating of the power system. These are more significant when synchronous starting is employed. These problems and disadvantages of a synchronous (line) starting can be eliminated using a static frequency changer. When the load torque is high, a very high starting torque is required. This high value of starting torque is achieved with a very high value of starting current, e.g, five times full load current may be required. In synchronous starting using static frequency converter (s.f.c), the required starting torque can be obtained at relatively low value of current (1.5 times full load value). The acceleration starts from zero speed and the stator frequency is gradually increased so that rotor is always in synchronism with the stator field. The voltage dips on the mains are automatically eliminated.

In asynchronous starting the energy stored in the rotating parts $\left(\frac{1}{2}JW^2\right)$ calls for increased thermal rating of the rotor due to difference in the speeds of rotor and stator field. When this type of starting is used for synchronous motors driving turbocompressors or blowers, very high rotational energy is stored and hence the rotor of the motor is heavily thermally loaded. But the rotors of salient pole motors have limited thermal capacity. Starting by means of s.f.c. is suitable here.

Another advantage gained by s.f.c. is that one unit can be used to start several motors if speed control is not required (Fig. 7.12). The s.f.c. brings the motor to rated speed and the motor is synchronised to mains and the motor continues to run at synchronous speed. The s.f.c. can be used to start other motors.

When used for starting purposes on s.f.c. should be designed according to

i. number of startings required
ii. starting time
iii. starting torque

If the load torque of the compressor can be by disconnected by some means during starting the converter becomes very compact.

A static frequency converter can be used to provide a frequency which gives operating speeds of 6000 rpm at power ranges of 10 MW. The motors are available for this speed. Direct driving without gears is possible. The s.f.c. used performs both starting and running at high speed.

The s.f.c. can also be used for speed control. The speed can be varied over a wide range (0–base speed). Above base speed there is flux weakening. The speed control varies the delivery rate of the pump or blower. The power consumption by this method of regulation is low. A constant delivery rate can also be achieved by running the motor at constant speed.

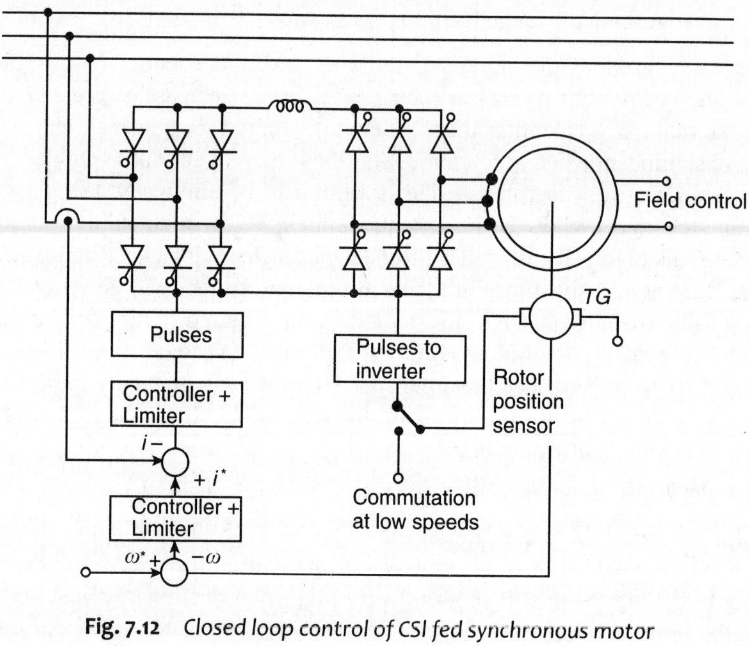

Fig. 7.12 *Closed loop control of CSI fed synchronous motor*

To summarize, an s.f.c. eliminates starting problem, makes a motor run at a speed higher than base speed, avoids power transmission equipment such as gears, and permits speed control over a wide range. An s.f.c. provides favourable start-up and acceleration so that the disturbances on the power system are least. The s.f.c. also prevents a synchronous motor from delivering fault or surge current when short circuit occurs on the mains. This saves switchgear.

When a synchronous motor uses s.f.c. for speed control, some reactions occur on the network:

i. ***Power factor:*** Although the machine side converter receives its reactive power from the motor, the line side converter receives its reactive power for commutation and control from the mains. The line power factor is poor as the firing angle increases. The mains power factor deteriorates as the speed is varied. The power factor is low at low speeds. At the rated speed only a reasonable power factor is possible.
ii. ***Harmonics:*** The mains and motor current are rich in harmonics.
iii. Twelve pulse circuits (Fig. 7.11) may be employed to eliminate lower order harmonics and consequent pulsating torques. However the higher order harmonics are present and they have negligible effect. This also has the advantage that the outlay of thyristors decreases as two converters feed the motor.
iv. The torque pulsations may cause torsional vibrations. These may be prevented by suitable shaft design.

The system has a high degree of reliability. The availability of thyristors at reduced prices and higher voltages and current ratings makes the s.f.c. very popular. Modular construction is also possible, in which repair and fault finding are easy and quick. In the case of applications where the interruption of equipment would cause very grave consequences, the principle of redundancy may be used. Extra equipment such as thyristors may be used which permit the continuity of service in case of failure.

In short, a synchronous motor with static variable frequency converter using thyristors is becoming very popular. The s.f.c. functions as starting equipment as well as unit for speed control. This is used with compressors, blowers and pumps.

7.11 SUMMARY OF THYRISTORISED AC AND DC DRIVES

A detailed discussion of dc and ac drives has been provided with respect to industrial applications detailed in the previous sections. An evaluation of dc and ac drives leads to the following conclusions.

At present the converter fed ac drives are competing with converter fed dc drives. The highest installed capacity and highest speed of operation are limited by the commutation of the motor. The commutation of a dc motor deteriorates when supplied from a converter, due to ripple content of the armature current as well as possible discontinuous conduction. The speed control of an ac motor over a wide range in a smooth manner has become an easy task, with the

Table 7.3 *Thyristorised AC Drives: A Swnmary*

Drive system (1)	*Type of Converter (2)*	*Direction of rotation (3)*	*Regeneration (4)*	*Power factor (5)*	*Max Speed (6)*	*Max. of conv. freq. (7)*	*Speed range (8)*	*Field weakening (9)*	*Application (10)*
Cycloconverter fed synchronous motor (cycloconverter induction motor)	Line commutated cycloconverter	2	2	Lagging, similar to converter fed dc motor drive		1/3 to 0.5 supply frequency	0 to (0.33 to 0.5) n_s	Yes	1. Gearless drive for large tube and bowl mills 2. Mine winders 3. Reversing rolling mills 4. Ship propellor 5. Large reciprocating compressors
Wound rotor induction motor with sub/super synchronous converter cascade	Line commutated	2	2	Lagging to leading		1/3 to 0.5 supply frequency	0.9 to 1.1 n_s 0.5 to 1.5 n_s	No	1. Network interconnecting converters 2. Flywheel converters 3. Drivers with high ratings in the low speed range
Wound rotor induction motor with sub/supersyn-chronous converter cascade	Indirect converter DC Link converter Line commutated	1	1	Lagging similar to converter fed dc motor		Supply frequency	$1 - 0.7\ n_s$ $1 - 0.3\ n_s$	No	1. Centrifugal pumps 2. Blowers
Converter fed synchronous motor	DC link converter Line/machine commutated	2	2	Lagging similar to converter fed dc motor		120 Hz	$0.05 - 1\ n_s$ $0 - 1\ n_s$	Yes	1. As start up converter for large synch. machines 2. Boiler feed pumps 3. Turbocompressors 4. Continuous rolling mills

Cage induction motors with voltage source inverter	DC Link converter with dc voltage link forced commutation at motor end with pulse duration modulation. Non-controlled at line side.	2		Approx. 0.85 – 0.92 lagging over complete speed range		200 Hz	$0-n_s$		Diverse drives primarily where several motors connected to the same converter are running at coordinated speeds. up to approx. 2 MW.
	Reversible converter at line side.	2	2						
	Controlled at mains end forced commutation	2	2	Lagging to leading					At present traction drives.
Cage motor with current source inverter	DC Link converter with dc current link forced commutation at motor end	2	2	Lagging somewhat worse than converter fed dc drive		120 Hz	(.05 to 1.0) n_s $0-n_s$	Yes	Pumps blowers test-beds up to 2 MW. Open.end.drives

Column 3: '2' indicates continuous change over to other direction of rotation without switching. '1' the possibility of changing direction of phase switching.

Column 5: Lagging to leading indicates reactive power absorption/supply possible acting as a motor and as a generator preferably with $\cos\phi = 1$.

Column 6: Max. speed limited by converter in case of converter fed syn motor and by motor in other systems. Max. speed decreases as power increases.

Column 8: Top values are preferred. Supersynchronous braking possible.

Table 7.4 Comparison of performance of Electrical Drives

S.No.		*Ward Leonard*	*Static Ward Leonard*		*Rotor resistance control*	*Armature voltage control*	*Slip energy recovery*	*Stator frequency control*		*Cyclo-converter CSI*	*Sychronous motor*		*Cyclo-converter*
			Chopper	*Phase controlled*				*PWM*	*Square wave*		*Cyclo-converter*	*Self control*	
1.	Speed range	1 : 10	1 : 20	1 : 20	1 : 2	1 : 5	1 : 5	1 : 00	1 : 20	1 : 20	1 : 4	1 : 20	1 : 4
2.	Speed regulation	below 5%	below 5%	below 5%	poor	poor	below 5%	below 5%	below 5%	below 5%	below 5%	nil.	nil.
3.	Constant power load	above base speed	above base speed	above base speed			suitable	suitable above base speed	above base speed	above base speed	above base speed	suitable	below
4.	Constant torque load	below base speed	below base speed	below base speed	suitable	suitable	suitable	below base speed	below base speed	below base speed	below base speed	below base speed	below base speed
5.	Variable torque loads	suitable	suitable	suitable	suitable	not suitable	suitable	suitable	suitable	suitable	suitable	suitable	suitable
6.	Soft starting	possible	possible	possible		possible		possible	possible	possible	possible	possible	possible
7.	Frequent starts and stops	suitable	suitable	suitable	suitable	suitable	suitable	suitable	suitable	suitable	suitable	suitable	suitable
8.	Frequent speed reversals	suitable with regen-eration	suitable with regenera-tion	suitable with regenera-tion		suitable		suitable	suitable	suitable	suitable	suitable	suitable

9.	Periodic acceleration and retardation	suitable	suitable	suitable		suitable		suitable	suitable	suitable	suitable	suitable	suitable
10.	Better efficiency	Yes	Yes	Yes	No	No	Yes	Yes	Yes	Yes	Yes	Yes	Yes
11.	Accuracy of speed control	up to 1%	up to 1%	up to 1%	2.5%	2 to 5 %	less than 1 %	less than 1 %	less than 1 %	less than 1 %	less than 1 %	less than 1 %	less than 1%
12.	For operation in contaminated atmosphere	possible with reservation	possible will reservation	possible with reservation	Suitable	Suitable	suitable	suitable	suitable	suitable	suitable	suitable	suitable
13.	Maintenance	costly due to frequent maintenance.	frequent maintenance is required.	Frequent maintenance is required	less	easy	easy	easy	easy	easy	easy	easy	easy
14.	powerfactor	does not arise	good if diode rectifier is employed.	poor but can be improved	good	poor	poor but can be improved	good if diode rectifier is used	poor	poor	poor	poor	poor
15.	Braking methods	Dynamic and regeneration	Dynamic and regeneration	Dynamic and regeneration	all kinds of electric braking	plugging	all kinds of electric braking	dynamic and regeneration	dynamic and regeneration	dynamic and regeneration	regeneration	regeneration	regeneration

development and sophisticated control of thyristor power converters. With this possibility, the tendency to make use of simple, economical and robust squirrel cage motors as variable speed drives is increasing. Therefore in industrial applications the inverter fed induction motors have become real competitors to converter fed dc drives even in the range of speeds and ratings where dc drives are still applicable. Beyond this range, which is substantially wider, only ac drives are employed. The synchronous motor, when self controlled, has properties of a dc motor and is free from its hunting and stability problems. Further it has an advantage that its power factor can be varied by varying the excitation. When operated at unity power factor the inverter size decreases as the current to be handled is minimum. When over excited the motor operates at leading power factors and machine voltages can be used for the commutation of the inverter. The commutation circuit is not necessary. Accordingly the inverter configuration is simple. For high speeds and large power ratings synchronous motors are suitable. For low speed, large power reversible drives, cycloconverter fed ac drives are widely employed.

Therefore within the range of speed control where a dc drive or an ac drive can be employed, they satisfy the functional requirements of a drive. The choice between the two must therefore be based on some other criteria, generally economy. The other criteria such as environmental conditions, maintenance considerations, location of the motor and network reactions will also influence the choice of the drive system.

The features of dc and ac drives along with their applications in industry are summarized in Tables 7.3 and 7.4.

Multiple-Choice Questions

7.1 A spinning motor in a textile mill should have
(a) a very good peak torque capability
(b) moderate starting torque and uniform acceleration
(c) high starting torque and peak torques
(d) a very good starting torque

7.2 To have uniform acceleration the motor should have
(a) flat speed torque curve
(b) a speed torque curve with a small difference between peak torque and starting torque
(c) very high starting torque
(d) very high peak torque and low starting torque

7.3 Dc drives in the steel rolling mills are slowly being replaced by
(a) variable frequency induction motors with torque and speed control
(b) variable frequency synchronous motor fed from a load commutated CSI
(c) variable frequency ac motors with regeneration capabilities
(d) variable speed induction motor with chopper controlled resistance in the rotor circuit.

7.4 A drive motor for continuous mills
(a) must be capable of two quadrant operation
(b) must be capable of four quadrant operation
(c) operates only in one direction, reverse rotation is not necessary
(d) must be a very high speed motor.

7.5 For applications in cranes
(a) differentially compounded motors are suitable
(b) cumulatively compounded motors are suitable
(c) dc shunt motors are suitable
(d) dc series motors are suitable.

7.6 A kiln motor in a cement mill
(a) must be a constant speed motor
(b) must be able to have creeping speeds
(c) need not have very large overload capacity
(d) has very low power requirement.

7.7 For fans and blowers in cement mills
(a) variable frequency induction motor may be employed
(b) three phase induction motor with chopper controlled resistance is suitable
(c) DC motors with field control are suitable
(d) the three phase slip ring induction motors with slip energy recovery schemes are very well employed with advantages.

7.8 To drive a centrifuge in sugar mills the most suited system is
(a) a solid rotor induction motor with stator voltage control
(b) a normal commercially available induction motor fed from a voltage controller
(c) a slip ring induction motor with slip energy recovery scheme
(d) a slip ring induction motor with rotor resistance control.

7.9 When an oversized motor is used the energy saving can be accomplished on light load portion duty cycles when
(a) the motor is controlled from a variable frequency variable voltage converters
(b) the motor is controlled from a constant frequency variable voltage converter such as three phase voltage controller
(c) the motor is fed from conventional speed control systems
(d) the motor is controlled on the rotor side

7.10 For a mine winder the suitable choice is
(a) an induction motor with stator voltage variation
(b) a self controlled CSI fed synchronous motor
(c) a dc motor with Ward Leonard control
(d) a squirrel cage motor with ac voltage controller

7.11 For centrifugal pumps the most suitable drive system is
(a) wound rotor induction motor with slip energy recovery schemes
(b) wound rotor induction motor with rotor resistance control using a chopper
(c) vector controlled three phase induction motor
(d) self controlled CSI fed synchronous motor.

References

7.1 Pillai, S.K., *A first course on electrical drives*, Wiley Eastern, 1982.

7.2 Mukherjee, K.K., *DC drives and controls and application in rolling mills*, Electrical India, XIV, p. 13, 1974.

7.3 Sarkar, B.P., 'Electrical equipment for main drives of reverse cold rolling mills', *Journal of IE (India)*, 56, Pt.E1.2, p. 57, 1975.

7.4 Wurgler, H.V., 'Electrical drives for rotary cement kilns', *Brown Boweri Review*, 58, p. 106, 1971.

7.5 Harverydrake, W., and H.L Prescott, 'Controlled torque starting of textile machinery', *IEEE Trans. on Ind. and Gen. Appl.*, IGA-7, p. 267, 1971.

7.6 Pabidri, M.S., 'Electric motors for textile industry', *Electrical India*, XVII, No. 24, p.35, 1977.

7.7 Wagle, A.V., 'Electrical Drives for coal mining', *Electrical India*, XIX, No. 16, p.17, 1979.

7.8 Stemmler, H., and A. Meyer, 'Variable speed converter fed three phase drives', *B.B.C. Review*, 1982, p.114.

7.9 Ternes, L., J. Bommeli, and K. Peters, 'The cycloconverter fed synchronous motor', ibid, p. 122.

7.10 Sharpenberg, H., and A. Streck, 'Subsynchronous converter cascade for large centrifugal pumps', ibid, p. 142.

7.11 Meyer, A., H. Schweickhardt, and P. Strozzi, 'The converter fed synchronous motor as a variable speed drive system', ibid, p. 151.

8 Microprocessors and Control of Electrical Drives

8.1 INTRODUCTION

Conventionally, control of thyristor converter fed electrical drives (both dc and ac) is performed using analog discrete components. In the drive control, conventional speed control systems have been improved continuously and gradually in terms of their control performance, as the control devices have changed from magnetic amplifiers to transistors and IC amplifiers. This hardware oriented control of drives becomes more and more complex with the increase in sophistications required, i.e. the requirements of high precision, performance and reliability. The various controllers, the control circuits for the power converters are realised using hardware oriented circuits. The control suffers from the drift and parameter variations of the components due to temperature and the improvement is rather limited. However, the control has the advantage that it is capable of processing the signals almost simultaneously.

Because of the involved complexity and the limitations of improvement in performance due to drift problems, several sophisticated methods of controlling ac motors, such as vector control of induction motors, did not become popular until the developments in the area of drive technology, which have been very rapid in the recent past. With these the control is possible with fewer components with the advantages of easy maintenance, economy, serviceability and general applicability. Also this rapid and fast development of digital technology has been the opening for new concepts of control using microprocessors and associated digital components, A/D converters, etc. These have resulted in compact control systems for drives which effectively replace the complex hardware oriented control using analog discrete components.

The digital processing of speed and angular position is more exact and less expensive than analog processing, which makes use of tachogenerators and position transducers. The control is free from drift and parameter variations. The main feature of digital control with microprocessors stems from the fact that it is mainly software oriented. When once the necessary algorithms are developed, the software can be used for different plants with only a few modifications. This is not possible with dedicated hardware control.

The drive technology using thyristor converter fed dc or ac motors, as expected, has become very popular owing to the developments in the area of thyristor technology, high speed digital systems, microprocessors and, microcomputers. The future has already begun for the use of microprocessors and microcomputers in the area of drives. Sophisticated controls of these drives with high performance and reliability, complicated hardware oriented controls, such as four quadrant operation of dc motors using dual converters, field orientation or vector control both in constant flux and flux weakening modes of induction motors and margin angle control of synchronous motors can be realised with software programs on the microprocessor and microcomputers with least possible hardware. This hardware includes the interfacing of the microprocessor to the system. Sometimes to reduce the burden on the microprocessor and to make it available to perform other functions effectively, the firing circuits with necessary pulse transformers are realised in terms of hardware. The sensor measuring the terminal quantities must be interfaced through proper A/D converters. The controls become economically feasible. Various controllers in the closed loop controls can be realised on the microprocessor by means of software programs. Optimised PWM strategies for the ac motor can be realised. Advances of new digital techniques make it possible to manage different operating conditions. Microprocessors can be very easily adopted in the parameter identification and adaptation problems using machine models. A slip energy recovery scheme for both super and subsynchronous operation using vector control can be realised using high performance processors.

Various sophistications in the control of synchronous motors have also become feasible. One such sophistication is on-line determination or estimation of the rotor position without any rotor position sensor on the shaft. The identification and state estimation processes involved are very easily implemented using microprocessors.

An attempt is made in this chapter to describe the role of microprocessor and its function in the control of electrical drive. The architecture and the types of microprocessors are not discussed here. A few applications of the microprocessor in the dc and ac drives are discussed to bring out clearly the stages involved in the design of the control system.

8.2 DEDICATED HARDWARE SYSTEMS VERSUS MICROPROCESSOR CONTROL

The microprocessor control of electric drives offers the following advantages when compared to the dedicated hardware control.

As has already been discussed, the use of microprocessors reduces the complexity of the system. The software supported control using microprocessors to perform the functions of controllers, feedback, decision making, etc., will eventually result in the least hardware making the system economically viable. The present day developments in this area will make the system cost effective.

The reliability of the control of the drives is higher with microprocessors than with dedicated hardware systems. The number of parts is reduced. Software programs are employed for various controls. The control equipment can be realised as a unified hardware.

The control is free from drift and parameter variations due to temperature while in operation using microprocessors whereas the control is affected by these when the control is implemented with dedicated hardware. The calculations are exact except for the errors in the A/D converters.

The information can flow in both forward and backward directions. This property is very important and has significance in the control using microprocessors. Proper shielding may minimise the EMI problems. The control may be extended as it is compatible with the host computer.

The speed detection schemes are completely digital, avoiding the errors of measurement. This improves the accuracy even at very low speeds. Control efficiency is more. The concepts of modern control theory can be used in the control.

A generalised hardware supported by a flexible software can take care of a particular application. The software will have possibilities of addition, deletion and upgrading.

Microprocessor control is capable of performing the control functions, such as decision making, complicated computations, etc. These enhance the power of the microprocessors and are not possible with dedicated hardware. The role of a microprocessor is to replace the logic circuits that control the firing angle of SCRS. In some instances it performs, in addition, the control computations. Since a control cannot be applied until it is calculated or computed, there is invariably a delay between the beginning of computation and application of control. In the case of SCR control, if the delay is too long the time corresponding to a particular firing time may have already passed. The controller must then fire at a different instant of time in the present cycle or wait for the correct time in the next cycle. A suboptimal control will result in both cases. Thus in order to avoid long delays and particularly with slow microprocessors, the control law, if to be implemented by a microprocessor, must be simple. Ideally determining the necessary controls should require only table look-up operations and any computations should be kept to a minimum.

Microprocessor control also permits several other functions, such as data acquisition, monitoring and warning, diagnosis, etc. The fault diagnosis of digital control systems is easier than that of a dedicated hardware system. It is easy to maintain the system. However, the microprocessor control has also certain disadvantages.

The communication between the microprocessor and the analog circuitry is accomplished by A/D or D/A converters. There are sampling or quantizing errors. These, however, can be minimised by increasing the bit size. Care must be taken to see that the signal resolution does not suffer.

The response is more sluggish and slower than that of a dedicated hardware. This is mainly because the dedicated hardware can handle the signals and process them almost simultaneously, without any time delay. The microprocessors on the other hand can process the signals only in a sequence or in a serial manner, causing a delay in processing. Higher sophisticated systems, using microprocessor control to perform different functions may exhibit stability problems. Special techniques are required to enhance the speed and assure stable operation.

The development of necessary software may be costly and time consuming. The cost may be justified depending on the size of the production.

The variables are not accessible for measurement using instruments. The parameters cannot be easily monitored or changed under operating conditions.

Modifications in the system are possible to overcome several of these disadvantages. Multiprocessor control increases the speed of the system. Resolution can be increased by increasing the bit size. A microprocessor based digital control system, which is priced at the same level as an analog control system but compares favourably with the latter in terms of controlability, reliability and available functions, has become practical.

8.3 APPLICATION AREAS AND FUNCTIONS OF MICROPROCESSORS IN DRIVE TECHNOLOGY

With the remarkable progress in the area of microelectronics, reliable and powerful microprocessors have come to be more widely used in the control of electrical drives. A few of the applications of microprocessors in the area of drives can be listed as follows:

i. Static Ward Leonard control of dc motor using dual converters
ii. Four quadrant control of dc motor using multiphase choppers
iii. Control of PWM inverter for ac motors
iv. Four quadrant operation of CSI fed ac drives
v. Four quadrant operation of cycloconverter fed ac drives
vi. Static motor starters

In variable speed drives using thyristor power converters and electric motors (ac or dc) the functions of a microprocessor are the following:

i. Generating and providing firing pulses to the converters
ii. Generation of necessary waveforms to feed the motors
iii. Nonlinear function generation. Estimation of feedback signals. Implementation of software supported controllers in the feedback control, such as current controllers, speed controllers, etc. The limiters are also implemented to limit the control variables to safe values
iv. Processing the measured signals, such as voltage, current and speed
v. Storing and processing the information of controlled quantities
vi. Estimation of feedback signals and computation of reference quantities which cannot be directly measured, such as torque and flux
vii. Identification and adaptation of variable parameters

viii. Adaptive control and optimisation
ix. General sequencing control
x. Monitoring and warnings
xi. Diagnostics and tests

8.3.1 Speed Detection

One important factor in the closed loop control of variable speed drives employing solid state converters is the speed detection. In the dedicated hardware control using analog components, the speed sensing is accomplished by means of tachogenerator. For control of the drive using a microprocessor, this analog speed signal can be converted to digital signal by an A/D converter. This method evidently does not possess the accuracy expected of a digital system. The speed measurement must be qualified by high resolution, high accuracy, fast response over a wide range of speeds and quick sampling. Such a measuring system can be obtained only by using digital techniques. Therefore, in systems employing microprocessors the speed measurement is carried out by means of a shaft encoder or pulse generator, which generates a train of pulses depending upon the speed.

The shaft encoder is a circular light aluminium disc mounted on the shaft with equidistant holes drilled along the periphery. On the opposite sides of the disc are placed a light source and photosensitive electronic device, e.g. a phototransistor. These are aligned such that the phototransistor gets light when there is a hole across. Light activated phototransistor generates a pulse. Thus a pulse train is produced. These pulses are not very sharp and must be shaped using suitable circuitry. These pulses are processed in a microprocessor to get the actual speed of the motor. There are two methods of determining the speed (Fig 8.1).

i. The pulses are counted in a given period of time. This is suitable for high speeds.
ii. The interval is measured between two consecutive pulses. Suitable for low speeds.

When the speed is determined using these methods, special care must be taken to see that the problems such as delay in detecting time at low speed are not there. Only with such precautions can low speeds be measured with reasonable accuracy. Thus speed control is possible over a widened range.

The high resolution of the speed sensing signifies the stable control over a wide range of speeds. The high accuracy of speed measurement improves the steady-state control accuracy. The detecting time in the measurement of speed influences the sensing of instantaneous speed. If the time is large the results give only average speed during the time. So, in order to sense the instantaneous speed in the transient and dynamic conditions this detecting time must be as small as possible. A larger detecting time increases the lag in the speed control system, making it difficult to realise stable high speeds.

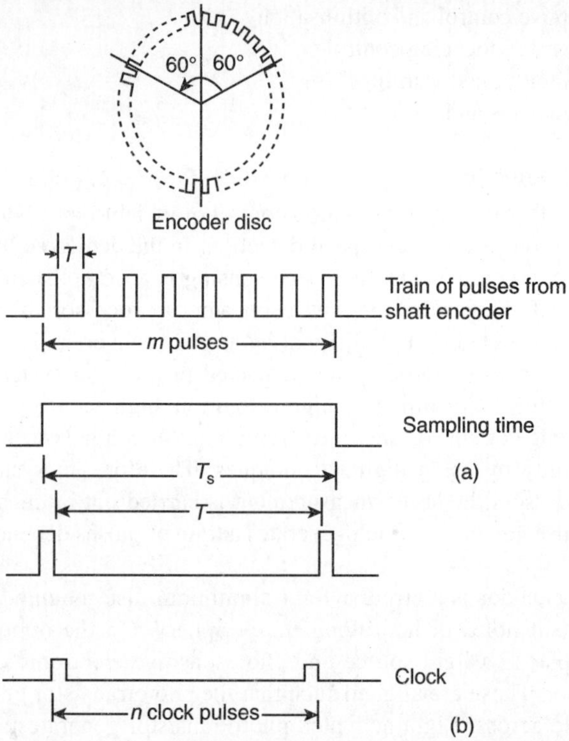

Fig. 8.1 *Processing of the encoder pulses to obtain the speed. (a) For high speed* $N = 6 \times 10^4 \dfrac{m}{MT_s}$ *M total pulses per revolution (b) For low speed* $N = 6 \times 10^4 \dfrac{f_c}{M_n}$

8.3.2 Gate Firing of the Converters

The firing pulse control unit constitutes the heart of any thyristor power converter. The gate pulses required by the thyristors are often derived from the digital pulses using a simple buffer unit. Analog firing controllers are used in systems with dedicated hardware. They are recommended for constant low frequency operation where the firing precision is not critical. They are simple.

However, in variable frequency applications, especially at high frequencies, the implementation of firing angle control using analog means is not appropriate. The firing pulses would depend highly on the precision and stability of analog components used, such as resistors, capacitors, linear and nonlinear devices responsible for time lags, etc.

The firing pulses cannot be realised with adequate symmetry. The all digital controller is free from these drawbacks. Further, a digital controller can be made programmable to generate accurate gate pulses in various configurations and to perform triggering of thyristors at accurately repeated intervals, independent of

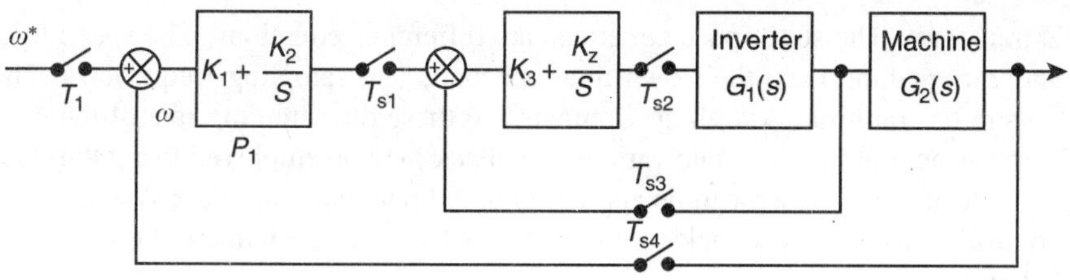

Fig. 8.2 *Representation of a speed control system with multirate sampling*

frequency with desired symmetry. It is also possible to alter the firing angle on line without affecting the symmetry of pulse train.

The requirements of a firing angle controller are:

i. A synchronised triggering output
ii. Thyristor bridge must provide the desired output. Suitable techniques such as linear ramp or cosine wave crossing may be employed.
iii. The transients and distortions of the supply voltage waveform should not have any effect on the firing angle delay.
iv. The firing angle must be varied in the range 0 to 180°
v. The firing angle control should be smooth and precise. The resolution of firing angle decides the sampling frequency when generating firing pulses.
vi. The scheme should have enough flexibility to accommodate variation in circuit parameters in order to obtain the best performance.
vii. The response of the controller must be very fast.
viii. The controller must have self regulating property when the amplitude of supply voltage varies.
ix. A unified or generalised hardware circuit which can be adaptable to any type of converter or cycloconverter operating under large variations of frequency.
x. The controller must be developed with low hardware content so that it can be integrated into a chip. This would reduce the cost and increase the potential field of application.

Depending upon the application, suitable software may be written to obtain the firing pulses. The method of obtaining the firing pulses to suit a given purpose, e.g., for dual converter will be discussed later.

8.3.3 Feedback Control

A closed loop control of a variable speed drive system employing thyristor power converters is normally a discrete time system when a microprocessor is implemented for control purposes. The system may be linear or non-linear. The power converter itself is a discrete time system due to switching of the thyristors. Just as a continuous system is represented by differential equations and Laplace transform, a discrete data system can be represented by difference equations and

Z-transform. The state space equations are difference equations. The speed loop and current loop have their own time constants and sampling frequencies. The system has multirate sampling. Compared to these the sampling rate of the converter is negligible. The state variable feedback may be employed using the state space techniques which can be applied to both linear and nonlinear systems. The controllers for such feedback control of a variable speed system are shown in Figs 8.3 and 8.4.

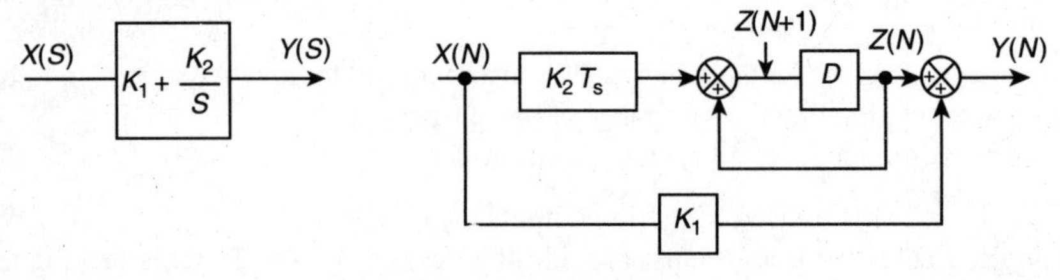

Fig. 8.3 *P.I. Controller*

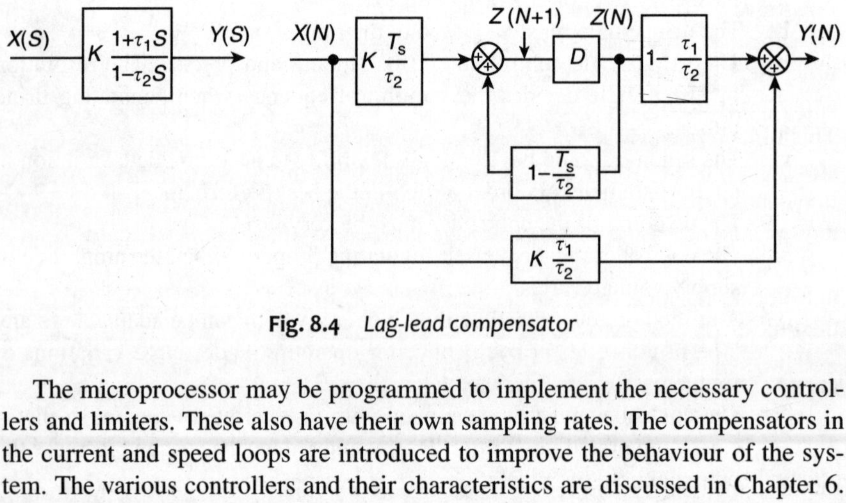

Fig. 8.4 *Lag-lead compensator*

The microprocessor may be programmed to implement the necessary controllers and limiters. These also have their own sampling rates. The compensators in the current and speed loops are introduced to improve the behaviour of the system. The various controllers and their characteristics are discussed in Chapter 6. These are normally PI, PID, or lag-lead compensators. In the hardware oriented system using discrete analog components, a controller is specified by its transfer function in Laplace transform. In digital control using either digital components or microprocessors, a controller is developed based on its time domain difference equations. The Laplace transfer function of an analog controller must be suitably transformed to time domain difference equations to develop an equivalent digital controller. A microprocessor based compensator has the following features:

i. It is cheaper than its analog counterpart.
ii. It offers advanced control capacity. Adaptive control is possible.
iii. It offers significant flexibility. The control parameter or even control algorithm can be changed.

iv. It is more compact and lighter than its analog counterpart. Its power consumption is also small.
v. It is highly reliable because the number of components is less and mainly supported by software. The noise is also less.

The digital implementation of PI, and lag-lead compensators is shown in Figs 8.3 and 8.4. The equivalent analog compensators are also shown.

The analog compensators limit the output to a safe value as desired due to the inherent saturation present. This limiting of the control variables is required in the closed loop control system so that the overloading of the components is prevented. The control variables are not allowed to have excursions beyond these limits. In a microprocessor based controller these limits are stored in the memory. Thus the limiting properties of a microprocessor based compensator are adaptable to any operating conditions by changing the values in the memory.

8.3.4 Function Generation and Linearisation

An important aspect of the closed loop control system is the use of non-linear functions, e.g. in the control of induction motors there exists a definite non-linear relationship between the stator current and slip frequency to maintain a constant air gap flux. The function generation can be easily carried out using a microprocessor. This task is very difficult to implement using discrete analog components. Multidimensional functions can also be generated on the micrfoprocessor. This function generation can be accomplished by means of a look-up table in the computer memory. If the microprocessor is sufficiently fast the function generation can be implemented using linear segment computations or curve fitting techniques. The necessity for function generation arises also in the dc motor control and synchronous motor control. In the former it may be required to represent the non-linear magnetisation curve and in the latter to represent the relation between the link current and margin angle in the constant margin angle control.

The linearisation techniques can be very well employed when a microprocessor is used in the control. The non-linear transfer characteristics of the system can be linearised using standard techniques of producing an inverse non-linear function in the processor. The gain of the system varies in the control of a dc motor in the field weakening mode as well as when the converter is in the discontinuous mode of operation. The reduction in the gain causes sluggish operation. The gain compensation in terms of firing angle as a function of I_d is a non-linear function. This is shown in Fig. 8.5. This may be linearised by the microprocessor. The inverse non-linear function for linearisation may be developed using function generation techniques discussed above, e.g., using a look-up table. The necessary corrections may be made to maintain the gain at a constant value.

During discontinuous conduction the converter voltage does not follow the cosine function. At a given firing angle the converter voltage is more if the conduction is discontinuous. A necessary correction to the firing angle is made to bring the voltage to a value corresponding to continuous conduction. A detailed discussion is given in the dc motor control.

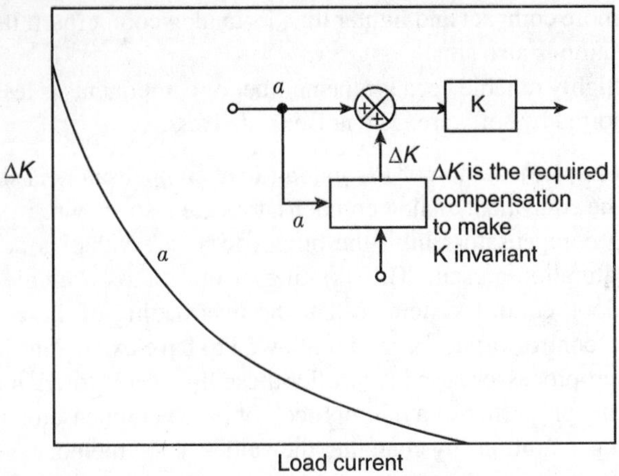

Fig. 8.5 *Non-linear gain compensation of a convener*

8.3.5 Processing the Feedback Signals

Some feedback signals can be determined by sensing. These are the signals of voltage, current and speed. These are analog signals when measured, and have to be converted to digital quantities by suitable A/D converters. The speed, however, can be measured digitally using a shaft encoder and A/D conversion can be avoided. These signals must be processed by the microprocessor for use in feedback loops. Further, the microprocessor should be capable of synthesising the feedback signals, such as torque, flux, etc., which cannot be easily measured. The microprocessor should perform the necessary arithmetic, including multiplication and division, while performing the function of estimating the feedback signals.

8.3.6 Implementation of PWM Techniques

The principles of PWM are employed in the control of inverter for the necessary voltage control in the inverter. These techniques are also used in the control of harmonics in the output voltage. The microprocessor can develop the PWM waveform using all the principles of PWM. These are sinusoidal modulation, or trapezoidal modulation. The necessary reference and carrier waves may be generated in the microprocessor by software programs to determine the points of triggering the thyristors. The technique of look-up table to generate the PWM waveform is very much suited for selective harmonic elimination. The computations to determine the firing instants in any of above methods become time critical at higher frequencies. Therefore at higher frequencies, a look-up table is desirable. The major application of microprocessor finds in the control of ac motors using PWM inverters.

8.3.7 Programmable Time Delay

One of the functions of a microprocessor in the control of variable speed drives is to provide an adjustable delay, e.g., delay angle in the gating of thyristors. This can be accomplished in a microprocessor by means of up or down counters. These are used to increment or decrement a digital word stored in the memory (periodically) until the memory location reaches the final count (in the former) or clears (in the latter). A pulse or an interrupt may be generated at the end of counting to execute the necessary event.

Digital filters can be implemented using a microprocessor. These correspond to analog transfer function and improve the performance of the system. The compensators explained before are simple digital filters. The microprocessors monitor various signals of the system and also give warning signals in case the variables exceed safe values. They finally issue a signal to shut down the system in case there is no response from the operator. The microprocessor may be programmed to provide the necessary protections against faults. Overcurrent protection, protection against single phasing, etc., may be provided with suitable software in the microprocessor.

A microprocessor performs the functions of data acquisition, sequencing of control for smooth transition from one mode of operation to the other, tests and diagnostics. It can be programmed to conduct evaluation tests on drive system and no load tests on the motors to determine the parameters. It may be supported by a powerful software to identify the faults. This software may be used if the system is down with a complex fault.

8.4 CONTROL OF ELECTRIC DRIVES USING MICROPROCESSORS

Before discussing the aspects of design of control systems for variable drives using a microprocessor a few of the application examples are discussed in the following. These are:

i. speed control of dc motors using dual converters.
ii. field oriented control of three-phase induction motor.
iii. speed control of synchronous motors.

The specific functions of the microprocessor in the above applications can be recognised after knowing the specifications of the performance. These serve in formulating the stages involved in the design.

8.4.1 Control of dc Drives Using Microprocessors

The dc motors fed from thyristor converters for variable speeds are being extensively used in general industrial applications. A dual converter, which is a combination of two antiparallel connected three phase/single phase bridge converters, provides a reversible dc drive with regenerative facilities. The response of the drive is fast. The speed control systems of the drive using the dedicated hardware with discrete components as well as a microprocessor, are shown in Fig. 8.6. Figure 8.6 clearly specifies the functions of a microprocessor in such a system. The scheme requires a number of parts or components and careful adjustment when based on analog techniques. However, by proper selection of counters to perform several jobs, the number of components can be reduced. The microprocessors

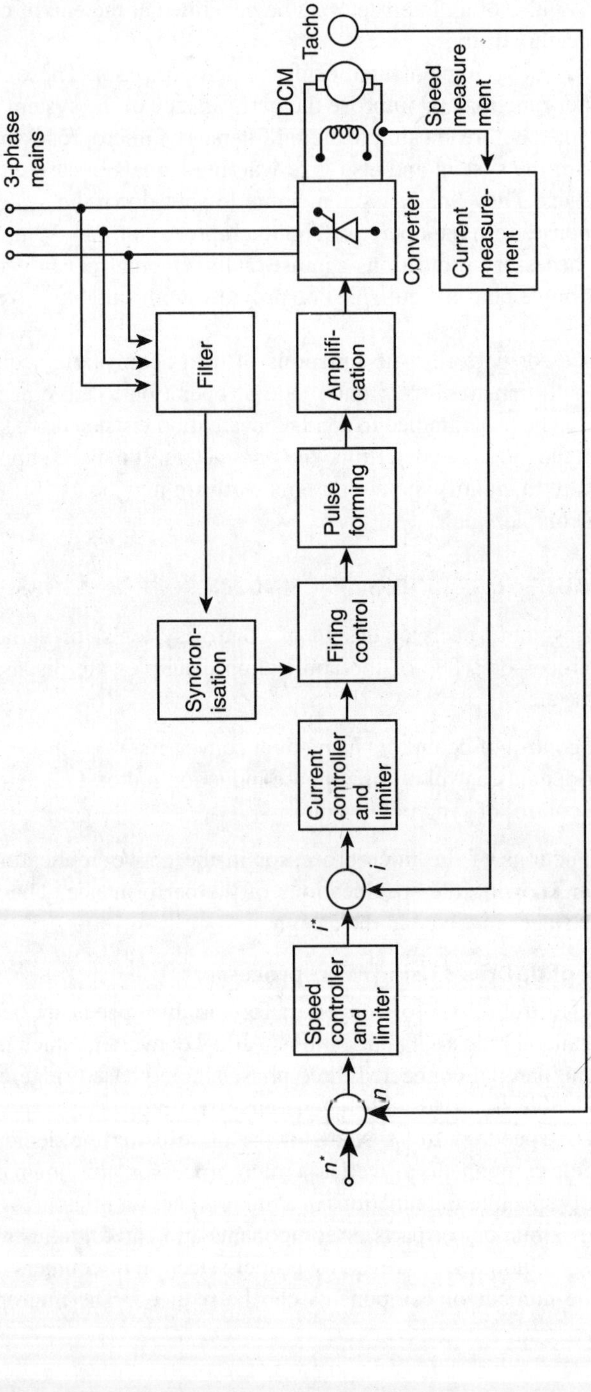

Fig. 8.6(a) *Block diagram of a thyristor converter fed separately excited dc motor using hardware*

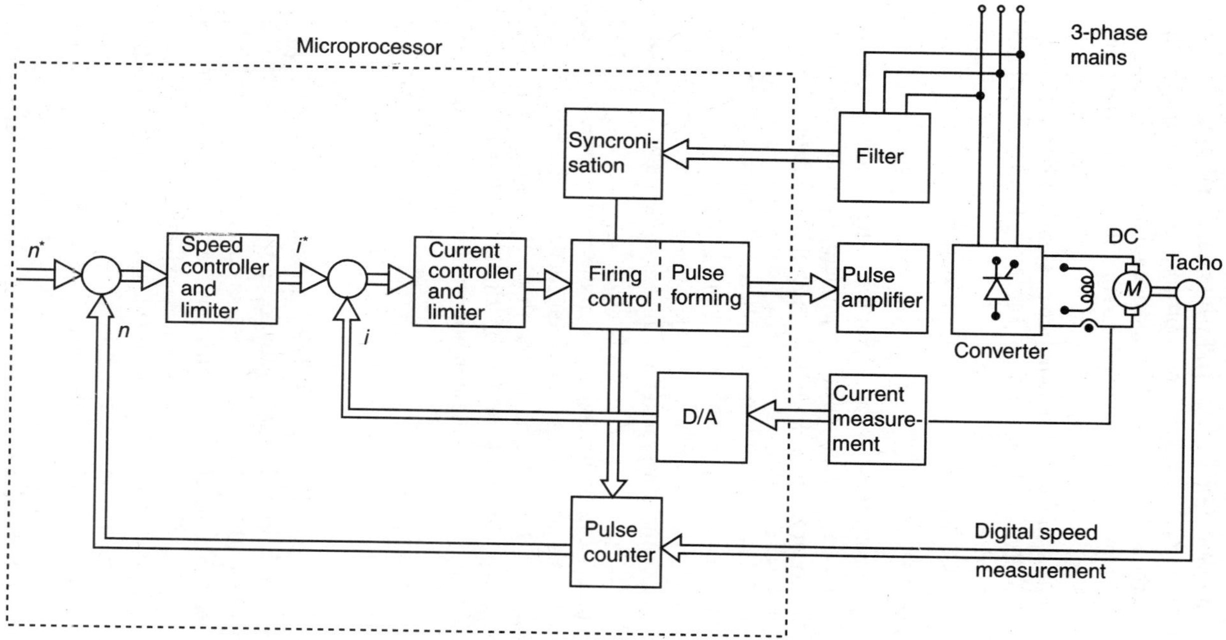

Fig. 8.6(b) *Block diagram of a separately excited dc motor drive using a microprocessor*

effectively replace these schemes for dc motors of high performance because of their varied functional capabilities. At the outset it may seem that the interface circuits between the microprocessor and the system tend to increase the overall cost and decrease the advantage of microprocessors. However the improvement in the functions, reliability, size of the control equipment, and rapid reduction in manufacturing costs are possible with the fast growth and developments of digital systems and A/D and D/A converters, to make the system economical and cost effective. Among the functional capabilities of a microprocessor are higher level reliability, availability, and serviceability, which are instrumental in the increase of productivity of a drive.

The following are the functions of a microprocessor in the control of a dc motor fed from a dual converter:

Speed Sensing As has already been explained, a digital speed encoder provides the information concerning the speed to a microprocessor. The train of pulses from the shaft encoder are processed in the microprocessor to estimate the speed.

Feedback Control The closed loop control here has an inner current loop and an outer speed loop. The necessary reference values are stored in the memory of the microprocessor. The current is measured and converted to a digital quantity using an A/D converter. After conversion the signal is fed to the microprocessor based control system. The speed signal is available in digital form. The necessary controllers and limiters can be implemented on the microprocessor, as has already been explained. The controllers implemented on a microprocessor are adaptive. These improve the performance and flexibility of the microprocessor. The system must be capable of taking care of variable gain during field weakening mode as decided by the operating condition to obtain the desired speed-torque characteristic.

A converter feeding a back emf load (as in the control of dc motor) operates in the discontinuous mode of operation under certain conditions of loading. The converter possesses a non-linear transfer characteristic with variable gain. The performance of the drive is sluggish. The non-linearity must be compensated by a proper feedback loop or linearising the operation using an inverse non-linear function. This function of the microprocessor must be supported by a suitable software. When the conduction is continuous the converter voltage is a constant, independent of load. In the discontinuous conduction, the angle of conduction and hence the converter voltage depend on the load. When once the discontinuous current limit is reached at a firing angle (Fig. 8.7), the converter voltage increases. To bring the value of V_d to the value corresponding to continuous conduction the firing angle must be increased by Δa. From the external characteristic of the converter the change in Δa as a function of I_d at a given firing angle a can be stored in a look-up table. The firing of the converter is controlled so that the gain of the converter is invariant. The non-linear characteristic of Δa vs I_d at a given firing angle is shown in Fig. 8.7, and the block diagram implementing the correction is shown in Fig. 8.7(b).

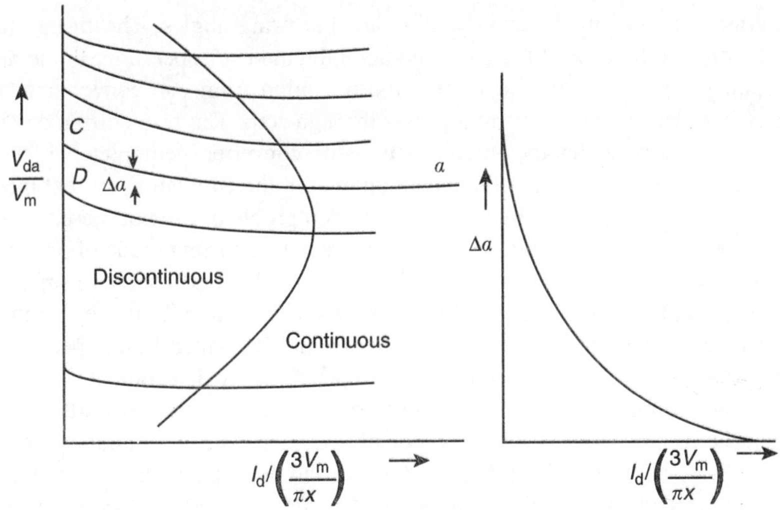

Fig. 8.7(a) *Non-linear compensation for α at the discontinuous conduction*

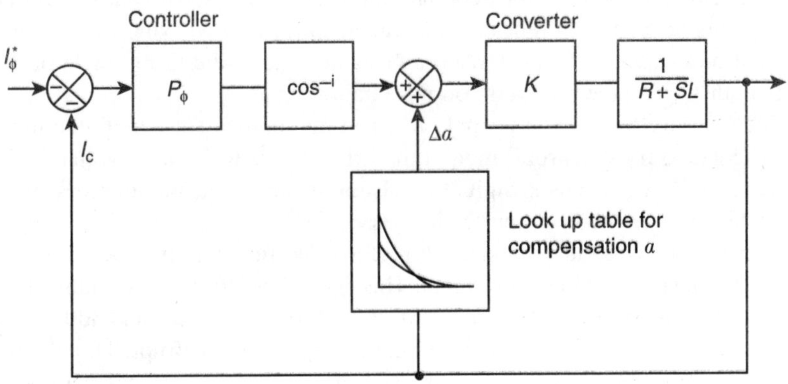

Fig. 8.7(b) *Implementation of firing angle correction at discontinuous conduction*

The feedback signals are limited by the proper limiters supported by the software of the microprocessor.

Gate Firing Signals A microprocessor is programmed to generate the firing pulses, depending upon the information available from the feedback signals. The digitised ac power signals are used to determine the firing delay with respect to the natural firing instant. The firing scheme must provide high dynamic performance to the drive, besides reducing the number of components.

The dual converter has two modes of operation. They are non-circulating current mode and circulating current mode. In the former only one converter is conducting whereas the other is in a non-conducting state. In the latter both the converters operate simultaneously with suitable firing angles. The firing pulse generation and the control logic depend upon the mode of operation. In the non-circulating current mode of operation the transition from one converter to the other takes place when the current passes through zero. The zero current persists for a short interval so that the outgoing (thyristor) converter is completely blocked before the pulses are given to the incoming one. In the circulating current mode, however, the current is continuous and transition problems are not there, as both the converters conduct simultaneously. The circulating current mode of operation is employed when zero current intervals cannot be tolerated. There are structural differences in the power circuits of these modes of operation. Similarly, software development for generation of the firing pulses must consider these aspects carefully. In the non-circulating current mode of operation zero detection of the current is required during the transition from one converter to the other. Also, there may be discontinuous conduction in this mode of operation. The program must be able to distinguish between the actual zero crossing and the discontinuous conduction. The firing circuit for non-circulating current mode is rather involved because of the features described above, whereas the circuit for circulating mode is simpler.

The digital firing angle controller should be programmable, should generate accurate gate pulses in various configurations discussed above, and should perform gating at accurately repeated intervals with the desired symmetry and independent of frequency. If must also be possible to alter the firing angle on line, without disturbing the symmetry of pulse train.

The interface between the supply line and microprocessor is required to provide the necessary synchronisation of the firing pulses with the ac voltages. This converts analog ac signals to digital signals for feeding to the microprocessor. The converter receives the firing pulses in intervals of 60° as per the firing sequence. The microprocessor may be made to perform the function of generating firing pulses by means of an interrupt signal. This signal has a frequency double that of the line. The interrupt signal occurs at 0°, 60°, 120°, 180°, 240° and 300°. A new firing sequence may be started at the falling edge of the interrupt. The hardware implementation and the signals are shown in Fig. 8.8. A microprocessor controlling the processes of a general servo system may be used to generate the firing pulse by going through the pulse generation program by means of the interrupt signal. After successfully completing the job the microprocessor will return to the normal function. The program should take care of performing the firing range selection, protection, etc. The firing angle range can be detected using the information of the line voltages of 1/6 cycle. This information must be sufficient to cover 0 to 180° in both the directions of current flow. This simplifies the system. The firing signals may be demultiplexed and amplified before they are fed to the thyristors. By this a saving of the hardware of the processor and I/O may be accomplished, which can be used effectively for other purposes. The necessary hardware and truth table for generating firing pulses are shown in Fig. 8.9. The gate firing program must handle both continuous and discontinuous conductions if the converter operates in the non-circulating mode of operation. The appropriate

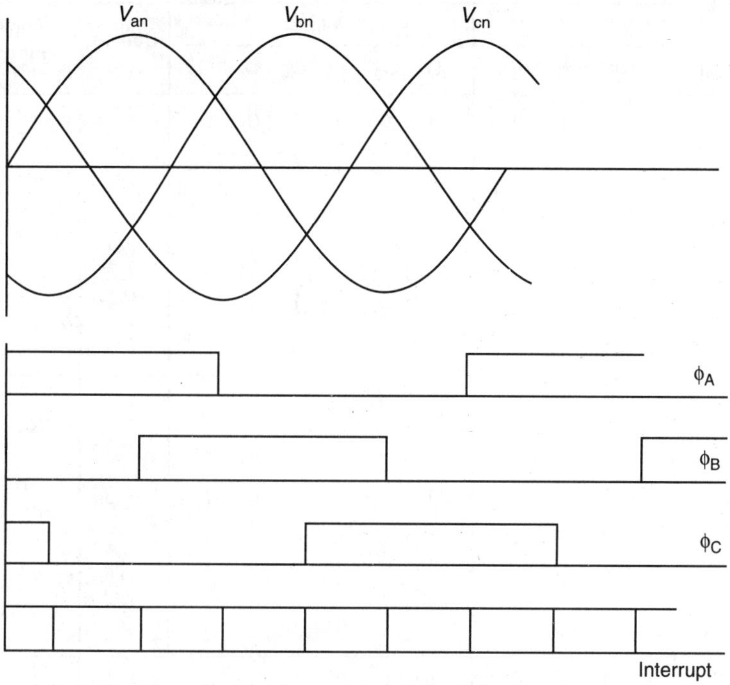

Fig. 8.8(a) *Generation of interrupt from digitised power signals*

converter must be fired depending upon the current direction. The discontinuous conduction must not be confused with zero of the current during transition, in order to assure the reliable firing of the converter. The necessary logic must be designed considering the following:

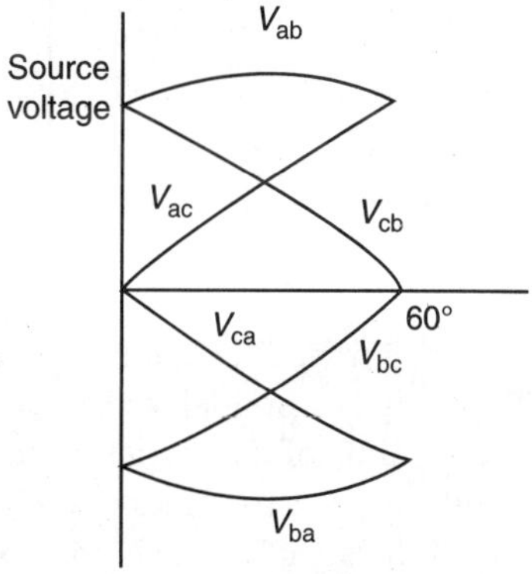

Fig. 8.8(b) *Power source voltage signals*

i. When the load current is positive converter A conducts and when it is negative the converter B conducts.
ii. If there is no discontinuous conduction sensing of either load current or thyristor conduction may be used.
iii. Zero current does not mean that transition should occur.
iv. When the load current is fetched to assess the condition of zero current it does not always mean that the transition should occur.
v. Gain compensation during discontinuous conduction or field weakening mode.

Inputs					Output						Source	Angle range
D_1	D_6	ϕ_A	ϕ_B	ϕ_C	Q_1	Q_2	Q_3	Q_4	Q_5	Q_6		
0	0	1	0	0	1	0	0	0	0	1	V_{ab}	0°–60°
0	0	1	1	0	1	1	0	0	0	0	V_{ac}	
0	0	0	1	0	0	1	1	0	0	0	V_{bc}	
0	0	0	1	1	0	0	1	1	0	0	V_{ca}	
0	0	0	0	1	0	0	0	1	1	0	V_{ba}	
0	0	1	0	1	0	0	0	0	1	1	V_{cb}	
0	1	1	0	0	0	0	0	0	1	1	V_{cb}	60*–120°
0	1	1	1	0	1	0	0	0	0	1	V_{ab}	
0	1	0	1	0	1	1	0	0	0	0	V_{ab}	
0	1	0	1	1	0	1	1	0	0	0	V_{bc}	
0	1	0	0	1	0	0	0	1	1	0	V_{ba}	
0	1	1	0	1	0	0	0	1	1	0	V_{ca}	
1	0	1	0	0	0	0	0	1	1	0	V_{ca}	120°–180°
1	0	1	1	0	0	0	0	0	1	1	V_{cb}	
1	0	0	1	0	1	0	0	0	0	1	V_{ab}	
1	0	0	1	1	1	1	0	0	0	0	V_{ac}	
1	0	0	0	1	0	1	1	0	0	0	V_{bc}	
1	0	1	0	1	0	0	1	1	0	0	V_{ba}	

Fig. 8.9 *Truth table for firing angle control*

The current direction can be sensed by the so-called current direction signals. These signals are delayed by a definite time period, normally greater than the turn off time of the thyristors, to make sure that the thyristor has turned off. The simultaneous zero values of the signals indicate that no thyristor is conducting. If one of the signals is zero, the converter with non-zero signal is conducting. The other converter cannot get firing pulse so that a short circuit is avoided. This logic is depicted in Fig. 8.10. The line voltages of l/6th cycle provide the information of the direction of current in a given range of firing angle.

To summarise, in a non-circulating current mode of operation the following points need consideration while developing software.

i. When the thyristors in one converter are conducting at any instant of time the other converter cannot be fired. During transition from one converter

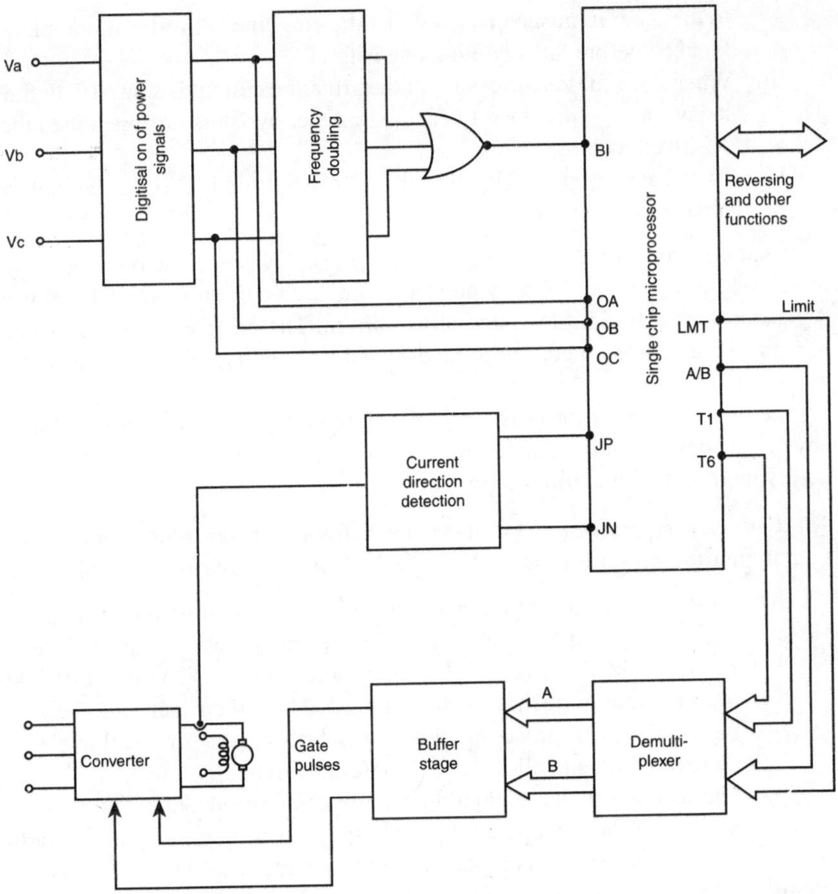

Fig. 8.10(a) *Block diagram of the control of dual converter*

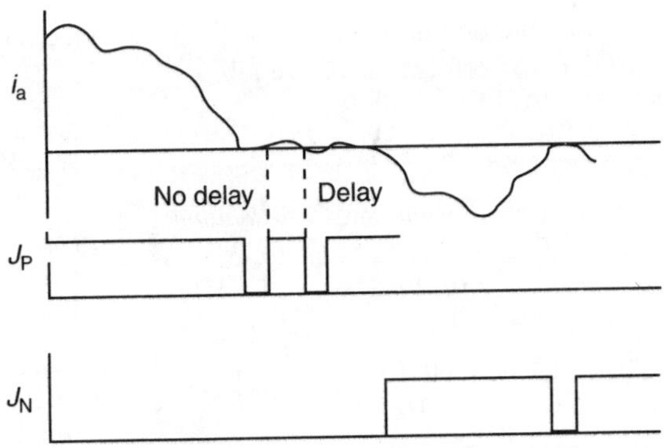

Fig. 8.10(b) *Current direction detection*

to the other it must be ensured that the outgoing converter is completely blocked before the incoming one is fired.

ii. When once the zero crossing of the armature current is sensed, a definite delay must be provided to ensure that the thyristors have regained their positive blocking capability.

iii. To arrive at proper logic, the state of load current is checked but not that of an SCR.

In some applications, such as servo drives, the delay cannot be tolerated during transition, however small it may be. The circulating current mode is employed. A reactor is used to limit the circulating current. The firing logic is very simple here, as there is no zero current to be detected and no discontinuous conduction at all.

The microprocessor must also provide suitable protections against failure of any of the phases, commutation failure, cross over condition, etc. The system reliability increases with multilevel protection.

Selection of a Microprocessor From the above discussion the criteria for the selection of microprocessor for the control of a dual converter may be derived:

i. The main function of the microprocessor is generating the firing pulses with least possible asymmetry. The asymmetry level is decided by the resolution of the firing angle. To get a resolution of the order of 1°–2° the sampling frequency must be 10–20 kHz while generating the firing pulse. The microprocessor chosen must be capable of small cycle and computing time to allow for high rates of sampling control.

ii. The accuracy in the calculation of current and speed loops. The sampling periods of these loops are however larger than those of the generating pulses. The error in computation must be limited.

iii. If extra accuracy is required in the speed loop the microprocessor must be capable of floating point arithmetic in order to handle large numbers. For this purpose a fast microprocessor with optimised programs may be required.

From the data available on microprocessors (Table 8.1) the 8 bit microprocessor would be sufficient to perform the above job.

Table 8.1

	Time constant	*Sample rate*	*Resolution*	*Accuracy*	
				8 bit	*16 bit*
Firing angle control		10–20 kHz	0.9°–1.4° 1.8°		7.8%
Current loop	10–20 ms	500–1000 Hz		0.37%	
Speed loop	100–200 ms	50–100 Hz		0.78%	0.3%

Some other applications of microprocessors in the area of dc drives are

i. chopper controlled four quadrant dc drives using separately excited dc motor or dc series motor.
ii. multiphase chopper controlled dc drives
iii. pulse width controlled dc drives
iv. The multiple firing schemes for improving the performance of dc drives, e.g., TRC and CLC.

8.4.2 Field Oriented Control of a Three Phase Induction Motor

The stator current of an induction motor has the functions of producing the required air gap flux (magnetisation) as well as developing the required torque to drive the load. An induction motor will have its operation similar to that of a dc motor if the stator current components (viz., flux producing and torque producing) can be separately controlled. This kind of control is possible in a separately excited dc motor where the torque and flux can be separately and independently controlled by varying the armature current and field current respectively. An inherent decoupling would exist between them, but for the effects of armature reaction. These effects can be eliminated by armature compensation. A perfect decoupling can be achieved in a compensated separately excited dc motor. This versatile control imparts a very good dynamic behaviour to a dc motor. Thus a high performance drive using an induction motor is achieved by attempting a decoupling of the stator current components. This principle is called field orientation control or vector control. This control improves the dynamic behaviour and a drive of very good performance can be obtained even at low speeds. The two components of the current are identified and they are oriented properly in orthogonal ordinates with respect to flux vector.

The current components can be oriented with respect to any of the three fluxes, viz., stator flux, air gap flux or rotor flux. The analysis shows that the dynamic performance of the drive is not up to the desired level if the orientation is carried out with respect to the stator or air gap flux. This poor dynamic behaviour stems from the delay of the torque in following the slip. A natural and efficient decoupling is possible if the orientation is carried out with respect to the rotor flux. This leads to a high performance torque control of the drive with a very fast response. The implementation of the principle of field orientation is illustrated in Fig. 8.11.

When the principle of field orientation was suggested, it did not receive the attention of the industry and was not very popular because of complicated hardware. With the developments in digital components and microprocessors by way of LSI, and miniaturisation of the components, this high performance control of induction motor is becoming popular. One main area of application and effective use of high speed microprocessors is the vector control of induction motors. This control requires

i. the exact information about the rotor flux.
ii. the precise adjustment of the stator current components according to the reference.

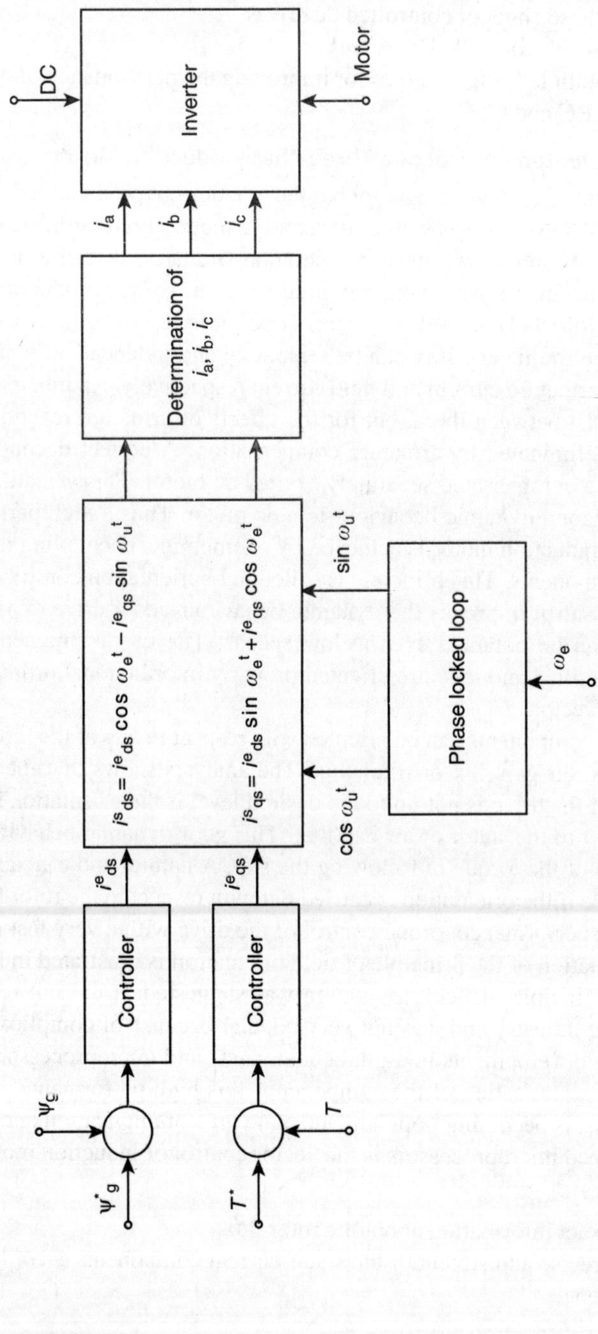

Fig. 8.11(a) *Implementation of the principle of field orientation*

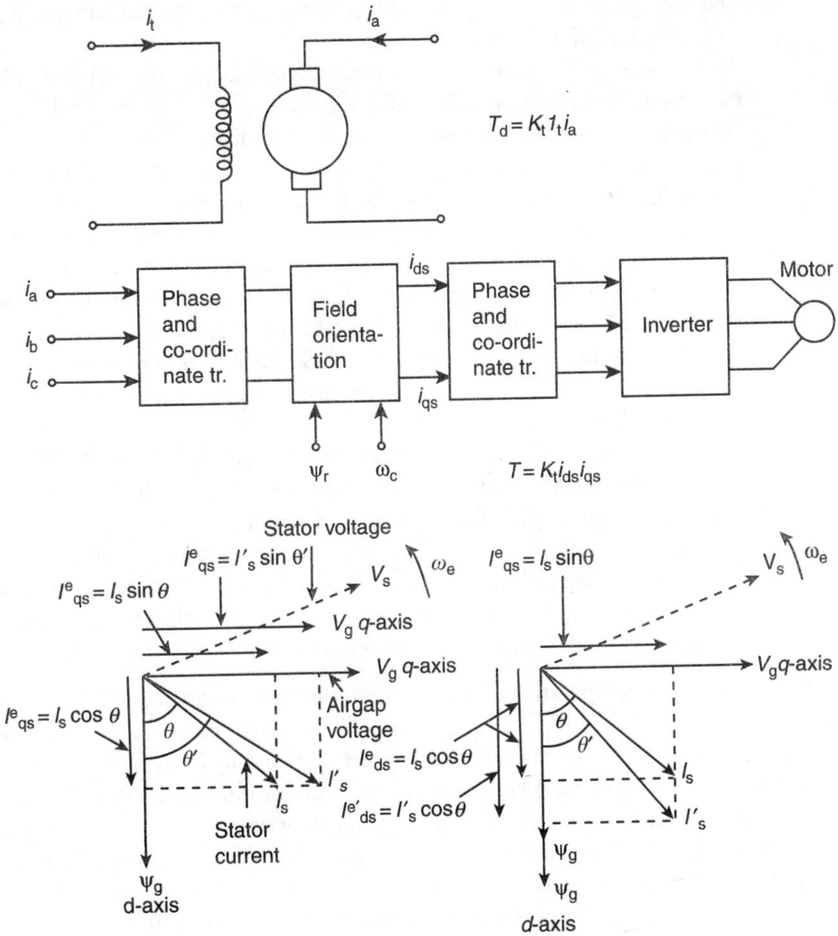

Fig. 8.11(b) *Illustration of principles of field orientation*

The functions of a microprocessor in the vector control of induction motor are as follows:

i. Processing of the signals obtained from the shaft encoder to determine the rotor speed and also the rotor angle. This rotor angle has to be used in the transformations from one frame to another.
ii. The flux estimation using the terminal voltages, currents and speed, based on one of the machine models.
iii. The computations with respect to phase and coordinate transformations to identify the two components of the current. After the necessary control these components must be transformed to provide the reference values for comparison with actual phase currents.

iv. The speed and current loops in the feedback control. The implementation of controllers in these loops.

v. To produce gate signals for both machine side converter and line side converter. The machine side converter decides the frequency whereas line side one decides the current/voltage. The firing signals to the line side converter are obtained in the same way as described for a dual converter. The firing delay is decided by the current/voltage required in the dc link. The firing signals to the machine side converter decide the frequency. The speed of the motor is added to the output of the slip controller to decide the frequency of the inverter output. The addition must be precise because a large quantity is added to a small one. The digital addition in a microprocessor is accurate. So the microprocessor must be capable of providing or generating the firing signals to the machine side converter also.

vi. *Data acquisition* The microprocessor must acquire the feedback signals in the digital form. A transfer of the data to CPU is required. A flow of the data both from and to the processor is required.

vii. *Limiting the control Variables* Linearisation of non-linear functions used in the control as well as non-linear behaviour of the converter during discontinuous conduction. The compensation of variable gain during discontinuous conduction and field weakening modes.

The mathematical operations include multiplication, division, addition and subtraction. The arithmetic processing unit of the microprocessor must be capable of performing these operations. The flux estimation may be done by numerical integration using the well known Simpson's rule or trapezoidal rule. But the results suffer from loss of accuracy due to truncation errors in eight bit processors. These errors lead to instability of operations. Use of floating point arithmetic with double precision improves the performance of the processor. This however requires a long time of computation. To reduce the burden on the microprocessor the flux estimation may be carried out by analog models external to the processor. The computed flux may be processed in the processor. The reference values of the currents and fluxes are developed by the processor. Depending upon the speed of the processor and its capacity to perform the above functions a boundary may be drawn between the local hardware and the microprocessor, to perform the functions. Sometimes multiprocessor control may be accomplished depending upon the speed and quality of control required.

Field orientation is possible with both voltage source inverter as well as current source inverter fed induction motors. The current source inverter is widely employed due to the simplicity of its power circuit. The features of field oriented control employing CSI are discussed here (8.12(a)). Sometimes a PWM inverter may be controlled suitably to provide the reference currents (Fig. 8.12(b)). This control can be accomplished using a microprocessor.

The flux may be measured directly using search coils or Hall probes. The analog signals are converted to digital ones for feeding into the microprocessor for

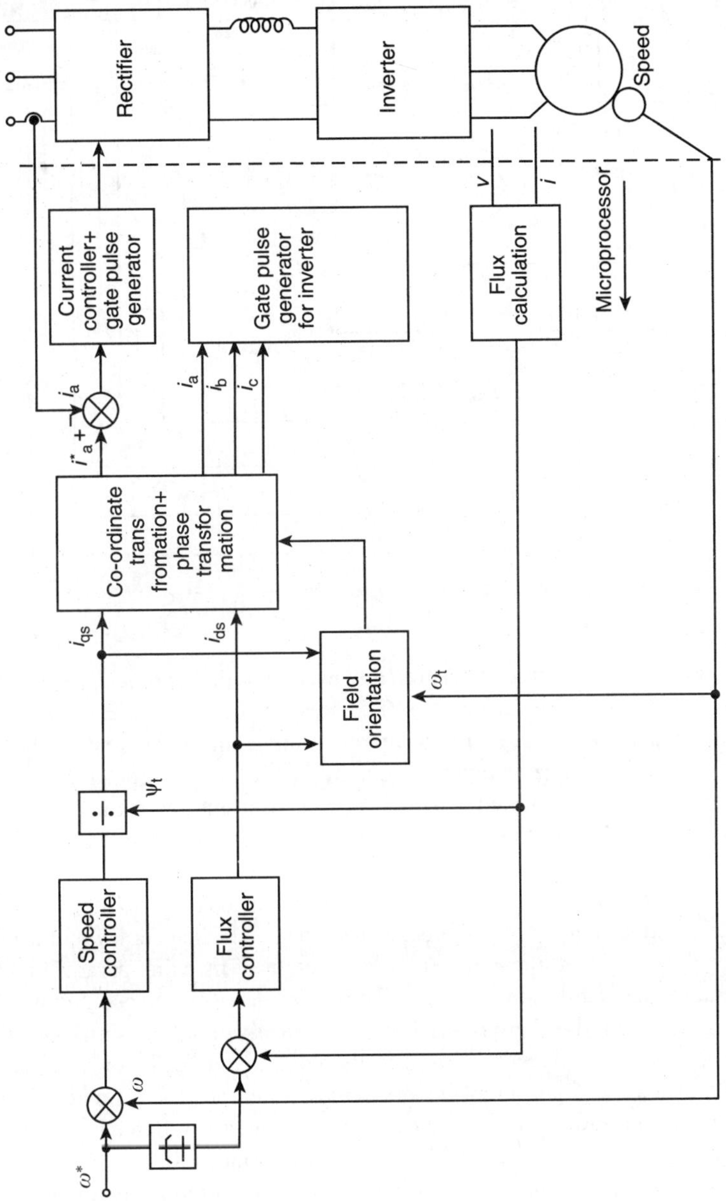

Fig. 8.12(a) *Field oriented control of a CSI fed induction motor*

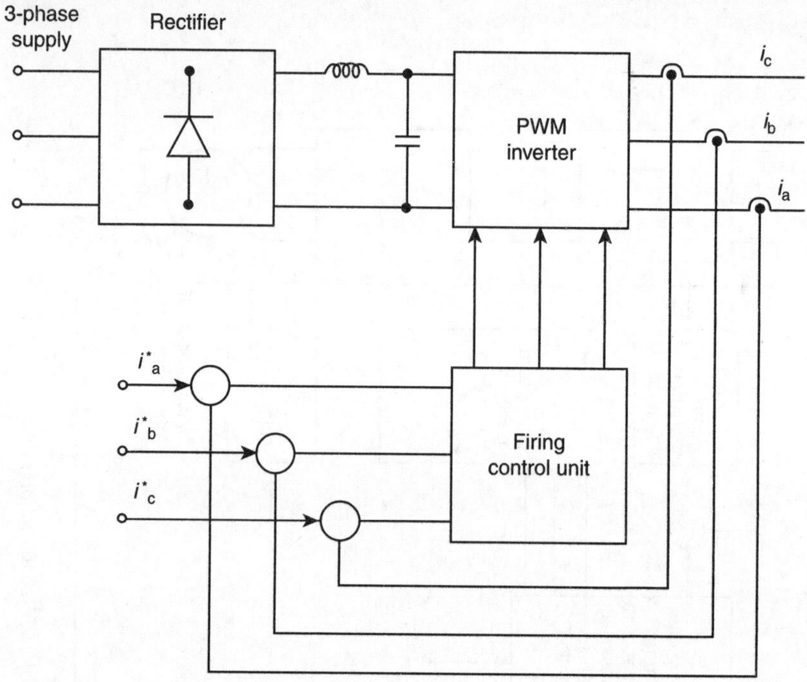

Fig. 8.12(b) *PWM inverter with current control*

further processing. This direct measurement, even though theoretically exact, has the following limitations or difficulties: The presence of sensitive Hall probes or search coils make the induction motor more sensitive but its inherent robustness is lost. The measured signal is superimposed by slot harmonics requiring filtering. The errors of measurement and A/D conversion make the results rather unreliable. Therefore, normally the computation of the flux using machine models is employed (indirect estimation).

These machine models for computing the flux make use of the motor parameters determined from no-load and blocked rotor tests. The inaccuracy of the models in the estimation of the parameters, the variation of machine parameters due to temperature and saturation must be exactly considered in the estimation. The accuracy gets impaired if there is any integration involved in the estimation. This happens if voltage and speed are used in the computations. At low speeds the integration introduces considerable errors in the estimation. The estimation using stator currents and rotor speed is free from these errors and reliable performance can be achieved over a wide range of speeds down to standstill.

To improve the performance the machine parameters used in the model must be as accurate as possible. Otherwise they can be corrected using a correction process, so that the exact coupling is possible. The correction process upgrades the parameter. The indirect flux estimation along with the so-called parameter identification or adaptation is used to obtain an induction motor with perfect coupling. The flux estimated from the actual sensed variables of current and speed together with the parameters, is compared with the reference value (determined from the reference quantities). The difference is used to correct the machine parameters entering into the calculations. The most influential parameter is the rotor resistance or rotor time constant and it is corrected until the required value of flux is given by the model.

The only solution for field oriented control with parameter adaptation is the use of microprocessors or microcomputers. Several techniques are available in the literature for parameter adaptation. One such scheme is shown in Fig. 8.13. The microprocessor has to perform the mathematical operations to estimate the flux both from the reference values as well as measured variables.

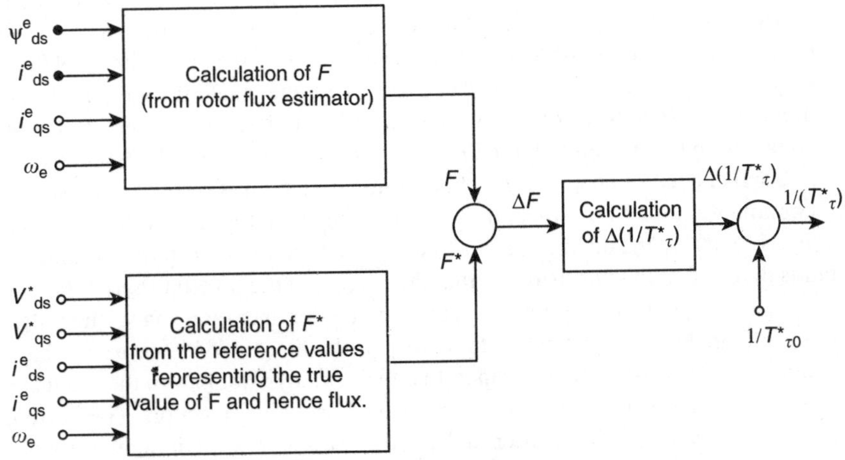

Fig. 8.13 *Adaptation or identification of rotor time constant*

The selection of a microprocessor for the above application may be based on the following considerations:

i. Resolution of the firing of the phase controlled rectifier. The resolution decides the asymmetry in the firing.
ii. The mathematical operations that can be performed. It should be capable of binary and decimal arithmetic, including multiplication and division
iii. The memory capacity required
iv. The internal clock generation
v. Software support for implementing the controllers, limiters, etc.
vi. The interrupt capability

Some other applications of a microprocessor in the control of induction motors are:

i. A microprocessor may be used to control the speed using a voltage controller

ii. It may be used to control the speed using slip energy recovery scheme. Vector control can be employed here also.
iii. It may be used for the control of slip controlled drive with flux and torque control in CSI fed drives.
iv. Can be used to implement PWM techniques to control the voltage and harmonics both in VSI and CSI fed induction motors.
v. It may be used for cycloconverter control.

8.4.3 Microprocessor Control of Synchronous Motor Drives

Variable speed drives employing synchronous motors are becoming very popular in industrial applications. They are an immediate solution for high power reversible drives and are becoming competitors to dc and induction motor drives. The synchronous motor operates at leading power factors when overexcited. The armature voltages can be used to commutate the inverter thyristors on the machine side. This is possible over a reasonably wide range of speeds. However, at very low speeds up to about 10% of base speed, the machine commutation is ineffective and therefore forced commutation is required. The inverter may be equipped with forced commutation circuit which operates up to 10% of the base speed, and above this speed the machine commutation takes over. The assistance in commutation may be provided by means of interrupting the dc link current. A thyristor across the dc link is used. When the commutation is required the current is diverted to this thyristor, so that the current in the inverter thyristors falls to zero. The line side converter is forced into inverter operation so that the polarity of voltage changes. After the thyristors in the machine side converter have attained their positive blocking capability, the next thyristor pair in the sequence is fired. The line side converter is forced back to rectifier action. The current is now transferred to the inverter and the thyristor across the inductance stops conduction automatically. One of the starting schemes must be employed. The machine commutation is possible only with current fed operation. The cycloconverter may also be used to control the speed of the synchronous motors. Machine commutation is possible here also. However, in the low speed range the line voltages may be used for commutation. The starting problem does not arise here.

The control of the inverter or cycloconverter feeding a synchronous motor can be accomplished using the position of the rotor with respect to stator. A rotor position sensor on the shaft senses the position of the rotor with respect to the stator and sends the firing pulses to the thyristors of the inverter. The six inverter thyristors are fired in a sequence once, by the time the rotor moves by two pole pitches or 180°. This provides synchronism between the frequency and rotor speed. The motor is said to be in self controlled mode.

The self control of the motor is also possible by sensing the stator induced voltages which effectively eliminates the mechanical rotor position sensing. In one cycle of stator voltages all the six thyristors are fired once in a sequence. Self control is possible with VSI, CSI and cycloconverter.

A synchronous motor in its self controlled mode has a performance similar to that of a dc motor, both in steady-state and dynamic conditions. It, therefore finds application where dc motors cannot be employed. It is also called commutator less motor (CLM).

When a naturally commutated converter operates in the inverter mode the overlap limits the range of firing angle. The upper limit of firing angle called inverter limit is decided by the turn off time of the thyristors and the overlap. The limit is set such that there is enough margin for (t_q) the turn off of the thyristor taking overlap into consideration, so that the commutation failure does not occur. A margin angle is defined as the difference between the lead angle of firing and overlap angle. A control is evolved to see that under no circumstances does this margin angle go below the turn off angle of the thyristor, so that commutation failure does not occur. This is called margin angle control. It has certain advantages:

i. Commutation failure is prevented.
ii. The maximum power output can be increased by simultaneous control of the field current to compensate for armature reaction. This improves the overload capacity of the motor.
iii. There is improvement in the power factor
iv. The torque pulsations under the load conditions are reduced. The ripple content of the dc current also decreases.

However, there are certain disadvantages. The upper speed is limited and hence the range of speed control is limited. To overcome this disadvantage, margin time control is employed, which improves the performance also. Microprocessor implementation of the synchronous motor is discussed in the next section.

Microprocessor Control of a Current Source Inverter Fed Synchronous Motor As discussed above, a drive system employing a CSI fed synchronous motor has the following features:

i. A four quadrant drive can be accomplished very easily.
ii. A self control, which synchronises the gating pulses of the inverter with rotor position, provides an improved steady-state and dynamic performance.
iii. The natural commutation using machine voltages is possible in a speed ranging from 10% base speed. At starting and low speeds forced commutation is required, which may be provided by additional commutation circuit of the inverter or by interrupting dc link current.

A typical block diagram implementation of a microprocessor for the control of a CSI fed synchronous motor is shown in Fig. 8.14. The system consists of a dc link converter which is made up of two six pulse bridge converters interconnected by a high smoothing inductance. The dc link inverter feeds a synchronous motor whose field may be controlled by a chopper or a phase controlled rectifier. In the normal operation, the line side converter operates as a rectifier and the

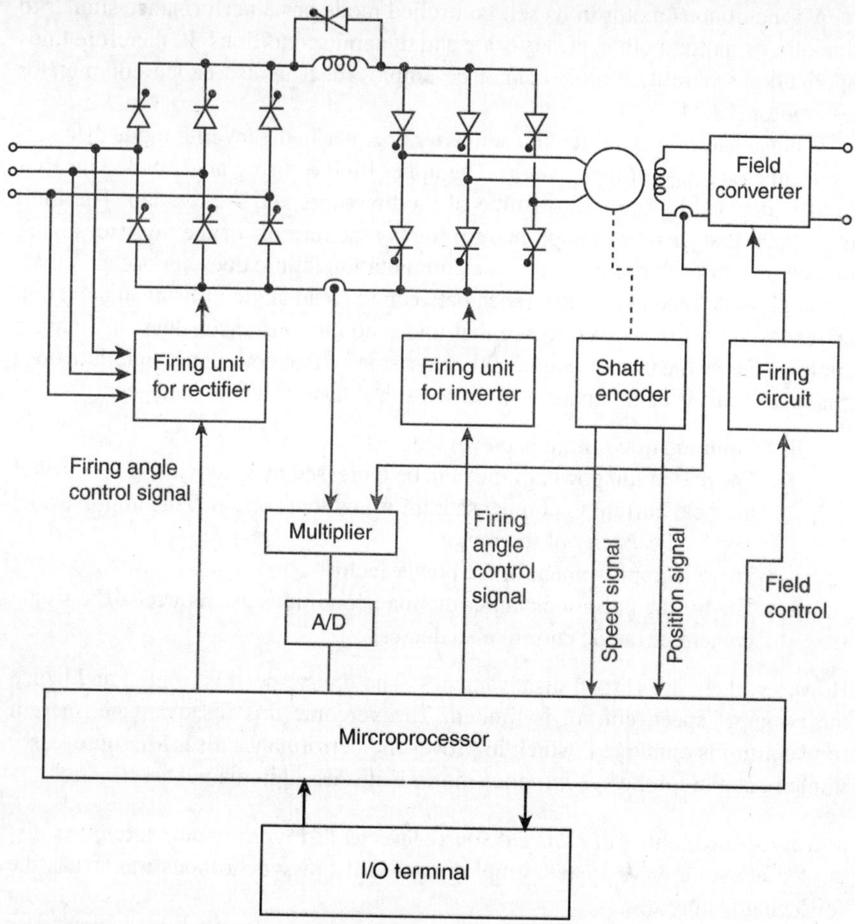

Fig. 8.14(a) *Block diagram of a microprocessor based synchronous motor control*

machine side one as an inverter. The control pulses to the rectifier are provided by the microprocessor with proper interface between the ac lines and microprocessor, to provide the proper interrupt signal, to synchronise the firing pulses with frequency and to vary the firing angle with respect to the natural firing instant. The synchronous motor is fitted with a shaft encoder. This is an aluminium disc with slots on the periphery (Fig. 8.15) mounted on the shaft. A combination of phototransistor and light emitting diode aligned across the slots provides the necessary train of pulses. These pulses are processed to obtain the rotor position as well as the speed. The forced commutation is available during starting until the machine accelerates to a speed where natural commutation can take over. The stator and field currents are also sensed and converted to digital form before they are fed to the microprocessor for further processing.

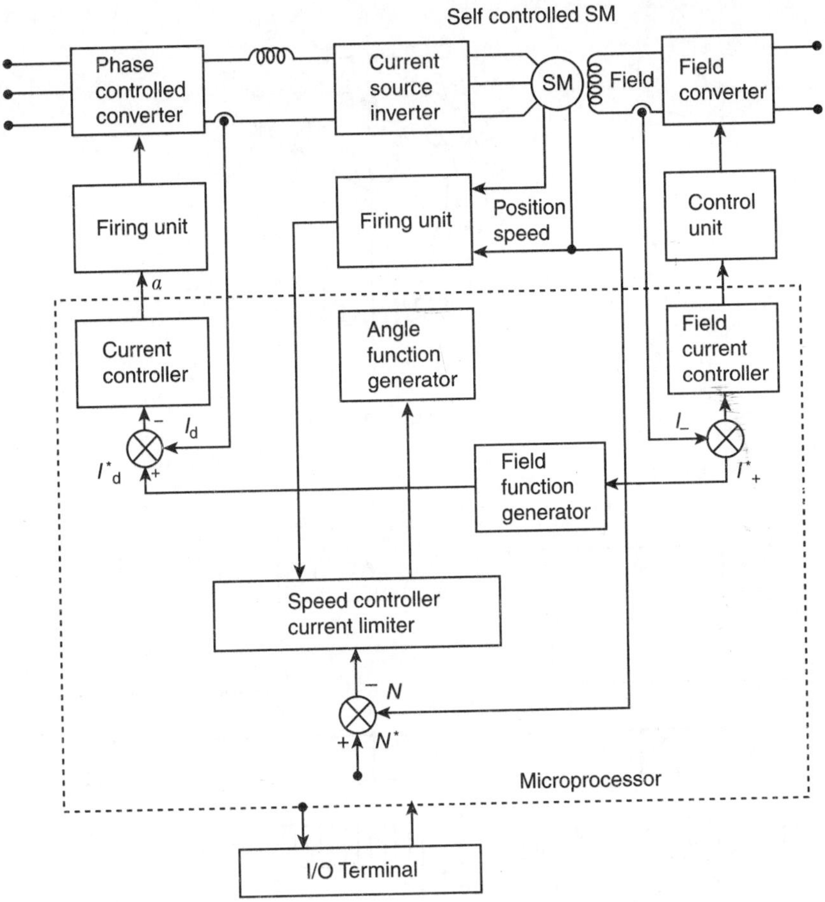

Fig. 8.14(b) *Microprocessor based control system of a CSI fed self controlled synchronous motor*

The functions of the microprocessor are now obvious and can be listed as under:

i. The microprocessor should perform the main functions of monitoring and control of the system variables to obtain the desired performance. The other functions are protection, diagnosis, and display.

ii. It must be supported by proper software to ensure the necessary commutation of the inverter at low speeds where the machine voltages are not effective.

iii. It receives the data concerning the system variables, stator and field currents, and processes them to issue the desired control signals to the rectifier and inverter to achieve the desired performance at all operating conditions.

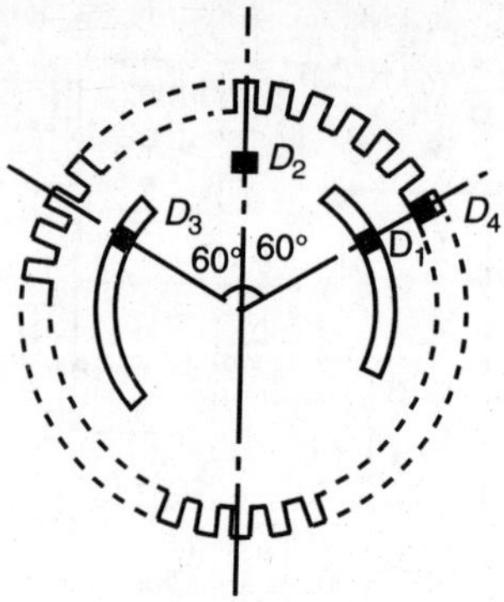

(a) Encoder disc

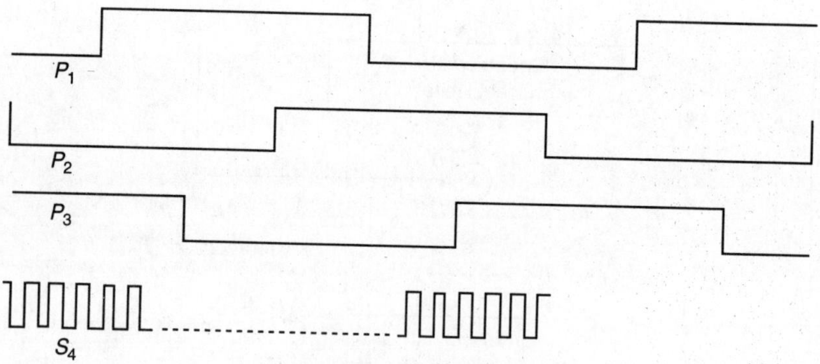

(b) Output waveforms

Fig. 8.15 *Shaft encoder*

iv. The link current control is obtained by controlling the firing angle of the line side converter. The firing angle control with proper synchronisation with line voltages and with respect to the natural firing instant may be obtained by implementing the method described above with the control of dual converters. Another scheme using phase locked loop is depicted in Fig. 8.16, the details of which are available in the literature.

v. The microprocessor receives the information regarding the rotor position and processes it to control the firing of the inverter.

vi. It must be software supported and have necessary hardware to accomplish the feedback configurations of the control. It must perform the generation of necessary feedback signals, necessary controllers, limiters and function generators using look-up tables. The controllers must be

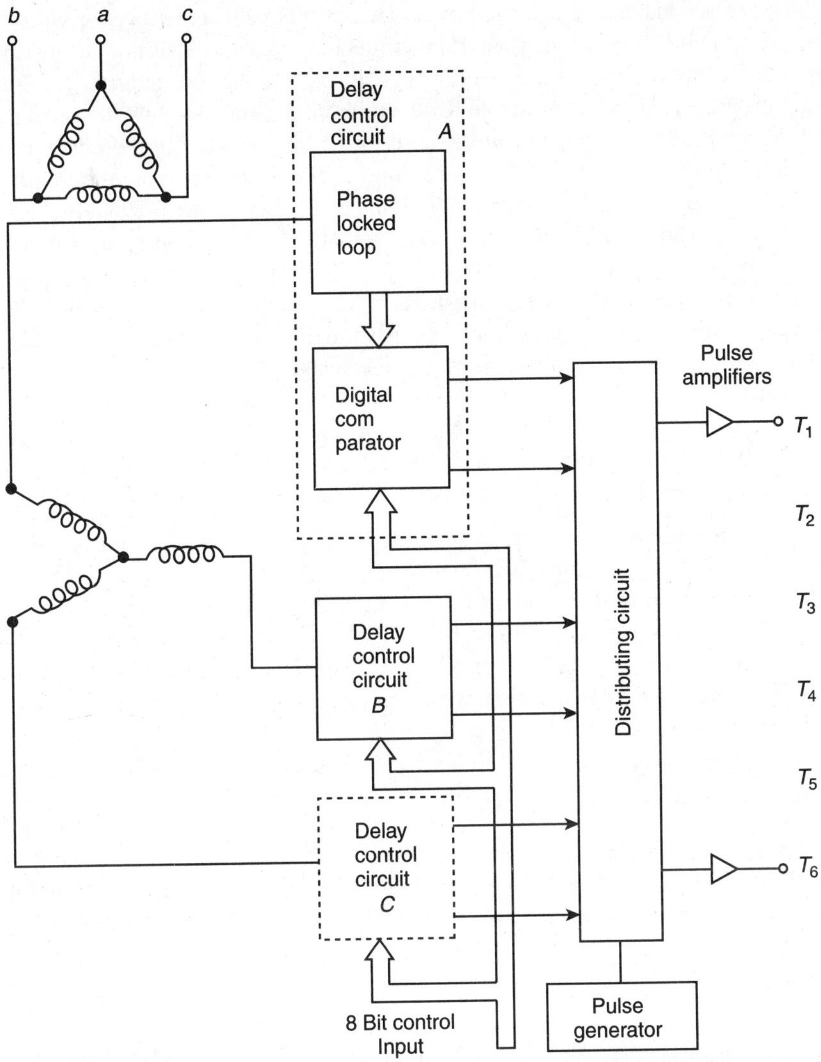

Fig. 8.16 *Digital firing circuit for the converter using phase locked loop*

software oriented so that they can be readily modified to suit a particular application.

vii. It processes the information from the rotor position sensor to determine the speed, which is one of the feedback signals.

Starting It has already been made clear that from zero to about 10% of base speed machine commutation is not effective. To accelerate the rotor to the speed where machine commutation can take over, forced commutation of the inverter thyristors is necessary. This can be achieved by an auxiliary circuit in the inverter with the facility of being cut off when natural commutation takes over. It is also possible by interrupting dc link current with a thyristor across the link inductance. Software programs should be available for the microprocessor to follow the sequence of operations. The second method is described here. On starting the inverter the thyristors are fired in pairs, as the quenching is obtained by making the link current zero. The information must be stored in the address registers. The initial rotor position is decided by the phase reference signals. A pair of thyristors is fired to obtain the maximum torque. At each change of state of the reference signals, commutation is executed by the microprocessor. The thyristor across the link inductance receives the gate signal. Simultaneously the line side converter firing angle is retarded beyond 90° to make it inverter. This changes the polarity of dc link voltage and decreases the link current to zero extremely fast. The current through the thyristor pair also becomes zero. The thyristors switch off. The line side converter is brought back to rectifier mode and the next pair of thyristors of the inverter are fired. The process is repeated till the motor accelerates to the desired speed.

i. The necessary software to perform the speed regulating and current regulating routines must be provided.
ii. A typical shaft encoder to provide the rotor position and to control the inverter firing angle must be provided. A typical shaft encoder is discussed separately as it can be used as a generalised hardware for any speed control employing microprocessor.
iii. During regeneration the firing angle control should be such that the line side converter operates in inverter mode and the machine side one as a rectifier.

Rotor Position Sensing A shaft encoder used for the purposes of both rotor position sensing and speed measurement is slightly different from the one discussed under the section speed detection. A typical sensor is explained in the following.

i. The shaft encoder used in the control of a synchronous motor is required to provide the triggering pulses to the converter, which are synchronised to the rotor position. The required firing angle delay must also be provided. These pulses must be available to the thyristor during starting employing forced commutation with link current interruption, as well as in the normal operation.
ii. It is required to provide the information to determine the speed.
iii. The firing pulses must be such that every 60° one thyristor is fired in sequence. The width of the firing pulses may be 120°. The delay angle information to provide the necessary phase difference between the voltage and current must be available.
iv. To be able to obtain the above information, the shaft encoder must give the necessary phase reference signals and one high frequency signal.

The shaft encoder is a light aluminium disc with a sufficiently large number of slots drilled at the periphery of the disc. The resolution of the firing angle delay depends upon the number of slots. A resolution of 1° may be obtained if there are 360 slots. A high frequency signal is obtained when the disc rotates through an optical sensor. This consists of a light emitting diode aligned with a phototransistor. Whenever a slot comes in between these two the transistor gives out a pulse. A train of pulses is obtained, which can be used as a high frequency signal to determine the firing angle delay as well as speed. The disc also has two more slots, the angle of which depends upon the number of poles, e.g., 90° slots for a four pole machine. Three sensors are placed with a phase shift of 120° electrical so that the sensors produce phase reference signals.

The waveforms of voltage and current during motoring and regeneration are shown in Fig. 8.17. The phase reference signals P_1, P_2, P_3 and the high frequency signal S_4 are also shown in the figure. These signals are processed to provide the firing pulses. The information of the mode of operation, i.e., motoring or regeneration is assigned to the first and the remaining delay information is assigned to the remaining seven bits of an eight bit digital word. The delay angle is a or $(180 - a)$ depending upon motoring or regeneration. The firing angle controller comprises three delay circuits, a thyristor address register and a pulse distribution circuit (Fig. 8.17). The digital control input is converted to the corresponding delay angle.

At the beginning of each half cycle of phase reference signal the delay information a is fed into the counter. This is converted to provide the required delay using linear digital ramp techniques. The pulses of the high frequency signal are used to increment or decrement the count. At the terminal count a pulse is generated, which operates as a clock to the D-type flip flops of the thyristor address register. The output signals having a delay of a with respect to phase reference signals are fed to a decoder, where they are combined to produce the modulated firing pulses for all the six thyristors. The pulse trains, shown in Fig. 8.17, are 120° wide and displaced by 60° from each other.

The firing of the inverter must take place during starting also, when there is no machine commutation. The thyristors are actually fired in pairs. For this the delay counters are disabled. The thyristor pairs to be gated are directly addressed by CPU through the thyristor address registers. The phase reference signals are available at the starting of the signals. The appropriate thyristors are fired to provide the torque in the required direction. At the change of state of the phase reference signals, the forced commutation is executed by the microprocessor in the following steps:

i. The thyristor across the link inductor is fired.
ii. The line side converter is forced into inverter operation. This reverses the dc link voltage and effectively makes the dc link current zero. The current in the conducting thyristors of the inverter is also zero, making them switch off.

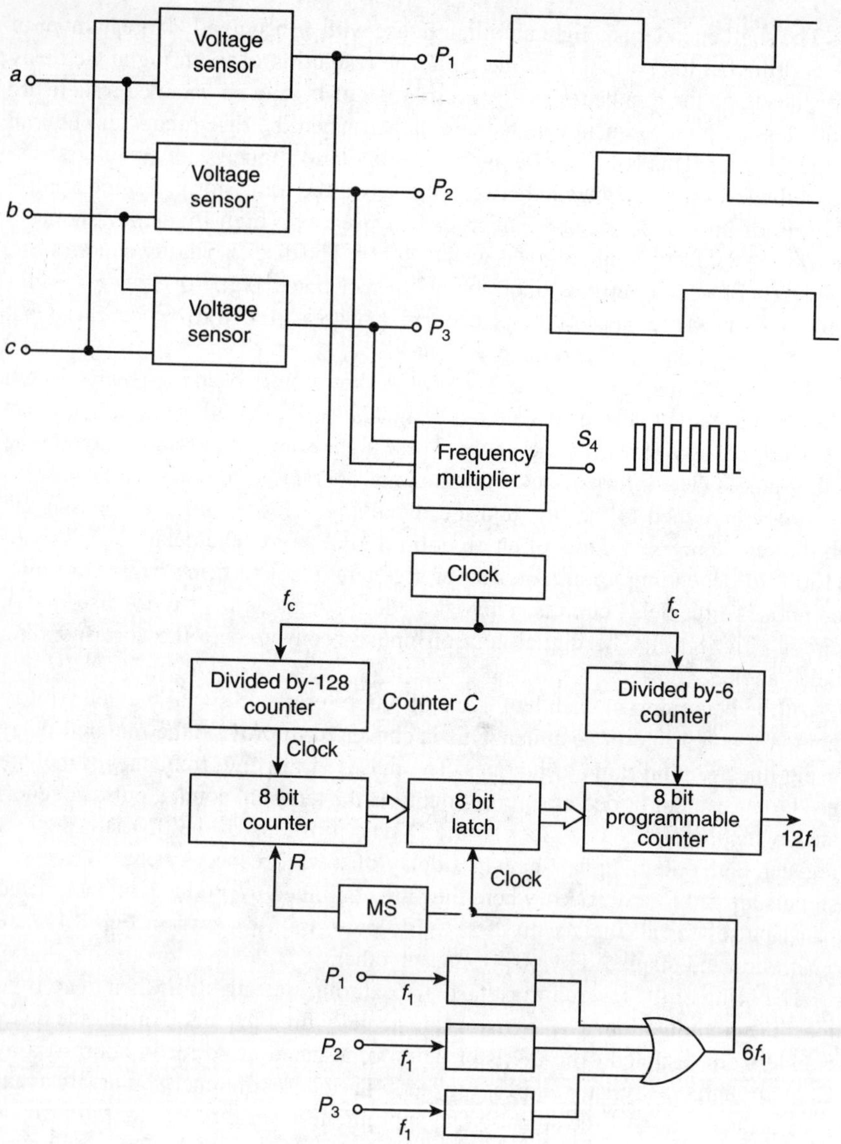

Fig. 8.17(a) *Terminal voltage sensing and frequency multiplication*

iii. The next pair of thyristors of the inverter is fired using the command from the CPU to the thyristor address register. The line side converter is brought back to rectification.

The necessary changes during starting are indicated on the schematic diagram of the firing angle controller. The forced commutation is continued till the motor accelerates to 10% of speed, where the machine commutation takes over.

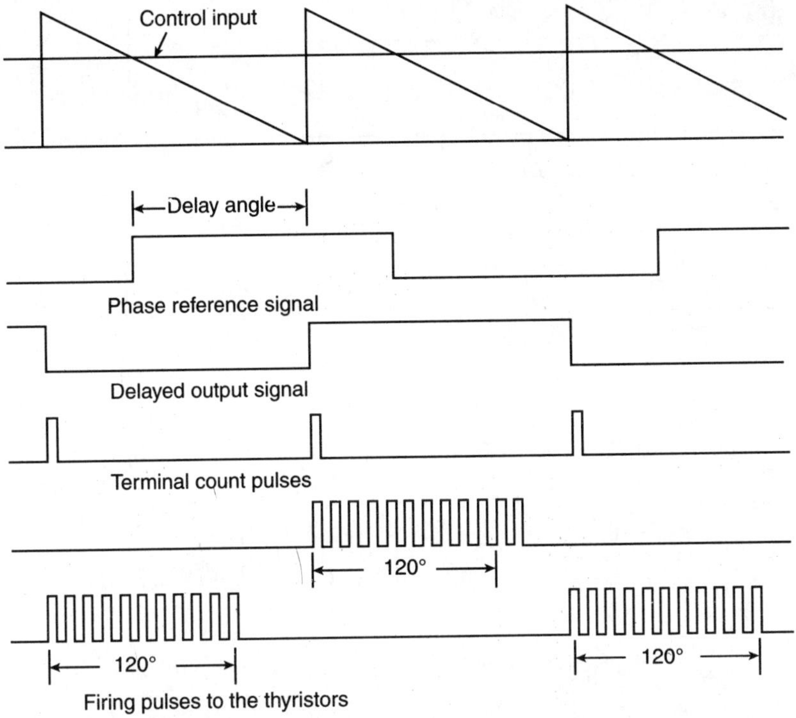

Fig. 8.17(b) *Firing pulse generation for motoring*

The high frequency signal S_4 can be used to compute the speed, as described previously, either by counting the number of pulses in a given time or by measuring the time between two pulses. The advantages and disadvantages of these methods have been already described.

Inverter Control Using Terminal Voltage Sensing The self control of the synchronous motor is obtained using the triggering pulses to the inverter which are synchronized with the rotor position. These signals are obtained by processing the phase reference signals P_1, P_2, P_3 and a high frequency signal S_4 obtained from a shaft encoder. This shaft encoder is a mechanical device fitted to the motor shaft. It therefore impairs the mechanical ruggedness of the system. The shaft encoder can be dispensed with, if the synchronizing of the trigger pulses with rotor position is accomplished by sensing the stator voltages. The thyristors can be triggered with minimum turn off angle to avail of the advantages of higher torque and improved power factor. The method can be implemented on a microprocessor used to perform the functions of monitoring and control. The control configuration and characteristics are supported by the necessary software, as described in the foregoing, so that the control strategies are adaptable to modifications without making any changes in the hardware. The only difference between the foregoing

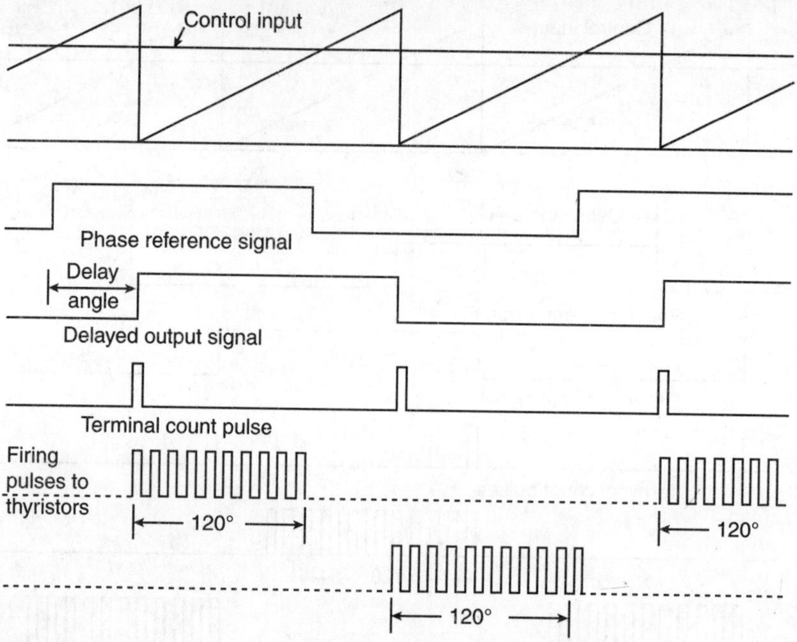

Fig. 8.17(c) *Firing pulse generation for regeneration*

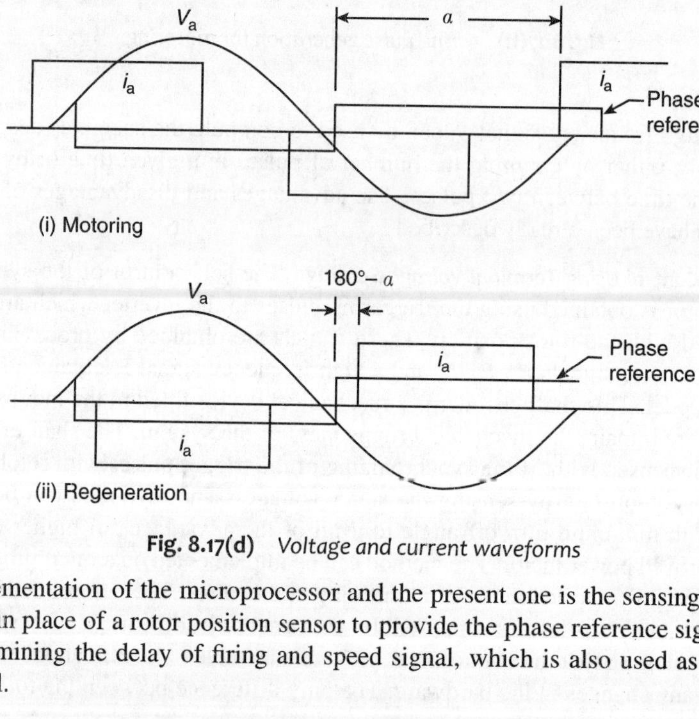

Fig. 8.17(d) *Voltage and current waveforms*

implementation of the microprocessor and the present one is the sensing of voltages in place of a rotor position sensor to provide the phase reference signals for determining the delay of firing and speed signal, which is also used as a firing signal.

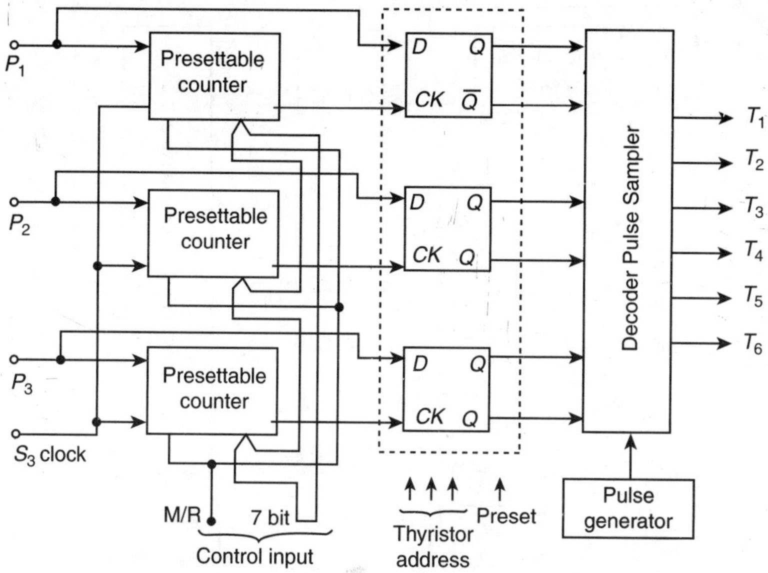

Fig. 8.17(e) *Schematic inverter firing angle controller*

Similar to the case of rotor position sensing (Fig. 8.17), signals identical to phase reference signals P_1, P_2, P_3 are obtained by detecting the zero crossings of the voltages (Fig. 8.18). The polarity of voltage between two phases is obtained. A frequency multiplier (Fig. 8.18) is used to obtain the high frequency signal S_4, similar to that obtained with a shaft encoder. The purpose of this signal is to provide the necessary delay angle in firing the thyristor so that the armature current has a definite phase difference with respect to the phase voltage, as well as to provide the information about the speed. The multiplication involved decides the resolution of firing angle. The signals P_1, P_2, P_3 and S_4 are processed in the same way as has been described in the previous section dealing with the shaft encoder.

However voltage sensing is not possible at standstill. The forced commutation routine must be involved to start the motor. A pair of thyristors addressed by the CPU are fired so that maximum torque in the required direction is developed. The pair of thyristors can be identified by measuring the induced voltages of the armature at the instant of applying the field voltage. The induced voltages depend upon the rotor position. The sign of the highest voltage is determined. This information is accessed and processed to identify the correct pair of thyristors to be fired. The starting process using forced commutation is implemented until the rotor accelerates to the speed at which natural commutation takes over.

Margin Angle Control of Synchronous Motors The commutation margin angle is defined as the angle measured from the end of commutation to the crossing of the phase voltage which was under commutation (natural firing instant). For satisfactory operation, without commutation failure, this margin angle must be greater

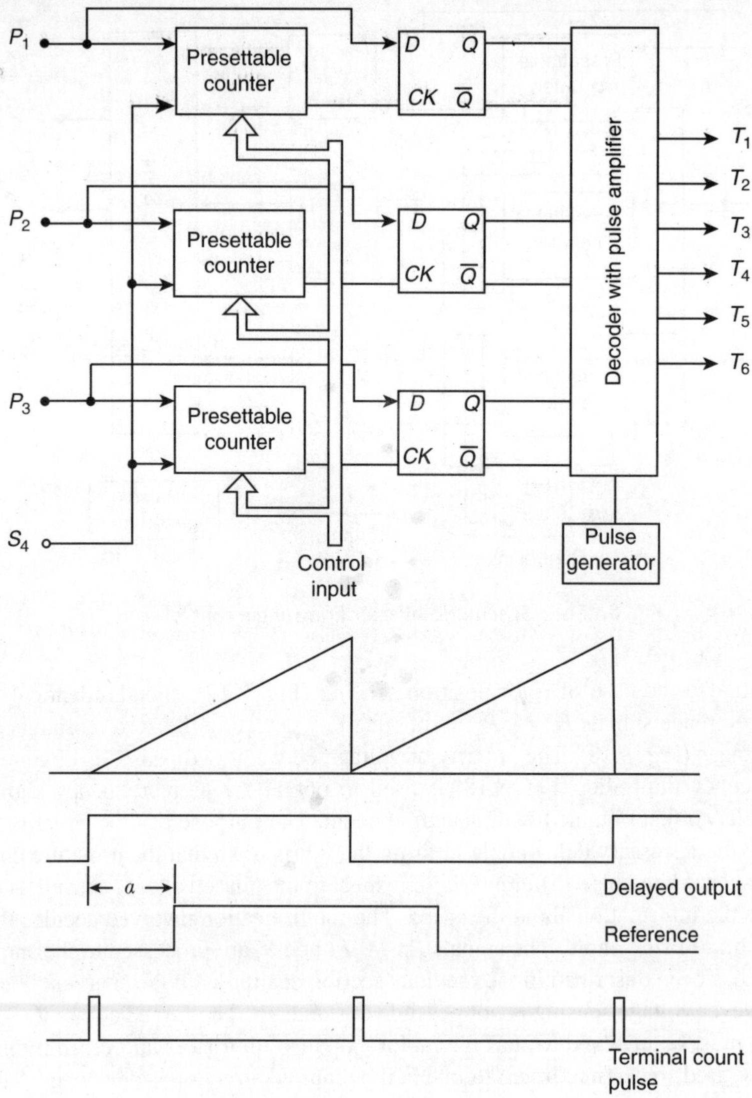

Fig. 8.18(a) *Inverter firing control circuit*

than the turn off angle (ωt_q) of the thyristors. In the constant margin angle control it is always observed that the margin angle does not go below a minimum value. Referring to Fig. 8.19 the margin angle is

$$\psi = \gamma - u$$

where γ is the lead angle of firing and
u is the overlap angle.

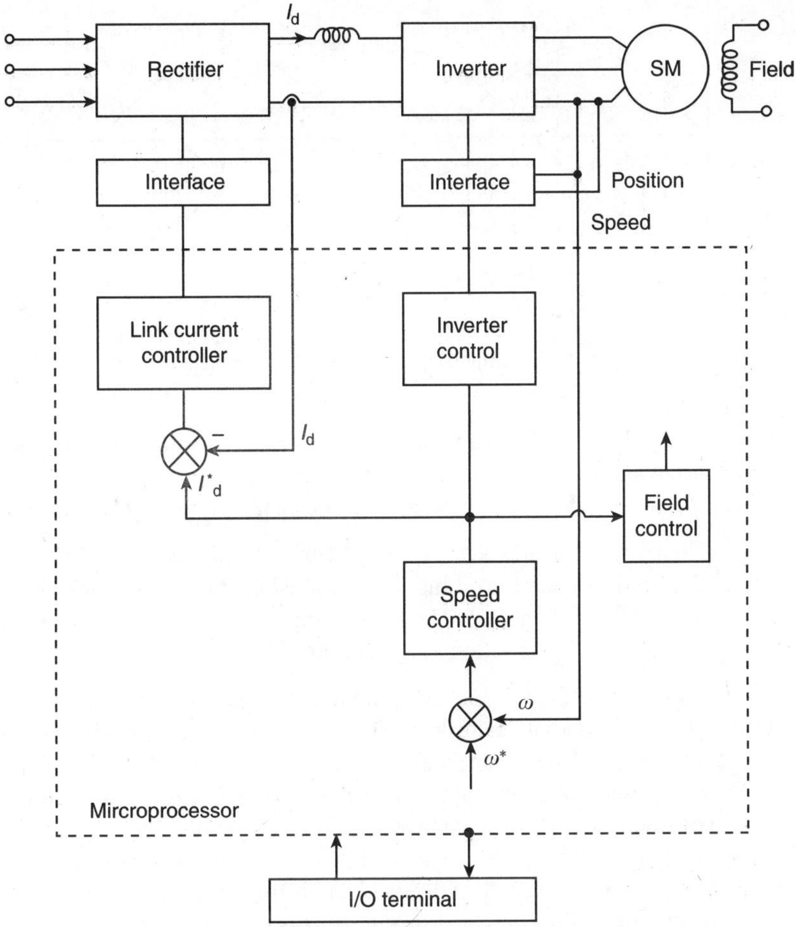

Fig. 8.18(b) *Microprocessor based control of a synchronous motor using terminal voltage sensing*

Among the factors, which influence the angle of overlap is the dc link current. Increase in the value of dc link current (I_d) results in an increase in u. Under transient conditions u increases. The value of γ also increases so that ψ remains the same. Obviously the performance improves as commutation failure cannot occur. On the other hand, in the constant lead angle control the margin angle decreases with increase in u, resulting in a commutation failure.

In the control, therefore, the firing angle changes such that the necessary lead angle of firing is available to keep ψ constant in case u changes. The overlap angle of the converter must be known and fed to the microprocessor. Microprocessor control of self controlled synchronous motor, discussed previously, may be used if the necessary changes are made in the firing of the converter, based on the

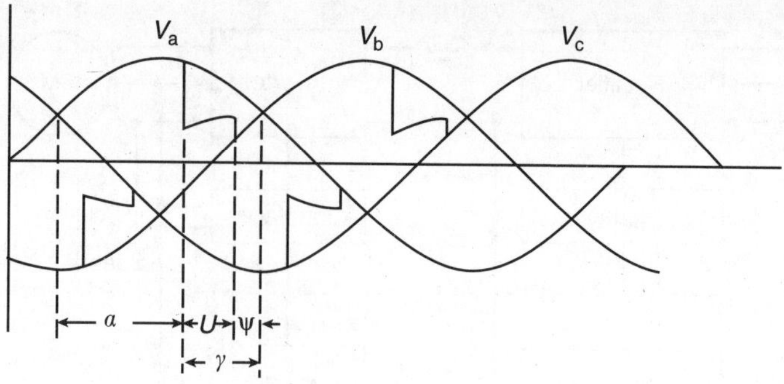

Fig. 8.19 *Definition to margin angle*

information available on the margin angle and overlap. The method of implementing the margin angle control is discussed in the following.

There are two ways of implementing the constant margin angle control:

i. The margin angle ψ is detected and controlled directly.
ii. Control is effected invoking the relationship between the link current I_d and the margin angle ψ. The margin angle is controlled using a function generator or a correction as a function of I_d.

In the first method the value of margin angle can be controlled to be constant at a value which gives a satisfactory transient and steady-state performance. However, the direct detection of margin angle is difficult over a wide range of speeds, due to ripple present in the machine voltage. This ripple causes difficulties in the detection of zero crossing of the voltages. The second method does not require the detection of margin angle and hence is very practical in the control of synchronous motors. The overlap angle is detected and is used in the control.

The method of control depicted in Fig. 8.20 is as follows. For a given value of margin angle, the relation between the lead angle of firing $\gamma = 180 - a$ and the link current is calculated. The relation is stored in the memory as a look-up table (function generation). This table is used by the microprocessor to correct the lead angle of firing in terms of I_d and u.

There are several difficulties with the margin angle control which require careful study, so that the control can be made advantageous. The constant margin angle control results in an intrinsic instability due to a voltage drop associated with overlap. Investigations show that the motor has unstable operation at heavy currents and high speeds. The motor can be stabilised by compensating the voltage drop due to overlap. A loop providing counter emf compensation corrects the voltage drop using the relation between I_d and loop gain in K_c in volts per rpm (Fig. 8.21).

The second difficulty arises from the transient behaviour of the margin angle, i.e., variation of ψ with I_d in transient conditions. The lookup table used for steady-state conditions may not be satisfactory under transient conditions due to

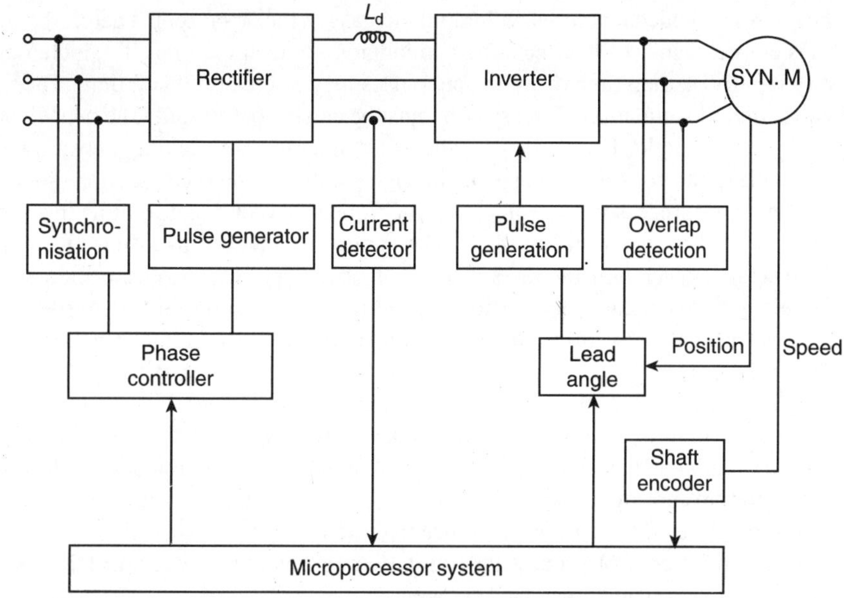

Fig. 8.20 *Implementation of margin angle control of a synchronous motor on a microprocessor*

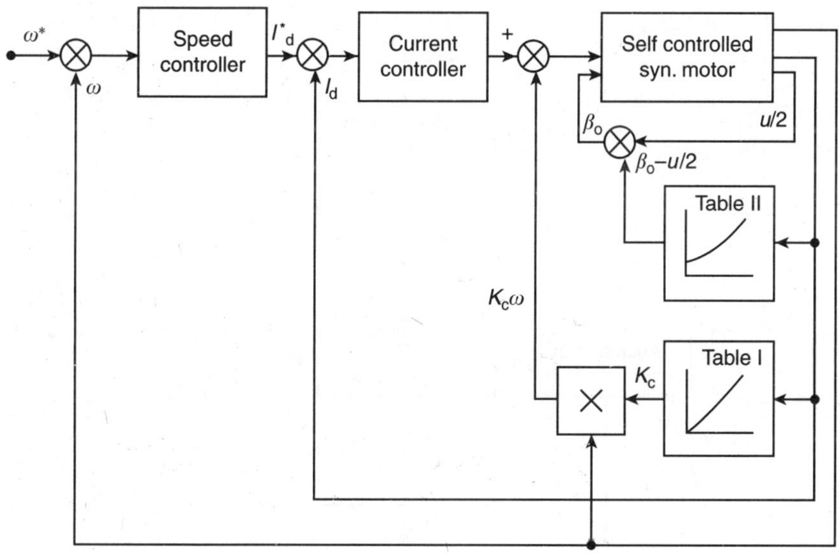

Table I Compensation for voltage drop.
Table II ($\beta_o - u/2$) v.d.c. current in the constant margin angle control

Fig. 8.21 *Block diagram of a self controlled synchronous motor with constant margin angle control* change in ψ. Commutation failure may occur. The margin angle depends upon the overlap u which increases with rapid increase in the dc current. Therefore, the transient behaviour of ψ affects the commutation, which may fail if I_d increases very rapidly. A long time delay in gate pulses may also cause this failure. Therefore, a digital simulation of the system must be carried out to study the behaviour of the margin angle. The reference value is set at a safe value taking the variation of ψ with I_d into consideration so that in any case there is no commutation failure.

The microprocessor control of a synchronous motor incorporating margin angle control is shown in Fig. 8.22. The control is achieved as a four quadrant drive with regeneration facilities. The error of the speed controller decides the mode of operation, e.g., acceleration, regeneration or motoring. If the speed error ε is greater than a set value n_2, the mode of operation is regeneration. The line side converter operates as an inverter reversing the polarity of link voltage. Accordingly regeneration is performed. This error also decides the acceleration ($\varepsilon > -n_1$). The acceleration is obtained with maximum current. The constant margin angle control is employed and care is taken to see that there is no commutation failure at the maximum current. In the region of errors $-n_1 < \varepsilon < n_2$ the normal PI compensation takes place with a current reference to current controller.

The current control processing routine comprises compensation of induced voltage, PI control and arc cosine compensation.

The margin angle control routine is shown in Fig. 8.23. This control takes place in the accelerating mode. When the drive is in this mode and the transition index

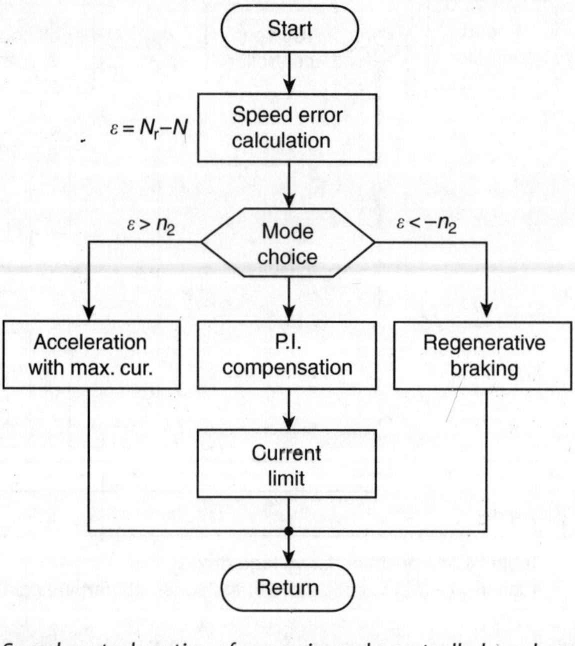

Fig. 8.22 *Speed control routine of a margin angle controlled synchronous motor*

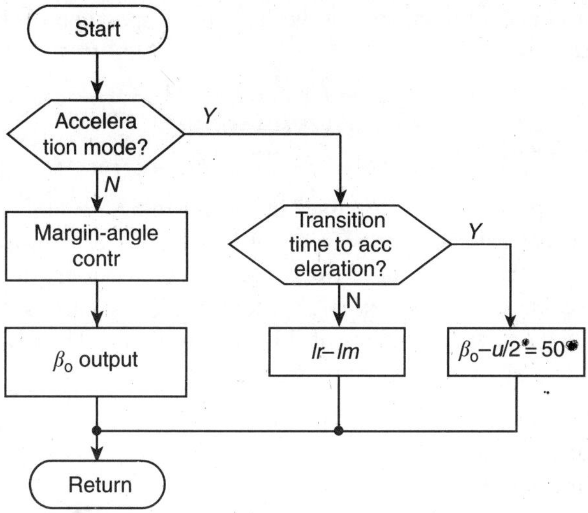

Fig. 8.23 *Flow chart of margin angle control*

is set, the margin angle is maintained at a proper value so that the commutation failure due to rapid increase in I_d is prevented.

The speed control routine in PI mode is performed with steady-state ψ control.

Selection of Microprocessor The criteria for selecting a microprocessor for controlling a synchronous motor discussed as above, are more or less the same as those discussed in the examples of dc motor and induction motor. The resolution of firing angle, instruction and cycle time, interrupt capability, and arithmetic operations, including the capacity to support multiplication and division, are several of the criteria based on which the selection can be made. The type of microprocessor selected should have sufficient memory capacity. It may be selected based on the number of bits, memory access time, interrupt I/O interface, kinds and number of instructions, reliability, cost and productivity. The future needs for development also may be considered. The local hardware design must consider the capacity of the microprocessor chosen. In a single microprocessor control, the boundary between the software and hardware is decided by the complexity of the functions of the microprocessor. The necessary interfaces are required in the form of A/D converters or others to convert the measured analog signals to digital quantities for processing by the microprocessor. The necessary software supports may be provided to generate firing pulses. If the capacity is limited these pulses may be generated by the necessary hardware to save the processor time for other jobs.

The microprocessor may be employed to control a synchronous motor using a cycloconverter, PWM inverter, etc. The cycloconverter fed synchronous motor can be used for low speed reversing mills. Both machine commutation and line commutation may be employed. The PWM inverter operation provides the necessary voltage and harmonic controls, so that the low speed operation of the motor is smooth with least torque pulsation.

8.5 SOME ASPECTS OF CONTROL SYSTEM DESIGN OF MICROPROCESSOR BASED VARIABLE SPEED DRIVES

Based on the examples of drives discussed in the foregoing, general outlines may be drawn for designing the control systems of variable drives using a microprocessor. It is evident that it is rather involved and time consuming to design and develop a microprocessor based variable speed drive. The control system design starts once the detailed specifications of drive are available. The main stages in the design are

i. system analysis
ii. computer simulation
iii. hardware design
iv. software design
v. system integration, debugging and performance.

A brief account of these stages follows:

System Analysis In this stage the complete structure of the system is formulated. All the components of the system are specified and their functions are identified, e.g., the power circuit configurations analysis of their commutation to decide the optimum values of commutation circuit components, etc. The performance specifications also decide the type of converter to be employed. If regeneration is not required half controlled converters with advantages of improved power factor, reduced ripple, etc. can be employed. If a PWM inverter is employed the line side converter can be a diode rectifier if no regeneration is required. The analysis also gives the details of the voltage waveforms, feedback signals, their generation and estimation. It is also useful in writing the algorithms for control purposes.

Computer Simulation If the system is complicated with several control loops and variables, a computer simulation of the system may be useful, to arrive at the optimised system and at the variables at every stage of control.

Hardware Design In this stage a decision is made, before selecting a microprocessor, about the boundary between the local hardware and microprocessor software. The functions to be carried out by the dedicated hardware circuits external to the microprocessor and by the microprocessor itself must be clearly defined. If the microprocessor is sufficiently fast, it may be assigned to perform the functions necessary for the control. On the other hand, if it is slow the hardware oriented circuits may be used to cover the jobs of signal generation for the inverter, high speed protection, firing angle generation for the converter, etc. The microprocessor can be saved to perform the other jobs. While selecting a microprocessor the following points need consideration:

i. Suitability of the performance of the microprocessor for the intended purposes. This is identified by the bit size, operating time, memory capacity, memory access, etc. The resolution required determines the bit size of the microprocessor. The sampling time of the processor is a measure of the resolution. The bit resolution to sampling time is identified as the figure of merit.
ii. *Suitability of the functions of the processor for intended purpose:* The functional capacity of a microprocessor is identified by the kind and number of instructions, microprocessor control, I/O interfaces, peripheral LSI, number of microprocessor devices, etc. Direct multiplication/division capability, functional integration, etc. show the effectiveness of the microprocessor in the computation of feedback signals in the control.
iii. The reliability of the system under operating conditions and temperature.
iv. Cost of the equipment.
v. Software and hardware support.
vi. Possibility for the expansion of the system to meet future needs.

Depending upon the program and data size ROM and RAM memory size may be designed conservatively. EPROM may be preferred to ROM in the initial stages. In many cases a single chip microprocessor with built in memory capacity may be sufficient, if aided by external hardware, to perform a few of the control functions in order to increase the speed. In case of multitasking a careful partitioning of the tasks may be necessary. Suitable methods may be used to increase the speed of operation.

Software Design Many of the functions of the microprocessor are software supported. The implementation of compensators, generation of PWM signals of a PWM inverter, etc. are carried out by suitable software. This makes them adaptable to various operating conditions. This is an important stage in the design of the system. The design of powerful software actually improves the adaptability and reliability of the system. The principal factors that decide the effective design are the language, bit size, sampling time and the identification of the function of the microprocessor. The resolution required in the operation decides the sampling time and bit size of the processor.

Once the above stages are completed, all the components are systematically integrated and the system may be tested for its performance.

8.6 STEPPER MOTORS

8.6.1 Introduction

The microprocessors are employed for the control of stepper motors which are generally used as position actuators. Some other applications of these stepper motors are to drive floppy discs, numerical control of machine tools, X-Y plotters etc. The ease with which these can be controlled using microprocessor has made them very popular in the applications cited above. The speed and position control using the stepper motor can be achieved without expensive feedbacks. A typical stepper motor has several stator phases. The rotor is a toothed structure. It may be sometimes a permanent magnet. Energising the phases in a sequence by direct current imparts the rotor a discrete step motion. The resolution of a typical motor is good if only it has large number of teeth on stator and rotor. The phases of a stepper motor are excited by pulses, each pulse being capable of providing rotor motion in smaller angles. The angular rotation depends on the steps of the motor per revolution. If a motor has 400 steps/rev, the angle by which it moves in a step is 0.9°.

Electronic circuits are used to excite the stator phases from dc supply. Now a days, semiconductor devices such as thyristors and transistors are slowly replacing the mechanical and electronic controls using vacuum tubes. The necessary logic which was expensive even with the introduction of semiconductor devices and electronic circuits has become easy and economical using microprocessors and digital circuits. The microprocessor has actually made a stepper motor and its control economically feasible.

Open loop control of stepper motor is the simplest method accomplished by means of step command pulses obtained from an external source. The motor is excited by these pulses and is expected to follow every pulse. This kind of control of stepper motor is very attractive and has wide acceptance in applications of speed and position control. The method has economic advantages. However it has its own limitations. The response of the motor to a given input command may become oscillatory or even unstable in some speed ranges. The control of the motor is not very fast due to this behaviour. To make the motor widely applicable it is necessary to improve the performance. A stepper motor may fail to follow the pulse when the frequency of the pulses is high or the load inertia is very large. Because of these reasons the open loop operation is only limited. The open loop control of the motor is depicted in Fig. 8.24(a). For most of the application open loop control is suitable.

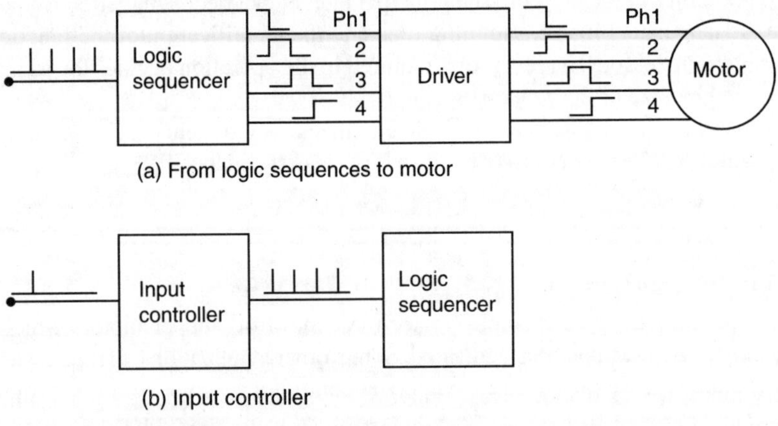

Fig. 8.24(a) *Open loop control of a stepping motor*

On the other hand a closed loop control of stepper motor is also possible where the switching of the motor takes place by means of the input pulse train using the position or speed feedback from the rotor. The closed loop control improves the performance of the stepper motor. The disadvantages of open loop control are not present. No step failure occurs, the response is quicker and the motion is smoother. The control is free from instability and capable of quick acceleration. A position sensor senses the rotor position and provides the information necessary for the control of the motor. A mechanical rotor position sensor with an optical encoder coupled to the shaft is used. The rotor position may also be determined using the machine voltages and currents. The motor position is monitored and a step completed. The closed loop control is employed when the maximum torque is required of a given motor and also when absolute step integrity is required. The cost of the system depends on the type of position sensor used. A closed loop control of the motor is shown in Fig. 8.24(b).

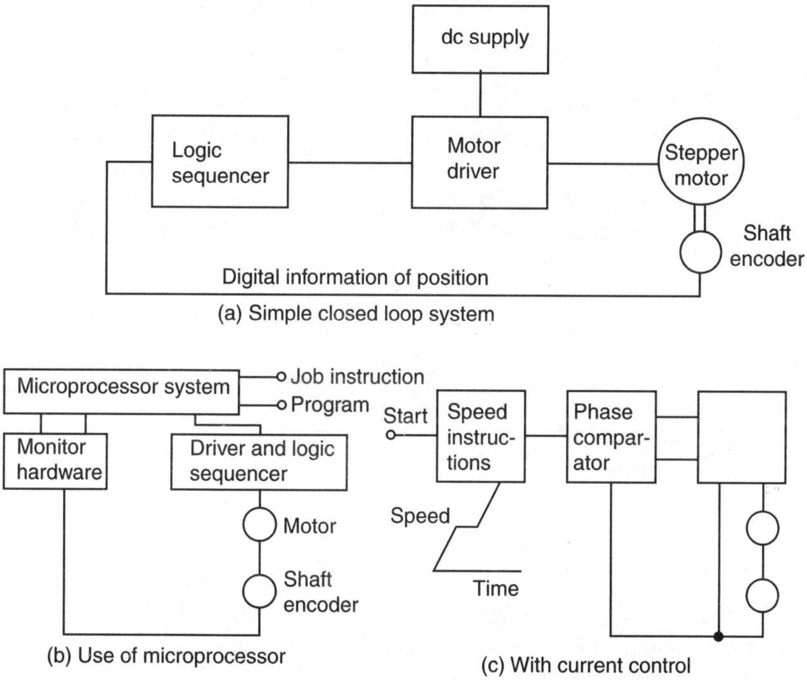

Fig. 8.24(b) *Closed loop control of stepper motor*

8.6.2 Applications of Stepper Motors

The stepper motors find applications in situations where two or more motors are required with independent or interactive running. The former is the simplest case. Every motor has its own control circuit. The open loop controllers are available commercially. The controllers may employ microprocessors.

The interactive control of the stepper motor is typically employed in places where the speed of the motors must be the same. One such example is the glass industry. If the speeds are not equal stretching of glass takes place. In the textile industry also the stepper motors are used in the interactive control to run at constant speed to avoid fabric tension and damage when fabrics are wound. Careful synchronisation of the running of the motors is required.

Another example of interactive control is in robotics. The control of the robot manipulator requires simultaneous control of several motors. Here the motors may not run at the same speed. The operation of one motor affects the other motors. Therefore there can be interactive control where no synchronisation is required, but a control is required to make the operation satisfactory with mutual effects of several motors.

The open loop control can be easily employed for interactive control of the motors also. The operation of the motors is however limited by the considerations of stability. When the load conditions are varying open loop control of several motors in interactive mode becomes difficult when stability is also to be assured. The possible solution is closed loop control with a common feedback control of the motors.

The synchronisation of two motors using single closed loop control strategy leads to instability. However a motor can be made to follow the velocity profile of the other with a suitable delay. This kind of operation however has limited applications. The closed loop control of robot motors in the interactive mode requires the knowledge of robot geometry. Microprocessors can be advantageously employed in complicated controls.

The stepper motors find application in computer peripherals. A few of the devices employing stepper motors are

i. Serial printers in typewriters or word processor systems
ii. Linear stepper motors to printers
iii. X-Y plotters
iv. Floppy disc drives
v. Numerical control of machine tools such as X-Y tables and index tables, milling machines, automatic drafting machine driven by a linear motor etc.

An X-Y table is a device operated by two motors working to move the object in two perpendicular directions independently. These are used to position the object for machining. In milling machines the movement of the object is required in three different directions which is accomplished by the stepper motors. Due to the oscillatory behaviour of the stepper motor the finish of the surface is not comparable to that obtained with dc servomotor control. The drafting machine is similar to the X-Y table. A separate minicomputer is used to control these motors. The oscillatory motion of the motor in the open loop mode must be given due consideration to have satisfactory control. Closed loop control may also be necessary.

The other applications of stepper motors are in space crafts launched for scientific exploration of planets. These are used to point the instruments such as T.V. cameras, ultra violet spectrometers at the desired objects. The scan actuator shaft position must be very precisely known to control the motors. This is obtained by means of feedback and telemetry potentiometers. The shaft and these potentiometers are geared together. The gear is highly precise with anti backlash.

8.6.3 Features of Stepping Motors in View of Application

For satisfactory operation of a stepper motor, it should have the following features:

- small step angle
- high positioning accuracy
- high torque to inertia ratio
- stepping rate and accuracy

Small Step Angle The angle by which the rotor of a stepper motor moves when one pulse is applied to the (input) stator is called step angle. This is expressed in degrees. The resolution of positioning of a stepper motor is decided by the step angle. Smaller the step angle the higher is the resolution of positioning of the motor. The step number of a motor is the number of steps it makes in one revolution. The stepper motors are realisable for very small step angle. Some precision motors can make 1000 steps in one revolution with a step angle of 0.36°. A standard motor will have a step angle of 1.8° with 200 steps for revolution. The step angles of 90°, 45°, 15° are not uncommon in simple motors.

High Positioning Accuracy The quality of a stepper motor is decided by its positioning accuracy. This accuracy of positioning is a significant factor. In any application the stepper motors are expected to rotate by step angle when a pulse is given as input. It should come to rest in a precise position. Care must be exercised in the design and manufacture of a stepper motor as the accuracy at no load depends on the physical accuracy of the rotor and the stator. The positioning accuracy depends only on the machine characteristics and the driving circuit, while other electronic parameters have no effect on positioning accuracy. The stepper motors are designed to have very high restoring torque when the rotor is displaced following load torque.

High Torque-to-inertia Ratio A stepper motor must move fast response to a pulse or a train of pulses. Fast response is also associated with quick start and quick stop of the motor. The motor must be capable of stopping at a position specified by the last pulse of the pulse train if the train is inhibited when the motor is running uniformly. A stepper motor must have a large ratio of torque to inertia so that it would fulfil the above condition.

Stepping Rate and Pulse Frequency The speed of a stepper motor is indicated by the number of pulses per second. Stepping rate of the motor is therefore used to indicate the speed. As the motor moves by one step for a pulse the speed can also be indicated by pulses per second as pulse frequency. The absolute rotational speed and stepping rate are related by $\underset{(\text{rpm})}{n} = 60\, f/S$, S = step number, f = stepping rate

8.6.4 Classification of Stepper Motors

The stepper motors are classified based on the construction and principle of operation. These are

i. Variable reluctance motors
ii. Permanent magnet motors
iii. Hybrid stepping motors
iv. Claw pole motors with permanent magnets

The stepper motors are used for linear motors also. These are also classified as 'variable reluctance motors' and 'permanent magnet motors'.

Variable Reluctance Motors The principle of operation of variable reluctance motors is based on the property of the flux lines tending to occupy low reluctive path. The stator and rotor therefore get aligned such that the magnetic reluctance is minimum. This type of motors being basic type, have a stator with different sets of winding which are excited by dc alternately. Each set may be called a phase. A simple motor may have six teeth on the stator. On each of the teeth there is a winding. The windings on the diametrically opposite teeth are connected in series and the resulting one is called the phase of the motor. Thus a motor with 6 slots will have three phases. The rotor on the other hand is simply made of magnetic material. It also has a slotted structure but no winding. This is capable of movement inside the stator whenever a phase of the stator is excited. Both the stator and rotor are made up of high quality magnetic materials having very high permeability so that the exciting current required is extremely small. A three phase variable reluctance stepping motor with four rotor slots is shown in Fig. 8.25. The principle of operation of the motor is explained with reference to Fig. 8.25.

The coils on the teeth AA′ are excited by the dc voltage by closing the switches S_1. The stator and rotor constitute the magnetic circuit and a magnetic flux is set up. The nature of the flux lines being to occupy the path of least reluctance tries to move the rotor to align with the stator as shown in Fig. 8.25. The rotor stops at a position where the reluctance is minimum and this position is said to be the equilibrium position. The next equilibrium position is obtained by exciting the next phase. The angle between two successive equilibrium positions is known as the step angle. The rotor moves by one step when a winding is excited. If any displacement is likely to occur in the rotor due to application of load or any disturbing forces, the rotor has a tendency to be in the equilibrium position. To accomplish this a torque is developed which is called the restoring torque. The external or load torque and restoring torque are of opposite nature. (Fig. 8.25), if one acts in the clockwise direction the other acts in the anticlockwise direction.

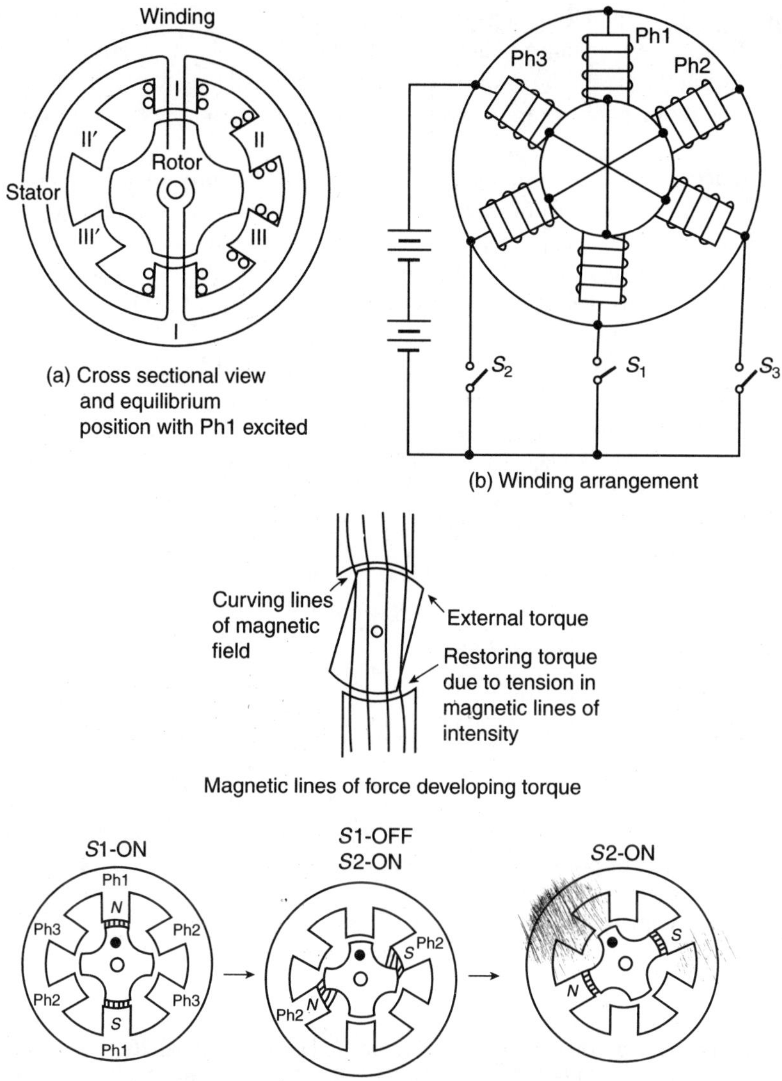

(a) Cross sectional view and equilibrium position with Ph1 excited

(b) Winding arrangement

Magnetic lines of force developing torque

Step motion when excitation is switched from Ph1 to Ph2

Fig. 8.25 *Variable reluctance stepper motor*

This restoring torque is due to the property of the flux lines to align the stator and rotor such that they have the least reluctive path.

Any switching operation to change the position is accompained by an increase in the reluctance which brings in a torque (or force) to seek the alignment of the rotor and stator, and the rotor is in equilibrium in the new position. Thus the rotor moves by one step or step angle at every new excitation. In the present example when S_2 is closed, opening S_1 the rotor moves by 30° in the acw direction. 30° is step angle in this case. The rotor will come to its original position after completion one rotor tooth pitch rotation which may take 3 steps. These are detailed in Fig. 8.25.

The accuracy of positioning which is a desired quality of a stepper motor requires a high torque from a given volume of rotor (which is normally small). To get the high torque capability a small air gap may be required. A small air gap means low reluctance and high flux providing a good torque. The restoring torque developed is also sufficiently high so that the displacement from an equilibrium position is rather small.

The resolution of stepper motor will be very high if the step angle is smaller. This can be made smaller by increasing the number of stator and rotor teeth. This depends actually on the stator phases and how the stator teeth are divided into phases. For example 12 teeth in the stator can be divided as 3 phases, or 6 phases. If rotor teeth are 8 in the former case the step angle is 15° and in the latter it is 7.5°. The step angle is therefore given by

$$\theta_s = \frac{360}{S} = \frac{360}{mN_r}$$

where m is the number of phases

N_r is rotor teeth.

To reduce the step angle the rotor teeth must be increased. The stator teeth is not a direct factor in determining the step angle or step number. A four phase motor with 50 rotor teeth has 200 steps/revolution or a step angle of 1.8°.

The rotor must be made to have low inertia, so that its torque/inertia ratio is higher for quick response of the motor.

Permanent Magnet Stepper Motors The rotor is made of a permanent magnet in permanent magnet motors. The stator is similar to that of variable reluctance motor. The coils on the stator teeth form the phases of the motor. The permanent magnet is cylindrical and housed in the multiphase stator. A four phase machine with cylindrical permanent magnet motor is shown in Fig. 8.26. The stator has four teeth around which coils are wound. The phases are excited by dc voltages using the switches. The excitation of phases in the sequence 1 2 3 4 moves the rotor in the clockwise direction. The step angle in the present case in 90°. To increase the step angle the number of stator phases and number of rotor poles must be increased. This method has limitations.

The rotor comes to rest at a fixed position even if excitation of stator ceases. The mechanism by which the rotor comes to rest is known as detent mechanism and the predetermined position is called the detent position. This is defined as the position at which a permanent magnet motor comes to rest at no load without excitation. A permanent magnet motor has the following disadvantages:

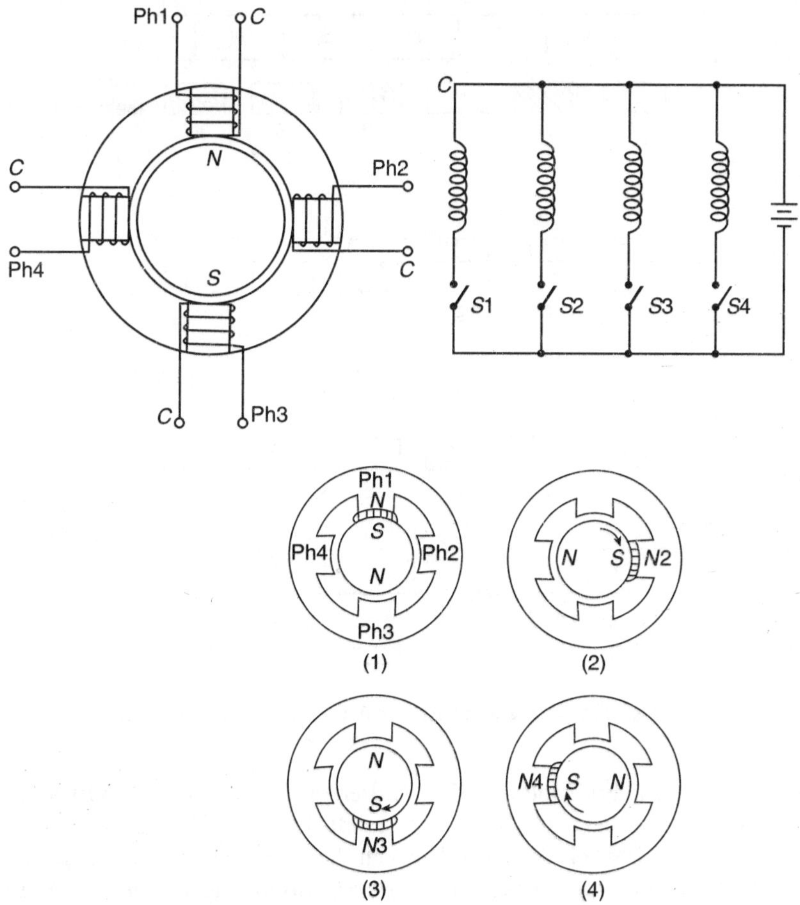

Fig. 8.26 *Permanent magnet stepper motor*

i. The motor is costly due to permanent magnets.
ii. The level of the magnetic remanance fixes the maximum flux density level. This being very low the torque capability of the motor is not high.

Hybrid Stepping Motors One main limitation of the permanent magnet motor is the step angle. It cannot be reduced because of the limitations on the phases of the stator. To increase the step rate and have a small step angle a hybrid motor working on the principles of variable reluctance motor and permanent magnet motor. The construction of stator of a hybrid motor is almost similar to that of a variable reluctance motor (Fig. 8.27). However there is a difference in placing the windings of the stator poles. In a hybrid motor two coils of different phases are wound on the same pole whereas in a variable reluctance motor only one coil of a phase is placed on a pole. In a hybrid motor the excitation of the coils cause different magnetic polarities.

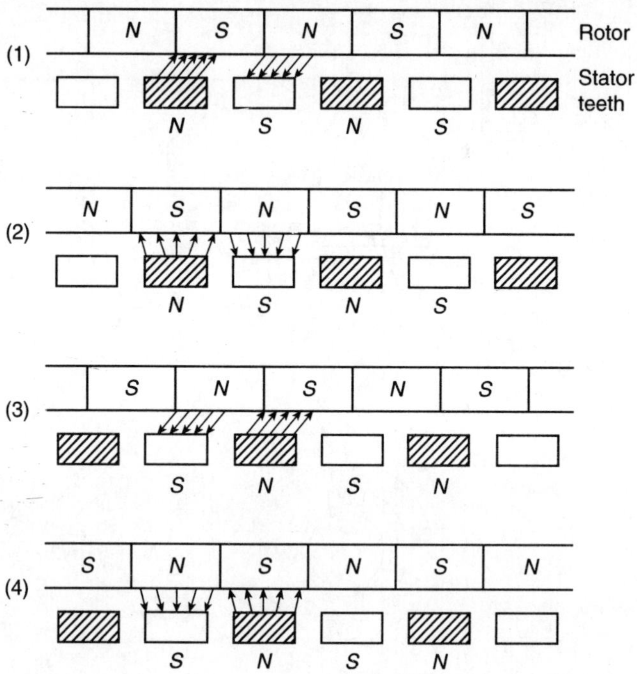

Fig. 8.27 *How steps occur in a claw-poled PM motor*

The rotor of a hybrid motor also has a different structure. The cylindrical rotor structure is magnetised to produce unipolar field. Each pole is covered by toothed soft steel. There is a misalignment between the teeth of two sections by half a tooth pitch. In some motors the stator core has misalignment. The toothed sections are made of silicon steel laminations. The stator coils generate heteropolar field.

The toothed structures of the stator and the rotor are designed to reduce the step angle in a hybrid motor. The permanent magnet is significant in producing a driving force. The torque is developed by the interaction of the two types of magnetic fields in the toothed structures in the air gap. Sequential switching on and off of the stator phases will result in the rotor motion in steps. The torque production and movement of the rotor is due to the superposition of the fields of stator phases and permanent rotor magnet. The superposition is such that the flux is strengthened on one side and weakened on the other side causing the motion of the motor. The rotor moves until the driving force is zero. At this point it has its equilibrium position. Multistack motors are used to raise the torque.

Claw Pole Motor A claw pole motor is also a permanent magnet stepper motor. It has the advantage of low manufacturing costs and has applications as the paper feed prime mover and head drive motor in floppy disc drives.

Multiple-Choice Questions

8.1 The centre of operations and control in a microcomputer is
(a) Memories
(b) Microprocessor unit (MPU)
(c) Address decoders
(d) Interfaces

8.2 The OR operation in a microprocessor is classified as
(a) arithmetic instruction
(b) data transfer instruction
(c) logical instruction
(d) decision making instruction

8.3 A subroutine in a microcomputer program is
(a) a set of branching instructions
(b) a set of logical instructions
(c) a set of transfer operations
(d) a special group of instructions that perform a commonly used specific task in a program

8.4 The CPU of a microcomputer typically contains a variety of storage devices called
(a) ROM
(b) RAM
(c) instruction decoders
(d) registers

8.5 A microprocessor responds basically to a listing of operations called
(a) high level language
(b) machine language
(c) assembly language
(d) to all the languages above

8.6 An assembler
(a) is a special computer program for translating from assembly language to machine language
(b) is a special computer program for translating from high-level language to machine language
(c) a unit used to connect several parts of μp
(d) is a computer program which translates machine language to assembly language

8.7 Interconnection of parts within a μp based system is called
(a) interrupt
(b) assembly
(c) interfacing
(d) programming

8.8 Keyboard is an
(a) output peripheral device
(b) input peripheral device
(c) output interface adapter
(d) input interface adapter

8.9 The 8085 microprocessor uses a
(a) +5 V power supply
(b) –5 V power supply
(c) +10 V power supply
(d) +15 V power supply

8.10 A dual converter is operating in non-circulating current mode. The microprocessor software must be capable of
(a) detecting discontinuous conduction
(b) sensing zero crossing of current
(c) identifying actual zero crossing of the current and it should not confuse with the discontinuous conduction
(d) identifying the discontinuous conduction and it should not confuse with the natural zero of the current.

References

8.1 Bose, B.K., *Power electronics and ac drives*, Prentice-Hall Inc., Englewood Cliffs, 1986.

8.2 Subrahmanyam, Vedam, *Thyristor control of electrical drives*, Tata McGraw-Hill Publishing Company Ltd., New Delhi, 1988.

8.3 Athani, V.V., and S.M. Deshpande, Microprocessor control of a three phase induction motor speed control system, *IEEE Trans. IECI*, IECI–27, pp. 291–298, 1980.

8.4 Buja, G.S., and P. Florini, 'Microprocessor control of PWM inverters', *IEEE Trans. IE*, Vol. IE 29, pp. 212–218, Aug., 1982.

8.5 Plant, J.B., S.J. Norma, and Y.T. Chan, 'Microprocessor control of position or speed of an SCR DC motor drive', *IEEE Trans. IECI*, Vol. IECI–27, pp. 228–234, 1980.

8.6 Tang, P.C., S.S. Lu, and Y.C. Wu, 'Microprocessor based design of a firing circuit for three phase full-wave thyristor dual converter', *IEEE Trans. Ind. Elec.*, Vol. IE–29, pp. 67–73, 1982.

8.7 Sathiakumar, S., S.K. Biswas, and J. Vithayathil 'Microprocessor-based field-oriented control of a CSI-fed induction motor drive', *IEEE Trans. on IE*, Vol. IE–33, pp. 39–43, 1986.

8.8 Gabriel, R., W. Leonhard, and C.J. Nordby, 'Field-oriented control of a standard ac motor using microprocessors', *IEEE Trans. on IA.*, Vol. IA–16, pp. 186–192, 1980.

8.9 Kenji Kubo, Masakiko Watanabe, Tsutomo Shamee, Kenjo Kamiyama, 'A fully digitalised speed regulator using multiprocessor system for induction motor drives', *IEEE Trans. IA*, Vol. IA–21, pp. 1001–1008, 1985.

8.10 Shmae, T., *et al.*, 'A microprocessor controlled high accuracy wide range speed regulator for motor drives', *IEEE Trans. IE*, Vol. IE–29, pp. 207–211, 1982.

8.11 Bowes, S.R., and M.J. Mount, 'Microprocessor control of PWM inverters', *Proc. IEE*, Vol. 128, 1981, pp. 293–305.

8.12 Oliver, G., V.R. Stefanovic, and G.M. April, 'Microprocessor controller for thyristor converter with an improved power factor', *IEEE Trans. IECI*, Vol. IECI–28, 1981, pp. 188–194.

8.13 Harashima, F., H. Nontoh, H. Taoka, 'A microprocessor-based PLL speed control system converter-fed synchronous motor', *IEEE Trans. IECI*, Vol. IECI–27, Aug., 1980, pp. 196–201.

8.14 Ishida, K., K. Nakamma, T. Izumi, and M. Ohara, 'Microprocessor control of converter fed dc motor drive,' *Conference Record, IEEE IAS*, 1982, pp. 619–623.

8.15 Bose, B.K. and H.A. Sutherland, 'A high performance pulse-width modulator for an inverter drive system using a microcomputer', *Conference Record, IEEE IAS*, 1982.

8.16 Le-Huy, H., R. Perret, and D. Roye, 'Microprocessor control of a current-fed synchronous motor drive', *Conference Record, IEEE IAS–79*, pp. 873–880.

Appendix A

ON THE ANALYSIS OF A CURRENT SOURCE INVERTER FED INDUCTION MOTOR

A method is presented for the analysis of a current-fed induction motor via state transition signal-flow graph technique. The input current waveform is neither stepped nor trapezoidal. It has a sinusoidal variation during commutation of the current in the inverter as well as in the phases of the motor. The equations developed are used to determine and compare the performance of the motor at two different slip conditions. At very large slips, it is found that the torque pulsations have larger amplitude owing to the departure of rotor flux wave from sinusoidal variation.

1 List of symbols

i_A, i_B, i_C	line currents of induction motor
i_a, i_b, i_c	phase currents of induction motor
I_-	dc link current
$i_{astr}, i_{\beta str}$	two axis components of stator currents in the stationary reference frame
$i_{arot}, i_{\beta rot}$	two axis components of rotor currents in the stationary reference frame
$i_{astr}(s), i_{\beta str}(s)$	*Laplace* transforms of $i_{astr}, i_{\beta str}$
L_{str}, L_{rot}	self-inductance of stator and rotor
M	mutual inductance between stator and rotor
p	pairs of poles
R_{str}, R_{rot}	resistances of stator and rotor per phase
s	*Laplace* operator
T_d	torque developed
$u_{astr}, u_{\beta str}$	two-axis components of stator voltage in the stationary reference frame
$\psi_{arot}, \psi_{\beta rot}$	two-axis components of rotor flux linkages
$\psi_{astr}, \psi_{\beta str}$	two-axis components of stator flux linkages
$\psi_{astr}(s), \psi_{\beta str}(s)$	*Laplace* transforms of two axis
$\psi_{arot}(s), \psi_{\beta rot}(s)$	components of stator and rotor flux linkages
ω_{rot}	rotor speed
ω_c	oscillating frequency of commutating circuit

A.1 INTRODUCTION

It has been well established during the last few years that the current source inverters posses a variety of advantages over the voltage source ones. Therefore, the drives utilizing current source inverters are becoming very popular. The output current waveform of a current source inverter is nonsinusoidal. When the commutation of the current in the inverter is instantaneous the waveform tends to be rectangular and it can be described as piecewise

* Dr. Habil Vedam Subrahmanyam is Assistant Professor at the Department of Electrical Engineering, Indian Institute of Technology, Madras/India.

Manuscript received: 23.7.1982

constant over a period. The analysis of current-fed induction motors operating on rectangular currents is discussed in detail [1, 2, 3].

The commutation of the current from one phase to the other of an induction motor fed from a current source inverter is sometimes assumed to be linear rather than instantaneous in which case the current waveform tends to be trapezoidal. A method for analysing the performance of a current fed induction motor using state transition signal flow graphs for trapezoidal currents has also been discussed [4, 5].

In practice, the commutation of the current in the inverter as well as in the phases of the induction motor is neither instantaneous nor linear. The current during commutation follows a damped sinusoid as the machine inductance and resistance form the commuting circuit along with the commutating capacitance. Neglecting the motor resistance the current variation can be taken to be sinusoidal during commutation.

The present paper aims at calculating the performance of a current source inverter fed induction motor with sinusoidal variation of current during commutation. The method developed using signal flow graphs has been used to obtain a closed form solution for the performance during interlude as well as commutating periods.

A.2 COMMUTATION PROCESS IN THE INVERTER AND STATOR CURRENTS OF THE INDUCTION MOTOR

To enable one to write the equations of the current during commutation, a brief description of the commutation process is given below. The following assumptions are made during the discussion:

—the diodes and thyristors are ideal switches,
—the link current is constant and ripple free, and
—the commutation overlaps do not occur.

The current source inverter feeding a three phase induction motor is shown in Fig. A.1. The inverter is of the auto sequential type according to *Ward* [6].

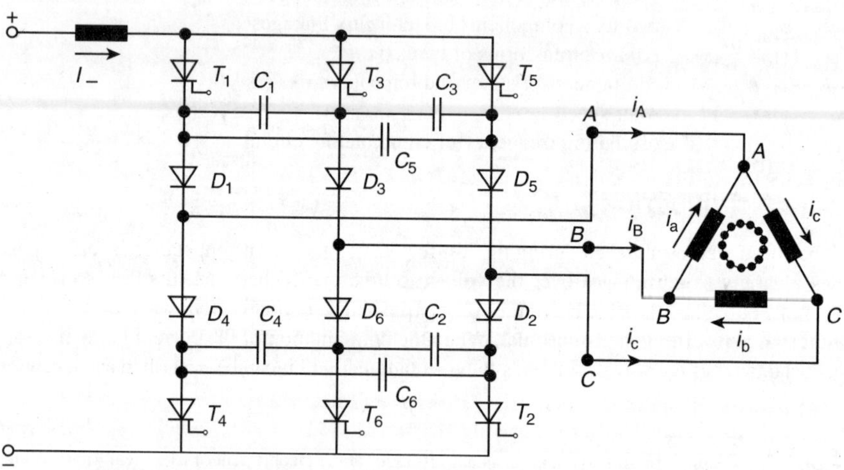

Fig. A.1 *Current source inverter feeding a three phase induction motor*

To start with it is assumed that thyristors *T*1 and *T*6 are conducting. The thyristor *T*3 is fired to commutate *T*1 at the instant $t = t_a$ Fig. A.2 shows the variation of current in the incoming as well as outgoing line feeding the induction motor. The commutation process initiated by firing *T*3 at t_a can be divided into three parts:

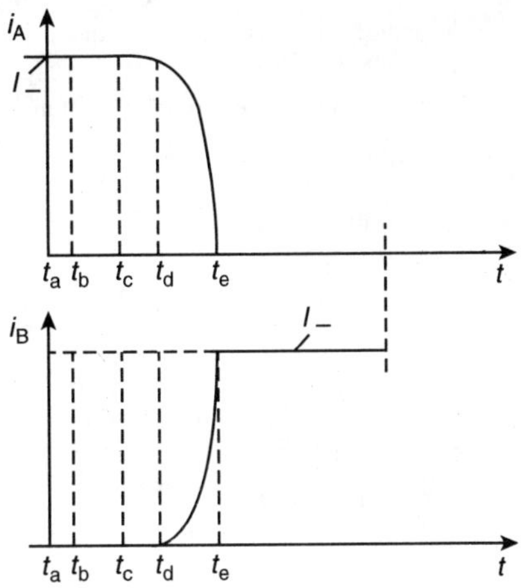

Fig. A.2 *Variation of line currents i_A and i_B of a three phase induction motor fed from a current source inverter during commutation*

—t_a to t_b is the period in which the current gets commutated from *T*1 to *T*3. This period is very small compared to the total commutation time and it can be neglected, i.e. it can be assumed that the transfer of current from the outgoing thyristor to the incoming one is almost instantaneous.

—t_b to t_d (t_a to t_d) is the first part of the commutation interval. During this period *D*3 is in blocking state. The capacitor voltages vary linearly and the machine currents remain unaltered. At t_d, *D*3 gets forward biased and starts conducting.

—t_d to t_e is the second part of the commutation interval (t_{de}). During this interval current is transferred from one phase to the other of the induction motor. The commutation of the current is influenced by the machine resistance, inductance and capacitance of the commutating circuit of the inverter. The current actually follows a damped

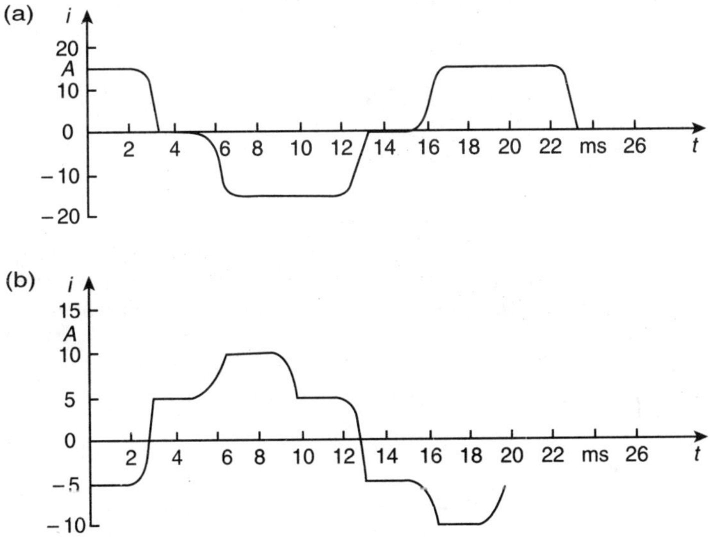

Fig. A.3 *Line (a) and phase currents (b) of a delta connected induction motor fed from a current source inverter*

sinusoid. But when the resistance is neglected the commutating circuit becomes a simple LC circuit and the current variation can be approximated by a sinusoid. Therefore the current can be written as

$$i_A = I_{-} \cos(\omega_c t) \tag{A4.1}$$

in the outgoing line, and

$$i_B = I_{-} [1 - \cos(\omega_c t)] \tag{A4.2}$$

in the incoming line. The phase currents can be determined depending upon the connection of the machine. The commutation is complete when al the current is transferred to the incoming line. ω_c is determined by L and C.

The current waveform at the input terminals i_A of the induction motor over a period can now be determined as shown in Fig. A.3 where the phase current for delta connection of the stator is given too.

A.3 MACHINE EQUATIONS AND FORMULATION OF SIGNAL-FLOW GRAPH

The performance equations of a three phase induction motor are well known. These equations derived with respect to stationary reference frame can be written as

$$u_{\alpha str} = R_{str}\, i_{\alpha str} + \dot{\psi}_{\alpha str} \tag{A4.3}$$

$$u_{\beta str} = R_{str}\, i_{\beta str} + \dot{\psi}_{\beta str} \tag{A4.4}$$

$$0 = R_{rot}\, i_{\alpha rot} + \dot{\psi}_{\alpha rot} + \omega_{rot}\, \psi_{\beta rot} \tag{A4.5}$$

$$0 = R_{rot}\, i_{\beta rot} + \dot{\psi}_{\beta rot} - \omega_{rot}\, \psi_{\alpha rot} \tag{A4.6}$$

$\psi_{\alpha str}$ and $\psi_{\beta str}$ in eqs. (A4.3) and (A4.4) are the direct axis and quadrature axis flux linkages of the stator whereas $\psi_{\alpha rot}$ and $\psi_{\beta rot}$ in eqs. (A4.5) and (A4.6) are the flux linkages of the rotor. These are given by

$$\psi_{\alpha str} = L_{str}\, i_{\alpha str} + M\, i_{\alpha rot} \tag{A4.7}$$

$$\psi_{\beta str} = L_{str}\, i_{\beta str} + M\, i_{\beta rot} \tag{A4.8}$$

$$\psi_{\alpha rot} = L_{rot}\, i_{\alpha rot} + M\, i_{\alpha str} \tag{A4.9}$$

$$\psi_{\beta rot} = L_{rot}\, i_{\beta rot} + M\, i_{\beta str} \tag{A4.10}$$

The stator current waveform is known in the case of an induction motor fed from a current source inverter. The analysis problem reduces to the solution of rotor equations only. The stator voltage can be calculated from the known values of rotor quantities and stator currents.

Substituting for $i_{\alpha rot}$ and $i_{\beta rot}$ in terms of $i_{\alpha str}$, $i_{\beta str}$, $\psi_{\alpha rot}$ and $\psi_{\beta rot}$ in eqs. (A4.5) and (A4.6) from eqs. (A4.7) to (A4.10) and rearranging the terms, we get

$$\dot{\psi}_{\alpha rot} = (R_{rot}/L_{rot})\, M\, i_{\alpha str} - (R_{rot}/L_{rot})\, \psi_{\alpha rot} - \omega_{rot}\, \psi_{\beta rot} \tag{A4.11}$$

$$\dot{\psi}_{\beta rot} = (R_{rot}/L_{rot})\, M\, i_{\beta str} - (R_{rot}/L_{rot})\, \psi_{\beta rot} + \omega_{rot}\psi_{\alpha rot} \tag{A4.12}$$

For a constant speed eqs. (A4.11) and (A4.12) become linear differential equations with constant coefficients which can be solved for $\psi_{\alpha rot}$ and $\psi_{\beta rot}$. Using eqs. (A4.9) and (A4.10) the rotor current components $i_{\alpha rot}$ and $i_{\beta rot}$ can be evaluated. The two axis flux linkages of the stator can be evaluated using eqs. (A4.7) and (A4.8). The instantaneous torque developed is determined using equation

$$T_d = 1.5\,p\,M\,(i_{\alpha rot}\, i_{\beta str} - i_{\beta rot}\, i_{\alpha str}). \tag{A4.13}$$

Equations (A4.11) and (A4.12) can be solved using the signal flow graph. The method offers a variety of advantages [4, 5]. Taking *Laplace* transforms of eqs. (A4.11) and (A4.12) we get

$$s\,\psi_{\alpha rot}(s) - \psi_{\alpha rot}(0) = K_1\, i_{\alpha str}(s) - Q_{11}\,\psi_{\alpha rot}(s) - Q_{12}\,\psi_{\beta rot}(s), \tag{A4.14}$$

$$s\,\psi_{\beta rot}(s) - \psi_{\beta rot}(0) = K_1\, i_{\beta str}(s) - Q_{21}\,\psi_{\alpha rot}(s) - Q_{22}\,\psi_{\beta rot}(s). \tag{A4.15}$$

The constants K and Q can be very easily identified from eqs. (A4.11) and (A4.12). $\psi_{\alpha rot}(0)$ and $\psi_{\beta rot}(0)$ are the initial values of rotor flux linkages. Equations (A4.14) and (A4.15) can be represented by a signal flow graph as shown in Fig. A.4.

Using the general gain formula derived by *Mason* [7] to obtain a functional relationship between the input and output variables expressions for $\psi_{\alpha rot}(s)$ and $\psi_{\beta rot}(s)$ can be obtained as

$$\psi_{\alpha rot}(s) = [1/(s^2\Delta]\,[K_1\,(s + Q_{22})\, i_{\alpha str}(s) - K_1\, Q_{12}\, i_{\beta str}(s) + (s + Q_{22})\,\psi_{\alpha rot}(0) - Q_{12}\,\psi_{\beta rot}(0)] \tag{A4.16}$$

$$\psi_{\beta rot}(s) = [1/(s^2\Delta]\,[-K_1\, Q_{21}\, i_{\alpha str}(s) + K_1\,(s + Q_{11})\, i_{\beta str}(s) - Q_{21}\,\psi_{\alpha rot}(0) + (s + Q_{11})\,\psi_{\beta rot}(0)] \tag{A4.17}$$

where $s^2\Delta$ is the characteristic equation which can be obtained as

$$s^2\Delta = s^2 + (2\,a_{rot})\, s + (a_{rot}^2 + \omega_{rot}^2) = 0 \tag{A4.18}$$

where $a_{rot} = R_{rot}/L_{rot}$; s_1 and s_2 are the roots of this characteristic equation.

As stated earlier, the input current waveform can be described as shown in Fig, A.3. For a portion of one sixth cycle the current is constant and during t_{de} it varies sinusoidally. From the known waveforms the α, β components of stator currents over one period are tabulated in Tab. A.1. The two axis components also are constant during interlude period and they are sine functions during commutation. Therefore *Laplace* transforms of the currents during interlude period are

$$i_{\alpha str}(s) = i_{\alpha str}/s, \tag{A4.19}$$

$$i_{\beta str}(s) = i_{\beta str}/s. \tag{A4.20}$$

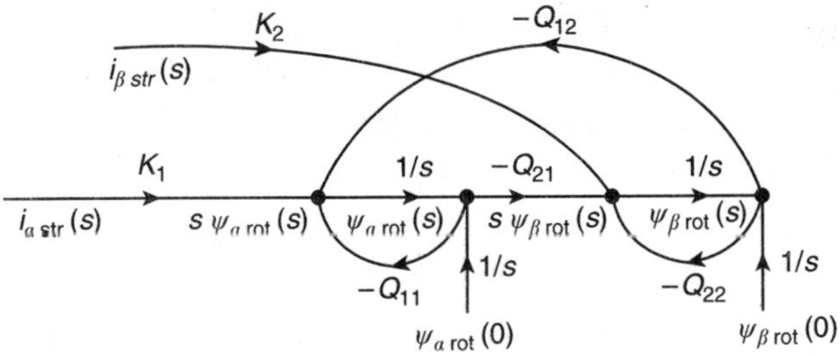

Fig. A.4 *Signal flow graph of a current fed induction motor*

Table A.1 *Expressions for the two axis components of the currents during interlude and commutating periods over a cycle (the machine is delta connected)*

	Interlude	*Commutation*
$i_{astr}\ i_{\beta str}$	$-I-/3\ -I\_/\sqrt{3}$	$(I\_/3)\,[1-2\cos(\omega_c t)]\ (-I\_/\sqrt{3})$
$i_{astr}\ i_{\beta str}$	$-I-/3\ -I\_/\sqrt{3}$	$(-I\_/3)\,[-2+\cos(\omega_c t)]$ $(-I\_/\sqrt{3})\ \cos(\omega_c t)]$
$i_{astr}\ i_{\beta str}$	$2I-/30$	$(I\_/3)\,[1+\cos(\omega_c t)]$ $(I\_/\sqrt{3})\ [1-\cos(\omega_c t)]$
$i_{astr}\ i_{\beta str}$	$I-/3\ I\_/\sqrt{3}$	$(-I\_/3)\,[1-2\cos(\omega_c t)]\ I\_/\sqrt{3}$
$i_{astr}\ i_{\beta str}$	$-I-/3\ I\_/\sqrt{3}$	$(I\_/3)\,[-2+\cos(\omega_c t)]$ $(I\_/\sqrt{3})\ \cos(\omega_c t)$
$i_{astr}\ i_{\beta str}$	$-2I-/30$	$(-I\_/3)\,[1+\cos(\omega_c t)]$ $(-I\_/\sqrt{3})\ [1-\cos(\omega_c t)]$

The time-domain solution for the rotor flux linkages during interlude period can be written as with k as running variable

$$\begin{bmatrix}\psi_{arot}\\ \psi_{\beta rot}\end{bmatrix}=\begin{bmatrix}A_{10}+\Sigma A_{1k}\exp(s_k t) & B_{10}+\Sigma B_{1k}\exp(s_k t)\\ A_{20}+\Sigma A_{2k}\exp(s_k t) & B_{20}+\Sigma B_{2k}\exp(s_k t)\end{bmatrix}\begin{bmatrix}i_{astr}\\ i_{\beta str}\end{bmatrix}$$

$$+\begin{bmatrix}\Sigma C_{1k}\exp(s_k t) & \Sigma D_{1k}\exp(s_k t)\\ \Sigma A_{2k}\exp(s_k t) & \Sigma D_{2k}\exp(s_k t)\end{bmatrix}\begin{bmatrix}\psi_{arot}(0)\\ \psi_{\beta rot}(0)\end{bmatrix}. \tag{A4.21}$$

$\psi_{arot}(0)$ and $\psi_{\beta rot}(0)$ in Eq. (21) are the initial values of flux linkages at the start of the 1/6th period.

General expressions during t_{de} for the two axis components of the currents are given in Tab. A.1. Their *Laplace* transforms will be

$$i_{astr}(s) = I\_[K_{a1}/s + K_{a2}\, s/(s^2+\omega_c^2)], \tag{A4.22}$$

$$i_{\beta str}(s) = I\_[K_{\beta 1}/s + K_{\beta 2}\, s/(s^2+\omega_c^2)]. \tag{A4.23}$$

Using these the time domain expressions for rotor flux linkages during t_{de} can be written as with k as running variable

$$\begin{bmatrix}\psi_{arot}\\ \psi_{\beta rot}\end{bmatrix}=\begin{bmatrix}A'_{10}+\Sigma A'_{1k}\exp(s_k t) & B'_{10}+\Sigma B'_{1k}\exp(s_k t)\\ A'_{20}+\Sigma A'_{2k}\exp(s_k t) & B'_{20}+\Sigma B'_{2k}\exp(s_k t)\end{bmatrix}\begin{bmatrix}K_{a1}I\_\\ K_{\beta 1}I\_\end{bmatrix}$$

$$+\left[\begin{matrix}\Sigma E_{1k}\exp(s_k t)+E_{13}\exp(j\omega_c t)+E_{14}\exp(-j\omega_c t)\\ \Sigma E_{2k}\exp(s_k t)+E_{23}\exp(j\omega_c t)+E_{24}\exp(-j\omega_c t\end{matrix}\right.$$

$$\left.\begin{matrix}\Sigma F_{1k}\exp(s_k t)+F_{13}\exp(j\omega_c t)+F_{14}\exp(-j\omega_c t)\\ \Sigma F_{2k}\exp(s_k t)+F_{23}\exp(j\omega_c t)+F_{24}\exp(-j\omega_c t\end{matrix}\right]$$

$$\begin{bmatrix}K_{a2}I\_\\ K_{\beta 2}I\_\end{bmatrix}+\begin{bmatrix}\Sigma C'_{1k}\exp(s_k t)+\Sigma D'_{1k}\exp(s_k t)\\ \Sigma C'_{2k}\exp(s_k t)+\Sigma D'_{2k}\exp(s_k t)\end{bmatrix}\begin{bmatrix}\psi'_{arot}(0)\\ \psi'_{\beta rot}(0)\end{bmatrix} \tag{A4.24}$$

t in eq. (A4.24) takes value from 0 to t_{de}. $\psi'_{\alpha rot}(0)$ and $\psi'_{\beta rot}(0)$ are the initial values of flux linkages at start of commutation.

The values of constants A, B, C and D in Eq. (A4.21) and A', B', E, F, C' and D' in Eq. (A4.24) are obtained using partial fraction expansion, for instance

$$A_{10} = K_1 Q_{22} /(s_1 s_2), \text{ etc..} \qquad \text{(A4.25)}$$

Using the symmetry relations the initial values and final value of rotor flux linkages over one sixth cycle can be related as (Fig. A.5)

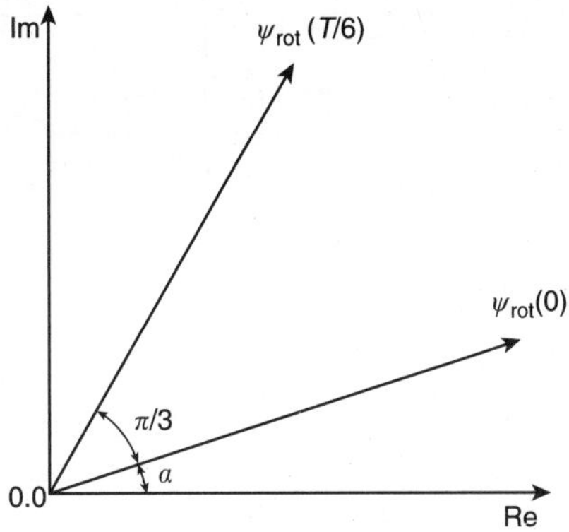

Fig. A.5 *Rotor flux vector at the beginning and end of one sixth cycle*

$$\begin{bmatrix} \psi_{\alpha rot}(T/6) \\ \psi_{\beta rot}(T/6) \end{bmatrix} = \begin{bmatrix} 0.5 & -0.866 \\ 0.866 & 0.5 \end{bmatrix} \begin{bmatrix} \psi_{\alpha rot}(0) \\ \psi_{\beta rot}(0) \end{bmatrix} \qquad \text{(A4.26)}$$

Equation (A4.26) is used to determine the initial values of flux linkages. Using these values the flux linkages over one sixth of one cycle are determined.

With the known values of rotor flux linkages the other quantities such as rotor currents, stator flux linkages, stator induced voltages and the torque developed can be determined.

A.4 RESULTS

The performance equation derived above via state transition signal flow graph are used to determine the steady state performance of a three phase slip ring induction motor the details of which are given in Table A.2. The steady state performance of the machine is calculated for two different slip conditions.

The locus of the tip of the rotor flux vector for one sixth cycle is shown in Fig. A.6 for two speeds of the rotor, one at 1460 min^{-1} and the other at 50 min^{-1}. The corresponding stator frequencies are 50 Hz and 3 Hz respectively. In both cases the link current is kept at 15A

Table A.2 *Details of the induction motor used for the investigation (rating 5.5 kW; n = 1410 min^{-1}; type: slip ring)*

	Stator	*Rotor*
Number of poles	4	4
Connection	delta	star
Voltage	220 V	170 V
Resistance per phase	0.98 Ω	0.90 Ω
Self inductance per phase	0.14 H	0.138 H
Mutual inductance between stator and rotor	0.132 H	0.132 H

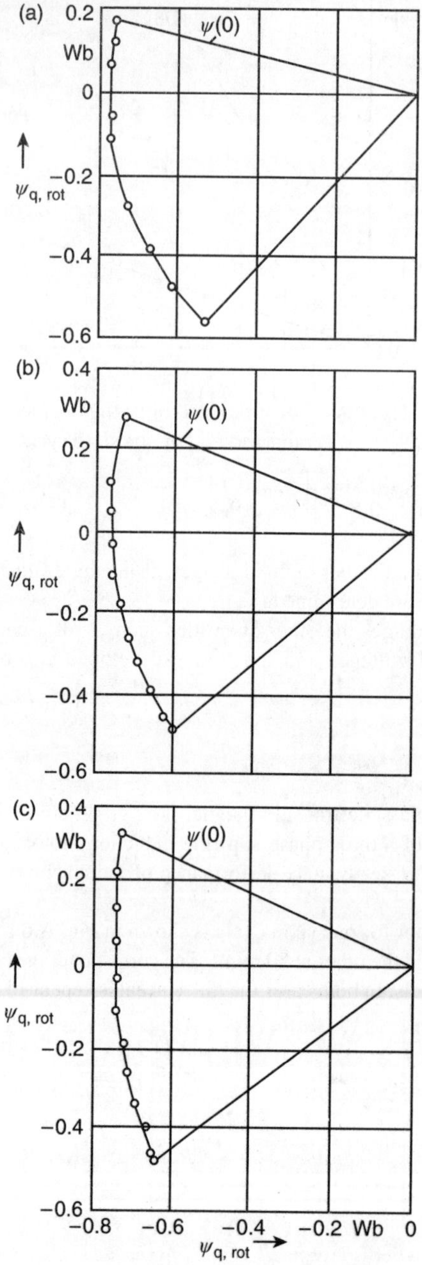

Fig. A.6 *Rotor flux vector over one sixth cycle at different frequencies (rotor frequency 1.33 Hz; link current 15 A) (a) $f_{str} = 50$ Hz; (b) $f_{str} = 10$ Hz; (c) $f_{str} = 3$ Hz*

and rotor frequency at 1.33 Hz. From the figure it can be seen that the locus is almost the arc of a circle at 1460 min^{-1} (low slip operation) whereas it deviates from the arc of a circle for operation at 50 min^{-1} (high slip operation). From this, it can be concluded that the rotor flux waveform is almost sinusoidal for the former operating conditions whereas it deviates very much from sinusoidal for the latter operating conditions. Therefore for a given rotor frequency and link current the harmonic content of the rotor flux increases as the frequency of operation decreases.

The induced voltages in the stator phases are evaluated from the known values of rotor flux linkages and stator currents (Fig. A.7). The waveform corresponding to 1460 min^{-1} is almost sinusoidal but distorted due to stator resistance and has spikes during commutation. At large slip operation a further distortion in the waveform can be expected because of the harmonics in the rotor flux waveform. The oscillogram of the voltage waveform is also shown in the figure. There is a very close agreement between the test and the calculated results.

Figure A.8 depicts the torque developed by the motor over one cycle for both the operating conditions described above. The steady state torque contains sixth harmonic torque pulsations. The amplitude of the torque pulsations is larger at larger slip conditions; here in this case at low frequency of operation. This can be expected because of the increased harmonic content of the rotor flux linkages at lower operating frequencies; or in other words the relative magnitude of the torque pulsations becomes larger as the slip increases for a given rotor frequency and link current.

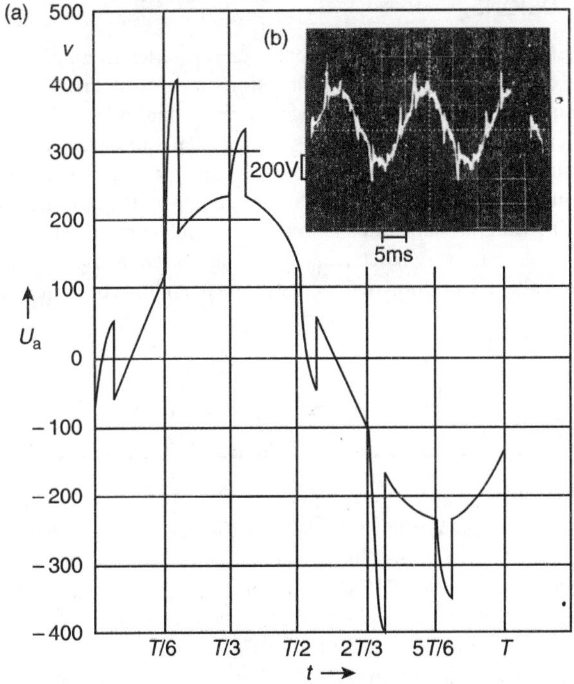

Fig. A.7 *Stator voltage waveform of a current fed induction motor (link current 15 A; speed 1460 min^{-1}; frequency 50 Hz) (a) Computed waveform (b) Oscillogram*

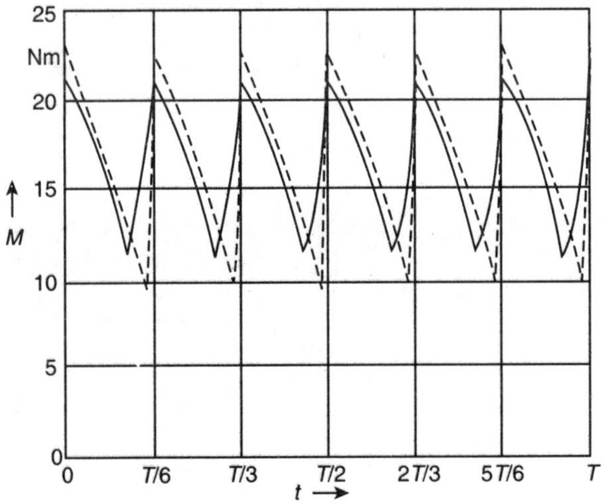

Fig. A.8 *Torque M developed by a current fed induction motor (I_{-} = 15 A)*
——— n = 1460 min^{-1}(f_{str} = 50 Hz)
------ n = 50 min^{-1}(f_{str} = 5 Hz)

A.5 CONCLUSIONS

An analysis of the commutation of a current source inverter feeding an induction motor shows that the current during commutation follows a damped sinusoid. When the resistance of the commutation circuit is neglected, it can be approximated by a sinusoid. With such an input current waveform which has a sinusoidal variation during commutation a closed form solution for the performance of the induction motor is developed via signal flow graph technique. These equations are used to compare the operation at two different stator frequencies with the same rotor frequency and link current. The comparison shows that the rotor flux deviates very much from sinusoidal variation at low frequencies. This in turn distorts the voltage waveform and increases the magnitude of torque pulsations.

References

A.1 *Naunin, D.*: Berechnung des Drehmomentverlaufs einer Asynchronmaschine bei Speisung mit Rechteckströmen. ETZ-A Elektrotech. Z. 90 (1969) no. 8, p. 179–182

A.2 *Klautscheck, H.*: Das Verhalten der Induktionmaschine bei Speisung über Strom-Zwischenkreis-Umrichter. ETZ-A Elektrotech. Z. 95 (1974) no. 5, p. 283–287

A.3 *Lipo, T.A.; Cornell, E.P.*: State Variable steady state analysis of a current controlled induction motor. Conf. Rec. IEEE IAS, 1974

A.4 *Subrahmanyam, V.; Subbarayudu, D.; Rao, M.V.C.*: On the utility of signal flow graphs in the analysis of current controlled induction motor. Conf. Rec. IFAC sympos. Düsseld., 1977

A.5 *Subbarayudu, D.; Subrahmanyam, V.*: On the steady state analysis of a current fed induction motor. Arch.f.Elektrotech. 61 (1979)

A.6 *Ward, E.*: Inverter suitable for operation over a wide range of frequency. Proc. IEEE 111 (1964) no. 8, p.1423–1434

A.7 *Dorf, R.C.*: Time domain analysis and design of control systems. London: Addison Wesley, 1963

Appendix B

LARGE POWER VSI DRIVES

A voltage source inverter possesses the following disadvantages as limiting factors restricting its application areas.

- Relatively poor efficiency; limited switching frequency due to high switching losses.
- Current harmonics leading to torque pulsatins.
- The maximum speed of operation is limited.
- Series, parallel operation of GTO's still poses a problem. With commercially available GTOs, the maximum power rating of a GTO inverter is around 2 MVA.

GTOs also require snubbers to protect against $\frac{di}{dt}$ and $\frac{dv}{dt}$. Energy stored in these snubbers (both $\frac{di}{dt}$ and $\frac{dv}{dt}$) has to be dissipated in the snumbber resistors.

These losses should be included in the snubber losses. However, the current during turn off should be considered but not the maximum value. These losses depend on the switching frequency. A study of these losses shows that the snubber losses form a larger part of the total losses. These increase in direct proportion to switching frequency. The upper limit on the nothing frequency is placed in view of these losses only. Maximum switching frequency with GTO converter is less than 300 Hz to limit these losses to a reasonable value. This is the case with RLCD snubbers with their disadvantages. Even the improved snubber circuits with reduced losses (about 60%) the problems are not solved effectively. The use of energy recovering snubber circuits lead to further improvement, however this is a very expensive solution.

Therefore, with the large power drives and their configurations the main objective should be to reach the maximum speed of operation with the lowest frequency, lowest possible current and torque pulsations.

The voltage control and harmonic neutralization in a voltage source inventer have been accomplished chiefly by the PWM methods. The research and development of the systems leading to efficient methods of modulation generating the pulse patterns of output voltage have been innumerable.

These methods are

- carrier frequency modulation techniques with sinusoidal reference
- space vector modulation
- offline optimisation of modulation using look up tables
- direct self-control techniques based on the signals of torque and flux

A variety of methods are developed where the devices used are IGBTs or MOSFETS and a very high frequency can be used as switching frequency. These are suitable for a voltage source inventer for small and medium power levels.

At large power levels, the GTOs are used. High switching frequencies render the system inefficient as discussed above due to high snubber losses. As has already been stated, the chief aim of these investigations is to have a VSI for large powers with least power loss, the lowest switching frequency and highest speed of operation. For this, a sinusoidal PWM

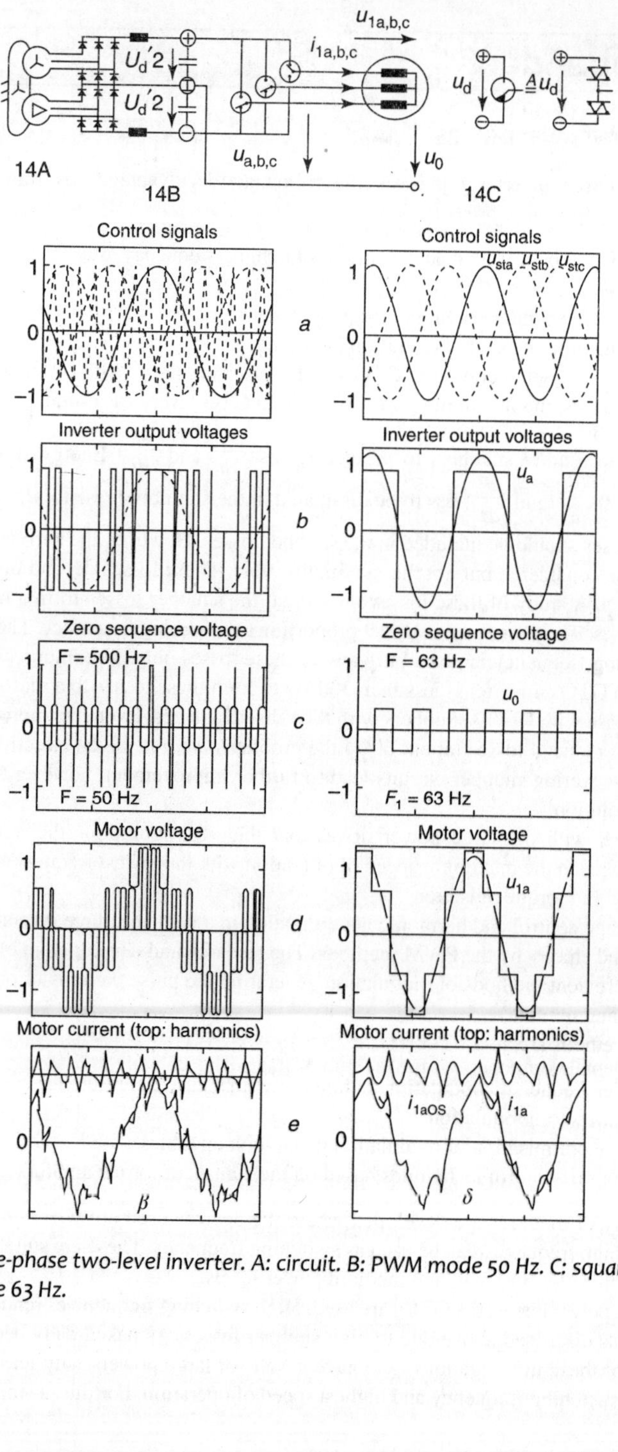

Fig.B1 · *Three-phase two-level inverter. A: circuit. B: PWM mode 50 Hz. C: square-wave mode 63 Hz.*

using triangular wave of high carrier frequency and sinusoidal reference is employed. The points of intersection of these waves are the instants of voltage reversal at the output of the inventer $U_{a,b,c}$. The carrier frequency is the actual switching frequency of the inventer.

Large induction motors have their magnetising current at 25% of the rated current of the motor. The rated current lags the applied voltage by about 25° to 30°. The harmonic equivalent circuit may be simplified and it can be assumed to be containing only leakage inductance. This simplification (leads or) results in the total harmonic (content) current independent of load. This harmonic current may be superimposed on the fundamental current, which is divided by the load. The harmonic current depends on the leakage factor; larger the leakage factor, smoother the current.

Considering from the viewpoints of high speed, low harmonic content, low switching frequency, the total harmonic component of the no-load current waveform can be taken as the representative. For large machines, the leakage factor is around 5%. This makes the THD to as high as 93%. By this, one means that the rms value of the harmonic content in the no-load current is 93% of the fundamental no-load magnetising current. It is rather difficult to get simultaneously lowest possible switching frequency and lowest THD. Taking for granted a switching frequency of 300 Hz and a THD of 0.93 as the upper limit, one has to strike a compromise between these two while going for high-speed drives with maximum reference frequency within the limits giving due consideration to normal frequency of operation.

When a VSI-fed induction motor drive is investigated taking these factors into consideration, two-level configurations (Fig. B.1a) for high power and high speed operation cannot be built with lowest THD and lowest switching frequency. Switching frequency can be greater than 300 Hz and THD in the range 0.93 for normal speed drives with a reference frequency of 50–60 Hz.

B.1 THREE-LEVEL DRIVES

Compared to two-level drives, these can be operated with twice the dc voltage and hence the rated power can be twice. A three-point change over switch (Fig B.2b) is placed in each phase of the motor. The output of these switches can be connected to positive, negative or zero potential of the input dc source. One three-level inverter can be regarded as an inverter which can be operated with two independent pulse patterns.

In the lower speed range, the inverter is operated in the PWM mode. Both the pulse patterns I and II (Fig. B.2) provide the same fundamental voltage. The harmonic content is however reduced by using staggered switching instants. The sinusoidal reference decides the fundamental component of the inverter voltage. The staggered switching is realised by means of 180° phase shift between the triangular control signals.

In the upper speed range, the inverter may be operated in the square wave mode. The pulse patterns here have a switching frequency equal to the reference frequency with a phase shift to provide or control the fundamental voltage (proportional to frequency).

Three-Level VSI Shows the Following Improvements

i. For normal speeds with reference frequency at 50 or 60 Hz, the switching frequency of 300 Hz reduces the THD of the motor current. Torque ripple decreases. The PWM mode can be extended to reference frequency close to normal value of 50 or 60 Hz as the case may be.
ii. For high-speed operation, this frequency can be increased to 130 Hz without exceeding the switching frequency beyond 300 Hz and without using field

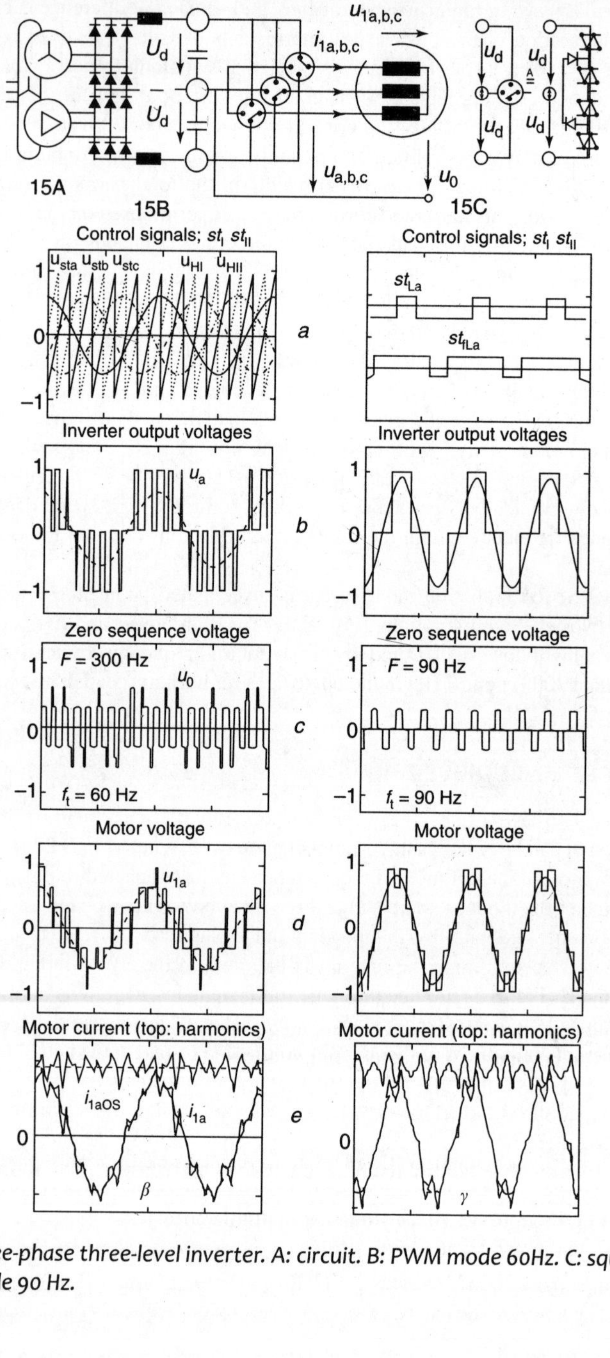

Fig.B2 *Three-phase three-level inverter. A: circuit. B: PWM mode 60Hz. C: square-wave mode 90 Hz.*

weakening. The control method should be changed at $f_1 = 65$ Hz from PWM switching mode ($f = 300$ Hz) to switching mode with fundamental frequency $f = f_1 = 65$ to 130 Hz.

B.2 LOW INDUCTANCE DESIGN

Taking into consideration the size of the components, the space needed for insulation and the increased distance of hot snubber resistors, the current loops may be regarded as having parasitic inductances. Even small leads may have such parasitic inductances at high frequency. The switching off of the current in these loops may result both *di/dt* and voltage stress. These may reduce the attainable rated power. Low inductance designs are therefore required.

A two-level inverter circuit with RLCD snubber is shown in Fig. 5 with four loops where currents are switched off with high *di/dt*. These loops must be kept as small as possible. Low inductance configuration is shown in Fig. 5. This is achieved basically by

—Low-inductance capacitors C_{dc} with sandwiched connections to the semiconductors.
—Snubber capacitors *c* and diodes (b_1, b_2)are placed as closely as possible to the GTOs.
—Snubber resistors *R* with low inductance design and sandwiched connections to other semiconductor elements.

B.3 CONTROLS

For good dynamic behaviour of the torque control and speed sensorless operation vector control using field orientation may be used advantageously both in low and medium power drives.

Large power drives use speed sensors. In most of the applications, high dynamic performance may not be required, e.g., pumps, fans and compressors.

High performance may be required in applications like steel mills along with regenerative speed reversal. For these applications, VSI drives are not competitive because they require additional convertor on the line side for regeneration. The FOC can still be used with VSI fed motors. But its advantages have not been found useful till now in this field.

B.4 FUTURE TRENDS

VSI-fed induction motors have increased potential for real high-power, high-speed drives. This potential is connected with the developments and advances in the semiconductor technology and also new motor–converter configurations:

These include the following:

i. Improved gate control of GTOs to reduce the turn-on and turn-off delay differences of individual GTOs. This may result in easier series–parallel operation of the units. Series–parallel operation with an additional GTO in each branch allows to introduce a new quality, fault tolerance.
ii. Higher power VSI at a justified cost. High-power, high-speed drives may be possible with VSIs at all power levels with twin configurations.

Looking further into the future, MOS-controlled high power GTOs (MCTs) may be expected with much simpler and economical gate control.

The introduction of IGBTs in the place of GTOs may open up a new trend in building cheaper and better inverters with reduced harmonics. IGBTs have specific advantages of higher switching frequency, simple gate control and perhaps no need for snubbers. It also offers high flexibility of operation and might also open the way for multiphase motor with high torque/volume ratio.

This transition is practicable for low and medium power drives. The transition at high power levels takes some more time till the developments of IGBTs to handle large power at high voltages and high currents as present-day GTOs do; series–parallel operation of GTBTs at the device level. It may be expected that IGBTs may be introduced at the low power end of large power drives.

B.5 CONFIGURATIONS OF VSI

Present-day GTO inverters have limited power and frequency capabilities, e.g., two-level inverter up to 2 MW/60 Hz and three-level one 4 MW/130 Hz parallel or series connection of VSIs may obviously increase the power rating but not operating frequency.

Developments of motors having different 3 phase wingdings fed from different VSIs. These have improved voltage waveforms impressed across the windings unlike CSI which have impressed currents. The magnetic coupling between the windings may cause interacting current harmonics due to voltage of both the VSIs. It is difficult to provide dc coupling between these coupled voltage systems to the extent that they do not produce high interacting current harmonics.

The power-factor limitations can be extended to twin configurations which are in conventional motor and inverter configurations. However, the existing VSIs and motors with conventional 3 phase stator winding with access to both ends of the phases may be used with two VSIs feeding on both the ends. The rated power can be increased to 8 MW, extending the frequency to 300 Hz (fundamental) develop high-speed motors (18,000 rpm). Low THD can also be combined at low switching frequency $F \leq t_{1max}$ without exceeding the maximum fundamental frequency.

Index